Recreational and Environmental Markets for Forest Enterprises

Recreational and Environmental Markets for Forest Enterprises

A New Approach Towards Marketability of Public Goods

U. Mantau
Institute of World Forestry
Department of Economics
University of Hamburg
Germany

M. Merlo
Department of Land and Agro-Forestry Systems
University of Padua
Italy

W. Sekot
Institute of Forest Sector Policy and Economics
University for Agricultural Sciences
Austria

and

B. Welcker
Institute of World Forestry
Department of Economics
University of Hamburg
Germany

CABI *Publishing*

CABI is a trading name of CAB International

CABI Head Office
Nosworthy Way
Wallingford
Oxfordshire OX10 8DE
UK

Tel: +44 (0)1491 832111
Fax: +44 (0)1491 833508
Email: cabi@cabi.org
Web site: www.cabi.org

CABI North American Office
875 Massachusetts Avenue
7th Floor
Cambridge, MA 02139
USA

Tel: +1 617 395 4056
Fax: +1 617 354 6875
Email: cabi-nao@cabi.org

A catalogue record for this book is available from the British Library, London, UK.

Library of Congress Cataloging-in-Publication Data
Recreational and environmental markets for forest enterprises: a new approach towards marketability of public goods / edited by
U. Mantau ... [et al.].
p. cm.
Includes bibliographical references (p.).
ISBN 0-85199-480-6 (alk. paper)
1. Forests and forestry—Economic aspects—Europe. 2. Forests and Forestry—Multiple use—Europe. 3. Forest products—Europe—Marketing.
I. Mantau, U. (Udo).

SD177 .R43 2001
333.75'13'094—dc21 00-063018

ISBN-13: 978-0-85199-480-2
ISBN-10: 0-85199-480-6

First published 2001
Transferred to print on demand 2007

Printed and bound in the UK by CPI Antony Rowe, Eastbourne.

Contents

The RES Team

Germany

Prof. Dr Udo Mantau, Project Co-ordinator. *Institute of World Forestry, Dept. of Forest Economics, University of Hamburg*
Berthold Mertens, Master of Forestry, Research Assistant
Borris Welcker, Master of Forestry, Research Assistant

Prof. Dr Ulrich Ramsauer, Associated Partner. *Institute of Public Law, Dept. of Environmental Law, University of Hamburg*
Birgit Malzburg, Master of Biology, Research Assistant

Italy

Prof. Dr Maurizio Merlo, Project Co-operator. *Department of Land and Agro-Forestry Systems, University of Padua*
Erica Milocco, Master of Forestry, Research Assistant
Paula Virgilietti, Master of Forestry, Ph.D., Research Assistant
Richard Panting, Bachelor of Forestry, Research Assistant

Austria

Prof. Dr Walter Sekot, Project Co-operator. *Institute of Forest Sector Policy and Economics, University for Agricultural Sciences Vienna*
Peter Plaimer, Master of Forestry, Research Assistant
Andreas Ottitsch, Master of Forestry, Contributing Author

The Netherlands

Ir. Kees Van Vliet, Project Co-operator. *ALTERRA, Green World Research*
Edwin Wiemann, Master of Forestry, Research Assistant
Arno Segeren, Master of Forestry, Contributing Author

Preface

The project 'Niche Markets for Recreational and Environmental Services from Multiple Forest Production Systems' (RES project) started long before it became a European Union (EU) project. Mantau, Merlo and Sekot have been working separately on different aspects to develop market solutions and strategies for various forest outputs, which are considered to be not or hardly marketable. Peter Glück knew about their efforts and initiated a meeting. It was the starting-point of the RES project.

The project was funded by the European Commission for 3 years, starting in March 1996. It was carried out by Udo Mantau (coordinator), Universität Hamburg, Germany, Maurizio Merlo, Università degli Studi di Padova, Italy, Walter Sekot, Universität für Bodenkultur, Austria, and Kees Van Vliet, ALTERRA (IBN – DLO), the Netherlands. The authors take sole responsibility for the content of the publication, which does not represent the views of the European Commission or its services. However, we would like to express our thanks to the EU for funding that made it possible to intensify the original ideas and to carry them out on an international level.

Within the application process, it was planned to document five cases in each country. After the project was started, the group found many more cases than originally expected. It was probably the most important decision in the project to commit ourselves to document a minimum of 20 cases in each country. We learned a lot from these innovative forest enterprises and we are thankful for the time and the valuable information they gave us.

The project was almost completed at the end of 1999 in several project reports. It was a big effort to select, summarize and harmonize each single report for this common publication. It would not have been possible to achieve this high standard without the outstanding commitment of Borris Welcker, who planned and carried out all the major and minor tasks of the challenges.

Since the basic idea of the RES project is to transform recreational and environmental forest functions into forest products, we had to transform the RES project into an RES product. We hope that this product/publication will help many forest enterprises to extend their options of economic actions and thereby extend the economic basis for sustainable forestry.

Udo Mantau
Hamburg,
June 2000

1

The RES Project – Introduction and Conclusion

1.1 Leading Ideas and Basic Assumptions of the RES Project

1.1.1 The economic situation of forestry

In many countries worldwide, the income received from forestry is insufficient (Mantau, 1999). On the one hand, free utilization of facilities in the areas of environment and recreation causes higher costs for the forest enterprises, while, on the other hand, timber prices are said to be too low. Furthermore, both issues are reinforced by the need to distribute an increasing number of cost-effective facilities and charges to be covered by proceeds from less and less marketable cubic metres of timber. The situation is additionally aggravated by political restrictions, as the economic freedom of choice of the forest enterprises is increasingly restricted.

The following, long-term, profound trends lead to the assumption that this development will continue:

- Products with a low degree of processing will increasingly lose their market share in favour of less developed countries (Ollmann, 2000).
- The trend towards economic cycles increases substitution processes and in particular affects small-sized wood (Umweltministerium Baden-Württemberg, 1995).
- Throughout Europe, timber growth increments are higher than cutting rates (Spieker *et al.*, 1996).
- Approximately from the beginning of the 1970s onwards, there has been a growing demand for recreational and environmental facilities (Ammer and Pröbstl, 1991).

Most of the proposals for a solution to the problem (e.g. job reduction, privatization) try to find a way out of the crisis by lowering costs and by increasing effectiveness. This can achieve the target, but the problem is both an issue of costs and a problem of offers. An economic entity does not provide products to meet a growing demand of society, or which supplies these facilities as forest functions free of charge to a larger extent than stipulated legally or covers demand only insofar as it occurs in the framework of joint production processes, not only abandons its economic future but will also be increasingly denied the right to act to the full range of its product potential, as others will make better use of production possibilities for covering the demand (Mantau, 1995).

Even worse is a policy which transfers the costs of infrastructural facilities of the forest on to timber prices. From the economic point of view, this not only promotes substitutive processes concerning raw timber, but also weakens competitiveness in subsequent markets (semi-manufactured and ready-made timber products). This leads to lower growth in these markets or even shrinkage, which means that increasing infrastructural costs will be charged to decreasing cubic metres of wood (Mantau, 1995).

At present, this situation develops spirally, causing the well-known problems in the financial standing of the forest enterprises and the forest economic situation of many countries worldwide. When an entire branch is shrinking to a healthy level, the respective companies lose part of their influence on the markets in which they were rooted and have failed to enter new markets.

1.1.2 The RES project

This is the starting-point of the European Union (EU) project 'Niche Markets for Recreational and Environmental Goods and Services from Multiple Forest Production Systems' (RES project; FAIR1 CT95-0743). In their application, the project partners[1] formulated the objective of the project, as follows.

The main objective of the research project is to develop market solutions and strategies for various forest outputs, which are considered to be not or hardly marketable. The phenomenon of 'public goods' (recreational and environmental goods and services of forest) is considered under dynamic conditions, in which marketability is rather a matter of product and framework development than of objective circumstances. The analytical framework is divided into five tasks:

- Strategies for product design and strategies for market transformation.
- Marketing strategies and training courses.
- Contracts, institutions and legal aspects (property rights).
- Management of multifunctional forests.
- Policy analysis and implication.

The development of new products and new markets requires quite different modes of action. Above all, practical solutions are demanded and, therefore, an interdisciplinary investigation approach is more or less compulsory. The development of a new product requires economical know-how, good planning abilities, the setting up of contract relationships within the given framework of action, etc. Although specific theories and abilities can be developed by partial approaches, practical success generally demands a heuristic mode of action.

In the present case, there is another problem to be borne in mind: in practice, the problem of marketability – namely, the lack of marketing of the recreational and environmental services (RES) of the forest – can be analysed, whilst marketing successes are observed quite rarely. Therefore, it is not surprising that the literature mainly deals with the determination of the quantitative extent of market failure. The RES project assumes that the RES of the forest could be marketed to a far greater extent if the instruments and the framework of action for this problem were developed more adequately. For the empirical work, this results in the necessity for a qualitative mode of action (case-studies).

Based on the theoretical knowledge of specific disciplines (economic theory, business administation, law, policy) and the available knowledge of specific successful case-studies, the instruments and the framework of action are progressively developed with a focus on the improvement of marketing possibilities. With a given low state of knowledge, progress is only possible with the invention of new hypotheses. Because of the state of the art, these hypotheses cannot (yet) be assured quantitatively. Nevertheless, their significance is not delimited by a few individual cases, but rests both on a broad empirical basis (98 case-studies) and on the logical combination of adequate problem solutions from literature deriving from other fields of application.

Within the task of 'strategies for product design and strategies for market transformation', the RES project connects with recent discussion in the economic literature. Present approaches are extended by the dynamic aspects of the project approach.

As aspects of active development of markets are seen as an important basis for a broader marketing of the recreational and environmental products of the forest, the task of 'marketing strategies and documentation/training of instruments' plays a significant role. Various instruments for the improvement of the marketing success are shown. At the same time it becomes obvious that success cannot be realized by just one solution but comes from the interaction of different economic and social framework conditions.

The task of 'contracts, institutions and legal aspects (property rights)' deals with the social framework conditions. Any exchange of goods postulates an action that is combined with a contract. Standard contracts are developed, the relationships between the contract partners are presented and analysed by the use of the theory of transaction costs. The influence of property rights is analysed on the basis of the German judicial system. As far as the right to access the forest is concerned, the comparison between the international judicial systems shows big differences.

The task of 'management of multifunctional forests' deals with the integration of new products into the existing company environment. The integration of a project within the business organization could be seen as a success factor (project management). Identification of the strengths and weaknesses of a forest company is a fundamental step in product development (potential analysis). Efficiency is a key issue of any RES activity, whereas accounting provides a framework for evaluating the profitability of different activities (business organization and accounting).

Finally, the task of 'policy analysis and implication' analyses the public acceptance of recreational and environmental forest products (RES) as well as the influential factors with regard to acceptance.

The RES project starts from the principle of an interdisciplinary investigation approach. Therefore, action opportunities are developed for the marketing of recreational and environmental forest products on the basis of qualitative case studies, along with a broad integration of theories and instruments as described in the literature. Furthermore, the investigation possibilities in an area of science characterized by a low state of knowledge are hypothetically extended.

The following outline reflects the leading ideas and basic concepts of the project. It does not claim to solve all of the marketing problems of public goods provided by the forest, though it can facilitate marketability by a far larger share of forest facilities than is assumed at present. The basic components for the expansion of marketability are product economics, market development, value transfer and behavioural changes.

1.1.3 Product economics – the complex structure of products

Goods display an inner structure of product elements and additional values. In the concept of the structure of a product as a homogeneous unit (black-box approach), many products do not seem to be marketable. Via the composition of product elements on various utility levels, exclusiveness and competition of goods and services can be considerably influenced.

- The feature of exclusiveness depends on the level of benefit of a product. In general, exclusiveness increases with the level of benefit. Therefore, exclusiveness cannot be defined for a complex product but only for its elements.
- Products are generated from a variety of elements. The product character (exclusiveness, competition) of these elements is different. From the variety of elements, marketing possibilities can be developed with the corresponding strategy.
- The value and marketability of a product to a large extent depend on dynamic changes due to changes in activities and environment.

These hypotheses and assumptions can be illustrated by the following example. In order to clearly illustrate the central idea, it may be indicated that the economic good 'hotel' on a higher level of benefit does not differ from the public good 'forest road' in its general economic value. For this aim, first the concept of benefit levels has to be defined.

A product is not solely a product in itself, but consists of several levels of benefit. The concept of benefit levels of a product is well known as a strategic concept in marketing literature (Kotler and Bliemel, 1992).

In general, the main benefit of a product is abstract in nature. When booking a hotel room, one expects to find sleep and rest at night. However, these features cannot be bought in this way. Only the generic product is marketable – in the case of the hotel, this will be represented by at least a bed and a receptionist's desk (Fig. 1.1).

When turning to the forest road, it becomes obvious that the main benefit for the hiker is recreation. In general, in the case of the forest, this can only be obtained if the hiker can have access to forest roads. In the case of the forest road, this generic product is not saleable, as it has become a public good by legal change because of the regulations of the German Federal Forest Law. As further reflections show, however, this does not exclude marketability. The theory of forest functions and the passive concept of joint products theoretically end on the level of the generic product. As joint products or in the wake of other facilities, only 'generics' can be produced, i.e. basic facilities. The attractiveness of a product, however, is mostly developed on the higher product levels.

Generally, consumers assume further additional benefits to be self-evident, which means that they would not consider buying the product or would only evaluate it as of minor quality. In the case of the hotel, this would be features such as cleanliness, towels, telephone and breakfast. This is the level of the expected product. In tourist areas, the recreationist will expect to find road signs and benches in the forest and views of beautiful landscapes. Already on this level, there are marketable product possibilities.

In industrialized economies, competition between enterprises in general takes place on the level of the augmented product (additional benefit). It has to be considered that there are different product strategies in different price segments, but it can be assumed that the choice of a hotel generally takes place on the level of the augmented product (swimming-pool, service, cuisine). Additional expense can lead to a considerable amelioration of market chances. The same is true for a forest track.

Finally, every product has a potential for development, i.e. product levels which exceed the present general product strategies. Some hotels already offer the organi-

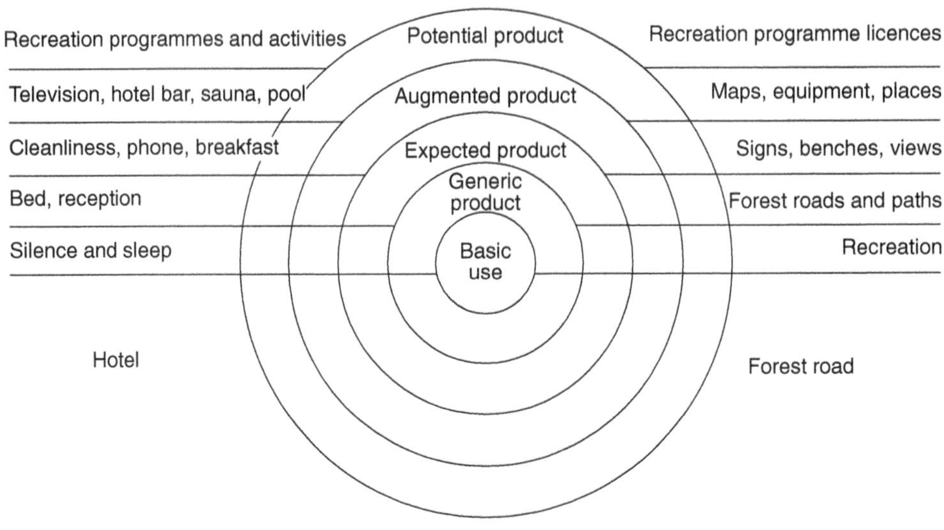

Fig. 1.1. Changing the character of rivalry and exclusion on different utility/product levels.

zation of adventure holidays or special leisure activities programmes. A comparable facility could be offered by a forest owner by combining different forest facilities in a leisure package offer (ecological or forest-cultural guided tours, hiking maps, utilization of sites, picnics and many more) submitted to the recreationist as the final consumer of the product. Therefore, it cannot be entirely excluded that a forest road and a hotel are identical on the level of the potential product – apart from the wish of the owner to actually make the offer.

So far, the concept of benefit levels has been described exclusively as a marketing instrument. Aspects of marketability have not been outlined up to the present. Mantau has described these aspects in different contributions (Mantau, 1993, 1994, 1999):

1. First, environmental and recreational facilities are basic benefits. As long as forest functions are marketed, market failure is always the case, as no product, not even a hotel, is marketable on the level of its basic benefit. This logical error is mainly caused by the theory of forest functions, which refers to forest facilities mainly in terms of the basic benefits.

2. Moreover, the product properties of rivalry and exclusiveness react entirely differently on the different levels of benefit. Via a design on higher levels of benefit, public generic goods can be transformed into marketable goods, i.e. rivalry and exclusiveness can be influenced.

3. The examples mentioned above, moreover, indicate that real products often represent a combination of goods (forest road, guided tour, accommodation). The structure of these combinations of goods concerning rivalry and exclusiveness also varies. Neoclassical economic theory is still based on homogeneous goods. This approach to goods (black-box approach) is one of the main reasons why theoretical trials by the neoclassical school to solve the problems, in large part, by measuring the assumed market value of fictitious goods subsequently correct 'market failure' Pareto-optimally.

4. Finally, dynamic aspects are important for the marketing of goods. Rivalry and exclusiveness of a product can be caused by environmental changes, as well as by activities of the participants.

When considering these four statements on the properties of goods, one also reaches other conclusions concerning the marketing of goods. All approaches that work with alternative categories like private goods and public goods lead to the conclusion that there is a clear borderline between marketable and non-marketable outputs. This is not the case. Rivalry and exclusion are not 'yes or no' properties. They vary in a continuous range from 'as if not existing' to 'abundant'. Different groups may be rivals for an attractive forest road close to a city, while a badly managed logging path is not frequented at all. Thus, product properties are not only determined by market conditions but can be changed, depending on owners' actions and demand-side changes.

Figure 1.2 illustrates the continuous structure of goods. The vertical axis represents rivalry, from low-level rivalry to high-level rivalry. The horizontal axis represents exclusion. It is assumed that exclusion is a matter of costs. It is very expensive to produce exclusion close to the zero point. In the other direction, it becomes less cost-intensive.

Therefore, the term non-marketability is misleading. There are goods with low marketability properties (low rivalry and high costs of exclusion) and goods with high marketability properties (high rivalry and low costs of exclusion). Every good to some extent has properties of marketability. However, these may not be enough for it to be successfully marketed.

The space of marketability can be separated into four strategic fields. Forest outputs in the first quarter with low rivalry and high costs have low marketability chances. If they are offered, legal regulations or state subsidies have to be applied. Forest outputs with high rivalry but also high costs of exclusion will utilize the high valuation by the users. It is very cost-

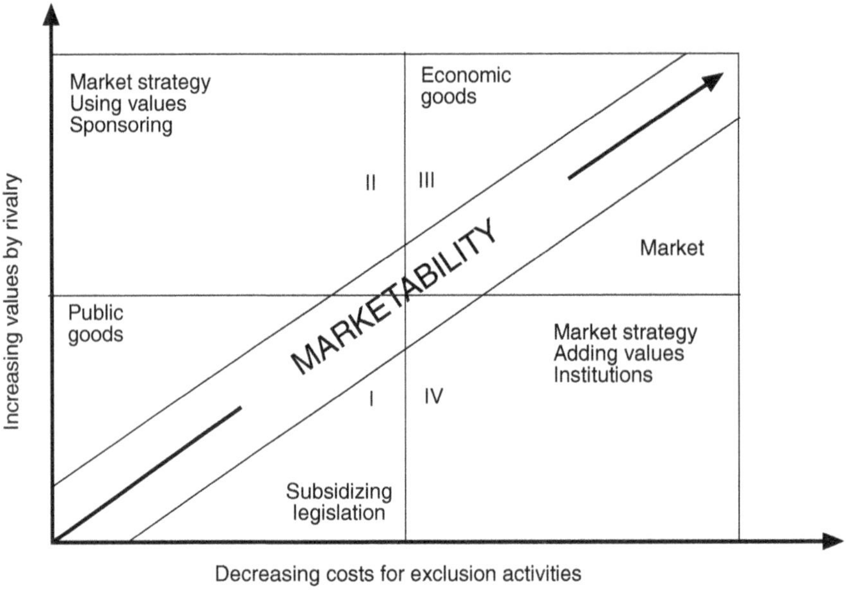

Fig. 1.2. Continuous structure of goods and strategic fields.

intensive to exclude somebody from looking at a very unique tree formation. However, marketing strategies that focus on value selling, such as sponsoring, can be successful in this segment. It can be profitable to look for a sponsor or to organize a club. Products with high rivalry and low costs of exclusion are, in general, called private goods. Normal economic strategies can be applied. Products with low costs of exclusion and low rivalry may also be marketed by adding value strategies or by institutions promoting more effective market organization.

However, it should be omitted to fall into new fixed categories. Products can be shifted from one segment to another, depending on changes in demand, preferences, property rights, owners' activity, etc. Thus Fig. 1.2 shows a general dynamic pattern of product properties.

On top of the dynamics of market conditions and private behaviour, there is product dynamics due to time. Figure 1.3 illustrates some examples. Looking at a private good like pulpwood in 1993, the prices after the big thunderstorms of 1990 (in Germany) were still so low that some landowners decided to leave the small-sized timber on the ground instead of preparing it for pulpwood (A1). The price of rivalry was lower than the cost of preparation and transport. In 1995, this situation changed rapidly (A2). Apart from such normal business cycles, there are structural changes as well. Let us assume that, as a result of technological inventions, energy will be overabundant in 2010. All small-size timber may have a low value (A3). If big energy shortages occur, all small-size timber may become highly valuable and the costs of exclusion will rise (A4).

An environmental good like a 'biotope' (B1) may have some value to the public but the cost of exclusion will be high. However, designing the 'biotope' as a product for a sponsor may make it more valuable (B2). The right of using the value of the 'biotope' for business advertising activities can easily be made exclusive.

In the year 1955, cycling was not seen as an important activity. Exclusion in economic terms was not practicable (C1). Mountain-biking has become a major recreation activity in our day. While foresters in Germany are discussing the problem of

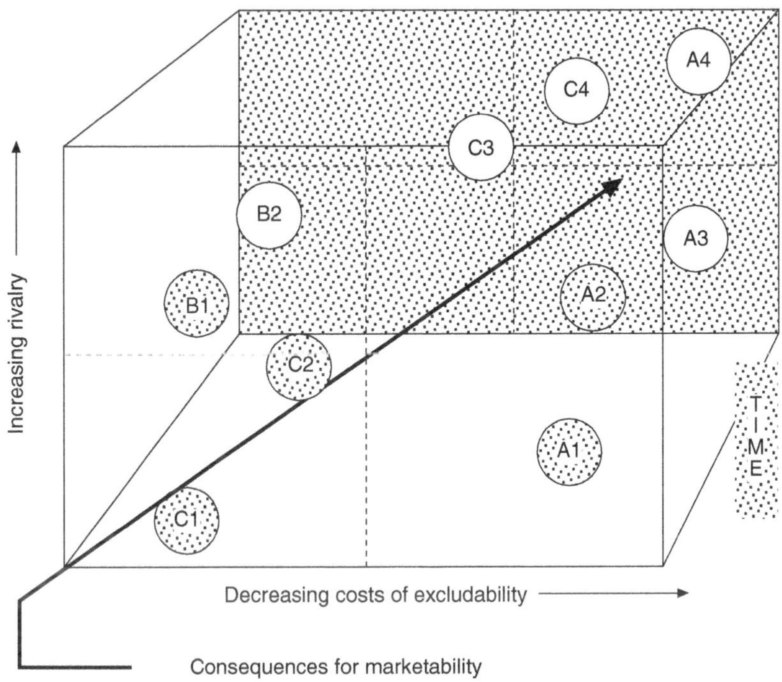

Fig. 1.3. Dynamic changes in the character of marketable goods and potentially marketable goods. See text for explanation of symbols.

whether or not the mountain-biker or the hiker has the right of access, the Austrian state forest has defended its right to exclude mountain-bikers from using forest roads as long as it does not receive any payments (C2). If forest landowners are successful in making forest roads attractive to mountain-bikers, this recreation may become even more valuable and the costs of exclusion will decrease (C3) because added value opens new ways to produce exclusion. If regional recreation becomes more attractive in the future, this will again increase marketability (C4).

When the product properties rivalry and exclusion are changed by market conditions, by behaviour and over time, there are many theoretical implications. Together with the dynamics to which all products are submitted with regard to their marketability, this indicates that a clear borderline between private and public goods cannot be drawn. For this reason, one should only differentiate between marketable and potentially marketable goods, instead of non-marketable and marketable goods. Otherwise, there is a danger of dropping future marketing possibilities or of stating non-marketability.

1.1.4 Non-material values in the marketing process

A variety of forest facilities are classified as non-material goods (scenery, environmental values). The design of non-material values can be used for enhancing marketability. Traditional concepts in forest economy formulate values in the form of forest functions, but when their importance is dismissed, they are withdrawn from the process of marketing. Values, however, can be integrated in the marketing process in different ways:

- Transformation of an already existing tangible product of one's own company by trade marks and certificates.
- Transformation of an already existing

tangible good of another company by sponsoring and licensing of trade marks.
- Establishing a product mix of non-material values and tangible product characteristics to form products and services.
- Persuading customers to pay voluntarily via prices and contributions.

There are many product structures that provide possibilities for overcoming non-marketability. A forest road is an example of a product that is integrated in different utility networks, such as the use of a forest road for logging, recreation or environmental education. Rivalry and exclusion are different in each of these utility networks.

More important for environmental goods is the difference in value levels. Values are becoming a more and more important factor in competition. There are already huge markets for values. Companies and their products are more and more valued by their contribution to society and ecology. Quite often they are not able to serve this demand with their simple products or they want to compete with more engagement. Forest landowners have a great potential to offer such value products. Figure 1.4 shows the difference in exclusion. A good like a 'biotope' may be public on the material level (nature). However, the right to advertise with the environmental value of this good can be sold exclusively to one sponsor. Sponsoring is just one way of selling values. Organizations like the Worldwide Fund for Nature (WWF) and Greenpeace are very successful in the marketing of values. But this leads us to another disadvantage of environmental forest products. It is not necessary to be the owner of nature to become a successful company in nature-value selling.

Values are invisible and non-touchable entities. It is important to 'materialize' values. The advertisement of a sponsor of a 'biotope' shows the 'biotope' in the background and the logo of, for example, WWF at the bottom. The price is paid for the logo and not for the 'biotope' The logo represents competence in environment. Whatever their usefulness, tasks which cannot be visually communicated by advertising will not be remunerated. They have to be visible for those who are willing to sponsor or choose the product that represents the nature value or who agree to pay for outputs that do not have to be paid for. Most forest landowners are 'no-names'. For the marketing of values, it can be put as simply as this: no name, no money!

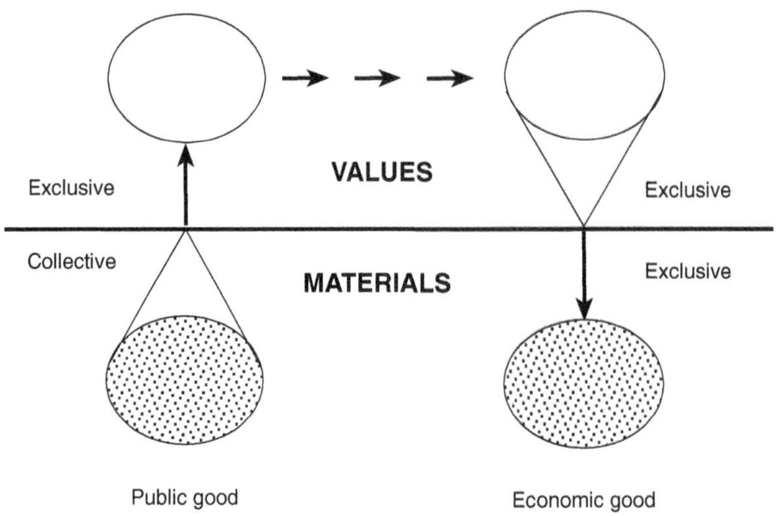

Fig. 1.4. Changing character of exclusion on different value levels.

1.1.5 *Active market development – the role of law, contracts and organizations*

Markets are complex entities of institutional and organizational structures. The formation of markets has to be shaped, i.e. can be influenced.

- Markets are not concentrated in one single spot but often result from a suitable network of intermediaries and channels of distribution (seminars for executives).
- Markets often start to function only when institutions of coordination and confidence in interactions are established (a network of bridle-paths).
- Markets consist of exchange transactions, and exchange transactions in turn require contracts and conditions of sale. Without business rules concerning financing and trade, the timber trade does not function. Each market transaction is related to overhead charges (transaction costs), depending on the company and the product environment, because of initiation, execution and control of market transactions. By lowering these costs (e.g. by sample contracts), marketability rises.
- Markets require a clear, trustworthy legal framework. Forest regulations came into existence under different political conditions and require modification. Even more important for forest economy is the circumstance that self-imposed limitations are often not based on a legal background.

The model of a perfect market has left considerable damage to the market's economic philosophy. The neoclassical concept of a perfect market has inflated the problem of public goods by excluding the possibility of exerting active influence on markets and their dynamics.

A product can pass through different kinds of distribution channels. In one case, it may be marketable, while in another it may not. In a distribution channel with low rivalry or a high degree of exclusion or high transformation costs, the product will fail. The best product is not marketable when it is not marketed in the right way. Thus, distribution channels have to be designed, as well as products (Fig. 1.5).

Every business is based on contracts,

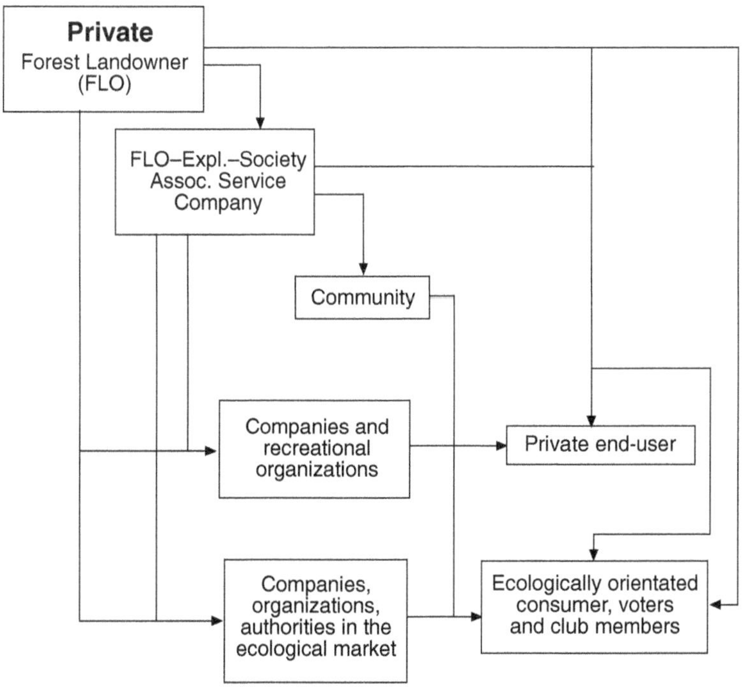

Fig. 1.5. The role of distribution channels for marketability.

Forest landowners have an excellent contracting system for timber selling but almost no commercial practice for the recreational and environmental outputs of forests. Contractual tools and trade usage are an important requirement for marketability.

Finally, there are many coordination problems for forest landowners wishing to market RES. Many of these services can only be offered on a large scale. Small-size forestry has to coordinate its supply. On the demand side, many consumers must be coordinated entities that can be dealt with. Totally new institutions may become important to make RES marketable.

In economic theory and in economic policy, the approaches of the newer transformational and institutional theory are increasingly discussed (Williamson, 1990), as they have already been formulated in the theories of the liberal school, partly using another terminology, for decades (especially in von Hayek, 1969). In this theory, it is accepted that the process of exchanging goods does not proceed automatically, but that the process itself leads to costs and has to be organized. However, it would exceed the framework of this book to reflect on this discussion.

Its significance can be illustrated by the following example. In most forest economic and forest products curricula, there is a lecture series on the timber trade. Within the framework of this weekly 2-hour lecture over one semester, a variety of subjects is dealt with which a forester or forest economist has to master, from issues such as classification and grading standards, sales procedures, national and international trade customs and measuring systems to legal trade regulations. These trade customs have been developed over decades and provide an organizational and institutional 'asset' of the timber economy for dealing in the timber trade. The lecture which, in more theoretical terms, could be called 'forest economic transaction and institution theory', is utilized to introduce only the substantial issues in timber trade. It has to be taken into account that these instruments are almost entirely lacking for the infrastructural facilities of the forest. But without these instruments timber would be difficult to market, which can still be seen in timber markets in developing countries. There is a great unfulfilled requirement to adapt infrastructural facilities (environment and recreation) to marketability.

Exchange transactions in markets require a legal framework at several levels (Fig. 1.6). The political state (distribution of the functions of power, constitutionality) and the economic conditions (property, contractual liberty) establish the framework in which legal regulations are set up (law of contracts, Federal Forest Law). The legal regulations define the participants' capacity to act. Frequently, the exchange transactions are so specialized (timber market) that the participants in the market expand the standard regulations (trade customs, norms, sales conditions and terms of payment). In particular, such trade customs and regulations are formulated by the exchange parties if by doing so the costs of the exchange, the transaction costs of opening, negotiation and control, can be lowered. Thus, there exist a multitude of regulations concerning settlement of the transactions, especially in the international timber trade. But, without their existence, complex exchange transactions would not take place, as risks and expenses would be too high.

At the next level of market organization, contractual knowledge is required. The offers for environmental and recreational facilities frequently have to be introduced in markets which are new to the forest companies. The development of contract patterns and the legal design of conditions of utilization, terms of notice, relations to forest law and traffic regulations, as well as issues of liability and insurance, to name only a few, are important preconditions for the marketability of the recreational and environmental facilities of the forest.

Finally, the operational abilities of the market participant are necessary to make goods marketable. The market participant can launch the offer on the market as a single entity, or may be required to do so in

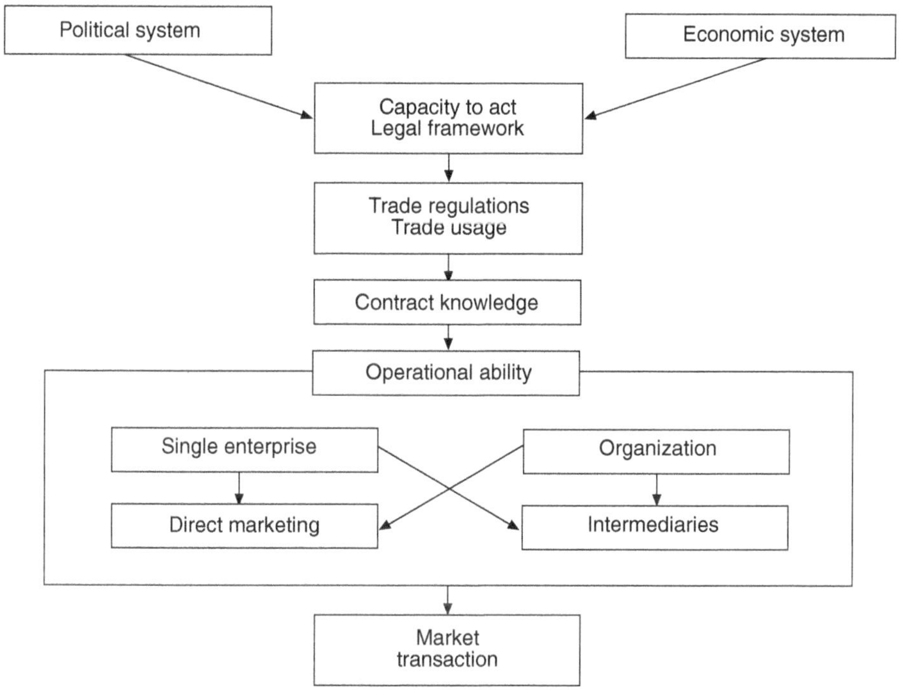

Fig. 1.6. Elements of marketing transaction.

the framework of a larger organization. The forest owner either possesses the knowledge and contacts required to execute the exchange in direct contact with the customer or he/she appoints the necessary intermediaries.

Each element of this complex market framework could be the reason for the non-marketability for a product. The neoclassical concept of 'market' is abstract and theoretical. A market comes into existence only by the development of the relevant system of rules and the participants' competence to act. Thus, frequently one may jump to the conclusion of 'market failure' although the reason is actually human failure, as the operational framework (politicians) or the participants (economic operators) have failed.

1.1.6 Behaviour – ability and willingness to act

As mentioned above, the human factor is an important element in the marketability of goods and services. A lack of marketing of forest products is frequently due to the human factor. From the broad spectrum of the human aspects of marketability, the most important are the following features:

- New markets and new business fields have to be operated by new, adjusted competences and instruments. In general, companies are used to this situation. In the case of forest management and the marketing of recreational and environmental facilities, however, this issue is an unfavourable phenomenon.
- Even if the capacity to act is provided, the willingness to act (motivation) has to be present. This involves overcoming barriers due to personal (professional training/tradition) and organizational (public forests) reasons. This situation is commonly known to companies. However, it has grave effects on forest management and the marketing of RES products, because forest landowners often have other than economic targets.

- The basic element of every market transaction is not complete information, but limited information, underlying preferences and emotions. Therefore, subjectivity overrules objectivity. In a competitive situation, the history of the product often has a greater influence on market success than the product itself. In the ideal case, the two aspects match.

The RES project has no influence on participants' willingness or disposal to act. It rather accepts the basis of private autonomy and with this the right to decide against marketing. However, if there is a willingness and the economic necessity of opening new markets, the RES project can contribute to marketing by expanding the instrumental abilities of the participants. This was a substantial reason for documenting and processing case-studies. Case-studies contribute to the understanding of the instruments offered by their relation to practice and indicate the relevance of models and instruments.

The integration of environmental and recreational facilities in the operational process of an enterprise is assigned great importance in the RES project. In the section on 'Management of multifunctional forests', aspects of potential analysis (PA) as a multifunctional management tool as well as aspects of project management (PM) for RES projects are outlined. Multifunctional management, planning and accounting are integrated in the economic background of the forest companies. Marketing instruments concerning the recreational and environmental facilities of the forest are developed according to the individual case.

The main hypothesis of the RES project can thereby be formulated, as follows:

> With the concepts described above concerning product economics, non-material values, market development and the human factor, the degree of marketability for recreational and environmental goods and services from multifunctional forests can be extended widely.

1.2 Summary and Recommendations

1.2.1 Economic strategies for transformation and product development

The main objective of the RES project is to develop instruments and solutions for a better marketability of recreational and environmental goods and services from forest outputs. Thus the methodological background is strongly related to institutional economics, business planning and marketing. However, the economic literature on the problem of public goods is broad and has a long tradition. In the contribution of Merlo *et al.* (1999), the linkage between the economic literature on public goods and the RES project is shown. The first part of this section is devoted to the characterization of the economic nature of public goods with reference to the environmental and recreational goods and services (ERGS) provided by forestry and the related environment. Theoretical backgrounds are briefly outlined according to public-finance approaches, such as taxation and pricing and the search for optimum provision of public goods. Excludability and rivalry criteria are employed for defining public and private goods.

ERGS are defined as public goods and services provided by forestry and the related environment before any transformation/development takes place. By a process of transformation (e.g. legal status, property rights, planning permissions, contractual agreements, etc.) and/or by product development (provision of complementary/additional goods and services, marketing promotion, changes of existing contracts, etc.), RES products (goods and services provided by forestry) are created.

The different character of the RES case-studies is documented by country. Products are offered both by private and by public estate proprietors. Environmental products are mainly offered on a big scale, whilst recreational products are quite often also found in small areas. Within the last 4 years, the introduction of RES products has increased in all four countries. However, a great proportion of products with a long tradition (before 1980) was identified, particularly in the field of recreation. Various specific conclusions were drawn from the differences of the participating countries. General patterns were identified that could give more empirical evidence for the mostly theoretical discussion of public goods.

The experience is such that both the approaches and means (institutional and market) are often used or needed in order to achieve remuneration of ERGS through RES products. The first institutional approach forms the base for transformation: e.g. new regulations are introduced for using the ERGS. However, the status of the ERGS is not dramatically changed. The core of the development process is given by the application of market-based techniques, taking advantage of the existing complementarity between RES products and ERGS. In fact, in most cases the presence of additional goods and services is necessary to create a market for the ERGS through the RES products.

In other words, what can be firmly stated is that the remuneration of ERGS is often quite indirect, through the development of complementary/additional RES products supported by some institutional changes of ERGS. The possible level of complementarity between ERGS and RES products is variable. There is a high complementarity between the pure environment and the footpaths to gain access to the environment, while it is lower with restaurants and shops. At the limit, when complementarity is very high, the ERGS and the RES products overlap almost completely. However, in almost all cases, the RES products are paid for in the market, and not the ERGS.

The case-studies show that each transformation/development path is different. However, a simplification model applied by Merlo *et al.* (1999, unpublished results) allows the following conclusion.

In general terms, the transformation/development paths can be divided into sequential stages, all represented in the investigated cases. The transformation starts with the creation of a certain excludability that causes some changes in rivalry, due to congestion effects (for shared goods) and to the different intrinsic characteristics of RES products (for both shared goods and private goods). In the following stage, the RES product is further developed (through marketing techniques) in order to respond better to customers' demand. If congestion is creating problems (as sometimes is the case for recreational products), measures to reduce it are implemented. In some cases, other technical or structural modifications to improve service quality are undertaken, without consequences for excludability and rivalry.

In particular, the following selected issues were found to be important to achieve market remuneration of ERGS through the transformation/development into RES products:

- The type of product is particularly important. In general, those products linked to recreation can more easily find market remuneration, while the environment itself is not so easily transformed into a market good.
- The transformation/development of RES products, when the environment component is more important, is usually conditional upon large estates. It is therefore not surprising that environmentally based RES products mainly appear on public land or in various forms of common properties.
- A remarkable number of RES products based on recreation (camp-sites, sports facilities, etc.) do not need a large land base, though they exploit the surrounding forest environment.
- Increasing management costs and other economic pressures often encourage the establishment of RES products. Opportunity exploitation is, however, the driving force supporting transformation/development.
- The main strategy behind development is the 'complementarity' between ERGS and the RES products, and the provision of structures necessary for recreation and access to the environment.
- Failures are not uncommon and can sometimes be due to property rights

violation as well as lack of demand and/or poor management.

- Concerning the relationships between traditional forest products, such as timber and RES products, the case-studies essentially referred to forestry, where competition was not so acute as it could have been in the case of agriculture.

The contribution of Merlo *et al.* (1999, unpublished results) starts with an overview of the main literature with regard to the economics of public goods. A theoretical framework has been developed to systematize the transformation/development activities of the case studies. The main transformation/development paths were documented and divided into sequential stages on the basis of the theory of public goods, on some basic assumptions of the RES project and on an empirical basis. Thereby it was possible to add some more differentiated and empirical-based views to the theory of public goods. Furthermore, the importance of dynamic changes was shown for any kind of transformation.

Mertens and Welcker (1999) develop a special typology for marketing and contracting analysis. They use the degree of non-materiality of a product and the customer integration in the product developing process to group the case-studies. The investigation showed the following results concerning the degree of customer influence on the development of RES products: in 53 cases, it was a mainly autonomic product and, in 36 cases, the product developed in an integrative way, which means that customer participation was necessary to develop the product. Concerning the degree of non-materiality, the systematization showed the following results: material products (four RES cases); material products with additional non-material components (13 RES cases); non-material products with additional material components (21 RES cases); non-material products (59 RES cases). The combination of both criteria (customer participation and non-materiality) was helpful for deriving general statements concerning possible marketing instruments and organization and contract structures.

1.2.2 Marketing tools for the development of RES products

Marketing strategy and potential for RES products

The development of marketing tools for recreational and environmental goods and services and its competent application is most likely the key for marketability. In the contribution of Welcker (1999), marketing tools as described in different marketing textbooks were related to the comprehensive empirical material of case-study interviews. By explorative research, innovations and general marketing theory, a specific marketing theory was developed for recreational and environmental goods and services for multifunctional forest management.

Since every offering company, every product and every product environment are different, this marketing process needs to be strictly orientated towards individual cases. Thus, it is not possible to draw common conclusions about which instruments have an impact on which factors and which combinations always lead to the best results. External influences, product design and the factor 'management and organization' have a determining influence on the success of an RES product. Active marketing was said to be decisive for the success of the product in only a few cases. This initially points to the special significance of product policy in the early stages of the marketing process. It is recommended to take the possibilities of marketing – particularly beyond the stage of product development – into more profound consideration.

RES products are often simple in their structure and, with regard to their demand for special staff, harmonized with the given company conditions. Thus the planning process depends to a great extent on personal factors and special circumstances. The availability of physical production factors as well as staff and know-how increases the prospects of success of a new product. Many RES products are in the early stage of their life cycle. For RES products in the introduction stage, there is the danger that they are not carried further due to insufficient financial benefits. It is recommended not to assess the success of RES products in comparison with traditional products (wood), as most of the products do not become profitable unless they are in their mature phase.

RES products are, to a large extent, traded on new, unsaturated markets. Available competitive products are mostly successful and have often been established in the market for some years. A significant part of RES products competes with alternatives that are free of charge. Successful RES products remain competitive through additional benefits. They are not strongly affected by competition and the partly positive impact is not regarded as highly as the negative impact. It is strongly recommended that public forest proprietors take into consideration the consequences of their action with regard to the offer of private forest owners.

RES products are often developed and offered on the basis of information on customer wishes which is not obtained systematically but presumably partly obtained by chance. Customer orientation with regard to RES products is usually reached by measures of product policy. The design of the facilities offered, additional offers, measures of infrastructural design in the forest, service, flexibility and care in forest management optimize the customer orientation regarding recreational offers. Concerning market segmentation, niche markets for RES products must not be limited too closely. Segmentation should be seen by forest companies as a chance to differentiate themselves from competitors by their special competence concerning their forest, enterprise and products. For a clear positioning of their company, it is important to establish and transmit a succinct corporate identity (CI).

As a basis of strategic thinking, a forest company should have a clear view on the 'portfolio' of its present and potential business units. The potential utilization of attractive business fields through RES products means a lateral diversification for many forest companies. Therefore, the risks and difficulties of the necessary diversification need to be recognized and managed before the necessary investments are actually carried out. The difference between this area and the trade of raw timber as the usual business of a forest company has to be realized.

Marketing tools for RES products

Many examples of using marketing tools successfully are introduced, along with applied examples, by Welcker (1999, unpublished EU Report). Starting with the important aspects for a marketing mix for recreational and environmental goods and services, it can be shown that all these aspects are already used in the case-studies, even if the intensity varies widely. The aim of product policy concerning RES products is, first of all, to turn forest facilities into marketable products.

Product packages, additional services and the creation of a shopping event are important additional benefits for tangible RES products. Innovative benefits can be added to RES recreational products on several levels. The special value of the landscape and an adequate forest management are the basis for the first, fundamental level. On this level, especially, simple recreational products are added value – for example, by building facilities for sports and resting in the forest. On the second level, there are additional organizational offers, with guided tours, events, additional forest facilities and all-inclusive programmes lasting for several days. The success of many complex offers for recreation and accommodation is based on such innovations. RES offers on a third level can be turned into 'luxury events' with an individual atmosphere. The creation of atmosphere has become the most important additional value concerning accommodation offers.

Advantages should result from a horizontal and a vertical distribution structure for all concerned parties. In particular, intermediaries promote RES products because of special advantages from this business. The forest company should take these economic or political advantages into account and use them actively for its own purposes. Thereby, the distribution system becomes an important success factor. Recreational forest products, for example, have positive effects for the regional tourism, on the one hand, and profit from the know-how of horizontal distribution partners from the tourism business. Environment conservation contracts are sometimes arranged by non-commercial organizations. Informal contacts are also important for the success of RES offers.

Prices can be changed flexibly without generating direct costs. The prices of initial offers should be determined under consideration of the company's own production costs, the willingness to pay and the prices of substitutes. Environmental-economic valuation techniques are only partly suitable for finding a price for RES products. Prices need to be orientated towards the chosen marketing strategy. RES products are mainly offered at market-orientated prices. However, there are, for a certain part of the products, purely cost-orientated prices or prices stipulated by the customers. Time-related price distinctions are in most cases supposed to compensate for fluctuations in demand and are therefore an important marketing tool, especially for services. Seasonal prices are an important factor of success for many RES products. It is recommended not to regard the price too much as a reaction of the market. Within the broad field of services – in comparison with mass products such as industrial wood and normal logs – creative possibilities with regard to the degree and the mode of the price increase are much bigger.

The relationship between marketing strategy and pricing can be shown by the extremely high variability of types of payment for RES products. Characteristics of every product influence the optimum type of payment. For material RES products, the calculation is mostly made according to the degree of consumption. It is possible, though, to agree upon basic amounts with institutions. Whereas for directly performed RES, it is particularly the performance duration that serves as the basis of the price, there are entry tickets and licences for utilization, which are especially suitable for permanent offers. Single negotiated RES recreational offers are often remunerated via bundle prices. In the case of offers for an overnight stay, tangible aspects, such as water consumption, are often calculated separately. Concerning integrative contract products, it is preferred to come to voluntary agreements as to type of payment, though donations are possible in this case. For those RES products which are only marketable in a limited way due to considerable legal restrictions, earnings can nevertheless be obtained on a voluntary basis.

Communication is inevitable for the long-term existence of forest companies. At present, advertising for RES products is done mainly through advertisements and leaflets. The expenses of advertising via electronic mass media are covered only in a few cases. Advertising is mainly suited for those RES products supplied as consumer goods. Complex recreational offers in the forest often utilize advertising strategies, because of their economic potential. In general, advertising is not sensible for RES products offered to institutions in the form of a contract. With the sponsoring of products, advertising for a product switches to advertising with the product.

Public relations (PR) are a means of communication preferably utilized by forest companies. However, they are rarely used for targeted sales support of certain RES offers. Public relations by third parties create the impression of high credibility. PR are extremely important for RES products, that were formerly obtained free of charge. The utilization of the internet is an innovative form of public relations, as well as of advertising, for certain RES products. Personal communication is mainly characterized by the direct exchange of impact and reaction between offerer and customer. Such a contact exists with all RES products that have been worked out together with the customer. Especially for RES-contract products, this is an essential basis. Personal contacts with selected customers are often an important factor of success for complex recreational products as well.

Free publicity and outsourcing mean the execution of a positive communication for a product or an enterprise by a third party. The special advantage of free publicity lies in the high credibility and the fact that it is free of charge for the company. Above all, RES recreational products often benefit from free publicity. Regional tourism institutions have a special significance with regard to this.

Brands as a recognizable identification mark of quality are particularly important for non-material products. Sponsored products require a plain name in order to be used by the sponsor. The generally high significance of brand creation as a success factor in contrast to the relatively few indications of their use in RES case-studies. This could indicate that the significance of brands has not yet been recognized by many forest companies.

1.2.3 Contracts and organizations as a basis for market development

One of the basic assumptions of the RES project is that products are not homogeneous entities and static phenomena. Their properties of exchange can be changed by product structures and the contractual and organizational framework. The main objective of the work of Mertens (1999) was to analyse product structures and their impact on contracting and business organization in niche markets for RES of forest enterprises.

The explanatory value of the main approaches of institutional economics, such as property rights theory, principal agent theory and transaction cost theory, for analysing contracts is discussed and transaction cost theory is chosen as a theoretical basis. The marketability of a product depends on the cost of realizing the transaction, such as costs for initiation, agreement negotiation, management of organization and controlling. This part of the study analyses the transaction costs in recreational and environmental goods and services from multifunctional forestry and develops a contract solution. Thereby a contribution to decreasing transaction costs is made. Other important aspects are the organization of the business (e.g. by a middleman) and the risks and liabilities. Several product groups are analysed concerning their cost aspects. Many examples are given for the explanatory relevance of transaction cost theory.

Offering tangible products, such as drinking-water or electricity, forest enterprises have completed individual written contracts with organizations because of the long-term perspective of the exchange process and because of the specific investments of the contract partners. When offering tangible products with additional non-material value to individual users such as Christmas trees, mushrooms, game meat, certified wood, etc., no individual contracts have been completed, as the tangible product part is a traditional product and therefore standardized contracts written down in the Civil Code provide a suitable framework for these exchange processes.

Forest enterprises do not complete detailed written contracts when offering accommodation possibilities, because general regulations in the Civil Code and juridical decisions are detailed enough to secure the exchange process. Furthermore, offering accommodation possibilities is a very common business. Standardized application forms are used to diminish the costs of dealing with numerous individual users. The special non-material value (forest surroundings) which is offered by forest companies is not secured by contract with the tourist but by market competition.

For seminars, the combination of different rights, tangible products and services in a product package makes the recreational and environmental quality of a certain forest area marketable. Forest companies cooperate with business consultants, as in general they do not have contacts with companies that want to train their personnel or have the relevant knowledge. Personal contacts with business consultants help to establish their cooperation and therefore those forest enterprises that do not have these contacts face a market barrier. These contracts are often frame contracts, which define the obligations very generally, because the product itself combines numerous non-material parts with credence qualities that cannot be defined in detail. Forest enterprises cooperate with business consultants or other offerers of seminars, because they do not want to face the liability problems of a tour operator.

Environmental services of forest outputs can only be sold as sponsoring products if combined in a product package containing different additional parts, such as communication, information, choice of a suitable product, etc. By differentiating single, well-described environmental products instead of offering the environmental quality of the forest in total, the forest enterprise has the possibility of offering several sponsoring products. Sponsoring contracts are long-term business relationships with high investments by both contract partners. Thus, the different obligations have to be regulated in a very detailed manner or, if this is not possible, the trust potential of the contract partners has to be very large because of personal contacts.

The possibility of offering use contracts concerning the recreational use of forests depends on the legal situation in each country. If the law requires permission of the forest landowner for certain recreationists in general (as for mountain-biking in Austria), it is easier to offer such use contracts. But an offer of additional value (e.g. the right to mark forest roads to design a network, the offer of insurance or the offer of road maintenance) is almost as important for marketability as the legal situation. Regarding use contracts concerning forest roads, it is very important to define the subject of the contract in detail (with a map as part of the contract) and to regulate use conditions, such as use period, right to put up signs, obligation to remove rubbish, etc. The forest landowner should try to transfer the responsibility for traffic safety to the contract partner as far as possible.

The various possible RES products are sold with different contractual and organizational structures. Numerous RES products dispose of a high degree of non-materiality, adding value and customer integration, and therefore differ greatly from the product timber. Due to these differences, forest enterprises have to face relevant initiation costs before contract completion. Furthermore, Mertens (1999, unpublished EU report) documents that certain characteristics of exchange processes have an impact on contract design and can be very well described and structured in terms of transaction costs theory. It was possible to develop some general suggestions for forest landowner organizations to support the offer of RES products.

Several regulations can be relevant according to the individual situation. Nature conservation law and forest law, as important laws regarding the use of forests, have been relevant in almost all cases. Transaction costs for information about legal circumstances seem to grow the more complex and distinct from normal forestry the offered RES products are and the more the forest enterprise is involved in developing the product. The observation that more than half of the contributors used external know-how when developing their products shows that offering RES products is no normal forestry business and that these information costs are of considerable importance when starting business in these niche markets.

Exchange partners are not arbitrarily exchangeable. They depend to a greater or lesser degree on each other and therefore are interested in instruments that secure contract fulfilment. In the case of these exchange processes, individual contracts with individual security instruments are negotiated. In the case of non-material product components, the customer does not have complete knowledge regarding the relevant product qualities. Whether the customer experiences the promised atmosphere of the Christmas fair or not is subjective and cannot be verified by jurisprudence. Rules that regulate the consequences in case the promises are not fulfilled cannot be designed in this case. Contracts regarding the exchange of products with non-material product components, therefore, have certain tendencies to be incomplete.

The missing physical comparability of RES products with a large amount of non-material parts results in low market transparency (e.g. in the case of the offer of seminars). A high degree of non-materiality of RES products, combined with high internal complexity, means a growing importance of the price as indicator for product quality, as well as references and images. As the offerer cannot present visible products, communication policy becomes much more important (quality of advertising, public relations). It takes considerable time to build up a certain image and customer confidence.

In the case of products with a high degree of non-materiality, it becomes difficult and produces considerable transaction costs to define all obligations of the offerer in a detailed contract. The uncertainty of both contract parties grows the more non-material parts a product implies, and therefore incomplete contracts are used. In this case, security instruments apart from juridical regulations become more and more important (e.g. image, confidence, personal contacts).

Customer integration has consequences for contract initiation. When a forest landowner develops a certain product together with and according to the special wishes of a customer (e.g. sponsoring products), he/she sometimes has difficulties in stating his/her contribution to the product development and in demanding a reasonable price. The more specific the products are, the lower the market transparency becomes. This might result in disadvantages when the forest landowner has to negotiate with a very powerful potential customer (e.g. in the case of nature conservation contracts with a public body or of contract negotiations with a big-drinking water company).

When offering products that can be characterized by the term product business, variable transaction costs to negotiate contract conditions do not arise, because legal regulations serve as contract conditions. In this case, transaction costs for the development of the respective regulations have been produced by the legislator and must be regarded as integral transaction costs for the relevant business. This example demonstrates that state authorities can influence marketability by providing an effective transaction framework.

1.2.4 Management of multifunctional forests

Project management (PM) for RES projects

Once the decision for the introduction of a new RES product is taken, effective and efficient coordination of all persons/institutions involved in the process of project realization is important to secure the success of the project. Plaimer (1999a) applied the methods of PM for RES products. Since he was the project manager himself for the Hebalm Beachball Camp of a forest landowner in Austria, he combined the theoretical and practical aspects of PM in a very target-orientated way. The integration of a project in the business organization could be seen as a success factor. In the course of the RES research, 98 cases were documented which, to a greater or lesser degree, are typical and individual projects. However, in about 60% of all cases, the process of establishing the RES activities was integrated in the daily business and fewer than 40% of all interviewed forest landowners defined a separate project.

What is daily work and when does a project start? For the judgement of a project, specific criteria can be defined. Such criteria are, for example, contents, duration, unusual features, complexity, meaning, risk and the costs of the project. The importance of PM is underlined by the researched case-studies. About a quarter of all respondents declared that organizational aspects would be possible improvements for establishing a new RES product.

For example, along with the development of new products or the introduction of new services, a clear allocation of tasks, the delegation of responsibility (motivation), clear aims and priorities, faster project realization, efficient use of resources, an early recognition of conflict potentials and other aspects have to be dealt with. Plaimer (1999a) combines the methods of

PM with the experience of the 98 case studies. Thereby, the usefulness of PM for large and small projects becomes evident and PM is easy to apply for RES products.

The advantages and disadvantages of different organizational forms are discussed and selection criteria for suitable organizational forms are presented. As shown in the project tasks, 'marketing' and 'contracting' intermediate experts or external project team members are quite frequently integrated in RES projects. Possible organizational forms for the integration of external persons and institutions are demonstrated.

The PM manual represents the simplest form for the arrangement of the project contents. Without large expenditure, projects can be planned in an efficient way and clearly laid out and are easily feasible. The tasks of all persons involved in the project can be coordinated and controlled. An example of such a project handbook summarizes all aspects of PM in a very comprehensive, practical way and makes it easy to set up one's own handbook for any project. Additional essential success factors can be arranged with this instrument: the acceptance and identification of all persons and communication between persons involved.

After the presentation of PM planning tools and their relevance for RES projects, the handbook demonstrates the use of definition of project boundaries and objectives. Examples are given for context analyses that can prevent the project from getting into external conflicts. Time schedules and check-lists are laid out and milestones are determined. The project organization is demonstrated and the needs of resources are defined. This contribution of Plaimer (1999a) on PM will make it easier for future projects to secure the success of a product development.

Potential analysis (PA) – a multifunctional forest management tool

Besides the implementation of new management methods (e.g. PM), it is essential to analyse various infrastructural performances of forests as well as relevant resources related to any RES activity in a more detailed way. The PA could be defined as an investigation of success factors with regard to expected development strategies (prediction). The contribution of Plaimer (1999b) presents related aspects of information management and the tasks and methods of PA. For the assessment of recreational and environmental forest functions, practical procedures of how to identify strengths and weaknesses are exemplified with projects such as a mountain-bike route, acquisition of water resources and the establishment of forest apartments.

For strategic information management, it is helpful to define specific business units. The essential features for a strategic business unit (SBU) are to fulfil a specific market task, to be organizationally independent, to cover a relevant market segment potential and to be relatively independent of the decision-making process in other business units. In comparison with other small and medium-sized companies, the internal information source takes on a greater meaning in forest companies. More than 40% use internal information sources. This circumstance could be explained by still existing forest management plans (forest maps, inventories, other existing plans). A variety of examples of external data sources, such as data banks, are documented. Since RES products often perform in markets that are completely different from those of traditional forest activities (wood production, hunting), outsourcing plays quite a big role.

An example of PA is given for the Austrian Federal Forest Administration (ÖBf AG). It starts with a situation and development description, followed by an internal analysis, and finally determines strategies and activities. For different company sizes environment potentials are discussed more generally. How can strengths and weaknesses be operationally identified? Checklists for recreational and environmental parameters are presented. Within the environmental field, valuation methods are quite often developed for political reasons, while, for product development needs, a complete checklist of possible offers is often a greater help than a complete model with indicators and criteria. However, the field of environmental product development is more sensitive.

The examples demonstrate that every project is different and needs a specific adaptation of theoretical methods. For a mountain-bike route, technical parameters (e.g. length of the course, drainage system, signposting) were documented. On the environment side, an assessment of use conflicts and possible sources of danger was explicitly carried out. In the context of the acquisition of water resources, the legal situation and possible changes were an important part of the strategic information management. On the other hand, the acquisition of internal water resources and quality had to be analysed and potential danger had to be identified. The determining factors for the establishment of forest apartments could be subdivided into general aspects, factors related to forest houses/apartments and factors related to the regional/global market situation. The criteria used for classification of farmhouses in Austria are almost a complete checklist. The three examples give a lot of impressions for practical potential analyses, but specific adaptations have to be made to each project.

Multifunctional forest management planning

The allocation of land-use priorities may be performed either by ranking land uses in accordance with their importance for the planning individual or group or by applying a weighing system, as shown in the section on 'Potential analysis'. A more complex approach is the use of geographical information system (GIS) technology. The contribution of Ottitsch (1999) applies GIS technology to RES planning. By linking objective physical information and subjective goals of the manager/landowner, as well as external influences, potential conflicts are identified. Multipurpose land-use planning is about visualizing the different angles from which a landscape is viewed by different potential users. By combining these different views, it is also possible to assess where different requirements will interfere with each other.

By more closely identifying the spatial distribution of land-use potentials, land-use managers are also able to detect niches in which to allocate specific uses without any production loss for the main production goal.

The use of GIS technology is nowadays widespread in land-use management planning, and methods are discussed widely in a broad range of literature. Thus, only the applied aspects for RES product development and planning are presented in this contribution. Yet the introduction of new planning tools itself will not solve any land-use problems or lead to success in marketing new products or services. Some elements of participation in planning may thus be rather seen as implementing techniques to assess the users' preferences or public relations in order to offer new products.

The project 'Land-use Analysis Achenkirch' was launched to design an integrated land-use concept. Nature conservation, agriculture, protection against disasters (torrents and avalanches), tourism, forestry and hunting were the land-use forms considered in this project. A temporal segregation-based approach was suggested for the improvement of hunting activities in the area, as well as lower conflict potentials between hunting and other land-use activities, specifically tourism and timber harvest. The contribution by Ottitsch (1999) on some applied aspects of GIS technology and the 'Achenkirch' example may increase the interest in using GIS for multifunctional forest planning in the context of RES product development. Concerning the technique itself, it is necessary to refer to the relevant literature.

Business organization and accounting

Accounting provides a framework for evaluating the input and output of economic activities in monetary terms. As Sekot (1999) stated in his contribution: 'Any RES project can and should be regarded as a kind of business undertaken to better achieve the goals of the forest owner.' Accounting tools are thus prerequisites for monitoring and controlling the profitability, as well as the efficiency, of RES projects.

This section first addresses the question of business organization. Roughly two-thirds of the RES case-studies were fully integrated in a forest enterprise. Thirteen projects, however, were run by a separate company and, in 16 cases, the RES business was organized as a cooperative effort of different companies. A formal delimitation of the RES activities from the forestry business may turn out to be necessary on legal grounds or just advisable in terms of, for example, liability or taxation. It is shown how the choice of the type of business unit triggers various consequences in different fields. Depending on the national legislation, such as tax law and trade regulations, respective indicators for a legal obligation to treat the RES activities as a separate business could refer, for instance, to items like the type or size of the business, the regularity of the activities and the real market revenues. It is noted how other aspects (cooperation, distribution, liability, creditworthiness) affect the choice for a RES-specific organization.

In eight out of 13 cases in which RES-based revenues contribute to more than 30% to the forestry income, a substantial investment has been made. The higher the financial risk and the contribution of the RES business to the companies revenues, the more cost-accounting is regarded as an essential activity. In 72% of the cases where no substantial investment was undertaken, cost-accounting was considered as being unnecessary. However, quite a significant number of forest company managers see potential improvements in this field.

The actual use of business accounting in the case-studies is documented. Checklists and calculation examples are presented. The contribution of Sekot (1999) does not aim to explain the instruments and methods of business organization and accounting. For this purpose, it is necessary to refer to the general literature. But the numerous examples and hints for specific aspects of RES business are a highly valuable source of information.

1.2.5 Delimitation of property rights

Economic activities can be described as an exchange of ownership titles. Property rights influence freedom of action and thereby product development, behaviour and market structure. It makes a difference for economic options if a forest estate owner is generally obliged to open his/her forest to the public, as in Germany, or as in the Netherlands, where receiving state subsidies goes along with the obligation to open the forest to the public. This example demonstrates that every analysis of property rights is linked to the national legal situation. Thus, the contribution of Malzburg (1999) focuses on the legal framework in Germany. It is an example for a property-right study in the field of recreational and environmental product development. The special interpretation of the legal situation has to be adapted, but the structure of the analysis can be used for other countries as well. However, the most important aspect – the right of access – has been investigated in all countries of the project partners involved. This part of the project was carried out by the Faculty of Law at the University of Hamburg under the supervision of Ramsauer.

For the first time, the relevant basic rights of the German constitution (GG) have been considered with regard to the frame conditions of legislation in the area of forest law. The most important basic rights in this context are Art. 2 I GG as part of the people's general freedom of action (*allgemeine Handlungsfreiheit*) and Art. 14 GG, defining the property rights of the forest estate owners. This has to be seen in the context of basic rights. It has to be noted that the basic rights are in relation to the state and usually serve as defensive rights (*Abwehrrechte*) of the citizens against interference by the state. Thus, forest visitors may not directly refer to the basic rights when the interference is by private landowners. In this respect, the citizens may only claim that the state has neglected its duty of protection. Accordingly, the state generally has to enable the citizens to practise recreation and mobility. But it is mainly left to the legislator as to how this duty is to be implemented. In this context, the legislator has many options and the actual regulation in the German Forest Law is just one of the possible options within the framework of the German constitution. Recreationists can claim interference in their basic right only if the state has acted insufficiently in a very obvious way, which usually is not the case.

The constitutional guarantee of property is characterized by private benefit of property (*Privatnützigkeit*) and freedom of disposition (*Verfügungsfreiheit*). But this does not mean in general that every matter or every good must be placed at private disposal. Private goods can be subject to administrative regulations and single rights can be excluded from private disposal without infringing the constitutional guarantee of property. As a result, the legislator has to delimit the different opposite interests (private and public ones) in such a way that every concerned position is preserved appropriately. Different reasons and types of compensation are discussed. In the context of the RES project, the following is most interesting. Where revenues had not been achieved by former use, compensation will not be paid. Even though this depends on the type of compensation, it can be generally concluded that the higher the income generated on forest land is, the greater is the price for restricting regulations in the public interest.

After a short description of the main structure of forest law in Germany, the relevance for RES products is analysed for access to forests, restrictions on specific forest uses, afforestation, construction of recreational facilities, restrictions on protected areas and intrusions in nature and landscape. Establishing enclosures for recreational facilities in the forest and the charge of entrance fees also need permission of the forest (or nature conservation) administration, as free recreation is restricted or excluded in particular areas. In general, access free of charge must be tolerated, even where additional services are offered (e.g. a cross-country skiing track)! If these permissions were given more freely when the possibility of free access is available close to the recreational facilities, these services would also be offered more often. This would indeed be in the public interest. Since the early 1970s when the forest law was launched, many things have changed both on the consumer side and on the side of the offerers.

The general right of free access for recreational purposes is not really a problem for marketability. People who are used to walking in the forest are potential clients for forest products and have positive attitudes to the forest. The problem is that the general right of free access can prevent the possibilities of offering products on an economic basis, because even in small spots the owner is not entitled to exclude people from free access. Thus, it should be possible to restrict the general right of free access for recreational purposes in the case of recreational offers as long as they do not infringe other laws and leave enough forests to exercise the general right of free access for recreational purposes.

The right of access describes the different starting-points of the countries for the development of RES projects. Therefore an example was chosen to give a comparison of the legal situation of Austria, the Netherlands and Italy with the German situation. Malzburg (1999) describes the difficulties of such an international comparison of legal systems. A problem in this context is that access to forests is not influenced just by a single legal rule or a single law, but rather, a large number of rules and laws of each country have to be taken into consideration. Furthermore, different kinds of legal interpretation apply in the different countries. Only Germany and Austria provide a right of access to the forest for recreational purposes. Whereas in Austria only walking is admitted, in Germany it is also permitted to ride on horseback and to cycle along forest lanes. In respect of the rights of the landowner offering RES products, an advantage of the Austrian situation may be seen in the fact that it is stated rather clearly by law which conditions have to be fulfilled in order to build up barriers for recreational facilities.

On the contrary, a right of access does not exist in Italy and the Netherlands. To this extent, the rights of forest landowners are less restricted. However, admissibility of RES projects mainly depends on specific provisions referring to particular uses. Public forests in Italy are run by regions or local authorities and are subject to their regulations on access. Usually, regional laws take account of public access to forests and admit (or do not prohibit) particular types of access. Regional laws cannot explicitly admit access to private forests, as this would infringe private property principles laid down in the constitution. Notwithstanding, it is assumed that access is accepted by the owner unless he/she has made use of his/her right to ban other persons from his/her land by enclosures.

The Dutch situation is very specific, as access is mainly determined by subsidies granted by the state. Similar subsidies do not exist in any other country concerned. The main consequence of restricting public access in order to offer RES products is the removal of subsidies, which is not directly a legal problem. The Dutch subsidy system merely does not encourage a landowner to develop RES projects that might conflict with public benefits.

In the area of recreational and environmental services, rather complex and specific products are to be developed which are usually not in conflict with the subjective right of recreationists. However, in some cases, the access right causes problems because it is difficult to exclude unauthorized people, e.g. on contract paths. But, in general, it cannot be concluded that RES products are easier to develop in Italy and the Netherlands, where no right of access is enacted. In respect of RES products, specific provisions imposing restrictions on particular types of forest use by forest, nature conservation and planning law are of major importance. As these provisions have not been investigated in detail in the other countries, a final statement cannot be made.

1.2.6 Public acceptance

Two research projects have been launched to analyse public acceptance. One was carried out in the Netherlands and one in Germany. Based on a study of the literature, Segeren and Van Vliet (1999) developed a conceptual framework for price acceptance studies. The concept has been used in two surveys, one for visitors of the forest areas 'Edese Bos' and 'Berg en Bos' and one for the general public. The research objective was to test the relevant factors for price acceptance. In Germany, a household study was carried out by Welcker and Mantau (1999) to analyse both the factors for price acceptance and the increment of the information basis for marketing strategies.

The research model by Segeren and Van Vliet (1999) made the conceptual framework operational. Personal variables (e.g. sex, age, education) and use variables (e.g. time, distance, expenditure) are the basic variables used in the model. Past use and past paying are important because they determine the accessibility of pre-existing information about the present fee. Furthermore, the price acceptance is influenced by contextual determinants (e.g. information on subsidies and facilities), immediate determinants (e.g. price, quality) and psychological determinants (e.g. values, beliefs, expectations). This conceptual framework has been used to design the questionnaires for on-site and household surveys.

Results show that the attitude towards all the different modes of payment is more negative in Edese Bos compared with Berg en Bos. This was to be expected because experience of paying is less in Edese Bos. Visitors are most negative about paying if they do not get something special in return. Results show that an average of 91% of the respondents in both areas agree or strongly agree that forests should be freely accessible for walking and cycling. An average of 49% of the respondents in both areas agree or strongly agree that a forest owner should be able to ask for money for additional facilities, such as mountain-bike and horse-riding tracks. Another conclusion is that, if the respondents know that their money is spent on additional facilities, their acceptance of paying is higher. Thus a customer-orientated information policy on the activities undertaken will reduce price resistance.

The results show a statistically significant relation between past paying behaviour and the attitude to paying. The results support the statement that 'recent payment at other similar sites will encourage acceptance of first-time payment'. Respondents with experience of pricing show higher support for pricing. Respondents with recent payment experience at similar sites also report higher 'fair' price levels (willingness to pay). In other words, their reference price has changed because of recent payment at other similar sites. From this, the importance of a dynamic approach becomes obvious. The higher the number of successfully realized projects is, the more the prices for RES projects, become 'normal'. For the same reason an assessment of actual marketing problems is indeed no convincing statement for the actually existing marketing chances.

The opinion that the government should pay for recreation in forests and nature is strongly expressed by visitors in both areas and by the general public in the household survey. There is hardly any support for the introduction of the 'user pays principle' for general visits. Public acceptance of the introduction of an entrance fee by private owners is somewhat higher compared with its introduction by public owners. But still most visitors disagree. The majority of the general public (household survey), however, does not mind paying when considering the possibilities for recreation in forests and nature. The majority of the general public, however, would rather pay if the money is used for additional recreational facilities.

The government should make clear what is basic (and thus subsidized) and what is not (and thus possibly subject to pricing). At present, the status of many activities is vague in the Netherlands, even an activity like cycling or the provision of benches. Private owners get a subsidy when the forest is opened to the general public, but there is no subdivision in subsidy levels for different activities and facilities.

The second acceptability study, by Welcker and Mantau (1999), deals with the behaviour of visitors in German forests and their attitude towards financing recreational and environmental goods and services. In a regional joint venture project with the forest administration of Schleswig-Holstein, a 'forest backpack' for forest visitors was developed, including information, experiments, games and other contents that are suitable for increasing the value of a forest visit. An example for market research on new products was included in the household questionnaire.

About two-thirds of the German population over 14 years of age use the forest as a place for recreation at least once a year. Though there are some shifts towards specific sociodemographic segments, it can be determined that forest visits are used for recreation by all parts of the population. On average, the forest is used for recreation purposes for 1–2 h on 30 days of the year. The main motive for these forest visits is 'quiet recreation' on foot. However, the forest is also regularly used for different athletic purposes, for playing (with children), for the purpose of nature studies and for collecting forest fruits by different sociodemographic population groups. There is a high information requirement in the population about the forest, its components and its social importance.

Forests where there is a liability to pay an admission fee are relatively unknown in Germany and therefore (also on account of legal impediments) only poorly distributed within the country. Nevertheless, a basic attitude towards financing specific forest services can be observed within the population, which appears to make marketing of these RES products quite realizable. Although free admission to the forest is important to about 90% of German forest visitors, a predominant proportion agrees with the financial support of the forest owners and states a readiness to pay for specific offers in the forest. It turns out that these individual forest services are respected by the population in a very differentiated manner. Therefore, with regard to their marketing possibilities, they have to be handled in a differentiated manner.

According to the point of view of the German households, specific forest uses by individual groups, such as riders or skiers, should be financed by these users themselves and, if possible, be deducted per use. Financing specific environmental services of the forest is predominantly imposed on the community of taxpayers, which supports market chances of corresponding cultivation/protection contracts with public authorities. The forest owner's duty is mainly seen only as the appropriation of the basic environmental and recreation services of his/her forest.

The huge information requirement of the forest visitors implies the conclusion that forest 'information products' would meet an absorptive market. According to the results of this study, a broad acceptance for specific customer-orientated forest recreation offers can be expected both within the society and by potential customers. The assumption that forest owners would have to produce and finance the entirety of all recreation services of the forest, along with the wood production, can, based on this study, be considered from the start with regard to more intelligent business and forest policy solutions.

In the questionnaire, the forest backpack, which should make the forest visit more attractive due to its various contents, was described as a product example. The households show a tremendous interest in the forest backpack. Of the forest visitors above 14 years of age, 63.5% were principally interested in buying the backpack. Even though the price for such a backpack is much higher than the average willingness to pay, the potential still remains high. The households only had a rough description of the backpack. They would expect to find such a forest backpack at the tourist information office (56.9%), the forest office/ranger station (44.4%) or their hotel/camp-site (21.2%). Families of higher income groups show a higher interest. However, marketing for very specific groups does not seem to be important, because the interest is high in all groups.

Evaluating the first experiences that were made in the sale of the forest backpack, the original idea to pay back receipts from selling the forest backpack to the regional forest landowners does not seem very realizable. In order to design the backpack attractively, one has to invest about 50 €(gross price). However, this is already very close to the limit of the willingness to pay. The executing company can indeed achieve interesting financial benefits, but at present, there is nothing to leave as a margin for the promotion of the forest. On the other hand, the forest backpack has taken over the function as a premium product for the offering company and therefore performs a significant signal effect for the company's service offer. Once again, this example illustrates that the willingness to act and the will to create are particularly important elements of marketability.

Note

1 University of Hamburg (Udo Mantau, Germany (DE), Coordinator), University of Padova (Maurizio Merlo, Italy (IT)) University of Vienna (Walter Sekot, Austria (AU)) and the ALTERRA Research Institute, Wageningen (Kees Van Vliet, the Netherlands (NL)).

References

Ammer, U. and Pröbstl, U. (1991) Freizeit und Natur: Probleme und Lösungsmöglichkeiten einer ökologisch verträglichen Freizeitnutzung, *Hamburg*. Parey, Berlin.
von Hayek, F.A. (1969) *Freiburger Studien*. Gesammelte Aufsätze, Tübingen.
Kotler, P. and Bliemel, F. (1992) *Marketing Management*. Stuttgart, pp. 621–660.
Malzburg, B. (1999) *Delimitation of Property Rights*. EU report for project FAIR/CT95-0743.
Mantau, U. (1993) Die forstliche Produktlücke – Infrastrukturleistungen marktfähig machen. *Allgemeine Forstzeitung* 4(93), 186–190.
Mantau, U. (1994) Produktstrategien für kollektive Umweltgüter – Marktfähigkeit der Infrastrukturleistungen des Waldes. *Zeitschrift für Umweltpolitik und Umweltrecht* 3(94), 305–322.
Mantau, U. (1995) Von der Waldfunktionenlehre zur Waldproduktlehre. *Holz-Zentralblatt* 26.
Mantau, U. (1999) *Introduction and Conclusions of the RES-Project*. EU report for project FAIR/CT95-0743.
Merlo, M., Milocco, E., Virgilietti, P. and Panting, R. (1999) *Economic Strategies for Transformation and Product Development*. EU report for project FAIR/CT95-0743.
Mertens, B. (1999) *Contracts and Organisations as a Basis of Market Development*. EU report for project FAIR/CT95-0743.
Mertens, B. and Welcker, B. (1999) *RES-product Typology for Marketing and Contracting Analysis*. EU report for project FAIR/CT95-0743.
Ollmann, H. (2000) *Holzbilanzen 1997–1998*. Berichte des Instituts für Ökonomie der Bundesforschungsanstalt für Forst- und Holzwirtschaft.
Ottitsch, A. (1999) *Multifunctional Forest Management and Planning*. EU report for project FAIR/CT95-0743.
Plaimer, P. (1999a) *Project Management for RES Projects*. EU report for project FAIR/CT95-0743.
Plaimer, P. (1999b) *Potential Analysis – a Multifunctional Forest Management Tool*. EU report for project FAIR/CT95-0743.
Segeren, A.H.J. and Van Vliet, C.J.M. (1999) *Public Acceptance of Pricing for Recreation in Forests*. EU report for project FAIR/CT95-0743.
Sekot, W. (1999) *Business Organisation and Accounting*. EU report for project FAIR/CT95-0743.
Spieker, H., Mielikäinen, K., Köhl, M. and Skoovsgaard, J. (eds) (1996) *Growth Trends in European Forests – Studies from 12 Countries*. Springer, Heidelberg, 372 pp.
Umweltministerium Baden-Württemberg (1995) *Verwendung von Durchforstungsholz und Altpapier zur Papierherstellung unter Berücksichtigung forstwirtschaftlicher Belange*. Stuttgart, 247 pp.
Welcker, B. (1999) *Marketing Tools for the Development of RES Products*. EU report for project FAIR/CT95-0743.
Welcker, B. and Mantau, U. (1999) *Public Acceptance of RES Products in the Federal Republic of Germany*. EU report for project FAIR/CT95-0743.
Williamson, O. (1990) *Die ökonomischen Institutionen des Kapitalismus*. Tübingen.

2
Case-study Analysis

2.1 The Theoretical Approach in the RES Project

What is theory? For the physicist and the chemist, a theory is a general description of principles from which specific predictions can be deduced by employing concrete data (von Hayek, 1969). The attempt of economic science to strictly follow this kind of theory implies that the object of analysis can be described in the same way as in natural science. A general structural difference between natural sciences and social sciences is the number of influencing factors. In a negotiation, the number of variables is many times higher than in an experiment in which, for example, the vibration of a panel is determined. Furthermore, the determination of the measurement of the specific variables (e.g. preferences, tactical behaviour) is extremely difficult or even impossible. On top of that, the variables change constantly and interact with each other.

Under these circumstances, theory in terms of natural sciences is only possible if the number of influencing factors is reduced to a minimum. This enables a strict logical, theoretical approach (deduction), following the tradition of constructivist rationalism (Descartes, 1596–1650). The problem of variable interaction is solved by the *ceteris paribus* clause. The dynamics of the system is totally neglected, as the conditions of the analysis are only static.

Within the framework of such an understanding of theory, it is almost impossible to develop solutions for new fields of problems. Therefore, the theoretical approaches are normally reduced to the calculation of the problem (market failure). Optimization calculations can indeed only include existing pieces of information. Nevertheless, problem solutions do expect innovative ideas in the form of technical or organizational progress. Instead of the perfect governmental intervention (Pigou) based on a theoretical parameter of comparison (perfect market model), the RES project implies market solutions under real conditions of information and action (see Streit, 1991, pp. 13–17).

Offering products in the field of environmental or recreational services is regarded as a lateral diversification, which means that the products as well as the markets on which they are sold are fairly new for forest companies (Sekot, 1997, p. 7). The main objective of this investigation is to analyse product structures and the institutional framework in niche markets for recreational and environmental services (RES) of forest companies. In this study, products are not interpreted as homogeneous entities and static phenomena. Instead of that, product structures and the organizational and legal framework for an exchange of these products between suppliers and users will be analysed (Mantau, 1997, p. 270, 1998, p. 45).

Within the RES project, two sources for the development of problem solutions are used: case-studies and existing theoretical knowledge from other fields of science (e.g. marketing, law). The case-studies were researched by a more or less standardized questionnaire. It was not the objective of the standardization to determine representative quantitative statements, but to distinguish connecting patterns between specific case-studies.

Therefore, some pattern of marketability could be shown in many cases, which, at the same time, appeared in several similar case-studies. By including theoretical knowledge of specific disciplines (institutional economics, business administration, law, policy), solution possibilities were evaluated hypothetically. The data evaluation is based both on a broad empirical basis (98 case-studies) and on a logical connection to problem solutions of other fields of application.

2.2 The Empirical Basis

When the RES project was started in 1996, the problem of the only vaguely developed marketing of environmental and recreational services was indeed known, but the number of possible case documentations was low and more or less unknown. Therefore, the first task was to identify case examples. Different approaches were carried out to identify case examples. Some cases were known from former studies. The project was promoted, partially by distributing survey sheets, along with several conferences. The survey sheets were also sent to forest enterprises and forest departments. The call for cases was also announced in the *AFZ-Der Wald*, a special forest magazine. Thus, more than 200 possible case examples were identified up to the year 1997.

Based on a theoretically preformulated product system (Mantau, 1995), the case examples were chosen in such a way that the diverse offer possibilities could be broadly determined. At the same time, the companies had to be ready to take part in the time-consuming determination of the case example. Furthermore, the case examples should meet the following premises:[1]

- There is a negotiation situation between the supplier and the demander.
- One or several (private law) contracts are made.
- The project focuses on financial benefits.
- The project develops new, creative products.

Along with these criteria, a clear definition between RES products and the field of contract nature protection was realized. Nevertheless, as a matter of fact, there are some unclear definitions on the borderline between RES products and nature protection contracts. The more a single project meets these four criteria, the more it is regarded as an interesting RES project. However, it was generally accepted that, according to state-specific projects, the same projects can be valued differently with the criteria.

The RES questionnaire was the core of the empirical evaluation. Beyond this sheet, as much documentation material (contracts, prospects, public relations (PR) articles, etc.) as possible was to be determined. At the same time, the questionnaire formed the main methodical basis, as the structure and the environment of RES products was prestructured theoretically. First approaches to work with an unstructured survey sheet were soon refused, as it was not possible to warrant the comparability of the data. Beyond that, the survey must be open to new and innovative matters. This was guaranteed by offering free space for individual notes within several fields of the structured questionnaire. Finally, several test interviews were carried out in every partner state (federal states within the Federal Republic of Germany (FRG)), in order to optimize the questionnaire formally and with regard to the content. All in all, the questionnaire addressed the essential objectives of the investigation:

- To operationalize the objectives with regard to the content.
- To coordinate the empirical survey with the partner states.
- To supply the different task steps of the RES project with empirical material.

- To prepare comparable evaluations for the formulation of a hypothesis.
- To find comparable action patterns between the 98 cases.

As a result of the empirical survey, the coordinator presented a case-study report. This report includes the statistical analysis of all questionnaires and a qualitative summary of each case-study. The qualitative summary included all the relevant information of a case. Thereby, every scientist involved in the project did not have just a complete statistical evaluation, but also a clear and detailed description of the case-studies. The following descriptions of case-studies are to be seen as a summary. The detailed description also contains, amongst other things, the background of the interviewer, confidential information, etc.

An essential objective of the RES project is to present marketing solutions within the general context of product development, marketing, contracting, organization, accounting, the legal context and the economic theory. The questionnaire is the core of these efforts. Within the framework of the case-studies, it supplied a huge amount of information deriving from forestry practice. Based on this, it was possible to elaborate these pieces of information – along with the inclusion of the current state of the art in the literature – for a more general application. Thereby, the RES project has succeeded in presenting the investigation object systematically within its entire range, in forming hypotheses concerning the marketing possibilities of environmental and recreational services and in creating extended mode of action possibilities for the forest enterprises.

2.3 The General Typology of the RES Case-studies

The case-studies investigated are described in the following sections, grouped under the headings of each country (Austria (AU), Germany (DE), Italy (IT) and the Netherlands (NL)). These RES products supplied or linked to forestry have been classified in three broad categories of recreational, environmental and traditional products, depending upon their characteristics and the forest functions involved. Occasionally, it was not possible to assign a case-study to a precise category, because of the complex structure of RES products. Therefore, the mixed categories of environmental/recreational (25 products), traditional/recreational (four products) and traditional/environmental (nine products) have been included in addition to pure recreational (42 products) and pure environmental (16 products), as shown in Table 2.1. Purely traditional forest/agricultural products could not be considered because a recreational/environmental component was required for a traditional product to be regarded as an RES product. Car test courses (AU 14) and military training (AU 04), though obvious market products, have not been considered as traditional

Table 2.1. Typology of RES products.

Type of RES products	Austria	Germany	Italy	Netherlands	Total
One category					
Recreational	9	10	16	7	42
Environmental	6	4	2	5	16
Mixed categories					
Environmental/recreational	3	10	8	4	25
Traditional/recreational	1	3	–	–	4
Traditional/environmental	–	1	4	4	9
Not classifiable	2	–	–	–	2
Total	21	28	29	20	98

products. They fall outside the traditional field of forestry and agriculture because, although they take advantage of the forest environment, the recreational/environmental component, in the strictest sense, is hardly identifiable. These two products have been defined as unclassifiable. The Dutch case of commercial activities in a forest (NL 03) was also difficult to classify but has been considered as recreational/ environmental.

Remarkably, the majority of RES products surveyed in the four countries had a recreational component and fell, at least to a certain extent, in the category of recreational goods and services (71 products out of 98). Case-studies of recreational activities ranged from fishing to skiing and related facilities, such as holiday houses and car parks. These RES products were mainly organized or structured activities, giving added value to the environmental and recreational goods and services (ERGS) and taking advantage of the forest environment. In many cases, it was found that some recreational activities could be carried out both within and outside the forest. This suggests that what is actually sold is not only the activity itself but also the enhancement of that activity due to its location within the forest environment, i.e. complementarity. This is discussed further in Chapter 3, where the relationship between ERGS and RES products is detailed, emphasizing how excludability can be achieved through the provision of additional services. However, it was also found that various other activities and facilities (e.g. guided walks and visitor centres) were dependent on the environmental component provided by the forest. Although they could take place elsewhere, their character would be different; for example, the ambience of a city centre car park is substantially different from that of a car park in a wood.

RES products with an environmental component (50 out of 98) included case-studies that were strictly dependent on the place where they occurred. Sponsorship provides a good example of this, where the real product is an environmental image helping the promotion and/or the business undertaken by the sponsor. Purely environmental products (16 out of 98) were represented mostly by sponsorships and other products closely linked to a certain specific environment image. Environmental products like 'Larch meadows' and 'Nature conservation of specific sites' (see AU 10, 12) were sometimes the objects of management agreements (*Vertragsnaturschutz*). Management agreements represent the first step towards the marketing of ERGS. In fact, they depend upon negotiation, though it remains a 'state pays' rather than 'beneficiary pays' approach.

Traditional products with a recreational/environmental component (13 cases out of 98) were marketable forest products normally produced by forest enterprises (like timber, mushrooms and game), which were transformed/developed to include a certain remuneration for the quality of the environment and for possible recreational activities where they were produced. These products have been considered as RES whenever they have an environmental/recreational component, or at least an image of such, shown by certification procedures or the assignment of particular labels, allowing the consumer to identify the product with the context where it is produced. Therefore, a strict link between the RES products and the environmental/recreational image quality (ERGS) was established.

2.4 Case-studies – Germany

DE01 Christmas fair Waldhütten forest and fish-pond management

A BRIEF DESCRIPTION. A private forest owner (190 ha of woodlands, in addition to fish-ponds) organizes a Christmas fair with some attractive events (camp-fire, mulled wine, soup, pony-rides, sales of Advent decorations and wooden handicrafts) in his forest. His main source of income from the fair, however, is from the sale of Christmas trees, although the Christmas fair establishes a sales-promoting programme by which the traditional product of Christmas trees has added value through the related activities. An additional RES product[2] offered by the enterprise is accommodation in holiday flats. Furthermore, riding holidays, the expansion of the marketing of fish and programmes for staff parties are planned.

PRODUCT DEVELOPMENT/MARKETING. The product is a further development of an already existing product, i.e. the sales of Christmas trees. The forest is utilized as the point of purchase to promote the sales of the trees. The target group of this offer is families. Though none of the competitors offer trees free of charge, there are many competitors on the market offering Christmas trees. Advertising in the market is done to a small extent.

ORGANIZATION OF DISTRIBUTION. The forest enterprise distributes the products directly to the consumer. There are no existing more extensive contracts, and problems with legal regulations have not been indicated.

INTERNAL ORGANIZATION OF THE ENTERPRISE/BOOKKEEPING. At present, financial targets have not been defined explicitly, but the owner is satisfied with the actual situation. The organization of the business is not complicated, as it is integrated in the daily business transactions of the forest enterprise and is not dealt with as a separate branch. Bookkeeping and cost calculation, therefore, are not very developed.

CONTACT WITH AUTHORITIES/RECEPTION. Contact with the local forest authorities was very restricted. The public reacted very positively to the offer.

DE02 Holiday flats Waldhütten forest and fish-pond management

A BRIEF DESCRIPTION. A private forest owner (190 ha) has redesigned plain accommodation for loggers, which are already in his forest, for utilization as holiday flats. The project can be described in terms of an RES product, as the surrounding forest is a decisive factor in the letting of the flats. Besides forest management, the forest owner gets profits from the management of fish-ponds. As additional RES products, the enterprise offers a Christmas fair (see DE01). In addition, holidays for riders, an expansion of the marketing of fish and programmes for staff parties are planned.

PRODUCT DEVELOPMENT/MARKETING. The product was newly introduced in the range of products of the enterprise. Although the houses already existed, it turned out to be very costly to redesign them for the new purpose. The idea was realized by the forest owner himself. The target group consists of urban families, youth organizations, riders and anglers. In this market segment, there is no competition free of charge, but there are numerous competitors offering similar products. Among other competitors, there is a holiday park with ten bungalows nearby. Advertising is used restrictively. Together with the letting of holiday flats, the product range comprises products such as fish, fishing permits or a Christmas fair.

ORGANIZATION OF DISTRIBUTION. The enterprise offers holiday accommodation in cooperation with tourist information. There are no written contracts concerning the rent of the accommodation yet. There have not been difficulties concerning the legal framework, but regulations dealing with industrial law and building laws had to be observed.

INTERNAL ORGANIZATION/BOOKKEEPING. A slight, though probably insignificant, conflict with timber production could occur. The aim is to gain an additional source of income. The break-even point is estimated at around 50% of the capacity.

CONTACTS WITH AUTHORITIES/RECEPTION. There have been only a few contacts with the tourist authorities, which supported the project. There have not been any statements on public acceptance yet.

DE03 Environmental information centre Forest authorities of Boeselager

A BRIEF DESCRIPTION. The so-called Naturerlebnis-Wildwald (nature experience deer-forest) is the business area of a private forest owner with a forest of around 2000 ha. The project consists of an environmental information centre developed from an already existing deer-park founded in 1970. Within the project, there are a wide variety of products offered which are related to the forest, such as guided tours through the forest, information material, events, a restaurant offering specialities in game and a forest shop selling forest products. As additional RES products, the enterprise offers contractual nature preservation, the letting of an area for motocross activities and the letting of a quarry to an archers' club.

PRODUCT DEVELOPMENT/MARKETING. The product's environment is very important for tourism, as the Ruhrgebiet is situated at about 45 min distance. The main target groups for the product are adults, families and teachers with their classes. The product is a further development of an already existing product (deer-park). Information on the forest was enlarged and a variety of additional services were offered. As well-trained personnel are an important factor, adequate measures for advanced education were organized. Demand is not restricted seasonally; however, the number of visitors is highest in autumn. Entrance fees vary according to the day of the week and the time. The offers are complemented by sales promotional offers, such as bad-weather tickets or field-glasses for visitors with annual tickets. At present, the brand name Naturerlebnis-Wildwald has not been registered.

ORGANIZATION OF DISTRIBUTION. The forest enterprise itself offers the product. Contracts exist between the forest enterprise, on the one hand, and the separate enterprise Naturerlebnis Wildwald GbR, on the other. A brewery acting as a sponsor is also involved in the offer. With reference to the product range, a number of regulations had to be observed.

INTERNAL ORGANIZATION/BOOKKEEPING. The financial target, profits, has not been entirely reached yet, 4 years after opening. The profit margin is at 100,000 visitors annually. However, at present, there are only around 80,000 visitors per year. The project was developed separately by a project team. Therefore, bookkeeping and cost calculation are of eminent importance.

CONTACTS WITH AUTHORITIES/RECEPTION. The product development was not promoted by special public programmes or measures on the side of the authorities. From the perspective of local nature preservation clubs and hunters' associations, the product has been welcomed. Previous customers criticized higher prices, but new customers are largely satisfied.

DE04 Organized hunting events [indication of the name not authorized]

A BRIEF DESCRIPTION. For about 30 years, a private forest enterprise has offered an organized hunting event for 1 day in a fenced shooting ground (600 ha), followed by a sophisticated dinner. The customers are hunters who are disposed and willing to spend the fees for a luxurious hunting day in an exclusive setting. In total, approximately 15% of the income of the forest enterprise is derived from the sale of RES products. Around 90% of the income from RES products is, in turn, derived from the offer of hunting permits in the hunting area outside the fenced ground and of the above-mentioned hunting day.

PRODUCT DEVELOPMENT/MARKETING. The know-how needed for the offer is mainly derived from the enterprise's own experiences. The service supplied to the hunting guests is highlighted – ranging from the specially designed descriptions of the different hunting stands for the amelioration of security and hunting success to the conservation of trophies. It is very complicated to get a similar product free of charge, and existing competing offers did not greatly affect the business of the forest enterprise. A marketing strategy is the tailoring of customer-orientated package offers, consisting of a high-quality hunting event, the luxurious surroundings (castle) and the meeting of interesting fellow-hunters.

ORGANIZATION OF DISTRIBUTION. About 10% of the hunting guests on average are procured by hunting intermediaries commissioned by contract. The rest are regular guests or are recommended by them.

INTERNAL ORGANIZATION/BOOKKEEPING. Competition between recreation, timber production and hunting has to be balanced permanently. The offer is recorded in a detailed manner in separate expense accounts in bookkeeping, a system that has turned out to be very useful for current economic supervision. The working effort for the organization of such a hunting event is considered to be very high at 6 man-months per year.

CONTACTS WITH AUTHORITIES/RECEPTION. In the contact with representatives of nature preservation associations and with authorities, it has regularly been necessary to constructively solve the problems and concern arising from fencing and high game density. These efforts have been successful, and the involvement of many neighbouring hunters without their own hunting preserve in the entire hunting organization has turned out to be very helpful in this issue.

DE05 Nature preservation contracts Borough of Gieboldehausen

A BRIEF DESCRIPTION. In this case, the forest owner is a borough owning 100 ha of woodlands, which result in 74% of the income. An area of 8.6 ha with lopped beeches is especially ecologically and culturally significant. The forest owner has negotiated two contracts with the Landscape Protection Administration Göttingen: a contract of lease (DM 250.- annually per ha) and a maintenance contract. The forest owner executes maintenance measures. Every 20 years, thick branches are cut. For this measure, the Landscape Protection Administration pays DM 35.- per tree. Maintenance is carried out by members of the borough, who are not paid for the job by the organization. However, they are allowed to keep the firewood in return.

DEVELOPMENT OF THE PRODUCT/MARKETING. Successful features of the product are the high value of the biotope and the intense contact with the interested Landscape Protection Administration. On the one hand, specially trained personnel are needed as it is difficult and, moreover, dangerous to prune the branches to a height of 2 m above ground. On the other hand, the trained personnel already were at the owner's disposal. There are similar forest stands in the vicinity, though the distinctly visible advantage of this stand is that the forest owners have a personal interest in preserving the stock in this manner. Information on the market potential was supplied by the Landscape Protection Administration. This organization also had the expertise for evaluating the offer. Another feature in price calculation was the alternative income possibility when turning the forest into an oak/beech high-forest stand.

ORGANIZATION OF DISTRIBUTION. The Landscape Protection Administration in this case functions as an intermediary for sales. The project is financed by means supplied by the Niedersächsische Umweltstiftung, which is a public foundation aimed at environmental preservation targets. Rules and regulations for forestry and nature preservation had to be observed.

INTERNAL ORGANIZATION/BOOKKEEPING. The product range is supervised by a forest facility. This is of eminent importance as it describes the special significance of the stock in view of nature preservation and culture. Therefore, there is a large potential of conflict involved as concerns timber production. There is no special bookkeeping for this project.

CONTACTS WITH AUTHORITIES/RECEPTION. There have been contacts with the local forest authorities and with the forest planning board of Lower Saxony in charge of the establishment of forest inventories. Both organizations supported the product because of their interest in nature conservation and because of the income facilities for private forest owners. The Landscape Protection Administration provided the expertise and is responsible for the payment of the financial support.

DE06 Direct marketing of game [indication of the name not authorized]

A BRIEF DESCRIPTION. Two private forest enterprises (in total 2800 ha forest area) owned by two brothers market game from their own forest. The game is processed for bacon, sausages and other products according to old family recipes of the forest owners and, subsequently, is sold at Christmas fairs from a mobile stall. Additionally, they use a portable barbecue and sell freshly grilled game. Around 10% of the forest income is derived from the sale of game. In addition to the product already mentioned, a camp-site in the forest is offered.

DEVELOPMENT OF THE PRODUCT/MARKETING. The specialities in game are traditional products. However, due to the proof of origin, they can reach a price increase of 10%. By the processing of the game (e.g. bacon, sausages), an already existing product was developed further. Moreover, illustrated recipes are supplied free of charge, and the sale on markets can also be regarded as an innovation. The idea of the product was developed by the forest owners themselves. There is some competition, though this does not affect business significantly. The target groups are individuals fond of game and, not so often, companies buying business gifts for their customers. Demand is restricted to 4–6 weeks before Christmas. Only a little advertising is done.

ORGANIZATION OF DISTRIBUTION. The enterprises referred to market their products direct. Sanitary regulations and commercial law have to be observed. One of the brothers has applied for a trade licence to separate this branch of business from the main business area (forest products).

INTERNAL ORGANIZATION/BOOKKEEPING. The target of making profits is reached. The process of product development was organized as a separate project. However, a special project plan did not exist. Four persons are involved in the processing and the sale of game, and the entire working hours total about 2 man-months annually. The processing capacity is around 200 units of red deer and boars, as well as 150 pheasants per year. For the distribution of the game, the brothers invested approximately 200% of the annual income in a cold storage facility and a market stall.

CONTACTS WITH AUTHORITIES/RECEPTION. The health authorities control the observance of the regulations for the sale of food. Regular and initial customers have reacted very positively to the offer and prefer processed game to buying game in units.

DE07 Riding permits [indication of the name not authorized]

BRIEF DESCRIPTION. A private forest owner (5100 ha of woodland) offers bridle-paths through his forest. The forest in question is the largest connected forest area in this region. Riders pay a certain fee for riding permits valid for a year. The forest enterprise takes care of the maintenance of the bridle-paths. Approximately 120 permits are sold annually, and about 60 riders use the bridle-paths without paying the fee. Less than 1% of the income of the forest enterprise is derived from the riding passes. In addition to the product mentioned above, the enterprise offers a butterfly garden.

DEVELOPMENT OF THE PRODUCT/MARKETING. The bridle-paths mentioned are offered in the vicinity of Hamburg, and there are many riders in the surroundings of Hamburg. The product is a further development of an existing product, as riding passes had already been sold for 30 years. The forest enterprise can utilize part of its infrastructure for such an offer. In the region, there are few bridle-paths in the forest, and conflicts often arise between hikers and riders. The separation of these groups of consumers was one of the reasons for the offer of bridle-paths. However, the forest enterprise has problems arising from riders using the paths without paying the fee, as control is difficult and the possible fine is very low (DM 20.-). Competition is low, as this offer cannot be compared with offers in other areas, where riding is free of charge. The forest enterprise does not advertise, and the manager is not interested in expanding the product, due to the problems with fee dodgers.

ORGANIZATION OF DISTRIBUTION. Every rider buying an annual pass has to sign an agreement. According to the manager, the only problem arising was the obligation to ensure traffic safety on the bridle-paths. It is being considered whether this part of the agreement will be subject to change. Other problems with legal regulations did not exist.

INTERNAL ORGANIZATION/BOOKKEEPING. The conflict potential with regard to timber utilization is low. However, there is a higher conflict potential from hunting and other recreationists, such as ramblers. The financial target of reaching a positive profit contribution could be fulfilled. The project is integrated in the daily routine, and there is no separate bookkeeping for this offer. The annual working effort is about 50 h for the project. An amelioration in bookkeeping is planned.

CONTACTS WITH AUTHORITIES/RECEPTION. There have not been any contacts with authorities during the stage of product development. The only contacts with private organizations are with riding clubs buying the riding passes. Customers criticize the product (bad condition of the paths), but accept it.

DE08 Water protection sponsoring in the forest Klimaschutz durch Wald e.V.

BRIEF DESCRIPTION. The Verein Klimaschutz durch Wald e.V. is a non-profit association founded by two foresters. The main aim is the increase in groundwater by turning pine forests into pine–beech mixed forests. Plantations are organized and paid for by sponsors. The organization acquires sponsors and organizes the PR necessary. Forest owners make their forests available to the project and by doing this get the plantations free of charge.

DEVELOPMENT OF THE PRODUCT/MARKETING. The non-exclusive service of the raising of water is advanced to the level of environmental responsibility, and this value is marketed via sponsorships. To reach this aim, the environmental facilities of the forest are supplemented by additional product components, such as information material or the invitation of the press. For tax reasons, the product is financed by donations. It was developed in 1995. The necessary knowledge was partly supplied by the forest authorities. The main target groups are enterprises and public institutions. The organization had to submit a report on the influence of the forest conversion on the groundwater level. Communication is via newspaper articles, radio and television broadcasts and participation in ecology fairs. The name and the logo of the organization can be considered as the brand name for the environmental contribution of the forest owner.

ORGANIZATION OF DISTRIBUTION. The product is distributed in cooperation with forest enterprises. For tax reasons, there are no written sponsoring contracts.

INTERNAL ORGANIZATION/BOOKKEEPING. Non-profit organizations must not pursue financial targets. Therefore, the forest owners receive the afforestation free of charge. The project was developed by a team. Working hours total around 140 h per sponsoring project. Bookkeeping and cost calculation are of low importance.

CONTACTS WITH AUTHORITIES/RECEPTION. Although the product was welcomed by the forest and nature preservation authorities, there were no incentives on the part of these administrations for the development of the product. Experiences with private organizations were varied. One of the organizations sought to put an end to the projects as there is allegedly only a limited sponsoring volume at their disposal. The public and the sponsors welcomed the project.

DE09 Specialized guided tours [indication of the name not authorized]

A BRIEF DESCRIPTION. The private forest (approximately 3200 ha) has been managed sustainably for about 40 years by now. For this reason, a good data basis on the conversion of a traditional method of harvesting, i.e. clear-cutting, of the high-forest stand to sustainable forest management is available. Therefore, other forest owners and foresters visit this enterprise to view and discuss the results. Formerly, such excursions were offered free of charge, but, for 3 years now, the visitors have to pay a fee. Often, a programme including a city tour and cultural activities is organized.

DEVELOPMENT OF THE PRODUCT/MARKETING. The target group of this product is a relatively limited group of specialists to whose needs the excursions are adapted. The fee had to be introduced because of the time-consuming effort and the increasing number of visitors. The idea for the product was developed by the forester in charge, who also supplies the knowledge required. It is very difficult to obtain a similar product free of charge, as the experience of the enterprise in question with sustainable forest management is significantly longer compared with other enterprises in the vicinity. Information on the market potential was derived from discussions with potential customers and from personal experience. The usual and adequate price has to be paid per hour. Package offers consisting of the excursion and additional city tours and cultural activities are intended to promote the product.

ORGANIZATION OF DISTRIBUTION. The enterprise markets the product direct without written contracts or intermediaries.

INTERNAL ORGANIZATION/BOOKKEEPING. Forest management is of great importance for the product, as it supplies the detailed information necessary for the discussion of individual forest management with the visitors. The financial target was to cover the expenses of the excursions, and this aim was more than achieved. The project is integrated in the daily routine, and there is no special bookkeeping done for it. The manager does not intend to expand the offer for reasons of lack of time.

CONTACTS WITH AUTHORITIES/RECEPTION. There have not been relevant contacts to administrations or to private organizations. The product, criticized in the beginning, is now being accepted.

DE10 Seminars for executives Forest-land farmer Joseph Spann

A BRIEF DESCRIPTION. Joseph Spann (silviculturist with a forest of 20 ha), in cooperation with a consultant, offers seminars for executives in his forest. In these seminars, the structures of a forest are compared with the structures of a company and its staff. The programme also includes a lunch and, subsequently, a discussion in the forest office. The product is offered by a private forest owner, who derives approximately 10% of his income from the forest, of which, in turn, 15% is derived from the seminars.

DEVELOPMENT OF THE PRODUCT/MARKETING. The setting of the product in a 300-year-old farmhouse (in which the lunch and the discussion take place) and the rural atmosphere play an important role in the product design. The idea and the development of the product were conceived by the consultant. The product combines the specialized knowledge in forestry of the silviculturist with the management knowledge of the consultant. As the successful features in the marketing of the product, the exchange of information, the special atmosphere of the setting and the fact that the silviculturist is able to present his personal experiences vividly are mentioned. The market situation of such a product is promising. Although guided tours through the forests are offered free of charge in the vicinity, there are no similar seminars for executives. The personal contacts of the consultant with companies have turned out to be very useful for the acquisition of customers. The price of the product varies between DM 200.- and DM 900.- per person, depending on the subject and duration of the seminar. Customers are invited personally or are informed of the offer by former customers.

ORGANIZATION OF DISTRIBUTION. The product is offered in cooperation with the consultant. There are no written contracts. Minor problems arise because of the obligation to ensure traffic safety. In most cases, the participants are insured by their companies, but in some cases they are not.

INTERNAL ORGANIZATION/BOOKKEEPING. A minor conflict arose between the offers of the seminars and hunting. However, the executives in charge are very satisfied with the financial outcome of the project. The development of the offer was explicitly organized as a project, though without a fixed time schedule. As the project is still in its initial phase, there are no special instruments for bookkeeping applied. The offer will be expanded in the future.

CONTACTS WITH AUTHORITIES/RECEPTION. The management in charge did not maintain contacts with administrations concerning the development of the product. However, the forest authorities in retrospect reacted positively because of the promotion on behalf of the forest and the interest of private forest owners. The opinion of the public on the product is not known, but new customers have immediately accepted the offer.

DE11 Survival and wilderness centre Freiherr von Poschinger Forst- und Gutsbetriebe

A BRIEF DESCRIPTION. The private forest enterprise (2400 ha of woodland) is part of an enterprise running an agricultural business with 100 ha, as well as a glass factory. The marketing of the product is organized as a separate branch of the business. The offer consists of outdoor events and seminars for executives. In charge of the product offered are the forest owner, the director of the forest administration and survival experts of the army, who cooperate on a part-time basis. The economic significance of the environment is very high in tourism, as the Bayerischer Wald National Park is in the immediate vicinity. Additional products offered are a deer-park, Christmas trees and guided tours through the forests for tourists.

DEVELOPMENT OF THE PRODUCT/MARKETING. The forest enterprise introduced this product as a new product development. The forest owner indicated the attraction of the scenery, the privately owned compact forest property and the trend to teamwork in enterprises as the important profit factors. The market situation for the product is considered to be favourable, as it is at present unique in the market. Market research of the product is done with the assistance of a student who has written her diploma thesis on the subject. The target groups are mainly companies. The price of such seminars for executives is fixed individually. A variety of promotional measures, such as advertisements in outdoor magazines, articles in journals or the regional newspaper and radio broadcasts, are utilized. A new brand name for the product has been developed, but has not yet been registered.

ORGANIZATION OF DISTRIBUTION. The product is marketed direct. For tax reasons, the offer was organized separately in the framework of a civil corporation. In some cases, there have been written contracts in the framework of the product offer. Sponsors are partly involved.

INTERNAL ORGANIZATION/BOOKKEEPING. The target of reaching a positive coverage during the developing stage has been fulfilled. There have been non-material advantages, such as the establishment of new business contacts and the development of new ideas. As reasons for backlashes, a lack of organization and experience was named. Bookkeeping supplies relevant information on the RES project. Further expansion of the project and the employment of a person in charge are planned for the future.

CONTACTS WITH AUTHORITIES/RECEPTION. There have not been contacts with administrative authorities concerning the development of the product. The product has been accepted immediately by the customers, as well as by the general public.

DE12 Holidays at the forester's Dahn and Schönau forest administrations

A BRIEF DESCRIPTION. The offer concerns a holiday package (including a day-trip and four trips lasting for half a day) consisting of guided tours through the forest with the forester. Transport to different locations of the excursion, guided walks, activities in the forest, such as plantation and timber harvesting, a lunch and a certificate at the end of the event are included in the offer. The package is arranged and marketed by the Dahner Felsenland tourist information office, which is a regional organization.

DEVELOPMENT OF THE PRODUCT/MARKETING. The area, with 70% forest, plays an important role in recreation. Numerous weekend visitors come from cities such as Mannheim, Kaiserslautern and Saarbrücken. When the product was first launched on the market, it was an entirely new concept. However, since then, there have been other suppliers of an almost identical offer. The main characteristics of the newly developed product design are its activity orientation and the fact that the forest itself is the focus of the programme. The product is either offered as a 5-day all-inclusive package or individually for single days. The necessary know-how was supplied by the forest administration and the tourist information office. The exclusivity and the positive image of the forester were mentioned as the profit factors. The market situation was initially considered to be favourable, as there had not been similar products in this region and as the infrastructure is extremely favourable for tourism. The target groups are walkers (40- to 60-year-olds) who are interested in nature and want to gain knowledge of forest economics. The season of highest demand is autumn. For promotion, advertisements, publications, television and radio broadcasts and competitions are utilized. The brand name '300 years in 7 days' has been newly developed, but is not yet registered.

ORGANIZATION OF DISTRIBUTION. The product is offered by the tourist information office, which in turn remunerates the services of the foresters. For this purpose, there exists a written contract between the tourist information office and the forest administration.

INTERNAL ORGANIZATION/BOOKKEEPING. The development of the product had been organized as a separate project. Due to too few participants, profitability could not be obtained, though the non-material targets could be reached. (Supplement in autumn 1999: The financial target of the project has not been reached up to now. For this reason, enforced marketing of the product in cooperation with the Dahner Felsenland tourist information office had to be ended. Subsequent cooperation with a professional travel agency did not lead to the expected financial success either and had to be finished also. However, there is still the possibility for individual groups to buy a guided tour, and this offer is still in demand.)

CONTACTS WITH AUTHORITIES/RECEPTION. Forest administrations and local tourist businesses welcomed the project. The reception of the general public is not known yet, but customers are enthusiastic about the product.

DE13 Christmas fair Alterfrade forest district, forest administration of Hamburg

A BRIEF DESCRIPTION. The Alterfrade forest district (540 ha of woodlands) belongs to the forest administration of Hamburg. Hardwood plantations, which also contain a large percentage of fir species, were established on areas formerly used agriculturally. Part of it is harvested for maintenance reasons and marketed with the label 'Forest product – free of chemicals'. In addition, at Christmas the enterprise offers events for companies and associations that want to give their customers or members a credit slip for cutting their own Christmas tree. The marketing of Christmas trees and decoration material contributes 90% to the income of the forest district.

DEVELOPMENT OF THE PRODUCT/MARKETING. Though the product is widely known, the concept of the production being distinctly based on the ecological aspect and the offer of additional events are new. The forester of the district was responsible for these developments. The profit factors mentioned are the uniqueness of the product, good promotion and support by nature preservation associations. Information on the market potential was derived from discussions with potential customers, the forest district's own experiences, market research and tests. The target group of the 'event package' product are mainly enterprises. When cooperating with companies, special demands are taken into consideration. In many cases, the companies combine the Christmas event with a visit to the neighbouring deer-park. A new brand name (see above) has been developed and registered at the German Patent Office.

ORGANIZATION OF DISTRIBUTION. The forest enterprise markets the product direct to the customers. Furthermore, there are informal contacts with eco-farmers. There are no written contracts. Tax regulations, in particular, have to be observed.

INTERNAL ORGANIZATION/BOOKKEEPING. As the area is run by a forest administration, the overall production target of the forest enterprise is of eminent importance for the offer. The financial targets of making profits were more than fulfilled. Besides, the motivation of the staff is an important target. Emphasis is laid on bookkeeping. Experience shows that the marketing mix could be ameliorated, as well as cost calculation and organization.

CONTACTS WITH AUTHORITIES/RECEPTION. The production of the Christmas trees without using chemicals was especially welcomed by the nature preservation groups. The product was immediately accepted by initial customers and the positive effect of the image of a product made without chemicals was an important point in promoting it.

DE14 Water protection forest Municipal undertakings of Hanover

A BRIEF DESCRIPTION. In Lower Saxony, the so-called *Wasserpfennig* ('water penny', an obligatory additional fee for water) is legally regulated. This *Wasserpfennig* is collected together with the bill for water by the waterworks and has to be utilized for activities in the area of water preservation. The case-study refers to the change of purely coniferous stands to mixed stands of conifers and hardwoods for the purpose of an increase of potable water and its financing via the extra fee. The forest administration of the Hanover municipal undertakings (1500 ha of woodlands) coordinates this measure in the framework of the pilot project named Pilotprojekt Fuhrberger Feld. In this project, forest areas of the municipal undertakings, private forest owners and the state are included.

DEVELOPMENT OF THE PRODUCT/MARKETING. The project idea is new. The director of the forest enterprise of the municipal undertakings was responsible for the development of the product, in which he was supported by the Klimaschutz durch Wald e.V. association. As profit factors, the conversion of the forest free of charge for private forest owners and a general increase in environmental consciousness were indicated. Information on the market potential was gathered by means of a comparison with similar projects established in agricultural areas. Target groups are water consumers, nature preservation groups on the administrative scale and forest authorities. Reports, publications and mutual events were executed to promote the project.

ORGANIZATION OF DISTRIBUTION/CONTRACTING. The forest administration of the municipal undertakings acts as an intermediary for the product and negotiates contracts with the individual forest owners interested in participating in the project. The drinking-water preservation law was mentioned as being very important for establishing the framework.

INTERNAL ORGANIZATION/BOOKKEEPING. The already existing forest institution was very important in the development of the product, but mapping of the drinking-water sources was said to be of at least equal importance. Financial targets were not pursued. The product was developed separately within a project team. Bookkeeping is done, which is considered to be an important issue. In the future, the project is intended to be accompanied by scientific research.

CONTACTS WITH AUTHORITIES/RECEPTION. Forest authorities especially welcomed the fees for environmental facilities of the forest, and the increase of biodiversity because of the forest conversion was highly valued by nature preservation authorities.

DE15 Organized tours through the forest [indication of the name not authorized]

A BRIEF DESCRIPTION. A municipal forest enterprise (3200 ha of woodlands, 25% of the forestry income derived from the marketing of RES products) organizes day-trips in the forest. The offer is always a package offer, including guided walks, overnight stay, food and transport. In some cases, the visit to the forest is combined with visiting a castle or with similar activities in the surroundings. The area is of average importance for tourism. The main target groups of the product are holiday-makers and day-visitors. Supplementary products are Christmas trees, decoration material and hunting permits.

DEVELOPMENT OF THE PRODUCT/MARKETING. The new aspect of the product is the package offer. It was introduced by the director of the forest enterprise. Aspects mentioned as profit factors are, for example, advertisements in the specialized press and mentioning the product on the enterprise's own internet site. The market situation of the product is said to be favourable as it is difficult to get a similar offer free of charge, and as the additional benefit is highly valued by the customers. Information on the market potential was given by potential customers and derived from discussions with the owners of hotels and restaurants. Institutional customers have to be contacted via the respective representative (companies – e.g. shop committee, administrations – e.g. heads of departments). Fixed charges and prices of similar offers were considered when calculating prices.

ORGANIZATION OF DISTRIBUTION. The product is either distributed by the forest enterprise itself with hotels as their cooperative partners or vice versa. There are no written contracts. They have been replaced instead by informal contacts to establish the product.

INTERNAL ORGANIZATION/BOOKKEEPING. A special description of the individual sights chosen for their particular interest was drafted. The financial target was to make profits. This target, though, was not quite reached. The product was developed within the daily routine, and lack of organization, experience and information was indicated as the reason for backlashes. The capacity of the enterprise concerning such an offer is not yet exhausted and, in the future, an amelioration of advertising measures is intended.

CONTACTS WITH AUTHORITIES/RECEPTION. The tourist information office supported the project. Acceptance from the side of the customers can be assumed. A positive effect is the increase in net proceeds, negative effects result from problems with local hotels that do not participate in the project.

DE16 Christmas fair Freiherr von Gravenreuth/Affing

A BRIEF DESCRIPTION. The manager of a private forest (1000 ha woodlands, 95% of the income derived from wood and 1% from RES products) initiated a Christmas fair in the castle-yard of the forest owner and persuaded various organizations in town (such as the youth club and the sports club) to participate. In this way, the sales of Christmas trees could be considerably increased.

DEVELOPMENT OF THE PRODUCT/MARKETING. The offer consists of a traditional product that has been developed further. The favourable acceptance in the market and the motivation of the local organizations are considered to be the aspects determining its success. Part of the income is dedicated to charity. The market situation is considered to be favourable as there are few similar offers in the vicinity and as the city of Augsburg is nearby. Information on the market potential was derived from personal knowledge of the region and from the manager's own experiences. The main target group is families with children. Carriage riding, a 'Punch and Judy' show and a living Christmas scene were set up to fulfil their needs. The product is marketed in combination with other traditional products, such as mead, mulled wine and game. Newspaper advertisements, the project's own publications, radio interviews and invitations to business connections are used for advertising and promotion.

ORGANIZATION OF DISTRIBUTION. The forest enterprise markets the product direct. All participants donate 10% of the turnover to charity. The different suppliers had to apply for permission to sell foodstuffs, and the observance of inspection laws concerning food was controlled by the health authorities.

INTERNAL ORGANIZATION/BOOKKEEPING. There is no competition with other forest uses. The aim of a positive profit contribution was reached, and there have been no backlashes up to the present. The development of the product was integrated in the daily routine. Bookkeeping is considered to be important.

CONTACTS WITH AUTHORITIES/RECEPTION. The experiences with the local forest authorities and the tourist organizations have, up to now, been positive. Customers accepted the offer immediately, and demand is increasing.

DE17 Events in the forest
County forest [indication of the name not authorized]

A BRIEF DESCRIPTION. The county forest refers to a forest of 640 ha commonly owned by several private owners. The forest enterprise owns a sheltered workplace, where in former times carpenters produced fences, benches and other products, but because of the slack financial situation the enterprise could not employ its own carpenters. At present, the director of the enterprise uses the hut as a location for seminars in the woods and for other events. These activities are organized in close cooperation with customers, according to their special needs.

DEVELOPMENT OF THE PRODUCT/MARKETING. The product idea was new to the manager. He derived information about it from the specialized press. The market situation for the product is said to be difficult as similar, though not as well-organized, events are offered by the public forest administration. Information on the market potential was derived from personal experience and discussions with potential customers. The main target groups are enterprises. The product is marketed adapted according to the special demands of these customers. Newspaper advertisements and their own publications were indicated as promotion measures.

ORGANIZATION OF DISTRIBUTION. In most cases, the forest enterprise offers the product on its own behalf. In some cases, however, it cooperates with a supplier of seminars in the vicinity. Good personal business relations have turned out to be advantageous in the development of the cooperation.

INTERNAL ORGANIZATION/BOOKKEEPING. The financial target of reaching a positive profit contribution has been fulfilled relatively well. The product was developed within the daily routine. The product is not considered in bookkeeping.

CONTACTS WITH AUTHORITIES/RECEPTION. Experiences with public institutions (forestry, nature preservation, tourism) have been satisfactory. Favourable features of the offer are the touristic promotion of the region and additional activities. The customers, as well as the general public, welcomed the product and immediately accepted it.

DE18 Outdoor events for enterprises Lahnstein forest administration

A BRIEF DESCRIPTION. A consultant offers a specialized seminar on team development for the executives of companies. The consultant is supplied the service by the federal forest administration (5000 ha), which in turn is remunerated by him. The programme comprises a guided tour in the forest, the utilization of a shooting-box, a carriage ride into the forest and a variety of outdoor activities in the forest. As an additional product, the sponsoring of horticulture is offered.

DEVELOPMENT OF THE PRODUCT/MARKETING. The natural surroundings and the significance of the area for nature preservation play an important role in the product offer. The recreational value of the area is relatively high, and there are numerous sights (castles) nearby. The product is a new development. Besides the guided tour, the forest administration also supplies material for activities in the forest, the right to commercial use of the forest paths and the organizational properties. The consultant is offered a product package. The idea for the development of the product and the know-how required are supplied, on the one hand, by the director of the forest administration and, on the other, by the consultant. The interesting tour, the atmosphere at the shooting-box, high customer orientation and the luxurious accommodation in a first-class hotel were indicated as profit factors. Target groups are larger companies. The consultant is in charge of the negotiations with the customers. Partners for negotiations are, as a rule, in the management of the enterprises or are representatives of the personnel department. The consultant developed a brand name for the marketing of such products, and the prices are calculated as all-inclusive prices.

ORGANIZATION OF DISTRIBUTION. The forest enterprise acts as a contractee of the consultant. There is a basic agreement between the consulting company and the Rhineland-Palatinate State Forest Administration. The cooperation was facilitated by contacts of the director of a forest district, who formerly organized the Camel Trophy. In the development stage of the product, there have been slight problems with nature preservation, but an exception could be made with a special licence.

INTERNAL ORGANIZATION/BOOKKEEPING. Conflicts arise with recreational facilities and with nature preservation, as the area is valuable for nature preservation. The financial target was the making of profit, which has been achieved relatively well. The development of the product was organized in the framework of a project. The capacity of 20 events per year is not exhausted yet and, for ensuring financial success, at least three events per year are required. Bookkeeping is assigned great importance.

CONTACTS WITH AUTHORITIES/RECEPTION. The nature preservation authorities especially underlined the conflict between the significance of the area for nature preservation and the organization of such events. The participants of the events were enthusiastic about the atmosphere, the teamwork and the outdoor exercise. The product is accepted by the general public.

DE19 Ski-tracks at the Taufstein Schotten forest administration

A BRIEF DESCRIPTION. The Schotten forest administration (a public forest of 4500 ha), in cooperation with the Hoher Vogelsberg national park, offers a high-quality ski-track network, consisting of a track for training and competitions and, parallel to it, a track for spectators. For financing the tracks, a sticker for the ski equipment is sold to the users of the tracks under the brand name Loipl. Together with the Loipl, a track map is sold. It is not obligatory for the users to buy the sticker, but the proceeds of the sticker are meant to contribute to the fixed charges of maintenance.

DEVELOPMENT OF THE PRODUCT/MARKETING. Throughout the country, the region is important for the training of cross-country skiers. There are distances of different levels of difficulty and a floodlit track for professional and amateur skiers. In the development of the offer, the aspect of direction of visitors was an important issue. The product is the further development of an already existing ski-track (contract of agreement without remuneration). The forest administration has ameliorated the track by the removal of trees at the forest margin which are exposed to extreme weather conditions. Staff members of the national park ameliorated the signposting. The idea of the fee was supplied by the consumers themselves. For testing purposes, the sticker was sold in hotels, restaurants and boarding-houses. Further promotional measures planned are articles in the regional press and television broadcasts. In the planning stage, also, is a wooden gate at the starting-point. At this gate, a track map will be supplied. The sponsors could be indicated on wooden plaques.

ORGANIZATION OF DISTRIBUTION. The track is offered in cooperation between the forest administration and the national park. However, there are no existing contracts for this cooperation. There is a legal problem as concerns the issue that the national park is not allowed, according to its statutes, to make profits. Theoretically, the proceeds derived from the sale of the sticker, therefore, would have to be deducted from the subsidies for the national park. As this would not ameliorate the financial situation, there is an agreement at present to deduct the proceeds of the sale of the sticker from the costs.

INTERNAL ORGANIZATION/BOOKKEEPING. The forest administration in charge supplies the following additional services: increased care to ensure traffic safety, longer hauling distances with regard to the tracks and seasonal limitation of hauling. The project has been carefully planned, and 3000 stickers have been printed. The aim is to gain a positive profit contribution.

CONTACTS WITH AUTHORITIES/RECEPTION. The reception of the product by the public has been positive in this context. Features of the product are the direction of visitors and the willingness of the users to pay for access to the track offered. The project conforms with the wishes of the regional politicians. As concerns the wooden gate, a conversation has been held with the nature preservation authorities, and the relevant licence has been applied for.

DE20 Sponsoring of recreational facilities Kassel forest administration

A BRIEF DESCRIPTION. The Kassel forest administration owns a total forest area of 8300 ha, of which 3600 ha are owned by the state. PR, environmental education, nature preservation and recreation play an important role for the forest administration on the edge of the city of Kassel. Recreational facilities and educational forest paths are sponsored by local enterprises and individual sponsors. In this way, the respective sponsor is considered to be the customer of the forest administration. The product in this case is the sponsoring project referred to. In this matter, the forest administration cooperates closely with the Habichtswald nature park and the joint board in the area of Kassel. Thus, educational forest paths, wooden bridges and benches and signposts have been financed so far. Furthermore, the forest administration offers a forest camp-site and guided tours.

DEVELOPMENT OF THE PRODUCT/MARKETING. The area is an important recreational area for the vicinity. Due to the costs, new sources of financing for the recreational facilities were searched for. Sponsoring in this case has turned out to be a very effective means to get the costs of recreational facilities refunded without charging the individual recreationist. The marketing of the product is based on direct contacts with potential sponsors. Price fixing is entirely orientated towards any costs arising.

ORGANIZATION OF DISTRIBUTION. The contractual agreements are negotiated individually. Contracting is a separate field of work and also separately organized. There are no legal problems.

INTERNAL ORGANIZATION/BOOKKEEPING. The forest administration plays a certain role in the product offer. In addition, there exists a map of forest functions. The financial target of a positive profit contribution has been more than fulfilled. However, the amelioration of the image has been valued even higher. The RES project as such is organized separately. Bookkeeping is not considered important in this context.

CONTACT WITH AUTHORITIES/RECEPTION. All of the administrative and private organizations contacted in this matter showed a positive attitude towards the product. Cost reduction for the public economy, the amelioration of recreational facilities and the increasing attractivity of the region have been enumerated as important aspects. The product has been welcomed by the general public, as well as by the sponsors so far. Demand is increasing.

DE21 A track for motor sports
City of Schlüchtern

A BRIEF DESCRIPTION. A municipal forest administration (1250 ha of forest) leases a smaller forest area to a local motor sports club for establishing of a track for motor sports. The public forest administration supports the offer, though proceeds are inconsiderable. In addition, open-air camping, barbecue and fishing permits are offered by the forest administration of Schlüchtern.

DEVELOPMENT OF THE PRODUCT/MARKETING. The reputation of Schlüchtern for motor sports has increased throughout the country due to the construction of the motor-sports track. Retailers and businesses dealing with tourism profit from this development. The development of the product is based on the needs of the local club for motor sports and the city of Schlüchtern wanting to promote this club. The initial contract between the parties was negotiated in 1972, and this contract was renewed in 1997. There are further tracks in the region, but because of high demand there is no real competition. For the loan for use of the forest area, the club only pays a nominal rent. The club in turn markets the track to individual customers.

ORGANIZATION OF DISTRIBUTION. Especially over issues of liability and landscape preservation, legal problems arose. They have been regulated in detail in the contract.

INTERNAL ORGANIZATION/BOOKKEEPING. The regular forest economy is impaired only to a minor extent, though there are considerable conflicts with nature preservation and recreation. The city mainly pursued non-material targets with the leasing of the area, which have been reached. In addition, the city gets DM 150.- rent per year.

CONTACT WITH AUTHORITIES/RECEPTION. From the point of view of nature preservation, the project is increasingly criticized. In particular, ecological damage and a negative attitude towards motor sports are feared. The users of the area are extremely satisfied, and the authorities involved have, in most cases, remained neutral, as they consider such a sports facility to be necessary. They also consider the aspect of direction of visitors, and they respect the city's right of ownership.

DE22 Sponsoring afforestation
Forstbetriebsgemeinschaft at the Fulda forest administration

A BRIEF DESCRIPTION. The owner of a small private forest in the forest administration of Fulda executed an initial afforestation project sponsored by a local car seller. The planning and execution of the afforestation were by the Fulda forest administration, within whose competence, as the local forest administration, is also the advising of private forest owners.

DEVELOPMENT OF THE PRODUCT/MARKETING. The idea of the product was developed by the sponsor himself. The Ford retailer wanted to sponsor a forest for PR reasons, taking the pattern from the Ford promotion campaign, and in this matter referred to the local forest authorities. They in turn passed the offer to two farmers willing to execute afforestation. Finally, the product was executed in cooperation with one of the two farmers. Thus, the marketing of the sponsored product was based on direct contact between the forest administration and sponsor. The price was entirely fixed from a cost-orientated viewpoint.

ORGANIZATION OF DISTRIBUTION/CONTRACTS. The forest administration and the FBG at the Fulda forest administration in this case acted as intermediaries for the sponsored product. There have been no written contracts and, according to the interviewee, there have been no legal problems either.

ANALYSIS OF THE POTENTIAL/BOOKKEEPING. The main aim of the project, as viewed by the forest administration, was to get the afforestation for the farmer on the best possible terms. Financing on the basis of sponsorship offers the advantage that the farmer maintains his own land usage with grants. Furthermore, public expenses are lowered.

CONTACT WITH AUTHORITIES/RECEPTION. The afforestation was carried out in the presence of the local inhabitants and of invited guests. Subsequently, there was catering, so that a party developed which the sponsor could utilize for promotional aims. Press coverage of the event was very positive.

DE23 Contract concerning cycling paths Nuremberg forest administration

A BRIEF DESCRIPTION. The Nuremberg national forest administration (11,300 ha) grants a contract of agreement to the city of Nuremberg concerning cycling paths. The city itself designs and maintains the paths. Liability and obligation to ensure traffic safety are also passed on to the city. Due to the contractual assignment of maintenance, liability and traffic safety, the city does not have to remunerate utilization. The forest administration, in addition, offers further contracts for productive use, such as facilities for traffic control, a golf club and a zoo, sportsgrounds, a greyhound race track and cross-country running.

DEVELOPMENT OF THE PRODUCT/MARKETING. The product is a recreational product marketed in a distinctly urban region. The forest areas of the forest administration surround the city of Nuremberg in a semicircle, which explains the intensive usage of the forest by recreationists. The first contracts of this kind were negotiated in 1963. An important aspect is avoiding conflicts, by separation of hikers, cyclists and riders. In the development stage of the product, the successful cycling concept of the city of Erlangen served as a model. Though, there is a broad range of cycling paths for which neither city nor consumers have to pay, the advantage of the cycling paths in the Nürnberger Reichswald is the good condition of the paths, their being in the vicinity of the city and the extremely good maps supplied, together with the appropriate signposts.

ORGANIZATION OF DISTRIBUTION. The city of Nuremberg is the contractee of the forest administration and offers the cycling paths to its citizens free of charge. The recreational value of the forest in this case precedes other forest functions (map of forest functions).

INTERNAL ORGANIZATION/BOOKKEEPING. The financial target of making profits in this case was reached. With the contractual transfer of the maintenance of the roads and the obligation to ensure traffic safety to the city, there are no further costs arising for the forest administration.

CONTACT WITH AUTHORITIES/RECEPTION. All of the authorities and private organizations contacted expressed their positive attitude towards the project. Features mentioned were an increase in the attraction of the region, the control of recreational traffic, support of an ecologically conscious means of transportation and an increase in security on the new paths. The offer was greatly welcomed by the public and the customers alike.

DE24 Utilization of paths by a riding-school Dinkelsbühl forest administration

A BRIEF DESCRIPTION. A farmer offers holidays for riders on his estate. In addition, he uses 950 ha of the surrounding forest area of the Dinkelsbühl national forest administration intensively and the entire area of 13,524 ha extensively. The commercial offer of holidays for riders is a recreational product mainly directed to families with children. The forest paths had already had to be repaired several times because of the damage caused by the horses' hoofs. For this reason, the forest administration has issued a contract for the commercial utilization of the roads in compensation. In addition, the Dinkelsbühl forest administration offers guided tours.

DEVELOPMENT OF THE PRODUCT/MARKETING. The product was mainly developed to restrict utilization to certain forest roads and to separate riders from other forest visitors. The price was calculated on the basis of a rent for the paths and of maintenance costs (depending on the extent of the roads). The riding-school could probably use the forest roads of private forest owners free of charge, but they prefer those of the national forest administration because they are in the vicinity of the riding-school and because of the beauty of the scenery and the good condition of the paths. At first, the riding-school was not willing to pay for access and could only be convinced after several periods of negotiation. The riding-school tried to evade the reproach of using the paths for commercial purposes by the foundation of a non-profit organization, but failed.

ORGANIZATION OF DISTRIBUTION. There is a contract between the riding-school and the national forest administration concerning the utilization of forest roads. The riding-school has to undertake the repair of damage to the roads. However, the obligation to ensure traffic safety rests with the forest administration. Independently of the riding-school, organized riding events have to be paid for, while individual amateur riders may use the paths free of charge.

INTERNAL ORGANIZATION/BOOKKEEPING. The product was not developed for financial purposes. However, important non-material targets were pursued, i.e. the decrease of conflicts with other recreationists and the increase in motivation of their own staff. As a positive side-effect, it was mentioned that, by directing the riders, hunting has been made possible again and thus the damage caused by game is decreased. In the future, the price for the utilization of the forest lanes will be raised.

CONTACTS WITH AUTHORITIES/RECEPTION. All of the public and private organizations had a positive attitude towards the project. Positive aspects mentioned were increasing the attractiveness of the region and the direction of visitors. Only the riding-school opposed the project in the beginning. However, it finally gained a profit by reports in the media.

DE25 Permission for downhill skiing Schliersee forest administration

A BRIEF DESCRIPTION. The forest administration (total size of the area 10,600 ha) has negotiated a contract of agreement with the community on the utilization of certain areas for downhill skiing (DM 0.04–0.10 m^{-2}, first contract). Part of the area had to be cleared before establishing the ski-run. In addition to the rent, the forest enterprise receives 2–3% of the turnover of the companies running the ski-lift (second contract). Furthermore, the forest administration also offers areas and buildings for film crews, as well as hunting facilities.

DEVELOPMENT OF THE PRODUCT/MARKETING. The product is mainly used by families and youths from Munich. Normally, they only stay for a day. In winter, there is no timber harvesting because of skiing. The product is a further development of contracts already existing since 1950. As there is no possibility of opening up new areas for skiing, the operators of the ski-lift are dependent on the areas of the forest administrations. There are agreements between the different administrations on the offer of ski-runs, and negotiations have been facilitated by sample contracts. Further expansion of ski tourism is not possible due to a provincial decision, dated 5 June 1984, referring to the preservation of alpine forests: 'Clearings in the alpine forest for new recreational facilities (for example for winter sports) or infrastructure on principle are not to be allowed any more.'

ORGANIZATION OF DISTRIBUTION. The forest administration has passed on the obligation to ensure traffic safety by several written contracts to the operators of the ski-lift and to the community. The marketing of the product is done via the two contracts described. There have not been any legal problems indicated.

INTERNAL ORGANIZATION/BOOKKEEPING. By the regional separation of recreational and preservation areas, the forest administration tries to avoid existing conflicts between the two branches. The target of product development was also making profits. In the future, prices for the utilization of the area will be increased.

CONTACTS WITH AUTHORITIES/RECEPTION. The interview partner did not comment on this issue.

DE26 Letting of ski-tracks Sankt Martin forest administration

A BRIEF DESCRIPTION. The national forest administration (18,500 ha of forest area) also runs areas in Austria (600 ha). On the latter areas, the forest enterprise has negotiated a contract of assignment for use with a community to install a cross-country ski-track. Maximum rent is DM 0.50 m^{-1} and the rent is graded according to the loss of use. The community offers the track to tourists free of charge. Furthermore, the forest administration offers paths for cycling, ski-tracks and hunting permits.

DEVELOPMENT OF THE PRODUCT/MARKETING. The region is part of the northern Alps, and the product is especially used by families on holiday and sportsmen of the region. By dispensing with timber production in winter, the needs of the skiers are taken into account. The product is a further development of existing forms of usage. It was mainly developed for reasons of directing visitors. A similar project in the region failed because of the resistance of private forest owners. A further aim was to transfer the maintenance costs to the tenants. Price fixing is also settled between the forest administrations of the region on the offer of such utilizations. Sample contracts and the mutual target of regulating ski tourism have facilitated negotiations with the community. The main advantage of the track mentioned is the beauty of the landscape.

ORGANIZATION OF DISTRIBUTION. The obligation to ensure traffic safety has been transferred to the community by the above-mentioned assignment for use. Furthermore, the community is responsible for signposts, maintenance and rubbish collection.

INTERNAL ORGANIZATION/BOOKKEEPING. With the development of the product, no financial targets have been pursued. Important non-material targets were the direction of visitors and the preservation of ecologically valuable areas. While the community wants to expand the offer, the forest administration is not interested in a further expansion of ski tourism. At the moment, the product is gaining profits as the forest administration does not have to cover costs.

CONTACTS WITH AUTHORITIES/RECEPTION. Tourist organizations have supported the product as it increases the attraction of the region for tourism. Farmers opposed the project, as they were concerned about a limitation of cattle-grazing in this area. The contracts are not known in public or to the individual cross-country skiers. Demand has increased recently, as downhill skiing is increasingly expensive, while cross-country skiing is still free of charge.

DE27 Youth hostel in the forest Sellhorn forest administration

A BRIEF DESCRIPTION. The Forstamt Sellhorn national forest administration (approximately 5000 ha of woodlands) runs a youth hostel in the forest for pupils. The school classes mostly stay for 2 weeks. During their stay, they have a fixed schedule, consisting of theoretical lessons held indoors and of guided tours and working-hours outdoors. The programme has already existed for 50 years. Accommodation is free of charge for pupils, as they participate in everyday tasks. The youth hostel at Sellhorn is one of 11 youth hostels in the forest in Lower Saxony.

DEVELOPMENT OF THE PRODUCT/MARKETING. The Lüneburger Heide nature preservation area surrounds the youth hostel and, with its heathlands, is very important for recreational purposes. The idea for the youth hostel was developed by national forest administrations and ministries of education and science, which, after the war wanted to dedicate a worthwhile task to young people. While at first working in the forests was foregrounded, later on plain lodgings in the forest were added, which were further developed into the present youth hostels. The main target groups are school classes. Spreading the word seems to be sufficient as product advertising. Demand at present is triple the capacity of the youth hostel.

ORGANIZATION OF DISTRIBUTION. There is close cooperation with the Schutzgemeinschaft Deutscher Wald, which receives the bookings for the stays and assigns accommodation. Many legal regulations have to be observed when running a youth hostel in the forest.

INTERNAL ORGANIZATION/BOOKKEEPING. All of the facilities of the forest are important in this area. The running of the youth hostel, however, is not in conflict with the other forest facilities. Non-material targets are more important than financial targets. Therefore, cost calculation and bookkeeping provide only little information. Nevertheless, the youth hostel needs its own management.

CONTACTS WITH AUTHORITIES/RECEPTION. The attitude of the public and of the different authorities towards the youth hostel is very positive. In total, the commitment to the environmental education of young people is highly valued, but also economic side-effects on public transport, stores, etc., is evaluated positively. The youth hostel is very popular with its visitors – an attitude that is also influenced by low prices.

DE28 Riding and hunting trips [indication of the name not authorized]

A BRIEF DESCRIPTION. The owner of a large private forest (5500 ha of woodlands) offers four comfortable holiday flats in combination with hunting, riding and fishing facilities. A large percentage of the customers are hunters from the surrounding cities. The product is very successful, as it gains considerable profits. Important profit factors are the location of the forest area, the name and reputation of the forest owner, the equipment and the ambience of the flats, the considerable stock of game without any fencing and the know-how and the motivation of the management. Around 25% of the total proceeds gained from forestry result from RES products.

DEVELOPMENT OF THE PRODUCT/MARKETING. The product area is considerably important for nature preservation and recreation. The product in its combination of accommodation and riding, hunting or fishing permissions is new to the forest owner. He invested a considerable amount in the development of the holiday flats to establish a standard above average. The product is based on his own ideas, and its realization did not lead to legal difficulties. Due to the outstanding quality, competition is very low. The product is a typical example of a successful product mix of the individual offers mentioned above. The price is calculated according to accruing costs and in comparison with similar offers. Demand exceeds supply.

ORGANIZATION OF DISTRIBUTION. The product is marketed directly to the customers.

INTERNAL ORGANIZATION/BOOKKEEPING. For the development of the product, maps of the forest facilities could be utilized. The main target of profitability was reached, to an even higher degree than planned. The setting up of the project within the daily routine caused no problems. Financing was done by the owner's own resources.

CONTACTS WITH AUTHORITIES/RECEPTION. A conflict arises with the forest administration in charge, which regularly reduces the high shooting rates applied for by the forest owner. In general, citizens often regard a project such as hunting very critically. The customers are highly satisfied and regularly come again.

2.5 Case-studies – Italy

IT01 Mushroom-picking permits
Consorzio Comunalie Parmensi

A BRIEF DESCRIPTION. Mushrooms are a traditional forest product and establish an important source of income for the inhabitants of the Val di Taro. Business with permits for the gathering of mushrooms began in 1964 and has continued to the present, despite some legal changes.

DEVELOPMENT OF THE PRODUCT/MARKETING. The significance of the area is low in tourism, as well as in nature preservation. The product is derived from the ideas of forest owners, and legal regulations allow them to issue permits for the gathering of mushrooms. Such permits are also sold by neighbouring communities. However, there is only slight competition with the offer of the Consorzio Comunalie Parmensi, as theirs are the best mushroom areas to be found. Demand is restricted to the mushroom season. The permits are also sold in pubs. In areas with an FGO certificate (EU standard 208/92), the price is around 30% higher.

ORGANIZATION OF DISTRIBUTION. The consortium has negotiated a contract with the Comunita Montana, which allows the issue of permits. There is no restriction of forest economy by legal regulation or the marketing of RES products.

INTERNAL ORGANIZATION/BOOKKEEPING. A certain part of the forest is dedicated to producing mushrooms, and there is competition between the transformation of brushwood to high forest and the marketing of mushrooms. In the case of managing the forest as brushwood, mushroom production is complementary to timber production. The business target is to supply the community with a source of income to be able to finance social issues. There are problems with former inhabitants who have moved and now gather mushrooms without permission. Bookkeeping and cost calculation are considered not to be important. There are problems with the controlling of permits.

CONTACTS WITH AUTHORITIES/RECEPTION. Conflicts arose with the regional forest authority and with the Comunita Montana. In the customers' and the general public's concern, it took a certain time before the ticket system was accepted. Furthermore, it was necessary to introduce a price reduction for the local inhabitants.

IT02 Mushrooms with FGO certificate
Consorzio Comunalie Parmensi

A BRIEF DESCRIPTION. The high quality and the good reputation of the Borgotaro mushroom in recent years has led to unfair competition, as mushrooms from other areas were sold with falsified marks of origin. In order to protect the Borgotaro mushroom from mislabelling, an FGO certificate according to European Union (EU) standard 208/92 was developed. So far, however, the consortium in charge of the marketing of the mushroom is not yet working.

DEVELOPMENT OF THE PRODUCT/MARKETING. The product is a traditional product with strong relations to the local surroundings. The development of the product began in 1995, but, because of technical problems with certifying, the consortium is not working yet. Under licence of the FGO certificate, there is now the possibility to definitely mark the Borgotaro mushroom and to prevent the misuse of the mark of origin. The product idea was developed by the forest owners themselves. There are no competitors or complementary products, and there was no market research.

ORGANIZATION OF DISTRIBUTION. A consortium was formed for the marketing of the product (gathering of the mushrooms, packaging, distribution). Business law applies to the marketing of the product.

INTERNAL ORGANIZATION/BOOKKEEPING. A certain part of the forest facilities is dedicated to mushroom production. There is competition between the transformation of brushwood into high forest and the marketing of mushrooms. In cases where the forest is managed as brushwood, mushroom production is complementary to timber production. The targets of the project are higher profits, amelioration of the image of the valley and local development.

CONTACTS WITH AUTHORITIES/RECEPTION. Relations with local authorities are difficult. At present, there have been no other experiences.

IT03 Letting of an area for a golf-course Cansiglio

A BRIEF DESCRIPTION. A golf-course is offered in a forest area, which is composed of national and regional forest areas. (The questionnaire is answered from the viewpoint of the operator of the golf-course and not from the perspective of the forest enterprise.) Other offers in this region are cross-country and downhill skiing, guided tours, mountain-biking, agrotourism, historical gardens and regional agricultural products.

DEVELOPMENT OF THE PRODUCT/MARKETING. The area is very well equipped with recreational facilities, and the surrounding forest is managed as a nature preservation park. The golf-course was developed to cover the demand of tourists interested in such an offer. The area is fenced in and, in winter, also serves as an area for cross-country skiing. The golf-course was already established in 1958. Target groups are mainly families, young people and other tourists of the region. Brochures, adverts and information in the local press on national and international contests are used for promotional activities.

ORGANIZATION OF DISTRIBUTION. There is a contract of lease between the operator of the golf-course and the regional forest administration. When establishing the golf ranges, all of the measures of moving soil and clearing had to be permitted by the forest authorities. Furthermore, former grazing rights had to be cancelled, and the holders of these rights received compensation.

INTERNAL ORGANIZATION/BOOKKEEPING. The targets of the golf club are the promotion of the sport, a balance between expenses and income and local development. The golf-course is financed by internal means, though certain contests are additionally sponsored. Bookkeeping is said to be less relevant; however, cost calculation is done.

CONTACTS WITH AUTHORITIES/RECEPTION. Forest authorities and golf organizations have supported the project, while environmental preservation organizations (especially the World Wildlife Fund (WWF)) were against it. The project is welcomed by the users and the general public.

IT04 Letting of a car-park area La Regiole (mutual forest ownership)

A BRIEF DESCRIPTION. Val Visidende is a very popular recreational area and, on weekends, is overcrowded with tourists. At the beginning of the 1990s, the forest enterprise therefore decided to offer a chargeable car-park at the entrance of the valley and to prohibit parking free of charge outside the car-park. Subsequently, there have been great problems with the acceptance of these measures.

DEVELOPMENT OF THE PRODUCT/MARKETING. The valley is situated in a landscape of outstanding beauty. There is only one entrance to the valley. The forest enterprise decided to offer two car-parks to limit parking in free nature. The local community is in charge of the control, and unauthorized parking is fined. The complete closure of the access road to the valley in 1995 failed, as it is a public road. Although there is a car-park free of charge outside the valley, the chargeable car-parks are frequented as they are closer to the forest and to the picnic sites. Standard fees are charged for parking.

ORGANIZATION OF DISTRIBUTION. The forest enterprise itself offers the product. An insurance contract against damage and thefts has been negotiated. The forest enterprise had to apply for permission to be allowed to charge fees.

INTERNAL ORGANIZATION/BOOKKEEPING. Maps of the forest enterprise have been used for establishing the car-park. The main targets of the offer were the regulation of unauthorized parking and the making of profits. Bookkeeping and cost calculation are considered to be of minor importance. The establishment of the car-parks was organized as a separate business branch. Lack of experience and information is indicated as the reason for backlashes during the stage of project development.

CONTACTS WITH AUTHORITIES/RECEPTION. There have been conflicts with the community because of the introduction of a fee. The acceptance by the tourists has changed over the years. This was mainly dependent on the organization and quality of the offer.

IT05 Offer of ski-tracks and downhill skiing Cansiglio

A BRIEF DESCRIPTION. In a forest area consisting of a national forest and a regional forest (similar to a rural borough), cross-country skiing and ski-tracks are offered. The area is extremely well suited for cross-country skiing and the vicinity to the Venetian plain is an important feature determining success. (The questionnaire in this case was also partly filled in from the viewpoint of the operator of the ski-lift.)

DEVELOPMENT OF THE PRODUCT/MARKETING. The area is well equipped with recreational facilities and the surrounding forest is managed as a nature preserve. During the development of the product, the utilization of the ski-track free of charge was prohibited in 1995 and the payment of the fee was controlled. There are only a few possibilities for substitution in the region. Main target groups are schools and ski clubs. In most cases, an all-inclusive package offer with ski school and hotel accommodation is offered. Advertising is done through advertisements in newspapers, TV commercials and brochures.

ORGANIZATION OF DISTRIBUTION. Several contracts for productive use with the national and regional forest authorities regulate the loan of the area for downhill skiing, lifts and tracks. For the installation of the ski-lifts, a building permit had to be applied for.

INTERNAL ORGANIZATION/BOOKKEEPING. Conflicts between the offer and the timber production in this region were not indicated. The project was financed internally. It was not developed within a special project plan, and bookkeeping and cost calculation are considered to be of minor importance.

CONTACTS WITH AUTHORITIES/RECEPTION. There have been no problems with forest administrations or other authorities. Their attitude towards the offer remained neutral. Nature preservation organizations have a critical view towards the installation and the operation of ski-lifts. The customers in general are satisfied with the offer and the price.

IT06 Admission to Nevegal forest
Nevegal

A BRIEF DESCRIPTION. Nevegal is an important recreational area, in winter as well as in summer, in the vicinity of the Venetian plain. A smaller woodland area, which belongs to the organization for the promotion of tourism of the southern province of Belluno, is equipped with a number of tourist facilities. The facilities are run by tenants, and a contribution is demanded for access to the area.

DEVELOPMENT OF THE PRODUCT/MARKETING. The area is of great importance for recreation, as well as for tourism. The product was initially offered in 1996, and its success to a high degree is due to the multitude of different facilities (bar, restaurant, hiking paths, etc.). Furthermore, its geographical location and the beauty of the scenery are important. The diversity of the offers in a limited space is also unique in this region. Demand is limited to the summer season. All-inclusive package offers are sold to groups of recreationists, and the entrance fee is L. 2000 per person. No advertising is done.

ORGANIZATION OF DISTRIBUTION. There are contracts with a main tenant, who in turn has several different contracts with subleaseholders.

INTERNAL ORGANIZATION/BOOKKEEPING. The offer is very plainly organized. The forest owner and the leaseholders share the financing. Bookkeeping and cost calculation are considered to be of minor importance.

CONTACTS WITH AUTHORITIES/RECEPTION. There have been good experiences with recreational organizations. These customers are very satisfied with the product. However, there have been problems with some of the former users of the area, especially with the local inhabitants.

IT07 Sponsoring of museums
Cansiglio (public forest)

A BRIEF DESCRIPTION. The ethnographical and ecological museums are situated in the hilly region of the Cansiglio, and they are surrounded by forests. Both are run by the national forest administration. No entrance fees are charged. Management costs are partly financed by a cooperative sponsorship. In the area, furthermore, are various other offers, such as hiking, horse-riding, cross-country skiing, a golf club and guided tours (see also IT03).

DEVELOPMENT OF THE PRODUCT/MARKETING. The equipment of the area with recreational facilities is very good, and the surrounding forest is managed as a nature preserve. The museums were opened in 1964 and 1984 and, since this time, have been managed by the national forest administration. In 1996, COOP decided to sponsor a special diorama. The sponsorship officially is with the community in the mountains, as the national forest administration is not allowed to make profits. Similar products are offered in the Venetian region, but they are at a considerable distance. Information on the market potential in this case is based on the administration's own experience. Target groups for the museum are visitors on day-trips and schools. The offer is promoted by brochures. Furthermore, during schooltime there are lectures by forest officers.

ORGANIZATION OF DISTRIBUTION. There are no contracts concerning the sponsorship. The contractor is the community, as the national forest administration may not directly gain profits.

INTERNAL ORGANIZATION/BOOKKEEPING. The management of the museums is integrated in the daily routine of the national forest administration. The main targets are an increase in environmental consciousness, PR and motivation of the administration's own staff. The project was partly financed by the administration's own means and partly by the sponsorship mentioned above. Bookkeeping and cost calculation do not exist.

CONTACTS WITH AUTHORITIES/RECEPTION. Experiences with authorities, especially with the local community, were generally positive. The community acts as an intermediary for the project. Furthermore, there are contacts with an association of mountain guides, who use the museum as an additional excursion visit. Experiences with the general public and the customers are good, and the acceptance of the museums as an offer for ameliorating the comprehension of the natural surroundings and the history of the area is very high.

IT08 Mushroom-picking permits Cansiglio

A BRIEF DESCRIPTION. In the region of Cansiglio, the gathering of mushrooms has always been forbidden. However, in 1994, a regional law allowed the local authorities to sell permits for gathering mushrooms. In the example referred to, the local forest administration offers the permits directly for their area. The introduction of gathering permits mainly aims at controlling the gathering of mushrooms. As the area is not widely known for its mushrooms, the gatherers are living in the vicinity.

DEVELOPMENT OF THE PRODUCT/MARKETING. The equipment of the area with recreational facilities is very good, and the surrounding forest is managed as a natural preservation area. The product was developed to hinder illegal mushroom-gathering. Since the introduction of gathering permits, there has been a reduction in the number of fines due to illegal mushroom-gathering. The observance of the regulations is controlled by the staff. The main customers are private mushroom-gatherers. Demand is limited to the mushroom season. The price differs according to daily, weekly or monthly licences. Free licences are given to local inhabitants and to a mycological organization.

ORGANIZATION OF DISTRIBUTION. The licences are issued by the office of the regional forest authority. There are no restrictions of forest management.

INTERNAL ORGANIZATION/BOOKKEEPING. The forest facility is not relevant for the product offer, and no conflict with timber production was indicated. Besides the target of making profits, the main target was to regulate the gathering of mushrooms. Issuing of permits turned out to be more successful than prohibiting gathering entirely. It is stipulated by law that 70% of the proceeds gained by the sale of gathering permits have to be reinvested into the forest economy. The bookkeeping is exclusively for tax purposes.

CONTACTS WITH AUTHORITIES/RECEPTION. There are no further experiences with administrative bodies. The customers, as well as the general public, reacted positively to the offer.

IT09 Cross-country ski-tracks Val Gares

A BRIEF DESCRIPTION. In agreement with the community, a local ski club offers cross-country ski-tracks. The beauty of the landscape, the certainty of snowfall in the area and the good condition of the tracks contribute to the success of the offer. The area disposes of a favourable tourist infrastructure.

DEVELOPMENT OF THE PRODUCT/MARKETING. Formerly, the tracks could be used free of charge. However, parts of the tracks could not be used after heavy snowfalls as the tracks were not cleared with a snowplough. The planning of the regional development, as well as of the development of the city, had to be changed for the offer of chargeable tracks. Thus, private owners are prohibited to fence in their properties. Instead, they have to tolerate the establishment of the tracks. Staff control the authorization of the cross-country skiers. Though there are facilities for cross-country skiing in the region free of charge, the quality of this offer is higher. Target groups are the local ski school and individual skiers. Furthermore, the track is let for contests. In summer, mountain bikes can be hired. The price mainly covers current expenses. Local inhabitants may use the track free of charge. Brochures and signposts along the road are used for promotion.

ORGANIZATION OF DISTRIBUTION. There is a contractual agreement between the town and the ski club, whereby the town supplies the machines (e.g. snowploughs, snowmobiles), while the ski club is in charge of the maintenance of the track.

INTERNAL ORGANIZATION/BOOKKEEPING. The ski club is a non-profit organization and is not allowed to retain profits. The establishment of the track was financed by the community. There is neither bookkeeping nor cost calculation.

CONTACTS WITH AUTHORITIES/RECEPTION. Generally, the contacts with authorities, as well as with the general public and the users, were positive. The track is well prepared, which is the reason why this track draws more users than other tracks.

IT10 Guided tours through the national park Casentinesi

A BRIEF DESCRIPTION. The national park of Casentinesi disposes of a multitude of natural and cultural-historically interesting objects. An organization of guides offers tours through the park. This organization employs guides from the local community and is supported by the administration of the park aiming at promoting rural development.

DEVELOPMENT OF THE PRODUCT/MARKETING. The touristic infrastructure of the area is very good – especially the equipment with hiking paths and information facilities. In former times, the tours were offered by the national forest administration free of charge, but on a smaller area. At present, the tours are offered for the entire area of the park. The product was introduced in 1995. Profit factors are the professionality of the guides, the rising demand for environmental education and the sales promotion of the administration of the national park. The park administration still offers guided tours free of charge, but on a limited area. The market for guided tours was investigated according to the number of visitors of the national park, supplied by the administration. Special attention is given to schools and hiking groups (all-inclusive package offers). The price is fixed by the park administration, and the size of the group may not exceed 20 persons. Advertising is done in specialized magazines and on the internet.

ORGANIZATION OF DISTRIBUTION. There are agreements between the park administration and the organization of guides.

INTERNAL ORGANIZATION/BOOKKEEPING. The organization is a non-profit organization and therefore does not pursue financial targets. Non-material targets are to support sustainable development according to the criteria stipulated at the conference in Rio and an amelioration of the understanding of environmental contexts, especially by young people. The park authorities consider cost calculation and bookkeeping to be very important instruments for the administration of the park.

CONTACTS WITH AUTHORITIES/RECEPTION. The project was welcomed, supported and partly financed by the authorities. Acceptance by the general public and the consumers of the offer is also favourable.

IT11 A camp-site Stelvio national park

A BRIEF DESCRIPTION. A camp-site, formerly run by the community, is now offered by a private contractee. It is situated in the centre of a national park and is only accessible for groups. Due to the transfer to the private operator, improved control of usage has been obtained. In the valley involved, the touristic infrastructure is only moderately developed. This conforms with the targets of the park administration, which wants to develop considerable tourism in this area. In general, the area is rich in rare and valuable plants and animal species.

DEVELOPMENT OF THE PRODUCT/MARKETING. In former times, the camp-site could be used free of charge. It was badly organized, and the site was partly overcrowded. After renovating the site and prohibiting other camping facilities in the community, the site is now open to groups (schools, scouts) and sometimes to caravans. There is no other camp-site in the valley. Customers who already know the site often book again 1 year in advance. The tourist office supplies information to new customers.

ORGANIZATION OF DISTRIBUTION. The contract between the community and the contractee lasts for a duration of 5 years. No legal difficulties were indicated.

INTERNAL ORGANIZATION/BOOKKEEPING. The target of this offer is making profits, and its development was organized as an individual project. The project was financed by regional subsidies. Bookkeeping only serves tax purposes. In total, the project is evaluated positively, although a larger number of public facilities of this sort would be necessary.

CONTACTS WITH AUTHORITIES/RECEPTION. The contacts with the forest administration and other authorities were positive. The users of the camp-site welcome the offer, especially since the sanitary facilities have been ameliorated. The local community is not altogether in favour of the offer, as they are not allowed to let camping facilities privately any more.

IT12 Letting of a car-park area San Cassiano

A BRIEF DESCRIPTION. On the advice of the city of San Cassiano, the private owner of an already existing car park introduced a fee for parking. After a month, the former users of the car park had complained so intensely to the city and the tourist information office that the city itself ordered the owner to withdraw the fees.

DEVELOPMENT OF THE PRODUCT/MARKETING. The touristic infrastructure of the area is good, and the landscape is characterized by forests and meadows for animal grazing. Before the introduction of the fee, the site was used for relatively disordered parking. The idea of introducing the fee was initially to ameliorate the equipment of the parking lot with the proceeds received from the fees. The fees were introduced in the summer of 1996, and a keeper controlled access to the car park. As there are many parking spaces free of charge in this area, the fees were calculated solely according to costs.

ORGANIZATION OF DISTRIBUTION. There was no special organization of distribution. However, the forest owner had to apply to the city for permission to charge.

INTERNAL ORGANIZATION/BOOKKEEPING. The main target was to gain a net profit in order to reinvest it in the amelioration of the car park. An important non-material target was the regulation of unauthorized utilization.

CONTACTS WITH AUTHORITIES/RECEPTION. Cooperation with officials was at first positive. After the increase in complaints, however, the city ordered the end of the project. Former users were not disposed to pay for a facility which they formerly had free of charge.

IT13 Certificate of sustainability Val di Fiemme

A BRIEF DESCRIPTION. The Magnifica Comunita di Val di Fiemme joint forest enterprise is very productive, already has a mark of origin for its timber and at present is in the process of certification. The certification refers to the forest enterprise, as well as to the sawmill belonging to it.

DEVELOPMENT OF THE PRODUCT/MARKETING. The forest in question consists of very productive forest stands managed according to sustainability. Certification is done along the lines of the Forest Stewardship Council (FSC), and the forest enterprise was urged to certify by some of its large customers. While the former certificate only referred to the origin, the FSC certificate is to certify sustainable management. In 1997, it referred to a current project. The market situation (in 1997) was considered to be favourable due to an entire lack of competition by other certified enterprises in the region. A market advantage via the certificate is intended to be gained, especially by customers such as the producers of picture frames in countries with a high environmental consciousness. An increase in price due to certification is not expected.

ORGANIZATION OF DISTRIBUTION. There are no contracts concerning the organization of the distribution. Important regulations to be observed are the rules of the FSC.

INTERNAL ORGANIZATION/BOOKKEEPING. The area is run by a forest authority, which supplied data for very important certification. Financing of the project was made by the enterprise's own means. Bookkeeping is said to be important, though cost calculation is considered of minor importance.

CONTACTS WITH AUTHORITIES/RECEPTION. There was close cooperation between the authorities in charge and the enterprise that submitted the application for FSC certification. At this stage of the project, there has been little experience with nature preservation organizations, the general public and environmentally conscious consumers.

IT14 Cross-country ski-track Nevegal

A BRIEF DESCRIPTION. A ski club runs a cross-country ski-track and allows access to skiers against payment of a fee. The city is involved in the project, as it owns part of the area that is utilized by the track and as it supports the ski club. The cross-country ski-track can be easily reached from different cities.

DEVELOPMENT OF THE PRODUCT/MARKETING. The touristic infrastructure of the area is very favourable, and families, as the main target group of the ski-track, can also use a sledge road. The track was already established in 1972 on private property, with the permission of the owners, and mainly followed hiking paths. After the community had become owner of part of the area, it proposed the ski club should charge fees for the utilization. Subse-quently, free access was limited, and tickets were introduced in 1995. Staff controlled the authorization of usage. In the region, there are several chargeable offers of this sort. However, they are not considered to be considerable competition, as the demand for ski-tracks is rising. The main target groups of the product are ski clubs and families. School classes can use the track free of charge. Prices are calculated according to maintenance costs and the prices of comparable offers. The local community should actually pay the fee as well, but most people use the track free of charge.

ORGANIZATION OF DISTRIBUTION. At the moment, there are no existing contracts. However, a contract will be completed with the city in the near future. The track is officially approved by the ski federation.

INTERNAL ORGANIZATION/BOOKKEEPING. From the legal point of view, the sports club, as a non-profit organization, is not allowed to gain profits. Therefore, the coverage of costs and the promotion of sports activities among young people are the main targets. In 1985, the university championships took place on the track and, on this occasion, most of the infrastructural facilities were built, with the support of the university sports clubs and the city. Bookkeeping only serves tax purposes.

CONTACTS WITH AUTHORITIES/RECEPTION. In general, contacts with authorities were positive. The same is true for contacts with nature preservation organizations. Even users who had formerly used the track free of charge are of the opinion that it is necessary to pay for the good service. This is especially the case as the activities of the ski club promote tourism in the community.

IT15 Cross-country ski-track Falcalde

A BRIEF DESCRIPTION. The product is a ski-track, offered by an association of hotel owners. In 1996/97, the community offered the track free of charge, as there was a dispute between the association and the representatives of the community. In the following season (1997/98), the community reintroduced fees.

DEVELOPMENT OF THE PRODUCT/MARKETING. The touristic infrastructure of the area is very good. In particular, the facilities of downhill skiing are outstandingly favourable. At first, utilization of the track was free of charge, but the fee had to be introduced because of rising costs and an increase in demand. Although authorization is controlled by the personnel, the track is used without payment by about 40% of the users. The fee was introduced by the association of hotel owners in 1992. The main target group is families with children and beginners. The track is also partly used by ski clubs and schools. It is an additional offer in a touristically well-developed area. There are daily, weekly, monthly and season tickets. The utilization of the facility is free of charge for the local community. A leaflet distributed by the local tourist office is used for promotion.

ORGANIZATION OF DISTRIBUTION. There are no written contracts. However, there is a resolution of the district council stipulating that the association of hotel owners is in charge of the maintenance of the track, while the community supplies the necessary machines (e.g. snowploughs, snowmobiles). The owners have tacitly consented to the utilization of their properties without the existence of a formal contract.

INTERNAL ORGANIZATION/BOOKKEEPING. The association of hotel owners wants to gain a positive profit contribution from this offer and, most of all, to ameliorate the attractiveness of the community. Due to the conflict in 1996/97 and the take-over of the management by the city, damage to its reputation is feared. Bookkeeping and cost calculation are little developed and only serve tax purposes. In general, the project is evaluated positively.

CONTACTS WITH AUTHORITIES/RECEPTION. The contacts with the forest administration and the ski club were positive. Recreationist associations and nature preservation organizations were slightly more reserved. The users of the track are very satisfied and disposed to pay for the service, as they are of the opinion that, by this means, the maintenance of the track can be guaranteed. In the preceding year, despite the utilization of the track free of charge, many users complained about the lack of maintenance.

IT16 Car parks and picnic sites Barcis

A BRIEF DESCRIPTION. Lake Barcis is a place of popular resort for weekend travellers who come from nearby cities to the countryside. In previous years, the car parks in the area have been ameliorated, and picnic sites have been provided with playgrounds for children.

DEVELOPMENT OF THE PRODUCT/MARKETING. The touristic infrastructure is developing since a small camp-site has opened and since there is a walking path around the lake. The area is very valuable because of its landscape. The entirely unspoilt area is situated in the vicinity of a regional park and a nature preservation area. In 1982, the work for amelioration of the area (public lavatories, picnic tables, a playground) had begun and in 1988 a fee was introduced. Staff for controlling access was employed on weekends between 11 a.m. and 3 p.m. There are no similar offers in this area. The main target group is families. The place is most frequented between June and October. Prices are calculated according to the cleaning costs arising. The ticket gives car drivers access to several car-park areas beside the lake.

ORGANIZATION OF DISTRIBUTION. There is no special organization for distribution. There is only a contract with an enterprise concerning the cleaning. Building law had to be observed in the design of the area.

INTERNAL ORGANIZATION/BOOKKEEPING. The main targets of the project are to avoid unauthorized parking and to create new jobs in the region (a cooperative of local citizens is in charge of the cleaning). The project was financed by the community and the regional association on the basis of capital budgeting. The project is considered to be positive from the viewpoint of the people in charge. Ameliorations are planned concerning the service offered.

CONTACTS WITH AUTHORITIES/RECEPTION. There have been positive contacts with the regional association. In the beginning, the introduction of the fee was criticized by the public and users alike. After some years, however, the tickets were accepted. High-quality service has contributed to this development.

IT17 Access to the forest Val Cimoliana

A BRIEF DESCRIPTION. Val Cimoliana is a valley with a beautiful landscape in Friulane park. The road leading through the valley is an important access road to upper mountain huts and especially to a ledge of rocks, which is well known as a visitor attraction. The community charges a toll for using the road.

DEVELOPMENT OF THE PRODUCT/MARKETING. The landscape of this area is of outstanding beauty and well equipped with hiking paths. Fees for access have been introduced recently. However, in contrast with former times, the utilization of car parks and picnic sites along the road is now free of charge. The charge of fees was enabled by a resolution of the district council. The fee was introduced in 1993 on the advice of the park authority. The supply of car-park areas and picnic sites free of charge is considered to be an important factor of success. At the moment, on the one hand, the valley of Cimoliana is the only valley with an access fee. On the other hand, the other valleys in the area are at present declared to be protected by the nature preservation law. A fee is only charged on Saturdays and Sundays between 15 June and 15 September and every day from 1 to 20 August. The fee is calculated on the basis of maintenance costs, and the utilization of the road is free of charge for the communities situated in the park.

ORGANIZATION OF DISTRIBUTION. Authorized utilization of the road is controlled by personnel employed by the community, though paid by the regional park.

INTERNAL ORGANIZATION/BOOKKEEPING. Financial targets were not to the fore in the project, and the project was intentionally developed by a separate project team. In general, the project was considered to be positive.

CONTACTS WITH AUTHORITIES/RECEPTION. The contacts with the regional authorities are positive. However, the attitude of the general public has often changed over the years.

IT18 Riding-school Val Canzoi

A BRIEF DESCRIPTION. A riding-school offers guided riding tours in the valley of Val Canzoi, which is part of the Dolomiti Bellunesi national park. The riding-school utilizes the paths free of charge, and the forest owners do not share in the profit of the riding-school.

DEVELOPMENT OF THE PRODUCT/MARKETING. Due to limited touristic facilities, demand is mostly restricted to day-visits. The riding-school utilizes the private forest areas free of charge. The product was introduced in 1990. Similar offers do not exist in this area. The main target groups are young visitors fond of horse-riding. The offer is partly combined with agrotouristic offers. Further customers are an institution caring for disabled people and sometimes companies organizing staff outings. The price is calculated from current expenses and is fixed according to information on the market price of similar offers. Advertising is done with leaflets distributed via the local tourist information office.

ORGANIZATION OF DISTRIBUTION. There are no contracts with the property owners. The owner of the riding-school mentions regulations concerning the keeping of animals as important legal framework conditions.

INTERNAL ORGANIZATION/BOOKKEEPING. Financial targets are to the fore in the project, though bookkeeping and cost calculation are considered to be of minor importance.

CONTACTS WITH AUTHORITIES/RECEPTION. In general, experiences with authorities, nature preservation organizations and the general public were positive.

IT19 Mushroom-picking permits Val di Fiemme

A BRIEF DESCRIPTION. In the Val di Fiemme, the gathering of mushrooms is an important additional event for tourists. In cooperation with the regional association, all of the communities sell uniform permits for mushroom-gathering. These permits have often been subject to changes over the years. In the meantime, they can be acquired simply by remittance on the postal account of the regional association. The tourists are satisfied with the offer.

DEVELOPMENT OF THE PRODUCT/MARKETING. In 1991, the gathering of mushrooms was restricted by regional law to 2 kg day^{-1} for non-residents. Before the passing of this law, the latter had to pay a tax to be allowed to gather mushrooms. By an amendment of the law in 1991, permits for gathering mushrooms are obligatory for all mushroom-gatherers who live outside the province. At first, there were substantial complaints, but now the fees are accepted. The main target groups are tourists and professional mushroom-gatherers. Previously, the permits could be acquired at tourist information centres, from local authorities and at special outlets. From 1996 onwards, sales were made by mail order. Prices vary for daily, weekly and monthly permits. Inhabitants of the province can get the permits free of charge. Leaflets and posters are used for promotional purposes.

ORGANIZATION OF DISTRIBUTION. The individual communities of the valley and the regional associations have formed an association enabling the users to buy a permit valid for the entire valley. Furthermore, the permit can also be acquired on Sundays at the tourist information office of the valley, which participates in the profits. In this way, the acquisition of the permit is considerably facilitated.

INTERNAL ORGANIZATION/BOOKKEEPING. In the offer, financial targets are not to the fore. Among the non-material targets, regulation of the mushroom gathering is the most important. Cost calculation is considered to be very important for the distribution of the profits among the participating communities.

CONTACTS WITH AUTHORITIES/RECEPTION. Cooperation among the different regional authorities and with the suppliers is indicated to be positive. As to contacts with private organizations, experience has been positive with clubs of mushroom-gatherers but negative with an association of hotel owners. In the view of the latter, the permit is too expensive. At the time of the introduction of the product, there was considerable resistance by mushroom-gatherers living outside of the province. They complained about having to pay for permits which the inhabitants of the province get free of charge. Acceptance of the fees took some years.

IT20 Guided tours in the mountains Stelvio national park

A BRIEF DESCRIPTION. On the basis of an agreement with the administration of the national park, a mountaineering school offers guided tours in the park. According to legal regulations, only authorized mountain guides are allowed to offer tours.

DEVELOPMENT OF THE PRODUCT/MARKETING. The national park offers a landscape of outstanding beauty and of cultural–historically interesting features. The touristic infrastructure of the area is outstanding, especially concerning the equipment of the area with a network of hiking paths, information signs and other features. Previously, the guided tours were offered free of charge by the national forest authority, which is also in charge of the administration of the park. Since 1996, on the basis of a law admitting only authorized guides, the park authority offers guided tours in cooperation with the local mountaineering school. The market situation from the viewpoint of the mountain guides is considered to be favourable as in practice tours cannot be offered free of charge. The visitors participating in the guided tours only pay a percentage of the actual price to the mountaineering school. The rest is covered by park administration.

ORGANIZATION OF DISTRIBUTION. There are agreements for cooperation between the mountaineering school and the administration of the park, as well as between the school and a regional private tourist association, which is in charge of promotion and bookings for the mountaineering school.

INTERNAL ORGANIZATION/BOOKKEEPING. Within product development, capital budgeting has been carried out. Bookkeeping and cost calculation are indicated to be important. The evaluation of the project is considered positive in retrospect.

CONTACTS WITH AUTHORITIES/RECEPTION. The administration of the park intentionally tries to promote tourism compatible with nature in this area and thus exerts a positive influence on the project. Also the experiences with nature preservation organizations and with the association of tourism were positive. Overall, the additional offer was welcomed by the general public, and a considerable increase in environmental consciousness is visible.

IT21 Timber for the soundboards of musical instruments Val Canale

A BRIEF DESCRIPTION. The Val Canale area, which is, in large part, managed by the national forest administration, is well known for its outstandingly high-quality wood. For this reason, the forest administration applied to the Chamber of Commerce of the province, which has now registered the trade mark 'Val Canale timber for musical instruments'. This trade mark is transferred to some reliable violinmakers in the valley, who mark their raw timber for violins with this trade mark.

DEVELOPMENT OF THE PRODUCT/MARKETING. Timber for the manufacturing of musical instruments is a rare, very specialized product, and thus the market price of this wood is about 300% above the price of normal raw timber of good quality. The trade mark was applied for in 1994 to exclude unfair competition and to obtain advantages for the entire valley through its good reputation. The raw material of the violinmakers is marked by the trade mark, which has been the official mark since 1997, and the observance of the rules involved is safeguarded by officials of the national forest administration. Due to the limited resources in timber for musical instruments, the market situation is considered to be very advantageous. Furthermore, marketing of this particular timber is facilitated by personal relations between the staff members of the forest administration and the local violin-makers. The product is promoted at fairs and exhibitions, where product advertising is done in cooperation with the violin-makers.

ORGANIZATION OF DISTRIBUTION. Apart from the cooperation with individual violinmakers described above, there is no complex organization of the distribution.

INTERNAL ORGANIZATION/BOOKKEEPING. Relevant information concerning the marketing of the product can be derived from the forest facility itself. The main target is realization of net profits and the amelioration of the reputation of the entire valley. Bookkeeping is only done for tax reasons, and there is no cost calculation. A lack of cooperation between the individual forest enterprises is indicated to be the reason for setback. A further step will be the certification of sustainability for the entire area.

CONTACTS WITH AUTHORITIES/RECEPTION. Experiences with authorities and the general public are indicated as positive.

IT22 Seminars on forest ecology
Gola del Furlo

A brief description. In cooperation with the province and the University of Urbino, a centre for ecological and faunal studies is managed. The courses take place in a training centre situated in a beautiful landscape.

Development of the product/marketing. The product is new to the market, and there are very few courses of this sort in Italy. In the meantime, demand has increased so dramatically that a selective system has had to be introduced. The courses were first held in 1989. The project is scientifically supported by the university. Among other features, the quality of the knowledge transfer, the interrelation between theory and practice and the beauty of the landscape in this area are mentioned as profitable factors. The market situation is advantageous, as there is no similar product on the market. The main target group consists of persons who deal with nature and landscape professionally, as well as students and other persons interested in environmental issues. Among the institutions, public administration is also a customer of this product. The province covers half of the actual price to open the courses to students and beginners on the job, as well. In the framework of communication policy, posters are supplied to universities, administrations and nature preservation organizations, and advertisements are published in specialized magazines.

Organization of distribution. The province appointed the centre for ecological studies with the organization of the offer. The annual seminar schedule is also approved by the province.

Internal organization/bookkeeping. A basic target of the project was to establish an educational training programme of high theoretical and practical value. This target has been reached. The realization of the project was done by a special project team, whereas financing was supplied by the province of Pesaro. Bookkeeping is done only for tax reasons.

Contacts with authorities/reception. There are favourable contacts with local administrations and with the University of Urbino. Nature preservation organizations, such as the WWF, are supportive and contribute to some of the seminars offered. The general public supports the project for reasons of reputation, and the participants of the respective seminars are satisfied with the favourable combination of tutorials and theoretical knowledge.

IT23 Access to the forest
Bosco Fontana

A brief description. Bosco Fontana is a nature preservation area fenced in since 1976 in a densely populated region. It is very distinctly characterized by intensive agriculture. The public administration in charge of the area introduced an entrance fee for access to the area in 1985 in order to facilitate regulation of the large number of weekend visitors.

Development of the product/marketing. The touristic infrastructure in this area in the vicinity of some larger cities is quite advantageous, and the area can be easily reached by public transport as well as by individuals. The area has been protected by nature preservation law because of its importance for certain insect species. Before the introduction of entrance fees, the area was regularly overcrowded on Sundays. In the region, a visit to forest stands free of charge is not possible due to the general scarcity of woodlands. The entrance fees are exclusively charged on the Sundays between March and September. Children and local citizens can enter free of charge. The forest owner himself does not promote the product. However, it has to be assumed that the local tourist information office supplies information.

Organization of distribution. There is no special organization of the distribution, and legal regulations are only important with respect to activities that are allowed or prohibited in the area.

Internal organization/bookkeeping. The main target was to avoid overcrowding on weekends, but the management is satisfied with the financial outcome of the project. In the future, an amelioration of the service offered and a limitation on the number of visitors on Sundays is planned.

Contacts with authorities/reception. There have been few contacts with other public authorities concerning the project. Nature preservation organizations were in favour of the project, whereas it took some time before the entrance fee was accepted by the general public. Therefore, the fee was very low in the beginning to facilitate acceptance. Local inhabitants could enter free of charge.

IT24 Offer of a picnic site Val Canzoi

A brief description. A forest owner who lives in a well-known valley of a national park offers a picnic site on his property, with additional services (firewood for barbecueing, food, beverages), and thus utilizes the popularity of the area.

Development of the product/marketing. Demand is restricted to day-visitors, and the touristic infrastructure of the area in question is limited. The entire product offer consists of a picnic site with swings, public lavatories and a basic building, in which the forest owner offers traditional snacks. The product was introduced in 1993, and the idea and the know-how were supplied by the forest owner. Although there are also picnic sites free of charge, the offer is successful due to the good service, and the owner is of the opinion that similar additional offers could positively influence his own enterprise. Information on the market potential were supplied to the forest owner by potential customers, and the product offer has positive effects on the other products he markets. For promotional purposes, the forest owner has run television commercials on local programmes, as well as the distribution of folders as a communication strategy. In his opinion, the commercials were far more effective than the folders.

Organization of distribution. The forest owner markets his product direct. In this way, he only needed liability insurance. However, he had to observe regulations concerning a watercourse nearby, and it took 5 years to get permission for the sale of agricultural products.

Internal organization/bookkeeping. There are certain conflicts with hunting in the area; however, these are avoided by the fact that hunting largely takes place at another time of the year. The project is marketed within the daily routine. The main target is the diversification of economic activities. Bookkeeping only fulfils tax requirements, and there is no cost calculation. Project evaluation in retrospect is positive, apart from obstacles due to bureaucracy.

Contacts with authorities/reception. The forest owner was dissatisfied with the lack of support by the local administrations. In the beginning, visitors complained about the fee for the utilization of the picnic site, but the complaints soon ended because of perfect service and good organization.

IT25 Leasing of ski-runs Cortina d'Ampezzo

A brief description. Cortina d'Ampezzo is an area in the Dolomites well known for skiing. The forest in this region (16,000 ha) belongs to a group of about 800 long-settled families. A small amount of the forest (around 100 ha) is let to a ski-lift operator on the basis of a long-term contract. The transformation of forest land into ski-tracks, however, is not on a permanent basis in this case, and after expiry of the contract, the areas have to be reafforested.

Development of the product/marketing. Touristic demand in this area is very high and, for this reason, the touristic infrastructure is well developed. The product is based on a 30-year-old contract of lease with the operators of the facilities for skiing. No legal provisions have been changed, as the owners still dispose of the right to manage their property according to their wishes and to demand the relevant rent. The contract was negotiated in 1971, the same year in which the joint owners were accredited as the forest owners. The local community and some of the other owners have negotiated similar contracts. However, these contracts do not affect the present project. Prices within the contract are subject to inflation and, furthermore, relate to the price of the ski-passes.

ORGANIZATION OF DISTRIBUTION. Contracts were negotiated between the group of owners and the operators of the ski-lifts. The operator of the ski-lift has committed himself to properly managing an existing forest area as compensation for the construction of the facilities for downhill skiing.

INTERNAL ORGANIZATION/BOOKKEEPING. The project is integrated in the daily business routine. The main target of making profits is reached. Among the profit factors, it is mentioned that some members of the group of owners are at the same time members of the companies running the facilities for skiing. Bookkeeping and cost calculation are of subordinate significance, and there are no future plans for adaptations.

CONTACTS WITH AUTHORITIES/RECEPTION. Experiences with the forest administration and other authorities have been positive. The same is true for experiences with local nature preservation organizations. The operators of the ski-lifts are dissatisfied with the obligation to pay for the management of a comparable forest area. They are of the opinion that the forest area will in any case be reafforested after the contract expires.

IT26 Letting of huts in the mountains Cortina d'Ampezzo

A BRIEF DESCRIPTION. Cortina d'Ampezzo is an area well known for skiing in the Dolomites. The forest in this region (16,000 ha) belongs to a group of about 800 long-settled families. By changes in the documents at the land register, mountain huts that had been built by third parties on the property of the group of owners were declared their property in 1991. The group now lets the huts and the surrounding area.

DEVELOPMENT OF THE PRODUCT/MARKETING. The area is greatly frequented by tourists and therefore has very much developed as concerns touristic infrastructure. The product, on the one hand, was developed from existing facilities. On the other hand, the leaseholders only paid a very low lease to the joint owners before they were granted property rights. The precondition of the rent increase was the transfer of the property rights by the land register. In the area, there are also mountain huts of other owners, which are rented privately, but they do not substantially affect business. The rent is fixed according to location, ways of access and tourist demand, and is subject to the inflation rate.

ORGANIZATION OF DISTRIBUTION. The huts are offered directly by the joint owners. Contracts were negotiated between the joint owners and the individual leaseholders. The contracts explicitly state that tenants do not acquire property rights with payment of the rent. The construction of the huts had to be approved by the municipal authorities.

INTERNAL ORGANIZATION/BOOKKEEPING. The project is integrated in the daily business routine. The business is facilitated by the fact that some of the tenants are also joint owners and thus care for the maintenance of the huts. Investments within the offer were covered by their own means. Bookkeeping and cost calculation are indicated to be of subordinate significance.

CONTACTS WITH AUTHORITIES/RECEPTION. The present tenants were very dissatisfied with the raising of rent, as they had constructed the huts themselves and now had to pay higher rents. A positive side-effect of the project is that maintenance is now safeguarded.

IT27 Sale of Montella edible chestnuts Avellino

A BRIEF DESCRIPTION. In an area of approximately 3000 ha in the vicinity of Naples, edible chestnuts are produced. According to Italian categorizations, the area is classified as a forest and, for the production, a mark of origin according to EU Regulation 2081/1992 was granted in 1996. This infers a thoroughly prescriptive method of production.

DEVELOPMENT OF THE PRODUCT/MARKETING. The traditional product is very much linked to the area. This link is underlined by the mark of origin. The application for the mark of origin is a further development of an already existing marketing strategy for chestnuts. It is significant that only local producers can obtain the mark of origin and for this purpose have to register with the community. The proper usage of the label is controlled by an official testing laboratory for agricultural products and by local authorities. Apart from the certified chestnuts, other suppliers offer chestnuts without a certificate. These suppliers, however, draw profits from the good reputation of the entire valley. The personal market knowledge of the suppliers of certified chestnuts was especially important for the evaluation of the market situation. The chestnuts were partly offered in a product mix of truffles, wine, cheese and mushrooms, which positively influences marketing. The price of certified chestnuts is about 10–30% higher than that of uncertified chestnuts. Events, promotional shows and television broadcasts are used for communication.

ORGANIZATION OF DISTRIBUTION. Associations of producers distribute the product to a large extent. Important legal requirements are the regulations involved with certification.

INTERNAL ORGANIZATION/BOOKKEEPING. The project is integrated in the daily routine, and a net profit is gained. An important non-material target is the preservation of the rural culture of chestnut producers in this area. Investments in the context of the project were partly covered by the supplier's own means and partly supported by subsidies from the EU and the government. Bookkeeping is only important for larger enterprises. In the future, there will be a union between regions, which will be especially in charge of marketing activities.

CONTACTS WITH AUTHORITIES/RECEPTION. The project is welcomed by the authorities, nature preservation organizations and the general public.

IT28 Admission to a regional park Maremma

A BRIEF DESCRIPTION. An entrance fee is demanded for a regional park with interesting plant and animal species. Furthermore, the number of visitors is limited to 500 persons per day. There are strict rules concerning the access to roads. In summer some of the roads can only be used when booking a guided tour because of the danger of fire. Few of the paths can be used free of charge.

DEVELOPMENT OF THE PRODUCT/MARKETING. The park lies in the vicinity of Rome, and it is very popular for its beautiful scenery and the possibility of bathing at the beach. In total, there are eight hiking paths, for five of which a toll has to be paid. Bus transport to the starting-point of the roads has to be paid by the visitors as well. The construction of the park and the setting up of management planning were necessary for controlling the number of visitors and for introducing an entrance fee. The park has been open since 1978, and its main factor of success is its widespread reputation. The knowledge of the market situation for such an offer was supplied by members of the park authorities. The main target groups are schools, families and other tourists. Schools have to book in advance and to state that they are executing a programme on environmental education. The price is orientated towards the means of an average family and varies according to the facilities included (e.g. bus transport).

ORGANIZATION OF DISTRIBUTION. The park cooperates with hiking guides organizations, who are involved in guiding the visitors on the basis of an announcement and an agreement contract. The most important legal framework condition of the project is the management plan of the park.

INTERNAL ORGANIZATION/BOOKKEEPING. The main targets are to obtain a positive profit contribution and the regulation of the utilization of natural resources. Both targets have been reached relatively satisfactorily up to now. The project is integrated in the daily routine and, for the future, ameliorations concerning the centre for visitors and the direction of visitors are planned.

CONTACTS WITH AUTHORITIES/RECEPTION. Conflicts concerning competence arose between the forest administration and other authorities. The general public has been prepared for the payment of a fee. Many visitors select the chargeable paths, because they consider these paths to be more attractive.

IT29 Sponsorship in cooperation with the WWF National park in the Abruzzen

A BRIEF DESCRIPTION. In the Abruzzen national park, the preservation of the Abruzzen brown bear is financed by sponsoring of the WWF. Furthermore, the Italian company SNAM has agreed a contract of sponsorship concerning the preservation of the Abruzzen chamois.

DEVELOPMENT OF THE PRODUCT/MARKETING. The area is of high touristic importance because of the national park. The rarity of the animal species sponsored is important for the offer of sponsorships (40 brown bears and approximately 500–600 Abruzzen chamois). The offer of sponsorships is newly introduced. Subsequent to research on the biotope, substantial measures to ameliorate the biotope were financed by the WWF and the SNAM. For this reason, the WWF collected money from its members and also contacted external donators. The sponsorships began in 1990 and 1992. Important profit factors for the WWF were the rarity of the bear and the emotional effect of the animal sponsored. The main target of SNAM was the external communication of environmental consciousness. The know-how for sponsoring at the WWF was supplied by other fund-raising projects; with SNAM, the contact came from the enterprise itself. In the framework of its communication policy, the WWF informed its members of the project by mail and also utilized television commercials for the project. SNAM utilized the sponsoring activity for its promotional activities, and the logo of the sponsor (SNAM) was printed on all information materials distributed by the park.

ORGANIZATION OF DISTRIBUTION. There is a formal sponsoring contract with SNAM. Strict environmental directions in the national park are important legal framework conditions and are especially important for the sponsors, as these regulations safeguard the natural development in this area.

INTERNAL ORGANIZATION/BOOKKEEPING. There are different management plans concerning this area. The inventory measures concerning the quality of the biotope for the animal species mentioned were especially important. Non-material targets were important for the projects; however, the financial support by sponsorships positively influenced the grant of EU aid.

CONTACTS WITH AUTHORITIES/RECEPTION. There are positive experiences with the Ministry of the Environment. According to a law passed in 1991, sponsorships are supported as financial resources for national parks. Environmental and nature preservation organizations showed a very positive attitude towards the projects. The general public especially acknowledged cooperation with the WWF, and some of the neighbours assisted voluntarily in measures for amelioration of the biotope.

2.6 Case-studies – Austria

AU01 Opening of forest roads for mountain-biking Österreichische Bundesforste AG

A BRIEF DESCRIPTION. The federal forests in Austria open forest roads for the construction of mountain-bike tracks on the basis of a contract of usufruct concerning the right of utilization. The contractee is a non-profit organization founded to serve the above-mentioned purpose. The organization is mainly financed by the federal state of Oberösterreich (situation in 1997; later on, the federal state of Oberösterreich directly acted as a contract partner).

DEVELOPMENT OF THE PRODUCT/MARKETING. The touristic infrastructure of the area is excellent. The product area is partly situated in the Kalkalpen national park, ecologically a very valuable area and one that is also remarkable for its scenery. The area is easily accessible, and tourism has been promoted by different regional initiatives for a long time. The product has been newly introduced. The setting up of signposts and the hire of mountain bikes are offered as additional services. The ideas concerning the development of the product were supplied by staff members of the forest administration, the media and the specialized press. Knowledge of the market mainly was derived from personal experience. Target groups are day-visitors and tourists, but also schools.

ORGANIZATION OF DISTRIBUTION. The distribution is organized by the tourist information office. For political reasons, a non-profit organization acts as an intermediary, i.e. as the contractee of the Österreichischen Bundesforste AG. Sponsoring is done by the post office (signposts) and an insurance company (liability insurance, situation in 1997).

INTERNAL ORGANIZATION/BOOKKEEPING. The offer involves a potential conflict concerning hunting and timber harvesting. The product was not developed separately. It was financed by the organization's own means, and there are no separate bookkeeping and cost calculation. In the future, new tracks will be opened and the opening hours will be changed.

CONTACTS WITH AUTHORITIES/RECEPTION. Tourist organizations have supported the project in many ways. The public first criticized the fees for utilization of forest roads, but the toll has now been accepted. The main argument was that the Austrian national forests as public forest owners should offer the public roads free of charge.

AU02 Hire of mountain bikes Österreichische Bundesforste AG

A BRIEF DESCRIPTION. The Austrian national forest administration hires out mountain bikes as an additional offer supplementing the network of mountain-bike tracks.

DEVELOPMENT OF THE PRODUCT/MARKETING. The product area is identical with the one described in AU01. For the offer, 50 mountain bikes had to be purchased, an old building restored and some signposts set up. The mountain bikes were initially offered in 1995. The national park and the already existing mountain bike network contribute to the success of the offer. Competition is with retailers of sports equipment and other lenders of mountain bikes. Market research was not undertaken. The offer is restricted to 6 months per year, and it is demanded by hotels, schools and sports suppliers. An opening ceremony, publications, leaflets and other means of information supported communication.

ORGANIZATION OF DISTRIBUTION. The supplier of the mountain bikes is involved in the offer as a sponsor. A special liability and casualty insurance was required.

INTERNAL ORGANIZATION/BOOKKEEPING. The financial target of positive net proceeds has not yet been reached. The reasons indicated for this are a lack of information and resources and the lack of project management. The mountain bikes are hired out by a pensioner, and there would have to be

500 bikes hired out per year to realize a net profit. There has been capital budgeting for the purchase of the bikes. In the future, different channels of distribution will be used and, above all, advertising and promotional activities will be intensified.

CONTACTS WITH AUTHORITIES/RECEPTION. Positive contacts have been established with retailers of sports equipment. The general public and especially the consumers have welcomed the additional offer.

AU03 Water-powered electricity [indication of the name not authorized]

A BRIEF DESCRIPTION. The forest enterprise runs an electric power station and distributes the electricity via the public network against a charge. At the same time, the enterprise has been authorized to use the water as potable water. The water, however, has not yet been marketed.

As the enterprise has issued a confidence-note, further details cannot be passed on.

AU04 Military training camp Chorherrenstift Klosterneuburg

A BRIEF DESCRIPTION. In addition to many other RES products, such as electricity production, potable water and horse-riding, the forest enterprise leases a forest area for military training to the Austrian Army. The offer of RES products represents 37% of the total income of the enterprise.

DEVELOPMENT OF THE PRODUCT/MARKETING. The location at about 50 km from some barracks determines the success of the product. The product is further developed. Due to rising demand, a new contract has been negotiated to diminish conflicts with recreationists. The idea of such an offer was developed by the forest owner in cooperation with a solicitor in 1985. The market is favourable for such an offer, as it is difficult for the military forces to utilize similar areas free of charge. Personal experience and contacts have facilitated the evaluation of marketing perspectives. The rent is ÖS 1000 ha^{-1} $year^{-1}$ and the army in addition has to cover land-tax.

ORGANIZATION OF DISTRIBUTION. The army is the only contractee. However, a very detailed contract has been negotiated, containing paragraphs on periods of utilization, damage, timber harvesting, pedestrians and the closure of the area.

INTERNAL ORGANIZATION/BOOKKEEPING. The surroundings of the product area are very important for hunting and, in this respect, there is a high potential of conflict. Financial and non-material targets have not been pursued with the project. There is a temporary utilization relating to the hunting season and the closed season. Investments have not been required for the project, and trade-offs were considered not to be relevant.

CONTACTS WITH AUTHORITIES/RECEPTION. There have been positive contacts with the general of the barracks. Organizations of tourism and nature preservation remained neutral. There have not been any reactions in the general public. A positive, external effect is the limitation of training activities to a certain area. However, there are negative effects, such as an increase in traffic, as well as damage and noise in the area.

AU05 Ecological park Pretterhofer

A BRIEF DESCRIPTION. A former farmer and forest owner (110 ha) offers a deer-park. Additional offers are a restaurant, a children's playground and different theme paths. The park covers an area of 4 ha. The income of the enterprise is derived, as follows: 70% from the sale of timber and 30% from RES products.

DEVELOPMENT OF THE PRODUCT/MARKETING. The area is very popular with tourists and day-visitors. A visit to the park can be combined with a visit to other places for sightseeing. Target groups are families, schools and holiday-makers. The needs of the target groups are met by adapting the flexible opening hours and offers such as guided tours to their needs. The offer is a further development of an already existing offer, developed to compensate for the decreasing proceeds of agriculture. The costs for fencing and control staff are low and, as the main profit factors, the clear arrangement of the area and close contact with the animals are mentioned. Marketing chances were mainly evaluated according to the owner's own experiences. A product mix is of considerable importance, combining the product with additional offers such as horns, a restaurant, a playground and guided tours. Furthermore, close contact with teachers is considered to be important. Means of communication are leaflets, advertisements in a travel guide and the label 'eco-park'.

ORGANIZATION OF DISTRIBUTION. Only the game is marketed in cooperation with a butcher. Special liability and casualty insurances were required for the offer. Furthermore, animal safeguarding provisions have to be observed.

INTERNAL ORGANIZATION/BOOKKEEPING. The financial target of net profits has been more than fulfilled. Problems in the realization of the project arose because of a lack of motivation of the staff and due to miscalculation of the amount of time required. The capacity is 500 persons and eight guided tours per day. Investments for fencing, animals and car-park areas required formal capital budgeting. Cost calculation supplied only restricted information and will be ameliorated in the future.

CONTACTS WITH AUTHORITIES/RECEPTION. All of the contacts with public authorities were positive. The general public welcomed the offer. At the beginning, former customers who previously did not have to pay for access were sceptical. After 2 years, however, the product was accepted.

AU06 Letting of a fish-pond
Gnaser

A BRIEF DESCRIPTION. The owner of a small forest area with a pond belonging to the forest lets part of the pond to staff members of Philips AG.

DEVELOPMENT OF THE PRODUCT/MARKETING. The area is scarcely attractive for tourism. The product, therefore, is mainly directed to the staff of Philips living in the area. The staff members usually combine their visit with a barbecue party. Special wishes have been taken into account, in so far as bathing was prohibited. The rent is about 10% above the average market price because of the favourable location and the equipment of the pond. The product is a further development of an already existing product: the owner set up additional fences, divided the pond into two parts and executed afforestations. The present tenant submitted the idea to the proprietor of the pond. Due to its location and its good supply of fish, the product has an advantageous position in the market in comparison with similar offers. Demand is limited to the months between March and November, and the price is ÖS 34,000 year^{-1}. No advertising is done. The contact with the shop committee of the enterprise was important in the product offer.

ORGANIZATION OF DISTRIBUTION. The basis of the offer is a regular contract of lease.

INTERNAL ORGANIZATION/BOOKKEEPING. Financial targets concerning the offer have not been defined. An important aim of the enterprise was the motivation of its staff. The working expense of the project is about 3 man-months per year. In the future, the owner intends to promote the natural regeneration of the surrounding forest stand.

CONTACTS WITH AUTHORITIES/RECEPTION. Contacts with local and regional forest authorities are positive. Former users of the pond wanted to buy the area and thus problems arose initially.

AU07 Guided boat tours
Österreichische Bundesforste AG

A BRIEF DESCRIPTION. The Österreichische Bundesforste AG offers guided boat tours through the Donauauen national park. The income is divided up as follows: 70% of the income is derived from the forest economy, of which, in turn, 10% is derived from timber production, 85% is subsidies given to the national park and 5% is derived from the offer of RES products.

DEVELOPMENT OF THE PRODUCT/MARKETING. The area is of eminent ecological significance (lowland forest). Vienna and Bratislava are at a distance of about 50 km, and the area is very popular among day-visitors. Target groups are schools, families and companies. Specialized tours are possible as well. The forest enterprise has newly introduced the product, for which, for example, the purchase of boats and a minibus was required. The already existent offer of guided tours through the Eckertsau castle can now be combined with a boat tour. The boat tours were first offered in 1995. An important factor determining success is the area of the national park and well-trained staff. A certain amount of competition exists because of offers by the administration of the national park and by private enterprises. Contacts with the local school administration are important for cooperation with schools. Advertising is done in the local newspaper, on television, by leaflets and at fairs.

ORGANIZATION OF DISTRIBUTION. A liability insurance contract of ÖS 50,000,000 has been negotiated. A legal provision is the prohibition of timber harvesting (compensation of ÖS 3.6 million).

INTERNAL ORGANIZATION/BOOKKEEPING. The forest administration was an important source of information in the offer of the RES product. Non-material targets are especially important in the offer (the competence for the national park was not transferred). Internally, it has been stipulated that only trained personnel should be employed and that tours only take place on preselected paths. Lack of organization and information, as well as insufficiently trained staff, was indicated as the reason for previous backlashes. The labour expense for the project is about 50 man-months per year. The number of guided tours increased from 37 in 1995 to the capacity limit of 1500 in 1997. Substantial investments (50% of the expected turnover) became necessary. Cost calculation supplies important information, and trade-offs are considered to be relevant. Pricing, organization and the offer of specialized tours will be ameliorated in the future.

CONTACTS WITH AUTHORITIES/RECEPTION. There have been positive contacts with the administration of the national park and the tourism organization. The citizens of Vienna especially welcomed the project. In the beginning, hunters and the local inhabitants criticized the boat tours (disturbances, traffic, etc.), but have come to accept the offer. A positive side-effect is the direction of visitors. In the long run, however, the increasing number of visitors could have negative side-effects.

AU08 Fees for access to a forest road Colloredo

A BRIEF DESCRIPTION. The proprietor demands a toll for a private forest road (800 m).

DEVELOPMENT OF THE PRODUCT/MARKETING. Tourism is of great importance for the region, which is characterized by a multitude of ecological factors and particularities of the landscape (Sölktäler nature park). The number of day visitors is very high. Users of the road are mainly families and sportsmen. The fee was newly introduced by the enterprise. Entry is controlled by stop signs and a gate. The offer was developed according to an idea of the forest owner, and was initially offered in 1992 in cooperation with the city and the nature park. It is difficult to visit similarly developed areas free of charge. Additional benefits in this case are car-park areas and a restaurant. Analyses were done to evaluate the marketability, and the enterprise offers information and maps on footpaths to increase the acceptance of the offer. Some articles on the project have been published in the newspaper.

ORGANIZATION OF DISTRIBUTION. There is an agreement with the city of Kleinsölk, which receives 30% of the profits for the maintenance of the roads. A liability insurance contract has been negotiated.

INTERNAL ORGANIZATION/BOOKKEEPING. The financial targets have been more than fulfilled. Also the ameliorated cooperation with the tourist information office is said to be positive. The offer requires a labour expense of about 8 man-months per year. In order to be profitable, 2500 vehicles have to use the road per year. In 1996, 4500 vehicles passed along the road. Investments have not been required. Bookkeeping is considered to be important, while cost calculation is said to be less important than bookkeeping. Trade-offs and overheads were not included in the calculations. In the future, the director of the forest enterprise wants to intensify contacts and cooperation with the administration of the nature park.

CONTACTS WITH AUTHORITIES/RECEPTION. The forest enterprise, as well as the city of Kleinsölk, showed a positive attitude towards the project as a source of income. However, the tourist organization was critical of the offer, because they feared a discouragement of visitors. The general public welcomed the regulation of access for reasons of nature preservation and the maxim of quality instead of quantity. As a positive side-effect, the public acceptance of different RES offers is assumed.

AU09 Sponsoring of a primary forest [indication of the name not authorized]

A BRIEF DESCRIPTION. In a daily newspaper founded by a distant relative, a private forest owner found a sponsor for the preservation of the remnant of a primary forest (40 ha). The sponsoring project was facilitated by the cooperation of an association. This association, founded by forest owners, has already executed several sponsoring projects.

DEVELOPMENT OF THE PRODUCT/MARKETING. The ecological significance of the area is outstanding. This is also true of its recreational importance. The forest owner supplied the idea for this product development and utilized the know-how of the association. An important legal regulation is the prohibition of timber harvesting. Guided tours are offered in the area. Informal contacts and perseverance are indicated as factors determining success. In particular, the wishes of the sponsor concerning a buffer zone of 45 ha and regarding the prohibition of timber utilization were observed. A detailed description of the ecological significance of the area has facilitated the acquisition of the sponsor. The level of the contribution of the sponsor was fixed on the basis of lost timber utilization and the non-material value of the primary forest.

ORGANIZATION OF DISTRIBUTION. It is stipulated that the contract automatically ends if the newspaper has to file a declaration of bankruptcy. Exclusivity is also stipulated in the contract.

INTERNAL ORGANIZATION/BOOKKEEPING. Maps and forest inventories were important instruments for describing the product in detail. Conflicts arose with recreation and timber utilization, as both are now restricted or not possible at all. An important non-material effect mentioned is publicity. A lack of organization, experience and information is responsible for backlashes. The working effort is 4 man-months per year. In retrospect, the evaluation of the market potential and the organization of the project (milestones, etc.) could have been improved.

CONTACTS WITH AUTHORITIES/RECEPTION. The national forest administration has supported the project concerning contractual nature preservation. Reception by the general public was very diverse, and it will still take some years before the project is entirely accepted.

AU10 Protection of larch meadows Strickner

A BRIEF DESCRIPTION. A private forest owner (50% of his income is derived from subsidies, 45% from agriculture and 5% from RES marketing) gets public funds for the preservation of 2 ha of larch meadows.

DEVELOPMENT OF THE PRODUCT/MARKETING. The area belongs entirely (100%) to the Oberberger See natural reserve, and its significance in nature preservation is very high. The product has been newly introduced by the forest owner, who was supplied with the idea by the mayor of a neighbouring community. The agreement has existed since 1984 and has been developed in cooperation with the forest and agricultural administration. Features determining success were the traditional management of the area, cooperation between several forest owners (the association of larch forest owners) and general environmental consciousness. The contractee and sole target group of the product is the federal government of Tyrol. The owner gets ÖS 2700 $year^{-1}$ for the mowing of the meadow, and for the maintenance of the fence ÖS 500 per 100 m and per year, which is the balance for regular management of high-forest stands.

ORGANIZATION OF DISTRIBUTION. The owners of the larch meadows are organized in an association. The association was formed on the initiative of the mayor in 1983.

INTERNAL ORGANIZATION/BOOKKEEPING. An inventory, with aerial surveys and a map, was executed for the project. The RES product is complementary to hunting (grazing area). Financial targets were not pursued, and non-material targets have not been defined. The working effort is approximately 1 man-month per year. In retrospect, the estimates of the area could have been more precise.

CONTACTS WITH AUTHORITIES/RECEPTION. All of the relevant public institutions supported the project. The reception of the general public was diverse, as the acceptance of subsidies is at present not very high. It took 5 years for the project to be accepted. A positive side-effect is the increase in job security in the rural region.

AU11 Accommodation in luxurious holiday flats Malteser Ritterorden Ligist/Hebalm

A BRIEF DESCRIPTION. The Ligist forest enterprise offers ten holiday apartments in five timber houses (spruce house, fir house, larch house and two pine houses) situated in the forest. Additional RES products offered are mountain-bike tracks of different levels of difficulty, an area for winter sports equipped with a chair-lift and six ski-tows and a camp for beach volleyball. Fifty per cent of the income of the forest economy stems from the sale of timber and 50% from the sale of RES products.

DEVELOPMENT OF THE PRODUCT/MARKETING. On the one hand, touristic facilities, such as restaurants, hotels, etc., are rare in the area. On the other hand, recreational facilities are favourable, and in the vicinity is the city of Graz with its 300,000 citizens, which is the reason for the large number of day-visitors. The forest enterprise supplemented the existing offer of the area suitable for winter sports around a mountain-bike track and, finally, constructed the apartments to give visitors incentives for longer stays. The offer has existed since 1996, and, as its factors of success, besides the willingness of the owner to take risks and the employment of adequate staff, among other features, the combination with other services is indicated. In the surroundings, there are no similar competitive apartment offers. Information on the development of the product was supplied to the enterprise by the tourist information bureau and by conversations with former customers. School classes frequently accept the offer, and the contact with the directors of schools is considered to be important. In part, package offers are marketed, including the utilization of mountain bikes free of charge or including beach volleyball and courses in skiing.

ORGANIZATION OF DISTRIBUTION. There is cooperation with the suppliers mentioned above concerning the organization of package offers. However, there are no written contracts. Suppliers may use the infrastructure of the forest enterprise, but have to pay a fee for utilization. Concerning legal provisions, regulations concerning nature preservation were especially relevant and obstructed the construction of the holiday apartments in this area.

INTERNAL ORGANIZATION/BOOKKEEPING. Slight conflicts arose between the RES project and hunting. The financial target has not been entirely reached yet, as considerable means have been used for the market launch (capacity output is increasing annually). However, no reasons for backlashes have been indicated. The enterprise invested ÖS 15,000,000 derived from its own means and, for this purpose, capital budgeting was done. Bookkeeping and cost calculation are important sources of information, and trade-offs and overheads are considered as well. In retrospect, it can be stated that in the initial stage the respective marketing measures could have been more intensive and pricing more differentiated.

CONTACTS WITH AUTHORITIES/RECEPTION. Forest authorities and two boards of works have been involved in the planning of the project. Initially, one of the boards of works was against the project. The project, furthermore, was supported by the tourism organization. Interestingly, the local community criticized the design of the apartments as unsuitable (architecture of the roof and choice of colours of the five houses), though the customers are very satisfied.

AU12 Contracted nature preservation Neckenmarkt

A BRIEF DESCRIPTION. A joint forest of 580 ha contributes 30 ha to a nationally supported programme entitled 'Network of natural forest preservation'. The forest areas have been selected according to detailed checklists. The members of the forest cooperative have certain rights to obtain firewood from the forest, but do not receive further proceeds from the property. As an additional source of income, the owners market a quarry.

DEVELOPMENT OF THE PRODUCT/MARKETING. The area has minor significance for recreational purposes. However, protective measures are of importance. The project is a new one, as concerns the mutual ownership of the forest. The idea was supplied by the maintenance organization of the forest administration. As a factor of success, trust of the forest associates in the management of the forest enterprise and in the public support programme are mentioned. Important regulations are the directions concerning the dismantling of a forest road, prohibition of timber utilization and the duty to inform the Federal Research Institute of any extraordinary event or development (e.g. windthrow) in the area. The amount of funds is ÖS 117,000 $year^{-1}$.

ORGANIZATION OF DISTRIBUTION. A private planning office participated in the development of the product, analysed the product potential and informed the forest owner on the existing programme. The contractee is the state sector.

INTERNAL ORGANIZATION/BOOKKEEPING. The information supplied by the forest institution was especially important for the development of the project. Financial targets have not been defined. As a non-material target, publicity has been indicated. Bookkeeping and cost calculation do not play a role.

CONTACTS WITH AUTHORITIES/RECEPTION. The project was supported by local and regional forest authorities. In particular, the Federal Research Centre for Forestry and Forest Products supported the project by a public promotional programme. However, a private nature preservation organization opposed the project.

AU13 Network of ski-tracks Kirchberg forest administration

A BRIEF DESCRIPTION. Among the targets of the local tourist office was to link two ski areas with a cross-country ski-track. For this purpose, a joint venture was founded in the framework of a non-profit organization, which was in charge of planning and the execution of individual contracts concerning the right of utilization. Members of the joint ventures are the owners of hotels and restaurants and individual forest owners. The Kirchberg forest administration (4200 ha), besides other forest owners, negotiates contracts concerning the right of utilization for some of the areas with the association.

DEVELOPMENT OF THE PRODUCT/MARKETING. The ecological and environmental significance of the area is high, and it is important as a recreational area for winter sports. Forest owners and the tourism organization were responsible for the product idea. There are no competitive offers free of charge. The product is offered in hotels and restaurants and by retailers of sports equipment. Prices were calculated considering the investment, fixed charges, losses in production possibilities, information on market prices and the outcome of a survey among users. Advertisements in newspapers and television commercials are used as means of communication. A logo and a brand name ('Semmering-Wechsel-Panoramaloipe') have been created.

ORGANIZATION OF DISTRIBUTION. The development of the product and the product offer have been organized via joint venture. The respective responsibilities have been stipulated in a special contract. Sponsors are involved in the offer as well, and the joint venture negotiated a special insurance contract for the tracks.

INTERNAL ORGANIZATION/BOOKKEEPING. Maps supplied by the forest facility were necessary for the development of the product. Conflicts exist with timber production, with other recreational activities and with hunting. The financial target defined by the forest enterprise was to gain a positive profit contribution. The project was organized as a separate project. Miscalculations and a lack of information were indicated to be responsible for backlashes. Moreover, 2000 m of new forest roads had to be constructed for the project. Financial resources for this target come from EU funds, resources of the tourism organization and the means of communities and landowners. Cost calculation is not considered to be necessary. For the future, it is planned to build a snack-bar.

CONTACTS WITH AUTHORITIES/RECEPTION. All of the authorities in charge have supported the project. However, there have been difficulties with some politicians. In the beginning, hunters were very critical towards the project, but, after the establishment of a hunting preserve, they showed a positive attitude towards the project, because of the direction of visitors. As the ski-track touches one of the last important habitats of black grouse in Niederösterreich, biotopes have been established by clearings in two woodland areas situated at a distance for substitution. Thus, a consensus for future touristic projects has been reached regarding reconciliation between tourism and the preservation of species.

AU14 Test tracks for cars
Gutenberg forest administration

A BRIEF DESCRIPTION. The private forest enterprise lets an area as a test track for a car manufacturer. The contract has already existed for a long time and pricing is adapted to current developments. In this case, 85% of the proceeds of the enterprise are derived from the forest economy (1650 ha of woodland).

DEVELOPMENT OF THE PRODUCT/MARKETING. In the present case, the contract is a further development of a contract existing since 1955, which was negotiated at the request of the car manufacturer, for whom pricing has now been adjusted. The profit factors mentioned concerning the product were the short distances between the car company and the forest area and the quality of the tracks. When negotiating the initial contract, special demands of the car manufacturer concerning routing and the nature of the terrain were taken into consideration. At present, the rent amounts to ÖS 15,000 for 7 km. In the future, a rent increase to ÖS 250,000 is planned.

ORGANIZATION OF DISTRIBUTION. The product is offered on the basis of a lease contract.

INTERNAL ORGANIZATION/BOOKKEEPING. The maps and the descriptions of ecological particularities supplied by the forest institution were especially important for the development of the product. The planned financial target of net proceeds has not yet been reached. A lack of experience and information is held responsible for inadequate pricing. Bookkeeping and cost calculation exist, though they are not utilized as a source of information for the project. In retrospect, target fixing and consideration of the project in bookkeeping and cost calculation could have been more adequate.

CONTACTS WITH AUTHORITIES/RECEPTION. Nature preservation groups tried to stop the project, due to noise and pollution of air, soil and groundwater, and underlined the significance of the area for recreation. There has not been any experience with the general public.

AU15 Participation at a town fair (sale of game)
Traun'sche forest administration

A BRIEF DESCRIPTION. The forest enterprise participates in a town fair. A variety of activities are offered at the forest administration site and the surrounding area, such as airgun shooting, carriage rides and the sale of game). The events are supervised by staff members of the forest administration.

DEVELOPMENT OF THE PRODUCT/MARKETING. Touristic demand in the area is slack. So the product is directed towards the local community. Stalls and other materials necessary for the offer were hired. The town fair took place in 1995. Decisive in the success of the project were efficient organization, a tasty game meal and prizes for airgun shooting donated by sponsors. The product idea was developed in discussions with the town administration. For the sale of game meals, a certificate of health for each of the staff members was required. In the offer, the origin of the game was especially underlined. As communication channels, canvassing by word of mouth and advertising in the festive magazine of the community were used.

ORGANIZATION OF DISTRIBUTION. The staff of the forest administration attended the stalls in their spare time. No formal contracts were negotiated. However, insurance was required for the offer of carriage rides.

INTERNAL ORGANIZATION/BOOKKEEPING. A positive side-effect of participation in the event was the acquisition of new customers for the sale of Christmas trees in winter. The financial targets of the project (net proceeds) were more than fulfilled. The contact with potential customers was especially important among the non-material targets. The project was planned within the daily routine, and the offer was planned to meet the demand of 2000 visitors. There have not been substantial investments necessary, and the project was not considered in bookkeeping and cost calculations. For the future, ameliorations concerning the attractions offered are planned (a forest fair).

CONTACTS WITH AUTHORITIES/RECEPTION. All of the local authorities showed a positive attitude towards the project. The offer was welcomed by the general public and especially by the visitors at the fair. Some of the local restaurant owners, however, complained about competition.

AU16 Sale of spring water
Mayr-Melnhof'sche forest administration

A BRIEF DESCRIPTION. A private forest enterprise in the vicinity of Salzburg sells potable water to smaller water-processing companies and other private consumers. However, public waterworks are not among the customers. The catchment area of the sources covers about 500 ha, which belong to the forest enterprise (4800 ha in total).

DEVELOPMENT OF THE PRODUCT/MARKETING. The significance of the area for nature preservation and recreation is considerable. The product is a further development of existing usufructuary rights. In addition, water-pipes and containers have been built, and the project had already been developed by the demand of customers in 1970. As profit factors, the experiences with water management, increasing demand, the good quality of potable water and the interest of the owner in new sources of income were mentioned. There are similar offers by other forest owners in the region, establishing a certain amount of competition. Information on the market potential was derived from personal experience and from existing price-lists of other water suppliers in the region. Besides individual customers, the enterprise also has contracts with eight water-processing companies. The enterprise records difficulties in negotiations with these companies.

ORGANIZATION OF DISTRIBUTION. The enterprise itself markets the product. Informal contacts and canvassing by word of mouth have assisted in gaining new customers. An individual contract was negotiated with each customer. Concerning the legal framework, forest laws, nature preservation law and rights of water are important. By the contract, the risk is transferred to the customer.

INTERNAL ORGANIZATION/BOOKKEEPING. Financial targets have not been defined. Considerable investments were not required, and bookkeeping and cost calculation were not applied. At present, the offer gains a positive profit contribution. False valuations of the demand for potable water are held responsible for backlashes. Retrospectively, the analysis of the market potential, the definition of responsibilities, the marketing mix and the control could have been ameliorated.

CONTACTS WITH AUTHORITIES/RECEPTION. There have been positive contacts with the regional planning authorities, as the offer of potable water is in the public interest. Water consumers as well as the general public welcome the offer.

AU17 Mountain-bike concept in Tyrol Tyrol forest administration

A BRIEF DESCRIPTION. The countryside department of the Tyrol federal forest board developed a model for mountain-biking for all of the regional forest owners, containing, for example, sample contracts, insurance regulations and signposting. The target was to meet the rising demand for mountain-bike tracks.

DEVELOPMENT OF THE PRODUCT/MARKETING. Tourism is of great importance in Tyrol due to the beauty of the landscape and the large number of natural and cultural–historical attractions. The main aim is a standard model for the states, providing remuneration for the access to forest roads and alpine pasture footpaths. The remuneration is financed by federal and local resources, as well as by means of tourist organizations. Special demands of the target groups are met by the offer of different levels of difficulty and by distinct signposting of the tracks. The project idea was developed in 1997 by the countryside department of the federal forest board. The utilization of forest roads and alpine pasture footpaths that have not been especially dedicated to this purpose is legally prohibited in Austria, but cannot be avoided entirely. Market information has been supplied by tourist organizations. To enhance the acceptance for the placing of public finances, a research on the frequency of utilization of the forest roads was necessary. The fee for usage is from about ÖS 2 to ÖS 4 m^{-1} of forest roads annually. As communication channels, advertisements, the department's own publications, maps, events and the internet (www.tirol.gv.at/mountainbike) are utilized.

ORGANIZATION OF DISTRIBUTION. For the organization of the distribution, the role of the federal forest board as an intermediary has to be underlined. The forest administration assisted with the supply of sample contracts and insurance conditions and in the discussion with local authorities. As to the legal requirements, especially forest and nature preservation law, as well as regulations concerning the obligation to ensure traffic safety, had to be observed.

INTERNAL ORGANIZATION/BOOKKEEPING. For the development of the product, information supplied by the federal forest board (especially maps and the database on roads) was utilized. The financial and non-material targets have been fulfilled and, at present, a positive profit contribution is gained. Expansion of the network from the current 3500 km to approximately 4000 km and intensification of advertising, especially in Germany, are planned for the future.

CONTACTS WITH AUTHORITIES/RECEPTION. In cooperation with other authorities, the problem was that the authorities were of the opinion that the forest owners should allow access to the forest roads free of charge. Tourist organizations and private customers supported the project, but it took 2 years before the offer was accepted.

AU18 Toll for a forest road Draiachberg Cooperative

A BRIEF DESCRIPTION. A cooperative of 21 forest owners has introduced an access fee for the utilization of a jointly maintained forest road. Ninety per cent of the members of the association are farmers. On average, forestry contributes some 30% of the membership income.

DEVELOPMENT OF THE PRODUCT/MARKETING. The touristic and ecological significance of the area is high. It is well equipped with touristic facilities. Tourists, as well as local inhabitants, utilize the forest road. The product is a further development, as it could formerly be used free of charge. Therefore, it was necessary to put up signs and toll-houses and to ameliorate the maintenance of the road. Target groups were mountain-bikers and hang-gliders, as well as other users of the area. A special feature in this case is a restaurant on the peak of the hill, which is often frequented as an excursion site. The idea of the project was supplied by the forest owner and was realized because of the close cooperation and diplomatic skills of the principal. Two sports clubs are registered as the users of the road. Member lists are kept at the toll-house. The price was calculated on the basis of current expenses. A leaflet and an annual church service in the chapel on the summit of the hill serve as communication channels.

ORGANIZATION OF DISTRIBUTION. The project is offered with the assistance of the association of forest owners as an intermediate. Informal contacts are considered to be important; formal contracts do not exist.

INTERNAL ORGANIZATION/BOOKKEEPING. The financial target is gaining a positive profit contribution. A non-material aim is ameliorating the contacts to sports clubs. The project was developed by a separate project team. In order to safeguard road maintenance, 450 tickets have to be sold annually. Considerable investments were required. Cost calculation supplies important information on the project. The definition of targets, such as the number of visitors, and the analysis of the market potential were mentioned as necessary adjustments.

CONTACTS WITH AUTHORITIES/RECEPTION. Regional and local forest authorities, as well as the sports clubs involved, showed a positive attitude towards the project. After 4 years, customer acceptance is slowly increasing. A negative side-effect is the increase in responsibility.

AU19 Sale of spring water Austrian Federal Forest Administration (ÖBf AG), Gmunden forest enterprise

A BRIEF DESCRIPTION. The Gmunden forest enterprise of the ÖBf AG sells spring water to local waterworks and individual consumers. Other RES products offered are mountain-biking, bathing facilities, car-park areas, areas suitable for skiing and, training facilities for sled-dogs or dog carts on forest roads, as well as the utilization of forest roads, shelters and the like for the shooting of films.

DEVELOPMENT OF THE PRODUCT/MARKETING. The ecological significance of the area is very high and there are two nature reserves in the area in question. The sale of spring water was developed because of increasing demand. Prices in general are calculated according to consumption. With older contracts, a single or annual payment is made. For the construction of facilities to extract water from a nature preservation area, special licences were required. Other private forest owners also offer spring water; however, they have problems with quality control as their water extraction area reaches far beyond their property. The main customer of the spring water is the city of Gmunden, in cooperation with which the form of utilization has been developed. The water is transported to town by a water-pipe along the bottom of Lake Traun, owned by the ÖBf. The town additionally pays a fee for use of the bottom lake.

ORGANIZATION OF DISTRIBUTION. The forest enterprise originally supplied the product from the water-pipes directly to the consumers. However, the town community wanted to keep the extraction and derivation of water under its own supervision. As concerns the legal framework, especially the Austrian rights of water, forest laws and nature preservation law have to be observed. The forest enterprise had to gain approval from the relevant authorities concerning the extraction of water and the establishment of the necessary technical equipment.

INTERNAL ORGANIZATION/BOOKKEEPING. On the side of the forest facility, the maps supplied were especially important. The financial target (net proceeds) is not fulfilled as planned (compromise at contract negotiations). The basic research and the negotiations were integrated in the daily routine. The forest enterprise has not made any investments. The financial outcome did not entirely correspond with the initial expectations, but the form of the contract is finally acceptable to both sides. In the future, old water contracts will be renewed.

CONTACTS WITH AUTHORITIES/RECEPTION. Local and regional authorities were in favour of the project. The federal government of Oberösterreich, however, was against the project because of the fee for water from public forests. The general public and the individual consumers at first criticized the project, but accepted it about a year later.

AU20 Letting of ponds [indication of the name not authorized]

A BRIEF DESCRIPTION. A private forest enterprise lets fish-ponds with huts by means of licensing or letting. Other RES products are the letting of a military training area, the sale of potable water and the rearing of Norway lobsters. The sale of RES products represents 3% of the income derived from forest economy.

DEVELOPMENT OF THE PRODUCT/MARKETING. The area is highly significant with regard to ecology (occurrence of Norway lobsters) and it is popular with recreationists. Touristic demand and the offer of touristic facilities, however, are low. Target groups for the product are anglers from the vicinity. Their preferences (silence, emotional attachment to nature, being undisturbed at their own pond) are especially met. Businessmen often invite their busines partners for an afternoon of fishing. An important factor determining success, therefore, is its uncomplicated accessibility. The existing ponds were extended in the development of the product from 12 to 25 fish-ponds. Special attention was dedicated to an appealing design of river banks and the surroundings of the ponds. There are no similar offers in the area. Information on market potential was derived from the enterprise's own experiences and from discussions with customers for Christmas trees and game. Advertising is not required. An enterprise uses the offer as a recreational facility for its staff.

ORGANIZATION OF DISTRIBUTION. The forest enterprise markets the product directly. Informal contacts were helpful. Building permits were required for constructing the huts. For the design of new ponds, clearing permits had to be applied for.

INTERNAL ORGANIZATION/BOOKKEEPING. There is a slight conflict between hunting, timber production and the RES product. The financial target (net proceeds) has been more than fulfilled. Each of the ponds is managed as a separate project. Substantial investments were necessary, and simple calculation was done on behalf of these. According to the statements of the interviewee, bookkeeping and cost calculation only supply a limited amount of information. In retrospect, the construction of the ponds might have been done less expensively.

CONTACTS WITH AUTHORITIES/RECEPTION. All of the authorities involved were in favour of the project. The general public and the customers accepted the new offer immediately.

AU21 Offer of a recreational forest
Community of Milstatt

A BRIEF DESCRIPTION. The community of Milstatt has bought a forest area of 22 ha and turned it into a recreational forest. Access is free of charge and the offer is intended to be an additional offer for the community and its visitors. The project was supported by public funds.

DEVELOPMENT OF THE PRODUCT/MARKETING. The significance of the area for recreational purposes, due to many touristic facilities, and for nature preservation, because of fens, lakes and ponds, is very high. The offer has been newly developed. For this reason, the forest area has been supplied with huts, sports facilities, road signs and key maps. In 1997, the recreational area was opened. The know-how required was supplied by a forest consulting organization and by the nature preservation organization. Features determining success were close cooperation between the city council and the nature preservation organization and the granting of public funds. There are similar forest areas accessible for recreation, but the forest areas in question are closer to the community.

ORGANIZATION OF DISTRIBUTION. The community itself offers the product, and there are no contracts.

INTERNAL ORGANIZATION/BOOKKEEPING. A potential conflict between recreation and timber production/hunting is not indicated, as recreation is explicitly stated as the main target. For the project development, a special inventory concerning the adequacy of the area for recreation was carried out. The offer was developed as a separate project, for which considerable investments were required.

CONTACTS WITH AUTHORITIES/RECEPTION. Relevant contacts were established with local and regional authorities in charge of forestry, nature preservation and tourism. All of the authorities mentioned were in favour of the project, and also a nature preservation organization actively participated in the realization of the project. The offer was welcomed by the general public and especially by consumers.

2.7 Case-studies – the Netherlands

NL01 Utilization of potable water
Appelscha forest administration

A BRIEF DESCRIPTION. On the basis of a voluntary agreement, the waterworks recompenses a national forest administration for extracting potable water from a forest area. The income gained by the recompense represents around 25% of the entire income of the forest administration.

DEVELOPMENT OF THE PRODUCT/MARKETING. Already in the 1960s, the area was utilized for extracting potable water. The annual water extraction is approximately 6.5 million m^3 of water. The recompense is based on the calculation of reduced timber growth. The decision on payment of a recompense on the side of the waterworks was influenced by the fact that the area disposes of a lower risk potential than agricultural or built-up areas. In the end, it is less expensive to pay compensation than to clean polluted water. However, the director of the forest administration wants to consider not only reduced growth but also the influence of the utilization of potable water on the ecological value of the area as a whole. At present, though, this target has not been realized. Information on the amount of recompense is supplied by studies executed in another area. At the moment, the waterworks pay 0.02 guilders m^{-3} of water.

ORGANIZATION OF DISTRIBUTION. At present, there is only a verbal voluntary agreement.

INTERNAL ORGANIZATION/BOOKKEEPING. In the first place, the financial target was compensation for a reduction in growth. However, the extraction of drinking-water also has negative effects on nature preservation, as the composition of the species changes with increasing drought. In the future, a realistic contribution of the waterworks to nature preservation measures in this area is wished for.

CONTACTS WITH AUTHORITIES/RECEPTION. The water authorities in charge supported the project. There are no further experiences with the general public.

NL02 Fees for the recreational use of a national park
Province of Nordholland

A BRIEF DESCRIPTION. The province of Nordholland lets a nature preservation area to waterworks. On the one hand, the waterworks use the soil water, while, on the other, they have developed the area for recreational purposes and demand entrance fees.

DEVELOPMENT OF THE PRODUCT/MARKETING. The area is a landscape of outstanding beauty and attracts about 5 million day-visitors annually. The visitors can choose between 30 different hiking paths, and the area disposes of panoramic views and restaurants. The direction of visitors is obtained by different offers for hikers, mountain-bikers and riders. This was also the main argument for the introduction of the fee and the extension of the area. Although the authorization of the visitors is intensively controlled, a percentage (10–20%) of fee-dodgers could not be avoided. Additional offers, such as guided tours and information facilities, are very important for the acceptance of the fee by the visitors. Furthermore, visitors feel more secure in the presence of rangers. Although other areas are easily accessible free of charge, the extension of the area is unique. The area is utilized by about 125 sports clubs, which have partly negotiated special prices. Membership of a nature preservation organization may also reduce the fee.

ORGANIZATION OF DISTRIBUTION. The province of Nordholland has let the area to waterworks, which in turn market the area independently. Partly, retailers are involved: they get a commission on the sale of tickets. Ticket machines have also been introduced. Important legal framework conditions are regulations concerning water preservation.

INTERNAL ORGANIZATION/BOOKKEEPING. A detailed management plan has been set up for the area. The enforcement of nature preservation and recreation in this case are complementary. Financial targets have not been defined. The most important non-material target was the direction of visitors. The project was developed within the regular daily routine. A permanent topic for discussion was the wish of politicians for the introduction of an annual ticket allowing the utilization of all areas in the province. The working effort of the project is 200–300 man-months per year. Costs for entertainment, ticket machines and administration, as well as opportunity costs of 350,000 guilders in compensation for lost subsidies for the opening of the forest, are met by approximately 1 million guilders from entrance fees. At present, a positive profit contribution is gained.

CONTACTS WITH AUTHORITIES/RECEPTION. In the meantime, the province and the communities have accepted the introduction of fees. However, there are still irritations concerning the closure of public property. Some of the sports clubs consider the fee to be too high. At the moment, the general public remains neutral, due to the favourable equipment of the area with recreational facilities.

NL03 Licence fees for filming
Forest administration of the city of Amsterdam

A BRIEF DESCRIPTION. The city of Amsterdam owns about 420 ha of forest areas. These attractive landscapes are rented for commercials and other business events at fixed charges.

DEVELOPMENT OF THE PRODUCT/MARKETING. Special tariffs have been developed for the offer in question. As an additional service, the preparation and supervision of the different activities are offered. The product was introduced in 1990. Important factors determining success are the personnel involved and a valid system of tariffs. The market situation is favourable, as there is no similar offer in the area. Target groups are television and film producers, commercial photographers, wedding photographers and event managers of fashion shows. The individual demands of the customers are considered in the offer. At present, there are three kinds of prices, which are calculated on the basis of current expenses and with an extra amount for administration costs. No advertising is done.

ORGANIZATION OF DISTRIBUTION. The forest administration markets the product directly.

INTERNAL ORGANIZATION/BOOKKEEPING. The financial target is a positive profit contribution. Internally, it has been stipulated that the utilization of the area for environmentally dangerous targets or for advertising for environmentally dangerous activities will not be approved. The project was developed within the daily routine and there have not been any backlashes. Capacity is about one event per day. In 1996, there were 350–400 events. Bookkeeping and cost calculation supply important information and, at present, net profits are gained. In retrospect, the definition of targets, bookkeeping and project organization could have been improved.

CONTACTS WITH AUTHORITIES/RECEPTION. There have been no negative contacts with authorities. The city of Amsterdam supported the project. Reactions of the general public are very diverse and range from non-existent to positive.

NL04 Sales of 'eco-meat'
Het Drentse Landschap nature conservation organization

A BRIEF DESCRIPTION. A nature preservation organization with about 3700 members utilizes a certain cattle breed for grazing on nature preserves. The meat is sold on the market, via stock-breeders and local butchers (eco-butchers), as eco-meat. The income derived from the sale contributes to the preservation of the nature preservation area. Approximately 10% of the proceeds of the nature preservation organization were derived from the sale of eco-meat.

DEVELOPMENT OF THE PRODUCT/MARKETING. The main further development of the previous sale of meat is the utilization of special channels of distribution, by which the ecological quality of the product is made profitable. There are similar projects in the region, which, however, do not negatively influence the offer. Members of the nature preservation organization have to order the meat in advance, and then can collect it at a stipulated date at a certain slaughterhouse. As the regular market could not be convinced of the special quality of the meat, the organization was forced to cooperate with an ecologically conscious slaughterhouse. However, in this case there arose difficulties with a certain cattle breed (Limousins) because of lack of experience. Finally, the price of the meat marketed via the 'green channel' is approximately 10–12% above the usual market price.

ORGANIZATION OF DISTRIBUTION. The organization markets the products in cooperation with local butchers.

INTERNAL ORGANIZATION/BOOKKEEPING. Financial targets have not been explicitly formulated. Unity between the members was important. The sale of the meat to members supplies advantageous publicity and contributes to the unity mentioned. In 1996, 12 Highland cows, 100 Limousins and approximately 100 sheep were sold to the members.

CONTACTS WITH AUTHORITIES/RECEPTION. There have been contacts with authorities, one of which complicated and hindered the project. Private organizations were not always in favour of the project either.

NL05 Offer of a network of bridle-paths Country estate 'Den Trek-Henschoten'

A BRIEF DESCRIPTION. The private country estate (1400 ha of woodlands) is situated in a densely populated area in the centre of the Netherlands. A network of 40 km of bridle-paths is offered for a charge. Other RES products are licences for all kinds of events. The income is derived as follows: 40% from the forest economy, of which, in turn, 60% stems from the sale of wood, 30% from subsidies and 10% from the offer of RES products.

DEVELOPMENT OF THE PRODUCT/MARKETING. Touristic demand in the region is high. Target groups for the bridle-paths are riding-schools and riding-clubs, as well as individual riders. The area is partly used for the execution of events. The product is a further development of a previous non-regulated usage of the entire area for riding. By the construction of bridle-paths, the direction of visitors and thus the prevention of conflicts are possible. In 1995, the fee was introduced for the utilization of bridle-paths. Value is added by car-park areas. Important profit factors in the development of the product were the cooperation and discussion with all the groups and individuals involved. Riders can also exercise their sport in other areas, but the combination of extension, the quality of the bridle-paths and the variety in the landscape of the area are unique. Information on the market potential was derived from personal experiences and from discussions with customers. The product is mainly demanded in summer. However, the tickets sold are mainly annual tickets. The price is calculated on the basis of costs, in addition to a profit margin.

ORGANIZATION OF DISTRIBUTION. The enterprise markets the product directly. The responsibility in case of damage or accidents is with the riders themselves. There are contracts with riding-schools concerning the right of utilization.

INTERNAL ORGANIZATION/BOOKKEEPING. For the development of the product, the existing forest facility was important. In addition, the possibilities for bridle-paths were especially researched. The financial target is the gaining of net proceeds. The project has been developed separately. Lack of organization, motivation and information is the reason indicated for backlashes. The project required investments of approximately 250,000 guilders. From an economic viewpoint, the break-even point has just been reached.

CONTACTS WITH AUTHORITIES/RECEPTION. The authorities showed a positive attitude towards the project, and the official plan for the promotion of bridle-paths of the province, set up in 1980, has had positive effects on it. Private organizations and new customers welcomed the project. Though criticized by previous customers, it has been accepted in the meantime.

NL06 Environmental sponsoring North-Holland Landscape Conservation Foundation

A BRIEF DESCRIPTION. A nature preservation organization supervises nature preservation areas of approximately 1900 ha. Sponsoring is used as a source of financing. In addition, excursions are offered and snowdrops and turkeys are sold in a shop, together with other products. Furthermore, rowing-boats can be hired. The organization is financed as follows: 47% of the income is derived from contributions of the province of Nordholland and 44% is derived from membership subscriptions. Sponsoring and donations only contribute to about 1% of the total income.

Development of the product/marketing. It sometimes happens that certain sponsors apply to be sole sponsors. Up to now, this intention has been denied, with the sole exception of the VSB bank, which sponsors the Association of Nature Preservation Foundation at the Level of Provinces as the sole bank. In 1995, a manual was published on all nature preservation areas in Nordholland. This year 200,000 guilders of sponsor finances were acquired in total. There are attempts to coordinate individual sponsoring, which, up to recent times, has been mostly uncoordinated. For this purpose, the foundation differentiates between three kinds of sponsorships: sponsoring of material, sponsoring of projects and long-term partnerships. The extent of sponsoring ranges between 5000 and 10,000 guilders per project.

Organization of distribution. There are no written contracts so far.

Internal organization/bookkeeping. Until June 1997, sponsoring activities were integrated in the daily routine. From then onward, a staff member has been solely in charge of the acquisition of sponsoring on a part-time basis. If required, an external consultant is commissioned.

Contacts with authorities/reception. Other organizations in general reacted positively.

NL07 Paths for riding and mountain-biking Streekgewest Öostelijk Zuid-Limburg agricultural association/Vereniging Naturmonumenten nature preservation organization

A brief description. The agricultural association has developed a network of tracks for mountain-biking (60 km) and bridle-paths (44 km), which now, however, is supervised by a nature preservation organization. Passes are sold for the utilization of the facilities.

Development of the product/marketing. The project is a further development of existing facilities. The fee was introduced in 1990. In addition, the mountain-bike tracks were constructed in the same year. The main target of the product development is the direction of visitors. The income derived from the sale of passes is low. No marketing strategies, except for a leaflet on the tracks for mountain-biking, are applied. The fees for utilization are 10 guilders for an annual pass for the mountain-bike tracks and a single payment of 25 guilders combined with an annual contribution of 20 guilders for the utilization of the bridle-paths. Cooperation with clubs is intended as, in this case, virtually no costs for the sale of the passes will arise. Individual passes can be bought at the information centre and from the forest administration.

Organization of distribution. The mountain-bike tracks are maintained by a mountain-bike club (costs for material required are covered by the nature preservation organization). In return, the club gets a price reduction of 50%. There are no written contracts.

Internal organization/bookkeeping. Financial targets have not been defined. The main non-material target was the direction of visitors. Backlashes were especially indicated concerning the cooperation with riding-schools, which neither participate in the maintenance of the paths nor buy passes for all of the horses. The Naturmonumenten nature preservation organization spends 8000 guilders annually for the maintenance of the bridle-paths and 3000 guilders for the maintenance of the mountain-bike tracks. In total, a negative profit contribution is obtained.

Contacts with authorities/reception. Positive experiences were had with administrations in charge and the members of the mountain-bike clubs, who maintain the tracks. There have been negative experiences especially with riding-schools.

NL08 Sale of eco-hay National forest administration

A brief description. The national forest administration supervises vast nature preservation areas, which have to be mown to avoid overfertilization. The cut hay is sold as ecologically high-quality (certified) material to producers of organic feed (*eco-brok*). Approximately 1500 tons are also directly marketed to customers, such as hobby farmers and pony clubs, at local markets.

DEVELOPMENT OF THE PRODUCT/MARKETING. The hay was at first sold on the regular market, and the forest enterprise had to pay extra for processing. Due to the low content of proteins, the material was not very popular in the regular market. On the market for organic feed, profits are approximately 30% above the normal price. However, there have been problems with the separate drying of hay. For this reason, in the future mainly the processing of *eco-brok* is planned. According to the interviewee, there is no competition by other producers of eco-hay. In the future, cooperation with fodder producers is planned to expand the market activities to Germany.

ORGANIZATION OF DISTRIBUTION. The forest administration has supply agreements with different drying companies and producers of animal feed. The staff members of the administrative department in charge have some difficulties in keeping the contracts, as regional administrations market the hay on the local market. The separate stocking and drying of ecological cut grass is stipulated in the contracts. The utilization of chemicals and fertilizers is prohibited.

INTERNAL ORGANIZATION/BOOKKEEPING. The offer of eco-feed still leaves the forest administration with a negative profit contribution. In contrast to previous marketing on the regular market, however, the losses could be compensated in part by the working of the eco-market.

CONTACTS WITH AUTHORITIES/RECEPTION. The administration of the province contacted the national forest authority to market the cut grass of their areas in a similar way. However, in the areas in question, difficulties due to pollution arose.

NL09 Nature preservation weekend Vereniging Naturmonumenten

A BRIEF DESCRIPTION. The nature preservation organization makes a package offer concerning a nature preservation weekend. This weekend consists of a stay at a luxurious hotel, a slide show, a visit to an information centre, a bicycle tour, an ornithological walk, a book and an information package. The offer is intended to be an exclusive one, thus it is only offered once every 2 years.

DEVELOPMENT OF THE PRODUCT/MARKETING. The product area in question has an interesting landscape of 2500 ha (1400 ha of woodlands, 800 ha of dunes and 300 ha of agricultural areas). Approximately 10 years ago, the first regional association of the Vereniging Naturmonumenten started such a weekend programme. In the meantime, similar offers have been made by different regional associations. Demand is limited to spring and early summer. Advertising is only done in the magazine of the nature preservation organization.

ORGANIZATION OF DISTRIBUTION. Participants have to book the event at the hotel, and there is an agreement with the hotel that the nature preservation organization does not assume responsibility if there are fewer participants than booked.

INTERNAL ORGANIZATION/BOOKKEEPING. The financial target is solely the coverage of costs. The main target, which is very important, is the support of the nature preservation organization. For the organization of the seminar, a working effort of about 0.3 man-months per year is calculated.

CONTACTS WITH AUTHORITIES/RECEPTION. Demand exceeds supply, and the participants are very satisfied with the offer.

NL10 National park De Hoge Veluwe foundation

A BRIEF DESCRIPTION. The park area of about 5500 ha belongs to a private foundation (50% of woodlands, the rest being dunes and fens). The park is well known for its game species (red deer, moufflons, boars and does). Besides the beautiful nature, a variety of recreational and information facilities are offered. Another component of the park is also a famous museum of art.

Development of the product/marketing. The surroundings of the area are densely populated, and the park is attractive for regional citizens, as well as for day-tourists and holiday-makers. The foundation has existed since 1935, and the Kröller-Müller family contributed the area to the foundation on the condition that all of the facilities should be accessible with a package price. This price is at 8.50 guilders and about 600,000 tickets are sold annually. The park is fenced in, which allows a high game stock (an attraction) and better supervision of visitors. There are no similar offers in the vicinity. The product is distributed via tourist information offices, travel agencies and travel organizations. Entrance fees vary according to the number of vehicles, the ages of the visitors and the duration of the stay. In the future, a separate entrance fee for the museum is planned. Leaflets, advertisements and participation in exhibitions are used for communication purposes. For special occasions, individual offers are tailored.

Organization of distribution. As concerns the organization of distribution, the park cooperates with the tourist organizations mentioned above. The restaurant situated in the park is let to a tenant, and there is an agreement according to which the park has to transfer 30% of the income to the museum (which is a foundation on its own). For certain projects (for example, the underground museum), there is cooperation with sponsors.

Internal organization/bookkeeping. Special investigations were carried out in the park concerning its recreational aspects. The financial target is cost coverage. The annual proceeds are around Dfl. 500,000, of which 60–70% are derived from entrance fees. The non-material target is mainly environmental education. In total, 42 permanent and 40 freelance staff members are employed. The capacity of the park is approximately 700,000 visitors. The construction of the park was financed by the foundation's own means and by sponsoring. Bookkeeping and cost calculation are of eminent importance. At present, the park covers its costs.

Contacts with authorities/reception. The offer was welcomed by the general public. Elderly visitors, in particular, have a feeling of security because of the fences. Other forest administrations acknowledge that the park works without subsidies. Nature preservation organizations partly criticize fencing of the area, but the attitude of authorities towards the park as a whole is positive. However, there regularly arise difficulties with landscape planning, with nature preservation law and with issues of liability; these problems are mainly solved informally, as the members of the foundation maintain close relations to politicians.

NL11 Offer of 'eco-meat' Vereniging Naturmonumenten

A brief description. Besides forest areas, the nature preservation organization also runs grasslands in nature reserves. On these pastures, cattle are grazed for nature preservation reasons. The enterprises are certified as organic businesses, and they sell the cattle to fatteners and butchers. The sale of eco-meat in smaller quantities to the members of the nature preservation organization has failed due to a lack of demand.

Development of the product/marketing. Due to decreasing proceeds, the marketing has been adapted to the demand on the market for organic products. The special quality of the product is promoted. The idea and know-how are supplied by the landowner (nature preservation organization). The eco-meat has been marketed since 1988/89. Other enterprises producing organic goods influence the business rather positively, as they expand the market for ecologically conscious products as a whole. Market research was not done. The main target groups of the nature preservation organization are organic cattle-fatteners, households, members of the nature preservation organization and the regular market. There is also a degree of cooperation with hotels and the catering industry. The meat is marketed at a price of about 13.50 guilders kg^{-1}. Advertising has been done only once, in the member magazine of the nature preservation organization.

ORGANIZATION OF DISTRIBUTION. The enterprises directly market their cattle to organic fatteners, butchers and, on a limited scale, to households.

INTERNAL ORGANIZATION/BOOKKEEPING. There is a management plan for the nature reserve. Financial targets have not been defined. An important non-material target is the promotion of nature preservation. Due to certification, the enterprises are submitted to strict regulations concerning their production. According to the interviewee, bookkeeping and cost calculation are of minor importance. The enterprises are not profitable, and the marketing of eco-meat oscillates between positive and negative profit contributions.

CONTACTS WITH AUTHORITIES/RECEPTION. There have not been any contacts with authorities. However, the enterprise cooperates with other breeders of cattle, and this cooperation is said to be satisfactory.

NL12 Environmental sponsoring Utrechts Landschap nature conservation organization

A BRIEF DESCRIPTION. A nature preservation organization of the province of Utrecht, called Utrechts Landschap, uses environmental sponsoring for financing. Enterprises can become 'friends', for a contribution of 2500 guilders or 'golden friends', for a contribution of 10,000 guilders, of the nature preservation organization. This kind of sponsoring contributes 2–4% to the finances of the nature preservation organization. Other RES products are boat tours, guided tours and the offers of a shop selling environmentally conscious products.

DEVELOPMENT OF THE PRODUCT/MARKETING. The product is a further development of the regular sponsoring. So-called 'friends' or 'golden friends' get the following additional benefits for their commitment: a magazine free of charge (four times per year), a wall newspaper (four times per year), a festival (once a year in September), an annual report, a Christmas present, the possibility of organizing certain activities in the nature reserves, the possibility of using the country house once a year, the right of utilizing the logo of the nature preservation organization and a link on the internet homepage of the nature preservation organization. It is important to contact management executives of potential sponsors and, for this purpose, adequate staff are required. For this reason, the nature preservation organization employed a former IBM manager solely for this task. There is competition with other nature preservation organizations, especially with Naturmonumenten, which also acquires sponsors. Target groups are enterprises. Their wishes are considered in detail when designing a sponsoring concept. For the nature preservation organization, a label has been developed that can be used by the sponsors in the context of sponsoring.

ORGANIZATION OF DISTRIBUTION. The nature preservation organization itself offers the product. Informal contacts are very important in the acquisition of sponsors. There are no written contracts.

INTERNAL ORGANIZATION/BOOKKEEPING. The financial target is to acquire approximately 200,000 guilders through sponsoring. This target was reached for 1997. Besides, promotion is an important non-material target. Sponsoring has been explicitly organized in the framework of projects. According to the interviewee, there have not been any backlashes so far. The working expense is approximately 560 h annually plus additional assistance in the administration and distribution of the information material.

CONTACTS WITH AUTHORITIES/RECEPTION. Decisive profit factors, according to the statements of the interviewee, are the 'green attitude' of the sponsors contacted and an exact knowledge of what will be achieved. Contacts with the WWF have been very supportive, while contacts with Naturmonumenten were rather difficult, as this nature preservation organization works in the same sponsoring market.

NL13 Tree-crown path
National forest administration

A BRIEF DESCRIPTION. A national forest administration (3000 ha of woodlands) offers an information path visiting treetops. Other RES products are paths for hiking and horse-riding, educational nature paths, tours with a tramway hauled by horses and activities for children.

DEVELOPMENT OF THE PRODUCT/MARKETING. There is no comparable product in the vicinity. Considerable investments were required for the offer. Building permits had to be applied for, and the landscape plan had to be adjusted. The product has been developed in cooperation with the province and with local organizations for tourism. In 1996, it was introduced on the market. Additional know-how has been supplied by architects and other consultants. Profit factors are good publicity in the introductory stage, the uniqueness of the product so far and the willingness to constantly innovate. Information on the market potential was supplied by the organization of tourism. The peak in demand is reached in summer, and the product is marketed in a product mix with, for example, an information centre, playgrounds and a restaurant. Special arrangements are offered for organizations on demand by the operator in charge of the restaurant. There are approximately 140,000 visitors annually. The entrance fee is 3 guilders for children (single ticket) and 9 guilders for a season ticket. Adults pay an entrance fee of 5 guilders for a single ticket and 15 guilders for a season ticket. A special communication strategy has been developed for the project.

ORGANIZATION OF DISTRIBUTION. The treetop tour has been developed in cooperation with the organization for tourism of the province, the hotels and the catering industry without a formal supervising organization. The forest administration is in charge of the maintenance and cultivation of the treetop path; the tourism organization deals with the marketing. Informal contacts have been very helpful in the cooperation. Contracts exist only with the companies in charge of cleaning, with the tenant of the restaurant and with tourist guides.

INTERNAL ORGANIZATION/BOOKKEEPING. A special inventory was executed when selecting the location. The financial target (net profits) has been reached even more satisfactorily than planned. The working expense is calculated roughly at 60 man-months annually and the period till regaining the investment (1.5 million guilders) is estimated to be 5–7 years. This time span could be shorter with lasting success of the project. The project has been financed by means of the national forest administration and the province. Bookkeeping and cost calculation of the project are done separately.

CONTACTS WITH AUTHORITIES/RECEPTION. Up to the present, there have been positive experiences with all groups involved.

NL14 Nature camp-site
Quadenoord–Boosbeek country estate

A BRIEF DESCRIPTION. The owner of a private country estate (160 ha of forest, 70 ha of agricultural area) offers a plain nature camp-site on his property. From this camp-site, 40% of his income is derived. The camp-site comprises 50 places for day-visitors and 100 places for seasonal campers. About 2000 campers use the offer per year.

DEVELOPMENT OF THE PRODUCT/MARKETING. The camp-site has existed since 1926 and has been developed gradually. Target groups are pensioners and young families with small children. The campers combine their stay with cycling tours in the surroundings and sight-seeing. The average duration per stay is 3 days. A special licence from the community was necessary for the development of the site. There is competition because of other commercial camp-sites and the fact that farmers are now allowed to let up to ten places on their properties. Demand is mainly in summer. Advertising is done by leaflets and advertisements in tourist guides. A label has been developed, but not registered.

ORGANIZATION OF DISTRIBUTION. The owner markets the product himself. An additional intermediary is an association of the owners of country estates. By their membership fees, the proprietors acquire the official label of the association, an entry in the guide for accommodation and joint advertising at fairs and exhibitions. The camp-site offered cannot be expanded because of nature preservation regulations. In the Netherlands, proceeds derived from the forest economy are subject to special tax regulations. These regulations, however, do not apply to the camp-site and, on the occasion of a tax inspection, the owner had to pay tax arrears.

INTERNAL ORGANIZATION/BOOKKEEPING. The financial target of net profits is obtained relatively well. The investments for showers and public lavatories were relatively low (low standard) and had already taken place some time ago. Also the working expense is not high and is mainly limited to the control and cleaning of the sanitary facilities by cleaners.

CONTACTS WITH AUTHORITIES/RECEPTION. While one community was supportive of the project, another showed a negative attitude towards extension of the camp-site. There are contacts with an association of nature-orientated camp-sites, as campers have to purchase a 'nature camping ticket' there to utilize the camp-sites.

NL15 Storage of carbon dioxide (CO_2) [indication of the name not authorized]

A BRIEF DESCRIPTION. A private owner of a country estate (870 ha, of which 145 ha are forest) afforests an area of an additional 100 ha. The afforestation is financed by the Forests Absorbing Carbon Dioxide Emission (FACE) foundation. This foundation was established in 1990 by Dutch electric power stations. The target is the afforestation of approximately 170,000 ha of woodlands, of which 5000 are situated in the Netherlands, for the purpose of CO_2 storage.

DEVELOPMENT OF THE PRODUCT/MARKETING. The product was a newly developed product, which markets the CO_2-storage of the country. By cooperation with the foundation, the reafforestation costs of an area which the forest owner has to reafforest are in any case covered up to 40% by the foundation. The forest planning of the afforestation is done in cooperation with an association of forest owners, which is also the formal contractee of the foundation. A substantial factor in the realization of the afforestation was the nature preservation law, which obliges forest owners to leave at least 30% of their property as forest if they want to continue to make use of the exceptional tax regulation for forest owners. A large part of the area is reforested in the tropics. The cooperation was realized because of personal contacts between the head of the association of forest owners and a representative of the foundation.

ORGANIZATION OF DISTRIBUTION. The execution of afforestation, cultivation, inventorization and other duties is stipulated in detail in a contract of 20 pages between the association of forest owners and the foundation. Furthermore, the foundation demands a management plan for the next 30 years, the observance of which is controlled by the foundation.

INTERNAL ORGANIZATION/BOOKKEEPING. The basis of the project is the management plan set up by the association of forest owners over a duration of 30 years. In total, 700 Dfl. have been invested for reforestation. A considerable part of the investment was derived from national funds.

CONTACTS WITH AUTHORITIES/RECEPTION. Authorities reacted positively towards the project or remained neutral. A variety of national funds for initial afforestations granted by the authorities in charge were important for the afforestation project as well.

NL16 Environmental sponsoring Het Geldersch Landschap foundation

A BRIEF DESCRIPTION. The nature preservation organization mentioned above maintains approximately 9500 ha of nature preservation areas, and for this purpose acquires sponsors as an instrument for financing. The main sponsors are three waterworks, which finance, for example, the transformation of softwood into hardwood forests to increase groundwater.

DEVELOPMENT OF THE PRODUCT/MARKETING. The first step towards the development of systematic sponsoring was the decision of the management of one of the waterworks that all of the staff members should be members of the nature preservation organization. This was expanded to systematic sponsoring. When utilizing sponsoring as a means of communication, it is important for the sponsor to be mentioned in different contexts (for example, in the magazine for members of the organization). According to the interviewee, it is important that one particular person, who contributes an adequate attitude and background for the acquisition of sponsors, is in charge of sponsoring and is solely concerned with sponsoring, without fulfilling other duties. There are other nature preservation organizations in the region – for example, Naturmonumenten – who compete for limited sponsoring resources. Potential sponsors (target groups) are waterworks and other enterprises, but also the Lions and the Rotary Club.

ORGANIZATION OF DISTRIBUTION. The organization directly offers sponsoring at the regional level. At the national level, the organization cooperates with the association of nature organizations in the provinces. There are no formal sponsoring contracts. However, there are written contracts with the three waterworks.

INTERNAL ORGANIZATION/BOOKKEEPING. Especially concerning the sponsoring of the waterworks, the regional management plan was an important basis. The financial target of net profits was reached quite well. Non-material targets of nature preservation and the promotion of environmental consciousness are the main aims of the organization. Internally, there is a regulation that enterprises with negative environmental behaviour are not allowed to act as sponsors for the project. A lack of experience in the acquisition of sponsors and too great involvement in daily routine are said to be responsible for backlashes.

CONTACTS WITH AUTHORITIES/RECEPTION. There were no contacts with authorities. However, there are contacts with the association of nature preservation organizations in the provinces and with potential and active sponsors. As a positive side-effect, it is underlined that the public increasingly recognizes the work done by the organization.

NL17 Nature camp-site Schellerberg country estate

A BRIEF DESCRIPTION. The owner of a private country estate (40 ha, of which 10 ha are forests, 19 ha pastures and 10 ha nature preservation areas) offers a plain nature camp-site and group accommodation on his property (40 parking positions, 22 beds). Green twigs for decoration are offered additionally.

DEVELOPMENT OF THE PRODUCT/MARKETING. Target groups are young families, pensioners and backpackers. When groups book the offer, the individual demand of the group is largely considered. The visitors mostly combine their stay with bicycle tours in the surroundings. The average duration of the stays is 2 weeks. Investments were required as the camp-site had to be furnished with bathrooms and as a part of the former stables had to be adapted to the needs of group accommodation. According to the interviewee, the investments were of medium extent, considering the expected proceeds. The product was introduced in 1990 and adapted in 1993. The owner now offers the accommodation facilities as a member of the association (see also NL14). Profit factors are hospitality, development from scratch, steady growth and the utilization of every business possibility. In the region, there is competition by similar offers. The site is only open from April until September, as otherwise damage would occur due to rainfall soaking through the soil. Means of communication are leaflets, advertisements and joint advertising (LKC; Association of country estates with camping facilities).

ORGANIZATION OF DISTRIBUTION. At present, the owner is the sole supplier of the product. A contract with a travel agency did not turn out to be successful. Concerning the legal framework, building permits were important and the licence of approval for offering a camp-site. The latter does not exist at present.

INTERNAL ORGANIZATION/BOOKKEEPING. The financial target is net profits. The investments have to be regained after 3 years. A backlash indicated is that some of the houses let cannot be redesigned because of tenancy law.

CONTACTS WITH AUTHORITIES/RECEPTION. The contact with the authorities in charge is satisfactory. In some cases, authorities have denied complaints of the waterworks because of camping and a newly built oil tank, because the owner of the country estate had fulfilled all relevant instructions. However, there have been difficulties so far concerning the setting up of a public sign-post due to high costs.

NL18 Holiday huts in the forest
National forest administration

A BRIEF DESCRIPTION. The national forest administration lets 15 former forester's houses in the forest as holiday accommodation. Target groups are young families, pensioners and groups of up to six people. The houses are equipped relatively luxuriously, and they are deliberately targeted to a slightly higher price level.

DEVELOPMENT OF THE PRODUCT/MARKETING. The forest administration could use the existing houses, but the redesigning required investments of 125,000–150,000 guilders per house. The preparations of the project started in 1995, while the first season of renting the houses was in 1998. The product is unique of its kind. In this way, competition is not expected. Prices for the offer vary between 650 and 1950 guilders per week. The basis of price calculation was the investments. Since October 1997, the project has benefited from good publicity. A detailed report was published in a national newspaper and, furthermore, the forest administration has issued a lavish catalogue indicating all the houses offered.

ORGANIZATION OF DISTRIBUTION. The product is marketed by the forest administration itself, in cooperation with a travel intermediary. As to the legal framework, conversion of buildings from permanent flats into holiday accommodation was important. Before developing the offer, the approval of the community was applied for.

INTERNAL ORGANIZATION/BOOKKEEPING. Financial target are net profits. The investments were originally planned to be regained after 3 years, but it had already taken 3 years before the forest administration could get rent for the first houses. Backlashes were due to a lack of experience in a new market; the administration of the buildings from a distance and a lack of experience in positioning the product on the market. At present (1998), 15 of the holiday huts could be rented. A long-term target is the letting of 50 or even 90 huts. Together with an interior decorator, the plan for future uniform furnishing of the houses was developed. This furnishing will be part of an overall concept and will mediate the feeling of living in a forester's house for the tenants.

CONTACTS WITH AUTHORITIES/RECEPTION. Contacts with authorities mainly consisted of applications for the necessary licences concerning the transformation of buildings.

NL19 Shop offering environmentally conscious products Marienwaerdt country estate

A BRIEF DESCRIPTION. On the private country estate (of approximately 900 ha), the daughter-in-law of the owner has established a store that sells a variety of farm products and also forest products. She rented the farm and the store (partly for tax reasons) from the owner.

DEVELOPMENT OF THE PRODUCT/MARKETING. For the construction of the processing facilities, larger investments were necessary. In the future, facilities for making cheese are planned. For this aim, funds are applied for. The idea for the development of the product was supplied by the owner. The young woman managing the nature store has over 5 years experience from employment at another enterprise. In addition, she attended courses in bookkeeping, PR and marketing. There is no similar shop in the vicinity. Information on the market potential was derived from discussions with farmers in Great Britain, specialized magazines, trade fairs, conversations with neighbours and business consultants. Target groups, for example, are recreationists, gourmets and enterprises ordering their business gifts for Christmas. There are also cookery classes and open days for visitors offered to the customers.

ORGANIZATION OF DISTRIBUTION. The product deals with typical direct marketing. A licence had to be applied for the opening of the store. Regulations concerning hygiene and a rule that 70% of the products have to be derived from their own production have to be observed.

INTERNAL ORGANIZATION/BOOKKEEPING. The financial target of obtaining net profits has been obtained quite well. The offer has been organized as a separate project, for which the daughter-in-law is solely responsible. Part-time employees additionally cooperate in the production of the goods and their sale. The project required capital expenditure planning, and the means for the investments required were derived from the shop's own means. In the future, cheese manufacturing will be added, more of their own products will be marketed, and they are thinking about an information centre.

CONTACTS WITH AUTHORITIES/RECEPTION. Authorities contacted on this matter in general reacted positively. Contacts with a recreational organization exist in so far as the shop is mentioned in their magazine. In the beginning, the general public was reserved towards the owner of a large country estate. But this reservation has now ceased.

NL20 Offer of holiday flats Zelle country estate

A BRIEF DESCRIPTION. The private country estate (350 ha, of which 165 ha are forest) offers 12 holiday flats. To a large extent, the accommodation consists of renovated houses that were formerly let. An additional RES product is a nine-hole golf-course. The income is derived as follows: 40% from the letting of holiday accommodation, 40% from the golf-course and a further 20% from lease contracts, agriculture and the forest economy.

DEVELOPMENT OF THE PRODUCT/MARKETING. In 1950, the father of the present owner began with two holiday houses and, 5 years ago, the last house was renovated. Visitors mostly stay for longer than 2 days (80%), but 20% of the visitors only stay for a weekend. An increasing number of visitors bring with them their own horses, which can graze on the country estate. According to the interviewee, it is an important factor for success to offer what the holiday-makers expect. It is important to preserve the typical character of a country estate. Other suppliers influence the market positively, as this promotes the 'extensive' form of recreation. The golf-course forms an important part of the product mix; for 10% of the opening hours it is exclusively reserved for the visitors in the holiday accommodation. In addition, package offers (accommodation and golf, accommodation and cycling tours) are made. Organizations use the offer for the execution of meetings. Advertising is partly done by the country estate itself and partly in cooperation with the LKC (see NL14).

ORGANIZATION OF DISTRIBUTION. The enterprise offers the holiday accommodation direct and in cooperation with the tourist information. Furthermore, there is membership of the LKC and contracts with the operators of the golf-course. For tax reasons, an operating company has rented the area for holiday accommodation from the country estate.

INTERNAL ORGANIZATION/BOOKKEEPING. The financial target of net profits has been obtained. The owner himself fulfils 90% of the tasks required. According to the owner, it would have been better to approve the construction of more holiday accommodation in the past. Such an expansion is not possible any more, due to present forest law.

CONTACTS WITH AUTHORITIES/RECEPTION. Contacts with authorities (building permits) have not always turned out satisfactorily. However, there is satisfactory cooperation in the areas of tourism and recreation.

Notes

1 Minutes of 3rd RES Meeting, Padua, 3–4 October 1996.
2 For the sake of brevity, the term RES product in the following will refer to products in the environmental and recreational areas.

References

Mantau, U. (1995) Forest policy means to support forest output. *EFI Proceedings* 2, 131–146.
Mantau, U. (1997) Alternative ways of correcting market failure – product structures – a key to marketability. In: Roper, C.S. and Park, A. (eds) *Living Forests*. HMSO, London, pp. 267–274.
Mantau, U. (1998) Beispiele und Konzepte für marktfähige Umwelt- und Erholungsleistungen des Waldes. In: Glück, P. and Nießlein, E. (eds) *Wer zahlt für die gesellschaftlichen Leistungen des Waldes?* Schriftenreihe des Instituts für Sozioökonomik der Forst- und Holzwirtschaft, Vol. 30, Vienna, pp. 45–52.
Sekot, W. (1997) Multifunktionale Leistungserstellung, PR-Gag, Peanuts oder Wirtschaftsstandbein? *Österreichische Forstzeitung* 11, 7.
Streit, M.E. (1991) *Theorie der Wirtschaftspolitik*, 4. Werner-Verlag, Düsseldorf, pp. 13–17.
von Hayek, F.A. (1969) Economy, science and policy. In: *Freiburg Studies. Collected Abstracts.* Mohr, Tübingen, pp. 11–17.

3

Economic Strategies for Transformation and Product Development

3.1 Introduction

This chapter analyses the transformation and development of public goods (environmental and recreational goods and services (ERGS)) provided by forestry and the related environment into marketable private goods and services – what is defined here as recreational and environmental services (RES) products.

Throughout this chapter the following definitions have been adopted:

1. ERGS are mainly public goods and services provided by forestry and the related environment before any transformation/ development has taken place.
2. Transformation is a modification of ERGS, mainly concerning their institutional nature (including legal status, property rights, planning permissions, contractual agreements, etc.).
3. Development is a modification of ERGS, mainly concerning their economic nature (provision of complemen-tary/additional goods and services, marketing promotion, changes of existing contracts, etc.).
4. RES products are mainly goods and services provided by forestry and the related environment after transformation/development has taken place.

A questionnaire was used in the four European Union (EU) countries participating in the research – Austria (AU), Germany (DE), Italy (IT) and the Netherlands (NL) – to gather information from 98 actual case-studies. The same framework for the questionnaire, translated into each language, was used throughout and covered all aspects of the research.

This chapter in particular reports on the results of the section of the questionnaire entitled 'Strategies for transformation', focusing on the mechanisms of transformation/development. The main goal is the identification of the general methods or paths of transformation/development of ERGS into RES products. Indications are therefore deduced so that the main factors behind transformation/development can be outlined and their applicability to different cases/sites where market remuneration of ERGS can be implemented.

The first part of the chapter is devoted to the characterization of the economic nature of public goods with reference to ERGS provided by forestry and the related environment. Theoretical backgrounds are briefly outlined according to public-finance approaches, such as taxation and pricing, and the search for optimum provision of public goods. Excludability and rivalry criteria are employed for defining public and private goods. Mixed cases are particularly considered as being very relevant for understanding the institutional and economic nature of 'local public goods', amongst which many ERGS and the related RES products must be included.

Public goods are also related to externalities and market failures. Policy options to cope with this are, therefore, outlined. Mandatory approaches (obligation to provide ERGS) are compared with

Pigouvian internalization ('state pays' approach) and Coasian market creation ('beneficiary pays' approach). It is stressed how mandatory approaches have been much applied in the past under the so-called 'forest regimes'. Pigouvian policies are now an integral part of the common agroenvironmental policy of the EU. Market solutions, including the transformation/development of ERGS into market RES products, are investigated in particular throughout this research project.

The relationship between ERGS and RES products is analysed in the second part, underlining the various degrees of identification and complementarity in consumption and how this complementarity can be exploited during the transformation/development process. The large variety of ERGS and RES products investigated in the four countries are then outlined and classified in the third part. The situation before transformation is described in the fourth part, where the basis for transformation processes is examined, as are the changes in excludability and rivalry. The institutional changes and related market approaches are discussed and the complex setting of complementarity relationships between ERGS and RES products detailed. The transformation/development paths are then summarized in the fifth part in order to identify some general guidelines, which are aimed at assisting the definition of actions at both enterprise and policy levels, to make possible the support of the transformation/development processes in different circumstances, cases and sites. The aim is to show the links between actual situations of forest properties/enterprises and how improvements could be made to generate income and to gain financial sources for providing ERGS and conservation. Parameters, factors and policy tools that could help the transformation/development are discussed and then proposed. Considerations on possible actual situations, theoretical backgrounds and policies/strategies to achieve market remuneration of ERGS through RES products are summarized in the final part.

3.2 ERGS and RES Product Characterization

In addition to traditional market commodities, forests and the related environment provide, or produce, several ERGS. Examples include watershed management, pleasant landscapes, maintenance of rural lanes, footpaths and other services, habitats for various kinds of flora and fauna, grounds for sports and other recreational activities. These ERGS are generally perceived by our societies as public goods, as people cannot be excluded and rivalry is not greatly felt. The degree of excludability can, however, vary according to the property rights of each particular ERGS and is sometimes very different across Europe, as are the local customs and the transaction costs to enforce exclusion.[1] The degree of rivalry is equally variable according to each particular ERGS: e.g. a landscape visible from a public road does not generally have a high level of rivalry, while the access to a nature reserve can create congestion and rivalry. As is the case for many other public goods, divisibility and possible rationing are crucial factors characterizing ERGS. In certain circumstances, rationing is easily achievable (e.g. limitation to a certain group of users, such as a club), while sometimes it is impossible or transaction costs are too high (e.g. difficult enforcement of limitation/exclusion).

The issue of public goods and services provided by forestry and the related environment is far from being a recent one. Watershed management, hunting and shooting and villagers' rights to the various forest products (firewood, mushrooms, etc.) are old debated issues, in addition to recreation and conservation, which are now considered to be the most important ERGS. One must be aware that, along with history and economic development, ERGS have been perceived in various ways by society, philosophers, economists, forest managers and landowners. However, the view of ERGS as public goods and externalities has been, and still is, dominant. This concept is basic to environmental economics, a discipline trying to 'offer a framework within which to analyse the problems which we face in making choices about the environment in which we live' (Hodge, 1995, p. 3). When making these choices, policy-makers need to take into account environmental externalities and public goods, or 'non-priced values'. No account, or improper accounting, easily generates the so-called market failures. The idea of accounting for the total economic value (TEV) of natural resources, including forests and the related environment, has therefore been developed, and is now central to environmental economics. It is accepted that the market is able to encompass only a part (the private one) of the TEV of natural resources, whose main components remain public: i.e. non-market use value, option, existence and bequest values (Turner *et al.*, 1994; Ferro *et al.*, 1995).

Forestry-related ERGS have been recently linked to what the present study has called RES, or RES products, which are an integral part of private market goods. Markets and marketing provide the link between ERGS (which basically remain in the field of public goods externalities) and RES products, which are an integral part of private market goods. It is, therefore, important to investigate the institutional-economic nature of ERGS in relation to RES products. These relationships, which can vary from close identity to various different degrees of complementarity[2] in consumption, have been the leitmotif, or the core, of the present study, being the driving force behind the transformation/development of ERGS. In order to investigate the economics of the transformation/development of ERGS into RES products, the following subsection analyses the theory of public goods, including ERGS.

3.2.1 The development of the theory of public goods

Philosophers and, above all, jurists have considered the nature and provision of public goods since the Middle Ages. According to Thomas Aquinas's scholastic philosophy, land ownership is a concession that should be exercised in the manner of a service. What is referred to here as ERGS were to be provided to society by those entitled to land ownership, far from the Roman concept of complete *dominium*. Property was, therefore, conceived as the *potestas procurandi* and *dispensandi* (the duty to manage and conserve the land, providing various services to society), rather than the *ius utendi, fruendi* and *abutendi* (right to use or abuse the land). Quite clearly it was a conception aimed at the protection of the various rights of villagers and peasants, traditionally entitled to a series of public goods and services, including various ERGS, provided by forests owned by kings and feudal lords.

Later on, at the time of the formation of modern European states (16th and 17th centuries), the provision of public goods was considered to be part of the government's policy and administration. The taxation necessary to finance the state's organization (provision of public goods), particularly 'law and order' and the national defence, became a central issue. However, financial problems were seen as pertaining to law and philosophy rather than economics. De Groot and La Court intended the services provided by the state as a 'social protection' and Pufendorf defined taxes as the payment to the state for the social protection. Other philosophers of the 'natural law' school (Locke, Hobbes and de Montesquieu) held slightly different views but all stated that equality should be the cornerstone of the state/citizens relationship, whereas taxes were a duty in exchange for the benefits received. The link was to be seen in terms of utilities (costs and benefits) rather than mere monetary transactions. It must also be mentioned that a few 'moral scientists' and quasi-economists of the 16th and 17th centuries started to pay close attention to the theory of government, state organization and the underlying financial implications, including Petty and Hume in Britain, Mun in the Netherlands, Boisguillebert in France and Galiani in Italy.

A real financial theory, aimed at defining the essential public goods and services that the state should provide in a market industrial economy, was developed at a normative level by Adam Smith, who was initially a professor of moral philosophy, before becoming the founder of classic economics, able to assemble all existing economic wisdom. He stated that law and order, national defence, public works and education (Brosio, 1986, pp. 27–28) should be amongst the few essential public goods and services a state should provide. The financing of these services should avoid possible market distortions. In particular, whenever the benefit of the individual citizen was clearly identifiable, the price (cost) of the service should be paid (tax), while for other more general public services (completely indivisible), like national defence, a contribution should be made according to personal income.

It was not considered to be the state's role to provide environmentally related public goods, such as the ERGS. The environment was far from being an issue to be dealt with by state intervention and policies.[3] However, on further consideration of the experience of the 17th-century modern states that were beginning to develop industrially, it can be seen that attempts were made to privatize various ERGS, with a more general definition and assignments of property rights over forests and land. This policy was more pronounced in countries like Britain (enclosures movement) and Austria (Maria Teresa reforms). The basic idea was that agricultural/forest products as well as the various other rights over the land, including ERGS, were to be clearly defined and assigned to the landowner. Large tracts of Europe remained, and still are, outside the field of application of these 'privatization' laws and so, particularly in marginal land of mountainous regions like the Alps, common properties still exist.[4] However, the 1791 French Constitution acknowledged the public ends of private property, as opposed to the Roman law of full *dominium*. Therefore, ERGS could also be seen as part of the public scope of private ownership.

Nineteenth-century neoclassical economics paid little attention to the provision of public goods and the underlying financial implications. Although taxation and incentives were considered, they remained far from being part of a comprehensive theory of public finance. State expenditure was conceived by British neoclassical economists as something pertaining to the sphere of policy rather than that of economics. The institutional situation of Britain, the cradle of neoclassical economics, was such that, on the one hand, public expenditure was part of a well-established procedure (see the exchequer tradition consolidated by the Gladstone reforms of the state budget in 1866) and, on the other, public expenditure remained rather limited until the 20th century. Britain also had a rather unique system, with a mixture of feudalism and liberalism that made it possible, until the 20th century, to have, for instance, public goods and services, such as country roads and drainage (and in general what is here called ERGS), provided as *noblesse oblige* by the 'landed gentry' – the aristocracy and ruling class of an otherwise well-advanced industrial market economy.

According to Samuelson (1955) and Head (1962), the most important modern scholars of public-finance and public goods theory could be found in continental Europe, particularly Austria–Germany (Sax, 1883; von Wieser, 1914), Italy (Pantaleoni, 1883; De Viti de Marco, 1888; Mazzola, 1890; Barone, 1912) and Sweden (Wicksell, 1896; Lindhal, 1919). It has been argued that the interest paid to public goods and public finance, at least in Germany and Italy, was linked to the financial problems in building a modern unified state in the 19th century (Brosio, 1986, pp. 29–31). Economists paid attention to the two sides of the state budget: revenues from taxes and expenditure allocation. The provision/distribution and amount of public goods services were thought to be connected to consumers' preferences and utilities as well as their marginal costs. The exchange between private goods (taxes) and public goods should have been seen, according to Sax (1883), in terms of 'exchange value'. Mazzola (1890) stated that the final utility of public goods must be equal to their prices. No doubt, neoclassical economics and marginalism were central to this approach.

Indivisibility, or jointness, in supply and consumption was considered to be the main feature of public goods (Sax, 1883; Mazzola, 1890; Barone, 1912; von Wieser, 1914), where everybody is entitled to their use. The consumption by A does not affect consumption by B, so quantities must be summed up on the vertical axis of the demand schedule, as shown later on by Samuelson (1954). Utilities, however, can be different according to the individual's needs and tastes. Quantification of preferences for utilities was, therefore, the main problem. According to the above group of economists, given the impossibility of applying different market prices to public goods, the solution should have been left to the government. Taxation (prices), allocation of financial resources, expenditure and provision of public goods were part of government decision-making. If this political process was correctly undertaken, efficiency should be achieved and, as a consequence, in a democratic system, the government should be re-elected, or otherwise changed. It is rather clear that the founders of modern public-finance theory attempted to encompass within the same model the neoclassical equilibrium between marginal utility and prices and the real-world mechanism of public-finance decisions taken at a political level.

A well-known contribution to the development of financial theory is that of De Viti de Marco (1888, 1934) who advocated a 'voluntary exchange' where, reflecting the Smith approach, the 'prices' of public goods and services should be met through taxes paid according to income. Public-finance should work like a market, with marginal utilities of public goods and services provided by the state being equal to marginal costs. Taxation of individual net income should, therefore, solve the problem of financial charge distribution respecting the two principles of income capability and payment of the public goods and services according to the use. The 'voluntary exchange' theory was specified by Wicksell (1896). The idea that every citizen should pay according to the use of public goods and services as decided by the government, formulated by the first group of public-finance scholars such as Sax (1883) and Mazzola (1890), was criticized for being unrealistic. Public goods and services are often indivisible and the use by individuals depends upon the society's contribution. In addition, individuals do not show their preference (free-riding attitude), therefore, obligation and regulation become necessary. It was therefore remarked that the compulsory (top-to-bottom) approach to public finance, based on government decision-making, has an element of inefficiency (Wicksell, 1896). In order to minimize inefficiency, a 'quasi-unanimity' principle was introduced, where each tax/service should gain the approval of a large majority, varying according to the relevance of individual cases. This is called the 'just taxation' principle, where a majority approval should guarantee that taxes satisfy the needs. The Wicksell principle required an even distribution of income throughout the society.

The above theory was completed by Lindhal (1919), who proposed a model where Pareto optimality[5] could be defined with reference to the supply and price of public goods. The model requires that citizens correctly reveal their preferences (quantity of public goods they wish and are willing to pay for), while the cost of production of the public good must be constant. The applicability of the model has been questioned because the citizens' strategic behaviour implies that there is no interest in expressing preferences. 'Voluntary exchange' is impossible and collective (state) decision-making becomes necessary. It has been noted, for instance, that it is the relationship amongst social classes that determines public-finance equilibrium and that the domination of democracy by specific interest groups does not guarantee justice. Goldscheid (1917) meanwhile tried to explain the development of financial policies according to social class relationships. Public goods and finance theory has triggered further development of Wicksell 'quasi-unanimity' through other theories – e.g. 'voting' and constitutional and 'public choice', as expressed by Arrow (1951) and Buchanan and Tullock (1962). Also relevant is the Downs (1957) economic theory of democracy, which shows how politicians try to maximize votes, while people are usually only concerned about their own personal interests. However, competition among political parties should allow the achievement of preferences expressed by the average citizen.[6]

At a strictly economic level, the 'voluntary exchange' theory was reconsidered in the 1950s by the work of Samuelson (1954, 1955, 1958), the well-known 'trilogy' published in the *Review of Economics and Statistics*. The modern definition and theory of public goods is accredited to this author and his contemporary, Musgrave (1959). Public, social or 'collective consumption goods' are those 'which all enjoy in common in the sense that each individual's consumption of such goods leads to no subtraction from any other individual's consumption of that good' (Samuelson, 1954). However, public and private goods are considered as polar cases of a more varied situation, with many intermediate mixed goods (Samuelson, 1955).

Unfortunately, markets can fail, particularly where public goods are concerned. Three issues are used specifically to imply the idea of possible market failures and the need to adopt policies that are able to solve the problem (Head, 1974):

1. Optimal provision of public goods where it is shown that Pareto optimality, defined in a world of private goods, can be radically altered when one of the goods is public.
2. Decentralized market mechanisms can fail to satisfy the demand for public goods. A price for additional units available at nil marginal cost will exclude consumers and affect the Pareto optimum; however, the impossibility of exclusion implies undersupply of public goods.
3. Political provision of public goods becomes necessary because neither a conventional market regime of perfect competition nor a generalized market regime of voluntary agreements can be expected to approximate Pareto-optimal provision.

3.2.2 Excludability and rivalry criteria for defining public and private goods

Following the work of Samuelson and Musgrave several authors have started to define goods in terms of rivalry and excludability, two terms much discussed in economics textbooks and articles, though sometimes with various degrees of misunderstanding (Randall, 1987). Pure public goods[7] should be non-rival and non-excludable in consumption and so be fully available to the public. On the contrary, pure private goods should be fully rival and excludable. The two concepts of rivalry and excludability are shown in Fig. 3.1.

Rivalry means that consumption by an individual excludes all other consumers. This applies to private goods, such as 'food, clothing, housing, automobiles, and millions of other marketable goods. A hamburger eaten by A cannot be eaten by B. Thus, benefits are internalized and consumption is rival' (Musgrave and Musgrave, 1973, p. 53). Meanwhile, public goods are non-rival and so can be enjoyed 'in common in the sense that each individual's consumption of such goods leads to no subtraction from any other individual's consumption of that good' (Samuelson, 1954). It could be remarked that public-finance scholars have traditionally expressed the concept of rivalry in consumption in terms of divisibility, portioning and marginal costs for providing an additional unit. Stiglitz (1986) underlines the fact that pure public goods have two critical properties, it being neither feasible nor desirable to ration their use. The classic example of national defence illustrates these points well: everybody can benefit from defence as nobody can be excluded and a newborn citizen does not increase the cost of provision (complete non-rivalry).

Excludability means that the use of a certain good is restricted to those who are

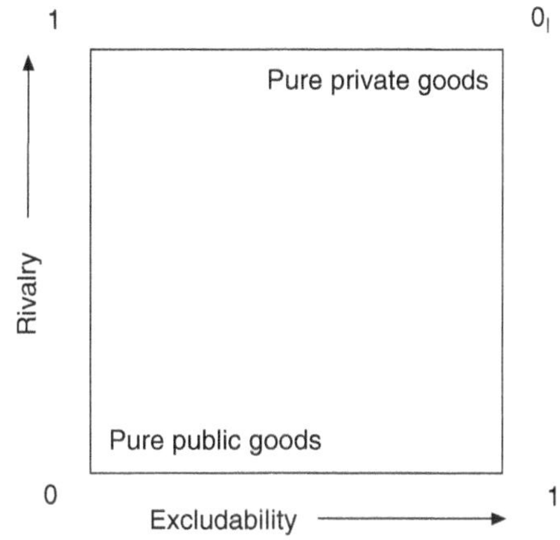

Fig. 3.1. Classification of goods according to excludability and rivalry: pure public goods and pure private goods.

entitled to it and generally pay a market price for the good. The concept can be related to rivalry, as excludability requires limitation. The simplest way to introduce limitation is by rationing (divisibility) of a certain good, making it possible for the market to operate rivalry: e.g. consumption by A excludes consumption by B. Exclusion is not, however, possible for many public goods as they must be jointly supplied and consumed, because divisibility is physically impossible: for example, public goods such as national defence and natural resources such as clean air, landscape, biodiversity, etc. The enforcement of excludability can also be linked to the property-rights regime and/or the transaction costs. It can be concluded that divisibility and property-rights make possible exclusion and rivalry, contributing to the creation of private market goods, a category clearly opposed to public goods.

3.2.3 The continuum from public to private goods: mixed cases

The diagram of Fig. 3.1 has been much used to distinguish public from private goods according to rivalry and excludability criteria. Excludability ranges from 0 (the minimum) to 1 (the maximum) along the horizontal axis and rivalry, again from 0 (the minimum) to 1 (the maximum), along the vertical axis. Examples of similar presentations can be found in Buchanan (1967, p. 188), perhaps the first author to propose the diagram. However, he put on the vertical axis the level of indivisibility (a proxy of rivalry) and on the horizontal axis the number of users (a proxy of excludability). In Fig. 3.1, therefore, it is shown that public goods are located at 0 and private ones at 0_I the so called polar cases.

Musgrave and Musgrave (1973, p. 54) also gave in their textbook a classification of goods explicitly using the terms rivalry (in consumption) and excludability (in supply). The result was a four-category representation (though not a diagram), as used by several subsequent textbooks on public finance. Stiglitz (1986) meanwhile puts on the vertical axis what he calls 'marginal cost and desirability of exclusion'. This concept is again that of rivalry: in fact 'non-desirability of exclusion' is explained as 'one individual's consumption does not detract from the amount that is available for others to consume', i.e. non-rivalry. The lower left-hand corner represents pure public goods like national defence or the natural environment, where exclusion is not possible and rivalry does not exist. The upper right-hand corner represents pure private goods, where rivalry and excludability are regulated by market prices.

Figure 3.1 quite clearly shows that pure public goods and pure private goods are polar cases, as termed by Samuelson (1955). The extreme cases given by pure public goods at the left-hand corner and pure private goods at the upper right-hand corner can help to understand their opposite economic nature, as argued by Samuelson (1955): 'doctrinal history shows that theoretical insight often comes from considering strong or extreme cases'. The actual situation is, in fact, much more differentiated and varied, as argued by various authors trying to criticize and complete Samuelson's model. Commenting on the diagram of Fig. 3.1, Buchanan (1967, p. 187) made clear that each good or service could be plotted in the diagram according to its degree of indivisibility (rivalry) and the number of users (excludability). Buchanan also introduced in the diagram three examples of mixed goods, including externalities as described by Pigou (1920). Incidentally, he had a dynamic view of goods, making clear reference to the fact that the group of users can be enlarged or restricted according to property-rights (Buchanan, 1967, pp. 190–191). He also showed that individual goods and services can move around the diagram according to rivalry and excludability, the allocation of hunting rights and their enforcement (Buchanan, 1967, p. 191) being the example given. Goods can be excludable but non-rival, and vice versa, but even more frequently there are goods that are partially excludable and/or rival and therefore are variously located within the diagram of Fig. 3.2.

'The real-world situation of a broad spectrum of economic goods on a continuum between pure public goods and private goods' (McGuire, 1987, p. 454) has been described by Buchanan (1965) in his economic theory of 'clubs'. In other words 'a club is an organization which offers a shared collective good exclusively to its members, defraying the cost of the good from member's payments' (McGuire, 1987, p. 454). According to Buchanan (1965), a 'club' can be defined as 'consumption ownership–membership arrangements'. The club goods category has thus been developed following criticism of the applicability of excludability to public goods (Cornes and Sandler, 1986). Once a club is created, excludability applies to non-members, but members can enjoy the good collectively, as is the case for public goods. Club theory applies where exclusion is possible; otherwise it has limited relevance. However, 'if the structure of property-rights is variable, there would seem to be few goods the services of which are non-excludable, solely due to some physical attributes'. Hence the theory of clubs is, in one sense, a theory of optimal exclusion, as well as one of inclusion (Buchanan, 1965).

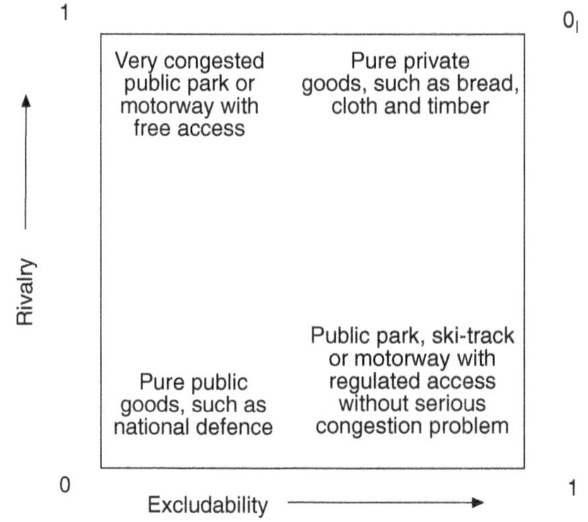

Fig. 3.2. Mixed goods: examples of impure public goods.

Another feature of many public goods conceptually linked to 'club goods' is their local nature. Consumers are selected on the basis of location, as is the case, to a certain extent, for the services provided by hospitals, fire brigades and several ERGS mainly used by local people. These local public goods have connotations of 'club goods', where exclusion and rivalry can first be decided by local government and then regulated by the market (a sort of voluntary exchange where taxes are given by membership fees), as suggested by Tiebout (1956). Meanwhile, citizens can 'vote with their feet' by deciding not to use a good if they are not satisfied with the resulting arrangements. The argument was developed again to show some limits of the public-good theory as developed by Samuelson. It was remarked that, for local public goods, optimum provision and efficiency can be achieved through specific cooperative arrangements based on local government organization or clubs (Brosio, 1986, p. 279).

The category of local public goods can be seen as opposite to more general public goods, according to the number of potential users. Brosio (1986, p. 53) has therefore added to the usual excludability/rivalry diagram a third dimension, as in Fig. 3.3, where at 0 public goods of local interest (e.g. an urban recreation park) can be found, while at 0_{II} are public goods of global interest (e.g. the ozone layer or greenhouse gases (GHGs) and related global warming). Therefore, within the box of Fig. 3.3, all goods can be located according to rivalry and excludability criteria, as well as local/global interests.

Remarkably, mixed goods of local as well as global interest are rather common where natural resources, such as forests, are concerned. Natural resources in general present various degrees of excludability (depending on property-rights and transaction costs for enforcing excludability), while rivalry is not felt up to a certain threshold. In addition, while recreation, landscape, etc., have a high local meaning, other types of ERGS, such as watershed management, can affect the life of a greater number of people – regions and nations. Other ERGS, such as carbon storage or biodiversity, have a global effect.

Forestry-related ERGS are often of primary concern at local, regional and global levels. In recent years, public goods and services of both local and global interest have been submitted to growing pressures.

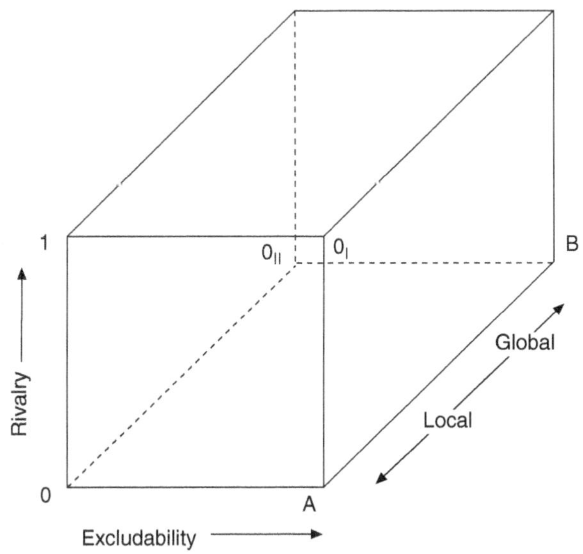

Fig. 3.3. Environmental/recreational goods and services according to their local/global nature.

Congestion has been felt for many natural resources and in various sites; therefore, rivalry in consumption has become a serious issue, as underlined by various studies, e.g. Cicchetti and Smith (1976). GHG emissions, potentially affecting climate change, have also been much studied, and considered by international summits, such as Rio 1992.

The theories of local public goods (Tiebout, 1956) and club goods (Buchanan, 1965) were formulated when considering specific services provided by the government. These theories have assumed momentum in the last decades regarding the environment and the related ERGS and RES products, such as those provided by forestry. Interestingly the classification of Randall (1987), which specifically referred to environmental products, goods and services, developed to a certain extent differently from that of Samuelson.[8] Randall classified four categories of goods, as outlined in Fig. 3.4: (i) non-rival, non-exclusive goods; (ii) non-rival, exclusive goods; (iii) rival, non-exclusive goods; and (iv) rival, exclusive goods.

It is evident that Randall's classification reflects previous analyses by other public-finance authors, such as Buchanan (1967), Musgrave and Musgrave (1973, p. 54) and Stiglitz (1986, p. 103). Categories (i) and (iv) represent Samuelson's polar cases of pure public goods and pure private goods, while categories (ii) and (iii) include mixed cases, like several ERGS. It can also be seen that categories (ii) and (iii) far from being in the extreme right-hand and upper left-hand corners of Figs 3.1, 3.2 and 3.4, are located within the diagram, because they are mostly partially non-rival and non-exclusive. Also of interest is the introduction of a 'congestible goods' category, where additional consumers may be added without any rivalry up to a certain threshold; 'eventually, however, the congestion of users sets in and the addition of more users reduce the utility of all users' (Randall, 1987, p. 177). In other words, the costs of congestion in wilderness and park recreation are felt, as shown by Cicchetti and Smith (1976).

3.3 The ERGS Provided by Forestry as a Case of Public Goods/Externalities

3.3.1 Possible relationship between timber production and ERGS

In practice, the various outputs of forestry could be represented as in Fig. 3.5, where the various segments of the production possibility curve (PPC) identify possible relationships between market output (timber) and ERGS. It is interesting to note

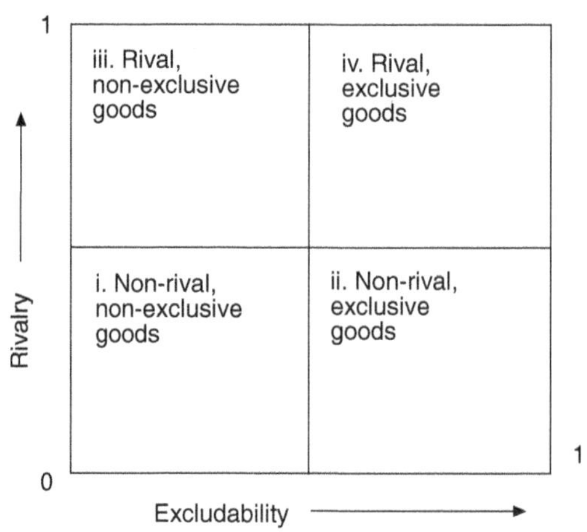

Fig. 3.4. Classification of goods in four categories according to excludability/rivalry criteria.

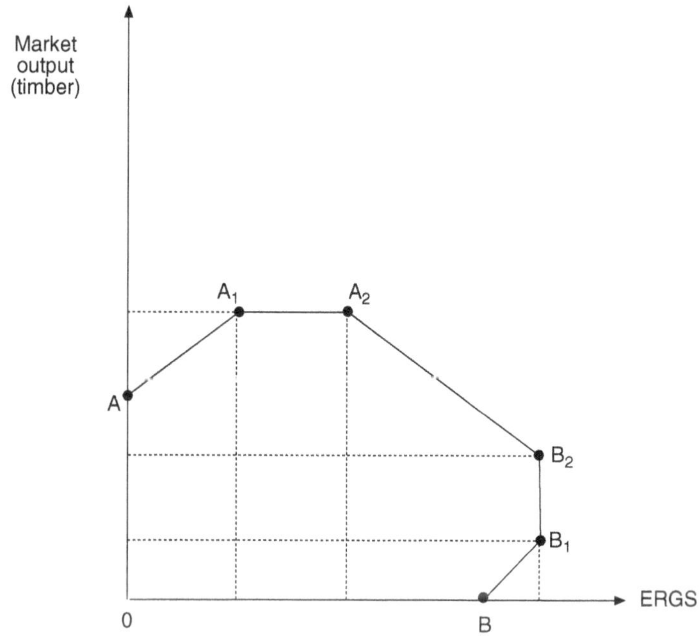

PPC segment	Jointness level between market output (timber) and ERGS	Economic nature of ERGS	Type of ERGS
A–A_1	Maximum complementarity	By-product reducing marginal costs of market output	Watershed management, erosion control, road maintenance, landscape
A_1–A_2	Complementarity	By-product without marginal cost for market output	Landscape quality, biodiversity
A_2–B_2	Substitution – trade-offs	By-product or main product while market output becomes by-product	Trees enhancing landscape and biodiversity, roads, picnic area, paths
B_2–B_1	Complementarity	Main product while market output becomes incidental by-product	Eco- and agrotourism
B_1–B	Maximum complementarity	Main/only product, with market output becoming by-product (and input to the production of ERGS)	Wildlife services, parks, recreational areas

Fig. 3.5. Production possibility curve (PPC) of market output (timber) and ERGS.

how, in addition to joint production, there is joint consumption of externalities (Bonnieux and Desaigues, 1998, pp. 22–24), and this certainly applies to both production and consumption of ERGS provided by forestry and the related environment.

Glück (1998, p. 219) uses the PPC, with reference to forest policy, in order to underline that 'the wake theory', an ideological justification of the primacy of timber production, worked satisfactorily as long as the demand for non-wood products and services did not exceed the external economies of timber production. When more of these other forest functions are required and, as a consequence, opportunity costs arise, more should be known about the relationship between timber and these products and services. This advocates the need for a better quantitative knowledge of the relationship between timber and the other forest products and services (ERGS), such as grazing, hunting, recreation, biodiversity and security.

3.3.2 From systematic to dynamic aspects of public goods

In the meantime, the overall 'forest production' has been reviewed according to public goods theory and environmental economics, identifying the private and public goods nature of the various products linked to forestry (market output and ERGS). Some authors have started to use the rivalry/excludability diagram, as shown in Figs 3.1 and 3.2 (e.g. Muraro and Merlo, 1987; Merlo, 1991), emphasizing how many ERGS belong to public and mixed-good categories. Also Cubbage *et al.* (1993), Mantau (1994) and Glück (1995) have provided a classification of the various ERGS, making a distinction between the economic nature of the forest products: public goods, private goods, club goods and common-pool goods (Fig. 3.6).

All the above contributions to forest economics and policy can be directly traced back to public economics, e.g. Buchanan (1967) and Musgrave and Musgrave (1973, p. 54), who reported a similar framework, and more particularly environmental economics (Randall, 1987). Buchanan in particular, as previously discussed, has given a dynamic view of the location of goods according to the property-rights regime. It was Mantau (1994) who formally introduced the marketability concept in the excludability/rivalry diagram by drawing an arrow from public to private goods (Fig. 3.7). This original idea has been empirically tested by the present study.

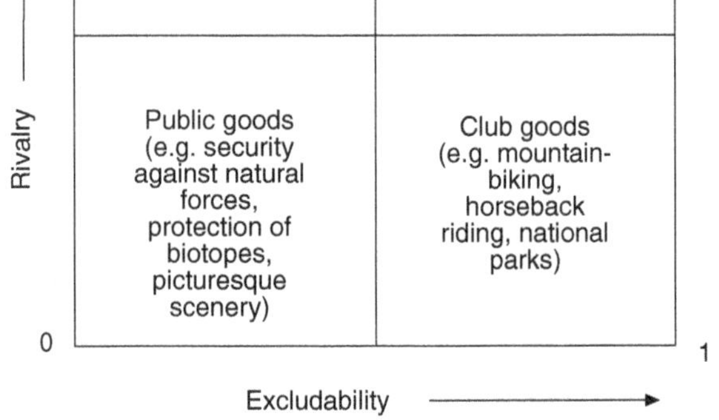

Fig. 3.6. Economic nature of various forest products (from Cubbage *et al.*, 1993, reported by Glück, 1995).

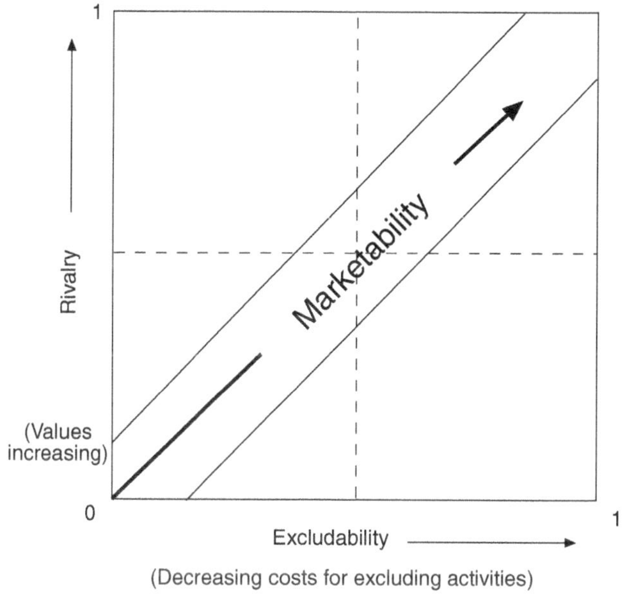

Fig. 3.7. The marketability arrow: from public to private goods (modified from Mantau, 1995a; original title: Continual structure of goods (in time and in creation) and in strategy fields for public goods).

3.3.3 The complementarity between ERGS and RES products

The analysis of policy options based on market approaches has shown that transformation/development of non-market ERGS into market RES products is mainly given by two approaches:

1. The first relies on institutional transformation from public to private goods, often through the modification of property-rights, resulting in changes in excludability/rivalry. This transformation is not always easy to realize, as social pressure, customs, etc., can prevent changes of excludability. The change must be accepted by the concerned parties and the transaction cost of enforcement must not exceed the net revenue from the marketing of RES products.

2. The second is based on market development techniques exploiting the relationship between the ERGS and what is actually sold in the market: the RES products. This relationship is often given by consumption complementarity; e.g. to enjoy the environment (the ERGS), one must use appropriate equipment (e.g. walking boots, mountain bikes) and infrastructures (e.g. footpaths, ski-lifts, restaurants). Quite clearly, the consumption of ERGS is linked with and is complementary to that of the infrastructure and equipment making up the RES products. This complementarity can also be seen in terms of 'added value', with the activity producing and providing the infrastructure, e.g. maintaining footpaths and restaurants, while the RES products are complementary and additional to the pure ERGS. The relationship of complementarity and the additional value of the equipment and infrastructure, therefore, represents a final step for making a market value for ERGS.

The experience is such that both the approaches and means (institutional and market) are often used, or needed, to achieve remuneration of ERGS through RES products. The first institutional approach builds the base for transformation: e.g. new regulations are introduced for using the ERGS. However, the status of the ERGS is not dramatically changed. The core of the development process is given by the application of market-based techniques, taking advantage of the existing complementarity between RES products and ERGS. In fact, in most cases the presence of additional goods and services is necessary to create a market for the ERGS through the RES products. Figure 3.8 exemplifies the transformation process, showing that the ERGS for which institutional status is slightly changed (continuous line) are enveloped within conventional complementary market goods, such as car parks, tourist and sport facilities (dotted line), and also traditional forest/farm products, whose quality and image is linked to the environment and the landscape (ERGS) where they are produced. In practice, exclusion is introduced through additional services, as stated by Mantau (1995a).

In other words, what can certainly be said is that the remuneration of ERGS is often quite indirect, through the development of complementary/additional RES products, supported by some institutional changes of ERGS. The possible level of complementarity between ERGS and RES products is variable, as shown by Figs 3.8 and 3.9: e.g. there is a high complementarity between the pure environment and the footpaths to gain access to the environment, while it is lower with restaurants and shops. At the limit, when complementarity is very high, the ERGS and the RES products overlap almost completely, as shown by Fig. 3.9. It is, however, almost always the RES products that are paid for in the market, not the ERGS.

Figure 3.10 outlines the different levels of complementarity between the ERGS (the pure environment) at the centre of the circle and the surrounding RES products. Mantau (1995a, pp. 131–145), with reference to the above complementarity, speaks of introducing 'exclusiveness through additional services'. An example could be the access to a nature reserve to be paid for by the consumer/visitor as a toll for using a footpath, whereas before transformation/development access was free but difficult because of the lack of footpaths. After transformation/development, the access was improved and paid for by the user, as the footpath provided by the developer

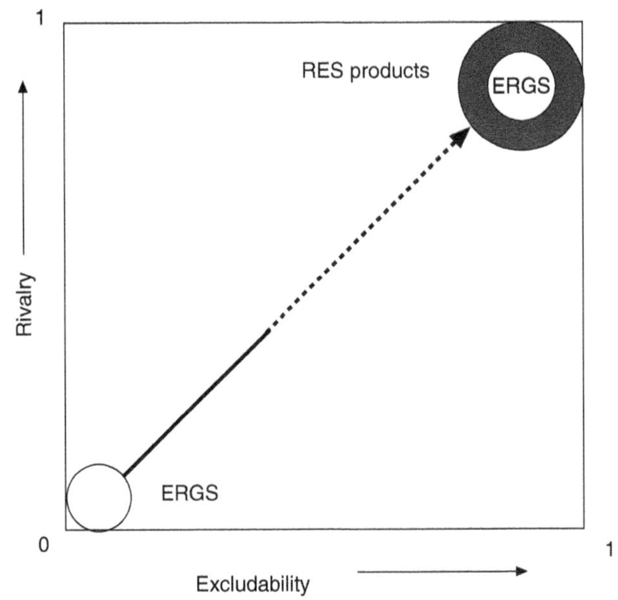

Fig. 3.8. Transformation/development paths from ERGS to RES products: the 'marketability arrow'. —— Excludability through modification of property rights: transformation; - - - - - excludability through provision of additional services (marketing techniques): development.

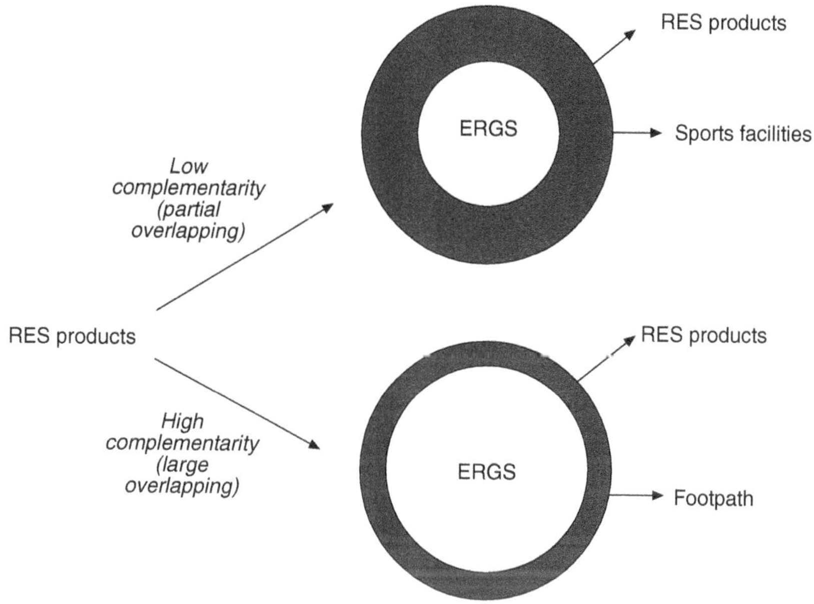

Fig. 3.9. Consumption complementarity relationship between ERGS and RES products.

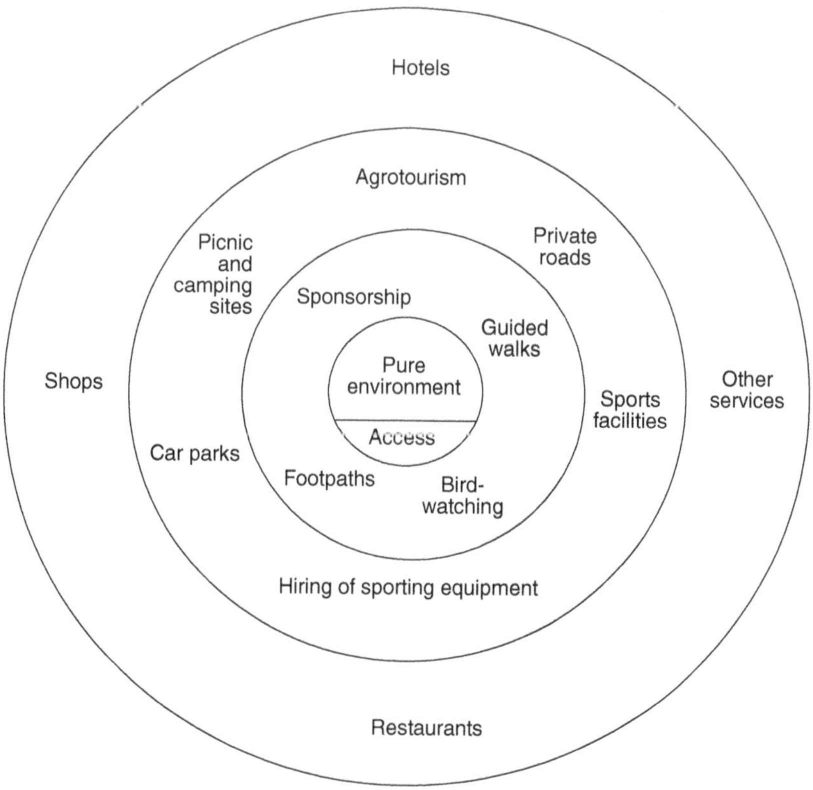

Fig. 3.10. Complementarity between the pure environment and the related consumption (elaborated from Mantau, 1995a).

must be financed. In other cases, the ERGS are rather different from the RES products paid for by the consumer, because the payments take place for specific facilities, such as camp-sites, holiday cottages, equipment and infrastructure in general, etc. However, the RES products paid for by the consumers take into account the attached ERGS (e.g. the attractive landscape/environment). In other words, people pay to use certain facilities or to have a product with certain characteristics.

3.4 Diverse Aspects of the Case-studies

3.4.1 Ownership and management

The land where RES products found market remuneration was in many cases publicly owned (40 cases out of 98), as shown in Table 3.1. In general, it was found that the public sector actively seeks to promote innovative exploitation of forest ERGS through RES products. For a large number of cases, however, RES products take place on private land. Distinct differences between the countries have been found, for instance, only two Italian cases had RES products that took place on entirely private land while in the other countries almost half took place on private land. These different types of ownership are of particular interest. Trusts and associations usually manage parks, nature reserves and RES products in Austria and the Netherlands, while in Italy traditional common-property ownership is particularly active in promoting RES products. Trusts, common properties and various forms of associations and non-governmental organizations (NGOs) can represent the link between the public and the private sector, because normally there are social goods and public participation in the running of the organizations responsible for ERGS transformation/development into RES products.

Common properties (which in many Italian alpine areas traditionally manage forests and grazing land) are often of the appropriate size for producing and marketing RES products. Economies of scale are particularly felt in these cases, even more than in traditional forest products. This can lead to the hypothesis that the production of certain RES products (particularly those environmentally based) can find more effective application in 'common/public land' of an adequate size with a legal status midway between public and private. Evidence of these situations was given by Pareto (1896) referring to Alpine commons: 'since it has undergone the trial of free competition for centuries, collective land tenure responds better [than other forms of land tenure] to certain specific needs, its destruction would mean a loss in the nation's total utilities'.

However, many case-studies, particularly in Germany and the Netherlands, were found which included recreational products that do not require a large area,

Table 3.1. Ownership of the land where RES products take place. (Question 1.1.6: Describe the ownership structure of the forest land base regarding the product.) Classification is documented case by case in the EU report.

Legal status	Austria	Germany	Italy	Netherlands	Total
Public					
Federal	4	–	7	4	15
State	–	10	5	1	16
Local authority	1	2	5	1	9
Common property	1	2	9	–	12
Trust	4	–	–	6	10
Private	10	12	1	7	30
Other	1	2	2	1	6
Total	21	28	29	20	98

such as camping, house renting and nature shops. It was also evident that the surrounding landscape and environment are a component of what is sold, even if they are not part of the enterprise or property managing the RES products.

The public-sector management of RES products was rather common, with 40 cases out of 98, while common property and trust enterprises are also widespread (22 cases). This situation could be due to the fact that public/common properties and trust enterprises often have social goals and demonstrate a particular interest in environmental/recreational products (Table 3.2). For example, organized guided tours often have an educational purpose or certain areas of land are managed only for conservation and recreation.

In general, public enterprises seem to prefer to receive a direct remuneration for their RES products. The payment for the entrance to public forests and parks, through, for instance, road tolls, is a direct way to gain remuneration purely from the existence of the forest and the people visiting it and does not require a large amount of organization. Common properties provide RES products with intermediate characteristics between the public and private sectors. They usually have a large land base like public enterprises, allowing efficient large-scale management of RES products. At the same time, common properties demonstrate an interest (particularly in Italy) in experimentation with innovative products. Similarly, trusts and other almost public bodies have sometimes developed quite sophisticated sponsorship systems to finance their activities of nature protection.

Private enterprises, meanwhile, produce all types of RES products (30 cases out of 98), with the largest amount in the 'pure' recreational product category. They also promote environmental/recreational and environmental/traditional products, where the remuneration for ERGS is quite indirect and can result in quality products or organized events that often require a large amount of organization. It is interesting to note that RES products managed by private enterprises usually use a smaller land area (e.g. for camp-sites, sports fields) than those managed by public/common properties/trusts (e.g. parks and nature reserves), which have an emphasis on the environmental component.

The managing status of RES products is shown in Table 3.3. Generally, there is a high coincidence between the forest owner and the RES product manager. The forest owner provided the management of the RES product in about half of the cases (52 out of 98), in both the public and the private sectors. This is very positive, since the owner of the forest resource is directly remunerated for the product or the service provided. In other cases, the management is provided by contractors, consortia and

Table 3.2. Ownership of the land where RES products take place: by product type.

	Legal status					
Type of RES products	Public	Common property	Trust	Private	Other	Total
One category						
Recreational	18	6	3	13	2	42
Environmental	2	3	4	4	2	15
Mixed categories						
Environmental/recreational	1	1	–	6	1	9
Traditional/recreational	17	–	–	3	–	20
Traditional/environmental	2	2	2	3	1	10
Not classifiable	–	–	1	1	–	2
Total	40	12	10	30	6	98

Table 3.3. Management status of RES products. (Question 1.1.4: Who is responsible for the product management?)

Type of manager	Austria	Germany	Italy	Netherlands	Total
Forest owner (private and public)	14	15	10	13	52
Contractor	–	2	4	1	7
Park authority	–	–	3	1	4
Consortium	–	–	4	–	4
Cooperative	1	3	1	–	5
Association	3	2	2	1	8
Local authority	2	4	–	–	6
Other	1	2	5	4	12
Total	21	28	29	20	98

park authorities, which are particularly common in Italy. However, the situation in Italy, which has arisen from the structure of the property-rights, is rather unique and contrasts with the other countries, as the forest owner is the RES product manager in only ten cases out of 29. The management of RES products is, therefore, more detached from the forest owners, although they provide the ERGS and support the related costs. RES product management by cooperatives, local authorities and associations of forest owners is more usual in Austria and Germany (Table 3.3). The RES products that fall in the category of 'other' are subject to a variety of legal status. In general this variety of RES product management reflects the need to adapt the business management to the legislation, which is rather different in the various countries.

3.4.2 New and existing RES products

The transformation/development of the RES products can be given by the creation of new products or by changes/adaptation to already existing products. Boundaries between 'new' and 'changes' are sometimes difficult to define. In general a 'new' product is a transformation/development implying the creation of new facilities (DE02, AU11), such as property-right modifications to allow mushroom-picking permits (IT01, 08) or the organization of an activity that was not in place before, like a particular event (AU15, DE18). The introduction of new labels and trade marks or sponsorship also constitutes creation of a new product.

A 'change' is more related to reorganization of existing products: for example, the adoption of a new payment system for the sale of drinking-water (AU16, 19) or advertising of the development of the traditional Christmas sales (DE01, 13). Another type of change could be in contract conditions; for example, for the case of a military training area (AU04), a new contract was signed with better-defined conditions, particularly for potentially damaging events. Both 'new' and 'changed' products were distributed quite evenly amongst the surveyed cases; 56 RES products were new, while 42 involved changes to existing products (Table 3.4).

Table 3.5 shows that since 1994 the introduction of RES products has increased in all four countries. The majority of purely recreational products were introduced before 1980 (Table 3.6), except for Austria, where recreational development took place mainly after 1994. Conversely, the introduction of environmental products and traditional/environmental products was mainly concentrated in the most recent years. In particular, for Italy the evaluation is undoubtedly due to EU Regulations 2082/92 on *appellation d'origine* (origin designation through labelling), which were once limited to cheese and wine and have now been extended to various forest products (IT02, 13, 21, 27). Finally, environmen-

Table 3.4. Transformation/development of RES products: distinction between new RES products and changes of existing RES products. (Question 3.1.1: Is the product totally new for your enterprise/management unit or is it a development of an existing product?)

	Austria	Germany	Italy	Netherlands	Total
New	13	18	16	9	56
Changes	8	10	13	11	42
Total	21	28	29	20	98

Table 3.5. Time of RES product's introduction. (Question 3.3.1: When was the product launched?)

Introduction	Austria	Germany	Italy	Netherlands	Total
Before 1980	2	6	5	5	18
Between 1980 and 1989	5	4	3	2	14
Between 1990 and 1993	3	2	8	5	18
Between 1994 and 1995	6	8	8	5	27
After 1996	5	8	5	3	21
Total	21	28	29	20	98

Table 3.6. Time of RES product's introduction according to RES product typology.

	Introduction					
Type of RES product	Before 1980	1980–1989	1990–1993	1994–1995	After 1996	Total
One category						
Recreational	10	4	11	11	6	42
Environmental	2	1	4	3	6	16
Mixed categories						
Environmental/recreational	5	4	3	6	7	25
Traditional/recreational	–	1	–	3	–	4
Traditional/environmental	–	3	–	4	2	9
Not classifiable	1	1	–	–		2
Total	18	14	18	27	21	98

tal/recreational products, including several innovative approaches, are often the result of the improvement of already existing recreational products. For example, the newly established network of authorized park tour guides (which charge a payment) offers a much better service than the previous rather limited tours, which were free (IT10). Also the improved organization and quality of excursions in a nature-orientated forest (DE09) followed a situation where the guided excursions were available free of charge, but were not so good.

There are many causes of development of RES products from existing products through 'changes', as shown in Table 3.7. Increasing costs of existing products or other economic pressures, such as decreasing demand and prices or unfair competition, have often been the cause for development through changes of existing products in all the countries. The growth of an already existing market demand represented the triggering for development of seven Austrian and German products. The problem of congestion control was often felt, and in two Dutch cases it was indicated as a major cause for product development. In several cases (NL05, 07) it has been, however, the awareness of new

Table 3.7. Causes of development of existing products.* (Question 3.1.3: If the product is not totally new, why have innovations been made in the old product?)

Cause of development	Austria	Germany	Italy	Netherlands	Total
Decreasing demand	–	2	1	–	3
Decreasing price	1	1	1	–	3
Increasing costs	1	5	5	2	13
Low profitability	7	5	–	5	17
Labour availability	–	3	–	–	3
Other	8	11	11	11	41
Total	17	27	18	18	80
Some examples of 'others'					
Increasing demand	2	5	–	–	7
Better techniques and infrastructure	2	1	–	–	3
Unfair market of high-quality products	–	–	–	3	3
Reduction and conflicts among users and fractions	–	–	–	2	2
Finding new sources of income	–	–	–	2	2
New ideas	1	1	–	–	2

*Multiple answers were included.

opportunities that initiated the development, such as legal innovation, reassignment of property-rights and market niche exploration.

3.4.3 Ideas, know-how, personnel, training courses and facilities

As shown in Table 3.8, the idea for the development of the RES products, in a substantial number of cases (67), was something endogenous and internal to the forest enterprises, even if the forest enterprises were not responsible for the subsequent management, as has clearly been the case particularly in Italy. It is remarkable how all types of enterprises are concerned with RES product development, whether private or public, individual or associated. This indicates that the pressures on forestry ownership/management have been coped with in similar ways in the different countries. Independently from financial pressure, imitation and external information have often played an important role. The same situation applies to the provision of know-how, which has developed internally to the enterprise in the majority of cases. However, total or partial external know-how was also common, as shown in Table 3.9.

In general, even if there was a high need for trained personnel to develop the RES products (Table 3.10), it was not necessary to organize specific training courses (Table 3.11), because personnel with the required skills were already available internally to the enterprise, or at least locally. This would suggest that RES product initiatives were not particularly difficult to organize. The highest need for specially trained personnel was for the development of environmental/recreational products, as they often have innovative contents and can be quite complex to organize. There was generally greater internal experience available for the development of recreational products, and generally there was no specific need for specially trained personnel (Table 3.12).

New facilities (e.g. buildings and sports facilities) for the product's development were needed in almost half of the case-

Table 3.8. Provision of the idea for development.* (Question 3.3.2: Who provided the idea of the product development?)

Provision of the idea for development	Austria	Germany	Italy	Netherlands	Total
Landowner/forest manager	14	20	19	14	67
Imitation	–	4	6	3	13
Mass media	1	–	2	–	3
Specialized press	1	–	3	1	5
Forest extension services	2	2	–	2	6
Agricultural extension services	–	–	–	–	
External consultant	–	1	–	–	1
Customers	4	8	1	1	14
Providers	2	–	–	–	2
Other	2	4	13	5	24
Total	26	39	44	26	135

*Multiple answers are included.

Table 3.9. Provision of the know-how for development.* (Question 3.3.3: How was the know-how supplied?)

Provision of the know-how for development	Austria	Germany	Italy	Netherlands	Total
Own knowledge	16	25	24	16	81
Public institutions	4	5	5	2	16
Forest extension services	5	9	–	2	16
Agricultural extension services	1	–	1	–	2
Other public institutions	–	3	3	–	6
Private consultants	–	3	1	3	7
Neighbours	1	2	6	–	9
Mass media	1	–	1	1	3
Specialized technical press	3	9	3	1	16
Providers	3	–	1	–	4
Other	8	8	4	7	27
Total	42	64	49	32	187

*Multiple answers are included. Not applicable to case NL03.

Table 3.10. Need for specially trained personnel. (Question 3.3.5: How necessary is it to have specially trained personnel in order to be able to offer your product?)

Need for specially trained personnel	Austria	Germany	Italy	Netherlands	Total
Very high necessity	5	10	8	8	31
High necessity	3	6	3	3	15
Low necessity	3	3	5	7	18
Very low necessity	1	1	5	1	8
No necessity	9	8	8	1	26
Total	21	28	29	20	98

Table 3.11. Training courses organized.* (Question 3.3.7: Were training courses organized for the personnel?)

Training courses	Austria	Germany	Italy	Netherlands	Total
No	17	23	17	14	71
Yes	3	4	8	6	21
Total	20	27	25	20	92

*Not applicable for case-studies AU04, DE10, IT01, 11, 22, 24.

Table 3.12. Need for specially trained personnel and typology of RES products.*

Need for specially trained personnel	Category of RES products					
	Recr.	Env.	Env./Recr.	Trad./Recr.	Trad./Env.	Total
Very high necessity	6	6	16	1	2	31
High necessity	4	3	3	2	3	15
Low necessity	9	1	4	1	3	18
Very low necessity	5	2	1	–	–	8
No necessity	18	4	1	–	1	24
Total	42	16	25	4	9	96

* Not applicable for case-studies AU04, 14.
Recr., recreational; Env., environmental; Trad., traditional.

studies (48 out of 98), as shown in Table 3.13. In the majority of cases these were not in place, therefore some kind of investment had to be made, although in general the investment was not extensive. The overall picture for personnel, capabilities, facilities and ideas for ERGS transformation/development into RES products is therefore rather positive: it appears that initiatives are not too complex and local/ enterprise means are generally available and sufficient. A conclusion could be that RES products are primarily developed according to locally existing conditions and opportunities.

3.4.4 Provision of additional services and means of payment

Table 3.14 outlines the fact that RES products are often sold with other similar products as part of a certain product mix or package developed in a certain area. Several cases indicated this, ranging from mountain-biking to hospitality in forest cabins (e.g. AU02, 11, DE02, 27, IT14, 16, NL09, 17). What clearly emerges from the survey is another form of consumption complementarity between different RES products, helping each other's consumption. The term 'area product-mix package' seems appropriate to express this situation.

The demand for RES products is, therefore, assisted by two different forms of consumption complementarity:

1. The complementarity between ERGS and RES products, as demonstrated in Section 3.3.
2. The complementarity that takes place between several different RES products available in one area.

These two types of complementarity are often connected, with locally available ERGS being transformed/developed into the so-called 'area product-mix package' of RES products (Table 3.14). This overall complementarity, made up of local or area interdependencies, was particularly found in Italian case-studies, where often several RES products were available together in one area as a group.

The consumption complementarity issue is much linked to the means of pay-

Table 3.13. Need for new facilities. (Question 3.1.6: If you need facilities, could you take advantage of already existing ones?)

Facilities	Austria	Germany	Italy	Netherlands	Total
No	12	16	12	10	50
Yes	9	12	17	10	48
Total	21	28	29	20	98

Table 3.14. Provision of additional services: the area product-mix package.* (Question 3.2.6: Were additional goods or services attached to non-marketable resources?)

Additional services	Austria	Germany	Italy	Netherlands	Total
No	10	7	9	7	33
Yes	11	20	20	13	64
Total	21	27	29	20	97

*Not applicable to one German case-study (DE27).

ments, as outlined in Table 3.15. Payment methods depend upon the characteristic of the RES products. Tickets and permits generally represent the payment method for club goods (car parks, access to recreational areas and parks, facilities for cross-country skiing, etc.), while billing according to consumption or duration is mostly used for pure private goods and services. Payment by ticket or permit was particularly prevalent in Italy. For German cases, payments were often also made through leases and various forms of contracts. In one case, a forest youth hostel (DE27), the payment was given in kind by the exchange of labour.

3.5 Empirical Evidence of Transformation/Development Strategies and Techniques: Institutional Changes and/or Marketing Tools

3.5.1 Quantification of excludability and rivalry

As already seen in Section 3.2, pure public goods are characterized by 'non-rivalry' in

Table 3.15. Means of payment.* (Question 3.1.7: What means of payment are used?)

Means of payment	Austria	Germany	Italy	Netherlands	Total
Billing according to consumption	4	7	4	5	20
Billing according to duration of service	8	8	2	5	23
Access cards	3	1	2	2	8
Tickets, permits	1	–	15	4	20
Subscription	2	–	3	1	6
Voluntary agreement	2	2	–	–	4
Donation	–	2	2	–	4
Other	7	15	6	5	33
Total	27	35	34	22	118

* Multiple answers are included. Not applicable to case IT07.

consumption and 'non-excludability' in their supply, the latter implying 'indivisibility' of benefits (Samuelson, 1955). Excludability means there is a boundary to free use, making the good 'divisible' between those admitted and those not admitted. Admission implies rivalry and, at least potentially, marketability. Rivalry and excludability, therefore, represent the two main criteria for classifying public/private goods and services, including ERGS and RES products. Throughout the survey, questions on excludability and rivalry[9] dealt primarily with the modifications before and after transformation/development of RES products. The following evaluation procedure was undertaken using the information from the questionnaires in order to categorize all cases in terms of excludability and rivalry before and after transformation/development:

1. A scale of five grades was adopted, both for excludability and rivalry (0; 2.5; 5; 7.5; 10), in order to plot ERGS and RES products in the public–private goods diagram.
2. Rivalry and excludability were evaluated independently, before and after transformation.
3. Before transformation/development, both excludability and rivalry were equal to zero when the RES product was new and the existing ERGS were freely used without rivalry/congestion. The exception was given by some cases where the public was not permitted to use the ERGS due to existing clearly stated property-rights (e.g. forest roads in Austria (AU08) or private property opened to the public (NL10)) and, therefore, excludability was complete, scoring 10.
4. After transformation/development of the RES products, excludability was evaluated according to the combined effects of actions undertaken in terms of institutional changes and marketing techniques. As far as possible, the two effects were considered separately; they are indicated, respectively, as a continuous and as a dotted line along the marketability arrow of Fig. 3.8. The valuation could be biased by subjective judgements, but the 'price' paid by the consumers for the RES products was significant for effective transformation/development. Whenever the RES product's 'price' was the only possible way to use the ERGS, excludability was valued at 10. If the various degrees of 'complementarity' between ERGS and RES products implied a reduction in excludability (for example, if paying the price for the ERGS could be avoided, e.g. voluntary contribution for a picnic area (IT06)), excludability was valued around 5–7.5.
5. After transformation, rivalry was rather difficult to evaluate. The concept is not so clearly defined in economic literature and rivalry was often interpreted as congestion by those completing the questionnaire. However, it is a more complex issue, linked also to physical divisibility and joint supply/consumption of ERGS and RES products. Moreover, the fact that only a certain level of excludability was introduced could imply the existence of a certain level of rivalry, due to the modified economic nature of the good – no longer completely public. Therefore, the introduction of excludability could result in higher or lower rivalry. For purely private goods, rivalry was valued at the highest level, scoring 10, when transformation was related to, or brought to, a fully rival good (e.g. product certification and house lease) where it is compulsory to pay a price for the RES product and the attached ERGS. For club goods and services, two types of rivalry were distinguished as either: (i) internal rivalry among people already entitled (i.e. club members) to use the good and/or services where congestion could be a problem; or (ii) an external rivalry among potential users. For example, for admittance to a park, there can be rivalry to obtain the entrance ticket if the number of admitted persons is regulated and demand is greater than supply (external rivalry). Once tickets have been issued, rivalry can exist among the admitted persons, due to congestion within the park – internal rivalry.

Again, the valuation of rivalry could be biased by subjective judgement. The following criteria have been adopted taking into account that, for club goods, rivalry, by definition, cannot score 10:

1. Rivalry = 2.5 means that there is a club good without any problems of rivalry among users (internal or external).
2. Rivalry = 5 is the 'standard' club good, where there can be accidental/temporary congestion problems (at some periods of the year, the week or the day).
3. Rivalry = 7.5 when there is a declared (through the questionnaire) problem of congestion (high internal rivalry) and/or a demand greater than supply (high external rivalry).

3.5.2 The state of ERGS (and potential RES products) before transformation/development

In most cases, ERGS before transformation (and the potential RES products) had the characteristics of a public good, with no excludability (everybody had, at least potentially, the possibility to use it) and no rivalry (Tables 3.16 and 3.17). This was sometimes due to the physical impossibility/difficulty of forbidding use or access to the ERGS. There were only a few existing pure private goods and services which did not require transformation, and in these cases the development of RES products dealt mainly with marketing, contracting and pricing activities.

Excludability before transformation/development

The few existing pure private goods and services (25 out of 95 cases) were mainly found in Austria, where the ERGS were already subject to some sort of exclusion before transformation/development. Because of existing property-rights, people were excluded from access to/use of the resources: e.g. use of forest roads, spring water and fishing. In some cases, this was due to a precise definition and assignment of property-rights: for example, in the Austrian legislation, water rights are private. In most cases, however, the ERGS which were to be exploited through RES products were non-excludable, or at least excludability was not clearly defined (70 cases out of 95). This also applied to traditional prod-

Table 3.16. Excludability before transformation/development.*

Excludability	Austria	Germany	Italy	Netherlands	Total
No (0)	7	25	23	15	70
Very low (2.5)	–	–	–	–	–
Low (5.0)	3	–	2	1	6
Relevant (7.5)	–	–	1	–	1
Very relevant (10.0)	9	3	2	4	18
Total	19	28	28	20	95

* Not applicable to case-studies AU10, 21 and IT07.

Table 3.17. Relevance of rivalry before transformation/development.*

Rivalry	Austria	Germany	Italy	Netherlands	Total
No (0)	14	22	14	12	62
Very low (2.5)	1	1	3	2	7
Low (5.0)	1	2	9	1	13
Relevant (7.5)	–	2	2	4	8
Very relevant (10.0)	3	1	–	1	5
Total	19	28	28	20	95

* Not applicable to case-studies AU10, 21 and IT07

ucts, such as those transformed into RES products by exploiting their environmental/landscape image. In general, before transformation, the image was that of a public good, as everybody could claim that a certain forest product came from a certain area.

Rivalry before transformation/development

Congestion was the main cause of high levels of rivalry before transformation, and was sometimes the driving force for transformation/development. As already stated, the concept of rivalry overlaps with that of congestion. In fact, a good is rival when the consumption by one individual takes away all the opportunities of consumption of the same unit of the good by other consumers. Congestion can be considered, in general terms, as 'the situation in which one individual's consumption reduces the quality of service available to others' (Cornes and Sandler, 1986). In this report the definition used is that of Randall's (1987) congestible goods: 'any good that can be enjoyed by many individuals but is subject to a capacity constraint and for which the fixed cost of provision far exceeds the marginal cost of adding additional users until the capacity constraint is approached'. From this definition, most club goods can be regarded as congestible. It is clear that congestion, diminishing the possibility of consumption by other people, causes an increase in rivalry. Rivalry, however, can exist independently of congestion, as long as a certain good is divisible and consumption by an individual excludes any other. As far as possible, rivalry due to congestion (internal and linked to indivisibility amongst people entitled to use a certain good) and to the good's intrinsic nature (external) were evaluated separately and summed up in the final assignment of rivalry level.

The concept of rivalry and the distinction between internal and external rivalry have clearly shown that the status of public–collective goods is applicable to a large number of ERGS that are transformed/developed into RES products. This concept has been used even though it can be misunderstood due to a certain level of ambiguity (which is still evident in the economic literature). Before transformation it was the rivalry concerning the ERGS, not the potential RES products, that was valuated, although the two identify themselves to a greater or lesser degree according to the level of their consumption complementarity (see Section 3.3, Figs 3.9 and 3.10).

Tables 3.16 and 3.17 show the status of the ERGS (would-be RES products) before transformation/development, using the usual excludability/rivalry diagram. It can be seen that the majority of ERGS fall into the category of public goods. The exception is Austria, where several ERGS (even before transformation) have high levels of excludability. This is due to their institutional status and property-right regime. The opposite situation occurs with rivalry. While Austrian cases generally show low levels of rivalry (which can depend on the already existing excludability), the other countries – Italy, the Netherlands and Germany – show a range of rivalry, from low to relevant (2.5–7.5). This may be caused by high pressure on forest resources (population density), accompanied by poor definition/assignment of property-rights.

The combined effects of excludability/rivalry define the public and the private status of ERGS (would-be RES products) of the four countries. Again, Austria appears different, as the ERGS generally have a low level of rivalry with a high level of excludability and the other countries have a higher level of rivalry and a lower level of excludability. This situation can certainly be related to forest tenure and to the fact that Austrian forest enterprises tend to be much more business-orientated than those in the other countries. The opposite situation occurs in Italy, where several ERGS still remain public, or largely public, and have high levels of rivalry/congestion in forest resources. They tend to be poorly regulated by excludability and have problems of definition and assignment of the property-rights of the ERGS. However, the transformation/development into RES products, as shown by the next section, proved to be less difficult than might be expected.

3.5.3 The state of RES products (previously ERGS) after transformation/development

Excludability after transformation/development

Excludability is a necessary condition for transformation/development. It can be obtained by modifying the institutional status of the ERGS and, therefore, the property-rights and/or less stringent rights, such as licences and permits. However, these changes, even if possible, are rather difficult to implement because they often require changes in the legislation and/or other institutional frameworks. Nowadays people can reject these changes where consolidated traditional rights are concerned, although in past centuries changes could be imposed, such as the UK enclosures and the Maria Teresa land reforms in Austria. Excludability can be established in a softer way by using marketing-based techniques. In this case, the consumption of ERGS is generally linked to additional or complementary goods and services (i.e. the RES product) and the ERGS are marketed with these additional features. The result is that the ERGS complementary to the attached RES products can be remunerated in the market, and existing rights are not usually dramatically changed. Certainly, excludability through marketing techniques can be achieved more easily where there is a high level of complementarity between the ERGS and the RES products.

Before transformation, many ERGS were open to the public free of charge, with no or limited excludability. The situation, however, was often rather unsatisfactory, because of pressure on natural resources, free-riding behaviour, poor management and lack of excludability and complementary services. Therefore, some sort of institutional changes, ranging from legislative to planning, administrative and contractual arrangements, have been introduced, capable of supporting substantial management of the ERGS and their transformation/development into market RES products. The situation after transformation is very different and various levels of exclusion are evident in most cases (Table 3.18). In all countries, there are, however, situations where the increase in excludability is not complete when the payment of the RES products is voluntary and not compulsorily linked to the use of the related ERGS. A marketing structure is set up to induce the customer to pay for the ERGS, through the RES products. An example is given by the organization of Christmas markets in forest areas (DE01, 13, 16). Before transformation, everybody could enjoy the atmosphere of the forest free of charge; after transformation, everybody can still go to the forest free of charge, but a 'trap' (i.e. the Christmas-tree market) is set to capture the consumer surplus. This market includes various attractions, such as a place to buy Christmas trees and other related goods. This so-called 'trap' tries to exploit the complementarity between the RES products and the ERGS linked to forestry by capturing the consumer surplus connected with the enjoyment derived from the forest.

According to the evidence given by the case-studies, after transformation/development (Tables 3.18–3.21) most RES products have assumed the status of private goods,

Table 3.18. Excludability after transformation/development.*

Excludability	Austria	Germany	Italy	Netherlands	Total
No (0)	–	–	–	–	–
Very low (2.5)	–	1	2	–	3
Low (5.0)	1	2	–	–	3
Relevant (7.5)	–	–	–	1	1
Very relevant (10.0)	18	25	26	19	88
Total	19	28	28	20	95

* Not applicable to case-studies AU10, 21 and IT07.

due to higher excludability; excludability is very relevant in 88 cases out of 95. The new status of private goods is, in most cases, due to both transformation (institutional changes) and development (marketing approaches).

In 42 cases out of 83 (Table 3.19), measures had been taken to prevent free-riding behaviour and to enforce excludability or to regulate access to the RES products. These measures have been adopted in more than half of the Italian cases and are also frequent in the other countries, including Austria, where one would have expected more consolidated property-rights. The adoption of these measures largely depends, of course, on the nature of the individual RES product. It is remarkable that, in all cases where measures have been adopted, they coincide with transformation strategies, resulting in relevant changes in excludability. Rather interestingly, transaction costs to prevent free-riding behaviour have been considered very relevant in only a few cases (Table 3.20); thus, it seems that 'normal' measures are usually sufficient. This fact can be related to the nature of the goods and services provided and paid for by the consumers, which are usually market products that they are already familiar with, e.g. car parks, guided visits, accommodation.

Rivalry after transformation/development

The level of rivalry depends on the intrinsic characteristics of the RES products and the associated ERGS. Modifications are linked to the change of the economic attributes, i.e. the creation of conditions of excludability because of institutional changes and/or the capture of ERGS within RES products through marketing techniques. Rivalry after transformation/development of RES products appears, in general, to be increasing (Table 3.21). However, it is difficult to identify a clear trend, as both increases (easier access) and decreases (regulated access) have been recorded. A clear-cut reduction of rivalry was evident only in three Dutch cases (NL02, 05, 07). Usually, after transformation, a certain level of rivalry can be ascertained in the majority of cases.

Rivalry also depends on congestion, since too many users decrease the enjoyment of the good/service. RES product development can, therefore, have different and opposite effects on congestion:

Table 3.19. Adoption of measures to prevent free riding.* (Question 3.2.4: Were actions to prevent free riding adopted (e.g. fences, forest wardens, ...)?)

Measures to prevent free riding	Austria	Germany	Italy	Netherlands	Total
No	7	16	9	9	41
Yes	13	3	17	9	42
Total	20	19	26	18	83

* Not applicable to case-studies AU03, DE05, 06, 10, 11, 15, 16, 18, 20, 22, IT18, 22, 29, NL17, 19.

Table 3.20. Relevance of costs to preventing free riding.* (Question 3.2.5: How relevant are the costs to preventing free riding?)

Costs to prevent free riding	Austria	Germany	Italy	Netherlands	Total
No	6	–	2	2	10
Very low/low	6	1	11	2	20
Very relevant	1	2	4	3	10
Total	13	3	17	7	40

* Not applicable to case-studies AU03, 04, 07, 09, 12, 16, 17, 21, DE01, 02, 04–06, 08–20, 22–28; IT03, 06, 07, 10, 11, 13, 18, 20, 22, 25, 26, 29, NL01, 03, 04, 06, 08, 09, 11, 12, 15–17, 19, 20.

1. The promotion of access and use, due to attracting a larger number of users, increases congestion.
2. The more regulated use of the resource, due to zoning the area and introducing entrance fees or permits, reduces, or at least controls, congestion.

It is interesting to note that, when rivalry was originally low, it tended to increase after transformation/development, while it increased very little or decreased when the level was already high before transformation. A low level of rivalry before transformation indicates that a certain ERGS is a typical public good with no congestion problems. In this case, the manager of the RES products is not concerned about congestion, and the goal of transformation is only to increase the use of the good and its 'private' components. Meanwhile, where a private environmental/recreational good with high rivalry before transformation is involved, the economic nature of the good does not change, and product development consists mostly of an increase in production efficiency. Finally, when there is a serious problem of congestion, transformation/development can often decrease congestion through market regulation, as shown by the experience of well-consolidated RES products.

When trying to summarize the two aspects of transformation/development of public ERGS into private RES products, it is clear that excludability was successfully achieved in the majority of case-studies (88 out of 95 (Table 3.18)). The situation with rivalry was rather different, due to the intrinsic nature of ERGS and the related RES products and the difficulties arising from divisibility and rationing, congestion effects, etc. Rivalry after transformation development was high in only 39 cases out of 95, remaining low and very low in 37 cases out of 95 (Table 3.21).

It can be concluded that in all countries about half the cases have shown full transformation/development into pure private goods when considering the criteria of both rivalry and excludability. In Italy, however, only a quarter of RES products were completely transformed into pure private goods with full excludability and rivalry. Rivalry was the criterion most difficult to achieve. In general terms, development mainly led to 'club goods', rather than 'private goods', due to the intrinsic characteristics of the goods and services offered, e.g. car parks, picnic sites and cross-country skiing.

3.5.4 Transformation/development success factors and failures

From Table 3.22 it is difficult to identify the most important success factors, since a large and differentiated number of possible factors have been identified. However, the quality of the supplied products is always considered important, while other factors can depend upon individual RES product characteristics and targets. The surrounding landscape has also been considered important, particularly in Germany. This evidence, which relates the RES products to the ERGS, has perhaps been overlooked in surveying other case-studies. Demand, again, is another important success factor, as well as the quality and the image of the RES product. From the survey it does not appear that specific factors, except the

Table 3.21. Rivalry after transformation/development.*

Rivalry	Austria	Germany	Italy	Netherlands	Total
No (0)	–	–	–	–	–
Very low (2.5)	4	6	4	1	15
Low (5.0)	6	5	5	6	22
Relevant (7.5)	1	2	13	3	19
Very relevant (10.0)	8	15	6	10	39
Total	19	28	28	20	95

* Not applicable to case-studies AU10, 21 and IT07.

Table 3.22. Success factors in the development of RES products. (Question 3.3.4: What in your opinion are the determining factors for the successful development of this product?)

Development success factor	Austria	Germany	Italy	Netherlands	Total
Product/service quality					
Good quality of service	3	8	7	5	23
Professionalism of personnel	3	2	–	3	8
Landscape/surroundings and area characteristics					
Attractive/exclusive surroundings	–	8	2	–	10
Recreational importance of the area	–	4	–	–	4
Closeness to the plain, major cities or popular places	–	–	4	–	4
Suitable area/period	–	–	–	2	2
Availability of resources	2	–	–	–	2
Easily controlled area	–	–	–	–	–
Existence of a demand					
Increasing demand	2	6	–	–	8
Presence of a demand	–	–	2	2	4
Popularity of mushroom-picking	–	–	4	–	4
Marketing					
Contact with client	3	–	–	–	3
Advertising/information	2	2	–	–	4
Good image of the enterprise	–	4	–	3	7
Public relations (for the sponsor)	–	3	–	–	3
Market orientation	–	–	–	2	2
Structural effects/additional services	2	3	4	–	9
Marketing activities	–	–	2	–	2
Other					
Teamwork between public and private organizations	4	–	–	–	4
Potential of national parks	3	–	–	–	3
No alternatives/lack of competition	–	–	3	–	3
Promotion of team spirit	–	2	–	–	2
Innovative product	–	2	–	–	2
Total	24	44	28	17	113

* Multiple answers are included – only causes indicated at least twice in a country are reported.

quality of the RES product, are always very important. Quite clearly success depends upon individual cases and situations.

When considering the crucial issue of competition between the ERGS transformed/developed into RES products and similar ERGS available free of charge, the survey shows that, in general, it was impossible to get an ERGS similar to the RES product free of charge in that area. Where free substitution possibilities existed, the new development has been possible because of additional advantages/benefits made available to the consumers or due to the peculiarities of the RES product itself. These benefits may include the quality of the RES product provided, the attractiveness of the landscape and the environment where the RES product is sold and the closeness of the new RES product to the places tourists usually visit.

Unfortunately, failures do occur. Managers of RES products were asked if they knew of any failures of products similar to those they were managing (Table 3.23). Twenty-five failed examples were

Table 3.23. Failures of similar cases. (Question 3.3.8: Do you know of any previous attempts to transform and develop a similar product that have failed?)

Failures of similar cases	Austria	Germany	Italy	Netherlands	Total
No	12	25	20	16	73
Yes	9	3	9	4	25
Total	21	28	29	20	98

reported, due to 48 various causes (Table 3.24), including problems with property-rights. In four Italian cases, this was considered to be the main cause of failure. Current property-right definition and assignment in Italy is less favourable to the development of RES products. Lack of financial resources was also blamed (one case in Austria and three in Italy) and so were organizational problems. The large range of reported causes does not give a clear indication of which aspects in the transformation/development process are responsible for the failure. However, property-right problems (except the Netherlands) and organizational errors seem to be the most important reasons.

3.5.5 Mechanisms of transformation/development

As already seen in the introductory sections and in the empirical analysis, the change from public ERGS to private RES products is due to changes of excludability and rivalry. These changes imply the transformation/development of ERGS into RES products. Transformation has been defined as a modification of ERGS mainly concerning their institutional nature (including legal status and property-rights, planning and permissions, contractual arrangements, etc.). Meanwhile, development has been defined as a modification of ERGS mainly concerning their economic nature, in particular the provision of complementary/additional goods and services, promotion, changing of existing contracts, etc. The roles of transformation and development are described separately in the following sections.

Transformation: the role of institutional factors

Institutional changes include modifications ranging from national/regional laws affecting property-rights to local planning. Other less important kinds of institutional

Table 3.24. Causes of failure. (Question 3.3.8: Do you know of any previous attempts to transform and develop a similar product that have failed?)

Cause	Austria	Germany	Italy	Netherlands	Total
Right to use not properly defined	1	–	2	–	3
Property-right problems	1	2	2	–	5
Lack of demand	2	–	2	1	5
Lack of financial resources	1	–	3	–	4
False perception of the poduct by the consumers	1	–	1	–	2
Organizational errors	2	1	2	1	6
Poor timing	1	–	–	–	1
Insufficient know-how	–	1	1	–	2
Insufficient advertisement	–	–	1	–	1
Other	7	1	7	4	19
Total	16	5	21	6	48

changes can concern the context in which transformation takes place, including regulations, standards, licences, etc. applicable to the RES products. Institutional changes can therefore be defined as follows:

1. Legislative changes concerning state and regional laws, often representing the first step towards transformation, creating the base for the RES product market niche: for instance, the Italian law on mushroom-picking or the EU regulations on product origin and certification.
2. Planning changes, though at a lower level, concerning regional and local authorities, which can again initiate the transformation of an ERGS into an RES product: for example, land-use and environmental planning, including zoning, delimitation of parks and protected areas, etc., where new opportunities for the establishment of an RES product are created.
3. Administrative changes, at an even lower level, being mainly local and including standards, licences and other regulations, making possible, again, the first step towards the establishment of an RES product on the basis of existing ERGS.

Therefore, institutional changes (or at least compliance with a certain administrative procedure) often represent the first preliminary step towards the transformation/development process. Notable differences do exist among the four countries: only a few German cases required a true legal change, while in several Italian cases such a change was required (Table 3.25). Conditions were more favourable to RES product transformation when large institutional changes were not necessary, because the existing situation (concerning property and use rights) already allowed, potentially, the market remuneration of the ERGS.

No dramatic changes of property-rights have taken place for the large majority of case-studies. In general, it has been a matter of adjustments/clarifications mainly linked to planning and administrative actions (Table 3.26): for example, local regulations, such as requesting a licence from the local authority for selling RES products attached to the related ERGS (Table 3.27). Minor institutional changes were needed in some cases when an RES product developer applied for a new contract according to organizational and legislative needs (Table 3.28).

It was also found that, if certain access regulations were already in place for the ERGS, it was much easier to develop the

Table 3.25. Occurrence of legal changes. (Question 3.2.1: Have there been any legal changes, such as access rights, right to exclude others from open use, limitations, prohibition to pick mushrooms, etc.?)

Legal changes	Austria	Germany	Italy	Netherlands	Total
No	12	24	10	12	58
Yes	9	4	19	8	40
Total	21	28	29	20	98

Table 3.26. Types of legal changes.*

Types of legal change	Austria	Germany	Italy	Netherlands	Total
Legislative	2	1	5	1	9
Administrative	7	2	15	4	28
Planning	2	1	3	5	11
Total	11	4	23	10	48

Multiple answers are included. * Not applicable for case-studies: AU01, 02, 06–08, 10, 12–15, 17, 18, 20, DE01–17, 19, 20, 22, 24–28; IT06, 07, 10, 13, 18, 20, 22, 24, 25; NL01–04, 06, 08, 09, 11, 12, 16, 17, 19.

Table 3.27. Need for a licence to implement the RES products.*

Licence	Austria	Germany	Italy	Netherlands	Total
No	13	22	13	9	57
Yes	6	6	16	10	38
Total	19	28	29	19	95

* Not applicable to all case-studies, i.e. AU02, 06, NL04.

Table 3.28. Existence of a contract activity.* (Question 5.2.1: Have you carried out any formal contracting activities regarding organization of the offer or distribution of the product?)

Contracts	Austria	Germany	Italy	Netherlands	Total
No	9	13	12	6	40
Yes	12	15	17	13	57
Total	21	28	29	19	97

* Not applicable to all case-studies, i.e. NL11.

RES products, as was the case for several Austrian and, to a lesser extent, German case-studies. However, this was not always so, as certain institutional changes were generally required to create the conditions for the RES product transformation/development. In the case of the entrance to a park, it is easier to ask for a payment when already existing regulations entitle someone (the park authority, the forest owner) to charge a payment. Otherwise a law becomes necessary to define or assign this right.

Development: the role of marketing techniques

As already discussed, consumption complementarity means that one good/service is preferably consumed jointly with another, as can be the case with cars and petrol, where a strong complementarity in consumption exists. The transformation/development of ERGS into RES products seems to be often based on such consumption complementarity: the enjoyment of ERGS involves the consumption of the related RES product. Complementarity can be very high where the RES product almost overlaps with the ERGS, or weak when the relation is less evident, as outlined in Section 3.3 (Figs 3.8–3.10). For every RES product, the complementarity level with the related ERGS can be estimated and quantified. The levels of complementarity between RES products and ERGS can be evaluated by using the scale and criteria identified in Table 3.29.

The list of the RES products with the related ERGS and the complementarity levels has been set up for all cases. Additionally the following categories of ERGS were identified: (i) recreation in the forest; (ii) water-quality protection; (iii) biodiversity conservation; (iv) environmental image/quality; and (v) carbon dioxide (CO_2) reduction. Each category has been related to the RES products according to the levels listed in Table 3.29.

Table 3.30 illustrates the fact that most of the RES products have a level of complementarity with the related ERGS of between 5 and 7.5. Quite clearly, transformation/development of the RES products has been easier where complementarity was high and well identified. Consumption complementarity between RES products and the related ERGS could also be seen as a way of exploiting the unique characteristics of a specific forest environment, avoiding standard products, which could meet competition in the market. This situation can also be related with the success factors described in Table 3.22.

In Table 3.31, the relationships between

Table 3.29. Criteria and scale for the assignment of complementarity levels.

Complementarity level	RES products
Strong (10.0)	Access Management agreements
Rather strong (7.5)	Sponsorship Traditional products with an environmental component Certificate products Footpaths Guided-tour visits Forest education centres/eco-museums Eco-museums
Weak (5.0)	Agrotourism Picnic and camp-sites Private roads Sports facilities Hiring of sports equipment Events in the forest
Very weak (2.5)	Hotels Shops Restaurants Flats

Table 3.30. RES product's complementarity with forest ERGS.

	Level of complementarity				
ERGS	Strong (10.0)	Rather strong (7.5)	Weak (5.0)	Very weak (2.5)	Total
Recreation in forest	4	16	40	4	64
Mushroom-picking	–	3	–	–	3
Recreational–environmental image	–	1	2	–	3
Environmental image	1	10	1	–	12
Environmental image/quality	–	3	2	1	6
Forest conservation (biodiversity)	3	–	–	–	3
Water-quality protection	2	4	–	–	6
CO_2 reduction	–	1	–	–	1
Total	10	38	45	5	98

institutional changes and market techniques based on complementarity between ERGS and RES products are investigated. When the levels of complementarity are high, there is no need for strong institutional changes; administrative changes are generally sufficient, so it is very rare for legislative changes to be necessary. In fact, these RES products largely overlap with the related ERGS (strong complementarity in consumption) and this relationship makes the RES product unique – it is not possible to separate it from the ERGS and/or to offer it elsewhere. For example, in the payment for access to a park, the RES product 'access' and the related payment could not exist without the park/environment with its unique characteristics.

The cases of RES products with weak complementarity emphasize the need of institutional changes based on administrative procedures and planning. In the case of RES products characterized by a weak

Table 3.31. Institutional changes for different levels of complementarity.

	Level of complementarity				
Change	Strong (10.0)	Rather strong (7.5)	Weak (5.0)	Very weak (2.5)	Total
Legislative change	1	7	2	–	10
Administrative change	2	9	17	2	30
Planning change	1	2	9	1	13
No change	6	20	17	2	45
Total	10	38	45	5	98

complementarity with the ERGS, the institutional changes are more related to the characteristics of the markets of these RES products (licences for selling products, as in the case of shops, and leases, in the case of houses, etc.) than to environmental legislation. Concerning the issue of product certification, the self-certification of a product does not require any institutional change, while the assignment of a label requires the existence of specific regulations.

3.6 Transformation/Development Paths: RES Products Relationship with Forest-site Conditions

3.6.1 Reference to country case-studies

From the previous section, it clearly appears that the transformation/development of ERGS into RES products involves both institutional and market-based approaches. According to each individual case-study, the two approaches can be applied alone or simultaneously, helping each other towards the marketability of the RES product. Both approaches, however, aim at increasing excludability, while rivalry may or may not be increased. For traditional products, which take advantage of the image of the environment where they are produced and the related techniques, the development paths due to marketing techniques are shown in Fig. 3.11 with a double-ended arrow.[10]

Figure 3.11, developed from Tables 3.16 and 3.17 (before transformation) and Tables 3.18 and 3.21 (after transformation), shows the transformation/development paths for the four countries. It is clearly demonstrated that transformation/development paths can be very different in terms of changes in excludability/rivalry; however, the 'marketability' arrow of Fig. 3.8 remains a common feature. It is also evident that transformation (concerning institutional changes, represented by a continuous line) and development (concerning marketing techniques, represented by a dotted line) can be applied either separately or together, according to the different case-studies.

The complementarity and synergy between institutional and marketing means is particularly evident for Italy and the Netherlands. In Germany, and particularly in Austria, the situation is rather different, because marketing means are more frequently applied, while support by minor institutional changes is less evident. Also the transformation/development paths appear to be similar and less dispersed in the excludability/rivalry diagram.

The differences in transformation/development paths, as outlined in Fig. 3.11, can be linked to Austrian and, to a lesser extent, German property-rights for ERGS connected with forestry, which are better defined and assigned by consolidated legislation. Traditionally, in Austria and Germany, forest owners are also more business-minded, as far as timber is concerned, than in the Netherlands or Italy. The lack of a business tradition in Italian forestry can also be seen in terms of 'late economic development', which has left a legacy of ill-defined and ill-assigned property-rights. The present situation, characterized by a rather affluent and environmentally sensitive society, makes it difficult to apply

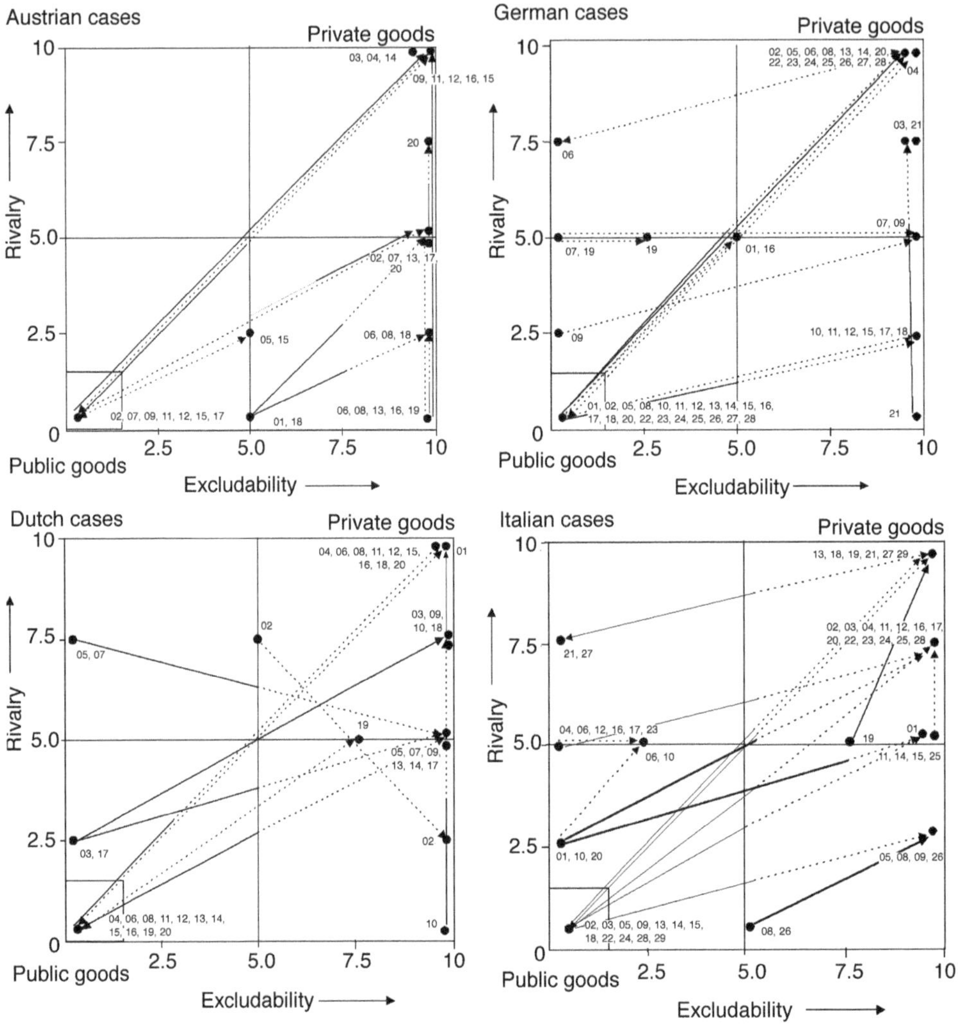

Fig. 3.11. Development paths for RES products by country. ——— Transformation through institutional changes; . . . . . . Development through marketing techniques.

'unpalatable' forest and land policies, which were easily applicable two centuries ago: consider, for instance, the Maria Teresa land reforms in Austria. Therefore, other more socially acceptable means have to be applied. This helps to explain the complicated mix of institutional/marketing approaches shown by the Italian and Dutch cases. These two countries also exhibit a common high pressure on forest resources determined by high population density.

3.6.2 Towards a synthesis of transformation/development paths

In order to find a synthesis of transformation/development paths, the usual excludability/rivalry diagram can be split into various levels of excludability and rivalry. The levels of excludability and rivalry from 0 to 2.5 have been separated to signify low/nil excludability/rivalry, while those from 7.5 to 10 signify very evident excludability/rivalry. This is shown in Fig. 3.12, where nine different areas have been iden-

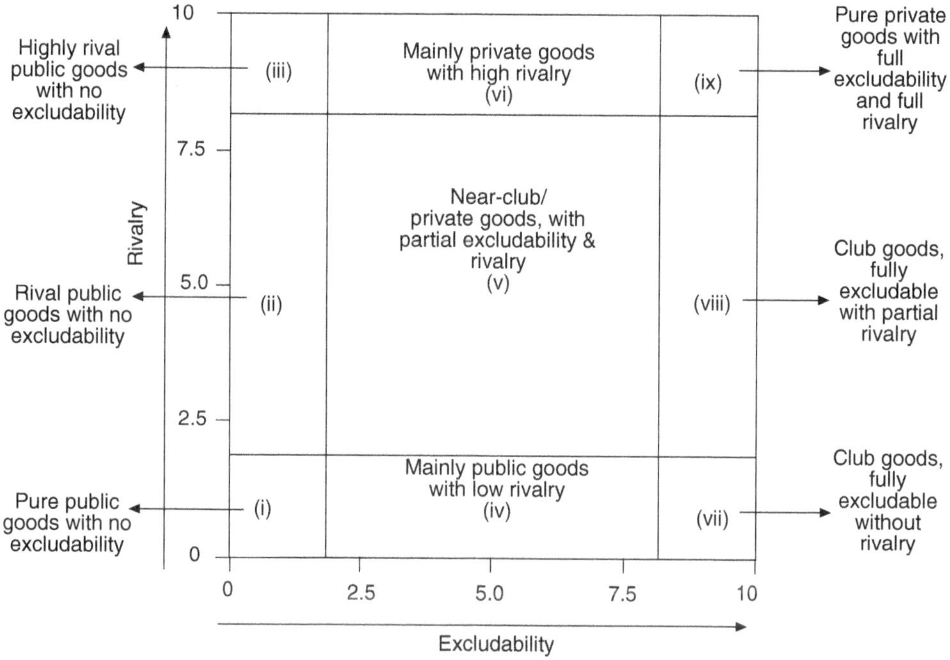

Fig. 3.12. Classification of various categories of public, private and mixed goods.

tified (five more than the four scarcely identifiable in the usual excludability/rivalry diagram):

1. 'Pure public goods with no excludability' or rivalry, e.g. a free-access forest/landscape without congestion.
2. 'Rival public goods with no excludability' while some rivalry is felt, e.g. a free-access forest/landscape with a certain congestion during weekends and in holiday time.
3. 'Highly rival public goods with no excludability', e.g. a free-access forest/landscape with serious congestion problems.
4. 'Mainly public goods with low rivalry', e.g. a forest/landscape where access can be forbidden without congestion being felt.
5. 'Near-club goods with partial excludability and rivalry', e.g. a forest where access can be forbidden, and the problem of enforcing exclusion is posed for the management because of congestion, particularly felt at weekends and in holiday time.
6. 'Mainly private goods with high rivalry', e.g. a forest where access can be forbidden, and the problem of enforcing exclusion is posed for the management because of high demand/congestion.
7. 'Club goods, fully excludable without rivalry', e.g. sports facilities where congestion is not felt; the implementation of exclusion creates a club good managed by members.
8. 'Club goods, fully excludable with partial rivalry' and congestion problems, e.g. sport facilities where the congestion is felt to a certain level so that members paying membership exclude non-members, while internal rivalry/congestion can be solved by the 'club management', on the basis of reservation or higher membership fees for peak hours/days.
9. 'Pure private goods with full excludability and full rivalry', e.g. a highly demanded forest where visitors must pay for a ticket; thus portioning and divisibility are implemented by the price of the ticket.

What emerges from the subdivisions of

Fig. 3.12 is the large central area of near-club goods where excludability/rivalry (between 2.5 and 7.5) does not succeed in creating a pure private good but which is no longer a public good. This central area must be overcome by transformation/development strategies and techniques; otherwise the management of the RES products risks remaining undefined and the cost of management is not sufficiently covered by the revenue.

The development paths for each country, as outlined in Fig. 3.11, can be grouped together to give six significant paths of transformation/development from ERGS into RES products. These six paths show general trends of transformation/development, which are illustrated in Fig. 3.13.

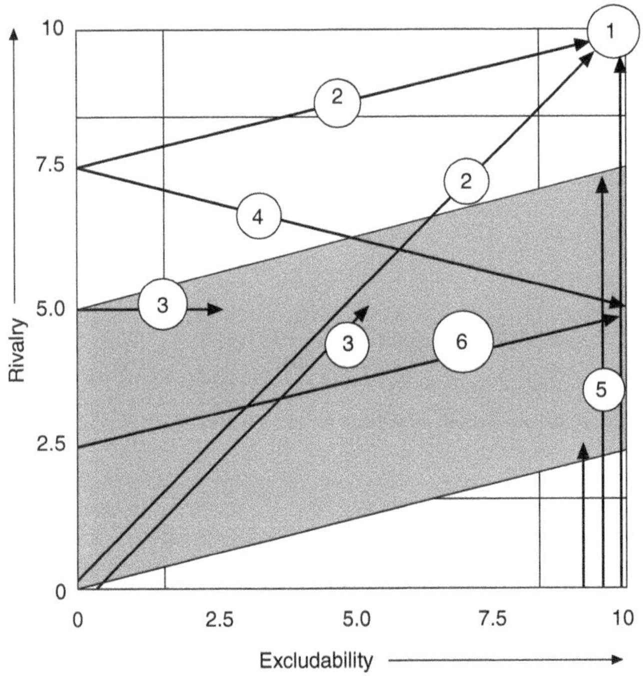

Fig. 3.13. The main transformation/development paths. Cases IT07 and AU10, 21 are excluded from the tables and so are not included in the paths of the figure. Cases IT08, 11, 19, 25, 26, AU01, 05, 18, NL09, 18 and DE03 do not exactly fit into these paths and so are also not included. To simplify the diagram, only paths that consist of two or more cases are shown; other cases that fit the general description of that path are written in italic.

(1) Relaunch of an already private (or club) good. In these cases, there was not any variation in the excludability/rivalry criteria (cases AU03, 04, 14, DE04, NL01).

(2) From a public good (pure or rival) to a pure private good (cases AU9, 11, 12, DE02, 05, 06, 08, 13, 14, 20, 22–28, NL04, 06, 08, 11, 12, 15, 16, 20, IT13, 18, 21, 27, 29).

(3) From a public good (pure or rival to various degrees) to a club good with partial excludability (cases DE01, 16, 19, IT06 + *AU15, IT10, NL19*).

(4) From a public good (partially rival) or a club good with partial excludability to a club good with a decrease in rivalry (cases NL05, 07 + *NL02*).

(5) From an excludable club good with low rivalry to a club or private good with partial or high rivalry (cases AU06, 08, 16, 19, 20, DE21, NL09, 10, 18, IT11, 25 + *AU13*).

(6) From a public good (pure or rival) to a club good with high excludability (cases AU02, 07, 17, DE07, 09–12, 15, 17, 18, NL03, 13, 14, 17, 19, IT 01–05, 09, 12, 14–17, 20, 22–24, 28). Path 6 consists of several paths, which are all within the shaded area.

The position of each path results from the various case-studies (as outlined in Fig. 3.16) when excludability/rivalry have been subjectively evaluated according to the scale 0, 2.5, 5, 7.5, 10, according to their position within the usual excludability/rivalry diagram of Fig 3.1 before and after transformation.

As already discussed in previous sections, each transformation/development path is different. However, the simplification of Fig. 3.13 allows the following common features to be pointed out:

1. Transformation/development paths always move from left to right and usually from down to up, as this is the consequence of the fact that the main purpose of transformation/development actions is to obtain private goods where remuneration is easier.
2. Some areas of Fig. 3.12 (e.g. iii, iv), characterized by high rivalry and low excludability, do not have paths to indicate that these position are far from the objectives of transformation/development. (Incidentally high rivalry and low excludability indicate such situations defined as 'free-riding' and 'tragedy of the commons'.)
3. The key issue for transformation/development is the increase in excludability, while changes in rivalry mainly result from this modification. Particularly in path 4, the goal was rivalry reduction, obtained through an increase in excludability.
4. Transformation/development usually leads to club and private goods, depending on the product's intrinsic characteristics. Path 3 is quite unusual; it corresponds to the use of voluntary contributions/donations as a means of payment, and it could be considered an incomplete transformation/development – the RES product remains in the central undefined area of near-club goods.
5. There is only one path 4 shown, which includes two Dutch cases, that goes between goods with partial excludability and club goods. More path 4s could have been expected, as it seems a logical further development of path 3 – rather common, with seven cases spread throughout the countries. The likely reason for this is that the change from voluntary donations to compulsory payments is often unpalatable to the public, and it is often unwise to disturb or upset people.

In general terms, the transformation/development paths could also be divided into sequential stages, all represented in the investigated cases. The transformation starts with the creation of a certain excludability (paths 2, 3 and 6), which causes some changes in rivalry, due to congestion effects (for club goods) and to the different intrinsic characteristics of RES products (for both club goods and private goods). In the following stage, the RES product is further developed (through marketing techniques) to better respond to customers' demand. If congestion is creating problems (as sometimes is the case for recreational products), measures to reduce it are implemented (path 4). In some cases, other technical or structural modifications to improve service quality are undertaken, without consequences for excludability and rivalry (path 1).

3.7 Conclusions: Perspectives for Remuneration of ERGS through Transformation/Development of RES Products

From the research supporting this chapter, the following issues were found to be important for achieving market remuneration for ERGS through their transformation/development into RES products:

1. The type of product is particularly important. In general, those linked to recreation can more easily find market remuneration, while the environment alone is not so easily transformed into a market good.
2. The transformation/development of RES products when the environment component is more important is usually conditional upon large estates. It is, therefore, not surprising that environmentally based RES products take place mainly on public land or various forms of common property.

3. In just under half of the cases, the management of the RES product is undertaken by various forms of associations or by contractors. Also around half of the cases took place on publicly owned land, which tends to have a larger land base than private land. However, there does not seem to be a strong correlation between RES product management and land ownership in terms of public or private.
4. A remarkable number of RES products based on recreation (camp-sites, sports facilities, etc.) do not need a large land base, though they exploit the surrounding forest environment.
5. Before transformation/development, excludability is, in general, very low, while rivalry varies from low to medium levels; transformation mainly increases excludability, so RES products are mainly provided as 'club goods'.
6. Increasing management costs and other economic pressures often encourage the establishment of RES products. Opportunity exploitation, however, is the driving force supporting transformation/development.
7. Some forms of institutional changes are generally needed (but only in a few German cases) for the establishment of the RES products, yet legal rights are not normally greatly affected. It is usually the case that minor administrative/planning changes are sufficient and these are granted by local authorities;
8. Excludability is certainly achieved after transformation; the concept, however, is applied to the RES products (market goods and services) complementary to ERGS and the environment in general.
9. Measures aimed at enforcing exclusion and preventing free-riding behaviour are also taken, though in general they do not call for high costs.
10. The main strategy behind development is the 'complementarity' between ERGS and the RES products, and the provision of structures necessary for recreation and access to the environment.
11. Failures are not uncommon and can sometimes be due to property-right violation, as well as lack of demand and/or poor management.
12. Concerning the relationships between traditional forest products, like timber, and RES products, the case-studies essentially referred to forestry; therefore, competition was not so acute as it could have been in the case of agriculture.
13. RES product development took place in many cases in areas of multipurpose forestry in high demand for recreation.
14. A positive correlation was generally found between the different RES products and the related ERGS and other products that were part of the same 'area product mix package', with the exception of hunting and shooting. It has also been proved, particularly in Italy, that RES products are substantially helped by the provision of packages of RES products available in one area.
15. The main shortcoming of any transformation/development strategies is given by the fact that those producing the forest-related ERGS are not always those managing the attached complementary goods or services, i.e. the RES products – sports structures, various facilities, hotels and accommodation, shops, etc. Compensation amongst economic actors at area level must, therefore, be considered, and this situation is particularly critical for Italy, where the forest properties, both public and private, are shown to be more detached from the RES product management, though they provide the ERGS and support the related costs.

3.8 RES Product Typology for Marketing and Contracting Analysis

3.8.1 Need and definition for a system of RES case-studies

In general, it is neither possible nor does it seem to make sense to list all possible products of forest enterprises, since the character of a product can change with time, depending on the appreciation of economic subjects (Sekot and Schwarzbauer, 1995). The authors of the present book therefore recommend defining every single product and relating it to the environmental and recreational services of forests, instead of establishing categories. It is, however, not very helpful to limit the view to single cases when trying to derive general statements concerning possible marketing instruments and organizational and contract structures. This investigation will rather describe the similarities and differences between different examples of RES businesses. These will be included in a typology of possible RES products in such a way that, on the one hand, it is open for new products and, on the other, it includes homogeneous product criteria, in order to derive type-orientated conclusions for marketing and organization and contract design.[11]

To begin with, the investigation has looked at the market-determining product attitudes of respective RES products and then at the parameters for demand necessary for marketing. The goal is to derive market structures that enable a sensible division of product groups for a comparative case analysis. Yet the typology derived in the following outline can only consider certain product features. If one wanted to consider all the properties of the product, the offerer and the market that influence the decisions of marketing, distribution, organization and contracting, one would inevitably return to a differentiation of individual cases. Therefore, the typology outlined below, which was reduced to certain properties of products and markets, mainly fulfils the following targets.

- Description of one of several possible, general and unlimited product typologies for RES products.
- Assistance in the search of certain product-dependent patterns of marketing and organization.

3.8.2 Basic approaches of product typologies

Mantau (1995a, p. 132; 1995b; 1997) recommends describing 'forest goods and services on a first level of reflection at first (...) and if necessary to [summarizing] them according to seven main groups for means of overview'.

1. Timber.
2. Other biological production.
3. Services.
4. Forest area.
5. Recreation.
6. Protection.
7. Aesthetics.

Thus, he remains at the first step on a descriptive level that was also used by other authors for the description of forest performances.[12] He opposes a description that already determines specific forest services in its connecting system by the choice of term (e.g. 'forest functions'). Descriptions are rather made in an open way, independent of any type of system (economics, politics), thus leaving space for development.

Ferro *et al.* (1995) distinguish the following types according to intended performances and conditions of competition in the production of raw materials:

- By-products (positive external effects).
- Joint products (intentional, coupled products).
- Main products (intended products in competition with raw-material production).

They go beyond a pure description and particularly include the production process in their establishment of categories.

Merlo *et al.* (1996) created a system of three different environmental goods and/or services provided by agriculture and forestry in mountainous areas:

1. 'Non-structured' environmental goods and services with direct remuneration (joint products).
2. 'Structured' (exploited through market transactions) additional/complementary services with respect to environmental goods and services with indirect remuneration.
3. Traditional quality products (certified) linked to the area and production techniques.

In marketing literature, it is common to differentiate between 'marketing of consumer goods', 'marketing of services' and 'marketing of investment goods'.[13] These areas, however, largely overlap each other. This is especially so because products generally contain goods, as well as services and rights, which can, on the one hand, be consumer goods and, on the other, investment goods. So Meffert and Bruhn (1995, pp. 5–6) assume a large 'common area of sectoral marketing' for different consumer goods, extending from consumer services and investment services[14] to systems and parts.

Mantau (1995a) define a flowing delimitation between certain product characteristics (compare Fig. 1.2). This flowing determination between public and private goods is very relevant for questions of product transformation (compare the previous sections in this chapter). It is of limited suitability, however, for the development of an open typology with common implications for marketing, organization and contract structures, because the investigated case-studies mainly stay in the lower right box of the Fig. 1.2. In the following, the authors therefore refer to an open typology of Engelhardt *et al.* (1993), which is suitable for a variety of uses, and they develop it further for RES products.

3.8.3 Parameters of the typology model used

Customer integration of RES products

Material goods, services and rights can be offered as products. Products normally consist of a combination of these three categories, and it is difficult to assign a certain product to only one category. Engelhardt *et al.* (1993) solve this problem by defining the term 'market object' (*Absatzobjekt*) as a package of different goods and services that are combined by the offerer to meet certain demands of the user and which are sold on markets. Depending on the degree of nonmateriality and customer integration, according to the authors four general product types can be distinguished (Fig. 3.14).

Products are defined as integrative products when the user is integrated in the production process. Engelhardt *et al.* (1993) also point out that a certain amount of customer integration always exists, because the customer has to express his wish to buy the product. Thus, the degree of integration is a continual feature. An example of a highly integrative product is management consulting. In this case, the demander has a high influence on the production process. Services of a consultant are discussed with the management of the enterprise and, in turn, the consultant has to rely on informa-

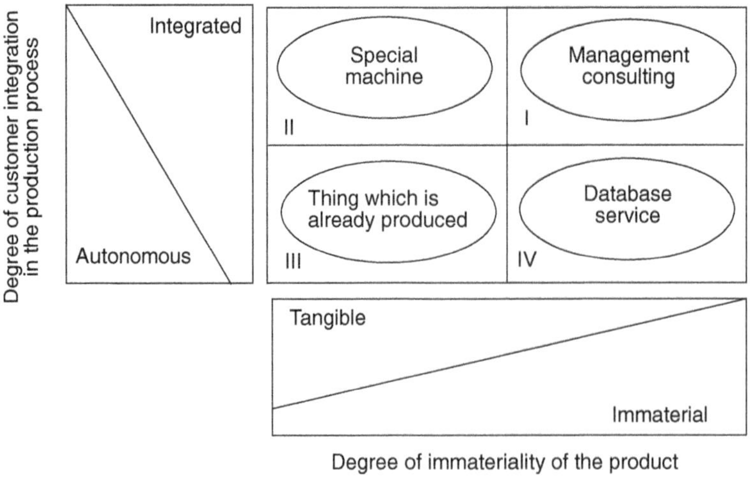

Fig. 3.14. Typology of products with four examples (from Engelhardt *et al.*, 1993).

tion (e.g. statistics) from the customer. RES products, which can be compared to this type, are use contracts that are set up according to the special wishes of the contract partner.

Products are defined as autonomic products when the demander has no direct influence on the production process. Autonomic products are characterized by standardized contract conditions. An example of a highly autonomic product is the offer of petrol at a petrol station. The customer has influence neither over the consistency of the petrol nor over the price. The only possible choice for the customer is to buy or not to buy the product. If the product properties, in contrast, are developed mainly by the integration of the customer into the production process, it is an integrative product.

The investigation showed the following results concerning the degree of customer influence on the development of RES products:

- In 53 cases, it was a mainly autonomic product.
- In 36 cases, the product developed in an integrative way, which means that customer participation was necessary to develop the product.[15]

In the remaining nine cases the product was developed in an integrative way between forest enterprise and offerer but was given to the end-user as an autonomic product. As a result, the description can either be 'offerer-orientated' or 'buyer-orientated'. This will be referred to later.

Immateriality of RES products

In addition, Engelhardt *et al.* (1993) describe a second dimension that is important for characterizing products – the degree of non-materiality of the product. Products are defined as immaterial when they are not tangible. Neither the information of a management consultant nor a database service is tangible. An example of a non-material RES product is the guided forest tour. In this case, the customer buys an intangible experience that consists mainly of information. Every product has a certain non-material part, according to Engelhardt *et al.* (1993), because they define the sales process (e.g. talking with the customer) as part of the product. According to this, the degree of materiality of different products is also a continuous feature that is not distinctly marked. Therefore, an operational instruction to categorize new products according to this typology is initially based on a distinct delimitation of product properties. If these product properties can mainly be evaluated materially, it is a tangible product. Examples of material products of forest enterprises are timber or the offer of drinking-water. In general, the customer can take away or dispose of a material product.

Even though the degree of non-materiality is not a sharp distinction,[16] the suggestion of Engelhardt *et al.* (1993) is helpful for finding certain product categories for products analysed in the RES study. Thus, besides distinguishing products by customer integration, the products are distinguished by four steps of non-materiality.

1. Material products (four RES cases).
2. Material products with additional non-material components (13 RES cases).
3. Non-material products with additional material components (21 RES cases).
4. Non-material products (59 RES cases).

In a single case, the product between forest enterprise and offerer consists of a completely non-material permission for utilization (*Nutzungsgestattung*) on a camp-site, the usage of which, however, is partly given as a material product to the end-user. This should be differentiated into an 'offerer-orientated' and a 'buyer-orientated' description as well.

Meffert and Bruhn (1995, p. 31) point out that 'specific marketing implications can already be derived' from the division according to the dimensions 'materiality' and 'integration of external factors' (participation of the customer with established traits). Nevertheless, different types of customers, product specifications and offerers can be distinguished in the case-studies in addition to these eight different product types and will be described in the following sections.

Customer characteristics of RES products

CUSTOMER TYPES. Regarding the customers of RES products in general, differences can be distinguished between individuals and organizations (institutional demand). While individuals always buy the products for consumption, organizations sometimes buy the product for investment purposes. The following list shows the different customer types of RES products:

1. Individuals. They demand the product for individual consumption purposes and are documented mainly for recreational products of forest enterprises (e.g. accommodation possibilities, recreational facilities, guided tours or use of forest roads, car-park areas or camp-sites).
2. Organizations
(a) Consumer organizations. They buy the product for purposes of consumption by their club members (e.g. horse-riding clubs, paragliding clubs, mountain-biking clubs) or by their employees (e.g. events, excursions, Christmas celebrations).
(b) Public institutions. They buy the product and put it at the disposal of the public (e.g. conservation contracts, protection of drinking-water resources).
(c) Groups with economic targets. They demand the product for investment purposes (e.g. sponsoring products, service for management seminars).

When analysing the case-studies, markets can be determined that are potentially accessible for different customer types, however: only for individuals (single customers), for individual customers and partly for organizations (institutions), for organizations (institutions) and partly single (individual) customers, or for organizations (institutions) only.

From Table 3.32, it is obvious that half of the products are mainly offered to private individual customers and half to institutions. Whether such a product is offered on a market characterized by private or institutional demand depends primarily on the product characteristics. The share of products purchased exclusively by institutions is 38%, which is high. The different organizational customer types are not further subdivided in Table 3.32.

PRODUCT BENEFITS TO THE CUSTOMER. Traded products can be divided into 'consumer' and 'investment products'. In a very general view, 'consumer products' can be regarded here as those products which are distributed to the 'end-user' and are not further processed. 'Investment products' include (also in general) all of those products that are part of another product and thus are distributed for further processing to institutional customers. These simplified definition approaches cover aspects of the demander and aspects of further usage. In the recorded cases, the following determinations of RES products appeared:

Table 3.32. Product shares with individual (consumers) and institutional (organizations) demand. (Question 4.3.1: Are there any institutions/organizations among your customers?)

Buyers	*N*	%
Only consumers (co.)	12	12.2
Usually consumers (co./in.)	38	38.8
Usually institutions (in./co.)	11	11.2
Only institutions (in.)	37	37.8
Total	98	100.0

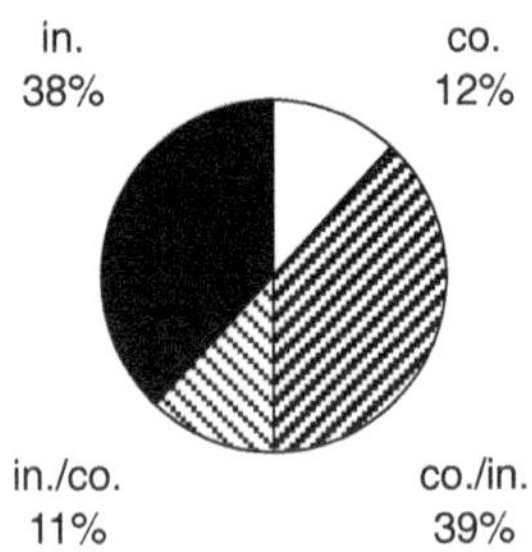

- In 42 cases, there was direct utilization as a 'consumer good' by exclusively or mainly private customers without further processing or passing to third persons. In an economic sense, such products can be considered as 'goods' as well as 'services'.
- In 37 cases, there was utilization as an 'investment good' by institutional customers, who further process it to a greater or lesser degree and pass it on to third persons, usually in a commercial way.
- In ten cases, utilization was as a 'common good' or as a 'good for common benefit', which is purchased by (partly public) institutions in order to pass it on to its own members, to third persons or to the public without further processing and generally without commercial aspects or even direct benefits.

The nine cases previously mentioned in which a distinction needs to be made between 'offerer-orientated' and 'buyer-orientated' can be divided into 'investment goods' on the first level and 'consumer goods' on the second level.

Additional selection criteria

However, some very different products need to be considered together according to the previous dividing criteria. Therefore, the typology was supplemented by some additional selection criteria for the description of RES product groups. In this way, a distinction was made with non-material products between complex, labour-intensive services and simple assignments of usufructuary rights. The advantages derived from a categorization according to product contents were utilized by comprising contracts concerning guidelines for increased environmental performances, agreed rights for advertising (sponsoring) and complex institutions for environmental education (teaching lessons, utilization of facilities, etc.) in their own product groups.

Also a more profound differentiation according to product-specific contents, such as drinking-water, overnight stays, guided tours or eco-sponsoring, is sensible and productive when examining certain issues. Therefore, in the following it is used where appropriate.

The most important feature of a typology dedicated to the further examination of certain marketing strategies concerning certain product groups is the required homogeneity of the types concerning the relevant categories. If the target of developing a typology that is all-inclusive with a manageable number of types and open for an unlimited number of new products is pursued simultaneously, it is necessary to integrate the important aspects of the product contents in a flexible framework. The typology developed here (division according to degree of integration, materiality and further usage, with additional features of product contents) is a possible stratification that seems suitable for the solution of the present problems.

Offerer of RES products

Concerning the goal of investigating new income sources for forest enterprises, the offerer should be a forest enterprise. Nevertheless, those products from RES of forests that are not offered by forest enterprises in the investigated case-study but could be offered by forest owners may be of interest as well. Therefore, the first distinguishing criterion is:

- The offering enterprise is (directly connected with) the forest landowner.
- The offering enterprise is not an organizational part of the forest ownership.

Within the group of forest owners it is crucial, for the sake of economic goals, flexibility and ability to pay, whether forest enterprises are:

- Private forest owners.
- Public forest owners.

This distinction can further be differentiated according to the degree of organization. It is mainly private forest owners who in this way profit from larger and thus more efficient management units.

Finally, the size of the offering enterprise can certainly have some significance for its supply opportunities. In this regard, a distinction can be made between:

- Small to medium-sized owners (0.1–1000 ha).
- Medium-sized to large enterprises (1001–5000 ha).
- Very large enterprises (over 5000 ha).

This distinction is rather rough, especially regarding the first group. It was, however, chosen in order not to differentiate too much between the offering enterprises. For German conditions (at present), these size levels correlate to the following division:

- Enterprises that can build up their own administrative structure only under specific preconditions.
- Enterprises that can have at least one 'forest ranger station' (*Revierförsterei*).
- Enterprises that possess their own administration (forest office) with several units (ranger stations).

In addition, forest enterprises as offerers of RES products have been documented in 14 cases. So the typology of products will be enlarged by the dimension 'number of offerers'. Combined with the division into the eight categories of non-materiality and customer integration, the typology of offers shown in Fig. 3.15 results.

'Cooperation of offerers' was chosen as an important dimension for the categorization, because it is an important difference whether a single forest enterprise offers a product or if a cooperation of enterprises does. When a cooperation of enterprises offers certain products, there are always organizational costs, which have to be added to the transaction costs, and therefore the total amount of coordination costs increases.

The higher amount of coordination costs compared with possible proceeds from selling recreational or environmental products could be one of the reasons why cooperations of small forest landowners as offerers of RES products have been documented very rarely in the case-studies analysed.

3.8.4 Typology of RES case-studies according to product groups

Explanation and consolidated description of RES product groups

On the basis of the classification parameters introduced before, the 98 case-studies

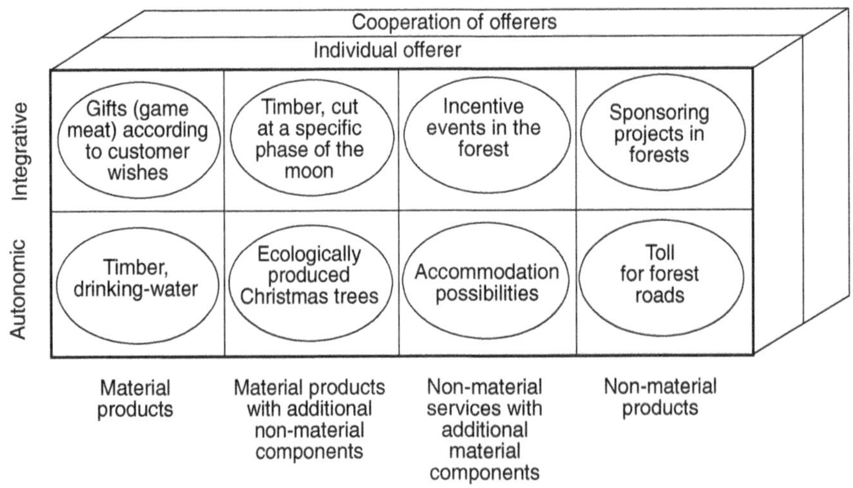

Fig. 3.15. Typology of offers.

can be divided into 13 product groups of comparable market structure and 19 subcategories of similar product contents. These groups are initially derived from the following seven product and demand characteristics that are important for RES offers:

- Degree of materiality (1: materiality of the main offer; 2: materiality of additional offer aspects).
- Degree of flexibility (3: autonomic/integrative product).
- Single buyers or institutions (4: main important buyers; 5: less important buyers).
- Use as 'consumer good', 'common good', 'good for common benefit' or 'investment (capital) good' (6).
- Service aspects, complexity (7).

Furthermore, the features 'content of the assignment of rights', educational aspects and product characteristics, such as certificates, accommodation, mushroom-picking, etc., are considered. The theoretical derivation of the product groups from these key parameters by a grouping matrix (yes–no structure) is outlined in detail in Welcker (2001). As a result, the product groups can be differentiated according to the market structures in Table 3.33.

Table 3.33. Product groups (PG) according to market structures.

PG No.	Comments on and examples of different market structures in product groups
1	Fixed, tangible products
1.1	Tangible investment goods (abbreviation: material, invest.) Market for material (tangible) products with environmental connection which are passed, by means of long-term agreements, to institutions for further processing. Product example: drinking-water
2	Fixed, tangible products with considerable non-material portion
2.1	Offer of consumer goods (abbreviation: material, direct use) Market for predominantly traditional, material (tangible) products – partly with recreational and environmental connection (non-tangible additional benefit) and connected with services – which are mainly passed, specifically determined, to single customers. Product example: Christmas market
2.2	Eco-investment products (abbreviation: material, eco-products) a Market for traditional, material (tangible) products with close connection to environment (non-tangible additional benefit) which are mainly transferred, specifically determined, to institutions for further processing and are rarely marketed to single customers. Product example: eco-meat b Assured specific properties through a certificate Product example: certified chestnuts
3	Fixed, partly tangible products
3.1	Complex recreational consumer products (abbreviation: rec., partly mat.) Market for complex, partly material (tangible) recreational products, which are generally passed on to single customers as specifically determined consumer goods but also (more rarely) to institutions, which pass the product to their members without further processing a Accommodation Product example: holiday apartments b Complex services Product example: hunting experience
3.2	Simple recreational consumer products (permits for mushroom-picking) (abbreviation and product example: mushroom-picking) Market for simple, partly material (tangible) recreational products, which are generally passed to single customers as specifically determined consumer goods but also (more rarely) to institutions, which pass the product on to their members without further processing

continued

Table 3.33. *Continued*

PG No.	Comments on and examples of different market structures in product groups
4	Fixed, non-material products
4.1	Complex recreational consumer products (abbreviation: fixed rec., complex)
	Market for complex non-material (non-tangible) recreational services, which partly include environmental and educational aspects and, as specifically determined consumer goods, are generally passed to single customers and also (more rarely) to institutions, which pass the product on to their members without further processing
	a Recreation facilities
	Product example: eco-park
	b Guided tours
	Product example: guided mountain tours
4.2	Simple recreational consumer products (abbreviation: recreation, simple)
	Market for simple non-material (non-tangible) recreational products, which, as specifically determined consumer goods, are mainly passed on to single customers and also (more rarely) to institutions, which pass it on to their members without further processing
	a Issuing of licences for utilization
	Product example: ski course
	b Issuing of licences for utilization and entry tickets with environmental relations
	Product example: forest entry
5	Variable, partly material products
5.1	Complex recreational products for direct use (abbreviation: variable, rec., complex)
	Market for complex non-material (non-tangible) recreational services, which include more or less environmental and educational aspects, as well as tangible components, and are generally passed on to institutions or (more rarely) to single customers as a variable, seldom fixed product for direct use
	Product example: guided forest tours
5.2	Complex recreational products as part of the offers of third parties (abbreviation: rec., complex, invest.)
	Market for complex non-material (non-tangible) recreational services, which include more or less environmental and educational aspects, as well as tangible components, and are passed on as variable product to institutions, which include this service and the associated rights in an offered product
	Product example: manager seminars
5.3	Forest education facilities (abbreviation: forest education centre)
	Market for complex non-material (non-tangible) educational forest facilities, which include more or less environmental and recreational aspects, as well as tangible components, and are generally passed on as a variable, seldom specifically determined product for direct utilization to institutions as well as (more rarely) to single customers
	Product example: forest youth hostel
6	Variable, non-material products, integrative contract products
6.1	Usufructuary rights (abbreviation: use contracts)
	Market for non-material (non-tangible) agreements on utilization rights, which are passed on in variable and, for the single case, determined form to institutions for further utilization
	a Rights for recreational utilization
	Product example: ski-run
	b Other utilization rights
	Product example: test track for cars
6.2	Nature and environmental preservation contracts (abbreviation: environmental services)
	Market for non-material (non-tangible) agreements for environmental protection, which are passed on in connection with public programmes for satisfying the needs of third persons or of the public in a variable or, in the single case, negotiated way to institutions only.
	Product example: contract for environmental protection

PG No.	Comments on and examples of different market structures in product groups
6.3	Sponsoring contracts (abbreviation: sponsoring) Market for non-material (non-tangible) sponsoring contracts on advertising rights, which are passed on as variably negotiated investment goods, mostly, in the single case, to institutions only a With strong environmental connection Product example: contract for environmental sponsoring b With strong recreational connection (partly given to patrons) Product example: contract for recreational sponsoring

The catogorization in Table 3.33, including the relevant case-studies, has been summarized again in the matrix of Fig. 3.16.

Differences of case classification according to goals of analysis

When the product groups were derived, nine 'special' case-studies were referred to. These need to be categorized into different product groups in accordance with the goal of the analysis.[17] Regarding the contract design and the organization of offers, the relation between the forest owner and the offerer of utilization is interesting. But, for a further investigation of marketing activities, the relation between this 'organizer' and the end-user has to be considered. Therefore, the respective cases were marked in the summarizing product-group matrix with the letters 'o' for 'offerer' for the first step and 'b' for 'buyer' for the second step and were presented twice. As far as the offering forest enterprises are concerned, clarification of the first step between forest owner and 'organizer' is certainly important.

Moreover, the following product groups in Table 3.33 can be considered together. When the contract design is focused on, in the groups of material products with non-material aspects (2.1 and 2.2) and integrative complex recreational products (5.1 and 5.2), there is no distinction between the distribution to the end-users and to further processors as customers. Concerning marketing activities, the product groups of integrative complex recreational products for end-users (5.1 and 5.3) can be viewed together.

When working out this typology and transferring it to RES case-studies, the basis for a differentiating consideration of comparable case-studies can be obtained (see Chapters 4 and 5 in this volume).

Notes

1 E.g. hunting rights and the related access belong to the landowners (strictly private) in the Anglo-Saxon and German countries while they are public and assigned by the state to individuals paying for permission in some Latin countries.

2 Incidentally, it may be noted that complementarity in consumption is at the base of many environmental valuation techniques, such as the travel cost (TC) and hedonic pricing (HP) techniques.

3 Except the scarcity of land, well considered by the Malthus and Ricardo theories of economic development.

4 Pareto (1896), the founder of modern welfare economics, has underlined the economic and social meaning of common properties, stating that: 'since it has undergone the trial of free competition for centuries, collective land tenure responds better [than other forms of land tenure] to certain needs ... its violent destruction would mean a loss in the nation's total utilities'.

5 The Pareto optimality criterion states that allocation of goods and services should allow somebody to gain but making nobody worse off.

Cooperation of offerers →
Single offerers →

	Integrative/flexible	Autonomic/fixed
Material products	**Key** • 100 = individual user • *100* = organization as user • **100** = organization/individual • [100] = cooperation of offerers	**Product group 1** **1.1 Long-term contracts concerning material products for processors** Drinking-water *316, 319, 401* Electricity *303*
Material products with non-material aspects	• 101 ff. = cases in Germany • 201 ff. = cases in Italy • 301 ff. = cases in Austria • 401 ff. = cases in Netherlands • Product group: 1.1, 1.2, ... • o209, o211, ... = **offerer**-orientated description • b209, b211, ... = **buyer**-orientated description	**Product group 2** **2.1 Products for individual users** Christmas trees, game meat, wood products 101, 106, 113, 116, 315, **419** **2.2 Products for processors** (a) Eco-products: **404, 408, 411** (b) Certified products: **[202], 213, 221, [227]**
Non-material products with material aspects	**Product group 5** **5.1 Complex recreational services for direct use (offer of own seminars/tours)** *109, 111, 115* **5.2 Complex recreational services for investment purposes (in cooperation with offerers of seminars)** *110, 112, 117, 118* **5.3 Forest education centre** *127, 222*	**Product group 3** **3.1 Partly material recreation products for individual users** (a) Accommodation possibilities: 102, 128, **b**211, 414, **417**, 418, **420** (b) Hunting, activity holidays, nature experience: 104, **311**, 409 **3.2 Mushroom-picking permits** [201], 208, **219**
Non-material products	**Product group 6** **6.1 Use contracts with organizations** (a) Recreation: *121, 123, 124, 125, 126,* ***o****203,* ***o****205,* ***o****206,* ***o****209,* ***o****211,* ***o****214,* ***o****215,* ***o****218, 225, 226, 301, 306,* ***o****313, 317* (b) Other forest usages: 304, 314, 403 **6.2 Contracts concerning environmental services in connection with public programmes** *105, [114], 310, 312, 415* **6.3 Sponsoring contracts** (a) Environment: 108, [*122*], 229, 309, 406, 412, 416 (b) Recreation: 120, 207	**Product group 4** **4.1 Complex recreational services for individual users** (a) Recreation facilities: **103, 305, 410, 413** (b) Guided tours: 210, **b**218, 220, 307 **4.2 Simple recreation rights for individual users** (a) Sports: **107, 119, b**203, 204, **b**205, **b**209, 212, **b**214, **b**215, 216, 217, 224, **302, 308, b**313, **[318], 320, 405, 407** (b) Access to the forest: **b**206, **223, 228, 321, 402**

Fig. 3.16. Matrix of product groups.

6 It is remarkable how, in most recent decades, economists and political scientists often find the environment at the centre of this kind of analysis; refer, for instance, to Nordhaus's (1993) approach to global public evils, such as greenhouse gas (GHG) emissions and possible climate changes.
7 Some authors prefer the term social goods (Musgrave and Musgrave, 1973, p. 52), others 'free' goods. Rather interestingly, other authors use the term public needs rather than public goods (Serpieri, 1922).
8 Randall (1987, p. 175) affirms that:

> In his classical article, Paul A. Samuelson defined a public good in exactly the same way as we defined non-rival goods. ... A public good is a good that is consumed without rivalry. However, Samuelson's paper and much of the subsequent literature implied that public goods are nonexclusive in addition to being non-rival in consumption.

9 Refer to the following points of the questionnaire: 3.1.8 Before transformation – was it possible to exclude people from the RES product? 3.1.9 What kind of rivalry was relevant in the consumption of the RES product before transformation? 3.1.10 If there was rivalry, how relevant was it? 3.2.2 How relevant is the change concerning the possibilities to exclude people from free use? 3.2.3 How relevant is the change concerning the rivalry among the customers?
10 This applies to all the countries that have been surveyed.
11 As a general goal of a typology in the field of marketing, Meffert and Bruhn (1995, p. 30) define the identification of specific types of performances possessing implications for marketing that are different among certain types but uniform within the same type.
12 See Sekot and Schwarzbauer (1995, pp. 7–10).
13 See, for example, Meffert (1986, p. 46).
14 By distinguishing between investment and consumer services, it seems to be admissible to categorize intangible services as investment products (see Scheuch, 1982, p. 57; Meffert, 1986, p. 44; Meffert and Bruhn, 1995, pp. 21–22).
15 'Integrative' RES products, which develop in cooperation with the user, can also be divided up in the following way. If regularly offered products, in this case services, are newly designed every time for the customer, they can be regarded as being variable. Products that are only developed once (e.g. contracts on a specific project) can be regarded as unique.
16 Meffert and Bruhn (1995, p. 30) underline the continuous character of special distinction criteria.
17 The cases referred to are IT03, 05, 06, 09, 11, 14, 15, 18, AU13.

References

Arrow, K.J. (1951) *Social Choice and Individual Value.* John Wiley & Sons, New York.

Barone, E. (1912) Studi di economia finanziaria. *Giornale degli economisti,* April–July, 313.

Bonnieux, F. and Desaigues, B (1998) *Economie et politiques de l'environnement.* Dollor, Paris.

Brosio, G. (1986) *Economia e finanza pubblica.* NIS, Rome.

Buchanan, J.M. (1965) An economic theory of clubs. *Economica* February, 1–14.

Buchanan, J.M. (1967) *The Demand and Supply of Public Goods.* Rand McNally, Skokie, Illinois, USA. Translated into Italian (1969) *L'economia pubblica: domanda e offerta di beni pubblici.* F. Angeli, Milan.

Buchanan, J.M. and Tullock, G. (1962) *The Calculus of Consent.* University of Michigan Press, Ann Arbor.

Cicchetti, J.C. and Smith, V.K. (1976) *The Costs of Congestion: an Econometric Analysis of Wilderness Recreation.* Ballinger, Cambridge, Massachusetts.

Cornes, R. and Sandler, T. (1986) *The Theory of Externalities, Public Goods and Club Goods.* Cambridge University Press, Cambridge.

Cubbage, F.W., O'Laughlin, J. and Bullock, C.S. (1993) *Forest Resource Policy.* John Wiley & Sons, New York.

De Viti de Marco, A. (1888) *Carattere teorico della economia finanziaria.* Pasqualucci, Rome.

De Viti de Marco, A. (1934) *Principi di economia finanziaria,* Book I, Part I. Einaudi, Turin.

Downs, A. (1957) *An Economic Theory of Democracy.* Harper and Row, New York.

Engelhardt, W.H., Kleinaltenkamp, M. and Reckenfelderbäumer, M. (1993) Leistungsbündel als

Absatzobjekte: Ein Ansatz zur åberwindung der Dichotomie von Sach- und Dienstleistungen. *Zeitschrift für betriebswirtschaftliche Forschung* 45(5), 395–426.

Ferro, O., Merlo, M. and Povellato, A. (1995) Valuation and remuneration of countryside stewardship performed by agriculture and forestry. In: Peters, G.H. and Hedley, D.D. (eds) *Proceedings of the XXII International Conference of Agricultural Economists, Harare, Zimbabwe.* Dartmouth, London, pp. 415–435.

Glück, P. (1995) Forest policy means for non-timber production in Austria. In: Solberg, B. and Pelli, P. (eds) *Forest Policy Analysis – Methodological and Empirical Aspects.* EFI Proceedings No. 2, p. 278.

Glück, P. (1998) The task of research in evaluation of multifunctional forestry. In: *II International Forest Policy Forum.* Centre Tecnològic Forestal de Catalunya, pp. 119–129.

Goldscheid, R. (1917) *Staatssozialismus oder Staatkapitalismus.* Azen Gruber, Vienna.

Head, J.G. (1962) Public goods and public policies. *Public Finance* 17(3), 197–221.

Head, J.G. (1974) *Public Goods and Public Welfare.* Duke University Press, Durham, North Carolina, p. 291.

Hodge, I. (1995) *Environmental Economics.* Macmillan Press, London, pp. 181–182.

Lindhal, E. (1919) Some controversial questions in the theory of taxation. In: Musgrave, R.A. and Peacock A.T. (eds) (1958) *Classics in the Theory of Public Finance.* Macmillan Press, London, p. 168. (Original title: *Die Gerechtigkeit der Besteuerung.* Lund.)

McGuire, M. (1987) Public goods. In: Eatwell, J., Milgate, M. and Newman, P. (eds) *The New Palgrave: A Dictionary of Economics.* Macmillan Press, London.

Mantau, U. (1994) Produktstrategien für kollektive Umweltgüter. *Zeitschrift für Umweltpolitik und Umweltrecht* 3, 305–322.

Mantau, U. (1995a) Forest policy means to support forest outputs. In: Solberg, B. and Pelli, P. (eds) *Forest Policy Analysis – Methodological and Empirical Aspects EFI Proceedings No. 2.* Meeting report, Working group S6.12-01, IUFRO/SNS, pp. 131–146.

Mantau, U. (1995b) Von der Waldfunktionenlehre zur Waldproduktlehre. *Holz-Zentralblatt* 26(95), 445, 451–453.

Mantau, U. (1997) Erschließung neuer Märkte durch forstwirtschaftliche Zusammenschlüsse. *Allgemeine Forstzeitung/Der Wald* 12(97), 636–639.

Mazzola, U. (1890) The formation of the prices of public goods. In: Musgrave, R.A. and Peacock, A.T. (eds) (1958) *Classics in the Theory of Public Finance.* Macmillan Press, London, p. 37. (Original title: *I dati scientifici della finanza pubblica.* E. Loescher, Rome, pp. 159–183.

Meffert, H. (1986) *Marketing*, 7th edn. Gabler Verlag, Wiesbaden.

Meffert, H. and Bruhn, M. (1995) *Dienstleistungsmarketing.* Gabler Verlag, Wiesbaden.

Merlo, M. (1991) *Economia ed Estimo Forestale Ambientale.* Patron, Bologna.

Merlo, M., Kuehl, G. and Ruol, G. (1996) Possibilities of additional income from environmental goods and services produced by agriculture and forestry in mountainous areas. In: *Proceedings of Cost Conference, Vienna 1996, Forestry in the Context of Rural Development: Future Research Needs.* Conference Report, p. 18.

Muraro, G. and Merlo, M. (1987) *L'economia del bosco come bene pubblico e privato (finalità multiple ed ottima gestione forestale).* Atti del XXVII Incontro di Studio CeSET, Tip. Baccini–Chiappi, Florence, pp. 45–80.

Musgrave, R.A. (1959) *The Theory of Public Finance – a Study in Public Economy.* McGraw-Hill, New York.

Musgrave, R.A. and Musgrave, P.B. (1973) *Public Finance in Theory and Practice.* McGraw-Hill, USA.

Nordhaus, W.D. (1993) Reflections on the economics of climate change. *Journal of Economic Perspectives* 7(4), 11–25.

Pantaleoni, M. (1883) Contributo alla teoria del riparto delle spese pubbliche. In: *Rassegna italiana.*

Pareto, L. (1896) *Corso di Economia politica.* Einaudi (Italian edition, 1949), Turin.

Pigou, A.C. (1920) *The Economics of Welfare.* Macmillan, London. Italian edition: *Economia del Benessere.* UTET, Turin (1948).

Randall, A. (1987) *Resource Economics – an Economic Approach to Natural Resource and Environmental Policy.* John Wiley & Sons, New York, p. 175.

Samuelson, P. (1954) The pure theory of public expenditure. *Review of Economics and Statistics* 36, 387–389.

Samuelson, P. (1955) A diagrammatic exposition of a theory of public expenditure. *Review of Economics and Statistics* 37, 350–356.
Samuelson, P. (1958) Aspects of public expenditures theories. *Review of Economics and Statistics* 40 (November), 332–338.
Sax, E. (1883) The valuation theory of taxation. Vienna. In: Musgrave, R.A. and Peacock, A.T. (eds) (1958) *Classics in the Theory of Public Finance.* Macmillan Press, London, p. 177.
Scheuch, F. (1982) *Dienstleistungsmarketing.* Munich.
Sekot, W. and Schwarzbauer, P. (1995) Methodische Ansätze zur Bewertung der infrastrukturellen Leistungen der Forstwirtschaft. *Schriftenreihe des Instituts für forstliche Betriebwirtschaft und Forstwirtschaftspolitik* 25, 325.
Serpieri, A. (1922) *Appunti di economia forestale ed estimo.* Dattiloscritto, Reggio Istituto Superiore Agrario e Forestale, Florence.
Stiglitz, J.E. (1986) *Economics of the Public Sector.* W.W. Norton, New York.
Tiebout, C.M. (1956) A pure theory of local expenditures. *Journal of Political Economy* 5(October), 416–424.
Turner, K.R., Pearce, D. and Bateman, J. (1994) *Environmental Economics: an Elementary Introduction*, Vol. VIII. Harvester Wheatsheaf, New York, pp. 1–328.
von Wieser, F. (1914) The theory of the public economy. In: Musgrave, R.A. and Peacock, A.T. (eds) (1958) *Classics in the Theory of Public Finance.* Macmillan Press, London, p. 190. (Original title: *Theorie der Staatswirtschaft*, 2nd edn (1923), Tübingen, pp. 292–302.)
Welcker, B. (2001) Marketing für Umwelt- und Erholungsprodukte der Forstwirtschaft – Qualitative Analyse und theoriegeleitete Konzeptionen auf Grundlage von 98 europäischen Fallstudien. Dissertation Universität Hamburg, zur Veröffentlichung vorgesehen bei: Sozialwissenschaffliche Schriften zur Forst- und Holzwirtschaft, Bd.2; Peter Lang Verlag, Frankfurt/M u.a., voraussichtlice 2001.
Wicksell, K. (1896) Ein neues Prinzip der gerechten Besteuerung (Intorno a un nuovo principio di giusta tassazione). In: *Finanztheoretische Untersuchungen.* G. Fischer, Jena.

4

Marketing Tools for the Development. of RES Products

4.1 Introduction

For decades, marketing has been of substantial importance in general economics. However, this field of knowledge is not widely considered scientifically in forest economy, which is traditionally orientated towards production. Furthermore, the recreational and environmental services (RES) of the forests still form a product potential which, up to the present, has only in a few cases been concretized by forest enterprises. In this respect, this chapter deals with two fundamental approaches to an economic field of activities which, to a large extent, is unknown to most forest enterprises. The questions arising, according to influential factors and as starting-points for successful RES marketing, are as follows:

- What is the definition of 'marketing'? Why is marketing necessary for a successful sale of products concerning the RES of the forest? Which of the characteristics concerning the environment, the enterprise, the product itself and the market could influence the marketing success of a product? How can these characteristics be evaluated, and what are the consequences for the enterprise?
- What strategies and marketing instruments are theoretically available for the promotion of the sale of RES products? Which of them were applied in the surveyed cases?
- What conclusions derive from the theory and the analysis of the case-studies for successful marketing of RES products?

The aim of this task of the RES project is to facilitate the forest owner's entry to RES marketing. For this reason, the aspects of marketing which could be derived from the interviews of the 98 case-studies were evaluated qualitatively without going into the details of marketing exhaustively. This method underlines some fundamental differences between the marketing of recreational and environmental facilities of the forest and the 'regular' products of our society, without aiming at supplying empirical proof for existing regularities.

On the basis of the literature and the case-studies, first assistance for further activities can be derived and subsequently taught to forest owners interested in RES marketing within the framework of an introductory training course in marketing.

4.2 The Strategy of Marketing

4.2.1 Basic significance of marketing for RES products

Success factors for RES products

As an introduction to the empirical side of the subject, primarily the following issue was pursued:

- Which factors were considered to be of determining importance for the successful development of their RES products from the viewpoint of the offerers?

The responsible interview partners were asked to enumerate up to four factors, conditions or measures determining the success of their product. Altogether, 92 of the interviewees stated 231 points, while six of the interview partners did not comment on the issue. The intuitively given comments were subsequently ordered systematically to be able to compare the content of the statements. This system was structured according to 31 factors which were subcategorized further in seven groups of similar factors (Table 4.1, Table 4.2).

This list distinctly shows the variability in the range of possible influential factors. It illustrates the assessment by the offerers, as follows. The factors assumed to be most important for the success of their products were the organization and the management of the offered product, closely followed by the scenic product environment, the sociocultural frame conditions and a considerable demand, as well as the general product design, the orientation towards the customers and the quality of the product. Furthermore, the legal and administrative frame conditions,[1] the existence of equipment and functional production factors and of already adequately trained personnel with the necessary know-how were considered to be of considerable importance for the success of the product.

It was possible to divide the seven groups of factors into two areas:

- The two categories frame conditions and market conditions contain factors which essentially cannot be influenced by the company itself and therefore have to be classified as passive factors.
- Above all, elements of an active marketing strategy are comprised under the headings of marketing management, product design, distribution and price policy, communication policy and general marketing.

However, it must be taken into account that, because of commenting on questions without given answer possibilities, the answers given were generally more complex and could be interpreted either as containing passive preconditions or as suggestions for an active formation. It is not possible, therefore, to establish a complete division of these areas for the purpose of a systematic evaluation. This turned out to be especially true for the area of product design and characteristics of the product, which contains the important factor customer orientation as well.

Table 4.2 illustrates the categorization of the factors of success into seven groups. From the observation of the groups of factors, it can be seen that, most of all, the frame conditions and the product design are considered for the predicted success of a product, followed by market conditions and marketing management. Active marketing measures from the areas of distribution, price and communication policy were mentioned far less than the passive factors as influencing success. The area of marketing instruments that reaches beyond product policy in total forms only 14% of all factors mentioned, while the area of passive influence factors comprises about 50% of all statements.

For subsequent consideration of the successful marketing of recreational and environmental forest products, it can already be stated that:

> External influences, product design and the factor management and organization have a determining influence on the success of an RES product. Active marketing was only in a few cases said to be decisive for the success of the product. This initially points to the special significance of product policy in the early stages of the marketing process. Furthermore, the philosophy and methods of marketing do not seem to be widely distributed in the area of forestry.

Basic concepts of markets and services

In general, the same connections and processes are valid for recreational and environmental forest products as for every other well-known product on the market.

Table 4.1. The factors of success as indicated in the RES case-studies.

Success factors	Indication frequency
Frame conditions	
Legal and administrative frame conditions	10
Sociocultural frame conditions	15
Scenic product environment	19
Already existing equipment and functional production factors	14
Already existing personnel and know-how	12
Geographical proximity to the customers	7
Already existing infrastructure	9
Market conditions	
Considerable/specific demand	15
No competition/sole offerer	8
Synergy with products of other offerers	1
Image of the offerer	7
Marketing management	
Management and organization	21
Market research	1
Product design and characteristics	
General design of the product	15
Innovation	2
Product mix	4
Customer orientation	16
Exclusivity	4
Quality	17
Age of the product	1
Distribution and price policy	
General sales policy	2
Horizontal organization of distribution	2
Informal contacts in distribution	4
Price policy	3
Communication policy	
General communication policy	1
Advertising	2
Sales promotion	1
Public relations	7
Personal contacts	5
Free publicity	4
General marketing	
General marketing	2

In the area of marketing, the enterprise, the customer, the offer, the demand, the price, the product, the production factor information[2] and, certainly, sales and management are of eminent importance.

Kotler and Bliemel (1995) define a market according to the existence of customers: 'A market consists of all kinds of potential customers with a specific need or wish, who are willing and disposed to satisfy this need or wish by means of exchange of goods or services.' Hence, from the viewpoint of marketing, a market is not something existent or something established automatically, but an opportunity to make use of actively. Mantau (1996b, pp. 103–107) is concerned with the definition of goods and markets from the perspective of new goods in the area of recreational and environmental properties

Table 4.2. Share of the different success factors in the factor groups. (Question 3.3.4: What are in your opinion the determining factors for the successful development of this product?)

Factor group	Indication frequency *n*	Indication frequency %
Frame conditions	86	87.8
Market conditions	31	31.6
Marketing management	22	22.4
Product design and characteristics	59	60.2
Distribution and price policy	11	11.2
Communication policy	20	20.4
General marketing	2	2.0
Missing	6	6.1
Total (basis 98 cases)	237	241.8

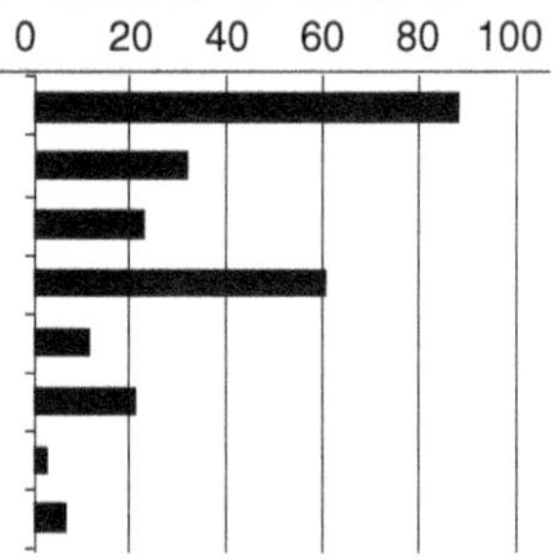

of the forests. On the one hand, the development from a basic service to an attractive object for exchange is the most important basis of a market for RES products. On the other hand, a market has to be established by formulating 'rules, internal institutions (trade customs, traditions) and exterior institutions (laws)' to be able to 'function'. There would not be a 'functioning' timber market if these preconditions were not given for, say, raw wood.

While the classical literature on economics and marketing is, above all, concerned with material goods, non-material services are prevalent in the area of RES products. Services are determined by their non-materiality, the efficiency of their provider and the integration of the external factor. Due to the non-materiality of services, they are typically characterized by the feature that they cannot be stored or transported. Potential service offers, such as a guided forest tour, will expire after the date indicated. Likewise, it is impossible to 'transport' the tour mentioned to the town of residence of the participant. ('Production and consumption of a service are effected simultaneously'; Meffert and Bruhn, 1995, p. 62.) From the non-materiality of services, among other things, the following conclusions can be drawn for their marketing: production and demand require intensive coordination and flexible planning of capacity on the side of the offerer and intensive management has to make it possible for the offerer to steer the demand at short notice.

The efficiency of the provider of a service directly influences the product. Therefore, it is necessary to secure, document and communicate certain specific service competencies. Finally, every service is determined by the external factor. By this term, it is understood that there is 'an object which constantly is at the disposal of the customer' which is the goal of the service in question. In many cases, this 'object' is the customers themselves with their individual properties and targets. This illustrates the considerable significance of customer orientation in the production process of the service and the difficulties of standardization of services (Meffert and Bruhn, 1995, pp. 61–68).

Basic concepts, definition and necessity of marketing

The definitions of 'marketing' are manifold and multilayered. In addition, they have undergone a continual development in the course of time. Thus, marketing was first equivalent to 'sales efforts' and then to 'comprehensive conceptualization' and, finally, it was increasingly understood along the lines of 'corporate policy'.[3]

For the present book, the multitude of definitions of marketing were concentrated in the following statement:

> The term 'marketing' describes all business activities concerning plan-

> ning, execution and control which are directed towards optimal exchange conditions on the basis of systematically gained knowledge of the market and consequent customer orientation by means of sales political instruments (product design, distribution, price design and promotion) for the products offered in present and potential markets. Marketing, therefore, is the expression of a comprehensive philosophy and conceptualization of management strategies and a process within socio-economic structures.

Among the widespread further possibilities of differentiating the term marketing, in the present chapter the differentiation between the marketing areas of consumer goods, capital goods and services (objects of marketing) will be underlined. A particular area with many contacts with RES marketing is the area of non-profit marketing, which is partly effected by public forest administrations to represent their institution and their social achievements.[4] Finally, differentiation between short-term 'operative' marketing and long-term 'strategic' marketing is advisable as well.

Marketing thus becomes a necessary part of enterprise philosophy and activities if supply exceeds demand of the relevant products. The question about the relationship between supply and demand for the products from the RES of the forest (RES products) referred to cannot be answered in general, as the term RES products is a collective term for a variety of goods and services in different markets (see Chapter 3, Section 3.8 on Typology). However, the following general statement can be made for new products:

- Initially, the demand for new products is limited, because of a restricted reputation – if the product design was not customer-orientated for a certain demand in a certain market.
- The total supply of a particular successful RES product will rise rapidly, as similar products of competitors will increasingly enter the market.

The sales concept, which is also called the 'traditional marketing concept' (Weis, 1997), focuses on active and, as a rule, aggressive marketing activities. The ('modern') concept of marketing on the contrary, aims at making sales activities almost redundant by developing product design, distribution terms and channels and communication of a product in such a way that the following preconditions are fulfilled: the product fits the wishes of the customers to a high degree, the channel of distribution is adapted to customer behaviour and the communication is understood by the customers and reaches them in their present needs.

The most important difference between the two concepts, therefore, consists of the fact that, with traditional marketing, an already existing product will be made a success in the market by using sales policy tools, while, with modern marketing, the development of a product is preceded by market research. Although some of the products in the area of recreational and environmental products of the forest which have to be supported are already developed products, a modern enterprise philosophy should generally be orientated towards the marketing concept, as the sales concept carries a variety of risks (Kotler and Bliemel, 1995).

Kotler and Bliemel (1995) enumerate and describe the following points as the 'four fundamentals of the marketing concept':

- The focus on the market.
- Orientation towards the customers and their wishes.
- Coordination of marketing.
- Profits for the enterprise from satisfied customers.

The orientation towards the customers has an especially large impact on the entire sales area of an enterprise, as a satisfied consumer, for example, of a new recreational product in the forest often has greater advertising effects by word of mouth than any advertisement.

Only an enterprise philosophy of marketing directed towards customers and the market in the long run secures the economic existence of forest enterprises. The development and the sale of new RES products is facilitated or even rendered possible by active marketing.

Market research as the basis for marketing decisions

Necessary information on the principles of the market is evaluated by market research as part of the situation analysis (Fig. 4.1).[5] Market research aims to gain and to evaluate information in order to apply it in a target-orientated manner in the marketing process. In this context, it is necessary to examine whether the information provides a minimum of accuracy, completeness, topicality and relevance. Market research is the basis for a customer-orientated and target-orientated management policy. However, market research involves a significant effort of time and financing. In order to avoid an unnecessary expenditure of these factors and, at the same time, wrong conclusions of serious consequence from research results, market research activities need to be planned precisely and conducted professionally.

The gaining of information can be divided into direct primary research (field research) and indirect secondary research (desk research). A typical example of primary research is opinion polls; secondary research is, for example, the evaluation of statistics or articles. Forest enterprises in general will entrust a suitably specialized company in the service business with the more comprehensive tasks of market research. Moreover, a permanent information system should be established within the forest enterprises to be provided with the most important current market data. By adequate qualitative and quantitative methods of analysis and prognosis, demand on the market and sales potential can be estimated.[6]

Orientation along the lines of marketing philosophy requires basic knowledge on the participants and on determining parameters of a market. Though market research for forest enterprises is normally conducted by third persons, basic knowledge about these instruments should be available. The benefit of gaining information has to be in an acceptable ratio to the costs of the survey. A permanent information system is also a cost-effective opportunity for forest enterprises to safeguard necessary decisions.

Within the field of RES products, most of all basic qualitative evaluations can be expected from results from secondary research. This is because of economic relations. The choice of the right technique of evaluation in any case should be precisely coordinated with the available information and the desired goal of information. The demand for the RES product of an enterprise corresponds to the market potential if the enterprise makes use of all marketing opportunities and if it is the only supplier of the product.

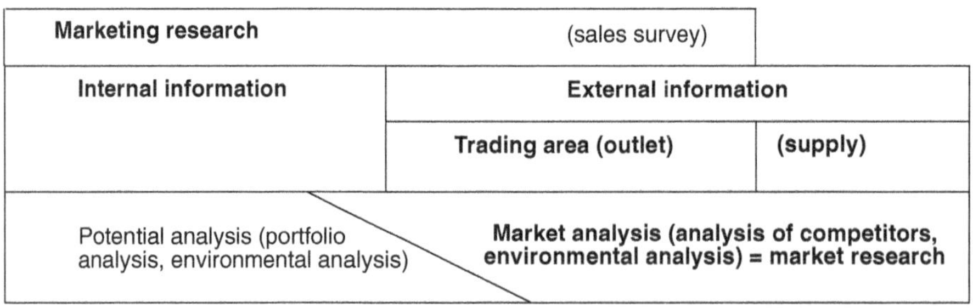

Fig. 4.1. Determination of the term market research (modified from Nieschlag *et al.*, 1991, p. 607, fig. 7.1).

What types of *ex ante* market investigation could be recognized in the RES case-studies? Has market research been required or were there other opportunities to determine the prospects for success of a product and the approaches of a marketing concept? Table 4.3 gives a survey. The table first shows that it is the manager's personal knowledge in particular which serves as source of information on the market potential of a product. Those customers with which a personal contact existed prior to product development were the second most important source of information within the framework of the RES case-studies, thus guaranteeing a close customer orientation. Similar products or exchanges with other forest enterprises (approx. 15%) were also important sources of information. Public institutions, such as communities or tourist offices, delivered suggestions and databases for new RES products in eight cases. In this respect, some partners of offers, such as the Worldwide Fund for Nature (WWF), the Schutzgemeinschaft Deutscher Wald (SDW) or hotels and restaurants, served as information sources as well.

Common techniques of market research, such as interviews, analyses, permanent market observations, tests, scientific studies and the outsourcing of market investigation, were summarized as 'market research' prior to product introduction. Such systematic surveys were, according to the interview partners, conducted in about 23% of the cases. Thus, market data, for example, were investigated and made available by an external consulting company or by studies and a diploma thesis on tourism. Sekot (1992, p. 25) states that, in the pioneer stage of RES products, 'what is missing is the investigation of market niches and, thus, the exploration of new potentials of success for forest enterprises further operating as private sector companies'. Though the author hereby refers very generally to the situation in forest enterprises, which is described as the 'forest production gap' by Mantau (1993), some parallels with the results of the RES case-

Table 4.3. Importance of market research as a source of information concerning the product potential. (Question 4.2.1: From which sources did you gain information about the market potential of your product?)

Information source about market potential	*n*	%
Personal knowledge	73	74.5
Customers	31	31.6
Similar products, forest enterprises	15	15.3
Tourist information office, municipality, region, etc.	8	8.2
Distribution partner	4	4.1
Market research		
General market research	6	6.1
Analysis and scientific studies	10	10.2
Tests	4	4.1
Permanent market observation	1	1.0
Systematic outsourcing	2	2.0
Print media, advertisements	7	7.1
Trial and error	3	3.1
Missing	2	2.0
Total (basis 98 cases)	166	169.4

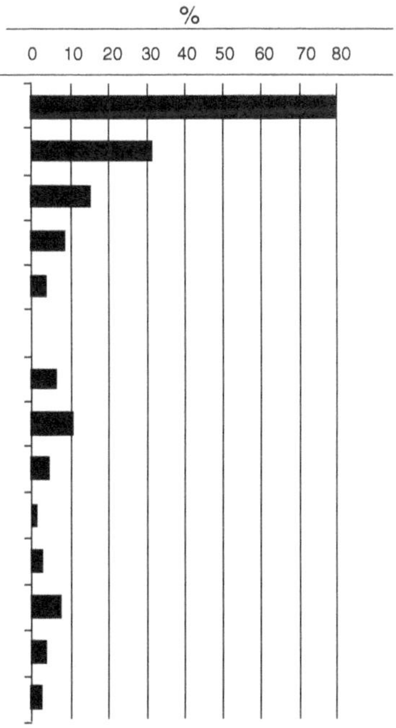

studies can be stated. Even in the investigated collection of almost 100 successful cases of RES business, the share of those cases in which market niches were actively explored is low.

When analysing this issue, it must not be neglected, however, that the recorded RES products were rarely capital-intensive and thus actually risky efforts. This is also indicated by the statement in three case-studies according to which the product was introduced and tested at the same time on the principle of 'trial and error'.

> RES products are often developed and offered on the basis of information on customer wishes, which is not systematic but presumably partly obtained by chance. Due to the mostly low financial expectations of forest enterprises concerning new RES products, investments for market investigation are minimized. Different information from their own enterprise or from neighbouring enterprises as well as from intermediaries or from public institutions can be summarized systematically. Thus, many offerers of RES products succeed in investigating the market cost-effectively.

FRAMEWORK AND MARKET CONDITIONS. An important framework condition for the marketing of RES products is the geographical location of the offerer and the enterprise environment. Moreover, the marketing possibilities of a product are influenced by a complex network of socio-cultural and legal–administrative framework conditions[7] which in the frame of situation analysis have to be evaluated according to their positive or negative influences on a new RES offer. Geographical and sociodemographical features can be utilized for the segmentation of a market and for the definition of target groups of potential RES customers.[8] Framework conditions are subject to dynamic changes, which should be considered in situation analysis as well.[9]

> The marketing of RES products is strongly influenced by a large amount of different formal rules and informal influences.

Market conditions are mainly defined by demand and competition.[10]

Competition in the case-studies. The evaluation of the general competitive situation, therefore, is a necessary feature of the situation analysis. Table 4.4 gives an overview of competitive payable offers in the RES case-studies. The result confirms the assumption that RES products are, to a high percentage, share-traded on new, unsaturated markets. Probably the low local flexibility which has already been mentioned contributes to this situation. In only fewer than 10% of recorded cases was there a highly competitive situation. In about 35% of the cases, there were few alternative products.

In order to evaluate these products qualitatively, a question was asked about the success and age of available competitive

Table 4.4. Existence of alternative offers for a price. (Question 4.1.3: Are there any similar products offered for a price in the region (influencing your product)?)

Similar products for a price	*n*	%
No	54	55.1
Yes, a few	34	34.7
Yes, a lot	9	9.2
Missing	1	1.0
Total	98	100.0

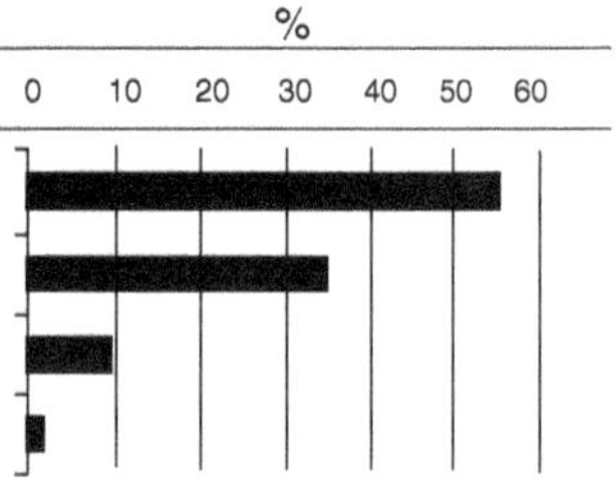

products. It turned out that almost half of the competitive products mentioned (44%) were successful above average and that low success of competitors was exceptional (5%). In 44% of the 43 case-studies concerned, the competitive products were said to be 'regularly' successful and, in three cases, no indications were made. Competitive offers were mostly older than 5 years (56%), while only 16% of the products were 3 years old or younger. A potential qualitative evaluation of competitors is, obviously, that established or older successful products are dominant. These products in particular give an impulse for the development of a similar product in order to gain market shares from competitors. However, the more saturated the market for such products is, the more difficult this becomes. Concerning the RES case-studies, the question therefore arises as to how the competitors affect the success of the enterprise's own products. This is considered after investigating an RES special case.

Competition of RES products with alternatives that are free of charge. Peculiar to RES products is the fact that potential customers of RES products have, in certain cases, the opportunity to choose alternatives that are free of charge instead. An example of such an opportunity would be a free guided tour in state forests instead of a guided tour in a private forest offered at a charge. Thus, the existence of such alternatives free of charge can be regarded as a special strength of the competitor – in this case, the state – on the market. In many cases, this situation probably causes another forest enterprise not even to try to introduce such a product. Thus, Mantau (1996a, p. 1275) states that no sustainable feasible market structures can develop if environmental and recreational services are offered by some forest enterprises free of charge. The question of competition free of charge leads to the results shown in Table 4.5.

It is evident that, at least in about one third of all cases, alternatives are available free of charge. For the user still to accept the new RES product and to pay for it, it is crucial for him/her that the price is (subjectively) reasonable in relation to the benefit received. Therefore, the question was asked, in the 33 cases in which there is competition free of charge, if RES products offer an additional benefit (Table 4.6). Table 4.6 verifies the assumption that there should be an additional benefit.[11] The advantage of competition free of charge is thus compensated for by the enterprise's own strength (e.g. the product design).

Mutual impacts with other offerers. Finally, it has to be evaluated whether comparable products negatively or – by certain synergy effects – positively affect the enterprise's own offer. Seventy per cent of the 49 enterprises that commented on this issue in the framework of the RES case-studies said that the offers of other enterprises did not

Table 4.5. Intensity of competition free of charge. (Question 4.1.1: How easy is it for the user to get a similar service/good free in the region (influencing your product)?)

Similar service free	*n*	%
Very easy	4	4.1
Easy	14	14.3
Difficult	5	5.1
Very difficult	10	10.2
Impossible	64	65.3
Missing	1	1.0
Total	98	100.0

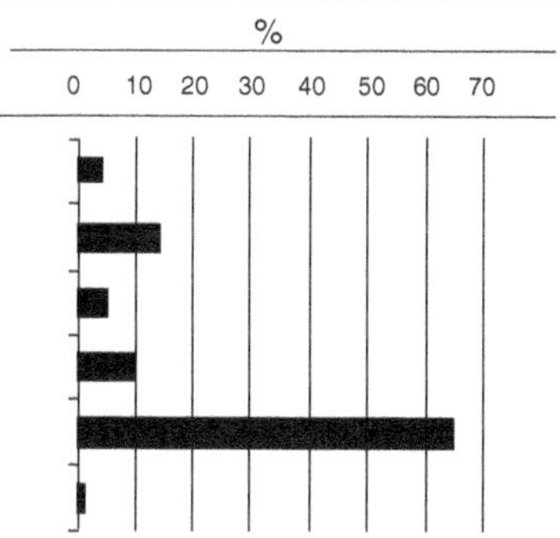

Table 4.6. Advantage of RES products over competition free of charge. (Question 4.1.2: If it is possible, how big is the clearly recognizable additional benefit of your product in comparison with the free service/good?)

Additional benefit	*n*	%
No	2	6.1
Small	0	–
Some	6	18.2
Big	16	48.5
Very big	6	18.2
Missing	3	9.1
Total	33	100.0

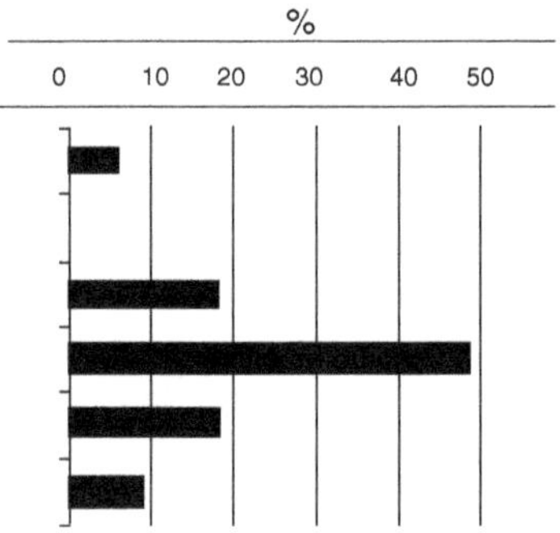

influence the marketing of their own products. While 20% stated a negative influence, 10% of the interview respondents mentioned positive influences of the competitors on their own business.

During the analysis of case-studies, this direct question was complemented by a comparison of the competitive analysis with some features of financial and intangible success of the RES products investigated. Thus, in particular, 'simple recreational products' (product group (PG) 4.2) are obviously influenced in a negative sense by alternatives that are free of charge, in so far as these exist. Positive mutual impacts with such competitive products in some cases appear with 'complex recreational products' (PG 3.1b, 4.1, 5.13 and 5.2). Regarding alternatives with charges, a slight tendency can be observed for the success of RES products to rise also with growing competition, even if this was not expressed in the interview. From complex competitive offers, agglomeration advantages could arise, as customers do not perceive such offers in passing, but search for them intentionally and, under certain circumstances, also conclude further purchasing transactions so that the 'effort is made worthwhile'. Moreover, distinct synergy effects and, most of all, the large share of RES products on which competition does not exert any influence, point to as yet unsaturated markets.

It becomes obvious that the competitive situation of RES offers has not yet reached the predominantly negative meaning that characterizes established products. Therefore, the area of competitive strategies will not be referred to in detail.

> RES products are, to a large extent, traded on new, unsaturated markets. Available competitive products are mostly successful and have often been established in the market for some years. A significant part of RES products competes with alternatives that are free of charge. Successful RES products remain competitive through an additional benefit. They are not strongly affected by competition. The positive impact that exists in some respects is not felt to be as important as the negative impact.

4.2.2 Strategic principles of the RES business

The utilization of marketing instruments, i.e. the optimal combination of the marketing mix of product policy, distribution, pricing and communication, is initially orientated along the lines of medium- and long-term strategic general principles.[12] These general principles are composed of different substrategies, which, for example, decide on the following points:

- Extent of market share (single-segment, multisegment, whole market).
- Primary characteristic of performances (low price, high quality, etc.).
- Attitude towards competitors (aggressive, defensive).

As has already become clear, the focus of RES marketing is at present still on the transformation of forest services into products, which frequently could be traded on as yet unsaturated niche markets. For this reason, the strategic considerations here will concentrate on product strategies, after a brief look at market segments and niche markets.[13]

> Marketing strategies are basic decisions containing different sub-strategies. Among the segmental strategies are the product strategies for RES products, which are currently of special significance.

Market segmentation, formation of target groups and positioning

The strategy of market segmentation and positioning aims at serving customer groups that are as specialized as possible, preferably as the unique supplier, thus optimizing customer orientation and evading competition.[14]

MARKET SEGMENTATION AND FORMATION OF TARGET GROUPS.[15] The data gained by market research generally allow segmentation of a potential market, according to different variables, into market segments with specific characteristics.[16] Kotler and Bliemel (1995) regard a market niche as a closer customer group than a market segment. If forest enterprises utilize special market niches for their RES offer, they have already conducted a close market segmentation.

Simple parameters for determination of market segments and resulting target groups for consumer goods are biological, geographical and sociodemographical factors, such as sex, age, profession, income, place of living and size of household. Deeper approaches, for example, are orientated towards psychological factors of motivation, perception, attitude and style of living. From these basics finally result praxis-orientated segmentations – for example, according to different opportunities for addressing the customer, the structure of preference and the resulting customer behaviour. Regarding the meat market, for example, it is reasonable for the marketing of eco-meat to categorize buyers according to their interest in ecologically processed products.

For the segmentation of markets for investment goods, the variables for consumer goods are often supplemented by further variables, such as the line of business, the customer size or the legal status. Since the product offer to institutions generally includes quite individual goods or services, market segmentation often happens in a more intensive way and small segments are established. The strongest segmentation is individual customer marketing.[17] Especially in the RES case-studies dealing with investment products (capital goods), a high percentage of one-customer segments could be found.

After establishing various market segments, potential buyers of a product can be categorized according to different criteria into economically interesting and less interesting target groups. In the Netherlands (NL) case of the Drents tree-crown path (NL13), among other subgroups of potential customers, one of the target groups that was considered to be especially important, and therefore was concentrated on separately, was children. Children greatly influence purchase decisions and play an important role in the leisure-time activities of families. Therefore, several permanent offers and special events for kids have been planned in the context of the tree-crown path.

POSITIONING.[18] By its own positioning via differentiation from comparable competitors, an enterprise describes its special competence and features.[19] The choice of means of differentiation should certainly correspond with the chosen market segments and target groups. Frequently, external frame conditions, such as the geographical location or certain inventory conditions of the marketed forest, offer the forest enterprises starting-points for effective positioning. However, positioning by a differentiated design of identity, such as for different cigarette brands, is quite uncommon in forest enterprises. There could be a great potential for success in this aspect, which is referred to later as corporate identity (CI).

STRATEGIES FOR WORKING THE SEGMENTS. Through its strategies of working the market segments, a company decides on those market segments to be selectively processed as 'strategic business fields' by means of marketing tools.[20] For (economically) small enterprises, as are most offerers of RES products, it is usually advisable to occupy certain market segments with certain products and to specialize in niche markets.[21] Specialization can be applied to a single market segment (a certain niche) ('strategy of concentration'), or the enterprise may specialize in several segments, on comparable products for different customers or on different products for similar customers.

From the RES case-studies, it follows that nearly all offers were developed for certain target groups. Moreover, many RES products are developed more or less substantially by the integration of the external factor 'customer' (PG 5.13–6.3). The strategy for working a market offered here is the governing idea of the 'Segment-of-one approach'.[22] This strategy individualizes every service to such a high degree that all of the customers, mainly organizations, get an individual result. Examples of this would be sponsoring contracts for certain components of a forest or use contracts which a forest enterprise sells to certain institutions, such as a community.

Finally, however, the following statement has to be made concerning the significance of segmentation strategies for RES products: Market segmentation continuously gains importance as the amount of markets and competitors and customer saturation are increasing. The recorded RES case-studies, however, describe many products at the introduction stage on not yet saturated markets. This is why differentiated marketing methods, such as the concentration on certain social market segments are, to some extent, of minor importance for these products.

> Every market segmentation needs to be conducted according to different objectives and reasonable criteria. In this regard, niche markets for RES products must not be limited too closely. Besides the usual suitable possibilities of differentiation of their own enterprise and their own products, offerers of RES products should aim at differentiating themselves from competitors by their special competence concerning the forest and in this way to position themselves distinctly. For a clear positioning of their enterprise it is therefore important to establish and transmit a succinct CI.
>
> Forest enterprises starting to introduce new RES products should first deal with concentrated strategies for certain target groups. In the beginning, they should try to pursue a strategy as pioneers on the market and to occupy niches. Often, small market niches are processed by a strategy of concentration. In the case of highly individual RES offers, the strategy can be adapted to a one-customer segment. The significance of market segmentation for RES products will increase.

Product strategies

DIVERSIFICATION. When introducing new RES products which have not been marketed yet by the enterprise concerned on new markets, the enterprise pursues a strategy of diversification.[23] On the one hand, forest enterprises in this case benefit from their excellent knowledge of forests as a central production factor of new RES products. On the other hand, problems result from little experience with new customers, with newly supplied markets, with suitable distribution chains and, to some extent, with those regulations which need to be considered. For forest enterprises it can therefore be recommended to make use of external experts concerning those products which are, as yet, 'far away from their own competence' (Ziesling, 1999b, p. 31). If corresponding customer groups of the same economic level are addressed, this is defined as horizontal diversification. This is the case, for example, with the selling of certified timber. Vertical diversification means offering products that are positioned prior to or after the process of product transformation. An RES example for this would be the offer of game meat prepared for cooking. In the past, forest enterprises have, in particular, tried to diversify on timber in a vertical direction. Despite the relatively favourable preconditions of the raw material basis, capital and know-how, attempts at processing raw wood in their own timber-yards have failed in most cases (Sekot, 1995, 1997).

Other types of diversification, such as the offer of guided forest tours to tourists, are summarized by the term lateral diversification.[24] The situation here is, in most cases, even more difficult, as experience or other success factors are not available from the core forest business concerning either the new products or the new markets. Therefore, it is quite risky to start a new business.[25]

THE PORTFOLIO APPROACH. 'Portfolio analysis' contributes to the design of product-orientated strategies which at the same time take into account the situation of further products, namely product groups subsumed as the strategic business units (SBU) of an enterprise. Based on the 'nine-field matrix' of the 'market attractiveness – business strength portfolio',[26] Fig. 4.2 shows a very simplified potential SBU portfolio of the use of sideline within a forest enterprise.

The term market attractiveness contains, among other success factors, the elements of market growth, condition of supply and environmental situation. For assessing the enterprise's own business strength, market attractiveness particularly refers to the relative market position, the relative production potential, the relative developmental potential and the relative qualification of employees (Nieschlag *et al.*, 1991, pp. 880–881). The data separately obtained for each SBU by situation analysis can be used for this evaluation.[27] Basic strategies of the forest enterprise chosen as an example could be strategies of investment and growth for the SBUs 'barbecue huts' and 'guided forest tours', strategies of consolidation and selection for the SBU 'forest seeds' and a strategy of skimming or liquidation for the SBU 'Christmas trees'.[28]

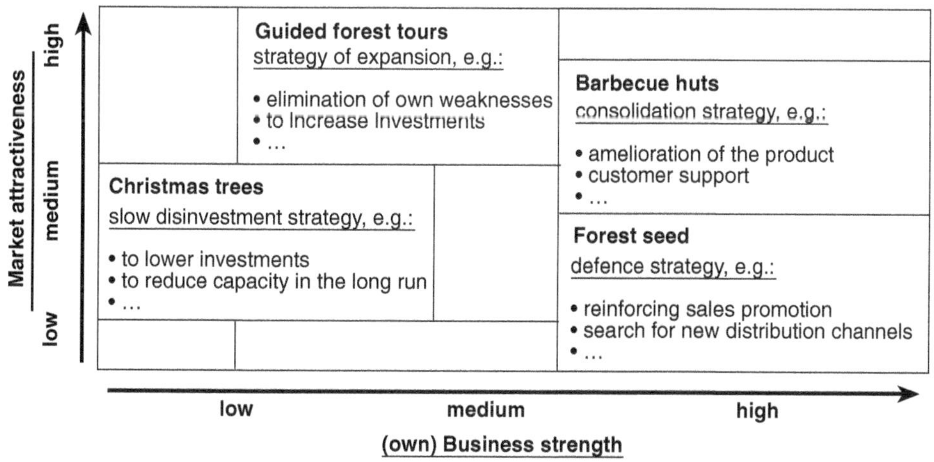

Fig. 4.2. Potential matrix of a peripheral usage portfolio of a forest enterprise (modified from Kotler and Bliemel, 1995, p. 103).

Besides the consideration of existing products a company should also include potential products in its strategic planning. As the basis of strategic management of a forest administration, Ziesling (1999b, p. 27) develops a comparable product portfolio, in which the existing and possible products of a forest enterprise are listed according to market attractiveness and enterprise competence. According to Ziesling, the traditional forest core competencies, such as raw wood production or forest side-utilizations, are not very attractive in comparison with future fields of operations, such as forest education, environmental protection, forest recreation, environmental research and forest tourism. However, at present resource expenditure is usually concentrated on those fields of operation which are not very attractive. This corresponds with the enterprise's advisable goal of not deviating too much from its own enterprise competencies because of economic risks. Thus, many attractive products of forest tourism are in business fields that cannot automatically be handled on the basis of a forest education. Without cooperation with external experts, there is a certain potential for failure inherent in individual areas of the portfolio that signify a lateral diversification for forest enterprises.

STRATEGIES THROUGHOUT THE PRODUCT LIFE CYCLE.[29] Marketing strategies change dynamically with market conditions and the particular product life stage.[30] In the beginning of a development is the product idea. An enterprise continuously developing market and/or enterprise innovations pursues a strategy of product innovation, which largely resembles the strategy of product development and the strategy of diversification. Innovative enterprises need to have an intensive market policy since new product developments entail great effort. There the success of market- or customer-orientated and initiated developments is generally higher than that of products which develop from scientific ideas or enterprise-specific production potentials (Nieschlag *et al.*, 1991, pp. 838–840).

Many of the RES products recorded in the case-studies are in the introductory stage. Possible strategies at this stage are illustrated in Fig. 4.3. Figure 4.3 shows four possible marketing strategies for new products, concentrating on the price and communication instruments of the marketing mix. According to this, a forest enterprise that wants to offer guided tours to tourists as a possible RES product could establish a strategic orientation. For example, it could assume the following preconditions for the offered tours:

Price	Sales support: high	Sales support: low
high	**1. Quick skimming of the market, if:** • (independent of market size) • *low name recognition* • *competition expected* • strong customer's wish • strong purchasing power	**2. Slow skimming of the market, if:** • limited market • *high name recognition* • *no direct competition* • strong customer's wish • strong purchasing power • no advantage for low-priced products
low	**3. Quick penetration of the market, if:** • broad market • *low name recognition* • *competition expected* • high sensitivity to price changes • high circulation digression	**4. Slow penetration of the market, if:** • board market • *high name recognition* • *no direct competition* • high sensitivity to price changes

Fig. 4.3. Four marketing strategies for the introductory stage.

- A broad market, since many tourists are interested (strategies 3 and 4).
- Low popularity, since it is a completely new product (strategies 1 and 3).
- A price-sensitive reaction, since similar offers were formerly obtained free of charge from the state (strategies 3 and 4).
- No expected direct competition, since only the forest enterprise has the opportunity to offer a forest tour in the particular area (strategies 2 and 4).

Therefore, the forest owner could decide on a strategy of market penetration (strategies 3 and 4) at medium speed. The price of tours would be kept at a relatively low level and sales support would be well balanced. In this regard, it would be necessary to take into account the fact that investments would pay rather slowly in the beginning but the benefit from the product could be kept for a considerable time.

The goal during the growth stage is mainly to maintain growth as long as possible.[31] Already at this stage (in the maturity stage at the latest), strategies have to be developed to render the product attractive to the customers in the long run and to differentiate the product from competitive offers. When considering abandoning an established product that does not show the usual increase of benefits after the growth stage has been accomplished, every enterprise should first take into account the effort of product development and introduction. Despite 'market-leaving barriers', such as capital appropriation or social obligations, at the stage of decline at the latest it may become necessary to eliminate products with a weak market performance.

If the product is maintained in the long run after leaving the growth stage, especially with service offerers, strategies of 'advantage by performance' (*Leistungsvorteilsstrategien*)[32] are advantageous. These aim at distinguishing the product from that of competitors by special measures of product policy. Continuous product improvements, guaranteed quality advantages, utilization of competitive advantages through establishing brands or an extension of the product range are all possible elements of this strategy. By these measures, in addition, new customers can be attracted by the product, preferably former customers of competitors, or the frequency of use can be increased. Basically all elements of the marketing mix should be adapted here. The enterprise should search for gaps in the previous mix; however, there are risks because of imitating competitors, particularly in price-cutting.

As a basis of strategic thinking, a forest enterprise should have a clear view on the 'portfolio' of its present and potential business units. Existing RES products need strategies that are adapted to their particular stage in the life cycle. Strategies of intensive growth on traditional markets are rarely possible for forest enterprises. The potential utilization of attractive business fields through RES products means lateral diversification for many forest enterprises. Therefore, the risks and difficulties of the necessary diversification need to be recognized and managed before the necessary investments are made. Customer-orientated innovations are generally an important factor of success.

4.3 The Operative Marketing Mix

The planning of marketing measures transfers the basic strategic targets to specific, 'operative' directions for action. In this regard, Nieschlag *et al.* (1991, p. 847) point out:

> The main aspects during this stage cover the selection, emphasis and design of marketing instruments, the determination of the activity level for single decisions, compiling them target-orientatedly to the highest possible level approaching the optimum marketing-mix, as well as the supply and allocation of financial support within the frame of budgeting.

Concerning the practical design of the marketing mix, consisting of product, distributive, price and communicative political instruments for RES products, crucial factors include, first, its properties, for example, as a consumer good for individual customers or as an investment product for institutional customers as a storable good or a service. Furthermore, among other things, the significance of the price for the particular kind of product (everyday product, luxury good) should be observed, what influence advertising exerts on the sale of these products and what effects the marketing tools have on other product markets, other corporate sectors or other target groups (spillover effects (*Ausstrahlungseffekte*)).[33]

The mix selected has to support itself and must not develop inconsistent effects. It would be useless, for example, to try to sell a product as a quality good while simultaneously pursuing a strategy of low pricing. Finally, a forest enterprise, when planning its marketing tools for RES products, should realize that, in general, different conditions apply to the marketing of RES products from those in the timber trade. Thus, for example the terms of payment are determined by the particular market structure, and the concepts of public relations (PR) and advertising for forest services should go well beyond the sales initiation in the timber trade (Sekot, 1995).

> On the basis of strategic decisions, the application of marketing tools is determined in a plan of measures in which, especially, the distribution of resources such as working hours and financial resources have to be planned. The difference between this area and the trade in raw timber as the usual business of a forest enterprise has to be realized.

4.3.1 Product policy

General basics of product policy

DEFINITION OF RES PRODUCTS. In general, products can be defined as follows: 'A product is everything which can be offered to (a person or on) a market for attention, acquisition, use or consumption that might satisfy a want or need' (Kotler and Bliemel, 1995, pp. 8 and 659). This definition comprises goods as well as services and similarly the cession of usufructuary rights. However, the 'RES products' discussed in the present chapter require an addition to the definition mentioned above: 'RES products are generated by the transformation of recreational and environmental services of the forest from their basic functional core benefit into marketable products.'

One of the core features of the RES research project is the transformation of the recreational and environmental functions of the forest into marketable RES products, for which customers are prepared to pay (see Chapter 3). With reference to this RES project task, ways have obviously been found to turn functional services into products by means of marketing strategies, as shown in Fig. 4.4.

Due to the significance of product design[34] as part of the marketing mix, its general theoretical basics are briefly outlined here.

MEANS AND TARGETS OF PRODUCT POLICY. For illustrating the principle of added value, a comparison will be drawn between two entirely different things, say a forest road and a chair, as simple everyday products. On the one hand, the producer of the chair tries to assign properties to his/her product

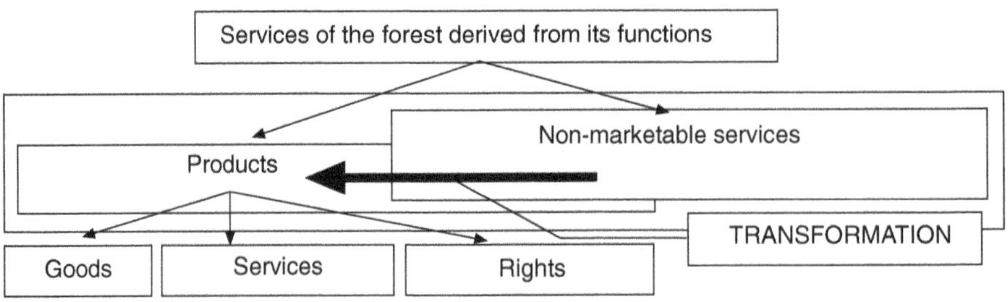

Fig. 4.4. The transformation into RES products.

at the design stage which will differentiate it from competing products in a way that will make a potential customer decide to buy exactly his/her chair. As the instruments of product policy, Weis (1997, p. 75) enumerates the following elements of design:

- Product variation, differentiation and diversification.
- Product programme and range of products.
- Quality, customer service and guarantees.
- Design and packaging.
- Branding and labelling.

Basically, the application of these instruments provides the customers with an additional benefit or conveys such value to them. This principle is valid for the development of entirely new products, as well as for the further development of already existing products to increase their attraction for potential consumers in opposition to alternative offers.

> Product policy, therefore, aims at a sole position ('minimonopoly'), in the broadest sense of the word, by limiting the possibilities of substitution for the customer, thus gaining an economic advantage.

Forest owners who want to market the use of their forest roads for recreational purposes, on the other hand, initially have to produce the marketability of their offer. This means that they have to overcome the non-marketability aspects of their public good. This, in turn, signifies that they have to create their forest roads as a product, for the utilization of which the consumer is disposed to pay voluntarily. This will only be the case if the visitors to a forest regard the usage of a certain forest road as an advantage, i.e. if this seems to be a more promising way of recreation than others of the same price or even free of charge.[35]

The model of 'five conceptual product levels' (Kotler and Bliemel, 1995, p. 660) is outstandingly suitable for the description of marketability through added value as a basic feature of product policy, which has already been described in detail in the introductory outline on the basics of the RES project (see Chapter 1). As has already become clear, products always consist of a variety of tangible and non-material parts, each of which by itself displays a different degree of rivalry and excludability. The task of product policy as part of the marketing mix, then, is to develop additional product features on the level of the 'augmented product' to obtain the minimonopoly aimed at for the product offered. This system, as a rule, is highly dynamic: some of the benefits of the augmented product level gradually turn into values expected by the customer, because of continual imitation by competitors. When considering the uppermost level, the level of the potential product, it must be taken into account, however, that the risk of failure increases with every further development, because of the increasing distance from the initial core competence.[36]

The marketing of recreation in the forest, therefore, is possible via the combination of the recreational function of the forest with the offer of additional benefits which sufficiently dispose of the properties of marketable goods: rivalry and excludability. A guided forest tour, for example, is composed of the forest with its manifold functions and the guided tour as a distinctly defined service. Participants can be excluded or included unequivocally and, because of words – though 'unrestrictably utilizable' in general – not being audible for every participant of a larger group, there is a certain rivalry concerning this service part of the product. A marketable product has thus been produced by product design and product composition from several components, each of them being more or less marketable on their own and more or less distinct in their partial benefit.

At this stage, the different product features of 'recreation in the forest' and the 'chair' meet again, as the chair also consists of marketable components (such as the wooden construction suitable for sitting on) and non-marketable components (such as the design or the advantageous environmental features of wood). Value addition in the meantime has become an indispensable sales condition, especially as concerns consumer goods, but increasingly so for capital goods also. For the distinction from additional products, which are considered to be part of the product mix, additional value in the present chapter is defined as follows: 'Additional value (benefit) results from the special features directly linked to the product itself or to properties which are inseparably connected with it and offered at one overall price.' This definition deliberately leaves open the assignment of marketable and non-marketable features concerning basic and added value of the product.

In concluding the comparison of the examples of the guided forest tour and the chair, there is a reference to the following feature: in general, marketing deals with products that require sales support, because of their position in a long-established, more or less saturated market. Marketing of products from the sector of RES of the forest, however, often deals with product placement in young, expanding markets, partly with high demand, as recent experience shows.[37] Moreover, offerers of RES products are often sole traders in the particular, in many cases regional, market. For this reason, their margin in further aspects of the marketing mix, for example in pricing, is quite considerable. For working these ready markets, first, the marketability has to be established, secondly, unstable preferences have to be overcome and, thirdly, working power and the necessary know-how have to be maintained (Mantau, 1997, pp. 636–637).

MEANS OF PRODUCT POLICY FOR RES PRODUCTS. In the design of RES products, some particulars arise. In many cases, for the RES offer, a primary clarification of property rights is required. In some circumstances, already at this stage, attempts are made for the marketability of forest products. Traditional forest products can often be turned into economically more attractive RES products by 'refining' (further processing). Game is a good example of the refining of traditional forest (sidelines of) utilization: it was formerly sold to game retailers intact and has recently been turned into an RES product, which means that the game is sold ready and prepared (*zerwirkt und küchenfertig*) to the consumers. Another example would be a downhill ski-run that was formerly used by agreement licensing (use contract) and nowadays is maintained as an open run. The strategy of refining and direct marketing of traditional products means that the profits of the processing company and of retailing are kept by the forest enterprise producing them. The relation between the costs of processing and the increase in proceeds determines whether this alternative is viable.

Forest services maintain a special significance in the field of RES products. Their characteristics have to be especially considered in product design:

- Non-materiality of the service, non-storability.
- Tight personal, temporal and spatial relation between the transactions.
- Great variability in execution.

Contrary to other tangible goods, non-material services have to express their added value by reinforcing tangible components perceivable by the customer before executing the respective service. However, these components are, above all, found in the area of distribution, such as personnel, place of performance or fittings of the agency or shop, in pricing or in the communication of information, brands, labels, symbols, etc. The attraction of a service to the customer in many cases depends on the personality of the provider. Instability in execution can be countered by strict quality-control practice (quality management). The feature of lacking storability requires special measures, as well, such as targeted variability in execution or, frequently, measures of pricing. Adaptability of organization of tasks by temporary employees or by increasing participation of the customers in the 'production' of the service can be interpreted in the lines of product design. Moreover, services such as guided forest tours can be made more attractive by customer-orientated flexibility in the time management of the service process (periods of transfer, settlement, waiting and transaction) (Kotler and Bliemel, 1995, pp. 711–715; Meffert and Bruhn, 1995, pp. 259–268).

Finally, many RES products depend on the voluntary formation of a contract as a fundamental aspect of production. Nature conservation on a contractual basis (*Vertragsnaturschutz*) is only one example in this area.[38] This product strategy, which was originally designed as flexible and openly useful for marketing purposes, in the meantime 'frequently serves only as the more acceptable alternative for compensatory payments derived from restrictions on title (*Eigentumsbeschränkungen*) or for promotional measures (subsidies)' (Mantau, 1994, p. 313). This statement is not intended to question the value of contracts concerning nature conservation for forest owners as such. However, the feature that has to be viewed critically from the perspective of economics is the development (back) to public programmes with fixed prices on the basis of additional expense and diminished proceeds.

The 'value hypothesis', i.e. the marketability of non-material values, which is important in the area of nature preservation contracts or of projects concerning 'eco-sponsoring' has already been outlined as a basic approach to RES projects (see Chapter 1). Therefore, it need not be outlined again in this chapter.

RES case-studies show a broad variability of very different products, which have already been described in detail in Chapter 3, Section 3.8, dealing with the typology for RES products. Moreover, on the basis of the interviews, answers can be found to the following questions concerning product policy:

- Is it an entirely new product or a further development?
- Is it a combination of a basic product and an additional benefit?
- Does the offer consist of a product mix?

In what follows, the strategies of product policy derived from the RES case-studies are outlined on the basis of the above questions. For this purpose, the instruments mentioned in the introduction of this section were divided into three groups: product innovation (variation of the product, differentiation and diversification), product mix (product line and range of products) and special product design (quality, customer support and guarantees, as well as product design and packaging). The area of brands and labels, which rather conveys the additional value of the product than produces it, will be referred to in the section on 'communication policy'.

> The aim of product policy concerning RES products is, first of all, to turn forest facilities into marketable products, for which customers are disposed to pay. For this purpose, a clear definition of property rights is necessary. From the refining (processing) of traditional forest products, i.e. from a strategy of vertical diversification, follows the development of some attractive RES products. Products are also developed by combining public forest services with marketable additional benefits or by marketing the exclusive value of forest properties. There is still a large market potential for environmental sponsoring in this area. Immaterial products, such as forest services, in the beginning have to be 'materialized', so that they can be evaluated by customers. Personal contacts within the enterprise and to the customers are especially important with regard to the offer of services. Voluntary agreements are an important basis of RES products.

Product innovation and product variation

The RES products covered by the case-studies have been categorized, according to defined criteria, as 57% new developments and 43% further developments of existing products (see Chapter 3, Section 3.4). In general, the distinction between product innovation, the new development of products[39] and the variability of products, the further development[40] is not clear-cut. It depends, among other features, on the definition of the product in question and the date of product development as the point of reference.[41] For this reason, both aspects fall under the heading of product innovation. This generalization seems to be permitted, as every variation is based on innovative alterations.

Innovative enterprises, which consider the development of products as their constant task, are, according to different research studies, outstandingly successful. However, product innovations always carry a certain risk, which, for example, is due to the costs arising from new developments or social and national limitations.[42] Therefore, comprehensive and careful planning is indispensable in the introduction of new products. Project management (PM), for example, is very well suited as a tool of product development and introduction (see Chapter 6, Section 6.1).

The transformation of RES of the forest into marketable products is essentially a process of innovation, which leads to the formation of new products or increases the attraction of existing ones.

> Product innovations require precise planning, because of the risk involved. Meticulous planning is a considerable factor for success. Many RES products are new products. RES products, like every other product, can undergo innovative changes in different ways.

INNOVATIONS CONCERNING TANGIBLE PRODUCTS. Material products comprise products such as consumer goods with a higher or lower degree of servicing (PG 2.1), on the one hand, and goods that are usually passed on to institutions for further processing (PG 1.1 and 2.2), on the other. With products such as game, Christmas fairs and Christmas trees or gifts (PG 2.1), there is frequently a high degree of product refining involved, such as with prepared or even grilled meat and with other processed products, such as bacon, sausages or jam. Package offers can also be counted among such offers, for example gift boxes, business and advertising gifts or the organization of an enterprise celebration in the forest.

Besides the product processing mentioned above, additional value is of high importance for this group of products. Thus, the following aspects of product design were mentioned:

- Shopping event at the Christmas fair.
- Good meals, shooting contests.
- Playgrounds and other special offers for children, such as a living Christmas scene, pony-riding or a Punch and Judy show.
- Transport for elderly visitors.
- Organization of a country fair and of an open-door invitation at Christmas-time.
- Programmes for companies to organize their Christmas party, offers for programmes concerning forest ecology for companies.

From these aspects of products, the strong interrelations between material and non-material components become very distinct. The shopping event becomes an integral part of purchasing a product, such as a Christmas tree, which one could get – maybe even more cheaply – in the car park of the supermarket nearby. The relation of the products with their environmental quality and the scenery as an important additional value is a purely non-material aspect. Especially as concerns one case of a Christmas fair in Germany (DE13), there is a reference to the protected ecology label for the Christmas trees offered. By the market advantage, the expensive production without the utilization of chemicals can be refinanced.

Ecological investment products (PG 2.2a) and, to some extent, also material goods (PG 1.1) combine environmental and tangible benefits. Besides the additional value, which more or less distinctly has to be communicated to the consumer, there are examples of refining even in such examples as 'eco-meat' or 'eco-hay' (NL04, 08 and 11). However, above all, distribution strategies, such as new distribution channels, but also new technologies and processes (water power in Austria (AU03)) and certainly the limitation of certain land-use strategies, materials and chemicals in the marketing of drinking-water were considered to be innovations of greater importance. Frequently, these processing regulations are guaranteed by a certificate (eco-meat/eco-hay, NL04, 08 and 11).

Products with origin labels (PG 2.2b) are very distinctly based on the principle of value addition, which in this case is guaranteed by a certificate. A variety of certificates are granted by different authorities, assuring different standards. Besides, all of the products of this group are offered in more or less processed form. Another additional value for the consumer can be illustrated by the case of certified chestnuts in Italy (IT27): 'Natural footpaths in the chestnut groves in which cattle graze and one can walk (and enjoy the peaceful atmosphere).'

In addition, aspects of marketing reaching beyond product design were mentioned in different questionnaires on tangible goods.[43]

INNOVATIONS CONCERNING RECREATIONAL PRODUCTS. In the vast group of recreational products, the first area is PG 3.1a, which almost exclusively comprises accommodation, offered to the individual consumer as fixed, partly tangible products. Product innovation in this area is mostly concerned with establishing an atmosphere, which, as a value-adding feature, will add to the attraction of a stay. A facility design orientated towards the target group, additional offers concerning leisure time, which can either be used in the framework of a package offer or can be acquired exclusively in a product mix, and the special, generally peaceful and natural, (forest) environment are parts of this atmosphere. The following features were mentioned as being characteristic of additional benefit:

- Playgrounds, a variety of different sports equipment and facilities, a schooling ring for ponies and a barbecue.
- Christmas fairs, a dinner with smoked fish ('smoked fish evening').
- Catering.
- Excellent possibilities for horse-riding, hunting and fishing, bridle-paths and hunting maps, box stalls for the guests' own horses.
- Individual group accommodation and special arrangements.

The complex services in the recreational sector with, in most cases, a distinct relation to the environment and frequently with aspects of learning (PG 3.1b, 4.1, 5.13 and 5.2), in view of their product design, have many similarities and therefore can be considered together. Guided tours through the forest, in the mountains or on horseback, deer- and eco-parks, as well as complex services comprising entertainment, education, boarding and often accommodation are characteristic examples of such services.

Starting from recreation in the forest as the basic product, there can often be found several levels of additional value. On the first level, there is the establishment of a forest environment orientated towards the target group by designing car-park areas and playgrounds, information tables or centres, road signs, different instructive paths or paths for ramblers or for horse-riding or mountain-biking. These basic infrastructural product parts are described in detail in the 'communication plan' for the tree-crown path (NL13). In particular, the area of traffic planning, consisting of parking and transport facilities for visitors, has been acknowledged as the main task of this recreational product. The benefit for the visitors on this first level of additional value consists of a feeling of security, the possibility of getting information, a clear choice of routes and the freedom to pursue their own interests (PG 3.1b and 4.1). But also 'wild' and natural woodlands without much human interference are important basics of certain products, such as 'outdoor events'.

On a higher level, guided tours in forests, in mountainous areas or on horseback are added. The organization involved – especially as concerns tours lasting several days – and, most of all, the specialized information represent the additional value on this level. The additional value of information and attention is also present in specialized programmes for local young people and naturalists (guided mountain tours, IT20), children's action days (tree-crown path', NL13)[44] and group packages such as manager seminars and outdoor events (a variety of seminars and guided forest tours, DE10, 11, 15, 18). In these cases, finally, material aspects, such as food, accommodation, transport, special equipment or a certificate on the participation in the vacation programme (holidays at the forester's, DE12), are included as parts of the product, thus representing additional values. Depending on the aim of the programme, events and leisure activities are offered free of (extra) charge, when they are intended to expand the forest experience as such. Products entailing great expense and energy in this respect are, for example:

- National park (NL10) with a broad range of natural scientific, leisure and cultural offers (information centres, an underground museum, a well-known museum of art, bicycle-hiring free of charge, a hunting lodge, maps and leaflets).
- Holiday weekends (NL09): a very comprehensive and luxurious all-inclusive programme, including information, leisure activities, board and accommodation (slide shows, day-trips by bicycle, ornithological excusions in the early morning, a visit to the information centre, accommodation and meals in a first-class hotel).

The luxuriousness, the high quality of the offer and the inherent atmosphere can be a further level of additional values (hunting experience, DE04; luxury apartments, AU11; holiday weekends, NL09; PG 3.1b). Group packages can be turned into very diverse products adapted to the needs of a certain group by skilful negotiations in advance with representatives of the particular group. Beyond the environmental information and recreation facilities, the consumer in this way frequently benefits from group dynamics and self-knowledge (for example, forest youth hostels, DE27).

The simple recreational products (PG 4.2) are non-material offers, which in most cases are sold to individual customers. These products dispose of a low degree of service, but of a more or less advanced degree of cession (*Rechtsabtretung*). Characteristic examples of this group are ski-tracks, toll roads, entrance fees for (private) forests, car parks, and bridle-paths. The innovative additional benefit of these products mainly consists in an amelioration of infrastructure and, therefore, generally corresponds to the first level described for the more complex product groups above. In the case-studies, above all the following features were mentioned:

- As concerns ski-tracks, special maintenance is needed. Tracks have to be of various degrees of difficulty for different users. Adequate road signs are necessary and, in the case of DE19, even lighting is required, footpaths along the tracks and special silvicultural maintenance of the tracks.
- For bridle-paths and mountain-biking tracks, a well-marked road network adjusted to the needs of the target groups has to be established and maintained in accordance with the wishes of the representatives of the respective target group, and additional car-park areas are required.
- The additional benefits to realize when there are entrance fees into forest areas are the adjustment of paths and the tailoring of offers according to customers' demands, the zoning of the area, information and educational centres, playgrounds, sports equipment and resting places along walking paths and special walking routes, partly obligatorily guided in the case of IT28.

The general additional benefit supplied by all of the forests with entrance fees is certainly the special value of the scenery. Two quite complex offers in this group are the cases of access to a small forest (IT06) with playgrounds, public lavatories, a dance floor and entertainment programmes, as well as a car-park area (IT16), where the entrance fee already includes the fees for utilizing several car-park and picnic areas, playgrounds and well-designed walking paths around a lake.[45]

Product innovations in the likewise 'simple' group of 'mushroom-picking permits' (PG 3.2) are, to a certain degree, different from the cases indicated above. The sale of licences for gathering mushrooms is a combined product of special recreational services with the mushrooms gathered as a tangible component. Therefore, in the case-studies, the potential income, the better control of mushroom-picking and the pickers and the advantage of controlled marketing over a prohibition are not referred to as additional benefits, but as innovative features.

INNOVATIONS CONCERNING INTEGRATIVE CONTRACTING. Agreements of licensing (use contracts), rent and lease (PG 6.1) attain only with a few exceptions a noticeable degree of servicing but always have a high degree of cession. Within this group, the additional benefit for consumers necessarily has to be differentiated from the benefit of the contractee as the customer. The institutional contractees gain additional benefit only indirectly, from the benefit to their members.

These benefits, for example, comprise a customer-orientated selection and infrastructural design of the sports grounds agreed on in the contract, good preconditions for exercise (scenic, as well as infrastructural), visitor-steering effects of the agreements or the deference of the contractors on the needs of the customers within their own forest management. Exceptions to this rule include the road network for mountain-biking (AU01), in which the additional benefit of the agreement on roads for mountain-biking is the possibility of renting bikes from the same supplier, and the lotting of forest areas for commercial targets (NL03). In the latter case, the supplier not only offers the utilization of the forest scenery, the value of which, in this case, lies in its effectiveness in advertising, but also helps in the preparation and the execution of the events permitted.

Sponsoring (PG 6.3) occurs in the recreational as well as in the environmental sector. Basically, sponsoring is a sales transaction with special characteristics. Thus, an exclusive right of using a value or service for advertising and PR is marketed. This is normally done on the basis of a contract individually negotiated at a certain price.

This contract, for example, settles where, when and how the sponsor may use the recreational facilities for advertising (recreation sponsoring, DE20), or it contains agreements on the location and time of the sponsored environmental project and on PR, the publication of an information paper or the active participation of the public in afforestation, together with a subsequent public festival at the plantation area for the sponsor's publicity purposes (afforestation sponsoring, DE22). With environmental sponsorship, the environmental value of an object of nature is used for advertising. Therefore, an agreement on nature preservation regulations is frequently an important component or the basis of a sponsoring contract. In the case of IT29, the WWF participates as a sponsor in a project at a national park. Thus, members of this nature preservation organization get special certificates on their contribution to nature conservation. In three of the RES cases, nature preservation organizations as well as their projects are supported by sponsoring (NL06, 12 and 16). Besides the usual agreements on the utilization of nature preservation areas and facilities dedicated to the PR of companies, as well as the mention of the companies in whatever context they want, a broad range of additional benefits for sponsors are used. Thus, it can be assumed that there might be a certain advantage in the additional benefit of environmental sponsoring over other possibilities of sponsorship.

Similar features are true of environmental products (PG 6.2). In this case, as well as in the case of environmental sponsorship, contracts fixing the components of the product in detail are negotiated individually. They always refer to a certain area and fix agreements on activities (contract of nature conservation, DE05: felling of trees) and non-actions, as well as on the liability to render account (contract of nature conservation, AU12). In one of the cases (DE14), a mapping of the biotope was required as a kind of additional service in advance, before the contract on silvicultural measures concerning water conservation could be agreed. Furthermore, information on the project was passed on to private households as an additional benefit. In another case (contract of nature conservation, AU10), nature and culture preservation were combined: the form of pasture farming remunerated also includes historical barns and fences.

Product innovations of RES products occur in a multitude of representations. Material RES products are frequently found at higher levels of processing. Furthermore, product packages, additional services and the creation of a shopping event are important additional benefits for tangible RES products. Important factors for the success of these offers, finally, are marks of origin or 'eco'-labels, which guarantee certain environmental standards. Capital goods have to communicate this additional value to the consumers.

Innovative benefits can be added to RES recreational products at several levels. The special value of the landscape and an adequate forest management are the basis for the first, fundamental level. On this level, simple recreational products, in particular, are given added value, for example, by building facilities for sports and rest in the forest. On the second level, there are additional organizational offers, with guided tours, events, additional forest facilities and all-inclusive programmes lasting for several days. The success of many complex offers for recreation and accommodation is based on such innovations. RES offers on a third level can be turned into 'luxury events' with an individual atmosphere. The creation of atmosphere has become the most important additional value concerning accommodation offers.

Support services can add value to integrative contractual products on recreational uses or concerning forest conservation. Contracts of sponsorship regulate the exclusive utilization of the forest components' value level for advertising. All of the contracts are, to a greater or lesser degree, individual agreements.

The product mix

Kotler and Bliemel (1995, p. 667) define the product mix (range of products) as 'the entire range of product lines and articles which a supplier offers to the customer for sale'. In this chapter, the term 'product mix' is used in a somewhat more limited meaning, and therefore is approximately equivalent to the 'product system' (Kotler and Bliemel, 1995, p. 662). Thus, in the following section, the products which are 'offered by a certain enterprise in an objective context orientated towards a target group' are considered. They differ from the additional values establishing the additional benefit of the product supplied basically in that they can be divided up and sold individually. Generally, the product mix comprises complementary products, which means that the products supplement the essential offer in view of an optimum customer orientation. With the supply of a product mix according to this definition, the utilization of joint effects is included in the offer (Nieschlag *et al.*, 1991, pp. 214–218). The forest as such in general defines the design of the product programme in forest enterprises by the production factors supplied. It is in the nature of things that the 'policy of product programme', as described, for example, by Nieschlag *et al.* (1991, pp. 211–218), is quite limited by natural features for forest enterprises. Nevertheless, in a total of 72% of the cases researched, additional RES products were offered in the range of the suppliers.

A product mix of related products was offered in 48% of the RES case-studies. As Table 4.7, categorized according to the product groups, shows, this product mix can consist of traditional goods, of RES products or of a mix of both lines of products:

Especially in the area of material consumer goods (PG 2.1), but also in the area of recreational products (PG 3.1–5.2), there is a high percentage of products which are included in a product mix. None of the offers which are not in relevant direct contact with the RES product of the case-study referred to were considered. This is gener-

Table 4.7. Type of the offered product mix by product groups. (Question 4.2.7: Is the product sold in combination with other products (product mix)? Which products?)

		Trad. prod.		RES prod.		Both		No mix	
Type of product mix	Cases	*n*	%	*n*	%	*n*	%	*n*	%
Material offers for institutions (PG 1.1, 2.2)	11	1	9	0	0	1	9	9	82
Material offers for single customers (PG 2.1)	6	1	17	0	0	5	83	0	0
Complex recreational offers (PG 3.1, 4.1, 5.13, 5.2)	27	7	26	2	7	12	44	6	22
Simple recreational offers (PG 3.2, 4.2)	27	2	7	7	26	4	15	14	52
Contracts with institutions (PG 6.1, 6.2, 6.3)	27	0	0	4	15	0	0	23	85
Total	98	11	11	13	13	22	22	52	53

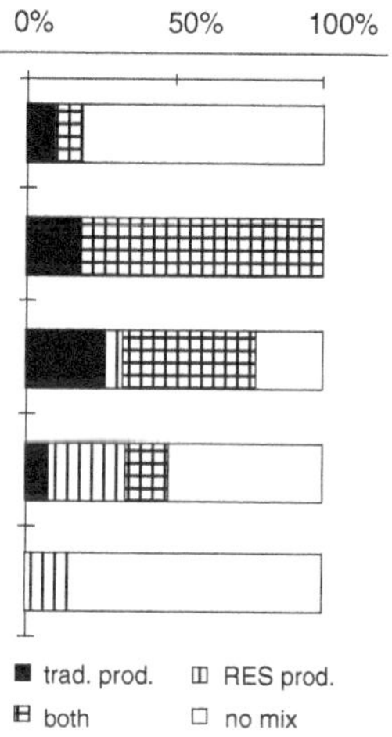

ally the case as concerns the sale of timber or semifinished wood products, traditional by-products, such as forest seeds or gravel, and some additional RES products.

PRODUCT MIX CONCERNING TANGIBLE PRODUCTS. The product group game, Christmas fairs and Christmas trees, gifts (PG 2.1) is characterized by an extremely high number of additional offers, concerning services as well as tangible goods. However, it is difficult to draw a line between additional services leading to an additional benefit, as described in a previous section, and the benefits of a product mix, for example, the sale of Christmas trees at a Christmas fair, as both the fair and the tree can be defined as a product in itself. In as comprehensive an outline of the product mix as possible, the following components arise:

- Traditional tangible components: food and drink, Christmas trees, firewood, flag-poles, game/specialities, fish, jams, honey, wine, beer, cookery-books, needlework.
- Services on the market: entertainment for children, such as pony-riding, games or a Punch and Judy show, transport for elderly people, round trips by carriage, shooting contests for hunters and other visitors interested.
- Further service offers: holiday flats and houses with a variety of additional offers, programmes on forest ecology in cooperation with enterprises interested, arrangement of a staff Christmas celebration.

These offers are certainly not all made in every case, but they serve as good examples of potential components. Material offers to institutions, as prevalent in PG 1.1 and 2.2, only in exceptional cases indicate a product mix. However, this would be possible, as shown by the example of the sale of drinking-water (NL01): besides the 'sale' of drinking-water, a building is let to the water-works company as the respective customer.[46]

PRODUCT MIX CONCERNING RECREATIONAL PRODUCTS. There are many components of a product mix comprised in the offers for overnight stays (PG 3.1a) as well. In general, these offers are additional offers to guests staying overnight, which can consist of tangible goods as well as services. Within this group, there are considerable differences in the extent of the offers.[47] As a summary, the following components were described concerning a product mix in this group:

- Traditional tangible components: fish, game, eco-meat, agricultural products, Christmas trees.
- Additional leasing and agreements: fishing and hunting licences, ponies for hire, box-stalls for horses, utilization of a barbecue, golf-course.
- Services: Christmas fairs, 'smoked fish evenings', individually organized arrangements for group vacations, organization of trips for handicapped customers or guided tours through a national park, riding lessons offered in cooperation with a riding-school, as well as a variety of offers made by the Dutch State Forest service in the case of NL18.

From this enumeration, the broad extent of potential product mixes can be easily seen. Thus, it is necessary to tailor one's offer exactly to the target groups of the main product.

Furthermore, extensive and numerous product mixes are found in the additional offers of complex recreational consumer products (PG 3.1b and 4.1).[48] Also in these cases, there are problems in defining the elements of additional value, as some of the facilities mentioned below, for example, are already included in package offers by the companies, although the individual customer has to pay extra for them. Frequently, especially for institutional customers (clubs and associations), a package offer is a combination of a main offer targeted individually to the respective group and of different additional offers. In the case-studies, different components of product mixes are mentioned:

- Traditional tangible components: restaurant (in the forest), accommodation and board for short vacations, as well as for guided tours lasting for several days, mountain guidebooks and maps, information packages, books, game, Christmas trees, firewood, a shop for forest products, souvenirs of all kinds, e.g. game bones.
- Additional leasing and agreements: holiday huts, camping, rental of rooms for celebrations, fishing licences.
- Services: educational events, guided natural-history tours, guided castle tours, package offers for tours guided by specialists, organized nightly tours on horseback and/or with board in country pubs, horse-drawn trams and various offers for exercise (summer and winter sports).

Integrative, partly tangible, recreational products (PG 5.13 and 5.2) are often sold to institutional customers. Therefore, particularly in this area, product mix and additional value are difficult to separate, especially as the offers are submitted in cooperation with the customer. The components frequently occurring in product mixes, as mentioned in the interviews, are:

- Traditional tangible components: lunch, drinks, wine, game, Christmas trees, twigs and other forest products for decorative purposes, honey, tree seedlings, firewood.
- Additional leasing and agreements: accommodation.
- Services: game parks, guided forest tours for companies and holiday-makers, music.

Similarly to material RES products, the extent of the product mix for institutional target groups is considerably smaller.

Product mixes offered in the group of simple recreational products (PG 4.2) are remarkably non-uniform. The products described in the case-studies are mostly of a more or less simple design, frequently with a low degree of service. Therefore, the effect of a product mix in this group is of special interest. In ten cases, there was no offer concerning a product mix. Nine of the case-studies indicated few additional offers:

- Traditional tangible components: Christmas trees, game, farmhouse catering.
- Additional leasing and agreements: hiring of mountain bikes at a car park and equipment for skiing on the ski-tracks,

ski-courses, accommodation, fishing licences;
- Services: guided tours, access to castles and libraries.

Finally, in five cases very comprehensive and multifaceted product mixes were described:

- IT06 (access to the forest): catering, organization of a light entertainment programme, recreational package offers with additional offers, lodging and overnight stays.
- IT28 (access to the forest): sale of information and souvenirs, horse-riding, canoeing, car-park areas on the beach, camping, guided tours, transport into the 'core' of the conservation area.
- AT13 (roads for mountain-biking): package offers with overnight stay, ski-lift tickets, transport offer for elderly customers, guided tours.
- NL02 (access to the forest): lease of a restaurant, information centre, camping, guided tours.
- NL07 (mountain-biking tracks and bridle-paths): a variety of offers, which in part could be used by the customers of the product (car-park areas, golf-course, camping, information centre).

The seasonally conditioned alternative utilization of RES facilities can be considered as a special kind of complementary goods; for example, two cross-country ski-tracks in Italy which, in the summer, are used by mountain-bikers, or, as in one case, skiing on an Italian golf-course in winter. In all three cases, the Italian providers do not own the forest concerned themselves.

PRODUCT MIX CONCERNING INTEGRATIVE CONTRACTING. When referring to letting, leasing, and agreements concerning utilization for recreational purposes (PG 6.1a), the additional offers to the institutional contractee will be observed, instead of the product mix offered to the consumer. The characteristics of the product (in general, individually negotiated separate contracts) lead to the assumption that an offer concerning a product mix will rather be an exception than a rule. This is confirmed by the case-studies: solely in the case of an agreement on a road network for cycling (DE23), licence agreements (use contracts) of different content are offered and marketed.

The agreements on environmental products (PG 6.2) are not supported by product-mixes. The only exception, in a sense, is the case of AU10, which contains non-exclusive agreements concerning pasture cultivation as the main product, but also building and the conservation of historical barns and fences.[49]

In the area of sponsorship (PG 6.3), there is no product-mixing in Italy, Austria and Germany.[50] The characteristic of sponsoring products as contracts individually negotiated with institutions supports this result. Another situation, however, is present in the Netherlands, where very complex structures of the sponsoring of different nature preservation organizations and projects are established. Sponsors, for example, are offered a mix of project-sponsoring, together with the sponsoring of a magazine (NL16). Another possibility is that they can get information, participation in guided tours, objects of art or facilities, which are also offered to private customers, in addition to the exclusive right usually granted to utilize a certain value of nature (NL06, 12). This points to the fact that market conditions for environmental sponsoring in the Netherlands are already harder than in the other three European countries.

> Due to different enterprise conditions, the range of product mixes offered by forest enterprises is quite limited. In spite of this, the offer of a product mix can contribute considerably to the success of RES products by compound effects. Tangible RES products for consumers and recreational RES products are frequently offered along with various other tangible goods. In most cases, the forest products offered are produced without chemical treatment, but are processed. A variety of other simple or complex services can also be marketed in the context of material and recreational RES products. Potential joint offers in the context of RES

recreational products are additional letting and use agreements.

The effect of product-mixing is especially strong when there is a direct orientation towards a certain target group. The offer of a product mix is often agreed on in individual negotiations with institutional customers in the framework of a package offer at a bundle price. Contracts are rarely offered in a context relevant to a product mix. However, this would be possible in different cases of recreational and environmental utilization of the forest.

Special product design

In the following, some of the product political tools are described which influence the appearance and the detailed design of the products offered to support their sale.

To a certain extent, packaging plays an important role in sales support, especially with consumer goods, depending on the motivation for buying the product. An exclusively rational buying decision, as with nails for example, cannot be influenced much by packaging, while an attractive package can be an elementary, integral part of a perfume or of similar products. In the area of RES products, such packaged consumer goods are rather exceptions than the rule, although the hygienic packing of game or an attractive, but ecologically conscious packaging of the so-called 'organic gifts' can increase sales to some extent. Furthermore, additional value can be established by convenient (technical) package designs. Besides, packaging provides an important means of communication from the possibility of placing brands and logos. The label can be considered an additional part of packaging and be used where appropriate (Kotler and Bliemel, 1995, pp. 700–702). One of the RES products utilizing this strategy, for example, is the Christmas tree, the label of which guarantees its ecologically sound production. In addition to the optical effect of packaging and the information provided, packaging determines the size of the respective sales units. In this case, the orientation of packaging to the quantity needed is an important part of product policy (Nieschlag *et al.*, 1991, p. 183).

It seems to be appropriate for material RES products that the customer can establish a relationship between the characteristics of form, colour and material and the forest eco-system.

The increase and securing of product quality, if possible in relation to an adequate guarantee which the customer can objectively evaluate, are an important part of product design.[51] Especially as concerns the service sector, the quality of the performance is an important element for avoiding the domination of a price war. As the quality of services is evaluated individually by the customers, individual customer orientation at this point is of determining importance. According to Zeithaml *et al.* (1992, in Kotler and Bliemel, 1995, pp. 722–723; Meffert and Bruhn, 1995, pp. 200 and 201), the most important aspects of quality concerning services are the following:

- Reliability.
- Obligingness, disposition to adaptation (*Reaktionsfähigkeit*).
- Competence through specialized knowledge and a trustworthy appearance.
- Sensitivity to individual needs.
- Appearance of personnel, equipment, set-up and leaflets.

Three characteristics of enterprises in the service sector with a strategy of quality are: high claims for themselves, an effective system of performance control and a high degree of job satisfaction among their staff.[52]

Finally, services accompanying a product, customer service and the granting of guarantees are important parts of product design as well. The issue of services as an additional benefit has already been outlined above. Customer service is different from guarantees in so far as it is offered without time limits (Nieschlag *et al.*, 1991, p. 219). It can develop a considerable effect on acquisition and is mainly of importance in the area of capital goods. Applications of business customer service, such as facilitating buying or giving advice (Weis, 1997, p. 248), could be also possible in the marketing of RES products.[53]

> In the marketing of some tangible RES products, customer-orientated packaging can contribute to the success of an offer. Forest services should aim at high-quality standards. Reliability, obligingness and competence of the personnel providing the service are of eminent importance. A counselling customer service can support the sales of RES products as well.

4.3.2 Distribution policy

Definition and tasks of distribution policy

The term 'distribution policy' comprises all decisions and activities that are related to the movement of a product from its place of origin covering the lapse of time and space to the final consumer.[54] This means that, in distribution policy, the relation between the production of the product and its consumption is observed. Distribution policy, like product policy, is a long-term marketing instrument. There are many relations and reciprocal effects between the selection of the distribution channel and the other marketing instruments.[55]

As a rule, in Central Europe, products are currently distributed via at least one and, in most cases, several marketing stages.[56] This presupposes a clear definition of distributor and customer in the distribution of a product and both distributor and customer have to be described in several clearly distinguished steps, if necessary. As a typical example of the area of recreational and environmental products, there is, for example, the cross-country ski-track, the utilization of which, as a first step, is subject to a lease contract between the forest enterprise and a sports club, and access to which, as a second step, is granted by the sports club to the skiers on payment of the membership fee.[57]

> The exact coordination of the distribution policy with the characteristics of the product and the customers is an important factor of success. Distribution is linked especially closely to all other segments of the marketing mix.

The stipulation of trade channels on the one hand, and the determination of the physical distribution, the product logistics on the other, can be distinguished as the main targets of distribution.[58] For this reason, it has to be determined how many and what steps there are between the forest enterprise and the final consumer (vertical distribution), whether formal or informal cooperation is included in the offer (horizontal distribution elements), if and how sponsoring is involved and, finally, the direct point of purchase of the product.

Distribution channels

GENERAL DESCRIPTION OF DISTRIBUTION CHANNELS. Stern and El-Ansary (in Kotler and Bliemel, 1995, p. 802) define the distribution channel as the entirety of cooperating organizations which help to ensure that the clients can use a product. Figure 4.5 gives an overview of the most important distribution channels according to Weis (1997, pp. 305–309).

Concerning different types of distribution, a basic distinction can be made between direct and indirect sales. Weis (1997, p. 310) defines direct sale as every distribution channel that does not directly involve any wholesalers. However, distribution helpers who only support the selling of products from the exterior can be associated with direct sales. With a typical form of direct sale, the management is responsible for sales, as is usual in forest enterprises. Also, market events, such as the selling of Christmas trees at the forester's house or timber auctions, are common in the forest sector.

The distribution of a product via intermediaries is referred to as indirect sales. If services are offered indirectly through intermediaries, they need to be 'materialized' – for example, by a ticket.[59] Distribution channels are distinguished by the number of intermediaries that help to distribute the product. Regarding the documented case-studies, the distribution channels listed in Table 4.8 have been found. Table 4.8 shows that direct sale is very common regarding the offer of RES products. In 74% of the cases, the offerers

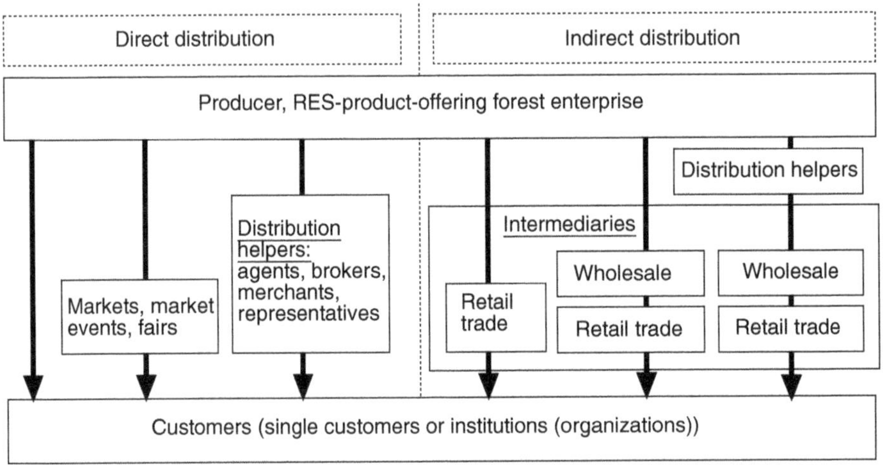

Fig. 4.5. Simplified system of different distribution channels.

use this distribution channel to offer their products. In the case of cooperation with intermediaries, these are mainly one-level distribution channels (39%). Multistage distribution channels are not used very often. In 18 case-studies, a mixed distribution was documented, where the offerers use both direct sale and cooperation with intermediaries.

After the fundamental decision on the distribution channel, it is decided how and what intermediaries and/or distribution helpers should be appointed as partners according to the strategy and the target, how many partners should be worked with and under what conditions and mutual obligations the distribution should be carried out together with the partner (Kotler and Bliemel, 1995, pp. 812–815). 'The essential reason for the participation of distribution partners is their higher efficiency concerning the comprehensive division of goods towards target markets' (Kotler and Bliemel, 1995, p. 803).[60]

Distribution helpers and intermediaries. Regarding market participants, namely institutions that fulfil certain tasks in the distribution process, a distinction can be made between distribution helpers and intermediaries. Distribution helpers support the distribution process without using distribution instruments of their own. Examples are banks, insurance companies and forwarding companies. Intermediaries use distribution instruments of their own in the distribution process. Wholesale or retail traders are typical intermediaries.[61] Forest enterprises may appear not only as producers but also as intermediaries or distribution helpers. Tasks that are undertaken by the forest enterprise in marketing

Table 4.8. Distribution levels in the RES case-studies.

Number of levels	*n*	%
3 levels	1	1.0
2 levels	4	4.1
1 level	38	38.8
0 levels (direct sale)	73	74.5
Total (basis 98 cases)	116	118.4

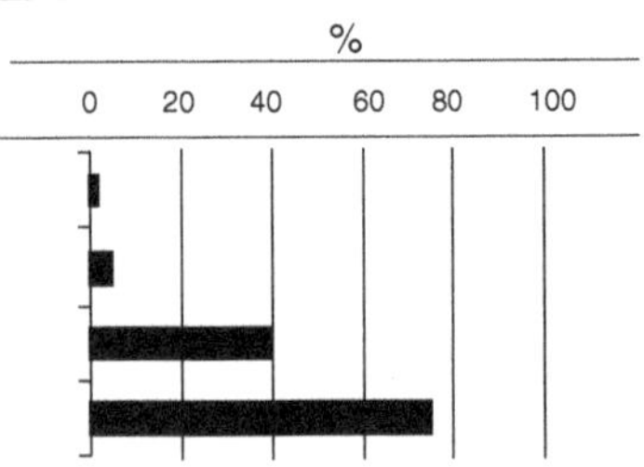

are totally undetermined and depend on its skills and readiness and on costs.

Types of participants in the distribution process and their tasks. When intermediaries are used to distribute RES products, these intermediaries fulfil, among other things, the following tasks:

- They solve structural problems by summarizing the offers of various smaller forest landowners.
- They are responsible for advertisement and PR services, which are not very common in forest enterprises.
- They present an image for offers that have important trust qualities.
- They offer professional help in questions of contract design, taxes, market investigations, improvement of production techniques.
- They help to lower tax payments.

As concerns legal regulations, third parties involved in a distribution process can be categorized into institutions acting on their own behalf and for their own benefit and representatives acting on appointment by and for the benefit of others. In the former group are tradesmen of mostly tangible goods (NL04, 08, 11, IT27). In the latter group, there are a variety of very different 'middlemen', the rights and duties of whom are stipulated by contracts, statutes and regulations of civil law.

Middlemen who are involved in the sale of RES products can act as independent organizations who work on the behalf of several offerers (28 cases), or they can be organizations that were founded, or at least influenced, by the offerers of RES products for a certain target (15 cases). Moreover, independent middlemen can be subcategorized into companies with economic targets (tradesmen, travel agencies, processing companies, etc.) and organizations with social targets (tourist information offices, nature preservation organizations, etc.). The latter are generally 'non-profit organizations'. Dependent middlemen are mostly organizations developed from horizontal cooperation. Thus the offerers can exert direct influence on their activities via the articles of association and their membership in the organization.

Advantages of independent middlemen in the distribution process. Intermediaries and distribution helpers have different motivations for acting on the behalf of the offerer. In Table 4.9 are listed the advantages that independent organizations can gain from participating in a distribution process.

DISTRIBUTION ORGANIZATION IN THE RES CASE-STUDIES. Table 4.10 gives an overview of the distribution organization of the RES products investigated. While the first answer to the question (see Table 4.10) points to an independent, direct sale of the RES product, the following alternative answers refer to different horizontal or vertical elements of the distribution organization. If the RES product is marketed via an organization or informally together with other forest owners, a horizontal distribution organization is prevalent (the second and third answers). If an intermediary is involved in the offer, distribution is organized vertically (the fourth answer). The

Table 4.9. Motivation for independent institutions to work as intermediaries.

Middleman	Motivation to work as a middleman and guarantees for successful work
Dealer or processor	Trading profit
Political organizations	Votes Politically determined objectives
Nature conservation organization	Image improvement Targets of the organization
Enterprises and organizations of tourism	Trading profit Commission for acting as middleman

Table 4.10. Organization of the offer in the RES case-studies. (Question 5.1.1: Who offers the product?)

Offer*	*n*	%
By the enterprise itself	69	70
By an organization	13	13
With other forest landowners	3	3
By middlemen	19	19
By another way	15	15
Total (basis 98 cases)	119	121

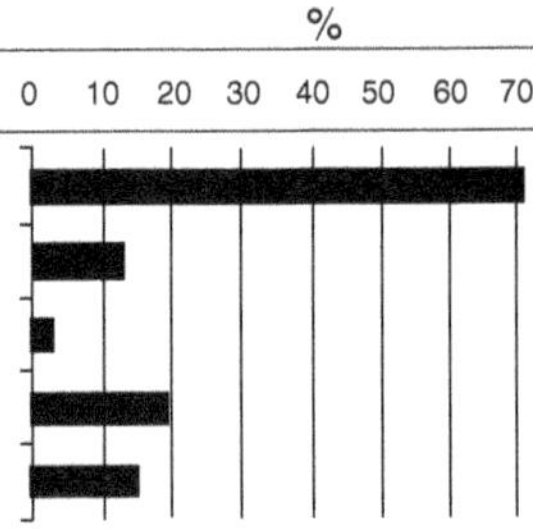

* Answer choice in full:
The enterprise offers its product by itself.
The product is offered as part of an organization (e.g. forest owners' association, purchasing cooperation ...).
The product is offered together with other individual forest owners without any formal organization.
The product is offered by middlemen (e.g. marketing bureau, tourist organization, municipality administration, ...).
Other forms [with the opportunity of a free input].

answer by 'middlemen' contains two forms of vertical distribution organization: the sale of a product via wholesalers and retailers and the sale in cooperation with intermediaries like tourist information offices, which cannot be primarily categorized as dealers.

Before referring more closely to the organization of an offer in which distribution partners are involved, the basic factors influencing the selection of distribution channels and the eminent importance of direct sale for RES products will be considered.

FACTORS INFLUENCING THE SELECTION OF DISTRIBUTION CHANNELS AND THE SIGNIFICANCE OF DIRECT SALE FOR RES PRODUCTS. Concerning the selling of RES products, direct sale is dominant in comparison with the common tendency towards indirect sale in the consumer field. Distribution channels are influenced by a complex system of factors. Table 4.11 summarizes the most important factors according to groups.

Factors connected with the enterprise.

1. Size of the offerer. Certain RES products, for example the offer of a cross-country ski-track or the offer of a bigger sponsoring project, need a minimum of forest area. In this case, intermediaries are able to solve structural problems by summarizing the offers of different small forest landowners. Examples:

- A middleman completes contracts with different forest landowners to offer a cross-country ski/mountainbike track (AU13, 17).
- A middleman completes contracts with different forest landowners to offer silvicultural methods for drinking-water purposes (DE08, 14).

2. Financial power of the offerer. This factor is connected with the size of the enterprise. It is important when the product offer needs certain investments (advertisement, packaging, accounting, certification, etc.). Example:

- offer of certified mushrooms by a consortium of forest landowners and dealers (IT02).

3. Distribution know-how of the offerer. Sometimes, the forest enterprise does not know the relevant distribution channels and distribution instruments for certain products. Therefore, cooperation with middlemen is sometimes necessary. Example:

- offer of management training in cooperation with business consultants (D10, 18).

Factors connected with the demand.

4. Image of the middleman from the customer's perspective. The image of the mid-

Table 4.11. Influencing factors on the choice of a distribution system (based on: Meffert, 1986, p. 421; Nieschlag *et al.*, 1991, p. 381; Kotler and Bliemel, 1995, pp. 810–812, 816 and 817; Meffert and Bruhn, 1995, pp. 320–322; Weis, 1997, p. 306).

Factors connected with	Some influence factors
Offering enterprise	Size of the offerer, financial power, available distribution potential (cost and benefit structure of possible market channels, opportunities of control of distribution channels as well as additional ways of modification, personnel, etc.)
Demand	Image and available distribution potential of potential market partners; structure of demand; strategic goals of customers regarding procurement (customer demand for distribution services concerning amount of marketed products, duration, local presence of offerer, diversity of supply and demand for supporting services; etc.)
Market competition	Competition, distribution channels of competitors, etc.
Product	Product type, product for consumption, product for investment, complexity, frequency of demand, age (need for explanation and for demand of an offered service, local flexibility of a service offer, etc.)
Social environment	Technical development, legislation (other conditions of environment, etc.)

dleman has great influence, especially when products are concerned which are characterized by trust qualities. In this case, the user cannot judge the quality of the product very well and the image of the middleman serves as a substitute for the missing impression of the qualities of the product. Examples:

- offer of environmental education with the help of consultants (DE10, 18).
- Cooperation with forestry-related environmental organization in the case of environmental sponsoring (DE08).
- Cooperation with WWF in the case of environmental sponsoring (IT29).

5. Behaviour of demand of the client. The client may search for certain products using intermediaries. In this case, neglecting these distribution channels would mean losing certain clients. This is very obvious with touristic products or with the offer of eco-products for example. Examples:

- offer of holiday apartments via a tourist information office (DE02).
- Offer of seminar programmes in the catalogue of an educational organization (DE17).
- Offer of forest apartments in the catalogue of a travel agent (NL18).
- Planned offer of eco-products in a supermarket (NL19).

Factors connected with market competition.
6. Distribution channels of the competitors. Not using the same distribution channels as a competitor can be both an advantage and a disadvantage for the offerer. When the offerer does not cooperate with the tourist information office offering holiday apartments, for example, he/she will lose potential clients.[62] But using distribution channels that differ from the regular ones can also be an advantage if new market sectors can be reached with these distribution channels. Examples:

- direct sale of Christmas trees via market events (DE01, 13, 16).
- Direct sale of guided tours with an additional programme via the internet (DE15).

Factors connected with the product.
Depending on the type of product (tangible goods, services or contract agreements), the potential distribution channels for RES products differ considerably. Regarding the industry for investments goods, direct sale is by far the most important. Besides, it is suitable when a small number of customers

are within reach of the producer. A direct distribution channel is also required if products need to be explained and are designed or produced interactively.

7. Complexity of the product. Specially designed products which cater for customer preferences and which are complex and have to be explained are mostly offered via direct sale. Examples:

- sponsoring of old-growth conservation (AU09).
- Special contracts on the use of a forest area to produce films (NL03).

8. Frequency of buying the product. Products which are bought infrequently are normally offered to the client directly. In the cases documented by this investigation, some products have been sold only once so far. If the customer is, furthermore, a large organization, the advantages of direct sale are reinforced still further. Examples:

- contracts for the sale of drinking-water (AU16, 19, NL01).
- Production of electricity (AU03).

9. Amount of rights and services in the product package. In general, RES products contain different components. Non-material parts, such as rights and services, are important product components. Pure material products and products with an important material product component have been documented in only 17% of the case-studies. Non-material products cannot be transferred to the customer, nor can they be stored. Only the promise of doing something (contract) can be offered. In most cases (60%), these promises take into account the special wishes of the client. The contracts are individually negotiated and completed. In only 40% of the promises are standardized contracts (e.g. tickets) used. One important function of the middleman – the presentation of products to the market – is not needed in case of the normally unstandardized RES products.

10. Life cycle of the product. Another reason for the lack of special distribution channels is the fact that the investigated products are fairly new. In 1998, a quarter of the products was no older than 3 years. But during the market introduction stage of the product, significant enterprise- and personnel-specific faculties are necessary to establish the performance capability of the offerer. When a new product is introduced, the customer is quite uncertain concerning its quality. Information about the long-term demand situation is not very profound, and experiences regarding the necessary distribution instruments do not yet exist. It takes time and labour (transaction-specific investments) to convince the customer of the product quality and to get information on the market situation. Normally, external intermediaries are neither willing nor able to make these investments and thus during market introduction direct sale is a common solution.

When the product has already existed for some time, the situation becomes easier for intermediaries. The uncertainty of the clients is reduced because of their experience with the product or with similar products which have now been developed. A certain standardization becomes possible and intermediaries have better information on the offerer and on the market situation for these products. During a later life-cycle stage, cooperation with intermediaries might become the better solution. In every early stage of the life cycle, products are usually not integrated within distribution channels. The introduction period is normally characterized by direct sales.

Forest enterprises only exceptionally have experience in the production of 'new forest consumer goods' and therefore they rarely have established relations with 'distributors', which in turn suggests the utilization of direct channels of distribution.

Factors connected with the social environment.

11. Technological development. The development of the internet is an example of the influence of technological developments on the distribution channel. Using the internet, offerers of RES products have new possibilities for direct sale. Example:

- offer of guided tours in the forest and excursions in the national park via the internet (DE15, AU07).

12. Legislation. The offer of holidays with the forester (DE12) is an example that shows the influence of legislation on the distribution channel. In this case, the tourist information office, and not the forest enterprise, is the offerer of the all-inclusive programme. This is due to legal circumstances (see Chapter 5, Section 5.3). The forest enterprise did not want to face the greater liability of a tour operator and this was one reason to cooperate with the tourist information office. Tax regulations also have an important influence on the distribution channel. Examples:

- offer of sponsoring projects in the forest via a non-profit organization to avoid income tax for the forest landowner (DE08).
- Foundation of a separate association for the offer of the RES product, because, otherwise, the forest enterprise would become a commercial enterprise and would have to pay business tax (DE03, 11).

Finally, the public develops, as a natural cross-current against what is progressively a 'supermarket culture', the desire to purchase individual, if possible 'organically reliable', consumer goods direct from small suppliers. The occurrence of such conditions can be utilized strategically by forest enterprises for the direct sale of RES products.

The above-mentioned factors both influence the choice of the distribution channel and are of varying importance according to the specific situation.

Many factors influence the choice of a suitable distribution system for RES products. The following hypotheses on the factors influencing the distribution of RES products can be derived.

Factors connected with the enterprise.

- The more necessary a certain land base for the offered RES product is and the more divided the forest property is, the more probable cooperation with intermediaries becomes.
- The more necessary a certain investment for the offered RES product is and the less financial power the single offerer has, the more probable cooperation with intermediaries becomes.
- The more necessary special distribution and marketing know-how for the offered RES product are and the less know-how the offerer has, the more probable cooperation with intermediaries becomes.

Factors connected with the demand.

- The more trust qualities the offered RES product has and the less trust capital the forest enterprise has in this niche market, the more necessary it becomes to cooperate with intermediaries who already have the necessary trust capital due to their image.
- The more similar to well-known products the offered RES products are, the more sense it makes to use those distribution channels the client is accustomed to when looking for similar products.

Factors connected with the product.

- The more complex the offered RES products are, the more integrated in the production process the customer is and the more seldom the product is sold, the more probable is direct sale.
- The more non-material parts (rights, services) the RES product package has, the more necessary direct sale becomes.
- The more innovative RES products are, the more direct sale is necessary and the older and better known the RES products become, the more intermediaries can help to sell RES products within special distribution channels.

As a summary, the following statements can be made.

> Direct sale without using any intermediate trade levels is predominant for RES products. But the sale of RES products is often supported by distribution helpers. Intermediaries solve problems of forest enterprises resulting from a lack of land-base size and knowledge of the RES market. This points to the assumption that forest enterprises have not had much experience with sales channels for RES products yet. For the sale of material RES products, it seems to be useful to use the trend for 'shopping on the eco-farm'.

Complex distribution organization in the RES case-studies

Complex distribution organization can include vertical cooperation with sales helpers and intermediaries and horizontal organization by cooperation and informal contacts, including the participation of sponsors. Therefore, it is necessary to appoint and to motivate distribution partners, to judge the distribution by efficiency and, if necessary, to modify the chosen distribution structure (Kotler and Bliemel, 1995, pp. 818–827).

VERTICAL DISTRIBUTION STRUCTURE. The distribution of products via one or several intermediaries can be designed quite effectively by efforts at vertical marketing.[63]

Distribution via intermediaries. Distribution via stages of intermediary trade is characterized by an increasingly dominating position of trade in the whole economic system.[64] Generally, the distribution via wholesalers is confronted with conflicts resulting from different goals.

The producer can handle these conflicts by four different means:[65]

1. Adaptation to the needs of trade.
2. Influence on the needs of trade by the 'pull effect' (demand pull).
3. Cooperation with trade partners.
4. Avoid trade.

In the RES case-studies, there was an additional possibility of evading commercial business by choosing a sales partner either from the area of public institutions or from complementary sectors of the economy, such as the catering trade. Concerning RES products, however, the complex, autonomous impacts and goals of trade with regard to the distribution of products are of minor importance. First of all, material goods for processors (PG 1.1 and 2.2) and mushroom-picking permits (PG 3.2) are passed to the final consumer via dealers in a significant way. In this connection, the following hints are mentioned:

- Distribution of licences partly via tobacco stores (IT01).
- Selling of eco-meat via butchers and dealers (NL04 and 11).
- Selling of certified timber via craftsmen (IT13).
- Forty per cent of the chestnut production is distributed via dealers (IT27).
- The forest enterprise sells its products itself and at the same time markets purchased products from other producers at its 'forest shop' (NL19).

Eco-investment products (PG 2.2), which are predominantly material and sometimes certified goods, are usually purchased by institutions. Regarding the aspect of simple methods of distribution at first, it could be decided to consider wholesalers as customers of the first distribution level and to view the offered products as a kind of 'investment good' for the dealer. However, this characterization of products with respect to their final utilization might not be appropriate, since these products do not go into a further production process.

Examples in which recreational products are at least partly marketed by intermediaries are as follows:

- Distribution partly via a travel agency (DE02 and NL10).
- Distribution of hunting licences via an agent for hunting opportunities (DE04).
- Ticket distribution, partly via dealers (NL02).

In the case of the Taufstein ski-tracks

(DE19), a sticker is sold at different places. It has to be stated, however, that the sticker itself is not the product but only serves as a 'payment vehicle' for the ski-track as a product. Travel agencies, hunting intermediaries and shops are typical examples of independent middlemen who pursue commercial targets and therefore are in the market as enterprises. Their main benefit in the distribution of RES products is in the utilization of the prevalent demand behaviour of potential customers and the utilization of existing know-how and 'trust in the competence of the offerer' as an asset (*Vertrauenskapital*).

Other examples of vertical cooperations. For many RES products, it is especially necessary to include organizations that cannot be primarily classified as traders in the vertical organization of distribution. Thus, for several recreational RES products, outlets under public law such as tourist information offices or (mountain) communities were mentioned as distribution helpers or intermediaries. Examples of this are as follows:

- offers of holiday flats via tourist infomation offices (DE02).
- Offers of seminar programmes via the catalogue of an educational organization (DE17).
- Offer of guided tours in a nature park (with an entrance fee) by nature guides who work for the park on the basis of an advertised contract (IT28).

Very complex systems of vertical cooperation can be found in the comprehensive recreational services with material aspects (PG 5.13 and 5.2). Here the chain of the service-determined products and product mix leads to the end-consumer via one or several, partly commercial and partly non-commercial institutions. For example, the stays at the Erhorn forest youth hostel (DE27) are distributed via the SDW as a non-profit association and outdoor events (DE10, 18) are offered via a consulting company.

Provided that direct contracts between a forest enterprise and a (institutional) contract partner are assessed clearly (PG 6.1–6.3), vertical distribution levels are the exception rather than the rule. If the offerer is related to the final user, however, the product 'contract' is on a vertical distribution level. Some peculiarities in which a middleman is between the forest enterprise and the contract partner are described in the following cases:

- The nature conservation organization WWF as intermediary for sponsoring products (IT29).
- The forest enterprise of a drinking-water organization serves as intermediary for private forest owners (DE14).

Also, vertical cooperation with non-commercial intermediaries mainly utilizes the established demand behaviour of the customers, on the one hand, and the know-how, image and confidence in the competence as an asset of the intermediaries, on the other.

Horizontal coordination of distribution. Horizontal coordination of distribution can generally be related to horizontal cooperation in the production process. Therefore, it is part of the product policy. The approaches for cooperation mentioned in the RES case-studies are summarized in Table 4.12.

A typical example of how the influences of a limited area and financial capacity can be overcome is the offer of certified mushrooms by a large consortium of forest owners and traders (IT02). Another interesting form of horizontal organization of an offer, in which considerable effects of synergy are utilized for the distribution of different products, are the events and markets described (in DE13, 16 and AU15).

First, typical recreational products are indicated in Table 4.13, such as the offer of a ski-track network by several forest owners (AU13). In this way, the necessary area size is secured by cooperation. However, it also becomes clear that horizontal cooperation, for example with organizations in the offer of management seminars (DE10), contribute the necessary specialized know-

Table 4.12. Horizontal distribution structures with the sales of material goods.

Supply of material goods	Cases
Certified mushrooms and certified timber by area-wide cooperation	IT02, IT13
Via events and markets	DE13
Christmas trees together with other suppliers	DE16
Game in cooperation with city administration	AU15
Chestnuts with certification of origin by a large cooperation of different offerers	IT27
Game, sold by two brothers	DE06

Table 4.13. Horizontal distribution structures with the sales of recreational products.

Supply of recreational products	Cases
Mushroom-picking permits, renting of parking areas, tickets for cross-country skiing, driving permissions on forest roads as result of a cooperation of several forest owners	IT01, IT04, IT08, IT12, IT19, AU13, AU18
Seminars and stays in the forest in cooperation with several independent partners	DE10, DE15, IT22, AU11
Camp-sites and weekend houses in cooperation with umbrella organizations (camp-site owner association, centre for group accommodation, hospital landowners, LKC (*Vereniging Gastvrije Nederlandse Landgoedenen en Kastelen*))	NL14, NL17, NL20
Nature conservation seminars and tree-crown path in cooperation with hotels	NL09, NL13
Seminars for managers on the basis of a frame contract with a consulting company	DE11
Financing of ski-tracks in cooperation between nature park and forest enterprise	DE19
Recreational forests by cooperation of owner and tourist institutions (hotels, restaurants)	IT06
Guided tours of nature by cooperation of park and mountain guides	IT20

how and confidence asset in the particular area. Besides, offerers can develop and expand this know-how more easily with the help of umbrella organizations. The 'communication plan' of the tree-crown path (NL13) explicitly considers business partners as institutions with which mutual offers can be developed to benefit from synergy effects.

With regard to sales of utilization rights (Table 4.14), difficulties arising from area size also have to be overcome for the permission for a network of tracks for mountain-biking (AU17), for example, and/or the utilization of the confidence assets of horizontal cooperation partners, for example in larger sponsoring contracts (DE08).

The multitude of potential vertical and

Table 4.14. Horizontal distribution organizations concerning the sales of utilization rights.

Supply of utilization rights for organizations	Cases
Lease of ski-courses and mountain huts by common forest property	IT25, IT26
Contracts for mountain-bike routes by coordination of several forest owners through forest administration	AU17
Sponsoring of recreation facilities in cooperation between natural park, forest enterprise and regional federation (*Zweckverband*)	DE20
Forest owner association as middleman for contracts for carbon dioxide storage	NL15
Products for environmental sponsoring by a non-profit association	DE08
Contracts for environmental protection via federation of forest owners	AU10

horizontal sales organizations becomes clear from the preceding sections. In the following, however, two special aspects of distribution policy will be outlined more closely.

INFORMAL CONTACTS AND PARTICIPATION OF SPONSORS IN THE PRODUCT OFFER. In general, informal contacts are of great importance for the distribution of products. This is especially the case for new products entering the market. Informal contacts with customers or with 'opinion leaders' can basically be orientated in a vertical as well as a horizontal direction. The case-studies referred to in this section mainly deal with contacts with cooperation partners. It is possible to expose some connections derived from the networks of informal contacts (mainly personal acquaintances) helping to establish the distribution method. In 50% of all investigated RES cases, hints about informal contacts could be found. It was noticed that a considerable share were mentioned for integrative recreational products (PG 5.13 and 5.2) and for sponsoring products (PG 6.3). Besides other opportunities, contacts could be realized in the following ways:

- membership of the tourist agency advisory board, the German Forestry Advisory Board (*Deutscher Forstwirtschaftsrat*), the regional Chamber of Commerce, the forest owner association.
- Membership of an organization for nature conservation.
- Meeting at a hunt or an outdoor rally or in the sports club.
- Other business.
- Family relations.

Efficiency of distribution can be increased by participating sponsors as well. This efficiency, however, is limited to special products of intangible value. The fact that sponsors are mentioned in almost one-quarter of RES case-studies indicates the high ideal value of forests. In this way, mainly non-material, autonomous recreational products (PG 4.1 and 4.2)[66] but also partly tangible, integrative products (PG 5.13) were supported by sponsors.

Concerning the participation of sponsors, the sponsoring products (PG 6.3) certainly have to be underlined. When regarding the services of forests as products, these sponsoring products would be offered 'with the help' of a sponsor in all investigated case-studies of PG 6.3.

Advantages should result from a horizontal and a vertical distribution structure for all parties concerned. In particular, intermediaries promote RES products because of special advantages from this business. The forest enterprise should take these economic or political advantages into account and use them actively for its own purposes. So the distribution system becomes an important success factor.

If an RES product is sold via intermediate trade enterprises, their marketing goals have to be included in the total marketing concept. The utilization of intermediate trade is a success factor for some special material and recreational RES products. Non-commercial intermediates are very important for most RES products. This is particularly valid as concerns recreational forest products with positive effects for regional tourism.

Horizontal distribution cooperation is an important success factor for many RES products. The offer of some RES products is not possible without large forest area associations. This applies especially to recreational forest products and use contracts, but sometimes also to material RES products. Material products can often be sold particularly effectively at market events in cooperation with other offerers. Recreational products profit from the know-how of horizontal distribution partners from the tourism business. Environmental conservation contracts are sometimes arranged by non-commercial organizations.

Informal contacts are also important

for the success of RES offers. In particular, recreational forest products and all contract products have an advantage from those contacts. A high non-material social value of forest can be used through the involvement of sponsors in offering the RES product.

The point of purchase (POP)

After the previous thoughts on the design for an optimum distribution organization, the (physical) point of purchase will be mentioned at the end of this section. The offerer basically has to decide where the optimum location for his/her enterprise will be.[67] As forest enterprises are stationary, in general this location cannot be chosen freely, however. Generally, the question of where to place the point of purchase goes hand in hand with the important question of marketing logistics.[68] This issue is especially important in the food sector but costs for logistics also account for a significant share of turnover regarding wood products, for example (Weis, 1997, table p. 351). Concerning material RES products, it is necessary to plan precisely how to organize the storage of products and the transport (either in self-direction or when run by others), as well as the lot size in order to achieve optimum customer orientation at minimum costs.

With the physical distribution of services, somewhat different problems arise, which already result from the non-materiality of these offers. When purchasing a service, a differentiation principally has to be made between the place where the service is effected and at the same time consumed, and the place at which the contract obligation, for example represented by an entrance ticket or a booking slip, is acquired. The in-depth survey on the opening hours of the Drents tree-crown path, as well as the infrastructural issues of parking and transport facilities for visitors of this recreational facility (NL13, sections 6 and 7 of the communication plan), outlines the significance and the connection between product logistics and product design concerning non-material RES products.

In the RES case-studies, statements were asked for on 'special places of demand' for the particular RES product. Besides the purely physical place, the answers contain revealing information on the organization of distribution.

In about 40% of RES case-studies, statements are made on specific locations from which it can be assumed that the forest enterprise itself is mostly categorized as the regular place of purchase. Special points of purchase for material products for further processing (PG 2.2), for complex recreational services for final consumers (PG 4.1) and for all integrative recreational products with material features (PG 5.1 and 5.2) were indicated especially frequently. Obviously, in the RES cases investigated, special points of purchase were scarce concerning contracts on sponsoring (PG 6.3) or environmental facilities (PG 6.2), as well as simple non-material recreational products (PG 4.2).

Sales points and special points of purchase for material RES products. Material consumer products, such as meat or gifts, can also be marketed in any supermarket. RES products, however, mostly use different methods. The enterprises interviewed made the following statements on their tangible products:[69]

- Direct selling from the farm or from the forest mainly concerns products such as game, Christmas markets and Christmas trees, gifts and certified (investment) goods, such as timber and chestnuts (PG 2.1 and 2.2b).
- Concerning the marketing of mushrooms with certified origin (IT02), there is the alternative of 'picking-your-own' besides selling ready-picked.
- Selling in a local shop was only done with timber of certified origin (IT21). Here, it has to be considered that distribution takes place on two levels, since selling of the raw wood is generally done in the forest.

In 71% of the material products for further processing (PG 2.2), specific locations of demand were mentioned. These statements point to either a well-established distribution system or at least at highly targeted selling locations. For example, a Christmas market is suitable for the marketing of game (DE06).[70] The demand for Christmas markets and Christmas trees (DE13) is mainly directed to the office of the forest district. In contrast to this, the demand for resonance wood (IT21) is directed to the local processors (carpenters).

Some of the interview partners gave a customer description instead of indicating the point of purchase.[71] Furthermore, often no clear distinction is made between the main sales areas (e.g. forest enterprise–restaurant) and further sales areas (restaurant–guest). In this way, the points of purchase in several cases can be understood as the first institutional customer for the product, on the one hand, and also as a place of resale to the final consumer, on the other. In detail, these were hotels and catering companies (NL04 and 11), organic farms (NL08), butchers (NL11) and local restaurants (IT27). This is an indicator that the offerers of RES products often do not yet consider the sales tasks as a differentiated tool of a strategy, but as a given result of the general sales effort.

SPECIAL MARKETS FOR NON-MATERIAL RECREATIONAL PRODUCTS AND CONTRACTS. Special locations of demand for recreational products (PG 3.1–5.2) were very often stated for complex consumer services (PG 4.1) and for all integrative recreational products with material aspects (PG 5.13 and 5.2). In general, these statements refer to places at which the contract obligation to execute a service can be acquired, as the service itself has to be executed on the spot.

In Table 4.15, the specific locations of demand mentioned in the case-studies are listed. The stated markets are categorized according to their institutional importance. Public (administrative) and tourist institutions (hotels, restaurants) are regarded as important groups. The group of other intermediaries is broadly spread rather than concentrated.

Table 4.15 indicates the significance of public institutions as markets for recreational forest products. Tourist information

Table 4.15. Special locations for demand for recreational forest products.

Market	Location of demand	Cases	
Public bodies	Tourist information agencies	14	DE02, 12, 17; IT10, 11, 19, 20, 28; AU05, 11; NL10, 13, 18, 20
	National park offices, office of mountain guards, natural park office	2	DE19, IT20
	Community, school office, schools	4	IT11, 22; AU02, 21
	Office of the forest district	1	NL05
Total		21	
Accommodation and restaurants	Hotels, boarding-houses	8	DE15, (17), 19; AU02, 05, 13, 21; NL13
	Camp-sites, tourist camps	3	NL05, 10, 13
	Restaurants	2	DE19, AU13
	Youth hostels	1	DE17
Total		14	
Retail trade	Sports shops	3	DE19; AU13, 02
	Motorway petrol stations, post office	2	IT19; AU11
Clubs	Horse-riding schools and mountain-bike clubs	1	NL07
Intermediaries	Travel agencies	1	NL10
	Office of an agent for hunting	1	DE04
Exhibitions	Exhibitions	1	NL13
Total		9	

agencies have the major share, but hotels and restaurants were also mentioned as important locations of demand. Recreational forest products therefore seem to be of regional importance for tourism, and synergy effects can be used for optimum distribution. Concerning other intermediaries, sports shops are important locations of distribution for sports-related offers, such as ski-courses or mountain-bike routes.

Commercial intermediaries, such as travel agencies, were only mentioned in two cases as special locations for demand. These institutions, which are very important in the 'normal' recreation business, still play a minor role in the offers of the RES case-studies investigated. Obviously, this method of distribution has been neglected and could be a new opportunity for improved distribution of specific RES products.

Since recreational forest products are often sold at the customer's place, it might be difficult to view the point of purchase and the customer separately. This is especially the case for the complex recreational products which are usually demanded by institutions (PG 5.13 and 5.2). In the following examples, the specific point of purchase resembles the customers of the respective recreational products:

- schools, clubs and groups (DE27, 17, AU02 and 21).
- Companies and worker organizations, planning excursions (NL10, DE15).
- Consulting companies (DE18).

While the demand for mushroom-picking permits (IT19) at tourist information agencies indicates the importance of these agencies for the distribution of such products, the same products might be offered at post offices. This points to a well-established distribution system. A customer will, however, only ask for a mushroom-picking permit at a post office if he/she knows that he can buy it there.

SPECIFIC POINTS OF PURCHASE FOR INTEGRATIVE CONTRACTS. The definition of integrative contracts includes contracts on the utilization of forests for recreational or other purposes, contracts on environmental services and sponsoring contracts (PG 6.1–6.3). In the case-studies, the following statements were made concerning specific locations for the demand of utilization rights for recreational purposes (PG 6.1a):

- sports shops, hotels, farms, tourist information organizations, sports clubs, companies (AU01 and 17);
- Car parks in the forest, at hotels, at tourist information offices (DE23).
- On the ski-tracks (DE25).

These statements refer to the locations of demand of the end-users of contract contents but not to the institutional contractees.[72]

Environmental products do not usually have end-users who can be directly determined. The institutional, mostly state, customers purchase them for the welfare of the public. As a result, no specific locations of demand are ascertainable. This is largely the case for sponsoring products, as well.

In general, the point of purchase for RES products cannot be chosen freely by locally limited forest enterprises. The highest flexibility in this regard exists for material products. However, material RES products are largely sold at the forest enterprise itself or on local markets. Multistage trade structures are rarely used. In the area of recreation, special points of purchase for the contract obligation exist particularly for complex recreational forest products. Public (especially tourist information offices) and commercial tourist institutions (especially hotels and restaurants) are important places in demand here. The utilization of these special markets contributes effectively to the success of RES products. Some products, in particular simple recreational forest products, are also demanded at the facilities of the retail trade. The point of purchase for contract products is generally is the forest enterprise itself.

The rare hints as to innovative points of purchase can lead to the following hypothesis: in many cases, offerers of

RES products do not yet view distribution tasks as part of a differentiated marketing strategy but as a given result of common selling efforts.

4.3.3 Price policy

General basics of price policy

The target of price policy as a part of the marketing mix is to optimize product prices and terms of payment in view of the marketing strategy selected.[73] It can be subcategorized into direct and indirect price policy. Direct price policy deals with the prices a certain product is offered at. Its subcategories are pricing, price changes over time and differentiation of prices according to temporal, local, customer-related and other aspects. The area of indirect price policy comprises means of price formation, such as discounts, conditions of delivery and terms of payment.[74] An important aspect of indirect price policy, moreover, is the optimal type of payment and special price offers connected with it in order to make a product more attractive by designing it as a compact offer.

The intensive use of price policy instruments points at an early stage of marketing.[75] Hence their application on a broad scale can be recognized in the RES case-studies as well.

Prices can be changed flexibly without generating direct costs. The intensive use of price policy means points at the early stage of marketing for RES products. Discounts and the conditions of payment can support the success of RES products. The strategy of a price battle is extremely risky for a small enterprise and should be avoided by forest enterprises.

Price determination

The basis of calculation applicable to determining the price of RES products is illustrated in Fig. 4.6.[76] Prices determined by the offerer according to economic aspects can thus be orientated along the lines of internal factors, such as product costs[77] and profit expectations, as well as external factors, such as the solvency and willingness to pay of the buyers[78] or the prices of competitors.[79] Generally, the optimum solution lies in the combination of different approaches for price setting. Thus, costs should expose the lower price

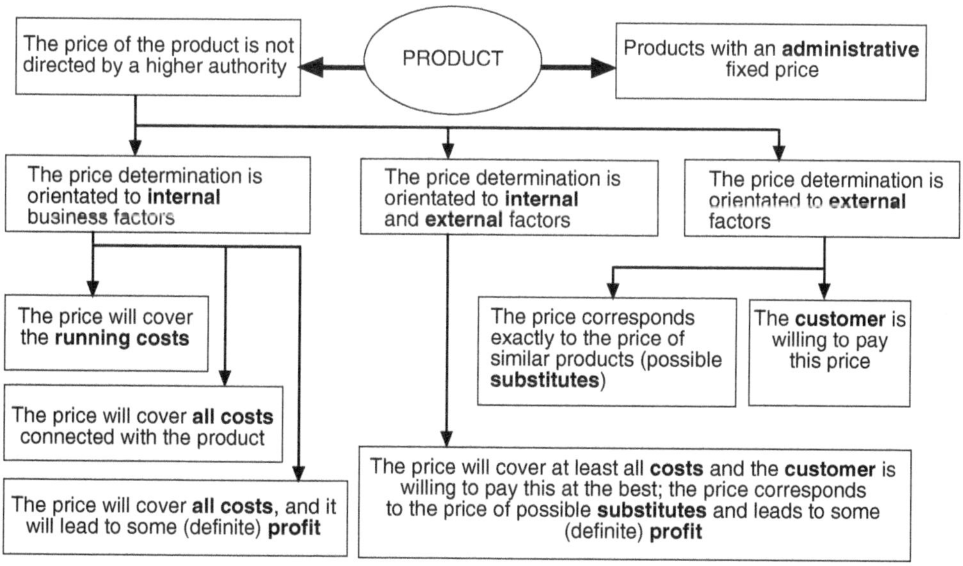

Fig. 4.6. Possible approaches of pricing.

limit, and willingness to pay (after making use of all suitable communication techniques) should be the upper price limit. The formation of a reference price from the prices of competitors' substitutes is in this respect, an important tactical orientation.

Infrastructural facilities of the forest in the area of environment and recreation have been subject to various evaluations according to environmental economic methods.[80] In the organizational and administrative area, these methods of evaluation can be used to derive values for developing theoretical balances or levels of compensation.[81] However, it is the task of marketing to positively influence valuation through strict customer orientation and to utilize prices not as the result of a market model, but as a sales instrument that can be designed actively. Therefore, environmental economic evaluation methods for pricing an RES product will not be referred to in further detail here.[82]

A product price should be stipulated on the basis of the marketing strategy selected and of product positioning considering the relations within the product mix. If a high price suggests to the customer a high quality (a 'price-orientated' purchase pattern in contrast to a 'price-conscious' purchase pattern), the price of a product also has a communication-political impact. Besides, an enterprise certainly needs to realize its financial goals through an adapted price policy (Kotler and Bliemel, 1995, pp. 748 and 749).

For the sale of consumer goods, a psychologically determined 'fine-tuning' of prices can be used effectively, thus exceeding the basic determination of prices. The most widely known instrument for this is keeping prices below an even figure, but also the formation of larger sales units and discounts for bulk purchases. In this way, customers perceive products for DM 1.99 to be cheaper than those for DM 2.00, and three cans of a beverage costing DM 1.45 as cheaper than a single can for DM 0.48. These measures of direct pricing policy, however, are scarcely important for the sale of goods and services to institutions, the purchasing processes of which are more rational.[83]

APPROACHES FOR PRICE FINDING IN THE RES CASE-STUDIES. Table 4.16 shows the basics of price setting within the RES case-studies. In about 32% of cases, the optimum method of price setting took a price that was agreed on externally as well as

Table 4.16. Basics of stipulating prices within the RES case-studies. (Question 4.4.2 (modified): Which factors were taken into account when setting the initial price?)

Price determination	*n*	%
Price to reduce the demand	1	1.0
Administratively fixed price	4	4.1
Purely cost orientation	15	15.3
Cost and profit orientation	10	10.2
Purely customer orientation	7	7.1
Purely competition orientation	1	1.0
External market price	23	23.5
External and internal determination	31	31.6
Spontaneous price	1	1.0
Missing	5	5.1
Total	98	100.0

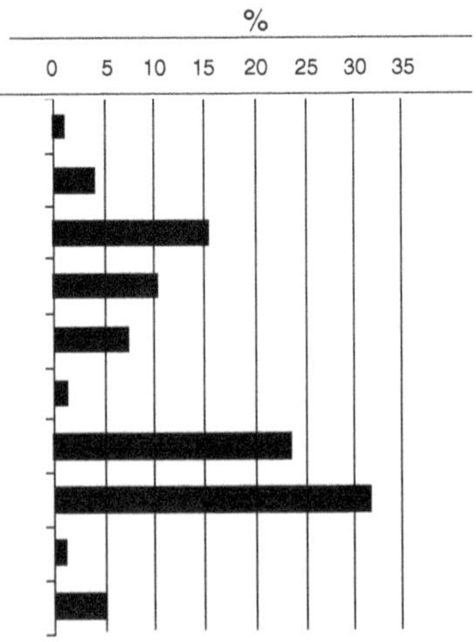

Table 4.17. Basics of stipulating prices according to product groups.

Price determination	Material products (PG 1.1, 2.1, 2.2)		Complex recreational products (PG 3.1, 4.1, 5.13, 5.2)		Simple recreational products (PG 3.2, 4.2)		Integrative contract products (PG 6.1, 6.2, 6.3)		Total	
Purely cost-orientated	1	6%	6	22%	5	19%	3	11%	15	15%
Cost- and profit-orientated	0	–	1	4%	4	15%	5	19%	10	10%
Purely customer-orientated	0	–	0	–	3	11%	4	15%	7	7%
Purely competition-orientated	0	–	0	–	1	4%	0	–	1	1%
External market price	8	47%	7	26%	4	15%	4	15%	23	23%
External and internal determination	4	24%	11	41%	7	26%	9	33%	31	32%
Spontaneous price	0	–	0	–	0	–	1	4%	1	1%
Price to reduce the demand	0	–	0	–	1	4%	0	–	1	1%
Administratively fixed price	1	6%	2	7%	1	4%	0	–	4	4%
Missing	3	18%	0	–	1	4%	1	4%	5	5%
Total	17	100%	27	100%	27	100%	27	100%	98	100%

internally. External market prices orientated towards the prices of the competition and the customer's willingness to pay play an important role (24%) as well. Together with the exclusively customer- and competition-orientated prices, almost two-thirds of all prices are externally orientated. Only about one-quarter of prices are cost-orientated. The remaining 10% of recorded cases show mechanisms of price finding that are not 'market-like': either administratively determined prices, 'spontaneous' prices, price for lowering demand or a price determination without a defined basis.

Table 4.17 shows the basics of pricing again, according to product groups. Pricing for material products (PG 1.1–2.2) was mainly conducted by the comparison with market prices of comparable products and therefore was mainly external, i.e. consumer- and competition-orientated. In five cases, consideration of internal costs was also mentioned. Concerning these costs, however, it cannot be stated whether these costs were not considered in the other cases or whether they were naturally regarded as at a lower price limit and were just not mentioned within the framework of interviews. In the case of water-power (AU03), with an administrative given price, and drinking water (NL01), with a purely cost-orientated price (loss of forest yield as a scale for compensation payment), special types of price setting can be observed.

Many of the recorded recreational products (PG 3.1–5.2) consist of services to a higher or lower degree. Meffert and Bruhn (1995, pp. 306 and 307) point at the particularly high share of fixed charges in the service area, which are mostly indirect costs. These can only insufficiently be related to specific products, usually according to a time-related distribution ratio. It is also very difficult to estimate the willingness to pay in this field, as the comparability between services and the integration of the customers in the production of the product are limited. Due to the degree of integration of service products, it is also difficult to determine uniform prices. Generally, this problem is solved by determining a single price after negotiation with the customer on the basis of price scales.

The recorded recreational products of product groups 3.1–5.2 frequently show an external and internal chosen basis of price determination or at least a strong orientation towards market prices of comparable products. However, in 11 cases,[84] the price is exclusively orientated towards production costs. The highest diversity of price determination can be found when looking at simple recreational products (PG 4.2). Here, a price for lowering demand and the strong cost orientation of many products are worth mentioning.

In the case of integrative contracts on forest utilization, environmental protection and sponsoring (PG 6.1–6.3), the share of market-orientated prices was much higher than the share of cost-orientated prices. The share of product prices (five cases) which were, to a large extent, fixed by the customers is relatively high, too.[85]

> The prices of initial offers should be determined under consideration of the enterprise's own production costs, the willingness to pay and the prices of substitutes. Environmental-economic valuation techniques are only suitable to a limited degree for the price finding of RES products. Prices need to be orientated towards the chosen marketing strategy. If entry prices are too high, though, they risk the success of a product, but subsequent price increases might be difficult to apply. Prices for consumer goods often make use of emotional price perceptions, whereas institutions judge prices in a more rational way.
>
> RES products are mainly offered at market-orientated prices. However, there are also purely cost-orientated prices or prices stipulated by the customers. This hints at 'non-profit offers' of RES products.

Time-related price changes, adaptations and discounts

This section describes all direct and indirect means of price policy that are related to time. For this reason, besides other issues from the RES case-studies, the question of the structures of payment and pricing of the offers was considered, as well as whether, and if so why, the price was stipulated after the introduction of the product.[86]

Price changes and price adaptions can basically happen actively, that means within the frame of marketing strategy, or as reaction, that means to compensate market changes.[87] Active price adaptations are long-term price adjustments, such as introduction prices for a limited period or the granting of discounts for the same service but at different times. Such temporal differentiations of prices occur especially in the service sector, on the one hand, to remunerate the disposition stability of the supplier by ordering longer in advance or in the sense of capacity utilization, with booking at short notice (last-minute booking). On the other hand, under this heading seasonal price differentiation also has to be understood, this being done to support supply in times of low-capacity output. Holiday houses that are rented at different prices depending on whether it is the main season or off-season are a typical example of seasonal pricing.[88] Reactive adaptations of prices can, on the one hand, be calculated on a long-term basis as a result of normal changes in the product environment, of which the adjustment of prices to inflation is the most important example. On the other hand, changes in price at short notice may be required as a reaction to a current reason, such as the change of a legal regulation or unexpected, profound market changes.

PRICE CHANGES IN THE RES CASE-STUDIES. Table 4.18 shows what price changes could be observed in the RES case-studies. Concerning price changes for material products, while price changes were not observed for consumer products (PG 2.1), prices for products for further processing (PG 1.1 and 2.2) were adapted to external conditions, partly as a 'normal' and partly as a 'concrete' reaction. In the case of the offer of electricity from water-power (AU03), there was even an active price change. This consists of seasonal prices for the electricity sold in winter, when the demand is high. In two cases of eco-products (NL08 and 11), as well as in the case of certified chestnuts (IT27), the price was adapted due to a specific occasion. Concerning the remuneration of drinking-

Table 4.18. Price changes within the RES case-studies.

Price changes	Material products (PG 1.1, 2.1, 2.2)		Complex recreational products (PG 3.1, 4.1, 5.13, 5.2)		Simple recreational products (PG 3.2, 4.2)		Integrative contract products (PG 6.1, 6.2, 6.3)		Total	
No changes	6	35%	11	41%	6	22%	18	67%	41	44%
Normally reactive	4	24%	3	11%	5	19%	2	7%	14	15%
Concretely reactive	4	24%	3	11%	11	41%	6	22%	24	26%
Active	1	6%	10	37%	4	15%	0	–	15	16%
Missing	2	12%	0	–	1	4%	1	4%		0%
Total	17	100%	27	100%	27	100%	27	100%	94	100%

water from forests (NL01), the possibility of offering a product at a certain price when it had formerly been free of charge was the specific occasion for price adaptation.

The introduction of a price for a formerly free offer as a special reactive price change is typical of some RES products. Table 4.19 gives an overview of the percentage of case-studies in which the product was formerly offered free of charge. This shows that about 25% of recorded products were formerly offered free. This concerns material as well as recreational products, and the following could be determined concerning their price changes.

In the field of accommodation and partly tangible complex services (PG 3.1), prices were actively changed in 50% of products. This can be explained by the seasonal dependence of these offers. In only one case – after a change of owner of a camp-site (NL14) – was a price change related to a specific occasion. In the case of mushroom-picking permits (PG 3.2), the specific occasion for the price adaption was the opportunity to offer the product for a charge when it had formerly been free of charge. The percentage of non-material complex services (PG 4.1) for which the price was actively changed is very high, too. An active price adaptation was, for example, mentioned for the eco-park (AU05).

Eighteen per cent (PG 3.1b and 4.1) and 11% (PG 5.13 and 5.2) of the complex recreational products are based on formerly free offers. These consist of two guided mountain tours (IT10 and 20) and forest excursions (DE09). In this case, it is worth mentioning that it was possible to demand a price without essentially changing the product. This indicates that the described guided forest tours are so unique – presumably due to their content – that demand is inelastic up to a certain price.

About half of the simple recreational products (PG 4.2) were formerly offered free of charge and therefore show an adaptation towards a special occasion or an additional active price change if seasonal prices are offered. This high amount is an interesting basis for the investigation of the necessary product transformations.

Concerning contract-based recreational products (PG 6.1a), in about half of the

Table 4.19. Share of RES products that were formerly offered free.

Formerly offered for free?	*n*	%
No	72	73.5
Yes	24	24.5
Missing	2	2.0
Total	98	100.0

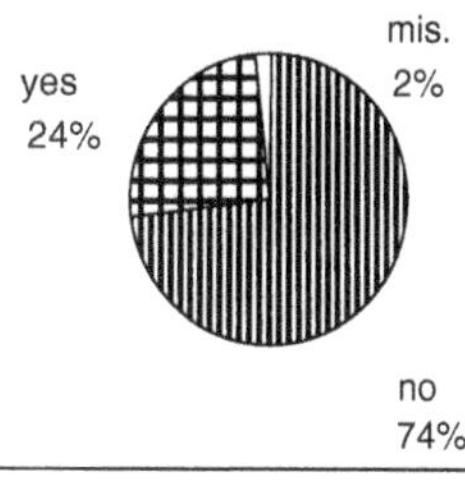

cases a price reaction is mentioned as a reaction to a specific occasion. These cases refer to products which were formerly being offered for free. In the range of environmental and sponsoring products (PG 6.2 and 6.3), almost no price change could be determined. For only one sponsoring case (NL12) was a normal price adaptation as a reaction mentioned. In none of the described cases was the product formerly offered free of charge.

TIME-RELATED PRICE DISTINCTIONS. A time-dependent price differentiation could be regarded as an active price change that returns periodically. In the following description, it will be made clear how often a time-related price differentiation could be determined in the 98 case-studies (Table 4.20).

First, it can be derived from Table 4.20 that time-related price distinctions are applied by means of seasonal prices only. The time relation of the term season is highly variable, though. Seasonal prices for holiday houses, leisure parks or guided tours, including higher weekend and holiday prices, are typical examples of a temporal differentiation of prices in the recreational area. Concerning those offers in which maximum demand is dependent on both season and time of the day (e.g. skiing courses), it is possible to develop very differentiated price systems, which, however, can turn out to be increasingly counterproductive with decreasing price transparency.

In the RES case-studies, such seasonal prices are mainly described with regard to offers of overnight stays (PG 3.1a) and complex recreational products (PG 3.1b and 4.1). Concerning these product groups, there are seasonal prices in about 50% of all recorded cases. Moreover, this instrument of price policy was applied to about 15% of simple recreational products (PG 4.2). The sale of energy from water power (AU03, PG 1.1) at seasonal prices has already been referred to above.

> Prices can either be changed actively or as a reaction to external occasions in the course of time. Thus, some RES products which were formerly offered free are nowadays marketed for a price due to changed frame conditions. Regarding the fact that it is nowadays possible to charge for guided forest tours were formerly offered free of charge indicates the high demand for this kind of product. Prices for RES products are only rarely actively changed. Time-related price distinctions are, in most cases, supposed to compensate for fluctuations in demand and are therefore an important marketing tool, especially for services. Seasonal prices are an important factor of success for many RES products. In addition to forest services, some material RES products can be offered at different seasonal prices.

Other price distinctions and discounts

The term price distinction in relation to time has already been introduced in the previous section. A general definition is as follows: 'A price distinction defines the charge of different prices for the same tangible or intangible performance' (Meffert, 1986, p. 386).

Table 4.20. Time-related price distinctions in the offer of RES products.

Time-related price distinctions	*n*	%
No distinctions	60	61.2
Seasonal prices	14	14.3
Missing	24	24.5
Total	98	100.0

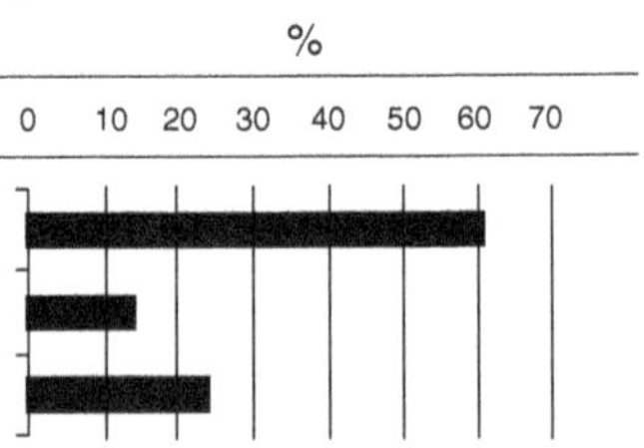

Price distinctions basically pursue two goals: on the one hand the utilization of different market segments at different prices and thus the greatest possible usage of market potential and, on the other, particularly in the service sector, the regular load of production capacities in order to avoid idle capacity costs (*Leerkosten*) (Corsten, 1988, in Meffert and Bruhn, 1995, p. 309). The first-mentioned goal is especially interesting for the field of sales support. Price distinction can be made according to the following criteria:[89]

- Differentiation according to time: this point has already been referred to above.
- Differentiation according to area: e.g. higher prices for selling in urban rather than in rural areas or for different agency areas; here cost differences are passed on or different levels of purchasing power or prices are made use of.
- Differentiation according to customers: single customers in contrast to institutional customers (trade discount), according to age, sex, social level (senior cards, discounts for locals).
- differentiation according to purchase amount: e.g. discounts for institutions, permanent cards, group cards.

When combining several groups of differentiation, overlapping segments have to be regulated conclusively (group tickets for senior citizens), to maintain price transparency.

Table 4.21 summarizes the extent to which such measures of price policy could be recognized in the RES case-studies. In almost half of the recorded case-studies, other (not time-related) price differentiations with RES products were mentioned. In about 20%, these marketing tools were applied quite intensively in different combinations. This concerns almost exclusively the material and recreational products and only in two cases integrative contracts.

The price distinctions most frequently used were made according to amount and customer. Since it does not seem to be reasonable to further differentiate the instruments, the discounts mentioned in the case-studies are included in the following description (Table 4.22). Prices differ according to the amount in barely 30% of all case-studies. Price differences dependent on the customer are made in about 30% of all case-studies. Price distinction according to area was mentioned in only three cases. In at least seven cases, approaches for discounts on services could be determined. Concerning the product groups and case-studies affected by this kind of price distinction, the following can be noticed.

Approximately 60% of material consumer

Table 4.21. Utilization of (not time-related) price distinctions according to product groups.

		Stressed		Mentioned		No		Missing	
Price distinctions	Cases	*n*	%	*n*	%	*n*	%	*n*	%
Material products (PG 1.1, 2.1, 2.2)	17	2	12	3	18	10	59	2	12
Complex recreational products (PG 3.1, 4.1, 5.13, 5.2)	27	7	26	8	30	8	30	4	15
Simple recreational products (PG 3.2, 4.2)	27	12	44	11	41	3	11	1	4
Integrative contract products (PG 6.1, 6.2, 6.3)	27	0	0	2	7	10	37	15	56
Total	98	21	21	24	24	31	32	22	22

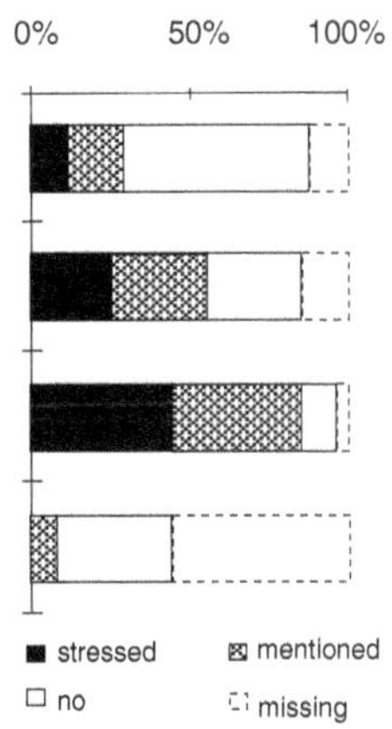

Table 4.22. Share of RES products whose prices vary according to amounts and customers.

	Yes		No		Missing		Total	
Price distinctions	*n*	%	*n*	%	*n*	%	*n*	%
According to the amount	27	27.6	47	48.0	24	24.5	98	100.0
According to the customers	31	31.6	45	45.9	22	22.4	98	100.0

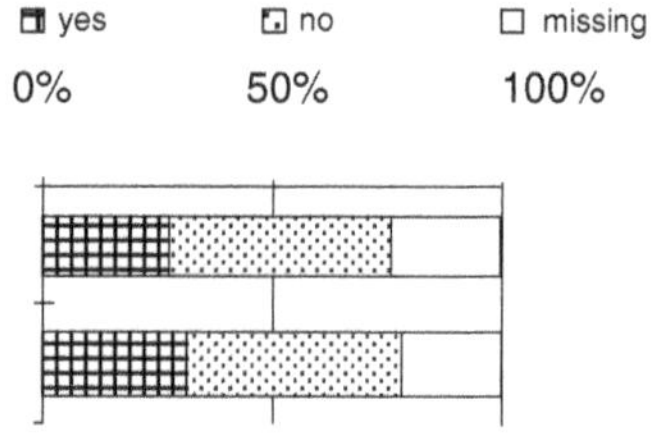

products (PG 2.1) and eco-investment products (PG 1.1 and 2.2a) are offered without further price differentiation or possibilities of discount. In the following five cases, such measures were utilized:

- Differentiated prices according to area and customers (dehydrating companies, feed suppliers) in the case of eco-hay (NL08).
- Prices vary according to customers (private customers, butchers) and amount in the case of eco-meat (NL04).
- Prices are only differentiated according to amounts in the case of drinking-water (AU19).
- Discounts on performance were mentioned in the cases of game (DE06), eco-hay and eco-meat (NL04, 08 and 11).

Thus, it becomes obvious that particularly the eco-products in the Netherlands work with these instruments of sales support.

In the field of recreational products, price differentiation is particularly and strikingly used in the following three product groups. In the group of complex consumer products (PG 3.1b and 4.1), prices differ up to 60% according to the customer and up to 55% according to the amount. In five cases[90] of this group, the price structure can be regarded as especially highly differentiated. In all of these cases, the price is differentiated according to the buyer as well as the volume of distribution. The second group is the mushroom-picking permits (PG 3.2). Here prices are differentiated either according to area and customers (IT01), to customers and amount (IT08) or, in one case, exclusively to amount (IT19). The third group, finally, includes the simple recreational products (PG 4.2). Here, price differentiation is mentioned in more than half of the case-studies according to customer or to amount. In ten cases[91] within this group, these two instruments are combined. In the case of the Taufstein ski-tracks (DE19), a discount on performance for middlemen can be recognized.

Of the group of offers for overnight stays (PG 3.1a), the case of accommodation and hunting in a count's forest (DE28) should be singled out. In this case, discounts according to amount and customers' performances are granted. Such a performance discount can also be claimed in the case of a forest youth hostel (DE27, PG 5.13). Thus, this instrument has been mentioned in three cases altogether.

However, fundamental price differentiation could barely be noticed concerning contracts on utilization (PG 6.1) as well as environmental and sponsoring products (PG 6.2 and 6.3). Here, a price differentiation according to the area can only be found in the case of mountain-bike courses (AU01).

Due to high fixed costs in forest enterprises, the importance of price differentiation arises in order to achieve permanent profit margins. Besides the time-related price differentiation, price differences according to customers and amount are particularly important for RES products. Price differentiation and discounts according to customers and amount are particularly applied concerning material eco-products, which are often purchased and further processed by institutions. For most RES recreational products, price differentiation according customers and amount is an important factor of success. This, however, is limited to those cases in which offers are not individually negotiated and designed. Contract products are hardly suitable for price differentiation.

Types and schedules of payment

Besides the type of payment for a product, the purely technical means of transferring money can become a component of price policy as well. Simplified payment transactions could thus have a positive canvassing impact, for example. An RES example is the payment for a mushroom-picking permit at any post office, as is the case in the Italian case-study IT19.

The type of payment will be understood here as the content aspect of payments for an RES product related to the basis of calculation. Table 4.23 describes the various types of payment that are used in the RES case-studies in a logical structure.[92]

The conceptual background of the categorization in Table 4.23 can be simplified, as follows. Calculations of money are based on costs through tangible consumption or consumed time and/or external parameters. Calculations for products or services can thus be variable or stipulated according to a fixed rate.[93] Products can be handed over, consumed or used either once or several times. A non-material proof of the money value (entrance ticket, etc.) frequently is necessary for the control of single usage of a non-material facility and it is always necessary in the case of repeat utilization. These aspects in turn can be combined optionally with one another and, thus, can be sold at bulk prices.[94] Moreover, the payment can be obligatory or voluntary (donation).

Often contracts containing several aspects of the above-mentioned basics of payment are completed with institutions on the utilization of complex performances. A special type of payment is services in return – for example, the help of pupils in forests in return for free accommodation and meals in forest youth hostels (DE27).

Table 4.23. Types of payment in RES case-studies

Variable calculation in the special case
Calculation according to degree of consumption
Calculation according to duration of service
Payment for defined single performance
Tickets and permits for single use
Payment of a fixed amount for a defined service
Payment for repeated or various performances
Duration permits (licences)
Tickets for several times of entry and various services
Bundle price for service-determined combinations
Pure bundling (all-inclusive)
Mixed bundling (divisible)
Payment via club membership
Voluntary acquisition of licences
Payment via subscription
Voluntary agreement
Lease
Donation
Sponsoring
Payment through quid pro quo
Combined payments
Individual contract and single calculation
Individual contract and profit participation
Duration of service and degree of consumption
Payment for services and club membership
Allowance for utilization and club membership

Table 4.24 shows the degree to which these types of payment are applied to the investigated products in the case-studies. The large variety of possible types of payment becomes quite obvious from Table 4.24. However, there is a basic connection between the type of payment and the product, since every type of payment is limited to certain products.

Concerning material products (PG 1.1–2.2), calculation and payment according to degree of consumption are dominant, as expected. In one case of a Christmas market (DE13), performances are basically calculated as a general bundle price. Moreover, particular conditions of payment were mentioned here. Such conditions of payment can be regarded as the

Table 4.24. Share of types of payment in the RES case-studies.

Type of payment	*n*	%
Calculation acc. to the degree of consumption	16	16.3
Calculation acc. to the duration of service	14	14.3
Tickets and permits for the single use	15	15.3
A fixed amount for a defined service	1	1.0
Long-term permits (licences)	5	5.1
Long-term (entry) tickets and various services	5	5.1
Pure bundling (all-inclusive)	4	4.1
Mixed bundling (divisible)	1	1.0
Voluntary licences	1	1.0
Voluntary agreement	8	8.2
Lease	3	3.1
Donation	2	2.0
Sponsoring	8	8.2
Payment through quid pro quo	1	1.0
Combined payments from		
Indiv. contract + single calculation	3	3.1
Indiv. contract + profit participation	2	2.0
Duration of service + degree of consumption	4	4.1
Payment for services + club membership	2	2.0
Allowance for utilization + club membership	2	2.0
Missing	1	1.0
Total	98	100.0

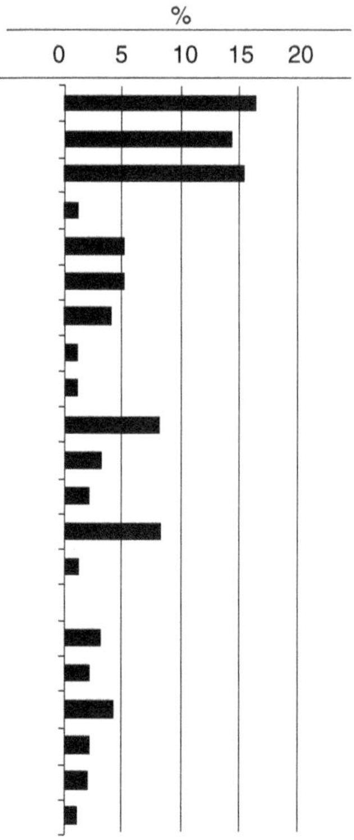

completion and refining of the type of payment. The share of RES case-studies in which particular conditions of payment appeared was approximately 4% in total. Concerning the sale of drinking-water (AU19), a long-term contract is the basis for single calculations.

Whereas for material products types of payment are characterized by their material value, recreational products are characterized by their non-material service orientation. Concerning the sale of services, there are different peculiarities of condition policy. Payment by instalments might be agreed for complex services. When a stipulated service is not used, agreements on the share of payments that have to be paid nevertheless need to be made. By specific conditions of delivery, it is easier for the customer to survey the service paid for. Agreements on guaranteed performance results – which need, of course, to be calculated according to their risk costs for the offerer – can also be an efficient instrument of targeted marketing (Meffert and Bruhn, 1995, p. 308).

The types of payment for recreational products in the RES case-studies are exceptionally numerous. In many cases, the payment was dependent on the duration of the service (PG 3.1, 4.1, 4.2, 5.13 and 5.2). In the groups of mushroom-picking permits (PG 3.2) and simple recreational products (PG 4.2), payment is dominated by single tickets, which, however, were also the type of payment for two complex recreational products for single users (PG 4.1). Permanent licences can be found in one case of a licence for mushroom-picking. In the case of non-material recreational consumer products (PG 4.1 and 4.2), permanent licences and tickets for different performances also play an important role. Concerning integrative recreational products (PG 5.13 and 5.2), package prices, which can be all-inclusive or divisible, are of significance. This is referred to in the following section. In addition, concerning the partly material recreational products (PG 3.1), combined payments can be found. These include, in one case, performance duration and club membership and, in two cases, performance duration and consumption. Here again the material character of offers for overnight stay is apparent (electricity, water, etc.).

Three single cases will be considered as special cases. In the case of the forest youth hostel (DE27, PG 5.13), performance by the forest enterprise for pupils is basically remunerated through the pupils' work in the forest. Varying prices for different user groups of this facility, moreover, results in special conditions of payment. The granting of cash discounts (*Skonti*) under certain conditions in the case of accommodation and hunting in a count's forest (DE28, PG 3.1) represents such a payment condition. In the case of the Taufstein ski-tracks (DE19, PG 4.2) use is voluntarily paid for via the purchase of a sticker for the skis.

As regards types of payment, contract products (PG 6.1) are certainly shaped by their product form – rents, leases and grants. Thus, voluntary agreements and lease contracts, with and without profit participation, are applied in particular. Types of payment in cases of environmental and sponsoring products (PG 6.2 and 6.3) are naturally quite restricted. Sponsoring determines the type of payment already by its product definition. Only in one case was the sponsored amount given as a donation, which was due to tax reasons. In the remaining cases, there are voluntary contract agreements, which, however, contain single calculations according to specific performances for some environmental products.

> Variability of types of payment for RES products is extremely high. Characteristics of every product influence the optimum type of payment. For material RES products, the calculation is mostly made according to the degree of consumption. It is possible, though, to agree upon basic amounts with institutions. While, for directly performed RES services, the performance duration is particularly fundamental to the price, there are entry tickets and licences for utilization, which are especially suitable for permanent offers. Single negotiated RES recreational offers are often remunerated via inclusive prices. In the case of offers for overnight stay, tangible aspects, such as water consumption, are often calculated separately. Concerning integrative contract-products, voluntary agreements are the preferred type of payment; donations are possible in this case. For those RES products which are only marketable in a limited way due to considerable legal restrictions, earnings can nevertheless be obtained on a voluntary basis. Conditions for payment, such as cash discounts (*Skonti*), can also be successful for RES products.

Bundle prices

Summarizing performances and unifying the different prices in bundle prices constitute a possible effective instrument of sales support. This especially concerns the sale of services and combinations of services and material products (Meffert and Bruhn, 1995, pp. 315–318). Potential customers, however, may react in different ways to such packages, so that a market investigation prior to their offer would seem to be very important. What is perceived as positive by the customer is the price advantage associated with the total purchase, as well as the fact that he/she is relieved from organizing and combining products. When purchasing complex combinations of services, the customer, moreover, can be sure that all components are harmonized with each other. What might be perceived as negative are the low price transparency and the feeling of having bought some inferior components as well.

Basically, an inclusive price can supply different advantages to the offerer of services, such as a more regular rate of capacity utilization (Meffert and Bruhn, 1995, pp. 315 and 316). Due to these advantages and the benefits for the customer, different price packages could be noted in the RES case-studies. In more than 31% of the recorded cases, they were at least partly established. The distribution of this third is quite uneven, though, as is shown by Table 4.25.

Concerning the material consumer products Christmas markets, events, game and gift articles (PG 2.1), in four out of six cases[95] product combinations are partly offered via bundle prices. This is not the case concerning the remaining material products (PG 1.1 and 2.2), which are mainly offered to institutional purchasers.

For the described recreational products, bundle prices are at least partly established in almost all groups. Whereas this was mentioned in only two cases concerning offers for overnight stay (PG 3.1a) and in only five cases concerning simple recreational products (PG 4.2),[96] it is common to develop bundle prices for complex recreational products (PG 3.1b, 4.1, 5.13 and 5.2). Five cases of these product groups[97] are always offered at bundle prices. Within these groups, only the case of an eco-park (AU05) does not apply this instrument of price policy.

Price bundling did not appear with pure cessions (*Rechtsabtretungen*).

From the perspective of the customer, bundle prices have advantages as well as disadvantages. When there is the greatest possible price transparency in an offer, the advantages mostly predominate. Bundle prices can be offered to the end-user for material RES products, but they are mainly used for RES recreational products. They are an effective factor for success, especially for offers with many complex single elements and with elements which are even negotiated individually with the customer.

4.3.4 Communication policy

Communication policy in general

Meffert (1986, p. 119) assigns the task of 'any conscious design of information on an enterprise which is aimed at the market with the goal of an influence on and control of current and potential buyers' behaviour' to communication policy as part of the marketing mix. This means that communication policy fulfils the areas of responsibility of information and motivation, in which the user of a product is informed about the objective qualities of this product and in which his/her subjective wish for the special product of the offerer is raised or strengthened. In contrast

Table 4.25. Bundle prices according to product groups.

Bundle price	Cases	Always		Partly		No		Missing	
		n	%	*n*	%	*n*	%	*n*	%
Material products (PG 1.1, 2.1, 2.2)	17	0	0	4	24	11	65	2	12
Complex recreational products (PG 3.1, 4.1, 5.13, 5.2)	27	5	19	16	59	6	22	0	0
Simple recreational products (PG 3.2, 4.2)	27	0	0	6	22	20	74	1	4
Integrative contract products (PG 6.1, 6.2, 6.3)	27	0	0	0	0	26	96	1	4
Total	98	5	5	26	27	63	64	4	4

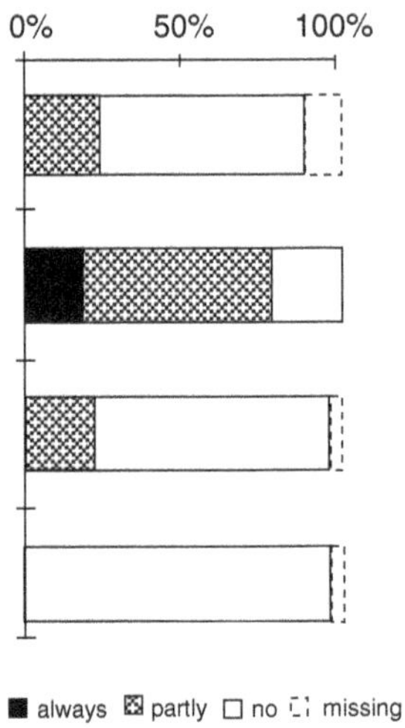

to the other three segments of the marketing mix, which are orientated towards the existing judgement and decision structures of the customers, communication is also intended to influence the internal psychological processes of the customer (see Hamm, 1991, p. 269). For this reason, it is necessary to consider the sociological aspects of the decision-making process in the framework of communication policy and the fact that the target hierarchy of communication policy contains not only economic but also operational psychographic targets.[98] Consequently, Nieschlag *et al.* (1991, p. 451) define communication policy as 'applied behaviour science, a social technique ... which takes use of knowledge about psychology, sociology and market research to fulfil business economical tasks'.[99] A further special characteristic of communication policy as the information process of the market is the fact that 'products and performances can be changed neither substantially nor functionally' according to Weis (1997, p. 363).

Any product must naturally have its own shape, a sales channel and a price in order to be in the position to be passed on. So product, distribution and price policy are most important in the rearrangement of products for markets which were scarcely known before now. On the one hand, these three elements theoretically suffice for the sale of a product when there is a high demand for it. On the other hand, a potential customer has to be aware of the existence of a product, so a minimum of information is necessary for every new product. Nevertheless, the 'function of motivation' of the communication policy has often been set back, particularly in regard to unsaturated markets. Besides the market situation, the type of product, the distribution channel and especially the particular target group influence the communication strategy. For RES products, this means that a distinct difference in communication policy can be expected, for example, between consumer goods and investment products, or that, by means of the significance of direct forms of communication in the area of services, it can be shown how performance and communication, which partly rely on each other, smoothly merge into one another.[100] Therefore, the details concerning the means of communication are non-uniform in the RES case-studies. Thus, the spectrum of communication tools mentioned stretches from 'nothing', which was indicated in 25 cases, up to the case of the tree-crown path (NL13), the communication policy of which is exceptionally diverse and intensive and is based on its own communication plan.

The most important sections of communication are advertising, sales promotion and PR, including sponsoring. Direct, personal communication is based on direct contact between supplier and customer resulting in a direct response, which frequently already exists with purchasing. This area, therefore, is the link to distribution policy.[101] A separate section in this chapter is dedicated to the development of labels and brands, as these are closely linked to product policy.

A special aspect of communication for RES products, which is often mentioned in the case-studies, is free publicity. Free publicity comprises the presentation of RES products in communications of third parties. Thus, the enterprise does not influence the presentation of the product directly.

> Communication is inevitable for the long-term existence of forest enterprises. The communication instruments for RES products always serve to inform the customers and, in the second step, generally to promote motivation. Communication pursues psychographic targets via the effective mediation of messages on the basis of behavioural science. Under certain conditions, forest enterprises therefore need professional assistance with regard to their communication policy. Communication policy strongly depends on the actual target group. It has to be perfectly harmonized with the other elements of the marketing mix. For RES products, direct marketing is frequently used. For services, direct communication forms are very important. For most of the RES products, the effects of free publicity can be used.

Kotler and Bliemel (1995, p. 913) subdivide the realization of an effective communication and sales support programme into eight stages, as described in the following list:

1. Indication of the target audience and its relation to the communication object. Who will be reached by the communication effort, what information and attitudes exist concerning the product and/or the enterprise?

> RES example (cross-country ski-track): cross-country skiers, living at or having a holiday within 100 km of the track; survey concerning knowledge and attitude regarding the offering forest enterprise and regarding the offered track.

2. Determination of the target effects of communication. This point is orientated towards hierarchical models of effects (*Wirkungsmodellen*) of communication (e.g. attention, interest, desire and action (AIDA) model); possible order of effects: knowledge – emotion – affection – purchase – satisfaction.

> RES example: the knowledge of the target group about the track will increase, the offer will lead to positive emotions and gain the affection of the target group, so an increasing number of customers use the track and become satisfied with this sports facility.

3. Design of the message. Elements of design are: the content, the motivations made use of (appeal), the structure, the kind of expression and the transfer medium of the message.

> RES example: presentation of the track and the offering forest enterprise, in a 'sporty–active' way, rationally founded and emotionally positively intensified, transmitted by reliable media.

4. Choice of communication channels. Possibilities range from direct communication face to face to any kind of multistage communication, using several media and/or opinion-leader positions.

> RES example: selection of preferably short, personal channels; use of famous sportsmen as opinion leaders.

5. Decision about the total budget for sales-supporting communication, depending on the financial possibilities, turnover or competitive situation; recommendable according to objective evaluation and tasks.

> RES example: determination of the number of potential customers who will be reached by the communication; this number leads to the total communication costs through multiplication.

6. Decision about budgets for the individual communication tools. According to the targets, the total communication budget is divided into advertising, sales promotion, PR and personal sale.

> RES example: 33% advertising, 10% sales promotion, 33% PR, 24% personal sale.

7. Execution and coordination of sales-supporting communication. Whereas the execution of the communication often requires the help of external service providers, the task of coordination remains with the enterprise as far as possible.

> RES example: communication for the track according to clear instructions of the enterprise through a PR agency; constant feedback between forest enterprise and service provider.

8. Controlling results. Possible through measures of accounting, calculation of key numbers and customer surveys, in particular in the first stage of the communication process.

> RES example: customer survey in sports equipment shops of the target area on Saturdays regarding recognition, use of the track and satisfaction with the offer.

Example of an RES communication plan

The communication plan of the Drents tree-crown path (NL13) is a temporally determined marketing plan with special consideration of the communication tasks. In the first of its three parts, the basis of communication is described. The second part deals with the utilization of communication instruments and communicative product parts, such as special events and activities, but also other aspects of product policy. In this case, the alternative significance of events in the marketing process, which, on the one hand, are part of the product mix and, on the other, contain aspects of sales promotion and direct communication, becomes clear. In the third part, there is an outline of which communication strategy is pursued to reach which target group. For this purpose, differentiated communication strategies of external and internal communication are described for the following target groups.

- Visitors (customers); potential customers.
- The press.
- Intermediaries, meaning persons who spread information to potential visitors.
- Sponsors and friends; business partners.
- Guides, i.e. external staff of the forest enterprise.
- Surrounding communities; the province in charge.
- The forest enterprise's own, internal employees.

In each section of these three parts, the targets described are assigned respective concrete measures. It becomes clear that the individual sections of this plan not only comprise the instruments of communication policy described below, but also refer to all relevant aspects of marketing with regard to the central task of communication.

Advertising

Advertising comprises all non-personal communication efforts of a supplier for a certain product or a product group via the utilization of media which charge for their service.[102] It consists of advertising means that are placed on or in advertising media and thus are introduced to the potential customer. Typical advertising means, for example, are advertisements, leaflets, posters or electronic commercials. The three most important media types used for advertising are daily newspapers, television and direct media (mailings),[103] besides other media such as magazines, advertising spaces in public places or radio broadcasts.

When selecting the advertising means and the advertising medium, emphasis must be laid on the highest possible efficiency. Thus, it would not be very sensible to advertise a local guided tour through the forest by TV commercials broadcast nationwide. The typical relation between expense and marketing effect has to be considered when deciding on the *budget* for advertising, as Fig. 4.7 indicates.[104]

Especially as concerns RES products with low turnover expectations, it has to be decided in advance whether advertising should be forgone, or whether a budget is available which surpasses the lower threshold of advertising effectiveness. The problem is that the actual level of the threshold budget is difficult to estimate, especially as empirical values are scarce with RES products. For medium-sized enterprises with complex communication needs, according to Nieschlag *et al.* (1991, p. 490), especially cost-effective types of direct marketing are advisable. This statement can be transferred to a variety of RES products. Advertising should be adapted to product features, as they supply a high potential for differentiation from similar products (profiling potential) and are difficult to imitate (stability), as the effect intended can only be reached after some time and involves high costs.[105] In practice, the complexity of effective advertising measures will, for the most part, persuade forest enterprises to commission advertising agencies and market research institutes for this task. The manager of the forest enterprise, though, should be familiar with some theoretical basics of advertising in order to be competent in discussions.

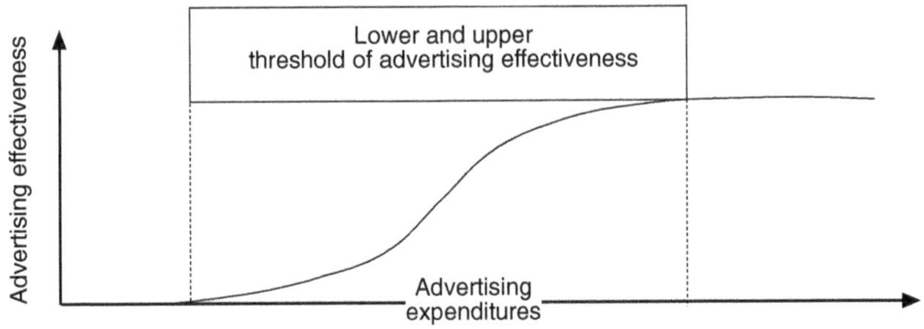

Fig. 4.7. Schematic development of the efficiency of advertising expenditure.

Advertising is a communication tool with high sales impact, but at the same time it is relatively cost-intensive. It does not establish a direct contact with selected customers. Expensive losses due to dispersal can be avoided by exact advertising planning. Advertising has to be innovative and creative to get the attention of potential customers and to have lasting effects. Suitable publicity media have to be selected according to their effects in view of the target. The advertising message should address the target group by introducing unmistakable features of the product. Immaterial services, for this purpose, have to be attributed to material features. When advertising RES products, the stipulated expense has to be sufficient to exceed the lower threshold of advertising effectiveness.

THE ADVERTISING EFFORT IN THE RES CASE-STUDIES. Table 4.26 indicates the share of cases in the RES case-studies in which the utilization of advertising is mentioned or even underlined and accentuated by several references and a variety of designs. It becomes clear that especially products from product groups which are mainly or exclusively acquired by individual customers are advertised. This is valid, for example, for material RES products (PG 2.1) or partly material (PG 3.1) and complex (PG 4.1) recreational consumer goods. But also integrative recreational products (PG 5.13 and 5.2) and sponsoring products (PG 6.3), which are generally sold to institutions, are, in part, intensively advertised. The data given for individual cases will be referred to further below.

Table 4.26 comprises all advertising means and media. For a more differentiated perspective on the advertising strategies utilized in these cases, see Table 4.27. This table refers to the share of RES case-studies in which the area of advertising by advertisements, leaflets or electronic mass media was utilized.

Advertisements in local and nationwide newspapers, magazines and journals were issued for about a third of the RES products researched. In another third of the cases, advertising by leaflets, comprising all kinds of leaflets and handbills, prospectuses, brochures and specialized catalogues, was mentioned. However, intensive advertising strategies were mentioned in only three of the cases. In seven RES cases, expensive advertising via electronic media, such as radio and television, was utilized. Only in five exceptions were RES products marketed through billboard advertising and posters. Further means of advertising, such as the issue of videos, postcards, printed T-shirts and other items, were also rarely used. In only four of the cases are there hints as to such measures. However, in the area of communication, especially measures rarely utilized give interesting hints as to how a message can be conveyed to the customer in unfamiliar and therefore eye-catching ways.

The advertising strategy most often referred to in the field of material products (PG 1.1–2.2) is the publication of advertise-

Table 4.26. Utilization of advertising according to product groups.

Advertising	Cases	Stressed n	Stressed %	Mentioned n	Mentioned %	No n	No %
Simple products for final users (PG 2.1, 3.2, 4.2)	33	1	3	18	55	14	42
Complex products for final users (PG 3.1, 4.1, 5.13)	23	4	17	14	61	5	22
Simple products for organizations (PG 1.1, 2.2, 6.1, 6.2)	29	0	0	5	17	24	83
Complex products for organizations (PG 5.2, 6.3)	13	1	8	7	54	5	38
Total	98	6	6	44	45	48	49

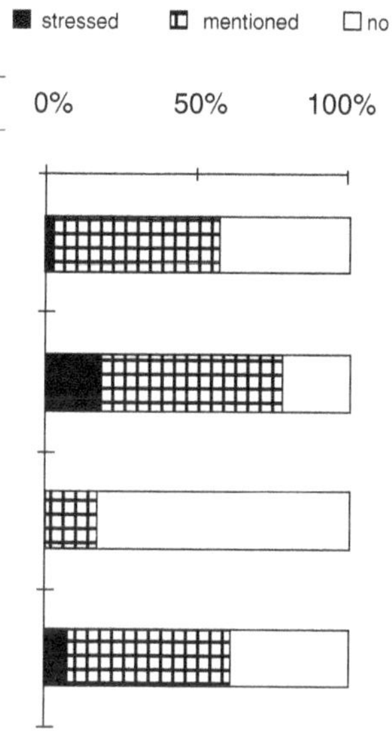

Table 4.27. Utilization of different kinds of advertising for RES products.

Type of advertising	Total	Stressed n	Stressed %	Mentioned n	Mentioned %	No n	No %
Advertisements	98	2	2	27	28	69	70
Leaflets, etc.	98	1	1	32	33	65	66
Electronic media	98	0	0	7	7	91	93

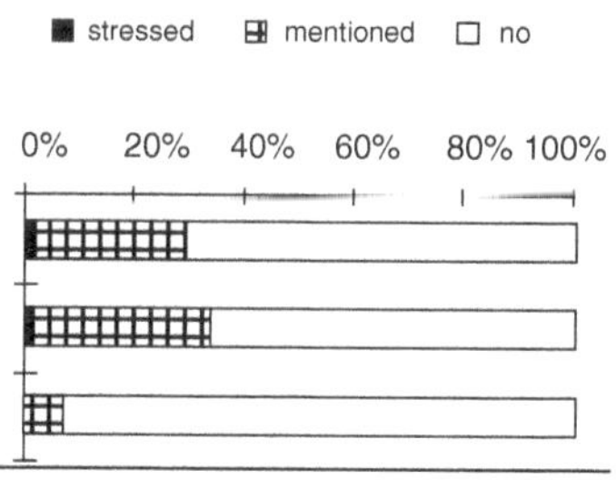

ments in mostly local, but also in some nationwide, newspapers, magazines and trade papers, as well as in club magazines. In general, the target group thus reached comprises the consumers of a product. Furthermore, handbills and brochures advertising the product in question are designed and distributed by shops and tourist offices. As expected, groups where the RES customers of forest enterprises mainly comprise institutions (PG 1.1 and 2.2), lower advertising efforts are indicated. In one case, the sale of drinking-water (AU19), advertising was even said to be counterproductive. In eight of the 17 cases concerning mostly material products, no advertising was done. In none of these cases were conspicuously intensive advertising activities described.

By far the most intensively worked and designed product groups with respect to communication policy are the partly material products and the complex recreational products directly distributed to the final consumer (PG 3.1 and 4.1). On the one hand, in only three out of 18 cases were there no indications of advertising activity of the enterprise. On the other hand, the advertising activities executed in the two product groups referred to in another three cases have to be described as strikingly intensive. Means of advertising were mainly advertisements in very different media, comprising club magazines for members, the local press, magazines and national daily newspapers. Moreover, leaflets and handbills, prospectuses, brochures, catalogues and other printed matter have been offered and distributed via tourist offices and shops. Besides the two very widespread advertising areas of advertisements and leaflets, advertising within these product groups also is done by posters, billboards, photos (cards), videos and spots broadcast on radio and even TV.

The case of the Drents tree-crown path (NL13) clearly shows the possibilities of advertising for complex RES products (sections 8/9 of the communication plan). The offer is promoted by various advertisements in the regional tourist guide and related publications, in special-interest magazines and in the brochure Drenthe Day Trip. Moreover, leaflets in three languages, posters and photos for external publications are available as advertising material. These advertising objects are distributed in approximately 1000 different places. For further support of the product, the introduction of its own sales point was planned in 1997. At the same time, the sales point will distribute advertising material and sell souvenirs (merchandising). A photographer supplies photos on demand according to a shoot-list which includes targeted material on the product. The production of videotapes is planned. As the most sumptuous advertising material, TV advertising spots were produced. These spots are shown nationwide and in the neighbouring countries. On the one hand, television advertising is characterized by addressing a large public, on the other, by its elevated costs. Television advertising for RES products, therefore, in general cannot be covered, although this means of advertising was considered to be very effective in the case-studies concerned. The intensive advertising for the tree-crown path (NL13) was made possible, first, by its being a sumptuous product targeted to a large potential public throughout the country and abroad, secondly, by its being owned by the National Forest Administration of the Netherlands and, thirdly, by the fact that advertising with the product and for the product is the task of the tourist administration of the province.

An exceptional case is the case of accommodation and hunting in a count's forest (DE28): in this case, a single advertisement concerning the offer published in a specialized magazine led to a demand that exceeded supply. As demand could not be covered any more, advertising had to be stopped.

Integrative recreational products (PG 5.13 and 5.2) are most often supplied to institutions. However, advertising activities for this group were, in individual cases, unexpectedly intensive, as already mentioned above. Although three out of nine cases did not indicate any advertising, advertising activities were very intensive in one of the cases in this group. The advertising activities mostly comprised advertisements and leaflets, too. The form and content of advertising indicate a similarity with fixed complex recreational products (PG 3.1 and 4.1).

As concerns simple recreational products (PG 4.2), which in turn are purchased directly by the consumer, however, only about half of the cases indicated advertising activities. In this group, leaflets seem to be more important than advertisements. In one of the cases, billboards along the road pointed to ski-tracks (IT09) and, in the case of the joint venture network of ski-tracks (AU13), the only one in this group where the advertising effort is classified as intensive, advertising through two television broadcasts was mentioned. In the group of mushroom-picking permits (PG 3.2), in one case (IT19) advertising was done by a poster.

The product line of use contracts (PG 6.1) comprises letting, leasing and licensing (utilizing) agreements, which are characterized by the fact that they are integrative and, in most cases, individually passed on to institutions (organizations). The fact that, in this group, advertising measures are mentioned in only one of the 13 cases underlines the strong dependence of the communication media on the product and its target group. Furthermore, in this exceptional case, many parts in the study dealt with the effects (of advertising) on the final user. Advertising means, therefore, are not targeted to the contractees, but to their customers as the consumers of the product. In a special case, the leasing of ski-runs (DE25), it was mentioned that the forest enterprise as the contractor carefully tries to limit advertising measures of the contractee in order to prevent the overuse of the forest by skiing.

Environmental products (PG 6.2) are analogous to the features mentioned above in relation to contractual products. Thus, it is not surprising that advertising is not done for such products. This, however, is different with sponsoring products (PG 6.3): in this case, in five out of nine cases, more or less intensive, but altogether very varied, advertising activities are outlined. At least in the case of sponsorship for a national reserve (IT29), which indicates an extraordinary variety of advertising aspects, advertising for a product and advertising with a product overlap, i.e. the content of the contract of sponsorship. In this group, club magazines, brochures, information sheets, stickers, posters and television commercials (IT29) are potential advertising media.

At present, advertising for RES products is mainly done through advertisements and leaflets. New strategies for advertising promise a higher impact for RES products. However, the expected proceeds of RES products in only a few cases cover the expenses of advertising in electronic mass media. Advertising is mainly suited for those RES products supplied as consumer goods. Complex recreational offers in the forest often utilize advertising strategies because of their economic potential. Advertising can lead to a non-utilizable excess demand for a limited RES product. In general, advertising is not sensible for RES products offered to institutions in the form of a contract. With sponsoring products, advertising for a product switches to advertising with the product.

Sales promotion and sales events

All special influencing to increase sales of a certain product or a product group by utilizing mostly short-lasting sales incentives can be understood in terms of sales promotion.[106] Such measures are generally directly associated with sales activity. Typical elements of sales promotion are, for example, incentives, competitions, selling aids and sales events. Measures of sales promotion can represent an important supplement of advertising, but they cannot replace advertising in respect of the establishment of an image and long-term customer loyalty. Like all instruments of communication, sales promotion also follows the process of communication from the point of view of the targets and strategies selected.[107]

Table 4.28 gives an overview of the extent of the measures of sales promotion applied in the RES case-studies. It becomes obvious that sales promotion was not regarded as very important for the recorded RES products. In the whole field of material products (PG 1.1–2.2), measures of sales promotion are only mentioned in the case of the eco-shop (NL19). In this case, however, this seems to be a central element of the communication strategy. Thus, an open-door day is organized once a year for the eco-shop of a farm, as well as exhibitions, rural fairs and cooking shows, with mulled wine offered free.

As concerns recreational products, the following is indicated. While measures of sales promotion have not been described for partly material recreational products (PG 3.1), this means of communication is still utilized by five of eight enterprises for the promotion of their non-material complex recreational products (PG 4.1). This is mostly done by exhibitions, exchanges and participation in trade fairs. The aspects of sales promotion involved in the case of the tree-crown path (NL13) have already been mentioned. Thus, the communication plan for this product deals with events in the framework of the touristic offers of the province, excursions offered by the forest administration, children's activities days, paper-chase rallies or integration in offers of third parties.

Table 4.28. Use of sales promotion for RES products.

Sales promotion	*n*	%
No	82	83.7
Mentioned	15	15.3
Stressed	1	1.0
Total	98	100.0

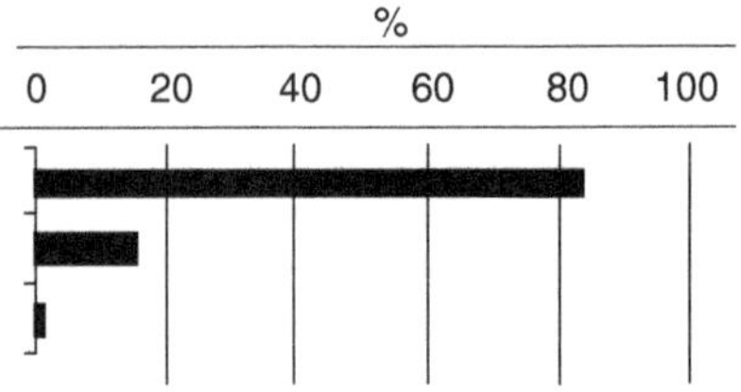

Within the cases of integrative recreational products (PG 5.13 and 5.2), one case uses a measure of sales promotion in such a way that a holiday package resembling the product forest adventure holiday with the forester (DE12) was offered as the first prize of a competition. Concerning simple recreational products (PG 4.2), measures of sales promotion were applied in two cases: in the case of IT04, mountain bikes are rented at special prices in order to encourage the use of a car-park area, and, in the case of NL02, an information centre serves as support of an entry ticket for the nature-protected area.

Concerning integrative contracts (PG 6.1–6.3), measures of sales promotion are only applied in three out of nine cases within the sponsoring products (PG 6.3). Here, participation at eco-fairs (DE08), the circulation of a list with potential sponsoring projects (NL06) and the particular mention of sponsors to club members (NL12) are described.

The relatively low levels of sales promotion in the RES case-studies are not inconsiderably due to the fact that this instrument of sales mainly supports classical advertising when the latter is no longer sufficient for the sales of surplus production in the area of consumer goods.[108]

> Sales promotion supports advertising measures by raising the attention of potential customers through special short-term incentives so that they will consider and/or purchase the product. For RES products, sales promotion in single cases is a suitable instrument of communication. It can be applied for all conceivable customer groups if the sales of a product requiring a more or less continuous demand are to be stimulated.

Public relations

With public relations, the enterprise itself is the focus of communication. In general, the addressees are the general public, as well as business partners, staff members, and special trend-setters who are important for the company. It can be assumed that information communicated through the channels of PR will get a higher level of acceptance when creating the impression of rendering an objective report than with advertising slogans (Kotler and Bliemel, 1995, p. 1021). Especially as concerns the promotion of RES products that are derived, through a transformational process, from services that were free of charge, it is of eminent importance to convince, above all, the public.[109]

The utilization of PR, as derived from the case-studies, is outlined in Table 4.29. PR has been maintained more or less intensively in about a third of the cases. This is valid for all product groups, with the exception of material capital goods (PG 1.1) and mushroom-picking permits (PG 3.2). PR was especially intensive in connection with certified material products (PG 2.2b) – for example, participation in shows and television programmes. The offer of certified chestnuts (IT27) in this context is very striking. Furthermore, integrative recreational products (PG 5.13 and 5.2) are related to outstandingly many PR measures, especially as concerns the offer holidays at the forester's (DE12). In detail, the following conclusions can be derived from the cases.

Characteristic PR tools of enterprises offering RES products are the publication of their own articles and other features in magazines, newspapers and professional journals. In some cases, lectures and radio interviews were indicated. Cooperation with the media was also established by keeping journalists informed through invitations and press releases. Some of the enterprises introduced themselves by participating in exhibitions open to the public. Thus, in the case of forest seminars (DE17), a joint event with the Association for German Forest Protection (*SDW*) was arranged. In Austria, as well, different events were related to the introduction and presentation of the respective products, for example an opening ceremony in the case of a mountain-bike road and mountain-bike hire (AU01 and 02), an event with the Ministry of Tourism and different sports organizations in the case of a concept for mountain-biking (AU17) and an annual event of the church free of charge in the case of a toll for forest roads (AU18). Such events are on the brink of sales promotion, because of their relation to a product. In the case of vacations (AU11), sports events bring the forest enterprise and its products to the notice of the public. A special form of modern PR is internet sites, which were mentioned by three enterprises (DE15, IT10 and AU07).

Finally, it has to be underlined that all communication tools mentioned with reference to environmental products (PG 6.2) can be attributed to PR. This is especially true in the case of water conservation (DE14), but also of carbon dioxide (CO_2) storage (NL15). Newspaper articles and other publications, such as club magazines, lectures, radio and television broadcasts and joint events, such as the planting of beeches, were among the measures utilized.

Here, it is shown that the type of product influences the suitability of PR as a communication instrument, especially the degree of utilization of special non-material forest facilities as product parts. While in approximately 53% of complex and eco RES products, including the environmental preservation and sponsoring contracts (PG 2.2, 4.1, 5.13, 5.2, 6.2, 6.3), there has been PR, this is only the case in 22% of the rather material and/or simple products, including licence agreements (PG 1.1, 2.2, 3.1, 3.2, 4.2, 6.1). The role of PR as advertising for the entire enterprise in this area of communication brings to the fore characteristics such as enterprise targets, type of organization or status of ownership, rather than the type of the single product offered.

As concerns the results of the interviews of the RES case-studies, it can be assumed that the area of PR has not been described conclusively, as the case-studies dealt with communication means in relation to a special product and not to the entire enterprise. Moreover, the borderlines between PR and advertising or sales promotion are not clear-cut. In this section, interviews and articles are categorized as public relations, although they can refer to special products or product groups. This seems to be sensible as, in this way, information of the public is brought to the fore. Therefore, a newspaper publishing an article or a broadcasting station broadcasting a radio interview will try to prevent the advertising character from getting in the foreground too distinctly.

Table 4.29. Use of PR in the RES case-studies.

Public relations	*n*	%
No	65	66.3
Mentioned	29	29.6
Stressed	4	4.1
Total	98	100.0

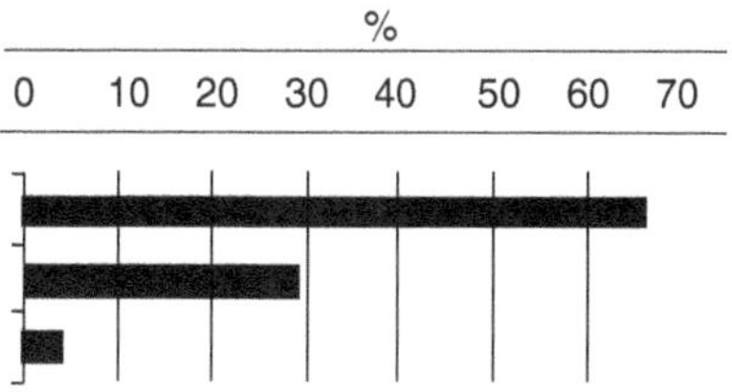

PR is a means of communication preferably utilized by forest enterprises. However, it is rarely used for the targeted sales support of certain RES offers. PR by third parties gives an impression of high credibility. PR is extremely important for RES products that were formerly obtained free of charge. RES products with far-reaching effects are frequently supported by PR strategies. Intensive PR measures can be effectively linked to the offer of certified, material RES products. The same is valid for integrative recreational products of the forest and for all of the environmental products.

The utilization of the internet is an innovative form of PR, as well as of advertising, for certain RES products.

Direct personal communication

Personal communication is directly linked with personal sale. It is mainly characterized by a direct intensive interrelation between the supplier and the customer. The customer is obliged to react to this direct contact.[110] These advantages place high requirements on the offerer as concerns ability and customer orientation.[111] While the selling of raw wood is generally done according to specific learned and practised habits, the selling of innovative RES products probably requires new selling skills and the mediation of technical competence in a hitherto unaccustomed market. Therefore, an intensive effort for new selling qualities of the forest management can be an important factor of success in the sale of RES products.[112] Especially as concerns non-material products, a personal sales contact is of eminent importance.

Table 4.30 makes it clear how far personal communication is mentioned in the case-studies. Concerning material offers (PG 1.1–2.2), personal contacts were indicated only in the case of a Christmas market (DE13), where invitations to business partners are mentioned. Concerning recreational products, there are different hints as to personal contacts for partly material and complex offers (PG 3.1, 4.1, 5.13 and 5.2), for example:

- Contacts with local schools.
- Personal talks with customers, personal invitation of customers.
- Personal mailings to regular customers.

This point is developed in more detail in the communication plan of the tree-crown path (NL13, PG 4.1). Visitors of the offer are free to indicate their names and addresses in the framework of recurring customer surveys or in the visitors' book and thus to be entered in a customer database. These regular customers are subsequently directly informed of events, novelties and activities concerning the product. In addition, the visitors' book fulfils an important function in the management of complaints.

Concerning the group of contractual products (PG 6.1), in many cases personal communication is regarded as natural and therefore it is not mentioned in particular. In the interviews, only in the case of the contract on permission for cycle paths (DE23), the statement 'contacts with the recreation club' was made. Personal communication is also of central importance for the sale of environmental and sponsoring products (PG 6.2 and 6.3). The following hints mentioned in some sponsoring

Table 4.30. Use of personal communication for RES products.

Personal communication	n	%
No	86	87.8
Mentioned	12	12.2
Total	98	100.0

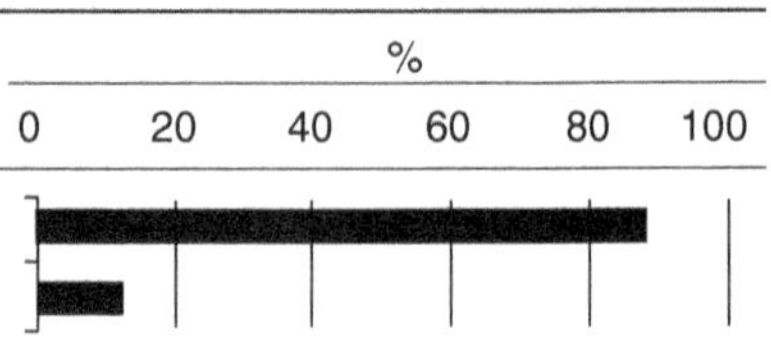

cases, therefore, can describe only a small part of personal communication aspects:

- Addressing potential sponsors by mail or personally (while playing golf).
- Meeting of enterprises interested in sponsoring projects.

The fact that, despite the significance of direct sale for RES products, there were few hints as to personal communication points to the fact that personal contacts between the enterprise and the customer are so self-evident that they have not been related to advertising and sales promotion when assessing the case-studies.

Thus, at least for all bilateral contracts on a single and unique object, it can be assumed that the personal contact is the central tool of communication.

> Personal communication is mainly characterized by the direct exchange of impact and reaction between offerer and customer. Such a contact exists with all RES products that have been worked out together with the customer. Especially for RES contract products, this is an essential basis. Personal contacts with selected customers are often an important factor of success for complex recreational products as well. For effective direct communication, particular personal attitudes are very important.

Free publicity and outsourcing

Free publicity as a communication policy advantage can result from the special characteristics of most of the RES products derived from the environmental and recreational services of the forest and from still unsaturated markets. This effect can be made obvious using a simple comparison, e.g. between a detergent and a nature reserve. While it cannot be expected that a certain detergent brand can be introduced positively to the public by third parties without some promotion from the detergent offerer, this is indeed possible for a reserve, as the latter can, assuming an attractive design and a certain uniqueness, achieve regional significance for accommodation business, the traffic business, retail businesses and some other branches of commerce. Based on the interest of these third parties, the significant product is frequently mentioned as an additional benefit for other products. In practice, such a combined communication is often carried out by public institutions – in the RES case-studies, most frequently by tourist offices. Thus, free publicity, normally containing aspects of advertising and PR, is in a close relation with outsourcing. Outsourcing is understood as the execution of an enterprise task – in this case, the communication task – by a third party. The decisive difference between free publicity and outsourcing is that outsourcing is normally directed by the offering enterprise and, furthermore, is cost-relevant for the enterprise. Nevertheless, the effects of free publicity cannot be integrated in a communication strategy as, by definition, they are not planned – otherwise it would be a PR instrument.[113]

Table 4.31 shows the use of free publicity for the RES products examined. Free publicity was more or less intensively described in 25% of the case-studies. For the recreational products of PG 3.1–5.2, there are more or less strong indications as to free publicity in 21 out of 54 cases, particularly distinctly in four cases.[114] In two cases (NL14 and 17), the nature camp-sites described are mentioned and recommended in diverse books, travel guides and catalogues. This observation is also valid for a picnic site (IT24).[115] In one case (IT28), various TV programmes and magazines mention the nature park that is offered as a RES product. Considering the great impact of electronic media, this should be of some significance for the product. A golf-course in the forest (IT03) is introduced in the media by broadcasting golf competitions.

Indications on advertising free of charge by pieces of information and brochures of the regional tourist office, to some extent also by the community or a regional park, can be found for six, mostly simple, recreational products from Italy.[116] In the case of

Table 4.31. Free publicity according to product groups.

Free publicity	Cases	Stressed		Mentioned		No	
		n	%	*n*	%	*n*	%
Material products for final users (PG 2.1)	6	0	0	1	17	5	83
Simple, non-material products for final users (PG 3.2, 4.2)	27	1	4	8	30	18	67
Complex, non-material products for final users (PG 3.1, 4.1, 5.13)	23	3	13	8	35	12	52
Material products for organizations (PG 1.1, 2.2)	11	0	0	1	9	10	91
Simple, non-material products for organizations (PG 6.1, 6.2)	18	0	0	0	0	18	100
Complex, non-material products for organizations (PG 5.2, 6.3)	13	1	8	2	15	10	77
Total	98	5	5	20	20	73	74

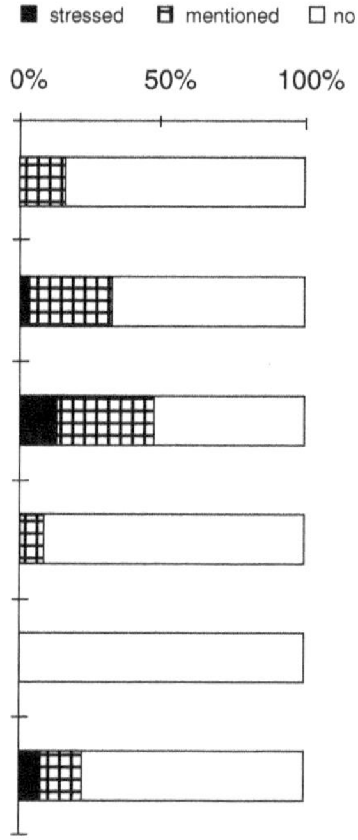

guided horseback tours (IT18), an announcement free of charge could be made. A further communication to be classified as free publicity is publicity by word of mouth. This effect is certainly an important component of the sales of any product and, furthermore, it is not typical of RES products above. For four German cases of quite different recreation products,[117] this effect can substitute for a broad range of targeted communication policies. This is similar to one case of camping (IT11), in which regular customers substantially contribute to the sales.

Some Dutch cases indicated free publicity for other products also, i.e. material products (PG 2.1 and 2.2) and sponsoring products (PG 6.3). The cases of the eco-shop (NL19) and eco-meat (NL04),[118] and two cases of organization sponsoring (NL06 and 16) can be stated here. Outsourcing of communication for recreational products to travel agencies is closely linked with distribution policy. It is mentioned in two cases (DE02 and 04) within the scope of the question about advertisement measures. Furthermore, the quite intensive communication policy in the case tree-crown path (NL13), as already mentioned, is not realized by the offering enterprise but by the province.

Free publicity and outsourcing mean the execution of a positive communication for a product or an enterprise by a third party. Free publicity results from positive side-effects of RES products because of their unique characteristics and their regard for the forest ecosystem. Outsourcing means that communication tasks are transferred with regard to a certain objective. Both are important for RES products.

The special advantage of free publicity lies in its high credibility and the fact that it is free of charge for the enterprise. Above all, RES recreational products often benefit from free publicity. Regional tourism institutions have a special significance with regard to this. Also for material organic products, free publicity is an important success factor. An important kind of free publicity is publicity by word of mouth, which, for all products, results from customers' satisfaction.

Names, brands and labels

As a matter of principle, every enterprise and every product has a name by which it becomes distinguishable in internal and external communication. In order to secure this value of recognition, these names should be designed to aim at the corresponding direction of communication and be determined as trade or brand names. The name itself should be characterized by a high value of attention and recognition; furthermore, it should be non-interchangeable, easy to pronounce and easy to remember and it should evoke positive associations (Kotler and Bliemel, 1995, p. 688; Weis, 1997, p. 224). Furthermore, a corresponding symbol can be developed to complement the name.[119] Finally, it is essential to protect these names and signs in the light of their uniqueness by registering them under the applicable laws. Thus, these brands become an important means of protection of property (Mantau, 1995b, p. 60). This development process and the protection of the result are combined with costs, which certainly have to be balanced against the result achieved. For most enterprises and most products with a long-term existence, the economic advantages will pay the expenses.

TRADE NAMES, PRODUCT NAMES, LABELS AND BRAND NAMES/TRADE MARKS. Names of enterprises can be protected as trade names. For RES products, the scope of marks for the producers' own products is quite important, namely brand names or trade marks. Names for all products of an enterprise, so-called umbrella marks, are frequently developed from the name of the enterprise or correspond to it. Especially as concerns services, due to their individuality, umbrella marks are frequently chosen, comprising the entire range of services of an enterprise under a single mark.[120] Therefore, the generating of names and symbols, as well as the safeguarding of these developments, can take place on several levels between the enterprise and the individual product. The RES case-studies consider the two stages of the product groups and the surveyed products. In a second step, it is investigated whether these names or labels were registered as true brands. An objective result can only be achieved if this second step is carried out. In the first step, the answer to the question 'Has a label/trade mark been created for the product?' also depends on the interpretation of the terms label/trade mark.

The most important economic advantage of trade-mark products results from customer preferences for well-known products, thus carrying lower risks once the decision to buy is made. The trade mark or the trade name of an enterprise then carries certain associations that are assigned to the enterprise. If these associations are positive, e.g. 'competence', 'kindness' or 'reliability', then a basis is formed for a successful enterprise communication. Labels, therefore, do not only represent an enterprise or a product, but are intended to symbolize the special value of a service and thus the fulfilment of customers' wishes. Trade marks can also be labelled as 'capital in the form of marketing efforts and the history of the enterprise and the product' (Mantau, 1995b, p. 60). Therefore, the establishment of brands is a particularly long-term marketing instrument.[121]

Regarding the economic sector of forestry, above all, the names of the forest owners can be found. Even if these names are more or less well known among the customers of raw wood, this would only be an exceptional case for the sales of RES products. This could be risky with regard to the simple rule of marketability of non-material goods, e.g. for sponsors: 'No name – no money!' (Mantau, 1997a, p. 275). Table 4.32 shows the proportion of products with labels/trade names within the total of investigated RES case-studies, categorized according to product groups. The case-studies showed that 37 products were sold under a name that was regarded as a label by the respondents of the case-study interviews. The question as to what kind of label was chosen according to each product was answered as shown in Table 4.33. Thus, in 22 cases new names were developed and in two to 14 cases a label was created for an entire product group. Nevertheless, Table 4.34 shows that only 11 labels were registered. The registration institutions for this are quite various. The following institutions registering the surveyed RES products were enumerated. It is important to state that some individual products were also registered by several institutions:

- European Union (EU) (in four cases).
- Government departments, e.g. German patent office, Munich (in four cases).
- Chamber of Commerce/Trade (in two cases).
- Local authorities (in two cases).
- Forest Stewardship Council (FSC) (in one case).
- Various (local) registrations (in one case).

As a summary, it can be stated that about half of the material products (PG 1.1–2.2) and most of the environment and sponsoring products (PG 6.2 and 6.3) have got a label/trade mark. Above all, the group

Table 4.32. Share of RES products with existing labels. (Question 4.5.2: Has a label/trade mark for the RES product been created?)

		Yes		No		Missing	
Trade mark/label	Cases	*n*	%	*n*	%	*n*	%
Simple products for final users (PG 2.1, 3.2, 4.2)	33	10	30	23	70	0	0
Complex products for final users (PG 3.1, 4.1, 5.13)	23	11	48	12	52	0	0
Simple products for organizations (PG 1.1, 2.2, 6.1, 6.2)	29	9	31	19	66	1	3
Complex products for organizations (PG 5.2, 6.3)	13	7	54	5	38	1	8
Total	98	37	38	59	60	2	2

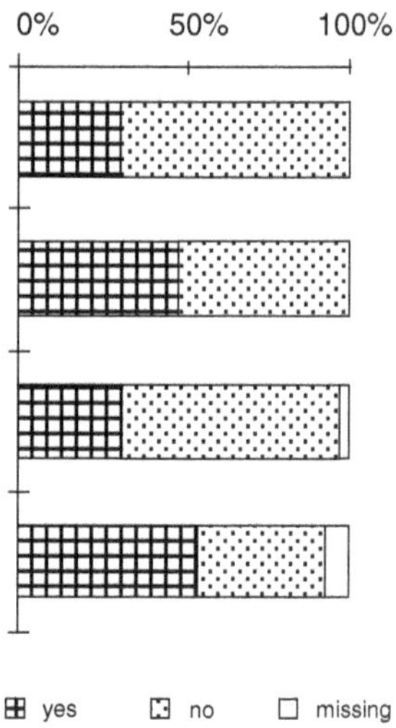

Table 4.33. Development of the label.

Label	*n*	%
New product label	20	54.1
Existing label attached	12	32.4
New label for a group	2	5.4
Missing	3	8.1
Total	37	100.0

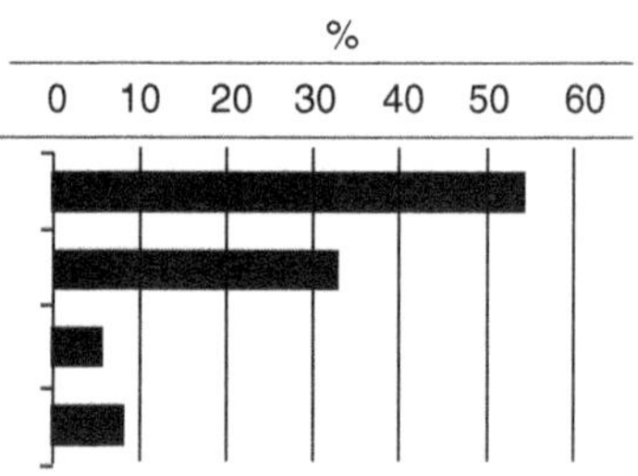

Table 4.34. Share of registered labels for RES products. (Question 4.5.3: Has the label been registered?)

Registered	*n*	%
Yes	11	29.7
No	26	70.3
Total	37	100.0

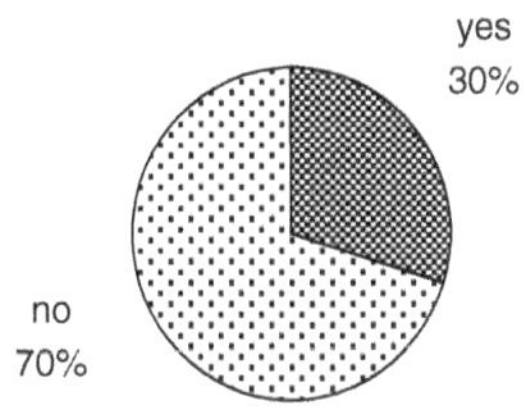

certified material products (PG 2.2b) is striking, in that they are particularly characterized by a label of origin. This label of origin is used as a substantial part of the brand name and, at the same time, a good protection of the name is guaranteed by its registration with official institutions. The fact that certified material products have been described exclusively in Italy indicates the particular significance of this marketing instrument in Latin countries. For example, the EU sign of origin 2081/91 is based on a product mark that has been established for a long time in Italy. It is quite possible that these labels of origin and certificates will increase in significance in these times of increasing demand for products whose origin and production should be entirely open to scrutiny and ecologically sound. Within this context are various eco-labels (EU Reg. 880/92) or forest certification following FSC standards (Merlo *et al.*, 1996, p. 9). Nevertheless, from the point of view of marketing, it must not be overlooked that certified standards only support the effect of sales as long as they appear as an exception within the entire offer. The quick reaction of the competitors on the market, normally at short notice, leads to a situation where only those enterprises that are not yet certified have disadvantages as far as their sales are concerned.

Within the group of sponsoring cases (PG 6.3), a high share of labels would be expected, as these are of some significance for the advertising of the user with the project. This is particularly valid for extensive sponsoring programmes. But existing labels are named in only four out of nine cases. However, out of the five environment products (PG 6.2), four are marked with a label but none is actually registered.

Labels of recreational products (PG 3.1–5.2) are mentioned for 37%, of which only three (6%) were registered. For the partly material and the complex recreational products (PG 3.1, 4.1, 5.13 and 5.2), some statements about labels can be found which partly describe the product itself, but in some other cases they consist, at least partly, of the name of the enterprise. The share of registered trade marks is low, as well as the share of labels for simple recreation products (PG 4.2), with five out of 24 cases. The name 'Loipl' as the voluntary-payment sticker for the Taufstein ski-track (DE19) is remarkable in this group. For the contract products (PG 6.1), no labels have been mentioned.

Finally, the slogan shows a close relationship to the product's name. Also the slogan should attract attention, be memorable, evoke positive associations and convey the centre of the advertising message as far as possible.

Trade and brand names/trade marks have a high significance in the business world. Nevertheless, they require certain non-interchangeable enterprise and product characteristics, and they produce costs that have to be covered by market advantages. These elementary basics of marketing have been neglected in the past by most forest enterprises characterized by timber production. Names for RES products quite often relate to the offering forest enterprise or to the region. RES products are rarely protected by trade marks/registered brands.

Brand names can be developed and protected for single products as well as for product groups or even for all the products of an enterprise. Beside the registration as a trade mark by the authorized government department, RES product brands can also be registered by various other institutions. Within the scope of the certification of an RES product based on a certain name, a basic brand creation may have already taken place. Brands as a recognizable identification mark of quality are particularly important for non-material products. Sponsoring products require a plain name in order to be used by the sponsor. The generally high significance of brand creation as a success factor is in opposition to the relatively few indications of their use in RES case-studies. This could indicate that the significance of brands has not yet been recognized by many forest enterprises.

CORPORATE IDENTITY.[122] Corporate identity is generated by connecting a name and a logo with an enterprise identity strategically and strictly developed internally and externally. For this aim, symbols are coined and features assigned to form a compact image and clearly defined targets, as well as a distinct positioning of the enterprise. In this way, the recognition value of an enterprise is safeguarded and customer loyalty can be established on the basis of reliability. It should be the declared objective of forest enterprises to create regional identities of this kind, by which competence in forestry, wood products, environment and recreation is communicated via corresponding performances.

A corporate identity (CI) serves the internal and external identification with the enterprise. An image of unity towards the outside and an internal bond between the whole staff is an important success factor in the business world. For generating a CI, team-orientated management methods and a complex integration of all the aspects of the enterprise are essential. Forest enterprises should use the opportunities for a CI and they should develop a clearly positioned regional institution, the competence of which for the forest finds broad acknowledgement. Impressive components of a forest CI can be developed from the unique characteristics of the forest and tradition. Nevertheless, each forest enterprise has to 'find itself' individually.

4.4 Summarizing Conclusions

4.4.1 General basics of successful RES marketing

What features contribute to the success of an RES product on the market?

The most simple basis for the sale of a product is the market demand. Nevertheless, in the highly developed national economies of Central Europe, the markets for most products are, to a large extent, saturated. Thus, the task of marketing is increasingly to create sale opportunities beyond the physical market demand by actively stimulating the desire for a product instead of passively waiting for action by the potential customer. From the RES case-studies, it becomes obvious that external influences, the design of the product and the organization and management have a decisive influence on the success of an RES product. On the contrary, active marketing that exceeds product design in only a few cases is judged to be decisive for the success of a product. Initially, this hints at the special significance of product policy in the early stages of a marketing process. Moreover, it may hint at the fact that the philosophy and the methods of marketing in the area of forestry have not been widespread up to the present and, therefore, are still of subordinate significance for some RES products, as many markets for RES products are not yet saturated.

In this work, it is shown that, by means of active marketing, the development and the sales of new RES products in any case can be promoted or even introduced, as an active, dynamic view of goods and markets can, in many cases, overcome the general opinion of the infrastructural services of the forest as public goods.[123] The basis for this is, above all, the strict orientation towards the customer, this being the core of the marketing philosophy. This philosophy leads to marketable offers and, therefore, secures the long-term existence of enterprises.

Marketing activities orientate to and fix their position in the system of objectives of an enterprise. The most important basis of any action is, therefore, in the first instance, an operational setting of objectives, which contain, besides economic targets, non-profit targets. Marketing supports the reaching of targets of forestry that are not related to income as well. On this basis, a marketing concept is developed, which is orientated towards a process of increase in value along the lines of the evaluation, value production and value mediation of the new product. Finally, the marketing concept is fixed in a written marketing plan. Thus, the objectives and measures of the marketing are understandable and controllable in the long run.

What factors influence marketing strategy and the marketing mix?

Selection of a suitable marketing strategy and the utilization of marketing instruments based on this are submitted to a variety of influences. Some of these influences were used for the formation of product groups:

- Tangibility of the product.
- Customer participation in product design.
- Target groups of the product (commercial or individual customers).
- Complexity of the product in view of the type (share of services) and the proportion (product grouping) of the offerer's performance.

Moreover, the marketing strategy is determined by characteristics that exceed product typology. Besides the superordinate target system, which is pursued with the offer and which is in reciprocal influence with other business environments, these characteristics include the following:

- Type of offerer (see, for example, the special social targets of public forests).
- Internal and external frame conditions of the project.
- Competitive situation for the product.
- Internal investment demand and the economic dimension of the project.
- Quantity of production.
- Age of the product on the market and in the individual enterprise.

The diversity of possible combinations of these forces leads to the statement that the individual aspects of a successful marketing strategy have to be defined and combined individually. Therefore, the following marketing instruments recommended for certain product groups can only represent a selection of measures which frequently can be sensibly implemented.

4.4.2 Securing success by the survey and use of internal and external influences suitable for the targeted objective

A comprehensive analysis of the situation, which should include all internal and external influences and finally lead to an opportunity–risk assessment, is the basis for all further marketing steps.

Environment, enterprise and product potentials as success factors

How do frame conditions and the characteristics of an enterprise and the product itself influence the product success?

The sale of RES products is strongly influenced by a great number of different formal rules and informal social influences. In particular, the limits of RES marketing according to law have to be considered. The effect of the influence of frame conditions on the sale of RES products depends on the characteristics of the offered product. The same may be stated for the influence of the ground property and the landscape environment as factors of success. Existing factual production factors, as well as available personnel and know-how, ameliorate the chances of success for a new product.

RES products have to be harmonized with the entire variety of products of the enterprise and with further claims on the forest. In contrast to timber sales, RES products play only a minor role in most forest enterprises as yet. For forest enterprises, decisions in marketing can be made easier by classifying their products into SBU. The various forms of life cycles of products complicate the judgement of the actual stage of life of existing products, but it is obvious that numerous RES products are at an early stage of life. There is often a danger that RES products will be dropped in the introductory phase because of insufficient financial results. Besides, RES products succumb to a seasonal influence.

Often, RES products are simply structured and harmonized with the situation of the offering enterprise concerning their need for specialized staff. A higher need for investments means a higher risk. Besides, RES products are often bound to their location or even unique, e.g. precious biotopes.

Frame conditions change over the course of time. Thus, certain trends affecting the sale of special products can be distinguished. The generally supposed trends for Central Europe point to the fact that RES products have indeed chiefly increasing chances.

Market conditions as factors of success

WHAT MEANING DOES MARKET RESEARCH HAVE FOR THE SUCCESS OF A PRODUCT? The philosophy of customer orientation already determines the objective to adapt new products in any case as closely as possible to the already existing or still to arise needs of the target groups. Thus, the preparing basis of any objective-meeting entry into the market should be a more or less comprehensive market enquiry. Within the course of such an enquiry, an assessment is made of all data important to the development of a marketing conception and collection of information about market participants, determinants and the state of the market. In this, the economic principles have to be kept in mind in order to hold the expenses of the market enquiry in the proper ratio to the economic meaning of the new product.

Market research is subdivided into the acquisition and analysis of new data (primary research) and the analysis of existing sources (secondary research). Within the scope of RES, basic qualitative analyses of secondary sources can be expected for economic reasons. Frequently, RES products are developed and offered only on the basis of non-systematically and probably accidentally acquired information as to customer wishes. Thus, investments in market enquiries are minimized, as forest enterprises mostly have low financial expectations of new RES products. A constant information system is a low-expenditure possibility for securing necessary decisions for forest enterprises. Therefore, a variety of information about their own and neighbouring enterprises and concerning sales people, distribution helpers, middlemen or public institutions is summarized systematically. Thus, a step towards the utilization of a low-cost market enquiry would be made for many suppliers of RES products.

The choice concerning the analysis operation should, in any case, be precisely harmonized with the available pieces of information and the desired information objective. A market potential determined in this way corresponds to the demand of the RES product of an enterprise if it takes all the marketing possibilities into account and, furthermore, is the only supplier of the product. More comprehensive market enquiries on higher expenses are normally carried out by specialized service providers. Nevertheless, basic knowledge on that topic should also be available at the forest enterprise.

HOW DOES THE COMPETITIVE SITUATION INFLUENCE THE SUCCESS OF RES PRODUCTS? An enterprise that draws public attention and is flexible will always have the best starting position in competition. RES products partly come into trade on new and unsaturated markets. For these RES products on small, open markets, competition can also be a factor for success. Existing competitive products are mostly successful and have often been established for some years. A special feature of RES products is that they sometimes compete with alternatives offered free of charge. By means of a recognizable additional benefit, successful RES products remain in competition. Competition has a relatively low influence on RES products. The partly existing positive influence is overlooked more easily than the negative effects.

WHAT SHOULD BE KNOWN ABOUT CUSTOMERS AND THEIR DECISION PROCESSES? For the successful introduction of RES products, applicable and positive information on potential customers is a basic premise. Customers' satisfaction as the 'highest good' of an enterprise is secured by the satisfaction of reasonable expectations and the steady quality of the product. Private consumers buying a product are motivated by needs and mostly emotionally based desires to satisfy their needs. The selection of perception and learning processes derived from positive and negative experiences determines the further actions of consumers. The various influences and influencing persons concerning a decision for buying and the phases of the buying process should be considered in a sophisticated manner in a marketing conception.

The buying behaviour of institutions is characterized by higher expenditures and, above all, it is determined rationally. If institutions intend to buy in the first instance the expenditures for negotiation are higher than if repeated buying is carried out. The binding of institutional customers to the seller is often tighter than on the consumption goods markets. Public buyers are normally bound into a tight legal frame of actions.

WHAT IS THE BEST WAY TO UTILIZE CUSTOMER ORIENTATION AS A SUCCESS FACTOR? Customer orientation of RES products is, above all, reached by product-policy measures. The design of offered facilities, additional offers, measures of the set-up of the infrastructure in the forest, service, flexibility and regard for the recreation offer within forest operations optimize the customer orientation regarding a recreational offer. The mediation of information has to be adapted to the customer group. Customer orientation is supported by regular customer enquiries. Individual contracts offer the possibility of a very close harmonization of the offer with the customer's need. Furthermore, instruments of price and distribution policy contribute to the customer's orientation. A great number of RES products do not enter into the particulars of individual customer's desires. This can be explained by the nature of the offer, on the one hand, while, on the other hand, the supplier may underestimate this success factor.

That orientation towards the customer has to be seen within the scope of economic restrictions is shown by the fact that maximizing the orientation towards the customer and neglecting the cost of production at the same time leads to acceptable but non-economic offers.

Strategic marketing decisions

Marketing strategies are decisions on principles. They consist of various strategy parts.

HOW CAN MARKET SEGMENTATION, THE SETTING UP OF TARGET GROUPS AND THE POSITIONING OF THE ENTERPRISE CONTRIBUTE TO SUCCESSFUL RES MARKETING? Forest enterprises starting to introduce new RES products should apply, in the first place, concentrated strategies aiming at certain target groups. Target groups are made up by market segmentation. Every market segmentation has to be carried out following diverse objective and suitable parameters, leading to special niche markets. Niche markets for RES products should not be defined too tightly. Chosen market niches are worked further by a concentration strategy. For particular individual RES offers, the strategy can be harmonized with a one-customer segment.

An individual positioning can be achieved by differentiation from other enterprises. Offerers of RES products should, beside the usual possibilities of differentiation, make efforts to differ from other suppliers by their unique competence in forestry. The setting up and mediation of a precise CI is of high importance for the clear positioning of an enterprise.

WHAT GENERAL STRATEGIC CONSIDERATIONS SHOULD BE TAKEN INTO ACCOUNT BY A FOREST ENTERPRISE? A forest enterprise should, as a basis of strategic considerations, consider the portfolio of its present and potential business units. Existing RES products need strategies that are adapted according to the particular life stage. For many forest enterprises, the potential utilization of attractive business fields in terms of RES products means a lateral diversification. Therefore, the risks and difficulties of a necessary diversification have to be recognized and worked out before investments are made. Forest enterprises should initially follow a strategy of being a market pioneer and niche-taker in the course of offering RES products. Thus, such an enterprise can turn out to be a market leader of the niche market, able to defend its position. Aggressive strategies and strategies of intensive growth on stable markets are rarely suitable for forest enterprises.

4.4.3 The formation of success factors by the marketing mix

The utilization of marketing instruments is fixed in an assessment plan on the basis of the strategical decisions.

Assessment of product policy

HOW DOES PRODUCT POLICY CONTRIBUTE TO THE SUCCESS OF RES PRODUCTS? Product policy in general aims at the establishment of a single position (mini-monopoly), the reduction of possible substitutions for the user and, therefore, the realization of an economic advantage. The product policy for RES products serves, in the first instance, the purpose of transforming the services of the forest into marketable products for which customers are ready to pay. Therefore, a clear definition and analysis of property rights is necessary. Attractive RES products can be generated by refining traditional forest products within the strategy of vertical diversification. Products are also generated by the completion of a public forest service with marketable additional services or by the marketing of the exclusive value level of a forest service. In particular, products concerning environmental sponsoring face a great market potential in this regard. Non-material products like forest services have to be 'materialized' in order to be evaluated by the customer. Within the offer of services, personal contacts within the enterprise and towards the customer are particularly important. Voluntary agreements are an important basis for RES products.

Figure 4.8 demonstrates the utilization of some product policy instruments that are important for RES products. Product innovation is an important factor of success. Nevertheless, innovations require precise planning because of the risks involved. Numerous RES products are new developments and, like all other products, they can also experience further innovative development in different ways. The product mix of forest enterprises is quite limited by different enterprise conditions. Nevertheless, the offer of a product mix orientated towards a target group can contribute considerably to the success of RES products via combination effects.

Measures of distribution policy

HOW DOES DISTRIBUTION CONTRIBUTE TO THE SUCCESS OF RES PRODUCTS? A distribution policy precisely harmonized with the product and the customer is an important factor of success. The distribution is particularly tightly bound to all the other sectors of the marketing mix. Many RES products are sold in a quite simple way. Nevertheless, multistaged distribution channels can indeed be sensible. Direct sale without using the intermediate trade dominates in RES products. In contrast to this, the sale of RES products is often supported by dis-

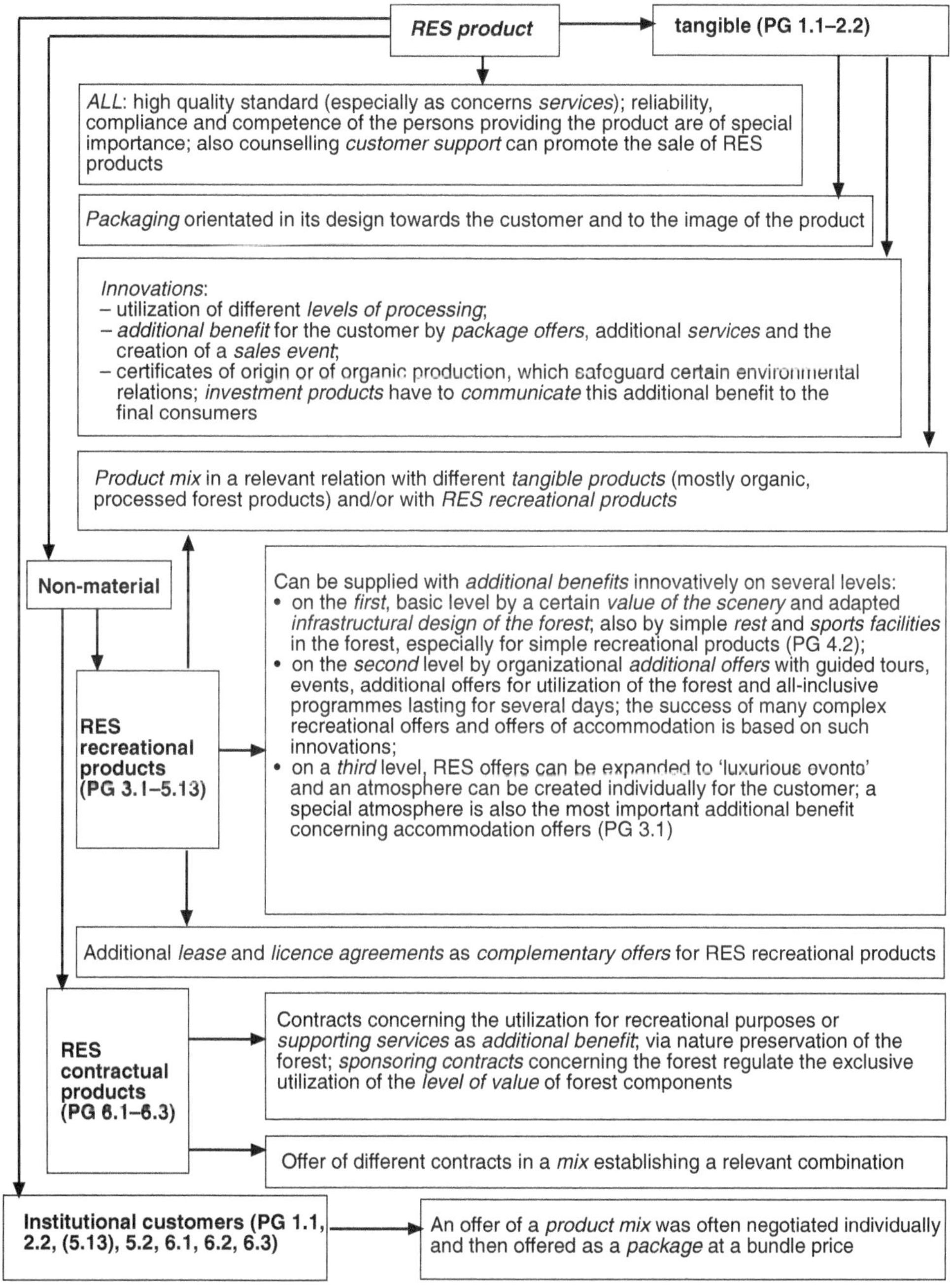

Fig. 4.8. The utilization of some product policy instruments for RES projects.

tribution helpers. These issues indicate that forest enterprises have little experience in trading RES products. In many cases, offerers of RES products do not see trading their products as part of a differentiated strategy but as a given result generated by general sales efforts.

Some of the distribution aspects

depending on the product type are represented in Fig. 4.9.

WHAT FACTORS INFLUENCE THE CHOICE OF A DISTRIBUTION SYSTEM FOR RES PRODUCTS? Cooperation with a sales intermediate (middleman) increases in significance with the following enterprise-related factors:

- In the case of increasing land requirement for the RES offer along with a fragmented forest property.
- In the case of increasing investment demand for the RES product, along with decreasing economic power of the individual forest owner.
- In the case of the increasing significance of a certain method of distribution and general marketing know-how, along with the decreasing know-how, of the individual forest owner.

The choice of distribution system is also influenced by demand-related factors. The greater the confidence capital of the RES product and the smaller the confidence capital of the forest enterprise on the corresponding niche market, the more significant becomes cooperation with intermediate salesmen, who, because of their image, already have the necessary confidence capital. The more similar the offered RES product is to already known offers, the more sense does it make to use those sales channels which the customer is accustomed to while looking for similar products.

While direct sale is important for quite young, innovative RES products, the

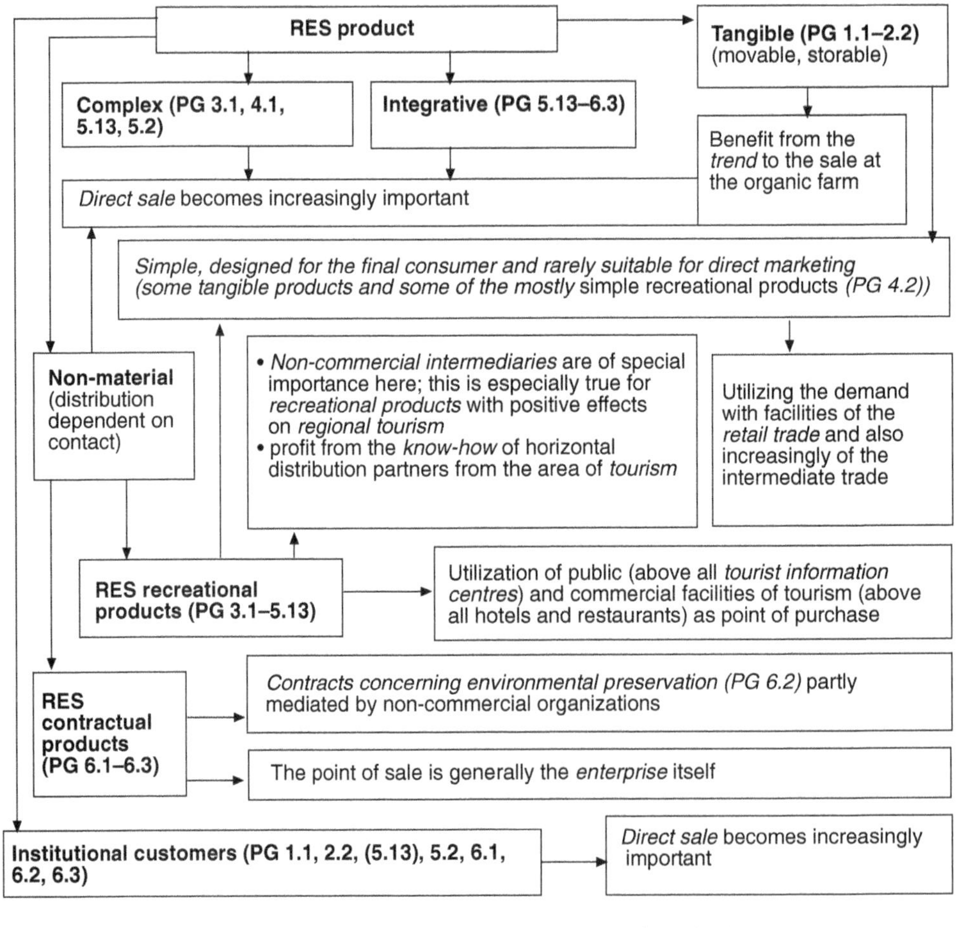

Fig. 4.9. Decision-making tools for the utilization of some strategies of distribution.

advantages related to the appointment of intermediate salesmen for older and better-known RES products increase with the sales figures. Intermediate salesmen can support selling the offer within the scope of special distribution channels. Finally, the intermediate salesmen overcome the problems deriving from not being well known on the RES market and from the lack of RES-market experience of certain forest enterprises.

HOW DO VERTICAL AND HORIZONTAL COOPERATION AND INFORMAL CONTACTS CONTRIBUTE TO THE SUCCESS OF RES PRODUCTS? Advantages for all participants should result from a horizontal or a vertical distribution structure. In particular, middlemen gain advantages from the sales of RES products they support. If these economic or structural advantages can be recognized and actively used by the forest enterprise, then this system is an important success factor. Within the scope of sales via intermediate salesmen, it is important to consider their marketing objectives within the overall marketing of a product.

Horizontal distribution cooperation with other enterprises and organizations is an important success factor in a number of RES products. Sometimes, only cooperation covering a great area, e.g. a whole region, makes the sales of RES products possible. This refers especially to RES recreation products and contract products, but also to some material RES products. These products can often be sold quite effectively at joint market events with different suppliers. By means of the participation of sponsors of the RES offer, the high social value of the forest can be used.

WHAT POINTS OF PURCHASE ARE PARTICULARLY SUITABLE FOR RES PRODUCTS? The sales point of RES products as the place of performance can rarely be selected freely, due to the fact that forest enterprises are stationary. Obligations to fulfil a service or tangible RES products, however, can be offered at various points of purchase (POP). These RES products are also sold mainly directly at the site of the forest enterprise or at local markets.

Measures of price policy

HOW DOES THE PRICE POLICY CONTRIBUTE TO THE SUCCESS OF RES PRODUCTS? Prices can be changed very flexibly without generating direct costs. The intensive use of price policy means points to the early stage of marketing for RES products. Many RES products are offered at individual prices. Discounts and the conditions of payment can support the success of RES products. The strategy of a price battle is extremely risky for a small enterprise and should be avoided by forest enterprises.

HOW SHOULD PRICES FOR RES PRODUCTS BE DESIGNED AND FIXED, AND WHAT TERMS AND MODES OF PAYMENT ARE ESPECIALLY SUITABLE FOR RES PRODUCTS? First price offers should be fixed in consideration of the enterprise's own production costs, the customer's willingness to pay and the prices for substitutes. Processes for environmental and economic evaluations are only partly useful for setting the prices for RES products. Prices have to be set with regard to the chosen marketing strategy. Price increases after the introduction can be quite difficult, but too elevated introductory prices may prevent the success of a product. Prices for consumption goods are often based on emotional perceptions, whereas institutions judge prices in a more rational way. RES products are mainly offered at market-orientated prices. In part, there are also prices that are only cost-orientated or determined by the customer. These hint at non-profit offers of RES products.

Prices can be changed actively or reactively due to external impacts, as RES products that were offered free before, now have a certain price to be paid due to the change of frame conditions. The great demand for products which were once free of charge and now have to be paid for quite obviously indicates their attraction. An example of this is guided tours through the forest. Prices for RES products are rarely changed actively.

Figure 4.10 represents some of the

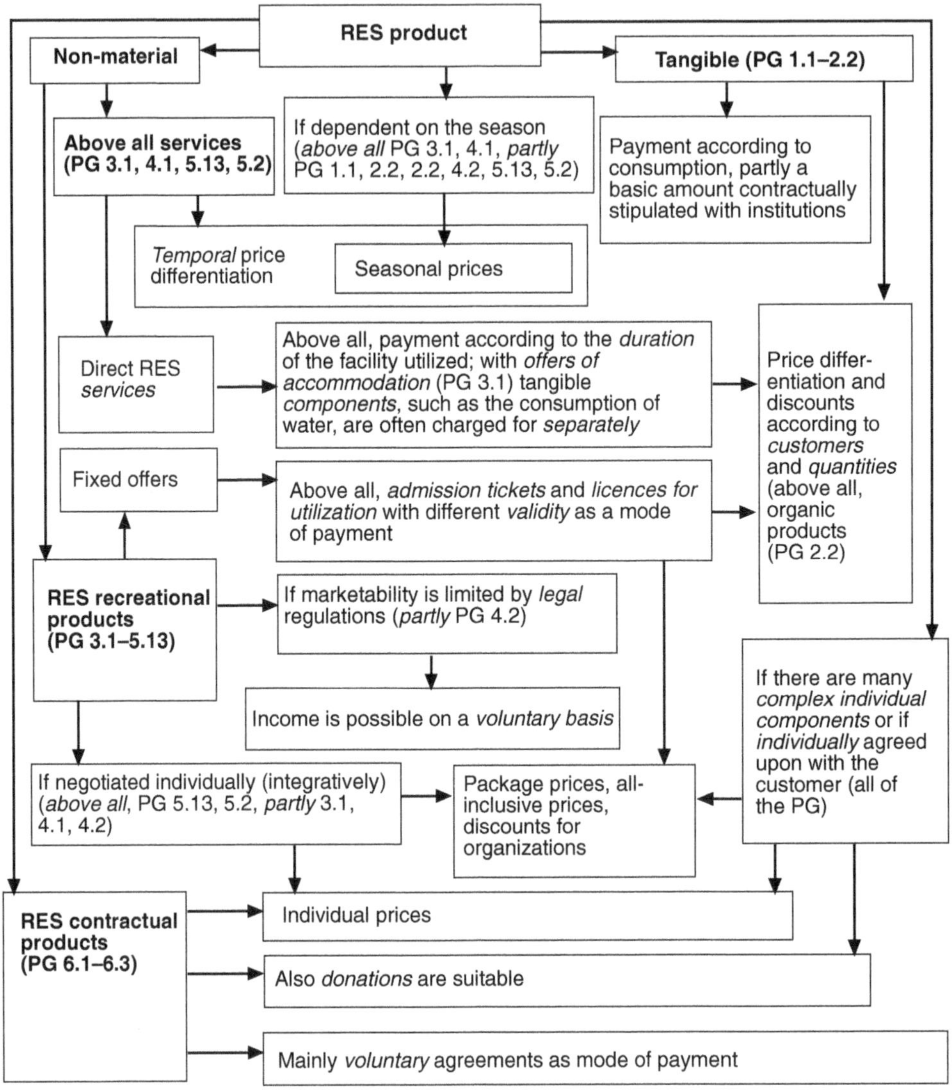

Fig. 4.10. Decision-making tools for the utilization of some price-policy instruments.

aspects of price policy depending on product types. Time-related price differentiation is mostly intended to balance fluctuations in demand, and therefore is an important means of marketing, especially as concerns services. Because of the high overheads of forest enterprises, the significance of price differentiation increases in order to reach constant contribution margins. Besides temporal price differentiation for RES products, mainly price differentiations according to customers and quantities are important.

The variability of the modes of payment for RES products is extremely high. The characteristic features of each product have effects on the optimal mode of payment. From the point of view of the customers, bundle prices have advantages and disadvantages. Mostly, the advantages dominate when there is the largest possible price visibility of the offer. Terms of payment such as discounts can also contribute to the success of RES products.

Measures of communication policy (Fig. 4.11)

HOW DOES COMMUNICATION CONTRIBUTE TO THE SUCCESS OF RES PRODUCTS? Communication is inevitable for the long-term existence of forest enterprises. The instruments of communication have to inform and motivate the customer on a behaviouristic basis. With RES products, the communication always fulfils its information task as well as its task of motivation up to a suitable degree. In order to send messages effectively and to obtain the desired effect, communication uses psychographic objectives and behaviouristic methods. Under certain conditions, forest enterprises need professional assistance with regard to their communication policy. Communication policy very strongly depends on its special target group. Thus, it has to be perfectly harmonized with the other elements of the marketing mix and should address target-group motivation as exactly as possible, while following the overall concept of the marketing mix, related to the product.

HOW CAN ADVERTISING, SALES PROMOTION AND PUBLIC RELATIONS BE USED FOR THE SUCCESS OF RES PRODUCTS? Advertising is an instrument of communication that develops a broad effect, along with relatively high costs. Direct contact with selected customers

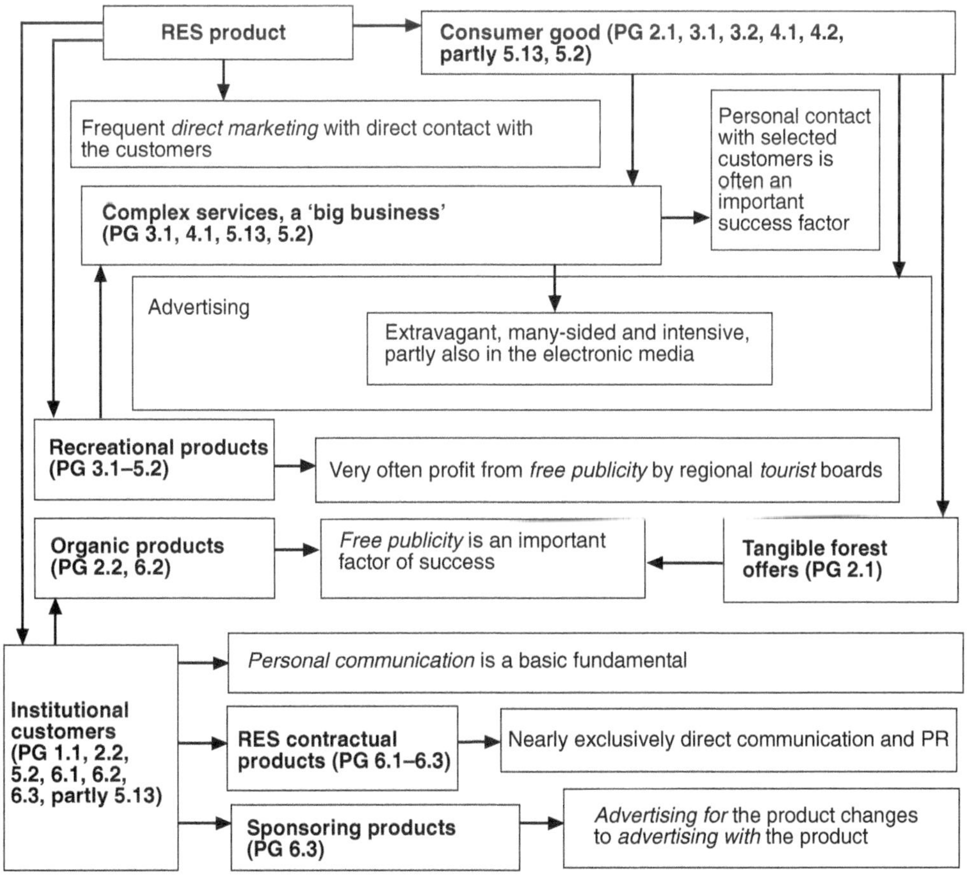

Fig. 4.11. The significance of some means of communication for RES projects.

is not established. By means of precise advertisement planning, cost-intensive coverage loss (*Streuverluste*) can be avoided. Advertising has to be carried out innovatively and creatively in order to be recognized and to show a permanent effect. The advertising media utilized should be selected according to their different advantages and disadvantages in view of the respective targets of advertising. The advertisement message should be precisely addressed to the target group with non-interchangeable specialities of the product. Non-material services therefore have to be materialized. If advertising is carried out for RES products, the planned expenses have to be sufficient to cross the lower threshold of advertisement effects (*Werbewirkungsschwelle*). RES products are currently advertised by all by announcements and brochures. New forms of advertisement also promise a greater effect for RES products. The expenses of advertisement by electronic mass media, however, is rather rarely covered by the profit expectations for RES products.

Sales promotion supports advertising measures by giving special stimuli and, for a short time, by directing the customer's attention towards the product. Sales promotion is, in some individual cases, a suitable instrument of communication for RES products. It can be used for all groups of demanders if the sale of a product is stimulated punctually, which frequently needs a constant demand.

PR is an instrument of communication that is preferably used by forest enterprises. Only rarely, though, is it used with regard to targeted sales support for special RES products. Executed by third parties, PR is characterized by high credibility. PR is particularly important for RES products that used to be available free of charge. Intensive PR can be combined effectively with the offer of certified, material RES products. This is also valid for integrative recreation products in the forest and for all environmental products. An innovative form of PR, as well as advertisement, for certain RES products is the use of the internet.

HOW CAN PERSONAL COMMUNICATION, FREE PUBLICITY AND OUTSOURCING BE USED FOR THE SUCCESS OF RES PRODUCTS? Above all, personal communication is characterized by the direct exchange of influence and reaction between the offerer and the customer. Such a contact exists for all those RES products which are elaborated together with the contracting party. Certain personal characteristics are particularly important for effective direct communication.

The communication aspects of free publicity and outsourcing, which are extremely important for RES products, mean the execution of a positive communication for a product or an enterprise by a third party. Free publicity results from the positive side-effects of RES products because of their unique characteristics and their connection with the forest ecosystem. The special advantage of free publicity lies in the high credibility and the fact that it is cost-free for the enterprise. An important form of free publicity is by word of mouth, which, for all products, results from customers' satisfaction. Outsourcing means delegating tasks of communication selectively to third parties.

HOW SIGNIFICANT ARE NAMES, BRANDS AND A CORPORATE IDENTITY FOR THE SUCCESS OF RES PRODUCTS? Names of firms and brands have a high significance in the business world. Nevertheless, they require certain, non-interchangeable enterprise and product characteristics, and they produce costs that have to be covered by market advantages. Therefore, these elementary basics of marketing have been neglected in the past by the mostly raw-wood-producing forest enterprises. Names for RES products quite often relate to the offering forest enterprise or to the region. RES products are rarely protected by brand names/trade marks.

Brand names can be developed and protected for single products, as well as for product groups or even for all the products of an enterprise. Beside registration as a trade mark by the authorized government department, RES product brands can also be registered with various other institutions. Within the scope of the certification of an RES product based on a certain name, a basic brand creation has already taken place. Brands as a recognizable identification (quality) mark are particularly important for non-material products. Sponsoring products require a plain name in order to be used by the sponsor. The generally high significance of brand creation as a success factor is in contrast to relatively few indications of its use in the RES case-studies. This could indicate that the significance of brands has not yet been recognized by many forest enterprises.

A CI serves the purpose of internal and external identification with the enterprise. An image of unity towards the outside and an internal bond between the whole staff are important success factors in the business world. For generating a CI, team-orientated management methods and a complex integration of all the aspects of the enterprise are essential. Forest enterprises should use the chances of a CI and they should develop into a clearly positioned regional institution whose competence regarding the forest finds broad acknowledgement.

Controlling and organization of marketing

WHAT ORGANIZING NECESSITIES ARE REQUIRED BY SUCCESSFUL RES MARKETING? Marketing activities should be controlled constantly, or at least after their application, with regard to their objective performance. Therefore, comparisons between the target and the actual effect are executed. As to the method, various processes, from a customer survey to a mathematical analysis of economic criteria are suitable. In this marketing field, forest enterprises can also be supported by external service providers.

An institutional marketing organization (marketing division) only makes sense in very large forest enterprises. Generally, marketing in forest enterprises is executed by the management. Nevertheless, the mode of thinking of the marketing in which the customer is the centre of all considerations has to be realized throughout every hierarchic level of an enterprise. The significance of conferring marketing tasks upon external service providers increases with the declining size of the forest enterprises. As this is not profitable for small forest enterprises, forest cooperations are particularly important as far as marketing is concerned.

4.4.4 Concluding remarks

This chapter is meant to be the basis of a development of differentiated marketing for products generated from the recreational and environmental facilities of the forest. For this aim, the fundamentals of RES marketing have been outlined. Furthermore, to reach this aim the context of different marketing textbooks was related to the comprehensive empirical material of case-study interviews on 98 examples of successful RES marketing.

By explorative research, innovations and basic approaches, as well as applied instruments of marketing, could be described mainly qualitatively and in relation to the product properties and frame conditions of the RES case-studies. Due to the selection and amount of case-studies, quantitative evaluations built upon this should not describe any continuous patterns. In contrast, the great variety of parameters and impact combinations of possible success factors could be demonstrated. These demonstrate the complementary, and under certain circumstances opposing, impacts of the use of different marketing instruments when developing factors for success. Likewise the impacts of different factors of success on the sale of a product can be either complementary or opposing.

Since every offering enterprise, every product and every product environment is different, this marketing process needs to be strictly orientated towards the single case. Thus, it is not possible to draw common conclusions as to which instruments have an impact on which factors and which combinations always lead to the best results. For every product, an individual concept of sales support needs be applied. The survey of factors and marketing efforts that has been developed in this chapter – which cannot be complete –

shows the opportunities and problems concerning the marketing of RES products. Finally, it is important to achieve a targeted, operationally planned and controllable concept of the marketing of the single product strictly according to the customer's wishes, considering all the properties that are relevant for this product.

Notes

1 The positioning of legal and administrative frame conditions as being of average significance concerning different success factors is surprising, when considering the significance this issue has in the discussion of forestry on marketing possibilities for RES products: see Bergen (1994), Blum *et al.* (1996) or Kissling-Näf (1998).

2 On the categorization of information as a production factor, see Kotler and Bliemel (1995, p. 179).

3 Weis (1997) accumulates a series of different definitions of marketing, among others the definitions by Gutenberg, Weinhold-Stünzi, Tietz, Köhler, Meffert, the American Marketing Association (AMA) and Kotler; the definition of the term chosen for the present chapter has been derived from the definitions by Nieschlag *et al.* (1991), Kotler and Bliemel (1995), Weis (1997, p. 19) and Meffert (1998).

4 The Forest Research Institute (*Forstliche Versuchsanstalt*) of Rheinland-Pfalz, for example, comprises in its Department E, 'Communication and Marketing', the offices of 'Non-profit marketing' and 'Profit marketing' (timber marketing) (Forstliche Versuchsanstalt Rheinland-Pfalz, 1997, p. 157).

5 Due to its significance for the economic success of enterprises and the variety of methods used in market research, this area has developed into a comprehensive area of knowledge on its own; for this reason, reference is made to the following sources, as examples: Henze (1994), Nieschlag *et al.* (1991, pp. 610–611, 627–636, 638–793 and 957); Kotler and Bliemel (1995, pp. 179–186, 190, 206–208, 278 and 394); Weis (1997, pp. 96 and 100–148).

6 See, among others, Nieschlag *et al.* (1991, pp. 793–809); Kotler and Bliemel (1995, pp. 192–206 and 394–416); Meffert and Bruhn (1995, pp. 97–99); Weis (1997, pp. 54–58, 96 and 149–165).

7 One central issue in the marketability of forest facilities is the limitation of property rights (see also Blum *et al.*, 1996, pp. 638–793 and p. 957); according to Barzel (1989, in Merlo *et al.*, 1996, p. 8), this follows from different regulations, habitual behaviour patterns and agreements; considerable differences which already exist on the level between the EU countries are clearly outlined by Merlo *et al.* (1996, p. 8); an insight into the legal frame conditions of the RES business in Austria is provided by the Österreichische Forstverein (1993, p. 49); for the offer of RES products, besides civil law, the areas of environmental law and trade law are of special significance (see also Nieschlag *et al.* (1991, p. 47); Kotler and Bliemel (1995, p. 260)).

8 See Kotler and Bliemel (1995, p. 280).

9 For the opportunities and risks that result from expected changes of the framework conditions, see Sekot (1992, p. 24); Österreichischer Forstverein (1993, p. 48); Glück (1995); Kotler and Bliemel (1995, pp. 241–274); Mantau (1995c); Meffert and Bruhn (1995, p. 121); Merlo *et al.* (1996); rights, drastically increasing qualitatively and quantitatively, concerning the infrastructural and social facilities of the forests result from various presumed trends.

10 As the most important market conditions, the 'five key parameters according to Porter' are frequently mentioned: strong rivalry within the segments by established and new competitors, by substitute products, by the increasing negotiating power of customers and by the increasing negotiating power of the supplier; these influences involve five risks that an enterprise should evaluate with regard to potential economic difficulties before concentrating its marketing effort on a certain market.

11 From the perspective of scientific theory, in this context the two cases in which no additional benefit could be found would have to be investigated more closely.

12 See, among others, Nieschlag *et al.* (1991, p. 833) or Weis (1997, p. 58).

13 On competitive strategies see, among others, Kotler and Bliemel (1995, pp. 468–469 and 597–615).

14 See Ziesling (1999b, p. 27).

15 On market segmentation and target groups, see, among others, Kotler and Bliemel (1995,

pp. 421–456), as well as Meffert (1986, pp. 243 ff.); Nieschlag *et al.* (1991, pp. 835–836); Meffert and Bruhn (1995, pp. 102–107).

16 In this context, it is important that these variables are measurable, separable, relevant as concerns purchasing behaviour, temporally stable and obtainable, that they dispose of an economically sensible size and profit potential and that they can be approached by feasible means.

17 See Kotler and Bliemel (1995, p. 423).

18 The considerations concerning positioning are mainly based on Kotler and Bliemel (1995, pp. 467–491).

19 Kotler and Bliemel (1995, p. 489) choose the following definitions:

> Positioning means the aim of an enterprise to design one's offer in such a way that it takes a particular and appreciated place in the customer's mind. Differentiating means the process by which reasonable differences are integrated into the design of a product offer in order to differentiate the own offer from the competitor's offer.

20 See observations on market segmentation strategies in, among others, Weis (1997, pp. 60–66).

21 Observations on the processes to select suitable market segments can be found in, among others, Meffert and Bruhn (1995, pp. 154–163).

22 See Meffert and Birkelbach (1992) in Meffert and Bruhn (1995, p. 189).

23 A classical approach frequently used concerning the general positioning of the enterprise's own strategy on markets and products is the matrix according to Ansoff (1966), in the four fields of which strategies are listed for the utilization of existing and new markets with existing and new products; the 'strategies of intensive growth' in the three remaining fields of Ansoff's matrix are still of minor significance for the offer of most RES products by forest enterprises; see, among others, Nieschlag *et al.* (1991, pp. 866–868); Kotler and Bliemel (1995, pp. 106–109); Meffert and Bruhn (1995, pp. 163–165); Weis (1997, pp. 69–71 and 194).

24 See Nieschlag *et al.* (1991, pp. 840–842).

25 Sekot (1995, 1997) outlines the problems that have to be acknowledged and solved by forest enterprises before utilizing a lateral diversification strategy: intradepartmental identification with the new business and motivation for this at all levels; acceptance of the business by the consumers and society; necessary internal and external communication; security of necessary investments and know-how; coordination with other enterprise tasks; possibly cooperation with external organizations, such as tourist organizations or other landowners.

26 See the nine-field matrix of General Electric and of McKinsey & Co., among others, in Nieschlag *et al.* (1991, pp. 879–884); Kotler and Bliemel (1995, pp. 102–105).

27 Thus, it becomes evident that the products and product groups of an enterprise should already be categorized into SBU (strategical business units) during the stage of situation analysis to facilitate the subsequent strategic decisions; thus, an enterprise can develop an individual concept consisting of targeting, strategy and tactical planning of measures, including the assignment of resources for the single SGE for each of the SGEs of its range of products; see Sekot (1997, p. 8).

28 See Weis (1997, pp. 485 and 488); also general standard strategies for different categories of SBU portfolios are briefly discussed here concerning the different marketing tools.

29 See Kotler and Bliemel (1995, pp. 568–585).

30 Weis (1997, pp. 178–181) gives a very concentrated description of product life stages and of strategic elements coordinated with the life stages; in particular, see the figure 'Examples for patterns of product life cycles of consumer goods' (p. 180).

31 See Meffert and Bruhn (1995, p. 131).

32 See Meffert and Bruhn (1995, pp. 171–174).

33 See, among others, Nieschlag *et al.* (1991, pp. 847–855) or Weis (1997, p. 75).

34 Nieschlag *et al.* (1991, p. 100) understand by the term 'product design' the development of a new or the modification of an existing product, in contrast to 'product eliminating', as a further fundamental means of product policy. Weis (1997, p. 173) replaces the term product design by 'product policy in the narrower sense of the word', which in turn is placed on the same level as programme and range policy, customer support policy and guarantee policy.

35 Such an alternative would certainly be to use this forest road, which for legal reasons must not be closed, without payment. The exclusion of unauthorized usage, i.e. dealing with the problem of free-riding, therefore, is an important part of RES product design.

36 See Meffert and Bruhn (1995, p. 273 and figures 5–8, p. 275).

37 See Eisele (1994, pp. 104 and 105); Merlo and Ruol (1996, pp. 34 and 35); Merlo *et al.* (1996, pp. 2f.), Ziesling (1999b, p. 27).

38 See Brabänder (1992) for contracts concerning nature conservation in forest economy.

39 The usual process of product innovation for the development of new products can be divided into different stages: generation of ideas, preselection of ideas (screening), development of the concept of the product and the temporary marketing strategy, economic analysis, product development, testing of the product and market testing, introduction of the product on the market, control of the market introduction; see Nieschlag *et al.* (1991, pp. 187–203); Kotler and Bliemel (1995, pp. 511–549); Meffert and Bruhn (1995, pp. 269–273); Weis (1997, pp. 194–238).

40 Possibilities of innovatively changing tangible or non-material products, for example, are ameliorations, which render the product more attractive for the consumer; modifications, which adapt the product to changes in customer behaviour; differentiations, to cover the demand of different target groups (Meffert, 1986, p. 364); see also Kotler and Bliemel (1995, p. 719).

41 The difficulties of an objective separation between product innovation and product variation (relaunch) are also indicated, in, among others, Nieschlag *et al.* (1991, p. 102) and Meffert and Bruhn (1995, p. 255).

42 See Kotler and Bliemel (1995, pp. 501–505 and the literature indicated there), as well as Nieschlag *et al.* (1991, p. 97).

43 In the case of PG 2.1, these have been instruments of distribution and communication (sale on weekly markets, outstanding personal service, cookery shows and tastings, recipes free of charge, promotional gifts and statements such as: 'The shop is indicated in specialized maps for hiking and cycling tours' and 'Management and marketing'); for PG 3.2 and 2.2b, as well, aspects of marketing were mentioned which exceed mere product design (simplification of payment and other tools of pricing or means in the area of communication, such as participation in shows and TV broadcasts to spread information on the healthy effects of edible chestnuts (IT27)).

44 The communication plan for the tree-crown path distinctly outlines the significance of constant further development of this product. Aspects of this product policy strategy were already outlined in the framework of the target definition and are described in detail in the following sections, most of all in 'Further development' and in 'Events and additional programmes'.

45 There are only a few hints of an additional benefit in the cases IT09, 12 and 15; in the case of AU18, a restaurant is mentioned as an additional benefit of levying a toll.

46 In the case of IT27 of PG 2.2b, certified chestnuts are also sold to individual customers as consumer goods; in this case, besides chestnuts, the range of products comprises mushrooms (among them truffles), wine and cheese, as well as holidays on a farm.

47 The cases of DE02, 28 and NL20 showed an outstanding variety of product mixes.

48 In the case of the tree-crown path (NL13), the communication plan deals intensively with various additional offers, such as excursions offered by the forest administration or events organized by the province, which is directly involved in the offer of the tree-crown path as an intermediary.

49 This cannot be called a product mix in the pure sense of the word, as it is indicated in the contract.

50 The facilities in the case of IT29 (information centre for visitors, museum of ecology, guided tours, environmental training, sale of environmental products) are not offered to sponsors, but to the visitors of the national reserve.

51 Nieschlag *et al.* (1991) describe the term 'product quality' as 'the basic suitability to satisfying customers' wishes constantly increasing'. The authors correctly underline the fact that certain (especially exterior) product designs can increase the benefit for the customer only for a limited amount of time if they are related to fashions and trends (Nieschlag *et al.*, 1991, pp. 177–183).

52 Concerning this paragraph, see, among others, Kotler and Bliemel (1995, pp. 721–729) or Meffert and Bruhn (1995, pp. 197–247).

53 See, among others, Nieschlag *et al.* (1991, p. 219); Kotler and Bliemel (1995, pp. 731–740); Weis (1997, pp. 238 and 247).

54 See the definitions by Meffert (1986, p. 421) and Hamm (1991, p. 207, referring to Nieschlag *et al.*, 1988, p. 367).

55 Such a network of distribution channels that results in acquisition effects can be utilized to establish customer loyalty to the offerer in many ways, and for the necessity and possible contents of operational targeting on the chosen distribution policy, see, among others, Kotler and Bliemel (1995, pp. 801 and 802) and Meffert and Bruhn (1995, pp. 319 and 320).

56 For the advantages of multistage channels of distribution, see Kotler and Bliemel (1995, pp. 806–808).

57 The issue of distribution levels oscillates between the survey of contracts and the organizational structure (Chapter 5), on the one hand, and RES marketing (this chapter), on the other, concerning the classification of some cases into product groups, e.g. in some interviews, the contract and organizational structure is described with regard to the connection between the forest owner and the offerer of a recreation facility and the marketing is described with regard to the connection between the offerer and the final user of the recreation facility. Thus, some cases are part of different product groups depending on the contract or marketing point of view.

58 See, for example, Weis (1997, p. 305).

59 See Meffert and Bruhn (1995, p. 321).

60 Nieschlag *et al.* (1991, pp. 54–76) critically investigate the term partnership between producer and intermediaries and also other areas of conflict in distribution. It can be viewed as a political task to maintain effective competition within the levels of distribution as well as to control the concentration of influence between distribution levels and to secure a suitable supply of consumer groups differently distributed regionally.

61 On the one hand, every organization that directly supplies products to the final consumers for their personal, non-profitable use, acts as a retailer (Kotler and Bliemel, 1995, p. 851). This obviously also concerns forest enterprises that market consumer goods. On the other hand, forest enterprises which market their products or services to retailers or to commercial users act as wholesalers (Kotler and Bliemel, 1995, p. 880).

62 Another example: all enterprises are almost forced to offer flights within the START system (a very common computer-based booking system for flights) of the travel agencies in order to be competitive.

63 Compare Kotler and Bliemel (1995, pp. 828–831).

64 With markets developing from production-orientated to buyers' markets, intermediaries who have been regarded as a 'necessary through-station' before have become very influential in the distribution range of producers producing goods for trade. Nieschlag *et al.* (1991, p. 367) point out that by using intermediaries, many areas of distribution policy have to be left to other companies 'which consider themselves not any longer to "fetch and carry" but usually as self-confidential, often too powerful business partners'. Marketing strategies of these intermediate stages, which can offer a specific product to the end consumer or refuse to pass it further, have to be considered very thoroughly in the distribution policy of every company. This is especially true for tangible goods (e.g. foods) (Hamm, 1991).

65 In general, basic disparities in targets between the producer and the trade result from the striving of both parties for the highest possible independence and profit from the mutual business; for the entire problem and further detailed target disparities, see Hamm (1991, p. 208) and the literature indicated there (especially table no. 7, pp. 209 and 210).

66 In the communication plan for the tree-crown path (NL13), potential sponsors are considered as an important target group and the respective communication strategies are described.

67 See also Nieschlag *et al.* (1991, pp. 373–378).

68 Kotler and Bliemel (1995, pp. 893–902) discuss this issue, together with technical questions of distribution; Weis (1997) considers this problem on pp. 351–359 of his book. Some hints on the logistics of distribution of services can be found in Meffert and Bruhn (1995) on pp. 325 and 326.

69 For some products, information on the customer area could be gained in the case-studies; this question, however, was almost exclusively answered for eco-investment products (PG 2.2a, NL04, 08 and 11). But there was also one determination from PG 1.1 (AU03) and are from PG 2.1 (DE06); in four cases, the customer area exceeded a radius of 50 km; in only one case, the selling of eco-meat (NL11), was the customer area limited to the vicinity with a radius of up to 50 km.

70 It is likely that, in the case of demand for eco-meat on the local market (NL08), the situation is similar.

71 This concerns the group of material investment products (PG 1.1), for example cases AU16 and 19, in which settlements are both customers and locations of demand for the offered water. This depends on the definition of settlement either as a gathering of human beings or as a group of houses.

72 The phrase 'private horse-riding school' in case DE24 describes this institution appearing as the customer of the contract. The rider, being the user of the product, asks for utilization opportunities at the school.

73 The contents of price policy are thus the remuneration of a service and the contract design related to it, so that it is also referred to as remuneration policy or contractual policy; see, among others, Hamm (1991), Nieschlag *et al.* (1991), Weis (1997) and Meffert (1998).

74 Regarding the different kinds of discounts and terms of delivery and payment see, among others, Hamm (1991) and Nieschlag *et al.* (1991, pp. 258 and 259).

75 See the historical review of the economy by Nieschlag *et al.* (1991, p. 235); if the price is handled in too flexible a way as an instrument for sales support, however, there is the danger of a disastrous price war; see, among others, Nieschlag *et al.* (1991, p. 240) and Weis (1997, p. 256).

76 See, among others, Hamm (1991); Nieschlag *et al.* (1991, p. 300); Weis (1997, p. 263).

77 Cost orientation is based either on absorption costing or on marginal costing; difficulties may arise from the independence of actual market processes and from the capacity output of production, as well as from the uncritical assumption of costs as a basis of calculation; for calculational approaches to cost-orientated price determination, see, among others, Nieschlag *et al.* (1991, pp. 304–321) and Weis (1997, pp. 264–267).

78 Customer orientation as the basis of pricing is based on the exploration of the demand function and of the customers' disposal to pay for the product; here, generally problems arise because of the fact that the investigation of the willingness to pay a certain amount methodically is extremely complicated and, furthermore, there may exist an internal dependence between the estimation according to the price as an indicator of quality (especially as concerns services); finally, it is also the task of marketing to change the willingness to pay; see, among others, Nieschlag *et al.* (1991, pp. 266–284 and 325–331); Kotler and Bliemel (1995, pp. 753–756); Meffert and Bruhn (1995, p. 306); Weis (1997, p. 267).

79 Due to its reliability, competitor orientation is an important part of market-orientated price formation; however, it does not suffice on its own, as the enterprise's costs are not considered and as the enterprise's own market potential is, under certain circumstances, not entirely exhausted; descriptive methods for competition-orientated price determination are described by Weis (1997, pp. 269–275).

80 The scientific branch of forest welfare economics (*Forstlichen Wohlfahrts-Ökonomie*) utilizes different methods of evaluation for forest facilities, such as the contingent valuation method (CVM) (1947), the travelling cost method (TCM) (1949) or the hedonic pricing method (HPM) (1966) and continuously optimizes them; see, for example, Bergen (1991, 1994); Bergen *et al.* (1992); Elsasser (1996); further hints on the subject can be found in, among others, Blum *et al.* (1996) and Merlo *et al.* (1996, p. 3).

81 Bergen (1994) mentions the possibility of judging the economic efficiency of a market solution for RES products by environmental economic methods; at the same time, he recommends these processes for determining maximum willingness to pay, while he considers the 'method of additional expenditure and diminishing return' to be suitable for the calculation of the price floor (see Bartelheimer and Baier, 1991).

82 See Mantau (1995a, pp. 140 and 141); also, among others, Meffert (1986) and Hamm (1991) show that the models in which price plays the central role in economic exchange processes are based on premises that cannot be fulfilled in the economic reality or which are even prevented by other marketing instruments.

83 See, among others, Weis (1997, pp. 280–283).

84 This refers to the case-studies DE17, 18, IT06, 09, 12, 16, 22 and NL05, 09, 10, 13.

85 This refers to the case-studies DE21, 24; AU14 and 20 (PG 6.1) as well as NL16 (PG 6.3).

86 This refers to the questions 3.1.7, 4.4.1 and 4.4.3 in the RES questionnaire.

87 The price is always strongly related to product image. For this reason, potential interdependencies between price changes and the rest of the marketing mix have to be considered; such interdependencies can be observed on Christmas markets for example: a price decrease of Christmas trees on 23 December, in order to sell the remaining trees, might be the incentive for repeat purchases on 23 December in the subsequent year!

88 With a time-dependent price differentiation, however, there is always the danger of an image loss as the customer is always orientated towards the lower price. Thus, it is more likely that the holiday house is considered to be expensive during main season rather than cheaper in the off-season.

89 See, among others, Hamm (1991); Meffert and Bruhn (1995, p. 309); Weis (1997, p. 278).

90 Cases DE03, IT20, AU05, NL10 and 13.

91 Cases IT03, 05, 09, 15, 28; AU02, 13, 18; NL02 and 05.

92 This categorization is mainly based on question 3.1.7 of the RES questionnaire of the case-studies, referring to the mode of payment.

93 Kotler and Bliemel (1995, p. 743) generally distinguish between 'individual prices' and 'unified prices'. Individually negotiated prices are of considerable importance with, for example, single contracts on lease or permission in the area of RES products. The bigger the circle of potential customers becomes and the lower the value of the single product and the degree of customer participation on performances become, the more likely it is that uniform prices are determined.

94 The all-inclusive and allocatable package prices, which also appear in the RES case-studies, are described by Meffert and Bruhn (1995, pp. 316–318) as pure bundling and mixed bundling.

95 Cases DE06, 13; AU15 and NL19.

96 Cases DE28 and NL20 (PG 3.1a), as well as IT05, 06, 28; AU13 and NL02 (PG 4.2).

97 Cases DE15, 18; IT10; NL09 and 10.

98 Some of these targets are: success of touch and contact (*Berührungs- und Kontakterfolg*) – minimizing of lost coverage costs (*Streuverluste*) in communication expenditures; attention and interest effects (*Aufmerksamkeitswirkung, Interesseweckung*) – penetration of the perception filter (*Wahrnehmungsfilters*) of the potential customer, activating his/her preoccupation with the product; creation of an emotional effect, a positive attitude – appealing to emotional motivation, changing unconscious needs into concrete desires as basis of a specific demand; information, memory effect – securing of durable information concerning the product; triggering of the purchase action – economically relevant realization of the demand; see Meffert and Bruhn (1995, pp. 284 and 285); further examples can, for example, be found in Hamm (1991, p. 273).

99 Nieschlag *et al.* (1991) describe the socio-economic basics of the science of publicity technique very extensively on pp. 452–478 of their marketing textbook; there are different models for the description of communication and its effects, all of which basically consist of the elements sender – message – medium – receiver – reaction (see, for example, Kotler and Bliemel, 1995, pp. 909–912); in certain areas the message is typically communicated and supported by opinion leaders (see, for example, Nieschlag *et al.*, 1991, pp. 474–477); a widespread model describing the effects of the advertising model is the AIDA model (attention, interest, desire and action); this simple four-part form has meanwhile been improved in many ways (see, for example, Meffert (1998, pp. 671–685); concerning fundamental factors of seeking attention, see also Mantau (1995b, pp. 50–56)).

100 See Meffert and Bruhn (1995, p. 287).

101 A direct link between distribution and communication is also called 'direct marketing'; the significance of direct sales for RES products has already been outlined; typical instruments of direct marketing, also feasible with RES products, are business letters, sales via catalogues, advertisements in newspapers and magazines, telephone solicitation or sale via the internet (see Kotler and Bliemel, 1995, pp. 908, 909 and 1089f.).

102 See Kotler and Bliemel (1995, p. 955); as concerns the typical properties of advertising, such as the range of dispersion, the psychological means, the feature of expendability or high costs, see Kotler and Bliemel (1995, pp. 939 and 940).

103 According to the sequence of the media-specific advertising expense, Germany 1993–1996 (Kotler and Bliemel, 1995, p. 974; Weis, 1997, p. 389).

104 See Kotler and Bliemel (1995, pp. 958–988); for the planning of the advertising effort, design, result testing of the advertising effort and possible approaches for optimization, see also Nieschlag *et al.* (1991, pp. 497–586); Kotler and Bliemel (1995, pp. 988–999); Weis (1997, pp. 380–422).

105 See Mantau (1995b, p. 57); on advertising aids for non-material offers, such as the description of material internal factors of performance (forester at a guided tour through the forests with beautiful scenery) or references by satisfied customers, see Meffert and Bruhn (1995, pp. 290 and 291).

106 See Hamm (1991) and Kotler and Bliemel (1995); on the typical properties of sales promotion, such as the direction of attention, effect of incentives or short duration, see Kotler and Bliemel (1995, p. 940).

107 See Kotler and Bliemel (1995, pp. 1006–1019); the instruments of sales promotion of RES products mainly concentrate on direct customers, since the intermediate trade and sales departments play only a minor role.

108 See Nieschlag *et al.* (1991, p. 492); Kotler and Bliemel (1995, pp. 1003–1006).
109 Kotler and Bliemel (1995, pp. 940 and 941) underline the following features as the special characteristics of PR: high credibility, as it does not support a certain product; the possibility of shows and exhibitions; for the processes of PR, see Kotler and Bliemel (1995, pp. 1024–1028).
110 See Kotler and Bliemel (1995, p. 941); note also the hint at the safeguarding of a long-term relation due to personal communication and the fixed-cost expenses due to sales personnel; for the requirements of the personal sale, see Kotler and Bliemel (1995, pp. 1069–1071); Meffert and Bruhn (1995, pp. 292–293).
111 Aggressive selling techniques often limit sales to one time for consumer goods and often lead to a failure of negotiations for investment goods. In contrast, a personal communication that is too cautious often seems to be a sign of insecurity and may lead to the customer's impression of a product of inferior quality.
112 Kotler and Bliemel (1995, pp. 1068–1084) deal with the basics of personal selling very concretely.
113 In the communication plan for the tree-crown path (NL13, PG 4.1), free publicity, besides PR, is integrated into the communication strategy towards the media; this can be derived from the slightly deviating definition of free publicity as free of charge, but nevertheless directable PR.
114 Cases NL13, 14, 17 and IT28.
115 Mentioned in a handbook about regional picnic places.
116 Cases IT06, 11, 15–17 and 23.
117 Cases DE09, 10, 27 and 28.
118 Within the course of introduction, the product was mentioned in the local press.
119 Methodical considerations concerning the creation of a name and a brand can be found, among others, in Weis (1997, pp. 224–225); information about the legal and technical interrelations of the creation of brands is offered, among others, by Kotler and Bliemel (1995, p. 679).
120 See Meffert and Bruhn (1995, pp. 276–283); Meffert (1998, pp. 793–808).
121 Concerning this paragraph, see, amongst others, Nieschlag *et al.* (1991, pp. 185 and 187); Kotler and Bliemel (1995, p. 678), Mantau (1995b, pp. 60–61) and Meffert and Bruhn (1995, pp. 276–283), dealing with the special conditions of service trade marks.
122 See, among others, Antonoff (1995, pp. 1157–1161); Mantau (1995b, pp. 61–63, 1995c, 1997, p. 637); Meffert and Bruhn (1995, pp. 302–303); Weis (1997, pp. 457–459).
123 See, among others, Mantau (1993, 1994, 1995a,c).

References

Ansoff, H.I. (1966) *Management-Strategie.* Munich.

Antonoff, R. (1995) Forstästhetik – der Begriff mit Zukunft. *Allgemeine Forstzeitung* 21(95), 1157–1161.

Bartelheimer, P. and Baier, M. (1991) *Belastungen der Forstbetriebe aus der Schutz- und Erholungsfunktion des Waldes.* Landwirtschaftsverlag (Schriftenreihe des BML, Reihe A, Vol. 399), Münster-Hiltrup. 149 pp.

Bergen, V. (1991) Monetäre Bewertung der Landeskulturellen Leistungen der Forstwirtschaft. In: Bergen, V., Brabänder, H.D., Bitter, A.W. and Löwenstein, W. (eds) *Schriften zur Forstökonomie,* Vol. 1. J.D. Sauerländer's Verlag, Frankfurt-on-Main.

Bergen, V. (1994) Vermarktung und Bewertung der Schutz- und Erholungsfunktionen des Waldes. *Holz-Zentralblatt* 5(94), 33–38.

Bergen, V., Löwenstein, W. and Pfister, G. (1992) Studien zur monetären Bewertung von externen Effekten der Forst- und Holzwirtschaft. In: Bergen, V. and Brabänder, H.D. (eds) *Schriften zur Forstökonomie,* Vol. 2. J.D. Sauerländer's Verlag, Frankfurt-on-Main.

Blum, A., Brandl, H., Oesten, G., Rätz, T., Schanz, H., Schmidt, S. and Vogel, G. (1996) Wohlfahrtsökonomische Betrachtungen zu den Wirkungen des Waldes und den Leistungen der Forstwirtschaft. *Allgemeine Forst- und Jagd-Zeitung* 5(96), 89–95.

Brabänder, H.D. (1992) Vertragsnaturschutz – eine Chance für die Forstwirtschaft? *Forstarchiv* 63, 41–45.

Eisele, F.-L. (1994) Bessere Wirtschaftsergebnisse durch besseres Marketing. *Forstwissenschaftliches Centralblatt* 113(94), 101–112.

Elsasser, P. (1996) Der Erholungswert des Waldes: monetäre Bewertung der Erholungsleistung ausgewählter Wälder in Deutschland. In: Bergen, V. and Brabänder, H.D. (eds) *Schriften zur Forstökonomie*, Vol. 11. Sauerländer, Frankfurt-on-Main.

Forstliche Versuchsanstalt Rheinland-Pfalz (1997) *Jahresbericht 1997.* Mitteilungen aus der FVA Rheinland-Pfalz No. 43/98, Trippstadt.

Glück, P. (1995) *Vermarktung forstlicher Dienstleistungen.* Vortrag vor der Vollversammlung des Verbandes der Akademiker der Österreichischen Bundesforste am 3. Dezember 1994 in Salzburg. *Internationaler Holzmarkt* 3(95), 18–21.

Hamm, U. (1991) *Landwirtschaftliches Marketing: Grundlagen des Marketing für Landwirtschaftliche Unternehmen.* UTB für Wissenschaft 1620, Ulmer Verlag, Stuttgart, 331 pp.

Henze, A. (1994) *Marktforschung: Grundlage für Marketing und Marktpolitik.* UTB für Wissenschaft 1792, Ulmer Verlag, Stuttgart, 512 pp.

Kissling-Näf, I. (1998) Grosser Wert und wenig Geld: Über die Honorierung von Waldleistungen. *Zeitschrift für Umweltpolitik und Umweltrecht* 3(98), 373–397.

Kotler, P. and Bliemel, F. (1995) *Marketing-Management: Analyse, Planung, Umsetzung und Steuerung*, 8th edn. Schaeffer-Poeschel-Verlag, Stuttgart, 1220 pp.

Mantau, U. (1993) Die forstliche Produktlücke: Infrastrukturleistungen des Waldes marktfähig machen. *Allgemeine Forstzeitung* 4(93), 186–190.

Mantau, U. (1994) Produktstrategien für kollektive Umweltgüter. Marktfähigkeit der Infrastrukturleistungen des Waldes. *Zeitschrift für Umweltpolitik und Umweltrecht* 3(94), 305–322.

Mantau, U. (1995a) Forest policy means to support forest outputs. In: Solberg, B. and Pelli, P. (eds) *Forest Policy Analysis – EFI Proceedings No. 2.* IUFRO/SNS – Meeting report, Working group S6.12–01, pp. 131–146.

Mantau, U. (1995b) Eine Strategie für forst- und holzwirtschaftliche Kommunikationpolitik. In: Bartelheimer, P., Gundermann, E. and Moog, M. (eds) *25. Forst- und Holzwirtschaftliches Kolloquium München.*

Mantau, U. (1995c) Von der Waldfunktionenlehre zur Waldproduktlehre. *Holz-Zentralblatt* 26(95), 445, 451–453.

Mantau, U. (1996a) Konstruktionsfehler im Nachhaltigkeitsdenken – Von der ergebnisorientierten zur handlungsorientierten Nachhaltigkeit. *Allgemeine Forstzeitung/Der Wald* 23(96), 1274–1278.

Mantau, U. (1996b) Öffentliche Güter und staatliches Handeln. *Forst und Holz*, 51(4) 102–107.

Mantau, U. (1997) Erschließung neuer Märkte durch forstwirtschaftliche Zusammenschlüsse. *Allgemeine Forstzeitung/Der Wald*, 12(97), 636–639.

Meffert, H. (1986) *Marketing.* 7th edn. Gabler Verlag, Wiesbaden.

Meffert, H. (1998) *Marketing.* 8th edn. Gabler Verlag, Wiesbaden, 1372 pp.

Meffert, H. and Bruhn, M. (1995) *Dienstleistungsmarketing.* Gabler Verlag, Wiesbaden, 627 pp.

Merlo, M. and Ruol, G. (1996) Naturschutz und Forstwirtschaft in Italien. *Österreichische Forstzeitung* 12(96), 34–36.

Merlo, M., Kuehl, G. and Ruol, G. (1996) Possibilities of additional income from environmental goods and services produced by agriculture and forestry in mountainous areas. In: *Proceedings of Cost Conference, Vienna, 1996, Forestry in the Context of Rural Development: Future Research Needs. Conference Report.* 18 pp.

Nieschlag, R., Dichtl, E. and Hörschgen, H. (1991) *Marketing.* 16th edn. Duncker & Humbolt Verlag, Berlin, 1064 pp.

Österreichischer Forstverein (1993) Multifunktionale Waldwirtschaft. *Österreichische Forstzeitung* 6(93), 48–53.

Sekot, W. (1992) Strategische Analyse der Dienstleistungsproduktion im Wald. *Österreichische Forstzeitung*; 6(92), 24–27.

Sekot, W. (1995) Produktion und Vermarktung Forstlicher Dienstleistungen aus Österreichischer Sicht. In: Jöbstl, H. and Zajac, S. (eds) *Marketing in Forest Enterprises. Proceedings of the IUFRO Working Group 4.04-02, Krakau, 29.8.–2.9.1994.* Warsaw, pp. 99–107.

Sekot, W. (1997) Multifunktionale Leistungserstellung, PR-Gag, Peanuts oder Wirtschaftsstandbein? *Österreichische Forstzeitung* 11(97), 7–8.

Weis, H.C. (1997) *Marketing*, 10th edn. F. Kiehl Verlag, Ludwigshafen, Rhein.
Ziesling, V. (1999a) Sind Landesforstverwaltungen zum strategischen Management fähig? *Allgemeine Forstzeitung/Der Wald*, 1(99), 26.
Ziesling, V. (1999b) Einführung eines strategischen Managements in die Forstbetriebe. *Allgemeine Forstzeitung/Der Wald*, 1(99), 27–31.

5

Contracts and Organizations as a Basis of Market Development

5.1 The Reason for Analysing Economic Institutions and How to Do It

5.1.1 Targets of the study

Environmental and recreational goods and services of forest outputs are discussed extensively in literature. Most studies describe these services as public goods, characterized by non-excludability and non-rivalry. Environmental and recreational goods and services are described as goods that are produced by forest enterprises but cannot be sold. Therefore, there is a market failure, and so the theory of external effects is often used to analyse the problem (Bergen, 1991, 1992; Elsasser, 1996b; Kissling-Näf, 1998).

In the field of environmental economics, many approaches to evaluating these effects have been developed and tested (Sekot and Schwarzbauer, 1995; Wibe, 1995) – especially in the field of recreational services of forest outputs (see Löwenstein, 1991; Merlo and Signorello, 1991; Elsasser, 1996b). The basis for a benefit-orientated evaluation is the view of the user. To analyse this, the travel cost method or the method of contingent valuation is most frequently used. Hampicke (1996, pp. 29–31) presents an overview of the results of relevant studies.

It is very difficult to obtain financial reimbursement for environmental and recreational services. Some foresters demand a correction of market failures by state subsidies, whereas others want to solve the problem by developing products in the field of recreational and environmental services (RES) (Brabänder, 1992; Sekot, 1992, 1994, 1997a,b; Mantau, 1993, 1994, 1995, 1996, 1997a,b; Bergen, 1994; Glück, 1995; Merlo, 1995; Merlo *et al.*, 1996; Blin, 1997; Schmidhauser, 1997). An overview of different economical approach to solving problems of external effects in forestry is presented by Hampicke (1996). In his opinion, storage of carbon dioxide, protection of drinking-water resources and nature conservation are the fields most suitable for financial reimbursement. He then discusses the possibilities for market solutions for recreational services. Although Hampicke presents a more differentiated view of recreational and environmental services of forest outputs than many other authors, he does not really present market solutions besides the well-known conservation contract. In fact, he states that relevant institutions summarizing the individual demand for recreational services do not yet exist (Hampicke, 1996, pp. 129–134).

Offering products in the field of environmental or recreational services is regarded as lateral diversification, which means that the products, as well as the markets on which they are sold, are fairly new for forest enterprises (Sekot, 1997b, p. 7). The main objective of this investigation is to analyse product structures and the institutional framework in niche markets for RES of forest enterprises. In this study, products are not interpreted as homogeneous entities and static phenomena (Mantau, 1997a, p. 270, 1998, p. 45), but instead the product structures and the organizational and legal framework for an exchange of these products between offerers and users will be analysed.

The following questions are fundamental to the investigation:

- What recreational or environmental products are offered by forest enterprises and what product structures can be recognized?
- How is the offer of recreational or environmental products organized in certain niche markets?
- What contract structures and connected legal regulations influence the transfer of the product from the offerer to the demander?
- Who are the cooperators with forest enterprises regarding the offer of RES and what are the contractual regulations for this cooperation?
- What explanations are offered by methods of institutional economics, especially transaction costs theory, for the cases analysed?
- Can hypotheses derived from a theoretical viewpoint be proved by certain cases? What are the general consequences for the offer of RES which can be derived therefrom?

The following main goals are pursued by this investigation:

- To describe practical market solutions and provide ideas and tips for forest enterprises to start their own RES-product business.
- To describe contract structures and their legal framework in detail and present ideas for completing contracts to the enterprises.
- To explain hypotheses of institutional economics (especially transaction costs theory) by case-studies and test their usefulness to explain market structures for RES products.

To achieve these goals, the assumptions and explanations of transaction costs theory, which are usually very abstract and general, will be made more operable and used to explain RES markets.

Since the investigation will combine practical and theoretical targets, a compromise between an explanatory description and a theoretical abstract style had to be found. In the opinion of the author, the best way to achieve this is to present the empirical results in a separate section. For practical use, they are arranged according to certain product groups. Initially, the main assumptions and explanations of institutional economics relating to the empirical results are described, and then the results are analysed by using the transaction costs theory.

Therefore, for each result, the practical market solutions and the theoretical background appear in the same section, enabling the results to be considered together with the theoretical background. General consequences are then derived by using the transaction costs theory, giving both practical and scientific interest.

5.1.2 Institutional economics and niche markets for RES products

In the last two decades, the theory of institutional economics has been increasingly applied. Thus theoretical solutions have been used to describe practical economic problems. Mechanisms or institutions are described as instruments for motivating and coordinating economic relationships. They are categorized into primary and secondary institutions (Dietl, 1997). Primary institutions are very basic, including, as they do, human rights, language and money. Since they are prerequisites for the possibility of living together in a human society, they are also fundamental institutions for economic relationships and constitute the very general framework. Dietl (1997) understands primary institutions as having developed unplanned within evolutionary processes and thus he refers to von Hayek's work. Von Hayek (1969, p. 79) criticizes the understanding of the social system and its fundamental institutions as a result of detailed human planning. He points out that human intelligence is limited and not able to plan a system of rules in detail because of information deficits. In his opinion, the development of abstract rules is the basis of progress in civilization.

Starting from this fundamental level, secondary institutions have been developed to coordinate human relationships in further detail. They exist on a (theoretically unlimited) number of different levels. These secondary institutions form a certain hierarchy, where institutions on the higher level constitute the framework for the next lower one. Derived institutions on lower levels are often very specific and correspond to special problems. According to von Hayek, it is not reasonable to change the whole system of institutions because no single human intelligence can foresee the consequences. Changing abstract rules is a slow experimental process, which begins at the lowest level of institutions (von Hayek, 1969, pp. 86–87). Private contracts as institutions on a very low level can be changed by the contract partners if required by the situation, but the framework of commercial law, for example, cannot be changed as easily.

The different approaches of new institutional economics deal with economic actions in a world where imperfect actors and people with limited rationality depend on each other in their economic activities (Picot, 1997, p. 53). The main sections are the property rights theory, the transaction costs theory and the principal–agent theory.

The main objective of the property rights theory is to describe the influence of restrictions of freedom of action due to traditions, laws, contracts – so-called property rights – on economic phenomena (Schüller, 1983, p. VIII). From the view of the property rights theory, economic actions are described as exchange of ownership titles connected with economic goods. The main target is to maximize the wealth of society by finding an optimum of property rights distribution. Regarding this target, the following main questions appear (Ebers and Gotsch, 1995):

1. What are the consequences of different distributions of property rights on the behaviour of economic actors and the distribution of economic goods?

2. How can the development, the distribution and the change of property rights be explained?

The property rights theory is mentioned as the basic theory in new institutional economics and important statements can be found in alternative approaches.

The principal–agent theory focuses on the economic relationship between principal and agent in agency relationships (Ebers and Gotsch, 1995, p. 195; Picot, 1997, p. 82). Typical topics are the relationships between employer and employee, owner and manager, supervisory board and executives, etc. It analyses agency problems that arise due to imperfect information and different risk behaviour of principal and agent. The main target is the development of motivation, control and information instruments within contracts to solve those problems. In doing this, the principal–agent theory focuses on the time before completing a contract. In an *ex ante* perspective, the theory assumes that the economic actors are able to anticipate the forthcoming agency problems in advance and are able to develop suitable instruments in the contract. From this perspective, the theory is not able to explain contract designs which explicitly take into account uncertain future developments (adjustment clauses, mechanisms for solving conflicts, control measurements, warranties, etc.).

The transaction costs theory focuses on the efficiency of alternative organizational structures and tries to explain why certain transactions can be organized in certain arrangements more efficiently than in others. Contrary to the above-mentioned theories, it also analyses the time after contract completion. It stipulates that economic actors act with limited rationality and cannot anticipate all future problems of their economic relationship before completing the contract.

A common basis of these three theories is the description of economic actions as contracting activities and the analysis of economic institutions. Though they are complementary scientific approaches, they focus on different topics and present different solutions for specific problems. Contrary to the property rights theory, the principal–agent theory and the transaction costs theory take into account the opportunism of economic actors and are able to explain their behaviour in different ways.

While the property rights theory focuses on the optimal development and distribution of property rights structures, the principal–agent theory and the transaction costs theory take the property rights distribution as a given framework for economic behaviour and concentrate on contractual arrangements. The advantage of the transaction costs theory is the consideration of uncertainty of contract partners and the explanation of incomplete contracts. Contrary to the principal–agent theory, it considers the total institutional arrangement that helps to solve transaction problems and not only the contract between principal and agent. The main differences are shown in Table 5.1.

The aim of this investigation is to analyse the organizational patterns that can be derived from the empirical data of the case-studies and to draw conclusions on the organization of niche markets for RES. To analyse this organizational problem in detail, the transaction costs theory seems to be most suitable because it focuses on the efficiency of alternative organizational structures in coordinating economic activities (Williamson, 1996, p. 2) and on contract relationships between market partners. The focal point for this investigation is not the property rights distribution itself or possible improvements but the institutional arrangements under the given property rights distribution. When compared with the principal–agent theory, the transaction costs theory offers a broader view on economic relationships, because it is not limited to principal–agent relationships, and it also analyses contract regulations that have been set up for the period after contract completion.

5.1.3 Transaction costs theory

Scientific studies carried out in the 1930s in the field of jurisprudence and economic and organizational theory provide the basis of the transaction costs theory. The main approaches originate from Ronald Coase, John R. Commons and Chester Barnard (Williamson, 1990, p. 2). The early results

Table 5.1. Main approaches in institutional economics (from Picot, 1997, p. 93).

	Property rights theory	Transaction costs theory	Principal–agent theory
Point of interest	Institutional framework	Transactions	Principal–agent relationships
Unit of investigation	Individual	Transaction	Individual
Criteria for efficiency	Sum of transaction costs and loss of wealth because of external effects	Transaction costs	Agency costs Costs of signalling Costs of controlling Loss of wealth
Conditions	Limited rationality Maximization of individual benefit	Limited rationality Maximization of individual benefit Opportunism	Limited rationality Maximization of individual benefit Opportunism Risk behaviour of the actors
Variable to change	Distribution of property rights	Institutional arrangements	Contracts

of these authors were developed further into an organizational theory by Oliver E. Williamson (Stahl, 1995, p. 85).

The transaction costs theory focuses on coordination processes both within and between organizations dealing with economic problems. The subject of investigation is the transaction, which is defined as the situation whenever a good or a service is transferred across a technically visible border. One activity ends and another activity begins (Williamson, 1990, p. 1). While describing the market process as an exchange of ownership titles between economic actors (see Streit, 1991, p. 71), the definition of the term transaction goes even further. As in Commons (1961), every new combination of property rights can be defined as a transaction (Meyer, 1983, p. 38). A property right is defined as the right of a (natural or legal) person to do whatever the person wants with this economic good within the legal framework. This right refers to the acquisition, use, charging for and transfer of goods (Streit, 1991, p. 72).

The main conditions of the transaction costs theory – for example, the methodological individuality, the consistent and static individual use function and the benefit maximization of the actors – are comparable to those of neoclassical economics. However, there are some differences. The transaction costs theory replaces, for example, the perfect information of market partners from neoclassical economics by conditions described as more realistic for economic behaviour (Stahl, 1995, p. 86). Thus, in particular, limited rationality and opportunistic behaviour, as described by Williamson (1990), are fundamental conditions for understanding the economy from a transaction costs theory point of view. These two main conditions are described in more detail in the following paragraphs.

Every economic action can be described as a decision about the use of resources to fulfil competing targets. It is the choice between different possible alternatives. In neoclassical theory, the economic actors are totally informed about all relevant circumstances when taking this decision. Von Hayek (1969) criticized this understanding and argued for the limits of rationality of human society. He pointed out that human intelligence is not capable of taking into account all relevant aspects concerning the development of a social system. Whereas von Hayek mainly analyses the social system as a whole, transaction costs theory concentrates more on economic transactions between individuals. Using the words of Simon (1986), Williamson (1990) describes human behaviour as 'intendedly rational, but only limitedly so'. Economic actors try to take into account all the information they have (intended rationality) but cannot take into account all information because they do not have it all (limited rationality). Thus, limited rationality describes information deficits of market partners for objective reasons (human intelligence is not capable of taking into account all aspects).

Another form of information deficit is that due to the subjective behaviour of market partners. It exists because market partners do not show their real preferences. Hiding their preferences, they try to reach economic advantages by deceiving their market partners and being cunning. In terms of transaction costs theory, this is called opportunism (see Williamson, 1990, p. 54).

The limited rationality and the above-described behaviour of economic actors are the reasons for the costs which arise when economic actors coordinate their wishes on markets or within organizations. Williamson uses Arrow's definition of transaction costs as the 'costs of running the economic system' and distinguishes them from production costs. In his opinion, transaction costs arise because of misunderstandings, communication problems and conflicts between the relevant economic actors (Williamson, 1990, p. 21). Because these costs can arise whenever property rights are combined in a new way, Meyer (1983, p. 36) defines them as the marketing costs of property rights.

As with Coase, transaction costs can be distinguished according to the *time when they arise* in the following way (Picot, 1997, p. 66):

- Initiation costs, e.g. costs of searching for information and obtaining information about potential transaction partners and their conditions.
- Agreement costs, e.g. costs because of the intensity and duration of discussions, contract formulating and agreement with contract.
- Management costs, e.g. organization.
- Costs for controlling, e.g. costs of securing agreements on deadlines, amounts and prices.
- Rearrangement costs, e.g. costs for contract rearrangement.

Every transaction produces the above-described costs, and Streit (1991) accordingly defines them as costs connected with the operation – or as running transaction costs. Running transaction costs normally arise because of the necessary communication between market partners about the specific exchange conditions. The other form of transaction costs, the costs for setting up an institutional infrastructure, e.g. trade rules or markets, are referred to as transaction costs connected with infrastructure. They are also called sunk transaction costs and have to be distinguished from the costs described above (Richter, 1994, p. 6). Even a contract law (e.g. the Civil Code) and the economic constitution are institutions that help to save transaction costs, although they cause transaction costs in their development stage.

Property rights are newly combined on markets as well as within organizations. Contracts are always the basis for these new combinations. From the transaction costs theory point of view, in principle there is no difference between a purchase contract between seller and buyer and a work contract between the entrepreneur and the worker. Regarding the place where costs arise, they can be distinguished further into transaction costs that arise on markets and organization costs, which arise within organizations (Schüller, 1983, p. 161). Costs that arise on markets are mainly costs for initiation, negotiating, management and controlling of contract economic activities. Internal organization costs are mainly costs for the control of the obligations that were stipulated by work contracts with the staff (Stahl, 1995, p. 91).

To make the arising costs operable, the following section follows the definitions of Streit (1991) and Schüller (1983). So all the costs that arise because property rights are newly combined will be defined as coordination costs and then be distinguished further into transaction costs and organization costs (Fig. 5.1).

The criterion for the organizational quality of economic exchange processes as seen by the transaction costs theory is the amount of transaction costs. The organizational structure with the lowest level of transaction costs is called efficient. The main objective of the transaction costs theory, therefore, is to find organizational solutions that diminish the transaction costs.

When trying to make the theory operable, there are some difficulties due to specialities of transaction costs. Although modern economic systems use more than half of their resources for exchange and agreement processes (Richter, 1994, p. 5; Picot, 1997, p. 7), the arising transaction costs often cannot be measured directly. The reason for this problem is the fact that such components as time and trouble over controlling contract fulfilment have to be considered, but cannot be easily measured. Although it is difficult to quantify transaction costs directly, the transaction costs theory is able to make suggestions for efficient organization. The theory stipulates that the transaction costs vary systematically with the characteristics of the transaction. When describing the characteristics of the transaction qualitatively, the approach draws conclusions on specific institutional arrangements that diminish transaction costs best. On considering the scientific approach, three steps can be distinguished.

1. Analysis of the transaction characteristics.
2. Conclusions about arising transaction costs.
3. Consequences for institutional arrangements.

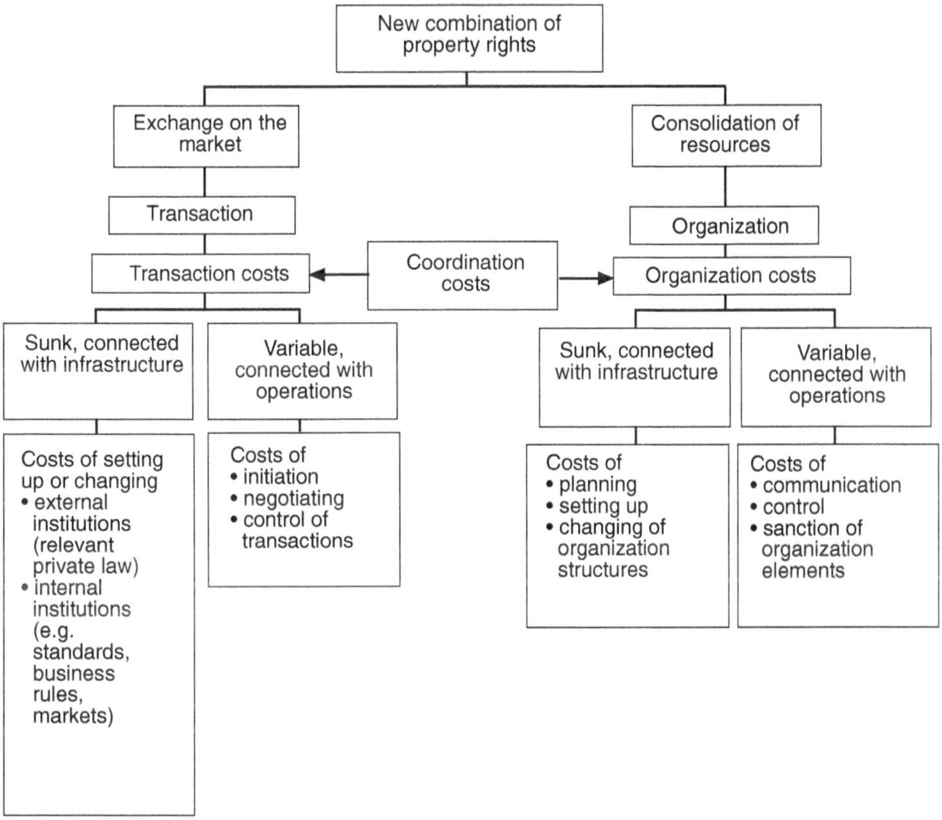

Fig. 5.1. Costs of new combination of property rights (adapted from Streit, 1991).

This investigation will follow these three steps. First, the empirical results from the case-studies regarding the legal and contractual organization of niche markets for RES products will be described. The arising transaction costs will be shown by examples from the case-studies. Then, the quality of the different transactions will be analysed in terms of transaction costs theory and the consequences for the institutional arrangement will be shown.

5.2 Contracts as Important Formal Institutions

5.2.1 Contract management

Contract management includes all those organizational steps that serve to design and realize a contract. Heussen mentions the planning of a contract, the design of a contract, the contract negotiation, the contract realization and the controlling of the contract. He describes 12 basic rules of contract management (Heussen, 1997, p. 24):

1. Distribution of risks. Risks and prognoses of risks are distributed by contract in a binding way: the more exact the risk expectations, the better the contract.

2. Orientation towards the result. Contracts have to serve the targets. Contradictory targets have to be negotiated and taken into account.

3. Procedures. Results always depend on procedures and conditions of negotiation (information, motivation, intelligence, place, time, structure, negotiation style).

4. Complexity and flexibility. Contract relationships are complex structures.

Reasons and effects are connected, and sometimes contracts can be neither planned nor designed perfectly. They have to be flexible to ensure the exchange process.

5. Rule of the magician. Everybody wants everything, immediately and free of charge.

6. Information and communication. Knowledge means power. Without information, it is impossible to judge certain situations; without communication, no contract relationship will develop.

7. Structure and documentation. The quality of ideas changes by the time they are structured (contract text) and documented (signature instead of verbal agreement).

8. Fantasy. Solutions sometimes develop from small ideas and can be destroyed, up until the end of the negotiation, by sometimes trivial items.

9. Opening new perspectives for those who are not persuaded. It is impossible to change someone, but it is possible to arouse interest for a solution, which he/she might accept if he himself/she herself participated in finding it.

10. The subconscious. Decisions are influenced not only by arguments, but always by the participants' feelings.

11. Balance of concessions. No final concessions before balancing the concessions of both parties. Different elements like money, goods, rights, fantasy and emotions can constitute the cost/use balance.

12. Keeping the overview. 'Invest confidence but control results simultaneously.'

Contracts only help to coordinate exchange activities, if they:

- point out common targets in a realistic way;
- stipulate the different interests, risks and conflict potentials of the two parties;
- develop rules as to how to behave in the case of conflicts;
- formulate sanctions and rules in a way that the contract realization is secured even in critical situations;
- are understandable for those persons who use the contracts.

How detailed contracts have to be to regulate certain things depends on the following (Heussen, 1997, p. 151):

- how detailed legal regulations are (in the case of very detailed legal regulations, the contract has to regulate only those things that are not regulated by law or which will be regulated deviating from the law).
- How far juridical decisions exist that have to be taken into account when designing a contract.
- How far other informal rules (e.g. commercial customs) secure the transaction.
- How far the transaction is secured by confidence that has been developed during a long business relationship.
- How complex the variables are that have to be regulated.

5.2.2 Contract negotiation

Contract negotiations will put in contact the possible exchange partners, they will exchange information, and they will regulate the exchange process in more detail. These targets can best be reached by negotiating in a way that is orientated towards facts and which tries to analyse risks and risk expectations and not to hide contradictory interests. Heussen (1997, p. 171) summarizes the perspectives and concepts of this kind of negotiation as follows:

1. Orientation towards targets and facts. Defining one's own target and not losing it during the negotiation. A position is not a target. Recognition of facts, even if it is hard. No erection of formal barriers.

2. Changing perspectives. Analysing the targets and perspectives of negotiation partners and trying to understand their motivation. There is almost always a common target: the contract will be completed and the conflict will be resolved.

3. Taking emotions seriously. Taking one's own emotions seriously and not suppressing the emotions of negotiation partners. Feelings always have reasons. One can recognize the emotions of others without losing one's own targets.

4. Looking for creative solutions. Looking for creative solutions together with the negotiation partner, because one cannot find the best solution when only having in mind one's own perspective.

The objective of any negotiation is a contract that can be signed by both parties. This target can best be reached by a constructive negotiation style characterized by open questions, which the other party is also invited to contradict in certain situations. One's own possible concessions can be discussed openly, but will not be suggested as final concessions. It is the exchange of arguments, not the exchange of positions that is the core of negotiating. The negotiation partners should request statements and explanations without judging them immediately. By explaining positions and arguments in a neutral way, searching for alternatives becomes easier. Summarizing of intermediate positions, looking for examples and defining agreement and differences can keep the negotiation going.

Statements and demands are not excepted, of course. Statements are useful if they concern positions accepted by both parties. Demands can be stipulated if they can be accepted by the other party or if, in case of refusal, the consequences are calculated (Heussen, 1997, pp. 179–180).

Heussen (1997, p. 188) mentions seven typical elements to be analysed if results are the goal of a negotiation:

- Interests. These are one's own interests, the interests of the other party and the interests of third parties. They are sometimes known and sometimes they have to be defined.
- Options. These are alternatives which every party or third parties have.
- Legal circumstances. The legal framework exists because of either other contracts or other relationships or it does not exist.
- Relationships. These are contractual or other relationships of the negotiation partners or third parties.
- Communication. These are communication paths that exist or have to be developed.
- Obligations. These are obligations that both parties or third parties are willing to fulfil or which they finally refuse.
- Alternatives. These are situations in which the parties cannot accept the existing options or do not want to accept them.

5.2.3 Contract law

Detailed and very specific contractual regulations from a juridical point of view differ in the relevant European Union (EU) countries (Austria (AU), the Netherlands (NL), Germany (DE), Italy (IT)), because every country has its own laws. Nevertheless, general contract designs are comparable in different countries. To give a comparative view on factors influencing contract design in general, conclusions from the transaction costs theory will be used (section 5.6). For practical reasons, detailed and specific contractual regulations will also be analysed. They will be analysed under German conditions in Section 5.3 in connection with the description of different product groups. By this structure, the investigation will serve the need of comparability within the EU countries, as well as the need of forest landowners for practical help. Furthermore, detailed legal regulations can be changed during economic development, whereas general factors of influence on contract design remain much the same.

For Germany, basic regulations on the organization of contractual relationships from a juridical point of view can be found in the German Civil Code. In the legal sense, every contract is an obligation from a legal transaction that entitles the creditor to claim performance from the debtor (§ 241 Civil Code). These obligations have to be differentiated from legal obligations and § 305 Civil Code stipulates that a contract between two parties is necessary to create such an obligation.

The most important legal aspects of contracts are summarized in the second book of the Civil Code, the 'Law of Obligations' (Fig. 5.2). Basic rules are written down in §§ 241–432 and particular contractual obligations in §§ 433–853. But, even between the basic rules, differences exist regarding the validity of the rules in general. Section one of the second book (§§ 241–304) deals with obligations in general, while section two (§§ 305–361) deals only with contractual obligations. Section 2 title 2 (§§ 320–327) specifies important rules for mutual contracts. Using the rules of the Civil Code to interpret contracts, one has to keep in mind that the general rule is always replaced by the specific one if applicable. Before using a more general rule, you always have to check if there exists a more specific one that applies to the problem. This is what is meant by the sentence: 'You have to read the German Civil Code backwards from the end to the beginning' (Medicus, 1995a, p. 18).

On a lower level of formal institutions can be found the 'standard-form contract conditions' (*Allgemeine Geschäftsbedingungen* (AGBG)). These general regulations are often applied when offerers want to complete the same contract with different contract partners on several occasions. Contract conditions which otherwise would have to be repeated in every contract are written down in a special document. § 1 AGBG (law about standard-form contract conditions) defines those conditions as contract conditions that are formulated for a number of contracts in general. § 2 AGBG says that the following requirements have to be fulfilled when using standard-form contract conditions.

- The user of standard-form contract conditions has to inform his/her contract partner clearly of the specific conditions.
- The contract partner has to have the possibility to get to know the conditions in a reasonable way.

In addition, the AGBG defines some contractual conditions that are not valid if they are part of a contract.

On the lowest level, we find the contract itself. This contract is a system of rules that has been set up by the contract partners themselves. The basis for the invention of specific rules is contractual liberty, which stipulates the following single principles:

- The freedom to decide whether to complete or not to complete a contract.

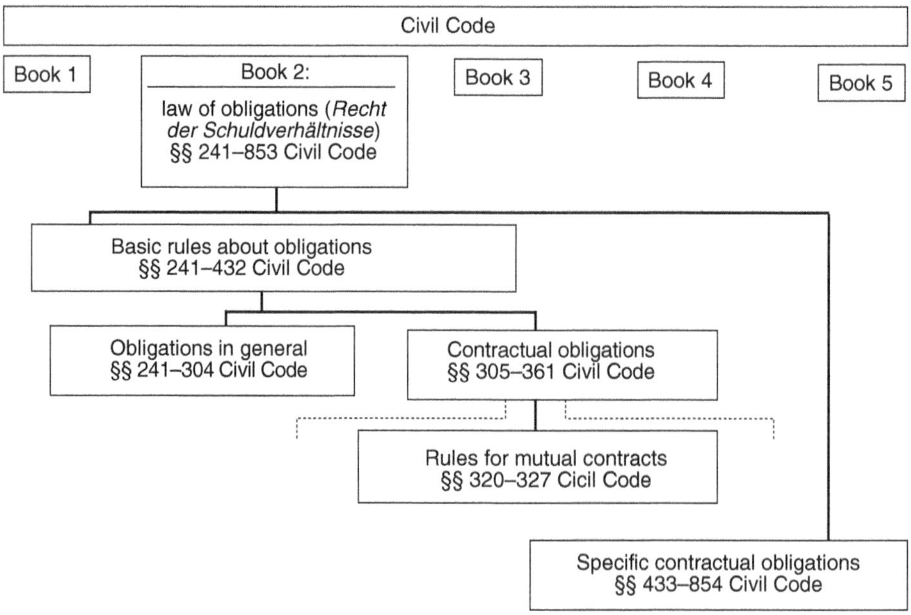

Fig. 5.2. Law of obligations.

- The freedom to choose every contract partner you want.
- The freedom to write a contract in the way you need it (only certain contracts have to have a special form).
- The possibility to differ from the legal regulations in the Civil Code.
- The freedom to set up an informal contract (if the Civil Code does not demand something else).

At the same time, the contract itself regulates exchange processes more specifically than other regulations. When completing a written contract, the contract parties set up rules that go beyond the legal rules and bind themselves to these rules. These rules are adapted to the individual purposes of the contract parties and can differ from legal rules in certain cases. Figure 5.3 shows the general legal framework for the coordination of exchange processes between offerers and demanders.

The completion of contracts is explained in §§ 145–156 of the Civil Code. The term contract is defined as a legally definite agreement that develops from corresponding declarations of intention of two or more persons. For the completion of a contract, a contract offer (§§ 145f.) and the acceptance of this offer (§§ 146f.) are necessary. For the completion of a contract, it is important to express the will for completion. It is not necessary to formulate the acceptance (exception: written affirmation of a merchant) (§ 346 Civil Code)) (Meins, 1993, p. 78).

The target of this investigation is to analyse, first of all, the last level of the formal institutions, the contracts themselves. In the case of written contracts, the regula-

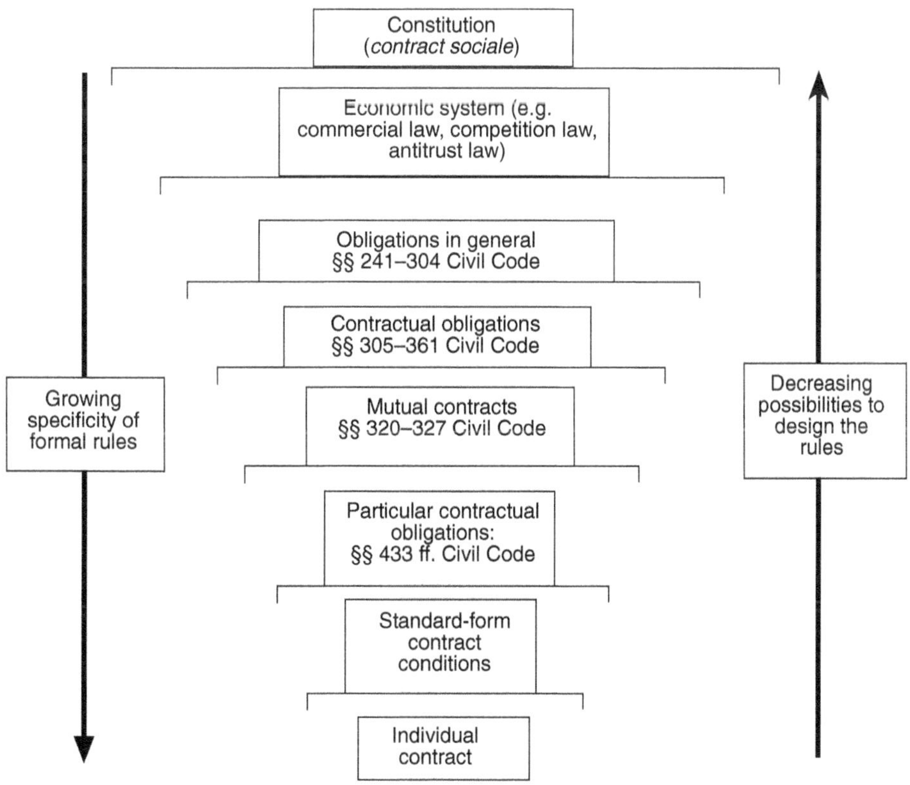

Fig. 5.3. Legal framework for the exchange of ownership titles.

tions are analysed, taking into account the above-mentioned regulations of the Civil Code. When written contracts do not exist between offerer and receiver, the regulations of the Civil Code are explained. This is because these regulations are the legal framework for exchange processes if the parties do not complete written contracts. Figure 5.4 shows the contract types used in the case-studies. Sometimes the contracts correspond very well to the contracts defined in §§ 433f. Civil Code, and sometimes the contract parties use a combination of different contract types. Besides written contracts, informal agreements also exist.

5.3 Product Structures, Distribution Channels and Contractual Cooperation in Different Niche Markets

5.3.1 Tangible products and tangible products with non-material components

In certain niche markets, forest enterprises offer tangible products. These are either environmental or recreational products themselves, or they have environmental or recreational components as additional value. Nevertheless, the general contract design is very similar in both categories. Therefore, these categories will be analysed in one section here.

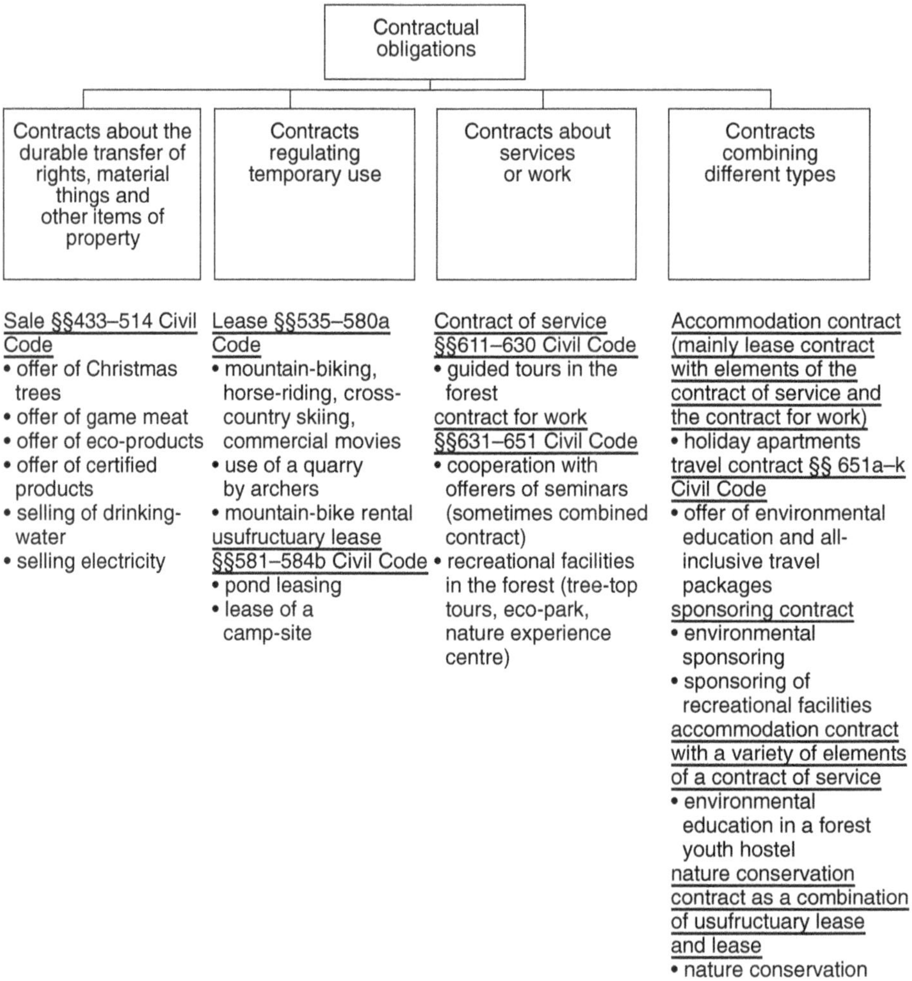

Fig. 5.4. Contract typology.

Niche markets with tangible products include the offer of drinking-water (AU16, 19, NL01) and of electricity production (AU03). In other cases, forest enterprises offer tangible products that contain non-material components as additional value. One example is the offer of Christmas trees (DE01, 13, 16). In this case, the non-material benefit is the organization of a Christmas fair with numerous attractions. Other examples are offers of traditional products with certificates representing environmental quality (IT02, 13, 21, 27), the offer of game meat directly by the forest landowner (DE06, AU15), the offer of mushroom-picking permits (IT01, 08, 19), the offer of eco-hay or eco-meat (NL04, 08, 11) or the offer of traditional farm or forest products at the forest enterprise (NL19).

The offerers of such products are the bigger private and public forest enterprises or associations of smaller enterprises. In general, direct sale is very common. In the case of the offers of drinking-water and electricity production, organizations are the contract partners. Regarding the offer of tangible products with additional non-material components, individual users as well as organizations are the contract partners. Neither user type has any direct influence on the production process. Table 5.2 lists the target groups that have been mentioned for the different products.[1]

The above-mentioned products are mostly sold directly to the user. Written contracts exist in those cases where organizations are the contract partners and offer the product to further consumers themselves. These are normally sale contracts (§§ 443f. Civil Code). This is also true of those cases where no written contracts have been completed (for Christmas trees, game meat, certified products, etc.). Legal regulations relating to sales contracts are normally well known to forest enterprises because of the selling of timber. Because of this, the contract design of these case-studies will not be explained in further detail in this investigation.

Conclusions

- When offering tangible products, such as drinking-water or electricity, forest enterprises complete individual written contracts with organizations because of the long-term perspective of the exchange process and because of specific investments of the contract partners.
- When offering tangible products with additional non-material value, such as Christmas trees, mushrooms, game meat, certified wood, etc., to individual users, no individual contracts are completed, as the tangible product part is a traditional product and therefore standardized contracts written down in the Civil Code present a suitable framework for these exchange processes. Furthermore, a long-term business relationship is not planned. When these products are sold to processing organizations, written contracts are completed, because these organizations have to have a certain security instrument regarding their obligations to their clients.

5.3.2 Accommodation possibilities in the forest

Product

The following description of the product structure uses the model of different use dimensions, as described by Mantau according to Kotler and Bliemel (1995, p. 660). According to the authors, the core benefit of a product is the fundamental benefit that is bought by the user but which has to be transferred into a generic product to be marketable. In case of a hotel, the core benefit that is bought by the tourist is recreation and sleep, and it cannot be sold without the generic product hotel room with a bed. But the customer always expects a certain standard product quality (e.g. that the room has a wardrobe to put clothes in). This dimension is called the

Table 5.2. Target groups for the offer of different tangible products.

Product	Target group
Christmas trees	Families with children, enterprises
Certified products	Families, restaurants in case of mushrooms, environmentally conscious consumers for chestnuts, carpenters in the case of certified wood
Game meat	Individual users of meat in general, enterprises
Mushroom-picking permits	Non-commercial mushroom-pickers
Eco-products	People who know the quality of the products, tourists, environmentally conscious meat consumers, enterprises that want to buy gifts for clients

expected product. On the subsequent level upwards, some products offer additional benefit to the user because of certain qualities (e.g. TV in the hotel room) which differ from competitive products. This product level is called the augmented product and is very important for realizing advantages compared with competitors (see also Chapter 3, section 3.8 on typology).

Transferred to accommodation possibilities in the forest, the main non-material product component is the lease of rooms or the lease of a forest area for camping. In addition, tangible products, such as gas, water or electricity, are offered. The main benefit of the product 'holiday apartment' or 'camping area' is recreation. The tourist buys this benefit when booking the room or the camping area. The holiday apartments and the camping area meet standards that are comparable with those of other offers. So there is no difference between the products of forest enterprises and the products of other offerers on the level of the expected benefit. Possibilities to differentiate are given on the level of the additional value, as the camping area and the holiday apartments are directly surrounded by the forests of the forest landowner. The additional benefit for the tourist consists of a certain kind of outdoor feeling. This outdoor feeling is strengthened by the fact that the holiday apartments are log cabins (AU11) or former forest-ranger stations (DE02, NL18). A variety of different additional offers have to be paid for separately (beach volleyball, mountain bikes, skiing, golf, bicycle trails, hunting, riding), but are an important additional benefit.

Contract design

Regarding the offer of accommodation possibilities, there are different possible contract designs (Fig. 5.5). Only one contract

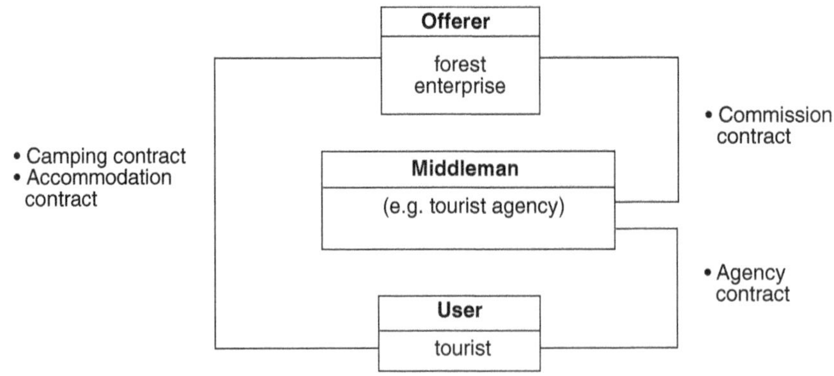

Fig. 5.5. Contract design in the case of accommodation possibilities in the forest.

between the offerer and the tourist is possible, as well as a contract between the offerer and the middleman and between the middleman and the tourist. In the following section, the contract between the offerer and the tourist will be explained in more detail.

HOLIDAY APARTMENTS. Offering accommodation possibilities on its own responsibility, the forest enterprise completes an 'accommodation contract' (*Beherbergungsvertrag*) with its customer. The offerer is obliged by the contract to offer one or more rooms for temporary use. Contrary to a lease, which is a 'continuous obligation' (*Dauerschuldverhältnis*), the temporary use is a typical character of the accommodation contract. This contract is not explicitly defined in the Civil Code. It is a combined contract, showing different single elements. Mostly additional services, such as cleaning, are part of the contract. Besides elements of the lease contract (§§ 535f. Civil Code), there are also elements of the contract for service (*Dienstvertrag* – §§ 611f. Civil Code) or of the contract for the delivery of work (*Werkvertrag* – § 651 Civil Code).

From a legal point of view, each single service has to be judged by those regulations which apply to the corresponding contract type. If there are problems of overlapping regulations, that type will be used which represents the core of the contract (Gitter, 1988, p. 177).

Like any other contract, the accommodation contract needs a contract offer and an acceptance of that offer. The following things are important in case the customer cancels his/her reservation:

- personal reasons (accident, illness) do not discharge the tourist from his/her obligations resulting from the accommodation contract (§ 552 Civil Code).
- The same applies to factual reasons (for example, lack of snow in the case of a winter holiday). These reasons are part of the risk the tourist has to take (§ 552 Civil Code applies accordingly).
- Regarding the amount of money that has to be paid by the tourist who cancels his/her reservation, the host has to take into account the value of the services saved (electricity, water, cleaning expenses – § 552 Civil Code) (Gitter, 1988, pp. 178–181).

Liability. The following regulations are relevant regarding warranty claims of the recreationist in the case of defects.

Claims resulting from a contract for work:

- removal of defects (*Mängelbeseitigung* – § 633 (2) Civil Code).
- Self-removal and compensation for the necessary outlays (*Selbstvornahme* – § 633 (3) Civil Code).
- Annulment of the contract (*Rückgängigmachung des Vertrages, Wandelung* – § 634 (1) Civil Code).
- Reduction of remuneration (*Minderung* – § 634 (1) Civil Code).
- Compensation for non-fulfilment (§ 635 Civil Code).
- Claim for cancellation (*Rücktritt vom Vertrag* – § 636 (1) Civil Code).

Claims resulting from a lease contract:

- claim for an appropriate condition of the facility used (§ 536 Civil Code).
- Release from payment of the rent (§ 537 (1) Civil Code).
- Self-removal and compensation (§ 538 (2) Civil Code).
- Compensation for non-fulfilment (§ 538 (1) Civil Code).
- Reduction of payment of the rent (§ 537 (1) Civil Code).
- Extraordinary cancellation (§ 542 (1), § 544, § 554a Civil Code)

The liability of the innkeeper for goods left in the room by guests (§§ 701f. Civil Code) is a speciality of the accommodation contract. In this case, an innkeeper is defined as a person who commercially accommodates guests. If the offerer only offers rooms without any additional service (cleaning, etc.), then only a lease exists and §§ 701f. does not apply (Gitter, 1988, p. 188).

The purpose of this regulation is to protect the guest from the risk resulting from changing guests and the possible unreliability of the cleaning personnel. Therefore, the above-explained regulations first of all apply to hotels, inns and bed-and-breakfast places. When offering rooms in a forest youth hostel (*Jugendwaldheim* – DE27), these regulations have to be applied accordingly. Whether the liability of the innkeeper is also valid for the lease of holiday apartments is not clear. (This is especially because the rooms of the holiday apartment are not accessible to the personnel during the time of the lease. In addition, the guest might be responsible for closing the doors and windows of the apartment carefully when leaving it.)

CAMPING. In this case, the tourist completes a contract with the forest enterprise according to the regulations for a lease contract (§§ 535 Civil Code). In addition to the area for putting up the tent or parking the caravan, the entrances, the roads and other facilities, such as showers or toilets, are rented too. Compared with the accommodation contract (*Beherbergungsvertrag*), the regulations of §§ 701f. about the liability of the host for goods of the customer left in the accommodation place do not apply in this case. The host only offers a place to accommodate (putting up a tent) and does not offer a room, which can be closed (Gitter, 1988, p. 207).

The completion of the contract does not have to follow a special form. It is necessary for a contract offer and an acceptance of this offer to be made within 5 days (Bartl, 1991, p. 98).

Liability. The offerer of the camping facilities can be liable in consequence of the camping contract (lease contract according to §§ 535f. Civil Code) and additionally in consequence of responsibility for traffic safety (§ 823 Civil Code).

Warranty claims of the tourist that are validated by the lease contract derive from the fact that the host is responsible for the camping area and for the facilities being appropriate for camping (see §§ 536f. Civil Code). Regarding the liability for traffic safety, see section 5.3.5. But, when leasing a camping area, the demands are lower than when leasing a room in a hotel. When using a camping area, the user has to be aware of certain dangers (irregularities of the ground, holes, stones, roots). If walking around, for example, during the night, he/she is responsible for his/her own safety (Gitter, 1988, p. 207). As regards dangers that result from facilities offered by the owner of the camping area, however, the situation is different. In this case, the offerer is responsible for the safety of his customer. Applicable regulations are §§ 836–838 Civil Code (liability for collapse of building) (Gitter, 1988, p. 208).

Cancellation. Regarding cancellation, the regulations of the lease contract apply (for details, see Bartl, 1991, pp. 99–104).

> Hypothesis: Forest enterprises do not complete detailed written contracts when offering accommodation possibilities, because general regulations in the Civil Code and juridical decisions are detailed enough to secure the exchange process. Furthermore, offering accommodation is a very common business. Standardized application forms are used to diminish the costs of dealing with numerous individual users. The special non-material value (forest surroundings) offered by forest enterprises is not secured by contract with the tourist but by market competition.

5.3.3 Seminars in the forest

Product

The results of the RES study show different possibilities for forest enterprises regarding seminars in the forest. One possibility is to offer only the forest area to an external offerer of seminars. The other possibility is to offer additional services. Regarding this possibility, two different common options have to be distinguished. The forest enterprise can be either the subcontractor of another offerer (DE10, 12, 17, 18) or the offerer of the programme itself (DE09, 11, 15, AU07, NL09).

In the case of being a subcontractor, forest enterprises cooperate, for example, with:

- management consultants in the field of management training;
- tourist agencies in the field of all-inclusive packages for tourists;
- education centres that organize seminars.

In the case of being the offerer, the forest enterprise offers, for example, the following products:

- guided forest tours (nature-orientated forestry) with a frame programme (restaurant, city tour, accommodation) for forest landowner organizations;
- team training for managers (accommodation, meals, training with outdoor elements);
- boat tours in the national park.

Cooperation with external offerers of seminars. In the examples mentioned above, the product offered by forest enterprises is generally a package of different product components (Fig. 5.6).

The forest enterprise sells this service package to other offerers of seminars, who use it as an investment good for the offer of an all-inclusive programme. In addition to the forest enterprise, the operators cooperate with different subcontractors (hotels, restaurants, bus enterprises). The product offered by the forest enterprise consists of components that are especially designed to meet the needs of the contract partner. These products are, therefore, characterized by high customer integration.

Offer at enterprise's own responsibility. In this case, the forest enterprise offers an enlarged product (Fig. 5.7), which contains additional services to the services mentioned above. The forest enterprise is responsible for cooperation with other subcontractors, such as hotels and restaurants.

Contract design

The contractual organization of seminars depends on whether the forest landowner offers a service to another organizer of the seminar or whether he/she offers such an all-inclusive programme him/herself.

In the first case, the forest landowner is legally the vicarious agent (*Erfüllungsgehilfe*) of the organizer (§ 278 Civil Code) (Larentz, 1994, p. 380). The relevant contractual relationship between the forest enterprise and the organizer of the seminar

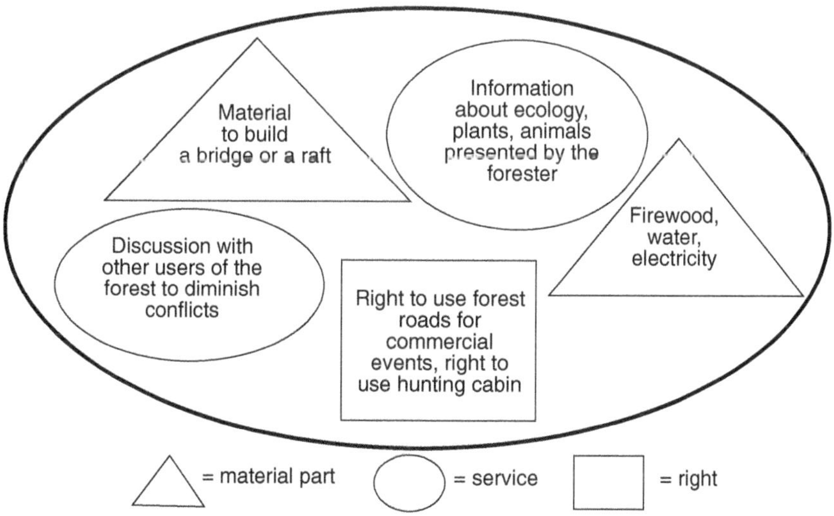

Fig. 5.6. Example of a service package 'forest'.

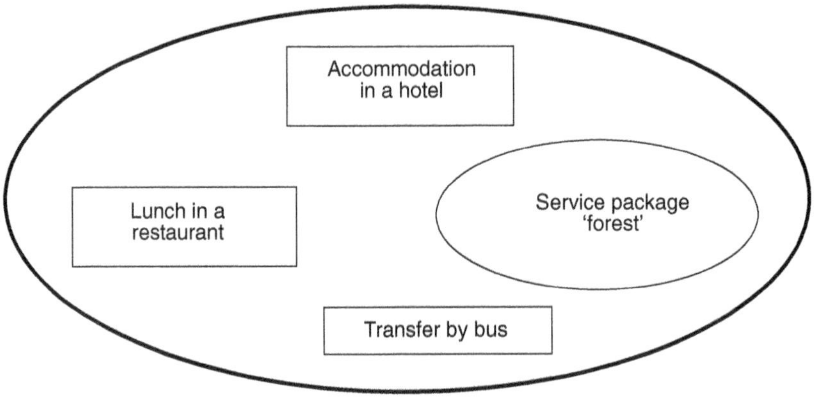

Fig. 5.7. Enlarged service package.

varies according to the services the forest enterprise offers (Fig. 5.8). This can be a combination of different contract types stipulated in the Civil Code. Elements of the contract for lease (§§ 535f. Civil Code – for example, lease of a hunting cabin), of a contract for work (§§ 631f. Civil Code – for example, transportation of participants), of the sale (§§ 433f. Civil Code – for example, sale of firewood) or of a contract for service (§§ 611f. Civil Code – for example, the offer of guided tours) can be part of the total obligation. Such contracts are called *Typenkombinationsverträge* (= contracts that are a combination of different contract types) (Larentz, 1994, p. 42). An important indicator for such contracts is the fact that the single obligations are owed in combina-

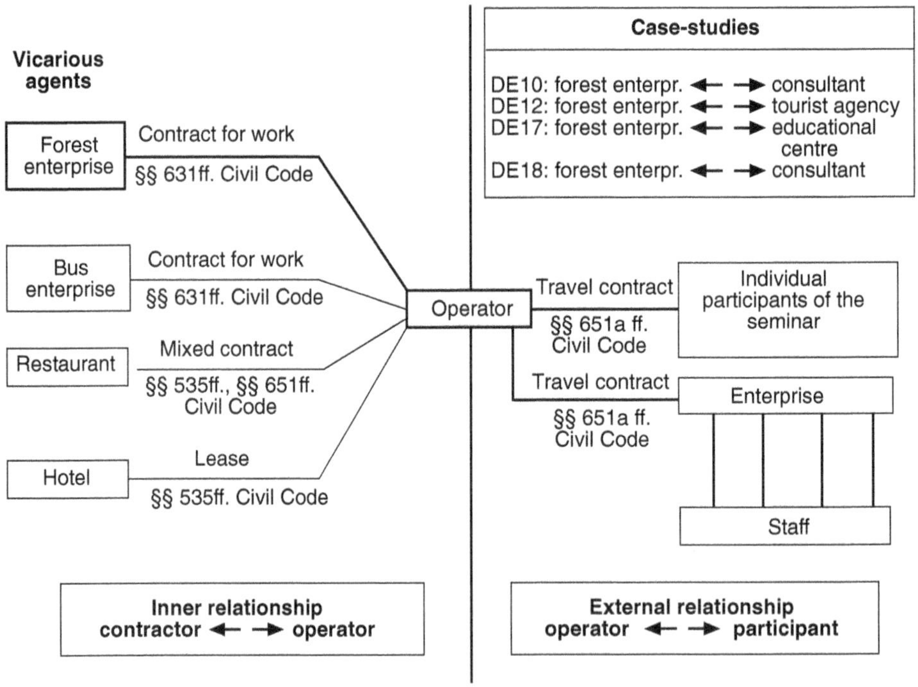

Fig. 5.8. Contract design in a case of cooperation with an external operator.

tion with each other (Medicus, 1995b, p. 271). This is the case if the creditor can only use the single services when they are offered together. In the examples analysed, forest enterprises offer a package of different services that have to be combined with each other to be of any use to the organizer of the seminar. For the organizer of the seminar, it does not make sense to rent a hunting cabin without being able to use the forest roads and the forest itself for the seminar. It is common sense that each single service in *Typenkombinationsverträge* contracts has to be judged by the legal regulations of the contract type which applies to that single service (Medicus, 1995b, p. 273). When organizing and coordinating the services is an important contract task for the forest enterprise, as Bartl (1991) points out, the forest enterprise has responsibility for the success of the total package and the regulations of the contract for work (§§ 631f. Civil Code) apply.

In the second case, the forest landowner is the organizer of the seminar and he/she offers an entirety of services (all-inclusive programme) for his/her clients. In this case, the contract (Fig. 5.9) contains certain elements of an all-inclusive journey (accommodation, meals, programme) and it is comparable to a contract type called *Reisevertrag* (travel contract), which is defined in §§ 651af. Civil Code. The seminars offered are not all-inclusive journeys as known from holiday catalogues; nevertheless, in this case the entirety of services (Larentz, 1994, pp. 380–381) is important and these seminars are on a par with study tours (Bartl, 1991, p. 127). Besides completing a contract with the participants of the seminar, the forest landowner completes different additional contracts with external cooperators (bus, hotel, restaurant) to offer the all-inclusive programme.

The cases where forest enterprises cooperate with organizers of seminars will be explained first.

COOPERATION WITH EXTERNAL ORGANIZERS OF SEMINARS. Written contracts have been completed in two cases. Regarding contract design, these are frame contracts, which serve as basic agreements on the cooperation between the forest enterprise and the organizer of the seminar. This is stipulated explicitly in the preamble of each contract.

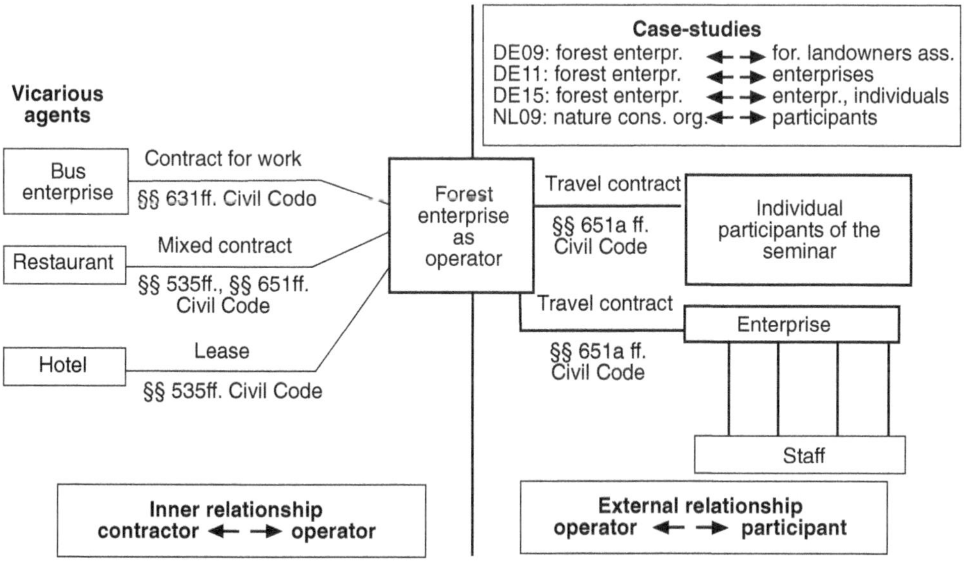

Fig. 5.9. Contract design in a case of forest enterprise as operator.

Example: [The operator] offers the programme ... Therefore, [he/she] cooperates with the forest enterprises ... The fundamental regulations of the cooperation between forest enterprise and operator are stipulated by this frame contract.

In addition, the contracts stipulate the following points:[2]

§ 1 Organizer of the all-inclusive programme.
§ 2 Contents of the programme.
§ 3 Obligations of the organizer.
§ 4 Equivalent considerations of the forest enterprise.
§ 5 Accounting.
§ 6 Liability.
§ 7 Other agreements.
§ 8 Term of contract.

§ 1 Organizer

Example: The [name of the programme] is a programme [of the operator], in which the forest enterprise is responsible for the practical management.
or: Operator and responsible contact person for the participants and the participating enterprises is [name of the operator].

The contractual relationship between the participants of the seminar and the organizer is defined as *Reisevertrag* (travel contract) in §§ 651af. Civil Code. The indication of the organizer has to be very clear, because he assumes liability for possible defects of the seminar and he/she is the person participants have to contact in case of defects (§§ 651c–651g Civil Code). In the above-explained examples, the contract partner of the forest landowner is indicated explicitly as organizer of the seminar. Therefore, he/she is liable for any defects of the seminar.

§ 2 Contents of the programme

Example: The contents of the programme are shown in the advertising leaflets. They have been agreed on by the persons responsible for the programme. The sequence of individual topics and the duration of the programme in total are stipulated by contract, as follows.
or:
[Operator]-initiatives can be:

- events
- projects
- patronages

(the contents are described in detail in the following text).

In this section, the contents of the programme are explained in more detail. In the first example, they are described as the topics of a programme flyer, which becomes part of the contract. In the second example, the basic forms of cooperation are described, and there is a notice that detailed contractual agreements have to be set up with each forest enterprise for each seminar (see § 5, second example). In principle, the contents of the programmes are described in a very general and basic way, which corresponds to the fact that these are frame contracts.

§ 3 Obligations of the organizer

Example: [The operator] is responsible for accounting. [He] organizes the advertisement activities and the distribution of the advertising leaflets. In addition, he offers to arrange accommodation for the participants. [He] organizes suitable transport to the places of activities. [The operator] arranges for accident insurance for the participants.
or:

- [The operator] develops a concept for the [name of the programme], as well as concepts for individual events.
- The development and implementation of the communication strategy and the placement of [name of the project] on the market are the responsibility of [the operator].

As concerns the obligations of the organizer, it is obvious that the organizer is responsible for the development and marketing of the all-inclusive programme. Again, the obligations of the organizer are described in more detail in the first example. This corresponds to the fact that, in the first example, the frame contract is the only written contract. In the second example, each partnership is itself stipulated by single contracts.

§ 4 Equivalent considerations of the forest enterprise

> Example: The forest service is responsible for the personnel and the material necessary to offer the programme. Every individual day of the programme is organized and managed by a staff member of the forest service. In case he cannot fulfil this obligation himself, the forest office is responsible for defining a representative who will manage the programme day.

This example, in particular, shows that the forest enterprise offers an entirety of services and their functional and temporal organization. In this case, it makes sense to use the regulations for contracts for work (§§ 631f. Civil Code) for the whole contractual relationship. It is important that the successful organization of the whole day and the activities are planned in detail by the forest enterprise itself.

§ 5 Accounting

Example:

- The total accounting is done by the office of the operator.
- The prices for the programme have been negotiated between the operator and the forest service according to the possibilities in the market. They are written down in the advertising leaflets.
- ...

or:

- Regarding individual projects, [the operator] completes a contract with the forest enterprise responsible for the area in which the project is offered. In this contract, the contents of the programme, organizational items and the payment of the forest enterprise are stipulated in detail.
- ...

In the explanation of accounting, it is obvious again that not the forest enterprise but the contract partner is the organizer of the programme. While in the first example accounting is explained in detail, this is left for individual contracts in the second one.

§ 6 Liability. When cooperating with organizers of seminars, different types of liability have to be distinguished. Regarding the relationship between the forest enterprise and the participant in the programme, the forest enterprise is the vicarious agent of the organizer. Accordingly, the forest enterprise is not liable towards the participants because of rules stipulated in the contract type 'travel contract' (*Reisevertrag*), but it could be liable because of regulations stipulated in § 823 Civil Code (duty to compensate for damage). Regarding defects of the programme, the participants, therefore, have to contact the organizer to enforce a claim that derives from §§ 651c to 651f Civil Code (travel contract).

In relation to the organizer, the forest enterprise is liable for defects of the services stipulated by the contract with the organizer. The liability for defects depends on the contract type which applies in the special situation. When the forest enterprise organizes the whole day itself (first example), its main obligation is the total organization and therefore the rules of the contract for work (§§ 631f. Civil Code) have to be applied. When the different single services can be more easily distinguished, each single service has to be judged by the legally stipulated contract type that applies to that service (Larentz, 1994, p. 44). The question of liability is also the responsibility of the forest enterprise regarding traffic safety. In the contract examples described here, the responsibility for traffic safety is stipulated by contract in only one case.

> Example: The forest enterprise is liable towards the operator and the participants of the programme only within the general liability for traffic safety, if there is no other regulation stipulated by individual contract.

The above-quoted agreement is not satisfactory concerning the needs of the forest enterprise, because in this agreement only the legal regulations are repeated. Instead of repeating the legal regulations, the forest enterprise should try to stipulate the responsibility for traffic safety by contract in detail (see section 5.3.5).

§ 7 Other general regulations. Regarding other general regulations, there are some interesting agreements on the protection of the programme concept and the obligation of the forest enterprise to cooperate exclusively with the special contract partner.

> Example:
>
> - For three years, [the forest enterprise] will not offer its own programmes or programmes with other partners which are comparable to the [name of the programme]. The forest service is allowed to make suggestions for such programmes. Events regarding touristic programmes and programmes which mainly consist of information about ecological knowledge are not affected by this regulation.
> - The right to grant a licence, the rights of the concept and name, as well as all labels belong to the property of the operator.

In the other cases of cooperation with organizers, the forest enterprises did not complete written contracts with the organizers but offered their services according to verbal agreements. Regarding the question of how far the regulations for contracts for work apply, the functional combination of the different services is important. The closer the functional context of services is, the more the regulations of the contract for work become relevant.

OFFER OF EDUCATIONAL PROGRAMMES ON THE ENTERPRISE'S OWN RESPONSIBILITY. As described above, §§ 651af. Civil Code are relevant for the contractual relationship between the forest enterprise as the organizer and the participants of the programme. The regulations of §§ 651af. Civil Code have been set up to protect tourists better in case of booking an all-inclusive journey. Besides legal regulations stipulated in the Civil Code, the standard-form contract conditions are very important for the contractual relationship between the organizer and participants in the programme. The contract must not stipulate regulations which put a disadvantage on the participant compared with the regulations stipulated in the Civil Code (§ 651k BGB). In addition, it is not permitted to violate the regulations of law about standard-form contract conditions (*Gesetz zur Regelung des Rechts der allgemeinen Geschäftsbedingungen*) (Bartl, 1991, p. 187). Especially if certain all-inclusive programmes are planned to be offered to individual customers, negotiations about each contract cost too much. In this case, formulating standard-form contract conditions is sensible. At the same time, standard-form contract conditions are only valid if the conditions have not been negotiated individually with the organizer.

An example of such standard-form contract conditions, which correspond mainly to the recommendations of the German Travel Agency Association dated 30 May 1994 (Bartl, 1991, pp. 255–264), is shown in the appendix. Certain sections are specified for the circumstances of forest enterprises.

> Hypotheses:
>
> - Success factors for the product 'seminars in the forest' are the actual development towards team training with outdoor elements, the interesting natural surroundings and personal contacts with cooperators.

- In the case of seminars, the combination of different rights, tangible products and services in a product package makes the recreational and environmental quality of a certain forest area marketable.
- Forest enterprises cooperate with business consultants as, in general, they do not have contacts with enterprises that want to train their personnel nor do they have the relevant knowledge.
- Personal contacts with business consultants help to establish the cooperation and are market barriers to those forest enterprises which do not have these contacts.
- If written contracts with cooperators have been completed, these contracts are frame contracts, which define the obligations very generally, because the product itself combines numerous non-material parts with credence qualities, which cannot be defined in detail.
- Forest enterprises cooperate with business consultants or other offerers of seminars, because they do not want to face the liability problems of a tour operator. If they offer the seminars themselves, they use standardized contract conditions, which correspond to the regulations described for the travel contract.

5.3.4 Environmental sponsoring

Product

The product basis for the sponsoring product is the environmental service of the forest enterprise. The enterprise offers, for example, nature-orientated silviculture, preservation of dead timber or protection of endangered species. But the environmental service alone is not sufficient to form a sponsoring product. A sponsoring product consists of more than just an environmental service. It is a package of different components, as Fig. 5.10 shows. Opportunities for environmental sponsoring in forests lie in the fields of afforestation, protection of endangered tree species, nature-orientated silviculture or cutting timber for energy purposes (Gundermann and Suda, 1996, p. 36).

Developing a sponsoring product involves at least the following three steps:

- choice of a suitable project.
- Choice of possible sponsors and contact with these sponsors.
- Contract design.

As the development of contractual solutions is a major goal, the following section will focus on contract design. Regarding the choice of a suitable project and the choice of possible sponsors, see Mertens and Schoop (1999).

Contract design

Depending on the market organization, there are different contractual relationships for the forest enterprise. When offering the sponsoring product on its own, the forest enterprise agrees on a sponsoring contract with the sponsoring partner. This contract is a combination of different contract types. When offering the sponsoring product together with different partners, the forest enterprise completes different contracts with these partners (normally contracts for work – §§ 631f. Civil Code). When the forest enterprise itself is only a partner of another offerer of a sponsoring product, the enterprise completes a contract with this offerer and not with the sponsoring enterprise. Depending on the kind of product, there are different possible contract designs. The following explanations concentrate on the contract between the forest enterprise and the sponsor. Therefore, the explained contract will be named the sponsoring contract.

Contrary to other contract types, such as an agency agreement (*Handelsvertretervertrag*), a broker's contract (*Maklervertrag*) or a contract for personal services (*Geschäfts-besorgungsvertrag*), there are no special legal regulations for the sponsoring contract. 'In principle, the completion of the contract does not have to follow a special form. Therefore, verbally

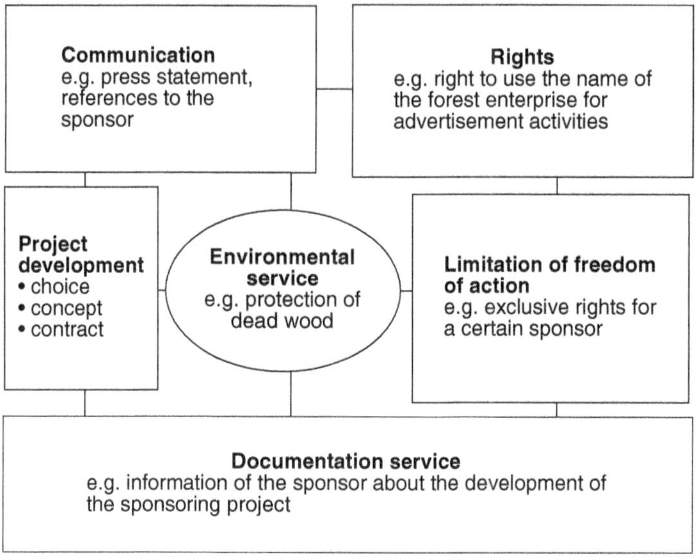

Fig. 5.10. Sponsoring product as package of different components.

completed contracts are valid for both contract parties, too' (Weiand, 1995, p. 39). But sponsoring projects in the forest are normally long-term agreements between the forest enterprise and the sponsoring partner. Especially when silvicultural activities with long-term consequences are the subject of the contract, written agreements become more necessary. The written contract can protect the contract parties from completing the contract in a rush, it might be necessary for the tax office and it is the basis of proof for possible disputes.

The minimum contents of the sponsoring contract are the contractual obligations of the sponsor and the equivalent consideration of the forest enterprise (Weiand, 1995, p. 39). Additional items that influence the contract are tax consequences for the sponsor and the forest enterprise and how the sponsoring project will be used in the public relations strategy of the sponsoring enterprise. When dealing with an advertising agency, other specialized services might become interesting. The following basic structure of the contract can be regarded as containing the core contents. It can be extended with additional items, depending on the economic importance of the sponsorship and the wishes of both parties to set up a detailed contract.

Preamble
§ 1 Obligation of the sponsor.
§ 2 Equivalent consideration of the forest enterprise.
§ 3 Position of the sponsor in comparison with co-applicants.
§ 4 General regulations (good behaviour, confidentiality, information).
§ 5 Contract period.
§ 6 Cancellation.
§ 7 Formal requirements.
§ 8 Law which applies, place of jurisdiction, location of debts.

Regarding sample contracts, see Zilessen and Rahmel (1991); Bruhn and Mehlinger (1992) and Weiand (1995). In the following, different sections of the contract are explained and details specific to sponsoring projects in forests will be explained if necessary.

Preamble. The target of the preamble is to describe the intentions of the contract parties and the activities planned for the contract period. The aim of the preamble is to give the basis for a better understanding of the following agreement.

> Example: The community of ___ owns a forest area near ____, which has a high natural and sentimental value. The area comprises about 14 ha and the core area has had the legal status of a reserve since 1986. With the common sponsoring project of the contract partners ... this area shall be preserved, protected where necessary and be embodied in the people's consciousness better.

§ 1 Obligation of the sponsor. The obligation of the sponsor can vary from payment in cash to a contribution of equipment to offering a service or may take any other form. To be sure, the contractual obligations will be formulated very precisely. The sequence of the written contractual obligations (first the obligations of the sponsor (§ 1) and then the equivalent consideration of the forest enterprise (§ 2)) is not mandatory. When the obligation of the sponsor is reduced to payment in cash, the obligations of the forest enterprise can be stipulated in § 1 of the contract, too (Weiand, 1995, p. 51).

The sponsor can deduct the payments entirely from tax if they are expenses of operation (*Betriebsausgaben*) or professional expenditures (*Werbungskosten*). Some requirements have to be fulfilled before the payments are accepted as expenses of operation. In the opinion of the tax authorities, normal advertising activities, a promotion plan and a control of success are indications for expenses of operation (for more detailed information, see Mantau and Mertens, 1997). The sponsoring contract is a mutual contract. Therefore, it will be clear that sponsors meet a certain obligation because they want to use the equivalent consideration of the forest enterprise in their public relations strategy (Weiand, 1995, p. 63).

§ 2 Equivalent consideration of the forest enterprise. The equivalent consideration of the forest enterprise can vary and the following example only describes some possible aspects:

> Example: The bank ____ pays an amount of __ to the ___, which is charged at the beginning of every year, totally __ during the contract term. This comprises:
>
> - the right to use the area ___ for advertisement activities;
> - maintenance and preservation of the area __ according to targets or environmental protection in the responsibility of the forest enterprise;
> - fixing the logo of the bank – if desired with additional text – on every two to three signs in the area ___;
> - maintenance of the signs, substitution in case of destruction;
> - in the case of events or dates with the press, the forest enterprise is obliged to give information on the sponsoring activities of the bank;
> - forest professional attendance at one event each year (__ man-days).

The equivalent consideration of the forest enterprise can be one of the following categories:

- the forest enterprise grants the sponsor a privilege without any further activities.
- The forest enterprise offers its own activities as an equivalent consideration.

Regarding the example above, the forest enterprise grants the sponsor the privilege of using the forest area ___ for its advertising activities. A second obligation is to offer certain services, such as the maintenance of the signs and participation at certain events. The differentiation between the alternative of granting someone a privilege and offering the enterprise's own activities as equivalent consideration is also made regarding tax regulations (Mantau and Mertens, 1997).

§ 3 Exclusiveness. Sponsoring derives from patronage. While altruistic targets

dominate the patronage, economic targets dominate the sponsoring activities. Sponsoring is used to document responsibility for the environment, to improve the image of the sponsoring enterprise and to reach the target groups better. An environmental service of forest enterprises sponsored by a variety of different sponsors at the same time would not be as effective for the public relations strategy of the enterprise as an exclusive sponsorship. Because of this, sponsoring contracts often define some kind of exclusiveness for the sponsor.

> Example: The [forest enterprise] covenants not to give any of the other reserves or old stands by a sponsoring contract to a competing bank for advertising reasons during the time from 1998 until 2002.

The transformation from environmental services to environmental products helps to differentiate between offers of the forest enterprise. The forest enterprise does not offer a global environmental service, which is not properly defined, but instead it offers specially defined single services. Regarding the example, these single services are:

- the preservation of tree monuments;
- the maintenance and preservation of a certain nature conservation area.

Further possibilities could be:

- removal of spruce in a certain forest area for nature conservation reasons;
- the preservation of a specific old stand of timber to support the black woodpecker.

By differentiating between single environmental products, the forest enterprise is able to complete exclusive sponsoring contracts for products defined in detail. While in this case the differentiation takes place on the side of the product offer, it can also take place on the side of the sponsor. As concerns the above-mentioned sponsoring contract between the forest enterprise and the bank, the forest enterprise agrees not to complete any sponsoring contract with another bank. A mixture of both possibilities is certainly feasible.

In any case, written agreements about the position of the sponsor compared with other sponsors are very important to the sponsor. Therefore, the contract parties often agree that the sponsored forest enterprise has to inform the sponsor on all communication activities.

§ 4 General regulations (good behaviour, confidentiality, information). Sponsoring projects in forest enterprises are planned as a long-term cooperation more than in any other area. This cooperation can be disturbed by a large number of things that were rather unexpected at the time of completion. Because neither party to the contract wants the sponsoring project to be a flop, normally some terms about good behaviour, confidentiality and information are included in the contract.

The following are very common regulations in all contracts.

§ 5 Contract period.
§ 6 Cancellation.
§ 7 Formal requirements.
§ 8 Law which applies, place of jurisdiction, location of debts.

Conclusions:

- Environmental services of forest outputs can only be sold as sponsoring products if combined in a product package that contains different additional parts, such as communication, information, choice of a suitable product, etc.
- By differentiating single, well-described environmental products, instead of offering the environmental quality of the forest as a whole, the forest enterprise has the possibility of offering several sponsoring products.
- Sponsoring contracts are long-term business relationships with high investments by both contract partners. Thus, the different obligations have to be regulated in a very detailed manner or, if this is not possible, the trust potential of the contract partners has to be very large because of personal contacts.

5.3.5 Use contracts with organizations

Products and distribution channels

To characterize the product more precisely, a description of a product within its different utility levels is very helpful (Kotler and Bliemel, 1995, p. 660). The core benefit of forest roads is the access. Horse-riding, mountain-biking and cross-country skiing are normally allowed only on forest roads. The infrastructure of forest roads is a fundamental resource for these sports activities. Being marked on a map, this forest road can be described very precisely and it becomes a potential market product. Mantau (1993) describes this situation as the level of the generic product. As described in the case-studies, no sales activity exists on this level – even under highly suitable legal conditions, as in Austria. All use contracts point out that, besides the right of access, the user has an important additional use. He/she has the right to mark the roads for his/her special purpose and thus has the possibility of orientating in the area much more easily. A closer look at the use contracts reveals that the product 'usage of forest roads' consists of different product components:

- the right to use forest roads for sports activities.
- The right to set up signposts in the forest (additional use).
- The right to design the road for special purposes (cross-country skiing).
- (Accident insurance for the users.)

It is obvious that the product 'usage of forest roads' often consists of product components that bring additional benefit for the user. These additional product components are the reason for the marketability of the product. The distribution of property rights and the demand determine the additional value the forest owner has to create to make access to forests marketable. Along with the difference in the distribution of property rights goes the difference in product components in different EU countries. Certain common aspects are shown in the following section.

Contract design

The contracts explained in this section are, in general, use contracts (*Gebrauchsüberlassungsverträge*) between offerer and demander (Larentz, 1986; Gitter, 1988; Fikentscher, 1991). It is common in this kind of contract for a good to be given to another party legally or economically for a limited time of usage. The most important basic types of these use contracts (lease, usufructuary lease, loan) are defined in Section Seven, titles 3 and 4, of the Civil Code. The part-time use of a thing against payment is defined as a lease (§ 535 Civil Code). If the lessor is bound to give to the lessee (during the term of the lease) the use of the object leased and the enjoyment of its fruits, in so far as they can be considered under the rules of orderly management, the contract type is defined as a usufructuary lease (§ 581 Civil Code) and, if the contract party does not have to pay anything for the use, the contract is defined as a loan (§ 598 Civil Code).

A common topic of the contracts explained in this study is the fact that certain forest areas or forest roads are dedicated for the use of the contract partner. Keeping the fruits (§ 581 (1) Civil Code) of these areas – for example, cutting the timber – is not in general allowed in the contracts. The recreational use of the area cannot be interpreted as a fruit of the forest area in a legal sense either, because, when using the area, the contract partner is normally not allowed to change anything (such as cutting trees, building roads, etc.).

The following contracts have to be considered legally as lease contracts. If there are no special contract regulations within the contract itself, the rules for lease contracts (§§ 535ff. Civil Code) have to be applied.[3]

The contract normally does not need to have a special form (Larentz, 1986, p. 215). A written contract is necessary when properties are rented for more than 1 year. When the contract parties have not observed the written form, the contract is regarded as completed for an undefined period. At the same time, a cancellation is not allowed earlier than 1 year after completion of the contract.

Nevertheless, a written contract is always more advisable, because obligation and equivalent consideration are clear and can be checked in case of disputes. The written form is even more important when the contract parties agree on contract conditions that differ from the regulations in the Civil Code.

The structure of use contracts will be explained with examples in the following section. Basic contract conditions derive from examples of the case-studies[4] but have been rearranged in a suitable way. In addition, the regulations of the Civil Code regarding lease contracts have been added where necessary.

The following contracts have been used from the case-studies:

- Contract for commercial horse-riding.[5]
- Contract for commercial coach tours.[6]
- Contracts for mountain-bike paths.[7, 8, 9]
- Permission to use a certain area for archery.[10]
- Cross-country skiing.[11, 12]
- Contract for a motocross course.[13]
- Network of bicycle trails.[14]

The above-named contracts are quoted as examples for certain usages of forest areas. If there are specific things to be aware of, these specifics are explained in more detail. The following general structure of these contracts could be recognized:

Premises.
Preamble.
§ 1 Subject-matter of the agreement.
§ 2 Obligations of the lessor.
§ 3 Use conditions.
  § 3a Time of use.
  § 3b Signs.
  § 3c Rubbish.
  § 3d Official authorizations.
§ 4 Liability.
§ 5 Insurance.
§ 6 Equivalent consideration of lessee.
§ 7 Contract period and cancellation.
§ 8 General regulations.

Premises. The first page of the contract (premises) shows the name of the contract and the address and name of the contract partners (Meins, 1993, p. 121). An example is given in the box below.

Preamble. The target of the preamble is to explain the goals and the planned activities of the contract partners in order to have a basis for the interpretation of the following agreements. In addition, it gives information on the underlying basis of the contract in a legal sense. Besides explaining the underlying basis of the contract, the preamble can be used to stipulate, first of all, the common interests of the two contract parties. This helps to create a mutual confidential relationship and to highlight the

Agreement on using forest roads	
Between	And
the forest landowner ... ,	...,
address: ____________	address: ____________
represented by ____________	represented by ____________
named lessor	named lessee

fulfilment of the contract (Meins, 1993, p.19).

Example: The association ... tries to start a dialogue between the different economic partners and to improve the cooperation at the point of intersection between agriculture and economy. On the basis of this, [the agreement] has the overall target of finding a reasonable solution for mountain-biking on forest roads by concluding a contract with ... . In addition ... a contribution to a positive economic development is intended with the contract.

§ 1 Subject-matter of the contract. The subject-matter of the contract is the core of every contract and should be explained carefully, as all other conditions depend on it in a certain way, and it often defines the main obligations of the contract partners.

Example: The lessor has the right to use the roads marked in red on the site plan, which becomes part of the contract.

Community:	Friedrichsruh
Name of the roads (forest area):	Sachsenwaldring, Zum Fuchsbau, Buchenberg
Size:	...
Number of the roads according to the road inventory:	Sa 3, Fu 5, Bu 2
Lot:	Abt. 324, Abt. 325, Abt.326, Abt. 327

The lessor opens the roads mentioned in the contract to biking according to the contract conditions. The lessee is entitled to use the area mentioned in the contract only for the above-described purposes and in a way which corresponds to the normal use. The contract parties know the boundaries of the contract area.

When renting a certain area, it is very important to describe the area properly (detailed description, size, number in the lot book) because otherwise subsequently disputes may occur. This description is often a detailed plan, which becomes part of the contract. In most instances, the contract text refers to this plan as an important attachment.

§ 2 Obligations of the lessor. The main obligation of the lessor is to dedicate the leased item for the use of the tenant during the term of the contract (§ 535 (1) Civil Code). When the forest enterprise only dedicates a certain area and does not offer additional services, the main obligation of the lessor is already explained in the subject-matter of agreement.

Besides the main obligation, 'assignment', the lessor has to ensure that the rented good is in an appropriate condition for the use stipulated, and has to keep it in this condition during the period of the lease (§ 536 Civil Code). Regarding the lease of forest roads for mountain bikes or riding, the offerer would have to maintain the forest road in an appropriate condition for mountain-biking or riding. The regulations regarding defects in the leased good – §§ 537–539 Civil Code – would apply accordingly.

Therefore, a non-warranty clause should be agreed on in the contract. When doing this, the offerer has to take into account § 540 Civil Code, which stipulates that such a clause is not valid if the lessor does not inform the lessee of defects that already exist when completing the contract. According to § 547 Civil Code, the lessor is normally obliged to compensate the lessee for any necessary incurred outlays on the leased good. Regarding the lease of forest roads for sports activities, the contract, therefore, should clarify in detail that the lessee him/herself is responsible for all outlays that concern the maintenance of the leased roads for the use stipulated in the contract.

Example: The contract area is handed over in the actual condition on which the contract partner is well informed. The condition corresponds to the standard of timber-harvesting roads.

Because of this, the traffic safety is strongly diminished compared with a public road. The lessor does not guarantee the correctness of the area size and the usability.

The lessor only maintains the contract area in a way that is necessary for his/her own economic purposes. For the special use, the lessee has to maintain the contract area according to prior agreement in his/her own responsibility. Outlays of the lessee using and maintaining the contract area are not compensated by the lessor.

According to § 541 Civil Code, the liability of the lessor also concerns defects in title (*Rechtsmängel*). 'If, through the right of a third party, the stipulated use of the leased good is wholly or in part taken away from the lessee,' the lessor is bound to allow indemnity.

Example: In the case any rights of third parties (for example, privileges of inhabitants) hindering the lessee totally or partly from using the contract area, the lessee releases the lessor from all warranty claims resulting from this. The lessor himself points out that he does not know of any other burdens being connected with the contract area than the things described in no. __.

When the lessor offers any additional service besides the dedication of the use (e.g. the maintenance and control of the roads), these services are defined in the part dealing with the 'obligations of the lessor'.

Example: The lessor will mark the roads mentioned in the contract in a way which does not harm the timber, and he will preserve these signs. The lessor is responsible for buying, fixing, maintaining and replacing the signs at the starting-point of the roads opened for mountain-biking. Prohibition signs for roads not opened, signboards and other signs have to be maintained by the lessor. At the starting-point of the road, the lessor has to put up prohibition signs with the following text: 'Exemption from prohibition for biking during the time 15.04. until 31.10. and from 2 hours after sunrise until 1 hour before sunset and for vehicles which have a licence at own risk.' The lessor will check the roads in reasonable time periods and remove obstacles, such as broken branches or fallen trees.
or:
The lessors will ensure that the forest roads leased for cross-country skiing will remain passable by the maintenance vehicles for cross-country tracks and are not closed because of any timber storage.

§ 3 Use conditions. The main obligation of the lessor in use contracts is the dedication of a good for use, and often no further services are offered. Normally, the use depends on certain conditions and is defined in detail. The following types of conditions could be recognized:

- rules concerning special use periods.
- Rules concerning the marking of the leased roads.
- Rules concerning rubbish in the leased area.
- Necessary official authorizations.

§ 3a Use period. An agreement on a certain use period can be helpful to minimize conflicts, for example, with hunters.

Example: The contract term is 3 years. During that time, the roads will be opened each year in the following time period:

May until August from 8:00 in the morning until 20:00 in the evening.
September from 8:00 in the morning until 19:00 in the evening.

§ 3b Signs. Often the lessee wants to put up certain signs to mark the roads for his/her purpose. Some remarks on the type of signs, type of attachment and removal at the end of the contract help to avoid disputes afterwards.

Example: Before opening the roads for public biking, the lessee has to ensure a suitable condition and marking of the road according to the special grade of difficulty. He has to maintain the road in that condition and to check it regularly at least once a month. He has to fix the mountain-bike rules on a signboard at the beginning of the trail in a visible way and without damage to the timber.

At the end of the contract term, the contract area has to be cleaned according to the orders of the lessor and has to be given back in proper condition. In case of instant cancellation, the lessor grants the lessee a reasonable time to clean the contract area and to give it back in a proper condition. The lessee has to pay the outlays for this on his own and the lessor is entitled to clean the area at the lessee's expense if the lessee does not fulfil his obligations.

Regarding use contracts for single sports events, a long-term relationship is often not planned and it can therefore make sense to oblige the lessee to deposit a security. The security will be repaid after restoration (removal of signs, removal of damages, etc.) of the former condition of the leased area.

Example: The lessee deposits an amount of DM 1000.- with the lessor until 05.09.199_. If there are no complaints about obligations not fulfilled, the amount will be repaid by 30.09.199_. Otherwise, this amount will stay with the lessor for security purposes and can be set off against outlays of the lessor.

§ 3c Removal of rubbish

Example: The roads mentioned in the contract and the surrounding area have to be cleaned at least once a year at the expense and in the responsibility of the lessee.

§ 3d Official authorizations

Example: The lessee himself has to ask for all official authorizations which are necessary to set up, maintain and use the contract object. The lessor informs the lessee of the fact that this contract does not replace any official authorization.

§ 4 Indemnification and liability. When dedicating forest roads and forest areas to the use of recreationists, different possible dangers and types of damage have to be taken into account and to be stipulated by contract:

1. Material damage caused by the lessee on the property of the lessor (damage to forests, forest roads, facilities, etc.).
2. Damage to persons (lessor, third parties) caused by the lessee.
3. Material damage or damage to person which the lessee suffers from because of dangers resulting from the forest.
4. Material damage or damage to person which the lessee suffers from because of dangers resulting from the leased roads.
5. Material damage or damage to person which the lessee suffers from because of dangers resulting from timber harvesting (broken trees, closed roads because of branches and harvest equipment, etc.).
6. Material damage or damage to person which results from the sports activities themselves (e.g. downhill mountain-biking).
7. Material damage and damage to person which result from natural phenomena.

Points 1 and 2 describe damage caused by the lessee. Points 3–5 deal with damage in the area of responsibility of the forest owner. Points 6 and 7 describe damage for which neither the forest owner nor the lessee is responsible. It is helpful to stipulate the different damage possibilities separately in the contract.

§ 4a Damage caused by the lessee. The lessor is, of course, interested in getting reimbursement for damage caused by the recreationist. In addition, he/she does not want to be liable for damage to third parties caused by the lessee.

Material damage.

> Example: The lessee is totally liable for damage to roads, trees and young plants, forestry, hunting and other facilities of the lessor which is caused by the lessee, his staff or other persons in the charge of the lessee.
>
> The lessee will be informed of damage to the roads or other damage by the lessor in the __ week. The lessee then has the possibility to remove this damage by __ . Afterwards, the lessor is entitled to repair the damage at the expense of the lessee.

§ 548 Civil Code says that the lessee is not responsible for alterations or deteriorations of the leased good which are brought about by the stipulated use. If the lessee is obliged to pay for such alterations or deteriorations (e.g. road damage that is caused by the stipulated use), this has to be written down in the contract, as in the example cited above. In the case of leasing roads for riding, damage is normal, and it is helpful to have a separate agreement for damage caused by the riding.

> Example: The lessee is obliged to reimburse the lessor for damage to forest roads caused by organized/commercial riding activities. This means a payment of:

a yearly lump sum of	__	DM
plus 9% value-added tax	__	DM
in total	__	DM

> Other damage is not included in this amount.

Damage to persons. The sufferer from damage to person can be the lessor, his/her employees and third parties. Liability for such damage can be stipulated by contract in the following way:

> Example: The lessee is totally and directly liable for all damage caused by the stipulated use or uses otherwise connected with the contract which concern the person of the lessor, his personnel and third parties. He has to fulfil all claims of indemnity of third parties, which – prior to his agreement – have already been paid by the lessor or which have been proved against the lessor by jurisdiction. In addition to that, he has to pay all expenses of the lawsuit connected with this contract. If third parties put up their claims against the lessor, the lessee has to be informed immediately.

§ 4b Responsibility for traffic safety. Forest enterprises often complain about a higher responsibility for traffic safety when forests are intensively used for recreational purposes (e.g. Thieme, 1993). When completing contracts about recreational use of forest areas, the lessor would do better to try to oblige the lessee to ensure traffic safety on the contract area. The target of the following section, therefore, is to explain the legal regulations about traffic safety in detail. Contractual possibilities to oblige the lessee to be responsible will be shown afterwards.

Development and structure of the responsibility for traffic safety. The responsibility for traffic safety is not described explicitly in the Civil Code, but it has been defined by jurisdiction. In a legal way, responsibility for traffic safety is a kind of case law, which developed step by step, beginning with the first judicial decision on 30 October 1902. In this decision, it was clarified that the rule of § 823 Civil Code ('A person who wilfully or negligently, unlawfully injures the life, body, health, freedom, property or other right of another is bound to compensate him for any damage arising therefrom.') does not only apply in cases where somebody injures someone else because of actions. It also applies when there is culpable neglect, if this neglect causes damage to the objects of an action named in § 823 (1) Civil Code. In other words, everyone who is responsible for a source of danger has to ensure that third parties do not get harmed by this. Gebhard defines the so-called *Zustandshaftung* (liability for the status), as follows: the person who is legally entitled to master a certain object is in a reasonable way responsible for ensuring that no danger results from this object (Gebhard, 1995, pp. 389–390). The term *Verkehrseröffnungshaftung* (liability because of opening to traffic) applies to the case where someone dedicates and maintains a road or a certain area for the common use of the general public. In this case, the lessor is responsible for any damage that is caused by bad maintenance or by not removing obstacles (von Gerlach, 1994, p. 488). The responsibility for traffic safety in general describes a duty to avoid dangers. The extent of this responsibility is not defined in detail but has to be judged according to the specific situation. The higher the danger (possible damage) and the lower the effort to avoid the danger, the greater is the obligation to avoid it (Gebhard, 1995, p. 390).

Responsibility of the forest owner for traffic safety regarding the recreational use of the forest. The difference between liability for the status (*Zustandshaftung*) and liability because of opening to traffic (*Verkehrseröffnungshaftung*) is important concerning the recreational use of the forest. In the following explanations, three groups are differentiated.

1. *The forest owner tolerates entering and using the forest on forest roads that have not been especially declared forest roads for recreational purposes because of forest law.* In this case, there only exists the described liability for the status.

(*a*) *Traffic safety within forest stands.* Most forest laws of the different German federal states allow free access to forests at a person's own risk for recreational purposes. This regulation was set up because the authorities did not want to oblige the forest landowner beyond the normal responsibility for traffic safety.[15] As stipulated in different judicial decisions, the forest landowner is liable only for atypical, not for typical, dangers. According to Klose and Orf (1998), typical dangers are, for example, broken tree-trunks and branches after snowfall and storm, holes in the ground and rotten trees within the forest. Atypical dangers are, for example, dangerous quarries, bridges, traps, insecure storage of timber, etc.

(*b*) *Responsibility for traffic safety on forest roads.* When private forest roads are used time after time for individual recreation and when these roads are not specially dedicated for recreational purposes, the forest landowner is only responsible for the general traffic safety, called liability for the status. The liability because of opening for traffic does not apply, because the forest landowner does not open for any traffic in a legal sense. His/her roads are used without any advertisement activities from his/her side (Geigel and Schlegelmilch, 1986, p. 353). Reviewing the main literature shows that there exist different opinions about the obligation of the forest owner to check the forest stands at regular intervals. While some authors think this regular check is more than the so-called liability for the status demands (Gebhard, 1995, p. 391), others compare forest roads with public roads and support a regular check (Frosch, 1990, p. 430).

Regarding dangers resulting from forest roads, the responsibility for traffic safety depends on the purpose of this forest road. For roads, there exists a rule which points out that their condition has to be accepted if it is sufficient for the purpose they have been built for. The main purpose of forest roads not dedicated to recreation is, certainly, the use for forestry activities. Irregularities of the ground, road holes, roots, etc., are typical for a forest road. The recreationist has to be aware of those possible dangers when he/she uses a forest road for recreational purposes.[16] Regarding atypical dangers resulting from bridges, barriers or other forestry facilities, the forest owner is responsible for warning recreationists early enough or for ensuring the safety of those facilities (Frosch, 1990, p. 430).

The same regulations apply for dangers that result from harvesting activities. In this context, it does not matter that the forest laws of almost all German states interdict the access to forest stands and the use of forest roads during harvesting activities. The person responsible for those harvesting activities has to ensure (with signs, barriers) that recreationists do not get harmed by these atypical dangers (Gebhard, 1995, p. 395).

2. *Responsibility for traffic safety when forest roads have been specially dedicated for recreation (riding, hiking) without any contractual agreements with the forest landowner.* In Gebhard's (1995) opinion, the responsibility for traffic safety depends on the intensity of recreational use in this case. A responsibility for traffic safety regarding typical dangers as described above would, therefore, not exist if the intensity of recreational use is very low. The situation would be different if special roads are recommended by newspaper articles, signs in car-park areas and other advertising activities. In this case, there exists a responsibility for traffic safety in terms of liability because of opening to traffic by the marking organization and, at the same time, a responsibility for traffic safety in terms of liability for the status of the forest landowner arises. So far, it has not yet been decided if the enlargement of traffic safety by dedicating forest roads for recreational purposes constitutes a claim for reimbursement on the side of the forest landowner or if this still falls below the social obligation of property (Gebhard, 1995, p. 391). In the opinion of the author, the liability of the forest owner cannot be extended in this case, because he/she does not open for traffic on his/her property him/herself. In so far as this is not yet clear in jurisdiction, the forest landowner should try to stipulate the use of his/her forest roads by contract with the responsible organization. In doing this, there are possibilities for assigning the responsibility for traffic safety to that organization.

3. *Recreational facilities (forest roads, picnic areas, car-park areas) which are offered by the forest landowner him/herself or which are leased to other organizations by contract.* In these cases, a larger responsibility for traffic safety exists on the side of the forest landowner. He/she is responsible in terms of liability because of opening to traffic because he/she him/herself dedicates certain forest roads for recreational use and he/she gets a certain compensation for doing this. This applies especially to facilities with numerous recreational uses, such as mountain-bike routes, bridle-paths, car-park areas, picnic areas, cross-country skiing trails and other recreational facilities. In this case, the forest landowner him/herself is responsible for the intensity of recreational use (Frosch, 1990, p. 429). Regarding dangers that result from the forest, a regular check twice a year is judged as being sufficient (Frosch, 1990, p. 431; Gebhard, 1995, p. 391). In the case of special dangerous situations (storm, snow, ice, etc.), an additional visual check may be necessary. If the visual check arouses any suspicion, detailed control is necessary. If the possible danger cannot be judged, the recreationists have to be warned early enough (Gebhard, 1995, p. 391).

It should have become clear by now that the responsibility for traffic safety will be extended especially in those situations in which the forest landowner offers recreational facilities by him/herself. The enlarged responsibility for traffic safety relates, first, to the forest and, secondly, to the forest road. Regarding dangers that result from roads that are specially dedicated to recreational purpose, there are large differences from those which are not specially dedicated to such purposes.

These roads not only have to meet forestry standards but the forest owner, as the owner of the forest road in a legal sense (*Wegehalter*) has to ensure that the use of these roads for the stipulated purpose is possible without any dangers (von Gerlach, 1994). When trying to stipulate the responsibility for traffic safety by contract, the forest owner has to keep in mind his/her responsibility for the road and for the forest.

Responsibility for traffic safety regarding the forest.

> Example: In addition to that, the lessee has to check the plants (whether forest plants or not) before opening the roads and during the opening period at least twice a year. He has to ensure the safety regarding

the stipulated use, and he has to remove visible dangers at his own expense.

Removing plants or changing the condition of the ground is subject to agreement with the lessor except in the case of prevention of accidents. If the lessee is not able to remove the danger instantly, he has to give a warning and, if necessary, close the road.

The lessee has to fulfil all claims for indemnity because of dangers resulting from the plants that surround the road. He has to compensate all indemnifications and all expenses for lawsuits or legal assistance of the lessor.

If the lessor did not agree to activities to prevent dangers, he himself is liable for any damage.

The lessor is obliged to repair the roads mentioned in the contract if it is necessary because of damage caused by natural phenomena (storm, rain, avalanches) or by harvesting activities. He has to repair these roads to a standard that is suitable for his purposes as a forest enterprise.

Responsibility for traffic safety regarding forest roads.

> Example: Regarding the roads mentioned in the contract, the lessor is not obliged to ensure traffic safety and the maintenance of the roads. Maintenance and the control of traffic safety will only be done in the way the forest enterprise requires. The road condition is the condition of roads for harvesting activities. Therefore, traffic safety is strongly diminished compared with public roads. The lessee agrees with the fact that he has to be more careful when using these roads. He uses the roads at his own risk. The lessee becomes a holder (*Halter*) of the road from a legal point of view and is obliged to maintain the roads for biking purposes according to the legal regulations for the holder of the road. This contract does not stipulate a liability of the lessee towards other users of the road (especially pedestrians, vehicles with the authorization of the lessor). The lessee will fulfil all claims of indemnity of bikers or third parties in connection with biking, when these claims are not covered by the insurance described in point __.

Responsibility of the forest owner for traffic safety. Though assigning the responsibility for traffic safety to somebody else, the forest landowner remains responsible. The responsibility is transformed from a responsibility to act to a responsibility to organize and control. If the responsibility is assigned to a financially weak person, the forest landowner can be accused of misorganization (Gebhard, 1995, p. 390).

Especially regarding this last fact, it is important for the forest landowner to get approval for liability insurance from the contract partner. How to stipulate this by contract is described in § 5.

Even if the responsibility for traffic safety is assigned to the contract partner, there still remains a remnant of responsibility. The forest landowner has to ensure safety during harvesting activities. Therefore, the possibility of closing forest roads during harvesting periods should be stipulated by contract.

> Example: The lessor is entitled to close the roads mentioned in the contract because of security reasons according to the legal regulations. He is entitled to do so for the duration of a danger from wildfire or because of harvesting activities. In addition, he is entitled to remove the above-described signs. During the time of closing the road, the owner of the road is holder in the legal sense.
>
> In this case, the lessee does not claim any compensation from the lessor. The lessor will restrict the barriers to the time which is necessary. If the barriers last up to 10 days and the number of days with barriers does not exceed 30 days in total, the payment is not reduced. If the barriers last more than 30 days in total, the payment is

reduced according to the days the roads have been closed to biking. The lessee is responsible for road closures because of natural phenomena that make the roads impassable.

Such regulations are necessary, otherwise the contract partner would be able to cancel the contract because of obstruction of the lease (§ 542 Civil Code).

§ 5 Insurance

Example: At the time of contract completion, the lessee will instantly effect an insurance that includes third-party indemnity insurance for the roads and liability insurance for the forests that surround the roads and from which dangers can derive. The insurance amount has to be at least as high as the amount insured known from personal car insurances.

§ 6 Equivalent considerations of the lessee. Normally, when paying a certain lease, compensation is the main obligation of the lessee. Besides this, other obligations (for example, the maintenance of the road) can be stipulated by contract. The payment can depend on the length of the road or the extension of the area. Another possibility is the participation of the forest owner in the revenues of the contract partner. Contractual agreement on how to calculate the value-added tax is necessary to avoid disputes afterwards. In addition, it is necessary to agree on a certain day of payment and to stipulate by contract how defaults of payment will be handled.

Payments

Example:
For the lease of the right described in § __ of this contract, the complications in forestry caused by the lease and the expected higher management expenditures, the lessee will pay a yearly amount of DM__ including value-added tax. This agreement is valid for 3 years.

The payment matures on 1 April each year for the following year. The amount has to be paid free of charge into the account no. __ at ___.

In case of delay of payment, accumulated interest of 4% above the normal accumulated interest of the bank has to be paid.

or:

The total receipts from selling cross-country tickets will be calculated from the number of tickets sold. This calculation will be done on 30 April every year by a representative of the lessor and a representative of the lessee. The expenditures of the lessee – proved by accounting – will be taken off the total receipts. From the surplus thus calculated, the lessor receives 10% (remuneration for using the roads).

In the case of long contract periods, regulations on adjustments of payment make sense. The forest enterprise should try to ensure the consideration in a constant value. This is possible with so-called 'escalator clauses', which compare the consideration automatically with a standard outside the contract. Regarding such 'escalator clauses', the contract partners have to bear in mind that official permission might be necessary (in Germany, the official permission of the Federal Reserve Bank (*Landeszentralbank*) is necessary (see Imbeck, 1997, pp. 413–414)).

§ 7 Contract period and cancellation. Regulations on contract period and cancellation are normal sections of long-term contracts. If nothing else is stipulated, the rules of the Civil Code will apply. A detailed written agreement is always important when specific wishes of the contract parties are to be considered.

Example: The contract starts on the __ and is valid until __ if it is not cancelled by one contract party. The cancellation has to be written and a notice period of 6 months has to be respected. The cancellation always becomes valid at the end of the year. If no contract party cancels the contract earlier, it will be extended for every following year automatically.

The lessor reserves a right for cancellation because of important reasons if:

- the contract partner contravenes the contract regulations;
- the contract area is needed for public or own purposes;
- no agreement on the adjustment of the rent is reached;
- the official authorizations have not been issued or public reasons prohibit the contract fulfilment;
- the contract area is sold or exchanged or the lease in perpetuity (*Erbbaurecht*) is sold.

Normally § 571 Civil Code ('purchase does not break lease') is valid in the latter situation.

§ 8 General clauses. General clauses are clauses about the completeness of the contract, fees, disputes and the place of jurisdiction.

Example: All expenses connected with this contract have to be paid by the lessee.

It is agreed that the district court __ is responsible in the first instance.

It is stipulated by this contract that, except this written contract, no other verbal agreements exist. Additional agreements, changes or additions to this contract will only be valid if they are written. This also concerns the agreement that the written form will not be necessary any more.

Conclusions:

- The possibility of offering use contracts concerning the recreational use of forests depends on the legal situation in each country. If the law requires permission of the forest landowner for certain recreationists in general (as for mountain-biking in Austria), it is easier to offer such use contracts. But the offer of additional value (e.g. the right to mark forest roads to design a network, the offer of an insurance or the offer of road maintenance) is almost as important for marketability as the legal situation.
- Regarding use contracts concerning forest roads, it is very important to define the subject of the contract in detail (map as part of the contract) and to regulate use conditions, such as use period, right to put up signs, obligation to remove rubbish, etc.
- Responsibility of the forest owner for traffic safety is always enlarged when dedicating a certain forest road for recreational use by contract. When completing a contract, the forest landowner should therefore try to transfer this responsibility to the contract partner as far as possible. He/she has to keep in mind that the contract partner must be able to fulfil his/her liability obligation in terms of financial possibilities (contractual obligation to complete an insurance).

5.3.6 Offer of recreational facilities

The products offered are all-inclusive programmes, which have to be paid by the user and which consist mainly of a non-material service. By paying for a ticket, the user obtains the right to use the facilities offered. Characteristics of the product are that the customer has no influence on the product realization. The product is offered with the help of standard contract conditions (tickets) and the offerer usually had to invest substantially to offer the product (e.g. building of facilities). Normally, additional components, such as restaurants and shops, are offered separately.

Contract partners in these cases are the offerer of a certain facility and the visitor or group of visitors who uses the facility. The relationship between offerer and user is a standardized contract. When buying a ticket or booking a programme, each user agrees with the contract conditions of the offerer. In a legal sense, a combined contract, with elements of the lease and of the contract for work, is completed in this case. More important than the single services are the presentation and organization of the all-inclusive offer. In the opinion of Bartl (1991, p. 134), the regulations of the contract for work (§§ 631f. Civil Code) apply to the contract between visitor and offerer.

There is a special feature regarding standard-form contract conditions: when tickets are bought at the ticket office, the offerer does not have to inform the buyer explicitly of the standard-form contract conditions. In this case, a notice at the place of contract completion meets the legal requirements (§ 2 I no. 1 AGBG). The visitor must have the option to read the standard-form contract conditions in a reasonable way. In addition, the offerer has to be aware of the fact that the contents of the programme and the catalogue are defined as guaranteed features of the offer. If certain items are changed, this might cause a defect and reduce the rights of the visitor. The offerer cannot avoid this by using standard-form contract conditions such as 'the programme is subject to change', because these clauses offend against § 10 no. 4 and § 11 no. 11 AGBG (Bartl, 1991, pp. 135–138).

The offerer of such recreational facilities is responsible for the safety of the visitors to a degree that exceeds the normal responsibility of the forest owner for traffic safety. Offering recreational facilities means opening to traffic in the forest. Thus, the offerer is responsible for atypical dangers resulting from these facilities. An exemption from this liability in the standard-form contract conditions with clauses like 'use at user's own responsibility' is not valid and offend against § 11 nos. 7, 8 AGBG. Therefore, all offerers of those facilities should have special insurance for their offers. In the case-study DE03 (environmental education centre), the forest enterprise enlarged the employer's liability (*Betriebshaftpflicht*) for the above-mentioned reasons.

Conclusions:

- When offering recreational facilities to individual users, it is most suitable to use standardized contracts. Using standard-form contract conditions saves costs for negotiating with any individual user and obliges the offerer only to observe the regulations of the law concerning standard-form contract conditions.
- Offering recreational facilities in the forest increases the responsibility for traffic safety and the offerer should therefore complete a special insurance.

5.4 Transaction Costs and their Sources when Offering RES Products

As mentioned above, the quantitative measurement of transaction costs is very difficult. Despite that difficulty, it did not make sense to investigate the exact amount of certain costs in a qualitative investigation like this RES study. The target of the study was to analyse different interesting case-studies and to find certain patterns regarding marketing, organization, contract design, etc. From a comparison of different transaction arrangements, the transaction costs theory tries to find systematic differences in the amount of transaction costs. The approaches of new institutional economics are, therefore, used to analyse the data from a more theoretical point of view.

Thus, the following section will point out transaction costs that arise in the process of offering RES products. Those costs encompass the single types shown in Table 5.3. As the exact amount of transaction costs was not asked for in the questionnaire, the answers to the questions in the questionnaire listed in Table 5.3 have been used as indicators to decide whether transaction costs arose or not. Nevertheless, it cannot be concluded that

Table 5.3. Types of transaction costs and correspondence to the questionnaire.

Theoretical category	Relevance in the questionnaire	Indicator question
Initiation costs	Costs for getting information about the legal regulations for certain transactions	5.3.2
	Costs for using external know-how for product development	3.3.3
Product development	Costs for getting information on market potential	4.2.1
	Costs for analysing enterprise potentials regarding specific RES products	6.1.2
	Costs for the planning process	6.2.7
	Costs for investment appraisals	6.3.2
Advertising	Costs for advertising activities	4.5.1
	Costs for development of a trade mark	4.5.2
Agreement costs	Costs for taking into account special wishes of organized user groups	4.3.2
	Costs for special activities to get acceptance for the offer	4.3.4
	Costs for formulating written contracts	5.2.1
Controlling costs	Costs for preventing free riding	3.2.4/3.2.5
Management costs (organization costs)	Costs for separate tax accounting	6.4.1
	Costs for financial accounting	6.4.2
	Costs for cost accounting	6.4.3
	Costs for training courses for the personnel	3.3.7

a case with transaction costs arising in only one category has less transaction costs in total than a case with transaction costs in three categories. Initiation costs and controlling costs will be explained as examples in more detail in the following paragraphs.[17]

5.4.1 Examples of initiation costs

The costs that arise before the transaction itself has taken place are, first of all, information costs (Streit, 1991, p. 80). Information costs arise from searching for information about the legal regulations for certain transactions and for information about potential contract partners and their special wishes and conditions (know-how for product development, market research). In addition, costs of the planning process (analysis of enterprise potentials, project management) and for information on the economic performance (investment appraisal) are part of the initiation costs arising. The costs encompass mostly labour and time to obtain the relevant information and to communicate with the relevant economic actors.

The following sections (and corresponding questions) in the questionnaire are relevant:

1. Costs for getting information about legal regulations (Question 5.3.2: Which fields of legal regulations did you have problems with or are significant regarding product implementation or realization?).
2. Costs for the assessment and use of external know-how (Question 3.3.3: How was the know-how supplied?).
3. Costs for market research (Question 4.2.1: From what sources did you gain information about the market potential of your product?).
4. Costs for analysing product-specific potentials (Question 6.1.2: Did you undertake any inventory of RES-related potentials (e.g. potential for recreation)?).

5. Costs for the planning process (Question 6.2.7: How was the process of establishing the RES business organized?).
6. Costs for an investment appraisal (Question 6.3.2: Has there been a formal investment appraisal?).
7. Costs for advertising activities (Question 4.5.1: What advertising activities do you implement (advertisements in journals/newspapers, your own publications, events, corporate actions, ...)?).
8. Costs to develop a label/trade mark (Question 4.5.2: Has a label/trade mark been created for the RES product)?).

Table 5.4 shows the relevance of certain initiation costs in the different product groups developed in the RES-project task 'Typology' (see Chapter 3, Section 3.8). The numbers represent case-studies in which the answers of the interview partners to certain questions indicate the existence of initiation costs.

Costs for information about legal regulations

Legal regulations are an important part of the institutional framework that influences economic activities. The regulations are written rules for the restriction of human interaction. They represent restrictions of possible activities predefined for numerous transactions and thus reduce the uncertainty of the transaction partners (North, 1992, p. 4). On one hand, they can reduce transaction costs, because market partners do not have to negotiate the rules again for every transaction. On the other hand, they cause transaction costs, because the offerers have to get information on the existing regulations when developing a new product (Streit, 1991, p. 81). In certain cases, it is necessary to apply for official agreement or to apply for exceptions from specific regulations. The costs for information on the legal framework arise before the transaction itself takes place. Therefore, they are initiation costs.

When asking about legal problems or important legal regulations from the offerer's point of view, those legal regulations will be documented which had to be taken into account by the offerers. At the same time, those parts of the legal framework are described where transaction costs arise in special niche markets.

The analysis is based upon results of the interviews in the case-studies. Due to the special circumstances of the investigated topic, the following open questions were used to ask about important legal regulations from the offerer's point of view:

- 3.2.1: Have there been any legal changes, such as access rights, the right to exclude others from open use, limitations, prohibition of mushroom-picking, etc.?
- 5.3.1: Are there any individual legal restrictions (timber production, access, ...) relating to specific forest functions?
- 5.3.2: Which fields of legal regulations did you have problems with or are significant regarding product implementation and realization?
- 5.3.3: If you have had problems with legal conditions, what have you done to solve those problems (e.g. modification of product, application for exemption, ...)?
- 5.3.4: Did the management unit have to apply for licences, permits, etc.?

In question 5.3.2, different legal areas, such as forest law or nature conservation law, trade regulations, tax regulations, liability regulations, property rights and other regulations, were given as examples by the interviewer to structure the answer and to help the interview partner. Regarding the results of this attempt to get information about the relevant legal circumstances, it has to be mentioned that, despite the quite impressive 76% of answered questions, the relevant regulations were explained in detail by the interview partners in only a few cases. Two reasons may be important for this observation:

1. Questions about legal circumstances were asked mainly in section 5 of the RES questionnaire, when the interview partner had already answered numerous questions about general circumstances, product development and marketing. Due to experiences in the interview, the interview partner was sometimes not able to concentrate on formulating detailed answers to open questions at this stage of the interview,

Table 5.4. Initiation costs of certain product groups.

Product group (total number of cases)		Number of case-studies with costs in the following categories* 1	2	3	4	5	6	7	8
1.1	Long-term contracts about tangible products (drinking-water, production of electricity) (4)	4	1	3	2	2	1	0	1
2.1, 2.2	Tangible products with additional non-material value (Christmas trees, game meat, mushrooms, certified timber) (13)	10	8	5	4	5	4	10	9
3.1	Partly tangible recreation products for individual users (camping, holiday apartments, hunting, nature experience) (9)	8	5	7	2	4	4	9	5
3.2	Mushroom-picking permits (3)	3	0	0	0	0	0	2	1
4.1	Complex recreational services for individual users (eco-park, nature centre, tree-crown path, guided tours in NP) (7)	7	7	6	3	4	5	5	4
4.2	Simple recreation rights for individual users (bridle-paths, fishing ponds, parking areas, picnic areas, access rights, etc.) (17)	17	12	11	4	7	9	14	4
5.1, 5.2	Complex recreational services for direct use/for investment purposes (manager seminars, holidays with the forester, etc.) (7)	4	5	6	6	3	1	7	4
5.3	Forest education centre (forest youth hostel, ecology centre) (2)	2	2	1	1	2	0	1	1
6.1	Use contracts with organizations (bridle-paths, mountain-bike trails, camping sites, etc.) (22)	10	8	7	2	3	5	6	1
6.2	Contracts concerning environmental services in connection with public programmes (storage of carbon dioxide, nature conservation, preservation of drinking-water resource) (5)	3	5	4	4	2	1	2	4
6.3	Sponsoring contracts (environmental sponsoring, sponsoring of recreational facilities in the forest) (9)	4	7	7	5	3	1	7	4
Total		73	60	57	33	35	31	63	39

* 1, Information on legal regulations; 2, know-how for product development; 3, information on the market potential; 4, analysis of enterprise potentials; 5, project management; 6, investment appraisal; 7, advertising activities; 8, creation of a label/trade mark.
NP, national park.

whereas questions with given answers, such as yes/no, possibility 1, 2, 3, ... were obviously easier to answer.

2. As the entire interview contains questions concerning tasks from product development to policy analysis, the duration of the interview was quite long and a detailed description of the respective legal conditions was not possible due to time restraints.

Despite the above-mentioned incompletenesses, the results of the interview show very distinctly that offering RES products has not yet become a regular business and that forest enterprises have to be aware of special regulations regarding the offered products in most cases.

PRODUCT GROUP 1. Long-term contracts concerning tangible products without additional non-material value. In the case of offering drinking-water, forest laws, nature conservation laws and regulations relating to drinking-water (AU16, 19, NL01) have been pointed out. For producing electricity by water-power (AU03), regulations for feeding the electricity into the public network have been very important.

PRODUCT GROUP 2.1/2.2. Tangible products with additional non-material value. Regarding these products, the offerers mentioned product-specific legal regulations. When offering game meat, chestnuts or mulled wine at Christmas fairs, health regulations (DE06, 16, IT27, AU15) have been important. When the activities exceed a certain limit, a business licence had to be applied for (DE06, IT27). For organizing markets, a special authorization was necessary (DE13) and, in some cases, permission for a bar (*Schankerlaubnis*) as well (DE13, 16, AU15). When offering certified products, the respective certification regulations had to be taken into account. Regarding the sale of eco-products in a farm shop, it was mentioned that 70% of the products have to come from the enterprise's own production (NL19).

PRODUCT GROUP 3.1. Accommodation possibilities in the forest. When offering accommodation possibilities, the interview partners pointed out the importance of trade regulations. (If the offer of accommodation possibilities exceeds a certain amount, trade regulations define that this is not forestry any more – DE02, NL20.) Furthermore, the offerers had to take into account forest laws and nature conservation laws, because offering accommodation possibilities mostly means a change in use of the forest area (AU11). When offering camp-sites, the enterprises had problems in extending the size of these areas, because of nature conservation laws (NL14, 17). When forest enterprises changed the use of buildings for accommodation purposes, they had to apply for special licences (NL17, 18, 20). When organizing hunting days, the forest enterprise had problems with hunting and animal protection law. Furthermore, permission for an exception to build a fence was necessary (DE04).

PRODUCT GROUP 3.2. When offering permits for gathering mushrooms, the Italian interview partners underlined the importance of regional laws concerning mushroom-picking (IT01, 08, 19).

PRODUCT GROUP 4.1. Recreation facilities in the forest and guided tours. In particular, forest and nature conservation laws had to be taken into account when building such facilities in forests (DE03, IT03, AU07, NL10). If the enterprise offers large facilities, tax regulations become more important (DE03, NL10). Animal protection regulations are very important when keeping wild animals for demonstration purposes (DE03, AU05). If a restaurant was part of the total offer, health regulations were important too (DE03).

PRODUCT GROUP 4.2. Use contracts with individual users. The products offered in this niche market are very different and accordingly the legal regulations regarded as important differ greatly. When offering bridle-paths or cross-country skiing trails, forest and nature conservation laws and, in addition, liability regulations were relevant (DE07, 19). This was the same in the case of offering forest roads as access to recreation areas (IT17, 28, AU18, 21). In a Dutch case, an application for exception from the free-access right was necessary (NL07). Regarding the offer of picnic areas and parking areas, special regulations had to be taken into account (IT04, 12, 24) and, when offering cabins in addition to fishing ponds, the forest enterprise had to apply for a building licence (AU20).

PRODUCT GROUPS 5.1/5.2. Seminars in the forest. Besides liability regulations (DE10–12), forest and nature conservation laws were pointed out by the interview partners. Sometimes, permissions for exception were necessary if something had to be built (DE18) or if the number of participants could have an impact on nature and wildlife (DE11). Tax regulations became relevant when the offer exceeded a certain amount. In one case, a special association was founded to offer the seminars (DE11).

PRODUCT GROUP 5.3. Environmental education centre. Only two case-studies document the possibilities in this niche market. While the offerer of a forest youth hostel (DE27) mentioned an authorization to offer food and beverages and had to take into account regulations on work with children, the Italian interview partner did not mention any regulations to be aware of.

PRODUCT GROUP 6.1. Use contracts with organizations. Again, forest and nature conservation laws are most important from the point of view of the interview partners (DE21, 23, 24, 26, AU01, 13). In the case of renting forest roads, liability regulations were pointed out (AU13).

Looking at the entirety of the examples, forest and nature conservation laws were considered as a very important framework in most cases. Trade regulations and tax regulations became important when the activities of forest enterprises exceeded a certain extent.

PRODUCT GROUP 6.2. Environmental service in a public programme. The name of the niche market indicates the importance of restrictions such as regulations for subsidies and nature conservation regulations (NL15). Because the environmental services of forest enterprises are mainly silvicultural activities in this case, forest and nature conservation laws have been indicated as important legal conditions (DE05, 14, NL15).

PRODUCT GROUP 6.3. Sponsoring. The respective legal conditions were very case-specific in this niche market. In the case of afforestation sponsoring (DE08), the statutes of the conservation area had to be taken into account. In the case of sponsoring a national park (IT29), the management plan of the park had to be taken into account.

Table 5.5 summarizes the different legal regulations that have been pointed out as important by the interview partners. This table is certainly not complete from a juridical point of view. It only shows those regulations which have been important in the opinion of the interview partners.

Conclusions:

- Several regulations (sometimes very specific) can be relevant, according to the individual situation.
- Nature conservation law and forest law, as important laws regarding the use of forests, have been relevant in almost all cases.
- Transaction costs for information about legal circumstances seem to grow the more complex and different from normal forestry the offered RES products are and the more the forest enterprise is involved in developing the product (e.g. product groups 4.1, 5.1/5.2, 6.3).
- A change in the legal frame conditions in favour of offering RES products can mean the extension of property rights on the side of the offerer or the restriction of property rights on the side of other economic actors. With 55% of the documented cases, official applications have been necessary or existing property rights have been changed.[18]

Costs for know-how during the product development

When offering environmental and recreational services, forest enterprises occupy gaps in the market in which they usually have had little previous experience. When developing RES products, the enterprises have to rely on external know-how (52% of the documented cases) besides their own experience (48%). Using external know-how produces transaction costs (informa-

Table 5.5. Costs for information on legal aspects.

Important legal conditions from the offerer's point of view	Product group* 1.1	2.1/ 2.2	3.1	3.2	4.1	4.2	5.1/ 5.2	5.3	6.1	6.2	6.3
Forest and nature conservation law	x	x	x		x	x	x	x	x	x	x
Area planning			x		x	x	x		x	x	x
Hunting law			x								
Animal protection regulations			x		x						
Drinking-water regulations	x						x				x
Tax regulations			x		x		x				x
Building regulations			x		x	x					
Trade regulations, regulations for registered trade marks		x	x		x		x				
Liability regulations					x	x	x		x		
Travel regulations							x				
Organization regulations (*Vereinsrecht*)											x
Certification regulations		x									
Regulations for subsidies										x	
Health regulations, bar permission, food regulations		x			x				x		
Highway code (*Straßenverkehrsrecht*)						x					
Regulations about feeding electricity into the public network	x										

*1.1, Long-term contracts about tangible products; 2.1/2.2, tangible products with additional non-material value; 3.1, partly tangible recreation products for individual users; 3.2, mushroom-picking permits; 4.1, complex recreational services for individual users; 4.2, simple recreation rights for individual users; 5.1/5.2, complex recreational services for direct use/for investment purposes; 5.3, forest education centre; 6.1, use contracts with organizations; 6.2, contracts concerning environmental services in connection with public programmes; 6.3, sponsoring contracts.
See text for further details.

tion costs). External information sources that have been used most often are public forest extension services (10% of the documented cases) and the specialized press (10%). These external information sources are often well known to forest enterprises and might be the first and easiest (with low information costs) step to getting the relevant information. Table 5.6 shows that a relevant and not very expensive information source is also the imitation of neighbours who already offer certain RES products. In the case of special information, business consultants, lawyers or other information sources, such as universities, have been helpful. Table 5.6 gives an overview of the relevant information sources regarding different product groups.

Conclusion: The observation that more than half of the offerers used external know-how when developing their products shows that offering RES products is not normal forestry business and that information costs are extremely important for starting business in niche markets for recreational and environmental services.

Costs for information on the market potential
Market research is one of the most important activities necessary before offering certain products, and the costs arising are initiation costs. Secondary market research (analysing secondary information sources, such as journals, newspapers, statistics) and primary market research (interviews, tests) can be distinguished (Erbach *et al.*, 1991, p. 85). Forest enterprises used both instruments in addition to their own knowledge. How far certain information sources have been relevant in certain niche markets is shown in Table 5.7.

Table 5.6. Costs to open up and use external know-how regarding product development.

Important legal conditions for product development from the offerer's point of view	Product group* 1.1	2.1/ 2.2	3.1	3.2	4.1	4.2	5.1/ 5.2	5.3	6.1	6.2	6.3
Own knowledge	x	x	x	x	x	x	x	x	x	x	x
Forest extension service		x		x		x	x	x	x	x	x
Agricultural extension service		x									
Private consultant		x	x		x		x				
Imitation of neighbours		x	x		x	x					
Mass media		x				x					x
Specialized press		x	x		x		x	x	x	x	x
Customers					x						x
Suppliers	x	x	x								
Lawyers											
Golf club					x						
Seminars					x			x			
Sports clubs						x					
Municipality									x		
Nature conservation authority						x					
Forest landowners association										x	
Tourist information office							x		x		
University							x				
Sponsoring agency											x

* 1.1, Long-term contracts about tangible products; 2.1/2.2, tangible products with additional non-material value; 3.1, partly tangible recreation products for individual users; 3.2, mushroom-picking permits; 4.1, complex recreational services for individual users; 4.2, simple recreation rights for individual users; 5.1/5.2, complex recreational services for direct use/for investment purposes; 5.3, forest education centre; 6.1, use contracts with organizations; 6.2, contracts concerning environmental services in connection with public programmes; 6.3, sponsoring contracts.
The following questions have been analysed for this table: 3.3.2: Who provided the idea of the product development? 3.3.3: How was the know-how supplied?

It is obvious that, in addition to the enterprise's own knowledge, discussions with potential customers and imitation of other examples serve as information sources. Whereas imitation of other examples has also been of considerable importance for gaining know-how during the development process (see previous section), discussions with potential customers have been more important for market research than for product development. Regarding recreational services, such as accommodation, recreational facilities and seminars in the forest, touristic information sources, such as tourist information offices, hotels, restaurants and the travel sections of newspapers, become important. In two cases, forest enterprises cooperated with schools or universities to get the necessary information. Economic analyses (which are quite expensive information sources) have been used only for those product groups where considerable investments have been necessary to offer the product (product groups 3.1 – accommodation possibilities – 4.1 – recreational facilities).

Costs for the inventory of RES-related potentials

Only one-third of the offerers used a special inventory of RES-related potentials to enlarge their knowledge. These inventories were necessary when offering drinking-water, because the hydrological conditions had to be analysed. When offering certified wood or certified chestnuts, special inventories for information about the product potential were carried out. In the case of recreational services, the inventories mainly analysed the suitability for

Table 5.7. Costs for information on the market potential.

Important legal conditions for product development from the offerer's point of view	Product group*										
	1.1	2.1/ 2.2	3.1	3.2	4.1	4.2	5.1/ 5.2	5.3	6.1	6.2	6.3
Own knowledge	x	x	x	x	x	x	x		x	x	x
Discussion with potential customers	x	x	x		x	x	x		x	x	x
Market research		x	x		x	x	x		x		x
Economic analysis			x		x						
Tests		x									x
Promotion		x	x			x	x				x
Limitation of neighbours		x	x		x				x	x	x
Trial and error									x		
Information from local authorities	x	x				x					
Tourist information office			x		x		x				
Schools					x				x		
Diploma thesis at a university							x				
Travel section of the newspaper							x				
Restaurants, hotels							x				
Private organizations								x			
Nature conservancy organization											x
Forest inventory agency										x	

*1.1, Long-term contracts about tangible products; 2.1/2.2, tangible products with additional non-material value; 3.1, partly tangible recreation products for individual users; 3.2, mushroom-picking permits; 4.1, complex recreational services for individual users; 4.2, simple recreation rights for individual users; 5.1/5.2, complex recreational services for direct use/for investment purposes; 5.3, forest education centre; 6.1, use contracts with organizations; 6.2, contracts concerning environmental services in connection with public programmes; 6.3, sponsoring contracts.
The following question has been analysed for this table: 4.2.1: From which sources did you gain information about the market potential of your product?

recreation of the respective area. For developing a nature education centre, for example, detailed descriptions of the recreational suitability were made by the forest landowner and his staff. As special 'education paths' are part of the offer, in the development phase all special features of the area which might be interesting for visitors had to be described and documented in detailed maps. The next step was a comparison of different path layouts regarding costs, natural features and direction of visitors. In the case of offering a tree-top tour, different possible forest areas were compared regarding accessibility, the possibility of building an information centre and nearby attractions. But transaction costs for analysing RES-related potentials also arose in cases with minor investments. In the case of guided tours, interesting places had to be chosen and, in the case of sponsoring, a suitable project had to be found.

Hypothesis: The market for recreational or environmental services of forest enterprises has just begun to develop. Only a few enterprises seem to define their activities explicitly as a new business field and, accordingly, only a few enterprises have started to develop products systematically (e.g. with a potential analysis before developing certain products). This observation corresponds very well with the fact that economic analyses or other more expensive market research instruments have not been used very often. Obviously, numerous products have been developed by chance or because a customer asked for a certain service.

Costs for the planning process

As known in other product developments, planning costs also arise in the case of development of RES products. The amount might be different depending on the product and the engagement of the enterprise. The use of project management methods[19] will serve as an indicator of planning costs. The results reveal that a special project team was responsible in only 38% of the cases. Milestones have been defined in 35% of the cases, and a written timetable was mentioned in 37% of the cases. It is remarkable that the above-mentioned costs arose in only one-third of the documented cases.

> Hypothesis: Even though a relationship between certain niche markets and the planning costs cannot be proved statistically, the investigation shows that such costs are relevant when developing RES products. A reason for the application of project management methods could be the amount of money that had to be invested. In 57% of the cases with project management, a higher investment was necessary. This is certainly not proved statistically, and a quantitative investigation would be necessary.

Costs for investment appraisals

Transaction costs because of investment appraisals have been documented in only one-third of the cases. Table 5.8 shows the results of the interviews regarding this question.

Investment appraisals have been regarded as important in those cases where the offerers plan a long-term engagement in RES activities and therefore tie up specific resources for a longer time period. Typical cases with investment appraisals are the offer of accommodation possibilities (DE02, AU11, NL18), the offer of complex recreational facilities (DE03, AU05, NL10, 13), the offer of seminars on the enterprise's own responsibility (DE11) and the offer of electricity (AU03). This observation corresponds very well to the observations regarding market research and potential analysis. The observation that only 41% of the offerers invested substantially in order to offer RES products could also be a hint as to the early development stage of the market for RES activities.

Costs for communication activities

The costs for communication activities are part of the initiation costs of the offerer, because they appear before the transaction itself takes place. Even if advertising is also necessary for business continuation, the costs arising can be seen as initiation costs, because, for continuing a business, advertising is used to initiate an unlimited number of sequential transactions. Advertising activities are implemented to communicate the existence and the qualities of the respective RES products and to point out which specific characteristics distinguish the product from the products of other offerers. On the one hand, the offerer has an

Table 5.8. Necessity for investment appraisals in the case-studies.

Investment appraisal	*n*	%
Yes, was necessary	24	24.5
Yes, but of minor relevance	8	8.2
No, not necessary	57	58.2
No, but could have been helpful	4	4.1
Missing	5	5.1
Total	98	100.0

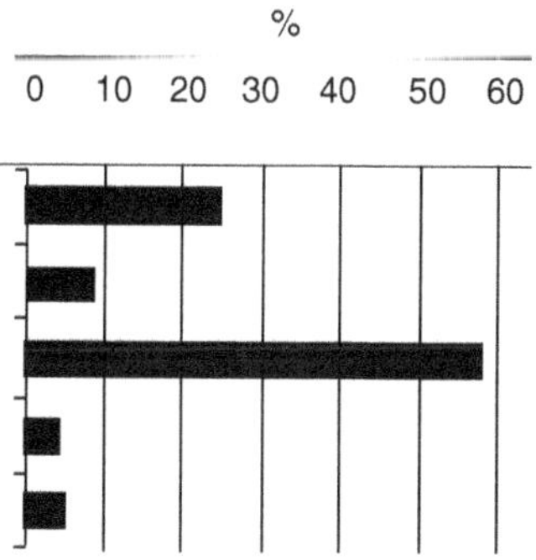

The following question has been analysed for this table: 6.3.2: Has there been a formal investment appraisal?

opportunity to underline certain strengths of his/her product by advertising, while, on the other, the user of the product uses the advertisement material as a help to choose a suitable product according to his/her wishes. In this way, advertising activities of the offerer diminish the information costs of the user. Reducing these information costs in the best way can be a strategic advantage, besides pointing out certain top qualities of the product.

Regarding the case-studies, 74% of the offerers carried out some kind of communication activities to give information about their products. Advertisements in newspapers or journals, brochures, handbills and other written information have certainly been most common. Some enterprises also use radio or TV spots, and even the internet has been used for advertisement activities. Table 5.9 presents the different advertising activities in different niche markets as shown by the empirical data of the questionnaire. The table shows that offerers of RES products have to face initiation costs because of advertising activities in almost all situations.

However, there are differences between certain niche markets regarding communication activities. In the case of offering tangible products, such as drinking-water or electricity to organizations (product group 1), no advertising activities were documented in the case-studies. Similar results were documented for most use contracts with organizations (product group 6.1). In this case, advertising activities (brochures, maps, etc.) were implemented by the contract partner of the forest enterprise in order to offer the product to the final user, not by the forest enterprise itself. Products that are offered under standard-form contract conditions, such as accommodation (product group 3.1) or access to recreational facilities (product group 4.1), were advertised by all kinds of communication activities.

Hypothesis: The difference regarding communication activities can be explained by the specific product profiles. When the product is sold to individual consumers for consumption purposes, it is mostly sold under standard-form contract conditions, and all kinds of advertising activities have been implemented to convince the user of the product qualities and to facilitate his/her information process. When a product is produced according to specific consumer wishes or for investment purposes, regular advertising activities, such as printed advertisements and TV and radio commercials, become less important. Often the customers initiated the product development, and personal contacts with potential customers in this way become more important.

Another indicator for initiation costs connected with advertising activities is the creation of a label or trade mark. About 38% of the offerers created some kind of label or trade mark for the offer of the RES product, and this corresponds very well with the general observations regarding communication activities.

The following section will show that transaction costs arise not only during the development and initiation stage of an exchange process but also after contract completion. A typical example of transaction costs that arise after contract completion are the costs for controlling contract fulfilment (Kieser, 1995, p. 187).

5.4.2 Example of controlling costs

In the case-studies, the interview partners were asked about measures they had to implement to ensure excludability and to prevent people from using the product free of charge (Question 3.2.4: Were actions adopted to prevent free riding?). These measures can be seen as measures to ensure contract fulfilment and were applied in 41% of the case-studies.

Table 5.10 shows that transaction costs to prevent free riding were considered to be very relevant or relevant for only 11.3% of the RES products and that normal measures, which are not very expensive, are in

Table 5.9. Costs for communication activities.

Communication activities	Product group*										
	1.1	2.1/ 2.2	3.1	3.2	4.1	4.2	5.1/ 5.2	5.3	6.1	6.2	6.3
Advertising											
Newspaper advertisement		x	x		x	x	x				
Advertisement in specific journals		x	x				x				
Articles in magazines, newspapers, etc.		x					x		x		x
Handbook on local picnic sites						x					
Travel guide, travel literature		x			x						
Ads in the association's journal							x				x
Brochures			x		x	x	x	x			x
Direct letter											x
Posters								x			
Stickers											x
Handbills		x	x								x
Signs along the road					x						
Radio commercials		x			x		x				x
TV commercials		x			x	x					x
Videos					x						
Sales promotion											
Road maps						x			x		
Open house to the public, cooking demonstration with free mulled wine		x									
Opening ceremony						x			x		
Events with the Ministry for Tourism										x	
Exhibition at fairs		x			x						x
Lectures							x			x	
Public relations											
Excursions with journalists					x						
Concerted actions with other parties						x	x			x	
Internet site							x				
Personal contacts											
Contacts with a travel agency		x	x								
Personal discussions					x				x		x
Cooperation with intermediaries			x								
Personal invitation, personal contact		x	x		x		x		x		x
Activities together with main client											
Communication plan											
Development of a communication plan					x	x					

* 1.1, Long-term contracts about tangible products; 2.1/2.2, tangible products with additional non-material value; 3.1, partly tangible recreation products for individual users; 3.2, mushroom-picking permits; 4.1, complex recreational services for individual users; 4.2, simple recreation rights for individual users; 5.1/5.2, complex recreational services for direct use/for investment purposes; 5.3, forest education centre; 6.1, use contracts with organizations; 6.2, contracts concerning environmental services in connection with public programmes; 6.3, sponsoring contracts.

general sufficient (38.7%). Measures to prevent free riding seem to have been very relevant in the case of product group 3.2, mushroom-picking permits, and product group 4.2, the offer of recreation rights in the forest. Mostly, fences and control personnel have been used to ensure contract fulfilment. Permits for gathering mushrooms, as well as recreation rights in product groups 4.1 and 4.2, are typically offered to individual users. Thus, standardized contracts, such as tickets, are

Table 5.10. Relevance of transaction costs for preventing free riding.

Relevance	*n*	%
Very relevant	3	3.1
Relevant	8	8.2
Low	11	11.2
Very low	10	10.2
No	17	17.3
Missing	49	50.0
Total	98	100.0

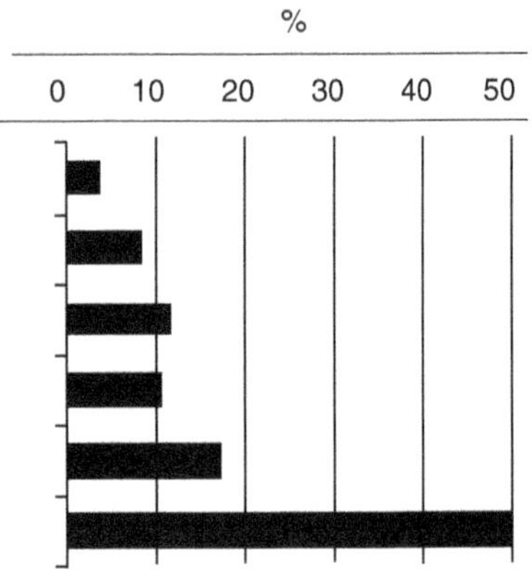

The following question has been analysed for this table: 3.2.5: How relevant are the costs for preventing free riding?

used and therefore transaction costs for negotiating a contract do not exist, but controlling costs to ensure contract fulfilment arise instead.

5.5 Transaction Qualities and their Influence on Contract Design

In the literature, transaction qualities such as specificity, uncertainty, frequency (Williamson, 1990, p. 59), internal complexity (Stahl, 1995, p. 95) and the strategic importance and atmosphere of the transaction (Picot, 1997, p. 68) are mentioned as important factors for the efficient institutional organization of transactions. These transaction qualities will be described in a general way before describing specific examples in the following paragraphs.

SPECIFICITY. The specificity of a transaction depends on the amount of specific investments the transaction partners have to make for the transaction. Specific investments are those which support the given transaction, but which can be used for other transactions only with a depreciation in value. Williamson (1990, p. 62) distinguishes the following specific investments:

- Specific investments connected with the location, e.g. investment in a cabin at a certain place, because of the offer of guided tours in the forest.
- Specific investments in permanent assets, e.g. buying of maintenance machinery for leased bridle-paths.
- Specific investments in human capital, e.g. organization of training courses for the staff to offer seminars for executives.

A transaction with high specificity is characterized by the dependency between transaction partners. The transaction partner who has invested in specific factors (e.g. the municipality which has bought signs to mark mountain-bike tracks) is at the mercy of an opportunistic transaction partner (e.g. a forest enterprise which will cancel the lease contract if the municipality does not pay a higher lease). In the case of specific investments, instruments to protect these investments against the opportunistic behaviour of the transaction partner are therefore necessary.

UNCERTAINTY. Uncertainty can derive, on the one hand, from missing information on the possible behaviour of the transaction partner and, on the other, from the missing knowledge about future developments. Uncertainty is normal in the case of the first cooperation with a transaction partner and in the case of innovative products. When dealing with a new market partner, common social values, targets and principles are missing. This is the reason for the development of control and insurance instruments in the case of new business relationships. While a forest landowner sells timber to a sawmill by verbal agreement, the same forest landowner will stipulate obligation and consideration by a written contract in the case of offering environmental services to a nature conservation organization.

FREQUENCY. The frequency of the transaction influences the development of special organizational structures. The more frequently a transaction takes place, the more important a special organizational structure to guarantee these transactions becomes. If the guided tours offered from time to time are to become a regular service, it might be valuable to set up a special management unit to offer these services.

INTERNAL COMPLEXITY. Internal complexity means the difficulty in describing, recognizing and evaluating single components of the transaction package. Complex transactions are characterized by the fact that the quality and quantity of goods and services transferred cannot be described in detail. Transactions that cannot be standardized, therefore, need institutional arrangements other than standard transactions. The selling of timber can be more or less standardized, while the selling of a complex environmental education seminar cannot be standardized as easily.

STRATEGIC IMPORTANCE. Strategically important transactions are defined by Picot as transactions where one transaction partner offers very specific services in order to attain market advantages compared with competitors (Picot, 1997, p. 70). Strategic importance is common in the case of innovations, which need insurance instruments in order not to be at the mercy of opportunistic market partners. An insurance instrument for RES of forest enterprises could be the development of a label or trade mark.

ATMOSPHERE OF THE TRANSACTION (Fig. 5.11). The legal (laws, regulations), social (customs) and technical (information and communication technology) conditions that are important for the transaction characterize the atmosphere of the transaction (Picot, 1997, p. 71). Depending on the atmosphere of the transaction, control or adjustment activities are more or less necessary during the transaction process. Therefore, different coordination structures have to be developed according to the atmosphere of the transaction. The lease contract is regulated in detail in the Civil Code, for example, while the spon-

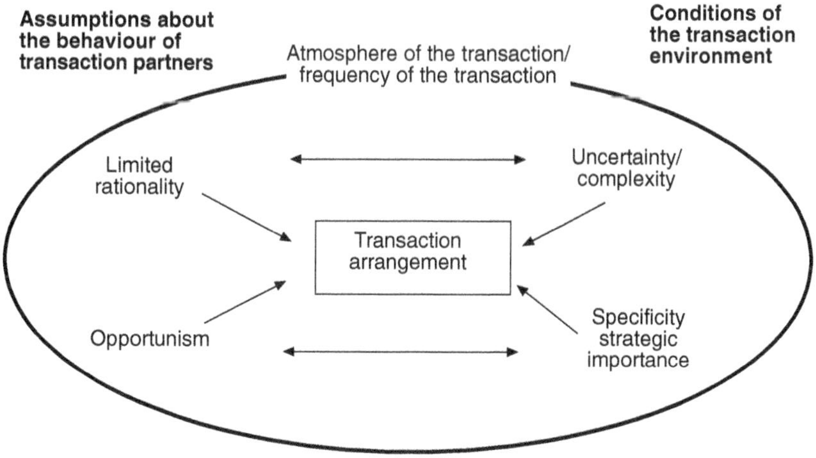

Fig. 5.11. Influence factors on transaction arrangements (adapted from Picot, 1997, p. 67).

soring contract is not regulated at all. Therefore, the negotiation of a sponsoring contract is normally much more time-consuming than the negotiation of a lease contract. Social and technical conditions exert an additional influence on the institutional arrangement.

In the following sections, examples of transactions in different product groups will be analysed using the above-mentioned indicators for transaction quality. Therefore, two columns are used. While the left column shows the relevance of certain transaction qualities for this example, the second column shows the consequences in keywords.

5.5.1 Example: accommodation possibilities

Accommodation possibilities have been offered in a number of cases (DE02, 28, AU11, NL14, 17, 18, 20) and these cases usually combine non-material parts, such as the right to use a certain building, with tangible parts, such as electricity, gas, water and, possibly, food. This example has been chosen as an autonomous product that is normally sold to individuals for consumption purposes.

RELEVANCE AND CONSEQUENCES OF SPECIFICITY

The contract partner (tourist):	
invests time and money to travel to the place of accommodation	→ Tourist asks for all information he/she can get (leaflets, postcards, catalogue, etc.)
has to apply for holidays at the workplace	→ Mostly a written approval for the booking is necessary → Contents of leaflets or catalogues are regarded as guaranteed qualities of the offer (see Section 5.3.6)
The forest landowner:	
invests in buildings, camp-site, etc.	→ Written application is required
has to reject all other applications once he/she has reserved an accommodation place for a certain tourist	→ Legal regulations if the tourist cancels the reservation (§ 552 Civil Code)

RELEVANCE AND CONSEQUENCES OF MARKET UNCERTAINTY

Market uncertainty is not very great because holiday apartments are very common products and prices can easily be compared by market partners	→ The offerers try to distinguish their own products by marketing activities (added value) from other products to increase market uncertainty and to realize price advantages

RELEVANCE AND CONSEQUENCES OF UNCERTAINTY ABOUT THE BEHAVIOUR OF THE MARKET PARTNER

This kind of uncertainty applies to both market partners, because neither the offerer nor the tourist normally knows the market partner before the first holiday. As the specific investments of the tourist are normally bigger than those of the offerer, uncertainty about the behaviour of the market partner is usually more important for the tourist	→ The offerers of accommodation possibilities try to reduce the uncertainty of the tourist by getting an independent certificate of hospitality → The offerers try to inform the tourists by all kinds of marketing activities → The offerers try to bind the tourists to their offer by gifts for tourists who have spent their holiday at the particular place several times (personal knowledge reduces uncertainty and saves transaction costs for the second holiday)

	→	The tourists ask friends about the place and the behaviour of the offerer

RELEVANCE AND CONSEQUENCES OF FREQUENCY

Accommodation possibilities are offered on several occasions to different tourists. They are long-term investments and will be used on several occasions.	→	Due to the frequency of the offer, standard-form contract conditions are developed and applied (booking formulas)

Strategic importance and internal complexity are not relevant in this case because holiday apartments and camping facilities are very common products.

RELEVANCE AND CONSEQUENCES OF THE TRANSACTION ATMOSPHERE

Legal atmosphere: Renting forest houses as holiday apartments is not a common business for forest enterprises and special regulations are normally unknown	→	Standard-form contract conditions have to be developed (transaction costs)
	→	Cooperation with a tourist agency has to be established
Social atmosphere: As the business is new, the forest enterprises do not have a network of clients yet	→	In former times, tourists have been regarded as disturbance factors in the quiet forester's life
Forest landowners are not very accustomed to dealing with tourists	→	Forest landowners have to get used to providing the service that tourists want

5.5.2 Example: cooperation with offerers of seminars

This example has been chosen to describe a mainly non-material product with high customer integration. Normally, the contract conditions are negotiated individually in this case.

RELEVANCE AND CONSEQUENCES OF SPECIFICITY

The contract partner: invests time and labour to develop the all-inclusive package responsible for the booking of a restaurant and a hotel discusses organizational matters with subcontractors is responsible for the production of advertising material is liable as tour operator	→	As the contract partner invests much more in the transaction than the forest landowner, the contract partner needs insurances for these investments
	→	Written contract in case of cooperation with market partners so far unknown (DE12, 18)
	→	Verbal agreement with personally known market partners (DE10, 17)
The forest landowner: invests time and labour to choose interesting forest locations for the guided tour invests time and money for specific know-how (educational know-how)		

Relevance and consequences of market uncertainty

Normally, the contract partner of the forest enterprise is better informed on the given market, because he/she developed the product and knows the market better than the forest enterprise	→	As the contract partner of the forest enterprise is better informed on the market, the forest enterprise has to face disadvantages regarding price negotiations (this was a fact an interview partner mentioned DE18)

Relevance and consequences of uncertainty about the behaviour of the market partner

The contract partner is not able to evaluate whether the service of the forest enterprise meets the professional standards (information on forest ecology, for example). The forest landowner is not able to measure the acquisition activities of the contract partner	→	Uncertainties on both sides are responsible for the fact that contracts are not completed for a long term but only as tests so far (uncertainty about behaviour of market partners can be solved by personal contacts sometimes)

Relevance and consequences of frequency

A regular cooperation is planned only in one case. All the other cases are test cooperations	→	No long-term contracts

Relevance and consequences of strategic importance

The strategic importance of the transaction especially influences the behaviour of the contract partner because he/she wants to attain market advantages by offering outdoor seminars in the forest	→	An agreement on exclusivity (DE18) exists
	→	The trust relationship between market partners (DE10)

Relevance and consequences of internal complexity

The internal complexity in case of offering seminars in the forest is very high, because the product package comprises considerable non-material services, which cannot be defined in detail, e.g.	→	Cooperation is only at the test stage so far
Well-designed guided tour in the forest Professional information Qualification of the forester to present an outdoor seminar	→	The product offer of the forest landowner has strong trust qualities and therefore personal contacts are very valuable to get the business started (In all documented cases of cooperation with offerers of all-inclusive programmes, informal contacts helped to start the cooperation)
	→	The forest landowner does not offer the seminars him/herself, but depends on the reputation of the middleman (consultant, tourist information office, educational organization)

RELEVANCE AND CONSEQUENCES OF THE TRANSACTION ATMOSPHERE

Social atmosphere: The forest landowner and the contract partner live in a different social atmosphere and did not have business contacts before	→	It is more complicated to start the business
	→	Contract negotiations are more difficult
	→	Is the reason for test contracts
Legal atmosphere: Regulations for tour operators	→	Is the reason for not offering all-inclusive programmes on the responsibility of the forest enterprise (DE12)
Liability for traffic safety	→	Written agreements on the liability for traffic safety are necessary
Income from RES products from a taxation point of view	→	Is the reason for founding a special association for the marketing of the RES product (DE11)

5.5.3 Example: environmental sponsoring

Environmental sponsoring is another example of an integrative non-material product which is sold to organizations and negotiated individually.

RELEVANCE AND CONSEQUENCES OF SPECIFICITY

Contract partner: develops an advertising strategy that is connected with the sponsoring project	→	The sponsor is entitled by contract to use the project for advertisement purposes (see Section 5.3.4)
	→	The forest landowner is contractually obliged to mention the sponsor in case of public relations (PR) activities in connection with the project (see Section 5.3.4)
The forest landowner: uses or is not allowed to use certain silvicultural methods with long-term consequences	→	Completion of long-term contracts (security of long-term financing)

RELEVANCE AND CONSEQUENCES OF MARKET UNCERTAINTY

There is low market transparency on both sides, because sponsoring projects are individually negotiated and contracts do not become public very often	→	Cooperation with a sponsoring agency (AU09)

Relevance and consequences of uncertainty about the behaviour of the market partner

The cooperation between sponsor and forest landowner has consequences for the image of both parties. Thus, both parties are looking for possibilities to reduce uncertainty about the behaviour of the contract partner	→	Either a detailed contractual regulation of obligation and equivalent consideration (see Section 5.3.4) or the public image of both parties (DE08, 20, 22, IT29) secures the fulfilment of the contract
	→	The forest landowner has to show indicators of his/her capabilities. Examples: Development of an interesting brochure (DE08) Summary of articles that have been printed in newspapers and journals (DE08) Cooperation with a well-known agency (AU09) Cooperation with an organization that secures the service of the landowner with its image (e.g. the World Wildlife Fund (WWF) – IT29)

Relevance and consequences of frequency

When sponsoring is a single project	→	The forest landowner does not develop special sponsoring concepts (DE22)
When sponsoring is used as a financial resource more often	→	The landowner develops brochures and sponsoring suggestions (DE08, NL06, 12)

Relevance and consequences of strategic importance

Being a communication instrument, sponsoring has strategic importance mainly for the sponsor	→	Exclusivity is secured in the contract by a certain exclusivity clause (see Section 5.3.4)
	→	An obligation to inform the sponsor of all PR activities is written into the contract

Relevance and consequences of internal complexity

The cooperation of contract partners in a sponsoring contract has characteristics of a long-term business relationship which is grounded on trust and confidence Not every obligation can be described in detail with reasonable costs The landowner has influence on the success of the sponsoring project with everything he/she does (in addition to what is written down in the contract) Contract partners cannot be changed	→	The offer of the forest owner has important trust qualities and therefore informal contacts are very important for the start of the business relationship. Examples: DE08: member of the board knows the sponsor from his job DE20: personal contacts between forest enterprise and sponsors DE22: sponsor knows about the targets of the local forest authority IT29: personal contacts with WWF personnel, NL 06, 16: personal contacts with the sponsors

RELEVANCE AND CONSEQUENCES OF THE TRANSACTION ATMOSPHERE

Social transaction atmosphere: Personal contacts facilitate the start of a business relationship	→ Examples above
Legal transaction atmosphere: Tax regulations	→ Are the reason for the foundation of special associations (DE08)

5.6 General Conclusions

5.6.1 Influence of product structure and transaction qualities on contract design and arising transaction costs

In contrast to other economic theories,[20] institutional economics takes information deficits of economic actors into account when analysing economic problems (see Section 5.1.2). It does not describe human beings as isolated individuals who decide on perfect markets with perfect information in accordance with the principles of perfect rationality and use maximization. Instead, it describes actions of individuals with limited rationality. These individuals do not interact on anonymous markets but with other individuals, who themselves show a certain behaviour, such as opportunism.

From this point of view, the decision-orientated behaviour of market partners is replaced by strategic behaviour to complete contracts with other economic actors. Initiation, completion and control of those contracts cause transaction costs, as described in Section 5.4 for RES products as well.

In the preceding sections, the decisive influence of transaction qualities such as specificity, uncertainty, frequency, internal complexity, strategic importance and transaction atmosphere on the transaction arrangement has been described from individual empirical observations in the case-studies. This influence will be described in a consolidated way in this section. It will be tested whether it is possible to assign the product groups that have been found in the case-studies to certain business and corresponding contract types in accordance with particular transaction qualities. By doing this, the present investigation mainly focuses on the transaction qualities specificity and uncertainty and tries to show their connection with the product qualities customer integration and non-materiality.

Specific investments in the business relationship of one contract partner always run the risk of being exploited by the other contract partner in an opportunistic way. As the investing partner cannot use these investments for other purposes or can only use them with less benefit, he/she becomes dependent on the opportunistic partner. When possible contract partners anticipate this dependence, there might be different consequences:

1. Contract partners try to develop security instruments within the contract which prohibit the contract partner from exploiting the dependence (the development of such instruments causes transaction costs (see Section 5.5)).
2. Very high transaction costs can make the contract completion much more difficult or can prohibit it entirely.

The amount of specific investments in a business relationship differs from transaction to transaction. The dependence of the investing contract partner on the opportunistic contract partner and the necessity to develop special security instruments for the exchange process vary accordingly. As a first step, therefore exchange processes can be distinguished into cases with negligible investments, on the one hand, and cases with considerable investments, on the other (Aufderheide and Backhaus, 1995, p. 53). Whereas standard contracts are completed in cases with negligible investments, contracts with individual security instruments are completed in cases with considerable investments.

But the criterion 'amount of specific investments' does not suffice to distinguish exchange processes taking place in real situations. If market partners had perfect information,[21] they would anticipate the future behaviour of their contract partner and all future aspects that might become relevant for the contract. The consequence would be a perfect contract, with detailed regulations taking into account all anticipated future developments. Looking at real transactions, it becomes clear that this condition is only hypothetical. Therefore, the transaction costs theory takes imperfect information and the resulting uncertainty of market partners into account and analyses transactions under much more realistic conditions. Different degrees of uncertainty can be distinguished with the help of the following two criteria:

1. Knowledge of all future aspects relevant for the contract at the time of contract completion.
2. Certainty about the fact whether those aspects will become relevant or not in future.

If both conditions are fulfilled at the time of contract completion, Aufderheide and Backhaus (1995) name the situation perfect knowledge and uncertainty does not exist. Perfect knowledge is a main condition for describing economic actions in neoclassical theory but it never exists in reality. Institutional economics acknowledges the much more realistic condition of imperfect knowledge of market partners. This knowledge can be imperfect in two ways. If contract partners know all aspects that might become relevant for the contract in future, but do not know whether these aspects will or will not occur, Aufderheide and Backhaus (1995) use the term risk. A higher degree of uncertainty exists when contract partners know neither the aspects which might become relevant for the contract nor whether they will occur or not. In this case, Aufderheide and Backhaus (1995) speak of incomplete knowledge or limited rationality. Thus, different degrees of increasing uncertainty are distinguished. Figure 5.12 shows the influence of increasing uncertainty on the contract design.

The first column of Fig. 5.12 shows the different degrees of increasing uncertainty. The second and third columns describe the criteria that have been used to distinguish these degrees of uncertainty. The fourth column shows the possibility of controlling the contract as another criterion to distinguish different contract designs. These different contract designs are described in column five. From the point of view of transaction costs theory, these contract designs mark five steps of the transition from market-led coordination to hierarchy-led coordination.

1. In the case of certainty[22] (the condition of neoclassical theory), perfect contracts[23] are completed. In this case, it is not necessary to control contract fulfilment because, in the case of perfect information, opportunism does not exist. Every contract partner is perfectly informed about the future behaviour of the other contract partner.
2. In case of risk[24] complete contracts[25] are designed which consider all relevant aspects. Obligations and equivalent considerations are regulated in detail. All aspects relevant for contract fulfilment are completely regulated and can be checked and implemented by jurisprudence. Complete contracts are normally related to one point in time and the identity of the contract partners is not very important. Preceding or following relations are not prerequisites for the actual obligation and the equivalent consideration.
3. In the case of incomplete knowledge, the contract partners are not able to take into account all future aspects relevant for the contract. In this case, obligation and equivalent consideration cannot be regulated in detail at the time of contract completion. Due to future developments which cannot be anticipated, the contract has to show a certain flexibility and therefore stays incomplete. Incomplete contracts can be distinguished further with the help of the criterion 'possibility of controlling and implementing the contract'.

- Sometimes, certain adjustment clauses

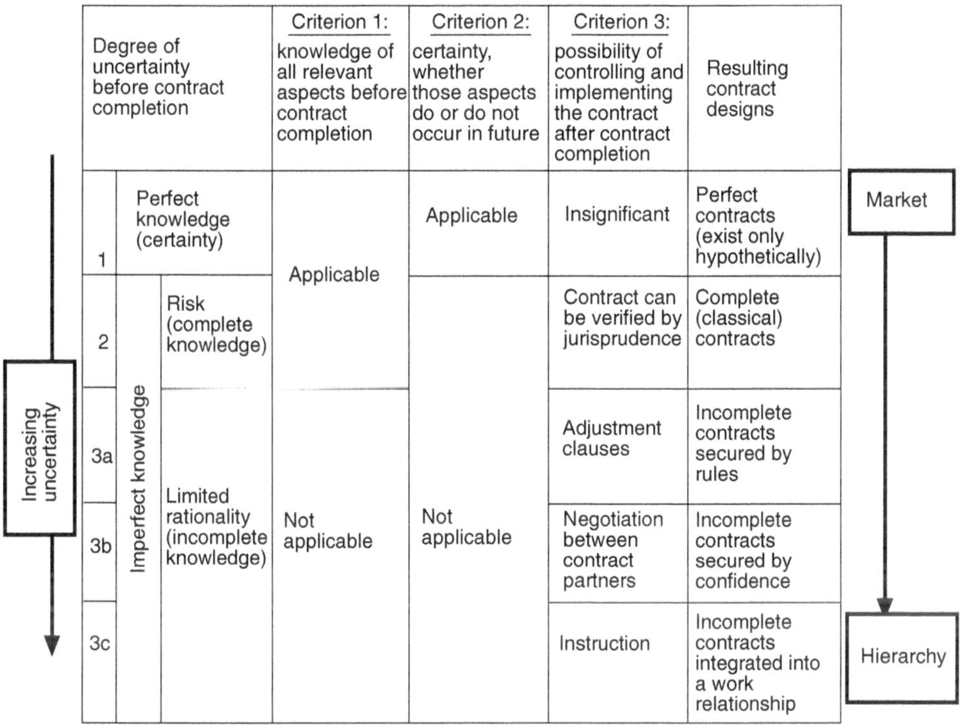

Fig. 5.12. Connection between degree of uncertainty and contract types.

exist in the contract or an independent third party is designated to decide in the case of problems. By doing this, the contract partners guarantee a certain flexibility of the contract without being obliged to negotiate an entirely new contract in case of small changes in a business situation. Such a contract design is a possible way of avoiding transaction costs which arise in case of problems with the necessity of new contract negotiations. Such regulations make sense if all relevant aspects cannot be regulated in the beginning, but contract fulfilment can be judged by an independent third party afterwards. In this case, the result is incomplete contracts that are secured by written rules[26] (adjustment clauses, designation of arbitrators).

- When obligation and equivalent consideration can only be judged by the contract partners themselves because of internal complexity, transactions take place within long-term partnerships (Williamson, 1990, p. 81). In these partnerships explicit contractual regulations are replaced by implicit agreements due to bilateral confidence. Personal knowledge, common values and self-obligations take over the function of adjustment clauses or the decisions of an arbitrator. Such transactions are merged into a long-term business relationship. Contract fulfilment cannot be judged by an independent third party because this arbitrator does not know the development of the business relation. In this case, the resulting contracts may be defined as incomplete contracts that are secured by confidence.
- When contract fulfilment cannot be judged and controlled by the contract partners either, because of information deficits, vertical integration becomes more and more probable. The cooperation of independent partners is replaced by cooperation within an enterprise. Integrated work contracts are designed and instructions guarantee contract fulfilment.

Thus the following three criteria to distinguish between exchange processes can be listed:

1. The degree of dependence of contract partners because of specific investments.
2. The degree of uncertainty.
3. The possibility of controlling and implementing the contract afterwards.

Using these criteria, the business types shown in Fig. 5.13 can be distinguished. Figure 5.13 was derived from the work of Aufderheide and Backhaus (1995) but has been enlarged by the distinction between contracts that can be controlled by independent third parties (adjustment clauses, arbitrators) and those which can only be controlled by the contract partners themselves (confidence, negotiation). Thus, the contract typology described in this investigation is more closely related to that described by Williamson (1990, pp. 77–81).[27] Contracts secured by rules are distinguished from contracts secured by confidence in order to take into account the importance of long-term business relationships in the real economy. Therefore, the 'long-term business relation' was added to the business types described by Aufderheide and Backhaus (1995).

The cooperation between employer and employee within an enterprise (vertical integration) is described as a contract relationship as well. In contrast with a long-term business relationship, this is not a cooperation of independent partners but rather the employee depends on the employer. Work contracts are the basis of cooperation.[28]

The basic connections between specificity, uncertainty, possibility of contract controlling and contract design have been described so far. They are also valid in the case of selling RES products. But, in accordance with the character of RES products, not every business type is evident. With reference to the distinction of products in accordance with the degree of non-materiality and the degree of customer integration

<table>
<tr><td colspan="3" rowspan="2">Degree of uncertainty and possibility of controlling the contract fulfilment</td><td colspan="2">Dependence because of specific investments</td></tr>
<tr><td>Negligible</td><td>Important</td></tr>
<tr><td rowspan="5"></td><td colspan="2">Certainty
(perfect knowledge)</td><td colspan="2">Business in neoclassical theory
perfect contracts</td></tr>
<tr><td>Risk
(complete knowledge)</td><td>Contract can be controlled by jurisprudence</td><td rowspan="2">Product business
complete or incomplete
standard contracts</td><td>Project business
complete individual contracts</td></tr>
<tr><td rowspan="3">Incomplete knowledge
(limited rationality)</td><td>Adjustment clauses</td><td>Combined business
incomplete individual contracts which are secured by rules</td></tr>
<tr><td>Negotiation between partners</td><td colspan="2">Long-term business relation
incomplete contracts which are secured by confidence</td></tr>
<tr><td>Instruction</td><td colspan="2">Vertical integration
contracts which are integrated in a work relationship</td></tr>
</table>

Fig. 5.13. Segmentation of exchange processes with the help of the criteria uncertainty, specificity and possibility of controlling contract fulfilment (adapted from Aufderheide and Backhaus, 1995).

(see Chapter 3, Section 3.8 on typology), two main criteria for assigning the sales processes of RES products to the business types mentioned become clear.

Autonomous products are characterized by the fact that the offerer produces the product for a variety of possible customers. His/Her investments in the production process are not specific because they have not been made for a specific customer. If the product is not sold to one particular customer, the investment is not lost, because the same contract can be completed with somebody else. Compared with integrative products, even the customer does not invest very much in the transaction (e.g. travelling to the offerer).

> Conclusion: The exchange of autonomous RES products (e.g. Christmas trees, mushrooms, game meat, accommodation possibilities, recreation facilities, standardized guided tours, lease of mountain bikes, riding tickets, car-park areas or picnic sites, etc.) is characterized by small specific investments by the contract partners. Specific instruments to secure contract fulfilment do not have to be developed. Instead, standard contracts (sales contract, lease contract, contract for work) are used. If necessary, legal regulations guarantee contract fulfilment (for example, the regulations in case of cancellation of an accommodation reservation).

The situation regarding integrative products is different. In this case, the forest enterprise develops products that are designed according to the special wishes of the customer. The forest enterprise invests specifically in the business relation (e.g. choice of specific excursion points for a seminar, implementation of specific silvicultural methods according to the wishes of a sponsor, acceptance of restrictions on forestry because of a mountain-bike trail). The customer makes specific investments as well in most cases (e.g. construction of a well to feed a fountain, production of advertising leaflets for the all-inclusive package of 'holidays at the forester's', planning of a PR strategy that takes into consideration the sponsorship, production of maps for a mountain-bike trail).

> Conclusion: The relations mentioned are characterized by the fact that the exchange partners are not arbitrarily exchangeable. They depend to a greater or lesser degree on each other and therefore are interested in instruments that secure contract fulfilment. In these exchange processes, individual contracts with individual security instruments are negotiated.

The connection between customer integration, specificity of the transaction and contract design has thus become clearer. High customer integration means that the transaction partner has to invest specifically in the transaction and to secure these investments by individual contracts. The following paragraphs will deal with the uncertainty and therefore with the question whether complete or incomplete contracts are used.

Complete knowledge about all aspects relevant for contract fulfilment or the possibility of getting this knowledge is a prerequisite for the use of complete contracts. The degree of non-materiality of the product and the degree of complexity of the transaction involved exert a decisive influence on the possibility of acquiring the knowledge needed.

Entirely tangible products already exist at the time of the exchange process. From a contractual point of view, the exchange of ownership titles is the centre of attention. Schade and Schott (1993, p. 17) define these products as 'exchange goods' (*Austauschgüter*). From the point of view of information, economic goods in general can be distinguished by the ability of the customer to check the product qualities relevant to the decision as to whether to buy a product. The following qualities can be distinguished:

- Search qualities, characterized by the fact that they can be entirely checked by

the customer before buying a product. The customer will stop searching for information when he/she reaches a level of information he/she subjectively judges as being sufficient or when the transaction costs of searching for further information become too high.
- Experience qualities, characterized by the fact that the customer can judge them only after the purchase of the product.
- Credence qualities, characterized by the fact that they cannot be judged by the customer either before or after the purchase of the product. This is due to the fact that the customer does not have the special know-how to be able to check the qualities and is not able to accumulate the necessary knowledge in a reasonable time.

A specific one of the qualities listed above cannot be assigned exclusively to one product, as certain products are normally a package of different product components (see Chapter 3, Section 3.8 on typology). Instead, products show a variety of these qualities in different combinations. As regards the example of ecologically produced Christmas trees (DE13), the pleasing form of the tree is a search quality, the freshness, or the time until the tree loses its needles, is an experience quality and the ecological production is a credence quality.

This example shows the decreasing influence of search qualities compared with experience qualities and credence qualities when the amount of non-material product components increases. Entirely non-material products do not exist at the time of contract completion. In this case, a promise of performance (*Leistungsversprechen*) (see Schade and Schott, 1993, p. 17) is the topic of the contract (e.g. the promise to guide participants of a seminar). In the case of totally non-material products, search qualities cannot be used for information purposes by the customer. As he/she cannot check the non-material part of the product before buying it, he/she relies on his/her own experience (e.g. in buying a Christmas tree at a Christmas fair) or on independent guarantees (e.g. certified eco-products or accommodation possibilities with a certificate). The two types – the total exchange good and the total promise of performance – can be seen as extremes on a continuum of exchange types in reality. Sometimes products resemble exchange goods and sometimes they resemble promises of performance. This is similar to the fact that products have both tangible and non-material parts (Weiber and Adler, 1995, p. 58).

Conclusions:

- In case of non-material product components, the customer does not have complete knowledge regarding the relevant product qualities. Whether the customer experiences the promised atmosphere of the Christmas fair or not is subjective and cannot be verified by jurisprudence. Rules that regulate the consequences in case the promises are not fulfilled cannot be designed in this case. Contracts relating to the exchange of products with non-material product components, therefore, have a tendency to be incomplete.
- When the specific investments of the customer are not very high, incomplete standard contracts are used most often. Contracts are more complete when the internal complexity of the transaction is not very high. This situation applies to use contracts with organizations which only sell a certain right to use the forest. Due to the low internal complexity of the transaction, it is possible to design detailed contractual regulations, which can be verified by independent third parties. Complete contracts are also possible for totally defined products which are sold on spot markets (e.g. timber).

Taking into account the above-mentioned connections, the business types when selling RES products can be summarized as in Fig. 5.14.

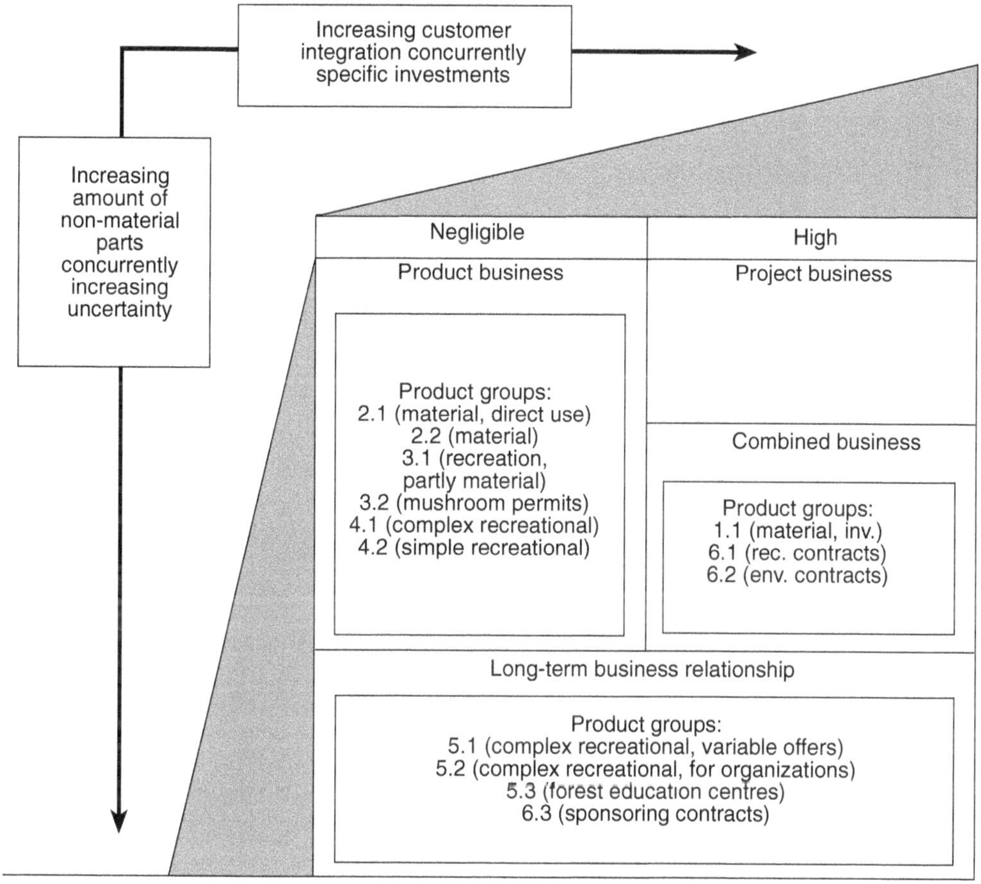

Fig. 5.14. Business types depending on product and transaction qualities. rec., recreational; env., environmental.

By assigning the product groups resulting from the case-studies to certain business types and corresponding contract types, the important main characteristics of the transaction and the transaction arrangement will be shown. A sharp distinction of the relevant business types and corresponding contract types will make the consequences more visible. In real business, an overlap always exists between these types. Nevertheless, the following general conclusions regarding the influence of non-materiality and customer integration on contract initiation and contract completion when offering RES products can be drawn:

CONSEQUENCES OF NON-MATERIALITY FOR CONTRACT INITIATION

Conclusions:

- The user cannot analyse the product quality physically and therefore relies much more on the credence and experience qualities of the product instead of search qualities.
- The missing physical comparability of RES products with a high amount of non-material parts results in low market transparency (e.g. in the case of the offer of seminars).

- As the user must choose the product before production, he/she is looking for substitutes for search qualities (e.g. references, image of the offerer, impression of the sales representative and of the communication material).
- A high degree of non-materiality of RES products, combined with high internal complexity, means a growing importance of the price as indicator for product quality (e.g. in the case of the offer of seminars and of guided tours in the forest).
- The forest landowner should observe the above-mentioned problems of the user when offering RES products with a high degree of non-materiality. As the offerer cannot present visible products, communication policy becomes much more important (quality of advertising, PR).
- Investment in a good image is very important for selling products with high credence qualities. Forest enterprises that want to offer these products should be aware that it takes considerable time to build up a certain image and customer confidence and that it takes only a few mistakes to lose this market advantage (long-term business relationship). If forest enterprises do not have the necessary image, it is often possible to secure credence qualities either by independent certificates (eco-labels, certificate of a certain accommodation quality) or by co-operation with middlemen who have the necessary image (e.g. cooperation with nature conservancy organizations in the case of sponsoring products, cooperation with business consultants in the case of offering seminars).

CONSEQUENCES OF CUSTOMER INTEGRATION FOR CONTRACT INITIATION

Conclusions:

- When the customer is very much integrated into the production process, it is difficult for him/her to judge the offerer's performance. Thus, when a forest landowner develops a certain product together with and according to the special wishes of a customer (e.g. sponsoring products), he/she sometimes has difficulties in stating his/her contribution to the product development and in demanding a reasonable price.
- The more specific the products are, the lower the market transparency becomes. This might result in disadvantages if the forest landowner has to negotiate with a very powerful potential customer (e.g. in the cases of nature conservation contracts with a public body or contract negotiations with a big drinking-water company).
- Due to customer integration in the production process, the offerer cannot influence the product quality independently of the customer. This might have the consequence of better excluding certain potential customers from the use of the product (e.g. by price policy) rather than producing dissatisfied customers.

CONSEQUENCES OF NON-MATERIALITY AND CUSTOMER INTEGRATION FOR CONTRACT COMPLETION

Conclusions:

- In the case of products with a high degree of non-materiality, it becomes difficult and produces considerable transaction costs to define the obligation of the offerer in a contract in detail. The uncertainty of both contract parties grows the more non-material parts a product has, and therefore incomplete contracts are used. In this case, security instruments apart from juridical regulations become more and more important (e.g. image, confidence, personal contacts).
- When non-materiality is combined with low transaction-specific investments by the customer (e.g. in the case of using a recreation facility in a

forest) the transactions can be defined as product business. Standard-form contracts are used and credence qualities are secured by market competition or sometimes by independent certificates.

- If a high degree of non-materiality is combined with a high degree of customer integration and accordingly high transaction-specific investments by the customers, the corresponding transaction increasingly resembles a business relationship. In this case, no standard-form contracts can be used and specific investments of both partners in the long-term cooperation serve as guarantees for contract fulfilment.

CONSEQUENCES OF DIFFERENT BUSINESS TYPES FOR THE AMOUNT OF VARIABLE TRANSACTION COSTS

Conclusions:

- In the case of offering products that can be characterized by the term product business, variable transaction costs mainly arise for the development of standard-form contract conditions, for communication activities and for the control of contract fulfilment. Variable transaction costs for negotiating contract conditions do not arise in this case, because legal regulations serve as contract conditions. In this case, transaction costs for the development of the relevant regulations have been produced by the legislator and must be regarded as sunk transaction costs for the relevant business.
- In the case of offering products that can be defined by the term combined business, variable transaction costs arise, mainly for negotiating and formulating written contract rules to ensure contract fulfilment (adjustment clauses, regulations about an arbitrator, definition of obligation and equivalent consideration). These transaction costs increase with the duration of the contract period. (Typical examples are long-term contracts concerning tangible products, such as drinking-water or electricity, contracts about environmental services in connection with public programmes or use contracts about the use of forest roads by organizations.)
- In the case of transactions that can be defined by the term business relationship, transaction costs mainly arise for the development of a certain image of the offerer, for PR to attract cooperation partners, and for the maintenance and further development of personal contacts. After starting a cooperation, variable transaction costs arise for information and communication processes between cooperation partners.

5.6.2 Suggestions from a contractual point of view for diminishing transaction costs of forest landowners to support the development of niche markets for RES products

Suggestions for supporting the activities of forest landowners in niche markets for RES products have to consider the broad variety of possible RES products and can therefore only be very general. Nevertheless, the following suggestions should be made according to the general business types described.

Conclusions:

- Regarding transactions defined by the term product business, transaction costs of forest landowners can be diminished by informing them of the special legal regulations of the relevant business field, by developing samples of standard-form contract conditions (as was part of this investigation) and by support in terms of advertising (e.g. development of labels, concerted actions regarding sales promotion or advertising). Forest landowner organizations, which build up special know-how or hire specialized personnel, can fulfil valuable tasks regarding these activities.

- Regarding transactions defined by the term combined business (long-term contracts), transaction costs of forest landowners can be diminished by the development of contract examples that consider the relevant regulations. Market transparency will be supported by describing successful contracting examples.
- Regarding transactions defined by the term business relationships, transaction costs of forest landowners cannot be diminished by developing standard contract examples. Nevertheless, it might be possible to design very general frame contracts, as has been done in two examples investigated (DE12, 18). As transaction costs to establish the necessary image and to justify certain performance capabilities play an important part in the total amount of transaction costs, forest landowner organizations can help to build up such an image by PR and cooperation with single landowners to arrange relevant contacts.

5.7 Summary

It has been shown that a variety of possible RES products exist, which can be characterized by certain main product qualities (see Chapter 3, Section 3.8 on typology) and are sold with different contractual and organizational structures (see Section 5.3). In general, products differ very much from the standard product timber, because non-material product parts, such as environmental quality or market atmosphere, are often used to support the marketability of tangible products by adding value. Numerous RES products dispose of a high degree of non-materiality and customer integration and therefore differ very much from the product timber. Due to these differences and because of the fact that forest enterprises offer new products on new markets (at least from their point of view), they have to face relevant initiation costs before contract completion. The relevance of these initiation costs when selling RES products has been described through examples derived from different product groups (see Section 5.4). Furthermore, it has been documented that certain characteristics of exchange processes and their influence on contract design can be very well described and structured in terms of the transaction costs theory (see Section 5.5). Methods of the transaction costs theory help to identify relevant product structures and transaction qualities and to assign to certain general business types the variety of possible transactions when offering RES products. Thus, it was possible to develop some general suggestions for forest landowner organizations to support the offer of RES products (see Section 5.6).

Notes

1 These target groups have been mentioned by the interview partners.

2 In the following section, the different topics are explained in detail. The empirical basis is the above-mentioned contract examples. Taking into account the wishes of those enterprises, their names are not indicated. For a better understanding, some passages of the contracts have been grouped and summarized under certain headings.

3 In contrast to this, conservation contracts, as explained by Moog and Brabänder (1992), are classified as usufructuary lease contracts because these contracts often entitle nature conservation organizations to cut timber, to fill up drainage channels or to carry out other activities to ensure the conservation goal. According to the authors, this corresponds more to a usufructuary lease contract than to a lease contract (Moog and Brabänder, 1992, p. 130).

4 Furthermore, sample contracts of the Austrian Federal Forests (ÖBF) and the Austrian Association of Agricultural and Forest Enterprises (HVLFÖ) have been analysed for this purpose.

5 Agreement of access between a Bavarian forest administration and a local riding club.
6 Licence agreement between the Allgemeine Hannoverschen Klosterfonds and a community.
7 Agreement between the association Initiative Land und Wirtschaft and the ÖBF concerning mountain-biking on forest roads.
8 A pattern contract drafted in cooperation between the Tyrolese Chambers of Agriculture, together with the Tyrol Federal Forest Authority, the ÖBF, the association of Tyrolese forest owners and the presidential department IV of the Office of the Tyrolese Government.
9 Agreement between the Bavarian Saalforstverwaltung St Martin with the association of cycling paths of the Pinzgau.
10 Contract of a forest enterprise and a local archers' club (indication of the name not authorized).
11 Licence agreement between the FBG (Forstbetriebsgemeinschaft) Westfeld, the Teilnehmergemeinschaft Westfeld-Ohlenbach, the FBG Oberkirchen and the centre for cross-country skiing Westfeld-Ohlenbach e.V.
12 Contract between the Bavarian Saalforstverwaltung and the local tourist authorities.
13 Contract between the club for motor sports in Schlüchtern and the City of Schlüchtern.
14 Contract between the forest authorities in Nuremberg and the City of Nuremberg.
15 Governmental printed matter 7/889, p. 29.
16 Compare the decision of the Kleve county court dated 28 May 1996, Holzzentralblatt No. 23/1998, p. 343.
17 Agreement costs and management are explained in the PhD thesis but will not be explained in further detail here.
18 Examples of the property rights extension of the offerer are all kinds of permissions (e.g. building permission, permission for afforestation, bar permission, business licence, exceptions from nature conservation law), the change in the area plan in favour of the product offer or a modification of the land registers to rent mountain houses. Examples of the restriction to property rights of other actors are the declaration of a certain area as drinking-water preserve, restrictions to the free-access right during the mushroom-picking season, prohibition of renting a private area for camp-sites, prohibition of fencing of private property as a basis for the offer of cross-country skiing tracks, prohibition of parking (as a condition for the offer of car-park areas) or the limitation of the access right of horsemen to horse-riding trails.
19 Question 6.2.7 in the RES questionnaire.
20 The microeconomic household theory simplifies the conditions of economic exchange, as follows:

- Complete knowledge of one's own preference structure.
- All economic actions are performed with the target of maximization of wealth (rational behaviour) and the use can be measured at least in an ordinal way.
- Complete market transparency (this means complete information for all market partners).
- Unlimited possibilities to process the information.
- No preferences regarding time, place or other conditions.
- No influence from other persons or experiences coming from other buying processes.

See Weiber and Adler (1995, p. 46).
21 In the case of complete information, the market partners know all future conditions relevant for the decision before contract completion. There is security whether those conditions become relevant or not. This is the central precondition for the exchange processes in neoclassical theory.
22 Aufderheide and Backhaus (1995) speak of complete knowledge in this case.
23 Agreements concerning behaviour in the case of different future conditions do not have to be in the contract, as the contract partners know all relevant conditions before contract completion and already take these future conditions into account at the present stage.
24 In this case, the contract partners know all relevant aspects at the time of contract completion but there exists uncertainty as to whether those aspects will arise or not in future (see Aufderheide and Backhaus, 1995, p. 54).
25 The contract partners are not sure whether those conditions will become relevant or not but they have, due to complete knowledge, contractual regulations for these possible future conditions, which can be verified by jurisprudence. This contract design is described by Williamson (1990, p. 78) under the term 'classical contract law'.
26 This contract design is named 'neoclassical contract law' by Williamson. This describes the situa-

tion when the 'classical' control mechanism, 'complete regulations and verification by jurisprudence', does not work any more. Adjustment clauses in the contract are used for contract fulfilment (Williamson, 1990, pp. 78–80).

27 See the reference to the work of Macneil (1974, 1978) in Williamson (1990).

28 In the literature, the typical long-term relationship is pointed out for business relationships, as well as for work contracts. Consequently, the corresponding contracts are named 'relational contracts' (see Richter, 1994, p. 84). Dietl (1995, pp. 569–585) distinguishes further between bilateral relational contracts and integrated relational contracts, highlighting the difference between autonomous contract partners in a business relationship and dependent partners in an employment relationship.

References

Aufderheide, D. and Backhaus, K. (1995) Institutionenökonomische Fundierung des Marketing: Der Geschäftstypenansatz. In: Kaas, K.P. (1995) *Kontrakte, Geschäftsbeziehungen, Netzwerke – Marketing und Neue Institutionenökonomik,* Verl.-Gruppe Handelsblatt, Düsseldorf, Frankfurt-on-Main, p. 53.

Bartl, H. (1991) *Reise- und Freizeitrecht,* 2nd edn. C.H. Beck, Munich.

Bergen, V. (1991) Monetäre Bewertung der Landeskulturellen Leistungen der Forstwirtschaft. In: Bergen, V., Brabänder, H.D., Bitter, A.W. and Löwenstein, W. (eds) *Schriften zur Forstökonomie,* Vol. 1. J.D. Sauerländer's Verlag, Frankfurt-on-Main.

Bergen, V. (1992) Studien zur monetären Bewertung von externen Effekten der Forst- und Holzwirtschaft. In: Bergen, V., Löwenstein, W. and Pfister, G. (eds) *Schriften zur Forstökonomie,* Vol. 2. J.D. Sauerländer's Verlag, Frankfurt-on-Main.

Bergen, V. (1994) Vermarktung und Bewertung der Schutz- und Erholungsfunktionen des Waldes. *Holz-Zentralblatt* 5(94), pp. 33–38.

Blin, B. (1997) Use it or lose it. *Österreichische Forstzeitung* 7(97), 34–36.

Brabänder, H.D. (1992) Produktionsorientierte oder absatzorientierte Forstwirtschaft? *Holz-Zentralblatt* 23(92), 365–366, 374.

Bruhn, M. and Mehlinger, R. (1992) *Rechtliche Gestaltung des Sponsoring.* Beck'sche Verlagsbuchhandlung, Munich.

Commons, J.R. (1961) *Institutional Economics. Its Place in Political Economy,* Vol. 1. Madison, Wisconsin. (1st edn, 1934.)

Dietl, H. (1995) Institutionelle Koordination spezialisierungsbedingter wirtschaftlicher Abhängigkeit. *Zeitschrift für Betriebswirtschaftslehre* 65, 569–585.

Dietl, H. (1997) In: Picot, A. (ed.) *Organisation: eine ökonomische Perspektive.* Schäffer-Poeschel, Stuttgart, p. 12.

Ebers, M. and Gotsch, W. (1995) Institutionenökonomische Theorien der Organisation. In: Kieser, A. (ed.) *Organisationstheorien,* 2nd revised edn. Kohlhammer, Stuttgart, pp. 185–235.

Elsasser, P. (1996a) *Struktur, Besuchsmotive und Erwartungen von Waldbesuchern: Eine empirische Studie in der Region Hamburg.* Arbeitsberichts des Instituts für Ökonomie der Bundesforschungsanstalt für Forst- und Holzwirtschaft, p. 5.

Elsasser, P. (1996b) Der Erholungswert des Waldes: monetäre Bewertung der Erholungsleistung ausgewählter Wälder in Deutschland. In: Bergen, V. and Brabänder, H.D. (eds) *Schriften zur Forstökonomie,* Vol. 11. Sauerländer, Frankfurt-on-Main.

Erbach, K.F., Follmer, F. and Rudnick, A. (1991) *Betriebswirtschaftslehre.* Winkler Verlag, Darmstadt, p. 85.

Fikentscher, W. (1991) *Schuldrecht,* 8th edn. de Gruyter, Berlin, New York, 883 pp.

Frosch, H. (1990) Die Verkehrssicherungspflicht des Waldbesitzers. *Allgemeine Forstzeitung/Der Wald* 17–18(90), 430.

Gebhard, H. (1995) Verkehrssicherungspflicht und Wald. *Agrarrecht* 12(95), 389.

Geigel, R. and Schlegelmilch, G. (1986) *Der Haftpflichtprozeß,* 19th edn. Beck, Munich, p. 353.

Gitter, W. (1988) Gebrauchsüberlassungsverträge. In: *Handbuch des Schuldrechts,* Vol. 7. Mohr, Tübingen, 541 pp.

Glück, P. (1995) Vermarktung forstlicher Dienstleistungen. Vortrag vor der Vollversammlung des

Verbandes der Akademiker der Österreichischen Bundesforste am 3. December 1994 in Salzburg, *Internationaler Holzmarkt* 3(95), 18–21.
Gundermann, E. and Suda, M. (1996) Ökosponsoring und Forstwirtschaft. *Forst und Holz* 2(96), 35–37.
Hampicke, U. (1996) *Perspektiven umweltökonomischer Instrumente in der Forstwirtschaft – insbesondere zur Honorierung ökologischer Leistungen.* Materialien zur Umweltforschung, Metzner-Poeschel, Stuttgart, 164 pp.
Heussen, B. (1997) *Handbuch Vertragsverhandlung und Vertragsmanagement: Planung, Verhandlung, Design und Durchführung von Verträgen.* O. Schmidt, Cologne, p. 924.
Imbeck, M. (1997) Basischeckliste für Austauschverträge. In: Heussen, B. (ed.) *Handbuch Vertragsverhandlung und Vertragsmanagement: Planung, Verhandlung, Design und Durchführung von Verträgen.* O. Schmidt, Cologne, pp. 413–414.
Kieser, A. (1995) *Organsiationstheorien*, 2nd revised edn. Kohlhammer, Stuttgart, Berlin, Cologne, p. 187.
Kissling-Näf, I. (1998) Grosser Wort und wenig Geld: Über die Honorierung von Waldleistungen. *Zeitschrift für Umweltpolitik und Umweltrecht* 3(98), 373–397.
Klose, F. and Orf, S. (1998) *Forstrecht.* Kommentar, 2nd edn. Aschendorff, Münster.
Kotler, P. and Bliemel, F. (1995) *Marketing-Management: Analyse, Planung, Umsetzung und Steuerung*, revised edition. Schaeffer-Poeschel, Stuttgart.
Larentz, K. (1986) *Lehrbuch des Schuldrechts*, Vol. 2, revised edition. Beck, Munich, p. 212.
Larentz, K. (1994) *Lehrbuch des Schuldrechts*, Vol. 2, 13th edn. Beck, Munich.
Löwenstein, W. (1991) Die monetäre Bewertung der Fernerholung im Südharz mit der Reisekostenmethode. In: Bergen, V. *et al.* (eds) Monetäre Bewertung der Landeskulturellen Leistungen der Forstwirtschaft. In: Bergen, V., Brabänder, H.D., Bitter, A.W. and Löwenstein, W. (eds) *Schriften zur Forstökonomie*, Vol. 1. J.D. Sauerländer's Verlag, Frankfurt-on-Main, pp. 162–168.
Mantau, U. (1993) Die forstliche Produktlücke: Infrastrukturleistungen des Waldes marktfähig machen. *Allgemeine Forstzeitung/Der Wald* 4(93), 186–190.
Mantau, U. (1994) Produktstrategien für kollektive Umweltgüter. Marktfähigkeit der Infrastrukturleistungen des Waldes. *Zeitschrift für Umweltpolitik und Umweltrecht* 3(94), 305–322.
Mantau, U. (1995) Von der Waldfunktionenlehre zur Waldproduktlehre. *Holz-Zentralblatt,* 26(95), 445, 451–453.
Mantau, U. (1996) *Vermarktung von Erholungs- und Umweltleistungen als wichtiger Bestandteil der Nachhaltigkeit.* Jahresbericht zur Tagung des Deutschen Forstvereins, Niederstein.
Mantau, U. (1999a) Alternative ways of correcting market failure – product structures – a key to marketability. In: Roper, C.S. and Pouk, A. (eds) *Living Forests.* HMSO, London, pp. 267–273.
Mantau, U. (1999b) Erschließung neuer Märkte durch forstwirtschaftliche Zusammenschlüsse. *Allgemeine Forstzeitung/Der Wald* 12(97), 636–639.
Mantau, U. (1998) Wer zahlt für die gesellschaftlichen Leistungen des Waldes? Beiträge zum forstpolitischen Kolloquium am 8. September 1997 in Klagenfurt anläßlich des 90. Geburtstages vom em. Univ.-Prof. Dipl.-Ing. Dr. Otto Eckmüllner. *Schriftenreihe des Instituts für Sozioökonomik der Forst- und Holzwirtschaft* 30, 45–52.
Mantau, U. and Mertens, B. (1997) Forstbetriebliche Umweltagenturen öffnen den Weg zur Vermarktung von Umweltleistungen des Waldes. *Allgemeine Forstzeitung/Der Wald* 17(97), 918–921.
Medicus, D. (1995a) *Schuldrecht: ein Studienbuch*, Part 2, 7th edn. Beck, Munich.
Medicus, D. (1995b) *Schuldrecht: ein Studienbuch*, Part 1, 8th edn. Beck, Munich.
Meins, J. (1993) *Die Vertragsverhandlung: Leitfaden zum Entwerfen, Verhandeln und Abschließen von Verträgen*, 2nd edn. Schäffer-Poeschel, Stuttgart.
Merlo, M. (1995) Der Privatwald in Italien: Strukturelle und institutionelle Änderungen vonnöten. *Östereichische Forstzeitung* 1(95), 15–18.
Merlo, M. and Signorello, G. (1991) Alternative Methoden zur Bewertung der Erholung in der freien Landschaft in Italien. In: Bergen, V., Brabänder, H.D., Bitter, A.W. and Löwenstein, W. (eds) *Schriften zur Forstökonomie*, Vol. 1. J.D. Sauerländer's Verlag, Frankfurt-on-Main, pp. 149–161.
Merlo, M., Kuehl, G. and Ruol, G. (1996) Possibilities of additional income from environmental

goods and services produced by agriculture and forestry in mountainous areas. In: *Proceedings of Cost Conference, Vienna 1996, Forestry in the Context of Rural Development: Future Research Needs.* Conference Report, p. 18.
Mertens, B. and Schoop, G. (1999) Wer sponsert den Specht? Vermarktung von Umweltleistungen des Waldes durch Sponsorships? *Forst und Holz* 54(12), 359–363.
Meyer, W. (1983) Entwicklung und Bedeutung des Property Rights Ansatzes in der Nationalökonomie. In: Schüller, A. (ed.) *Property Rights und ökonomische Theorie.* Vahlen, Munich, p. 38.
Moog, M. and Brabänder, H.D. (1992) Vertragsnaturschutz in der Forstwirtschaft: Situationsanalyse, Entscheidungshilfen und Gestaltungsvorschläge. In: Moog, M. (ed.) *Schriften zur Forstökonomie,* Vol. 3. Sauerländer, Frankfurt-on-Main.
North, D.C. (1992) *Institutionen, instituioneller Wandel und Wirtschaftsleistung* (transl. by Streissler, M.) (*Die Einheit der Gesellschaftswissenschaften,* Vol. 76). Mohr, Tübingen.
Picot, A. (1997) *Organisation: eine ökonomische Alternative.* Schäffer-Poeschel, Stuttgart.
Richter, R. (1994) *Institutionen ökonomisch analysiert.* Mohr, Tübingen.
Schade, C. and Schott, E. (1993) Kontraktgüter im Marketing. *Marketing, Zeitschrift für Forschung und Praxis* 1(93), 17.
Schmidhauser, A. (1997) *New Public Management in öffentlichen Forstbetrieben: Auswirkungen der Produktdefinition auf Struktur und Organisation, dargestellt an Beispielen.*Tagungsbericht zum Forstökonomischen Kolloquium auf Schloß Nienover, 15–17 September 1997.
Schüller, A. (1983) *Property Rights und ökonomische Theorie.* Vahlen, Munich, p. VIII.
Sekot, W. (1992) Strategische Analyse der Dienstleistungsproduktion im Wald. *Österreichische Forstzeitung* 6(92), 24–27.
Sekot, W. (1994) Multifunktionalität des Waldes: Schicksal oder Chance für die Forstbetriebe? *Holz-Kurier* 49(22), 18–20.
Sekot, W. (1997a) Naturvermögen Wald – Wohlfahrt durch Waldreichtum. *Österreichische Forstzeitung* 7(97), 37–40.
Sekot, W. (1997b) Multifunktionale Leistungserstellung, PR-Gag, Peanuts oder Wirtschafts-standbein? *Österreichische Forstzeitung* 11(97), 7–8.
Sekot, W. and Schwarzbauer, P. (1995) Methodische Ansätze zur Bewertung der infrastrukturellen Leistungen der Forstwirtschaft. *Schriftenreihe des Instituts für forstliche Betriebwirtschaft und Forstwirtschaftspolitik* 25, 325.
Simon, H.A. (1986) Rationality in psychology and economics. In: The Behavioural Foundations of Economic Theory. *Journal of Business (Suppl.)* 59(86), 209–224.
Stahl, D. (1995) *Internationale Speditionsnetzwerke: eine theoretisch und empirische Analyse im Lichte der Transaktionskosten-theorie.* Vandenhoeck und Ruprecht, Göttingen.
Streit, M.E. (1991) *Theorie der Wirtschaftspolitik,* 4th edn. Werner, Düsseldorf.
Thieme, F. (1993) Holzproduktion, Erholung, Naturschutz in einem stadtnahen Privatforstbetrieb. *Allgemeine Forstzeitung/Der Wald* 26(93), 1348.
von Gerlach, J. (1994) Verkehrssicherungspflicht im Straßenverkehr. *Deutsches Autorecht* 12(94), 488.
von Hayek, F.A. (1969) Arten des Rationalismus. In: Mohr, J.C.B. (ed.) *Freiburger Studien.* Paul Siebeck, Tübingen, pp. 78–79.
Weiand, N.G. (1995) *Der Sponsoringvertrag.* C.H. Beck, Munich, 39.
Weiber, R. and Adler, J. (1995) Informationsökonomisch begründete Typologisierung von Kaufprozessen. *Zeitschrift für betriebswirtschaftliche Forschung* 47(1), 46.
Wibe, S. (1995) *Non Wood Benefits in Forestry. A Survey of Valuation Studies.* UN-ECE/FAO Timber and Forest Discussion Paper, ECE/ TIM/DP/2. New York, Genf, p. 70.
Williamson, O. (1990) *Die ökonomischen Institutionen des Kapitalismus: Unternehmen, Märkte, Kooperationen* (transl. by Streissler, M.). Mohr, Tübingen.
Williamson, O. (1996) Transaktionskostenökonomik. In: *Ökonomische Theorie der Institutionen,* Vol. 3, 2nd edn. LIT, Hamburg, p. 2.
Zilessen, R. and Rahmel, D. (1991) *Umweltsponsoring: Erfahrungsberichte von Unternehmen und Verbänden.* Frankfurter Allgemeine Zeitung, Gabler Verlag, Weisbaden, 263 pp.

6

Multifunctional Forest Management

6.1 Project Management for RES Projects

6.1.1 Introduction

Increasing public demand on multifunctional forest effects and services represents new challenges for multiple forest management systems. European forestry should take changing values into account, and, more than ever before, there is a chance to establish various sources of income. Simultaneously, forest business is confronted with an increasing complexity of the economic environment. For new projects in the field of environmental and recreational activities, forest management has to consider also external resources (e.g. information).

The methods of project management (PM) secure a goal-orientated and structured strategy. They achieve an effective and efficient coordination of all persons/institutions involved in the process of project realization. The integration of a project in the business organization could be seen as a success factor. In the course of the recreational and environmental services (RES) research, 98 cases were documented which to a higher or lesser degree are typical and individual projects. However, in about 60% of all cases, the process of establishing the RES activities was integrated in the daily business and less than 40% of all interviewed forest landowners (FLOs) defined a separate project (Table 6.1).

In addition to the definition of an explicit project, more than a third of all respondents declared that different instruments of project management were applied and helpful in the realization process (Table 6.2).

In order to deal with an increasingly complex business environment, FLO should pursue a permanent development process able to take into consideration technological, economic and social changes (Hansel and Lomnitz, 1993, p. 1). The following sections will give an overview of a management instrument that should be used more often and more professionally by FLOs. The trend towards

Table 6.1. PM application related to the documented RES cases. (Question 6.2.7: What was the process of establishing the RES business organization?)

Organization	*n*	%
Integrated	58	59.2
Separate project	37	37.8
Missing	3	3.1
Total	98	100.0

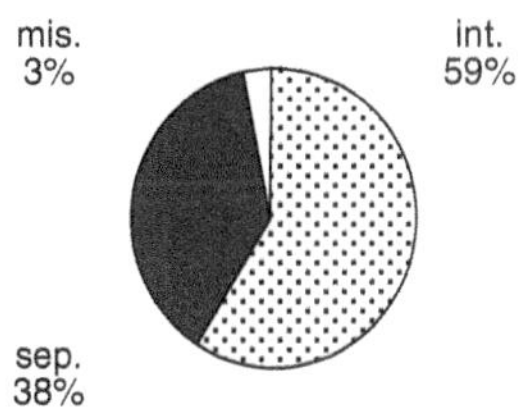

Table 6.2. Organizational items of PM applied in RES projects. (Question 6.2.7: What was the process of establishing the RES business organization?)

	No		Yes		Missing	
Organizational items	*n*	%	*n*	%	*n*	%
Responsible team	39	40	37	38	22	22
Milestones	42	43	34	35	22	22
Timetable	40	41	36	37	22	22
Checking	41	42	34	35	23	23
Total	162	41	141	36	89	23

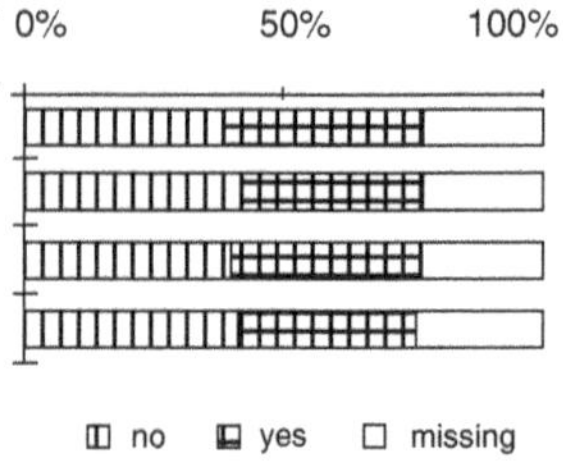

project-orientated work will be taken into account. Small changes in management structures or in the way of thinking (accepted by employees) can increase the potential for new achievements. Good working conditions should be available to promote the commitment, the motivation and the teamwork of each employee.

The instruments of PM represent helpful guidelines for working with innovations or managing new complex tasks. The realization and mastering of various tasks, with respect to the general expenses, mainly involves the internal company staff, and it is not only a job for internal and/or external experts.

With the application of PM, on the one hand a systematic way of thinking can be arranged (for example, by the definition of aims, by structuring the realization procedure, by temporal/spatial coordination of the conceptual formulations, etc.), while on the other it serves as an adaptive control of performance, costs and appointments, including parts of human leadership (communication, leadership, negotiations) (Zielasek, 1995, p. 6).

The implementation of most RES activities is a project in the understanding of PM. Consequently, PM could promote the successful establishment of an RES business. However, there is no unique or standard approach to hand. Depending on the scale and framework of the specific RES activity envisaged, a suitable PM concept has to be developed individually. In the following, the basic concepts and instruments of PM are introduced, thereby indicating how to adapt the general techniques for any specific case. Additionally, the discussion should provide guidelines for FLOs and forest managers on how to adapt and apply different PM instruments along with RES activities. As a result, any FLO should be able to describe a project in a PM handbook for better communication and controlling purposes. A practical application and an example of a PM handbook design is documented in Section 6.1.5 (related to a recreational project in a forest enterprise).

6.1.2 Projects and project management

Project and PM definition

For a better understanding, it is necessary to define the terms 'project' and 'project management'. Each project leader should be able to specify the borders and contents of a project.

A project is characterized by complex and, for the enterprise, mostly new and single tasks. The area of responsibility requires, to a varying extent, the cooperation of different subject areas (e.g. 'golden friends' of the Utrechts Landschap (Netherlands (NL)12): sponsoring, public relations (PR) and nature conservation tasks); this presupposes detailed project organization (with regard to the performance extent and responsibility of each team member, the appointments, the resources and the costs). Due to the uniqueness of a project, the management has to calculate a higher risk caused by the lack of experience. The challenges of uncertainty and indefiniteness must be faced (Gareis, 1991, p. 28).

'Project management' is defined in Fig. 6.1.

A project could also be simple and uncomplex, but in this case a project should be defined as a task or work package (WP) (defined later) and no PM is necessary for the realization process. Continuing procedures of routine should not be seen as a project in this context. The goal of the RES research is concentrated on typical projects (as defined above) and most RES activities fulfil this criterion at the level of forestry (see Chapter 2). However, the task (project or non-project) could be managed very individually. But the FLO should be aware that an explicit project requires special management tools for successful implementation. A strategy for distinguishing between (small or large) projects and non-projects is offered in Table 6.3. How many criteria could be defined with a degree of relevance (DoR)? On the one hand, non-projects could be felling, forwarding, afforestation, etc. and, on the other, forest management planning (FMP), chemical/biological control of insects (complexity, duration, importance, etc.) and RES could be defined as projects.

For the judgement of a project, the FLO can define specific weights and criteria. Such criteria are, for example, contents, duration, unusual features, complexity, meaning, risk and the costs of the project. These criteria could be evaluated, scale units determined and DoR estimated (see Table 6.3). The relative meaning or the value of the project is generated by the sum of weights times DoR.

The result can be used as general information (orientation help) in the planning phase, as a discussion basis during the decision process or for defining a task as a project.

Summarizing the specific characteristics of a project, they are complex, dynamic, relatively unique, goal-determined, important and risky tasks. For the realization process, a socially complex and temporary project organization (team) is necessary. A project often employs extensive internal human resources. Standard technologies and procedures cannot be applied, due to innovative, unique and non-repetitive activities (Gareis, 1994, chapter 4, p. 2). If the plots of the formulation are typical internal tasks, with a character of routine, with low risk and solved without interdisciplinary cooperation, then the concept is not considered as a project in principle.

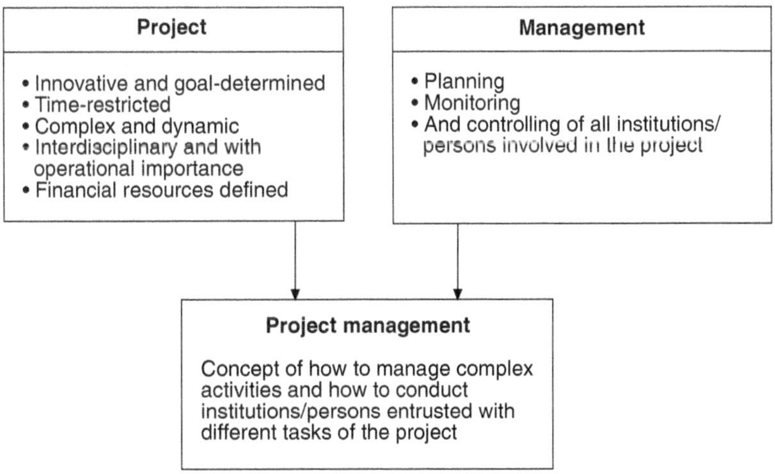

Fig. 6.1. Definition of project management (from Rinza, 1985).

Table 6.3. Example for a classification of relevance for PM (modified from Gareis, 1991, p. 29).

			Degree of relevance (DoR)				Weight × DoR
Criteria	Weight	Scale unit	1–2	3–5	6–8	9–10	
Contents	20	Number of tasks	Realization (e.g. to build a forest road for multifunctional purposes)	Planning and realization	Planning, realization and marketing	Planning, realization, marketing and organization of events	
Duration	10	Duration in months	Very short (0–2 months)	Short (3–12 months)	Long (13–18 months)	Very long (> 18 months)	
Unusual features	10	Task-related experiences	Internal know-how sufficient	Know-how partially available	Know-how scarcely available	Know-how not available; no experience	
Complexity	20	Number of internal and external team members (branches)	Very low (1–2 internal departments)	Medium (some internal and external departments)	High (many internal and external departments)	Very high (many external departments)	
Meaning, importance	10	Influence on daily business	Small (little additional employment)	Medium (relevant additional employment)	High (considerable additional employment)	Very high (extensive additional employment)	
Risks	20	Financial disadvantage by failure (Euro)	Low (up to 1000.-)	Medium (1000.- to 5000.-)	High (5000.- to 25,000.-)	Very high (more than 25,000.-)	
Costs for a forest road project	10	Investment (Euro)	Low (up to 20,000.-)	Medium (20,000.- to 200,000.-)	High (200,000.- to 700,000.-)	Very high (over 700,000.-)	
					Project value :		

DIGRESSION TO RES CASE-STUDIES. Referring to the general project definition stated above, the RES research implies an additional description of specific RES cases. The research team defined four relevant criteria for the extensive RES case-studies. Due to the goals of the project, each case should fulfil the following criteria:

- Negotiation.
- Contracting.
- Company profit.
- Innovation (new ideas, new products).

This definition was necessary so as to be able to evaluate and compare different projects in different countries at the same level.

PM starts out from two concepts. On the one hand, the leadership concept defines the tasks (WPs) necessary for the project and provides the required methods for the solution of these tasks. On the other hand, there is the concept of institution, which defines responsibilities for each task (carried out by ...) as well as the required organization. The organizational design should submit/secure an optimal integration of all involved project institutions in the existing enterprise.

Types of projects

The purpose of project distinction is used in particular to define common tasks for an economic PM design (Patzak and Rattay, 1996, p. 18). Projects can be differentiated by goals, scope and complexity, type of project owner and degree of repetition (Table 6.4). The judgement of the scope of the project must be defined internally by

Table 6.4. Distinctive features of projects.

Project targets	• New offer • Product innovation • Contracting or marketing project • Environmental protection • Organization of events • Rationalization, etc.
Branch	• Research project • Construction project • Industrial project • Tourism project, etc.
Project scope	• Small, medium, large
Type of project owner	• External projects: client comes from outside • Internal projects: internal client (department, management, owner)
Degree of repetition	• Pioneer project (unique) • Routine/standard project
Complexity	• Project of one department • Project of the whole enterprise • External institutions involved

various parameters, for example: number of the project employees, project duration, project budget, number of different disciplines involved in the project, etc.

From this differentiation of project types, various requirements and possibilities can be concluded for establishing PM methods. For routine or standard projects, PM methods can be standardized to a higher extent than in unique projects. Once developed, it would be possible to apply the instruments (checklists, guidelines, forms, standard breakdown structures, etc.) for succeeding tasks/projects in a cost-efficient way (e.g. pond management (Austria (AU)20): each pond was explicitly organized as a project and the same PM tools could be applied for several independent pond projects).

By the way, it is interesting that almost 50% of all forest managers interviewed apply internally defined codes of practice (Table 6.5) as guidelines or checklists for establishing/running the RES project.

Considerable differences exist between external and internal projects. External projects are initiated by external clients, so mostly very detailed ideas/contracts regarding targets, appointments, costs, etc. exist for the performances to produce (high goal reliability). Simultaneously, an essential controlling instrument is installed by the written specification (e.g. golf-course (Italy (IT)03), Tyrolese mountain-bike concept (AU17)).

Internal projects are usually formulated in an imprecise and verbal way. The tasks and the scope of the project are usually less clear. Written project definitions, from which project boundaries, goals, competence, etc., would be clearly obvious, are, in general, lacking. In many cases, a simple listing of WPs would make a project realization process more efficient and easier. Which PM methods have to be applied must be decided case by case. By doing this, it would be possible to estimate and optimize the project benefit with respect to the expected expenditures. As an example, afforestation sponsoring (Germany (DE22)) – a small project with a single character – could be mentioned here: on one hand, the project was integrated in daily business but several PM tools were applied and, on the other, the project evaluation indicated that press reporting could have been done in a better way. The result of a bad project organization could be documented impressively by IT04 (implementation of a parking area). The failures and consequences will be discussed later.

Table 6.5. Internally defined codes of practice. (Question 6.2.6: Are you applying any internally defined codes of practice serving as guidelines for establishing or running the RES business?)

Internally defined codes	*n*	%
Yes	47	48.0
No	49	50.0
Missing	2	2.0
Total	98	100.0

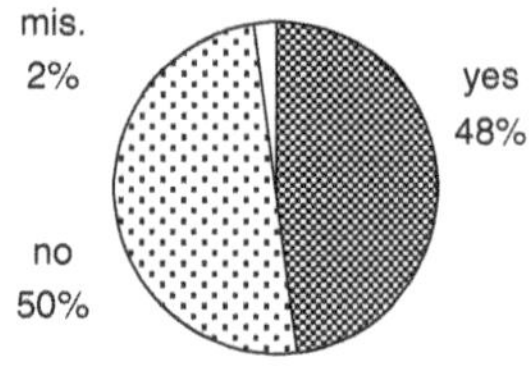

Those RES activities which are to be furthered by this research are typically internal projects. Hence, the first hint derived from PM is to explicitly formulate and document the main features of the projects.

Tasks and targets of the PM application

The application of PM methods guarantees a high quality of project performance by considering all possible problems that could appear during project realization (compare IT27 Montella chestnuts with DE08 Klimaschutz).

PM copes with several difficulties of new and/or complex formulations. The application of specific organizational forms and different methods for the support of cooperation and coordination are, in this connection, of considerable meaning. According to the literature, as well as drawing from the enterprise's own experiences gained along with a project, task structuring can be broken down, as follows:

1. Project planning.

- Definition of the project performance.
- Appointment planning.
- Planning of the project resources (personnel management, responsibility, communication, coordination, etc.).
- Planning of the project costs.
- Definition of project goals (Table 6.6).
- Risk analysis and countermeasures.

2. Project supervision and project control.

- Control of the project performances.
- Appointment control.
- Cost control.

Advantages to be achieved by PM

When considering PM tasks, various uses could be derived which arise from a consistent and professional application of the PM methods.

For example, along with the development of new products or the introduction of new services, the following are indicated.

- Clear allocation of tasks.
- Delegation of responsibility (motivation).
- Systematic and integral way of thinking.

Table 6.6. Non-material satisfaction. (Question 6.2.4: To what extent have your non-material goals been fulfilled so far?)

Non-material goals fulfilled	*n*	%
More than intended	23	23.5
As expected	61	62.2
Less than intended	3	3.1
Missing	11	11.2
Total	98	100.0

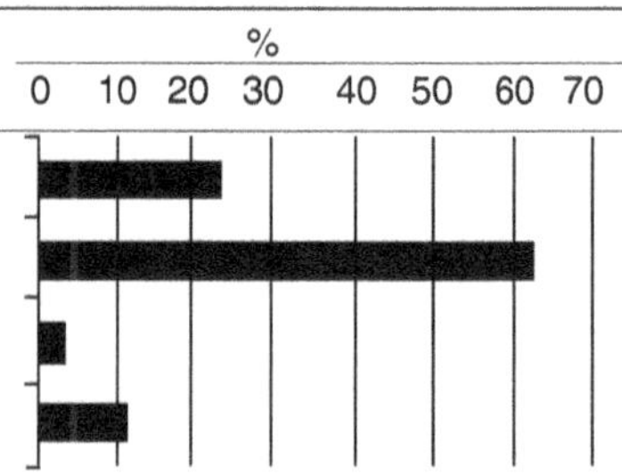

- Clear aims and priorities.
- Faster project realization.
- Efficient use of resources.
- Integral solutions.
- Optimal coordination and utilization of synergies.
- Early recognition of conflict potentials.
- Keeping of appointments and costs.

By the delegation of tasks and responsibility to project teams or single persons, the use of experience, knowledge, innovation, etc. is promoted. At the same time, effective 'management by motivation' is introduced.

With regard to the documented project experiences, the respondents defined several factors as responsible for drawbacks when establishing the RES project. More than one-quarter had problems with project organization (Table 6.7). Consequently, the relationship with other factors of drawbacks (motivation, training, etc.) must be considered. However, about one-quarter of all respondents declared that organizational aspects would be possible improvements for establishing a new RES product (minor and major hints are summarized in Table 6.8).

The increasing complexity of new types of problems requires efficient planning, supervising and control of the work arising. Caused by the exponentially decreasing influence on costs with progressive project period, the main attention has to be given to a precise appraisal of the expected costs (Pichler, 1992, p. 33).

The efficiency of the cost plan is decisively determined by the environment and the organizational integration of PM in the business. Pichler (1992, p. 39) concludes that it is most efficient to directly entrust the person in charge of the project with PM. Thereby the complete responsibility is transferred to the PM team in terms of a clearly defined project and its boundaries. In this way, the opportunity to be successful and to meet the goals is easier to obtain. Simultaneously, the necessary conditions must be created for leading the project. A project can only be efficiently and effectively conducted when the project leader is equipped with the corresponding competence (see next Section 6.1.3).

6.1.3 Project organization for RES activities

Conditions for the application of PM methods

The perception of projects as temporary organizations encourages the awareness that every project requires a specific organizational structure (Gareis, 1997, p. 4). Project-specific roles and tasks have to be defined, project-specific communication structures have to be agreed on, rules and norms for the cooperation in the project team have to be developed (see AU19: during the discussion of pricing, internal lack of information was finally responsible for a lower price being agreed).

Table 6.7. Extension of factors responsible for drawbacks. (Question 6.2.8: To what extent were the following factors responsible for any drawbacks the project has suffered so far?)

	Not at all		Average		Decisive		Missing		All
Factors of drawbacks	*n*	%	*n*	%	*n*	%	*n*	%	(100%)
Organization	40	41	20	20	8	8	30	31	98
Training of personnel	55	56	5	5	0	0	38	39	98
Motivation of personnel	52	53	8	8	3	3	35	36	98
Lack of experience	43	44	18	18	5	5	32	33	98
Misjudgement	47	48	14	14	5	5	32	33	98
Information	40	41	12	12	9	9	37	38	98
Lack of resources	48	49	9	9	5	5	36	37	98
Unsuitability of resources	53	54	5	5	3	3	37	38	98
Other	26	27	15	15	8	8	49	50	98

Table 6.8. Ideas for possible improvements. (Question 6.5.1: In the light of experience, what could or should have been done in a better way?)

	Minor		Major		No	
Possible improvements	*n*	%	*n*	%	*n*	%
Financial appraisal	9	9	12	12	77	79
Financial accounting	9	9	2	2	87	89
Cost accounting	8	8	8	8	82	84
Inventory of potentials	11	11	12	12	75	77
Organization	7	7	12	12	79	81
Marketing mix	9	9	7	7	82	84
Training of personnel	7	7	5	5	86	88
Motivation of personnel	11	11	3	3	84	86
Specification of goals	8	8	9	9	81	83
Definition of milestones	3	3	9	9	86	88
Definition of responsibilities	4	4	8	8	86	88
Controlling	6	6	10	10	82	84
Other	7	7	13	13	78	80
Total	99	8	110	9	1065	84

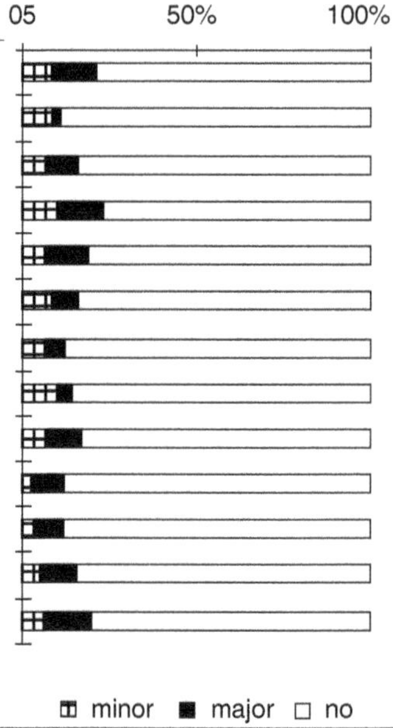

PROJECT-ORIENTATED PHILOSOPHY. In recent years, general project orientation and, with that, the application of projects for the realization of especially secondary processes (RES activities in forestry) can be observed (Gareis, 1997, p. 3). Project-orientated management strategies enable the enterprise to react fast and flexibly to various environmental influences. Due to the market situation, projects are currently started, carried out, concluded and broken off. By means of permanent control, the strategy is to secure the sustainable existence of the enterprise.

In accordance with Gareis (1993, p. 21), the RES-related management philosophy should consider the following principles:

- Project size: the FLO should define any activities of diversification explicitly as a project, independent of the project extent (see DE22, afforestation sponsoring, or AU03, hydroelectric power).
- Project contents: besides the handling and development of products, the areas of marketing, PR (see DE22: a better press report should have been made), staff, organization and investment should also be considered

The evolution towards a project-orientated enterprise has to be accompanied by several success factors. Identification of the personnel involved with the project may be supported by new ways of communication and specific planning and controlling techniques, as well as by the specification of project-related rules. The project culture can be documented in the project handbook (see Section 6.1.5). A very strong cultural element could be a project logo, which supports the identification of each team member and of the FLO him/herself with the project by 'symbolic management' (Tuman, 1994, chapter 11, p. 5). However, an enterprise logo could be used as a project logo as well, e.g. creation of a site on the internet (see www.forstverwaltung.de for the International Timber Exchange). The application of logos in the field of forestry should achieve a broader range in general.

Management by projects (MbP). Working on projects could be very attractive for employees because a rapid feedback of results is secured by the temporal demarcation of the project. The MbP provides new challenges for the team members. The solutions are not rigidly predefined, so that the management system promotes the creativity of each employee.

The motivation of the employee and the promotion of individual learning are the usual leadership strategies of MbP. An integrated way of thinking is promoted by teamwork and by definition of a complex and temporal limited project. A new pool of know-how is established in the enterprise (Fiedler *et al.*, 1994, p. 13).

The principles of MbP are as follows:

- finding out about possible success potentials.
- Acceptance of the complexity and autonomy of projects.
- New organization and communication systems (strategic business unit (SBU), see Section 6.2.2).
- New leadership understanding of management.
- Changes in regard to the staff and work planning.
- Promotion of internal business potential (innovation management).
- Use of project-orientated incentive systems and career planning.

At the level of farm forestry, MbP could help to divide several tasks into separate projects. For instance, the FLO of case-study AU05 (eco-park) is preparing additional services:

- an upgraded footpath (adapted to wheelchairs).
- New guided tours.
- Enlargement of the car-park area.
- Better restaurant service.
- The enterprise's own site on the internet.

Each task can be managed by MbP: responsibilities could be delegated within the family (Mr Pretterhofer, Jr: internet site; Mrs Pretterhofer: restaurant, etc.), the explicit definition of targets and boundaries would promote the realization process of each project and controlling tasks could be managed much more easily by concentration on one's own tasks.

Project team criteria. Working with methods of PM presupposes a homogeneous project team. One prerequisite is the availability of respective human resources in qualitative as well as in quantitative terms. Further criteria for organizing effective project teams are listed below (Thamhain and Nurick, 1994, chapter 19, p. 13):

- it takes additional effort for each team member to feel comfortable with the assignment. Although the overall tasks and objectives have already been discussed, it is essential for a better understanding to see individual tasks and the specific role in the overall teamwork.
- Individual task responsibility, accountability and organizational interface relations should be clearly explained. A simple work breakdown structuring (WBS) can facilitate a clear understanding of the project team organization.
- The project leader is responsible for the composition of the project team. It is important to assess individual capacities and abilities related to the requirements.
- Communication channels should be established as the project team is being formed. Regularly scheduled meetings are recommended and are particularly important when the team members are geographically separated.
- It is crucial for the management to provide a proper environment for the project realization. The extent of competence of the project leader and his/her relationship with the senior management is a critical factor for success.

Related to AU05, an example at a farm forestry level may be given. Mr Pretterhofer, Jr, is the leader of the internet site project. Relevant information (pictures, description) should be provided by each family member, especially related to their own project/business (e.g. Mrs Pretterhofer is responsible for information concerning the restaurant in general, etc.). The project leader should organize regular meetings (brainstorming within the family); he is responsible for the technical equipment, the web design and the current upgrading/updating. The responsibility for the internet site content could be delegated to project team members (Mr and Mrs Pretterhofer).

Project organizational forms

The perception of projects as temporary organizations encourages the awareness that every project requires a specific organizational structure. The structural organization describes the project institution that is engaged with the task of project realization and its integration into the still existing business. In more than one-third of all documented case-studies (best examples: NL12, 13, DE08, 14, 18–20, AU03, 11, 20, IT03), forest managers are still using PM tools, especially a separate project organization (see Tables 6.1 and 6.2).

There are a variety of possibilities as to how a project can be integrated in an enterprise. The correct incorporation of PM will promote, in particular, internal cooperation and coordination. In this way, the organizational risks can be minimized. This applies to extensive product innovations or rationalization projects of smaller size, as well as to various large-scale projects.

The division of formal competence between the project leader and the team members is of key importance for the performance of project tasks. The three traditional project organizational models (Madauss, 1991, p. 94) were developed for companies structured as a line organization:

- the influence project organization;
- the pure line project organization;
- the matrix project organization.

Within these models, the roles of the project manager and the project members, as well as the line manager, are defined. The various formal authorities are assigned to the initiation of work, the disposition of resources, the control of work quality, schedule and costs.

The following three subsections briefly describe these models.

INFLUENCE PROJECT ORGANIZATION (IPO). Related to RES projects, this organization is the most common application form at the level of (farm) forestry. The main object of the project leader is to fulfil coordination tasks related to the project without having any formal competence (Fig. 6.2). This kind of organization is recommended if the business management or the owner of the enter-

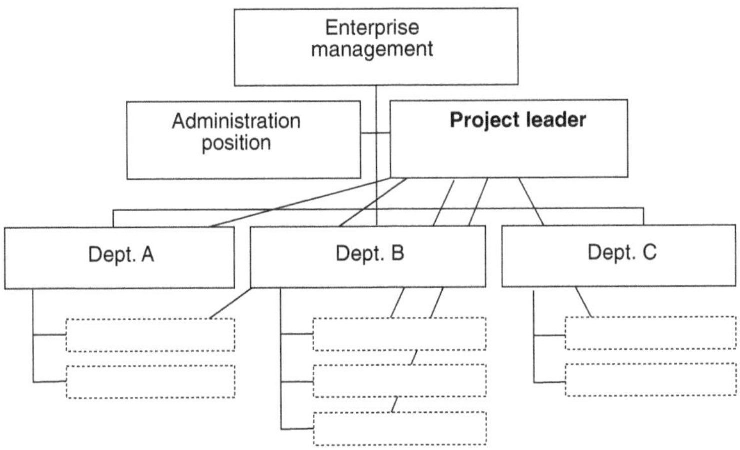

Fig. 6.2. Influence project organization (modified from Litke, 1995, p. 77).

prise wants to guide, control and influence the project (Zielasek, 1995, p. 18). The project leader organizes the project realization and informs the supervisor about the necessary measures to be taken. He/she is responsible for the information management, for the quality of the suggestions and for a proper reporting procedure. The decision and instruction authorization and thus the complete project responsibility lie in the hands of the supervisor (regarding performance, appointments and costs).

This organizational form makes the coordination of required project-related measures difficult. Successful realization depends on the project leader. The question is to what extent he/she is able to assert influence over the line management. This organizational form represents the weakest form of the project leadership. A decisive success factor, besides the project team, is the hierarchical level: where the whole project is located in the basic organization.

With the IPO, the structure of the enterprise remains unmodified. All project employees are left in their line jobs and are subordinated to their department leader.

Advantages of the IPO

- High extent of flexibility with regard to personnel use.
- No organizational restructuring necessary; therefore, simple integration in the basic organization.
- The problem of reintegration in the line job of the team employees is solved.
- Inexpensive to set up.

Disadvantages

- Process of decision-making ponderous.
- Weak position of the project leader.
- Conflicts of interest between the project employees and the respective department leaders.
- Identification of the project employees with the project aims is low.
- Strengthened control is necessary (by the decentralization of the project tasks).

The application of IPO is restricted to smaller projects. The daily business is not fundamentally influenced by the project.

RES related examples are: setting up a court for beach volleyball (see Section 6.1.5); forest guided tours; eco-sponsoring project (NL16, AU09).

Adapted to AU05, IPO could be organized according to Fig. 6.3. The project leader is dependent on detailed information regarding the four profit centres provided by Mr and Mrs Pretterhofer.

PURE LINE PROJECT ORGANIZATION (PLPO). For the purpose of the project, an independent department is formed (Fig. 6.4). The project leader has full responsibility for the department. This organizational form could be established in large forest enterprises with qualified and available staff.

For the duration of the project, the team members are exclusively responsible for their work in the project. The project leader disposes of all relevant resources (production factors, competence) and is responsible for the performance and attainment of the

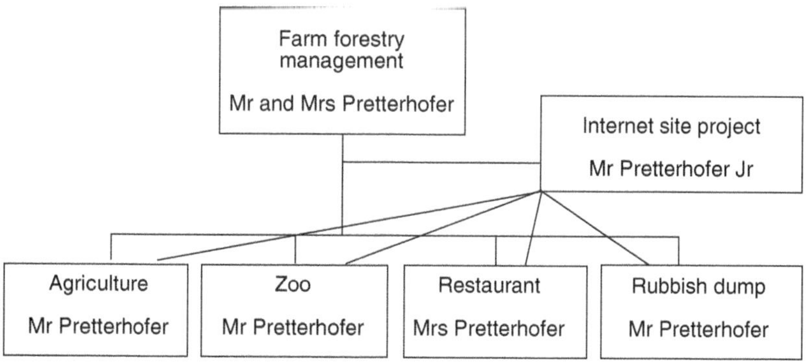

Fig. 6.3. IPO at the level of farm forestry (example).

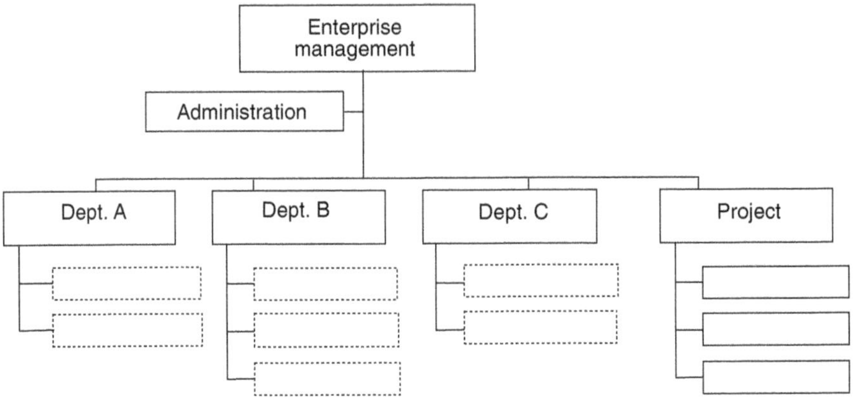

Fig. 6.4. Pure line project organization (modified from Litke, 1995, p. 76). Dept. A, B, C, forest districts.

project aims (Brandenberger and Ruosch, 1996, p. 48).

The difference of the PM department from the other departments of the enterprise lies in the temporal limitation. A separate line organization arises for the duration of the project.

After the project is finished, the internal team members are transferred back to their original jobs. Cooperation with external team members ends with the completion of a contract or its expiry.

This kind of organization is suitable for unusual projects of broad extent and represents the best solution to the project requirements.

Advantages of PLPO

- Integrative, goal-orientated leadership due to clear competence allocation (low conflict potential).
- Full concentration of the employees on the project by clear delegation of subresponsibility.
- Short decision procedure.
- Increased performance and high efficiency reduce the project duration.

Disadvantages

- Organizational restructuring causes high common costs.
- Overcapacities (continuous utilization uncertain) can occur because of the exclusive project allocation to the employees.
- Problems of integration into the project team and of reintegration in the respective old jobs (departments).
- Danger of business blindness and encapsulation by excessive project concentration via the line management.

RES related application examples are: construction of a hydroelectric power station (large project extent (see AU03)); forest guided tours in a national park (large project extent and large enterprise (see AU07)); selling of spring water (as documented in case-study AU19).

MATRIX PROJECT ORGANIZATION (MPO). By the MPO, the line management of the enterprise (vertical function structure) is superimposed from a second, horizontally project-orientated structure (Fig. 6.5). This organizational form is characterized by dividing the instruction authorization between the project leader and the line supervisors of project employees. Madauss (1991, p. 100) sees an optimal compensation of tasks and competence between the original business tasks and PM in this form of project organization.

As shown in Fig. 6.5, the project transaction for specific tasks is carried out by line management. The project leader defines the jobs and the time of realization

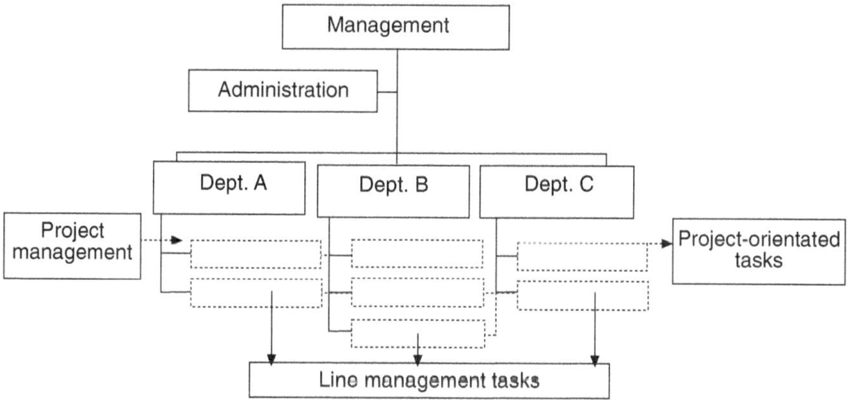

Fig. 6.5. Matrix project organization (modified from Litke, 1995, p. 78).

of the partial tasks. The main tasks of the project leader are to coordinate all activities and to do the controlling. An additional responsibility concerns meeting the economic frame conditions.

The line management leaders are responsible for the technical realization of the project tasks. Both the line manager and the project manager are responsible for the success or failure of the project.

The project team members are still responsible to their line manager. The project leader has the project-orientated technical instruction competence. Depending on project size, it could be very useful to share the working hours of the project employees between project and line tasks. For instance, at a 50% project utilization, the working week could be halved (Jenny, 1995, p. 103). The clear allocation of the working hours reduces the conflict potential between lines and project tasks (Hirzel, 1988, p. 261).

The concept of the so-called time-sharing organization recognizes this point explicitly. For a particular time period (days, weeks, months), an employee is fully integrated in the project. During the remaining time, he/she is doing the line management job. Accordingly, the technical and disciplinary instruction authorization is changing between the project leader and the line supervisor, which can cause many problems (Wischnewski, 1993, p. 11).

Advantages of MPO

- Flexible use of human resources.
- No difficulties resulting from change (workplace), specialized team members can work within their departments (exchange of thoughts with colleagues is possible).
- Optimal use of personal capacity by splitting the line and project tasks.

Disadvantages

- Intensive cooperation between project and department leader; high conflict potential with regard to the competence and the fulfilment of tasks.
- The dual load of the project employees requires independent and disciplined work.
- High requirements as to communication and information.

In spite of all the disadvantages, MPO represents in many cases the most effective organizational form. Taking into consideration the available scarce resources and economic aspects, MPO is often the only acceptable solution (Litke, 1995, p. 81). The project leader must be able to use internal know-how. The enterprise is also interested in promoting the technical knowledge of each employee.

However, it is the most flexible organizational form for projects. This kind of management method is used most often in enterprises that are operating with many simultaneous projects of various sizes.

With regard to RES-related examples, applying this organizational form depends on the internal business organization; it could be used for any project, independent of its extent, complexity, etc. (e.g. NL19, DE11, AU01, 03, IT11).

Selection criteria for a suitable organizational form

Every business has its own history, culture and structure. For the implementation of new management methods, it is important to take these special features into account.

Setting up a project organization requires careful planning and consideration of the specific features of the already existing organizational form. Although in general the owner decides on the organizational form, this decision should be made after discussion with the project leader and, if necessary, with the line manager.

The criteria for a suitable organizational form encompass the following aspects:

- structure of the business organization.
- Size and duration of the project.
- Economic meaning of the project.
- Necessity for interdisciplinary cooperation.
- Risk potential (with regard to attainment of goals, schedule and cost conditions).
- Disposal of resources.
- Experiences with PM.

The criteria listed in Table 6.9 could be useful for choosing an adequate project structure.

As already mentioned, the project documented in Section 6.1.5 (establishment of a beach volleyball camp in a forest enterprise) was realized with IPO, additional external project members being included. The decision on the organizational form was discussed according to Table 6.9: project meaning rather low, extent small, low risk, no special technology necessary, urgency normal, short project duration, low degree of complexity, coordination necessary (especially between the forest manager, the project leader and the forest staff), sporadic use of employees during the implementation; an additional argument for IPO: the project was initiated by a non-employee, who at the same time was the project leader (RES cooperator).

PM adaptation to RES-related forest demands

The organizational forms of PM mentioned can be used as the basis for any organiza-

Table 6.9. Selection criteria for a suitable organization form (modified from Jenny, 1995, p. 104).

Criteria	Pure line project organization	Matrix project organization	Influence project organization
Project meaning	Very important	Relevant	Rather low
Project extent	Very high	High	Small
Risk of realization	Very high	High	Low
Demand of technology	New	High	Normal
Urgency	Great	Medium	Medium
Project duration	Long	Medium	Short
Degree of complexity	High	Medium	Low
Need of coordination	Very high	Rather low	Medium
Temporal use of employees	Permanent	Part-time employment	Sporadic
Demands on project leader	Highly qualified (methodical and technical knowledge)	Good methodical knowledge	High requirement for personality
Coordination with line management	Low	Important	Important
Identification with the project	High	Medium	Low

tional modification. Every project requires the development of an adequate PM concept. Concerning the PM instruments, it is necessary to examine the extent to which the instruments can be efficiently used by forest owners (the expenditure in relation to the actual benefit).

Even for the smallest projects, it is very useful to prepare a written document of all essential project tasks and milestones. On the one hand, it is useful for purposes of self-control and, on the other, it is very helpful for all persons involved in the project, e.g. for a common project understanding (main targets of a project handbook).

FARM FORESTRY. At least with bigger RES projects, the forest owner will have to integrate external team members for project realization (Fig. 6.6).

After a rough goal definition, it is relevant to analyse the project environment and to identify possible institutions of cooperation. As a result, competent partners (specialists) could be integrated in the project team. The external team members, therefore, are under the control of the project leader (simultaneously the initiator and forest owner) and also responsible to their line management. The internal project employee (hunter, district ranger) is directly under the control of the project leader. After deciding on the team, the project leader creates a work schedule acceptable by everyone (delegation of responsibility, coordination of the project phases, costs, schedule, milestones, etc.) and is responsible for achieving the objectives defined. The project leader, who may have low technical and methodological knowledge, obtains an essential controlling instrument with project planning.

FOREST ENTERPRISE LEVEL.

1. An application of the essential project management instruments should be used at the forest enterprise level (separate business) to a higher extent. Starting from the classic forest organizational form (Sagl, 1993, p. 69), the structure of an implemented project organization could be designed in accordance with Fig. 6.7. This organizational form (a mixture of PLPO and MPO) appears to be practicable for the realization of any RES project in multifunctionally orientated forest enterprises (see AU03).

2. By using an external project leader (as in the project beach volleyball camp, which is described in Section 6.1.5), the IPO form could be used or modified (see Fig. 6.2).

For the successful implementation of RES projects, a specific project organization is an essential prerequisite. All persons involved are informed of their tasks and responsibilities. A well-structured and efficient organization maintains the interdisciplinary cooperation of internal and external project employees. The competence of the project leader, the image, methods/technical knowledge and support of the project funder/director are imperative for project realization.

6.1.4 PM planning tools relevant for RES projects

After the consideration of organizational aspects of how to integrate project-orientated procedures into daily business, the instruments of PM are introduced in this section. The reason and purpose of the explanation

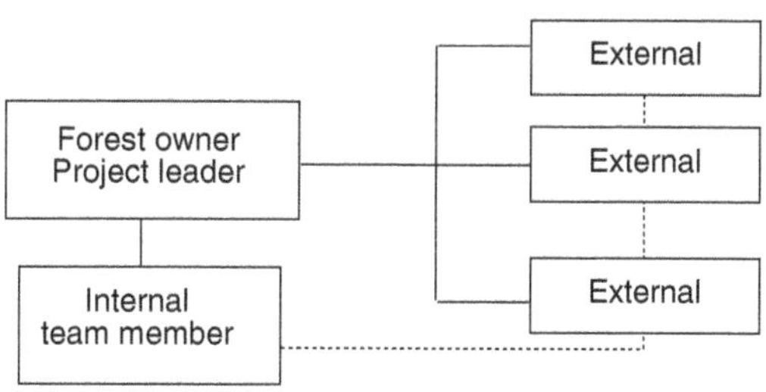

Fig. 6.6. FLO as project leader.

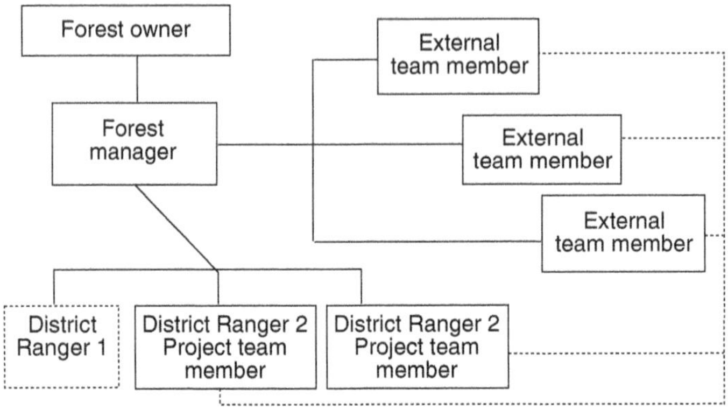

Fig. 6.7. Forest manager as project leader.

are to provide instructions for the setting up of a PM manual (PM handbook). Even without large expenditure, projects can be planned in an easy, efficient way, clearly laid out and easily feasible. The tasks of all persons involved in the project can be coordinated and easily controlled. The PM manual represents the simplest form for the arrangement of the project contents.

An additional essential success factor can be arranged with this instrument of administration: the acceptance and identification of all persons directly or indirectly involved in the project content by the representation of common and interdisciplinary coherence (Eid, 1986, p. 2). According to Table 6.7, many drawbacks were caused by the lack of information (DE11, 12, AU09, 13, IT17, etc.), which could easily be avoided by using a PM handbook.

Tasks and objective targets

Depending on the formulation and extent of the project, it could be realized within several weeks or several years. Detailed planning is the basis of PM, which helps to derive a specific structuring of the project. The work flow can be supervised, and problems can be recognized at an early stage. Finally, correction measures can be taken in time.

The planning consists of a systematic, future-orientated way of thinking and encompasses the decision on likely goals, measures, methods and ways concerning the achievement of objectives (Sekot, 1993, p. 36). The basic context and possibly also bottlenecks are shown by project planning. At the same time, alternative strategies for the project realization process should be developed and judged.

Detailed project planning should answer the following questions (Birker, 1995, p. 31):

1. What are the goal, object and contents of the project? What must be done? See cases NL03, DE04 and IT17, where goal specifications, etc., were missing.
2. Who can do it? What is the internal/external task distribution (project leader, project employee)? In project NL05, no project team was defined.
3. What measures are useful? Are there regulations, rules or special procedures that should be taken into account? For example, legal borderlines for dealing with drinking-water (NL01, DE14, AU16, 19).
4. How will the tasks be solved? What resources are available, what is absolutely required? See NL13 and IT12 for lack of potential analysis (PA) and specialist workers.
5. When will the various jobs be started or completed? See NL15, DE23, 24, AU07 and IT17, where several drawbacks were caused by bad time management (no schedule, no milestones, etc.).
6. Where are the various tasks to be carried out?

All results of the project planning process should be recorded in the form of written representations (charts, descriptions, etc.). These plans represent an essential communication instrument in PM. The application of adequate methods and techniques in PM cannot suffice to guarantee a smooth project procedure. It is also important to establish a personal culture of communication, which is essential for any business activity. The final prerequisite for success is the acceptance of all employees regarding the project and its planning tools. Therefore, project planning should provide the following items:

- A systematic procedure.
- A clear definition of all project goals.
- Definition of guidelines to recognize bottlenecks in time.
- Solution alternatives.
- Definition/allocation/structuring of WPs and the local and temporal processing (delegation of responsibility).
- Transparent and realistic time/cost calculation.
- Coordination of the project's internal and external procedures.
- Clear representation of the planning contents to promote communication.
- Specification of milestones to create a control mechanism
- Performance guarantee through defined aims.

Construction of project boundaries and project context

The determination of the project boundaries is the basis for project planning, controlling and the assignment of responsibilities. A holistic project perception ensures project quality (Table 6.10).

Project boundaries define what belongs to the project and the tasks/areas/goals that are not involved (Eid, 1986, p. 5). It is important to consider all goals of the enterprise and to take the special company philosophy into account (see, for example, AU11 Malteser Ritterorden: the forest apartments are adapted to handicapped persons because of their philosophy of social engagement). The delimitation of goals, which to a high degree are dependent on the motive for pursuing the project, endangers the quality of the project. For instance in an eco-sponsoring project, one should not only follow the goals of implementation, but also non-material goals. A second example is given by Balck (1994, chapter 2, p. 5): with the installation of a new coffee-maker an immediate usefulness is achieved, whereas this is not the case with the installation of a new personal computer (PC) system. Such projects do not end with software design and development and hardware and software implementation. The goals of organizational and personal implementation

Table 6.10. Dimensions of project boundaries and project context (from Gareis, 1995a, p. 4).

Dimension	Tasks	Brief description
Contents	• Project objectives • Project contents • Project budget • Relationships to other projects • Relationships to internal strategies	The delimitation is made by the definition of the project aims and by the description of the performance size and relevant project features
Time	• Start/end event, date of the project • Duration of the project • Pre- and post-project phase	The temporal delimitation is carried out by the specification of the start/end event of the project. Afterwards, time agreements can be assigned
Social	• Project name and logo • Roles in the project organization • Project-specific values • Relevant project environments	The social delimitation is made according to the definition of central roles in the project, e.g. project owner, project leader and employee, and by the definition of project-specific values

requirements for use must be equally considered (e.g. controlling aspects after project implementation: AU03, IT17).

Apart from the definition of the project scope, the construction of the project boundaries occurs by agreeing on the project start and the project end event, as well as agreeing on the project budget and on central project roles (see Section 6.1.5). For example, the project owner has to fulfil the following tasks (Duncan, 1996):

- Selection of the project manager.
- Arrangement of goals with the project team.
- Disposition of resources.
- Participation in the project start-up and close-down workshops.
- Strategic project control.
- Informing of the participants about the project context.
- Contribution to the solution of conflicts.
- Project marketing.
- Mutual feedback with the project team.

The construction of project boundaries makes a holistic project solution possible and supports cooperation with external institutions/persons.

By the definition of the project start and the project end and by the definition of the pre- and post-project phase, time is recognized as a key dimension within the context of the project. Information about the motivation for a project, as well as decisions before the formal project start, about costs and goals, are of great importance for the comprehension of the project goals and for the development of the project structures (pre-project phase).

The post-project phase gains importance in temporary systems. On the one hand, the post-project phase has to be planned and, on the other, expectations concerning actions and decisions in the post-project phase have to be considered again in the development of project structures (additional sponsoring relationships, further business organization, etc.). Considering the conclusion of a project leads to a specific identity and project culture. It develops temporary structures and temporary environmental relations.

Furthermore, the project context is derived from the contribution of a project to meeting the company's strategies, as well as from its relationships with other internal projects performed at the same time (see Section 6.1.5 on the point of influence on other projects). These relationships can be of a synergistic or conflictive nature. Interim results of a project can be a prerequisite for the further performance of a considered project.

Additionally, for a common project understanding, the results should be presented in written form (e.g. as shown in Section 6.1.5) and should be handed out to all project team members.

Project environmental analysis

The social context of a project is determined by its social environment. The relevant environment is constructed in a project environmental analysis (see Sections 6.1.5 and 6.2). Relevant for a project are those environments that could be expected to influence the success of the realization process. A differentiation can be made between an internal and an external project environment (AU13: cross-country skiing project, where many interest groups are involved).

By the explicit consideration of the project owner and the project team as internal project environments, it is possible to direct appropriate attention to these areas. The external environment can be defined by customers, suppliers, interested parties, media and cooperating partners, as well as by divisions of the project's own enterprise (especially relevant with delicate projects such as drinking-water production (DE14, AU19), cycling routes (AU01) and cross-country skiing (AU13, IT09). Project environments are to be differentiated in so far as different strategies and measures for each environment are required. The dynamics of projects results in interventions in relevant environments. Examples of interventions in relevant environments are new legal conditions imposed by the authorities, changes in the scope of work by the clients, bankruptcy of suppliers, an unexpected media reaction or demoralization of the project team.

REGULATION OF THE MEANING OF RELEVANT ENVIRONMENTS FOR THE PROJECT AND DISTANCE BETWEEN THE PROJECT AND RESPECTIVE ENVIRONMENT

- The degree to which a relevant environment is represented will represent the meaning of the environment. The significance of an environment for the particular project depends on its ability to influence the project. The borderline separating the project from its environment determines the intensity of the interaction between the two.
- The analysis of the relevant relations and the intensity of the interactions can be represented in list form (see Table 6.11, with regard to Section 6.1.5) or graphically (Plaimer, 1997b, p. 12).

Concerning the Volleyball Association and two tourist associations, it was not possible to get any relevant information or subsidies. They were not interested in cooperating, for various reasons (competition between indoor and beach volleyball, lack of money, small size of the project, restriction to merely local effects, etc.).

The analysis should present the 'big picture' for the project team. It is worth mentioning that valuation is sensitive to time. During the phase of project realization, the meaning and intensity of interaction are in a state of permanent change.

ANALYSIS OF PARTICULAR RELATIONSHIPS

- Important actions and decisions in the projects are guided by expectations of the environment. The quantification of a relationship between a project and its environment can be carried out by the description of the mutual expectations (see Section 6.1.5).
- The advantage of these procedures is that the project employees know each other and their specific expectations. Potential conflicts between the project and the relevant (selective analysis) environment can be recognized in time. Due to this feature, concrete strategies and measures can be derived for successful project realization.

The implementation of an environmental analysis should be organized in a project start-up workshop by the project team. Setting up a common point of view should be stimulated. At the same time, representatives of relevant environments (suppliers, customers, etc.) could be called in. If demanded, the extent of the documentation can be enlarged to provide more detailed results (mutual expectations, potentials and conflicts, strategies and measures, etc.).

Project performance planning

WORK BREAKDOWN STRUCTURING (WBS). The WBS is the principal item of PM; it is the basis of successful project planning. By structuring the scope of work, the overall project is segmented into practicable and controllable WPs. It is a relatively stable PM method, a central communication instrument and the basis for the project documentation system. As a basic principle, it is an alternative to network analysis

Table 6.11. Analysis of meaning and distance of relevant environments (example).

Environment	Meaning (1, meaningful; 5, insignificant)	Distance (1, very near 5, very far)
Project owner	1	2
Project team	1	1
Supplier	2	2
Sponsors	2	2
Banks	2	2
Austrian Volleyball Association	4	4
Tourism Association	4	5

(NWA). In contrast with WBS, NWA is mainly used by extensive and complex projects. Both project structure and schedule are integrated in NWA. Thereby tasks which depend on each other can be presented in a better way.

Processing of individual project tasks is increasingly transferred to external specialists/institutions. Dividing a project into WPs is essential. The project manager can transfer the different project tasks and delegate responsibility in a simple way (Madauss, 1991, p. 177). Additionally, WBS is an important controlling instrument and easy to handle. WBS does not represent a schedule, procedure, cost and resource plan. It is the common basis for all project planning and organization activities and, therefore, the general acceptance of all persons involved in the project must be achieved.

Creation of WBS. Setting up WBS should be done by specific, simple rules in close cooperation with the project team members. Practical experience for structuring a project is an essential success factor. However, each project calls for a modified, completely new or at least varied solution. The following procedures can be distinguished by the strategy of structuring (Patzak and Rattay, 1996, p. 166).

Top-down method. The project name and the content are fixed on the first level. On the second level, the project is subdivided into subprojects after specification of suitable structuring criteria. Again, criteria must be defined for dividing the subprojects into WP (third level; Fig. 6.8). The degree of detail must be in tune with the WPs and depends on the available level of information (rule of thumb: as few as possible and as detailed as necessary).

Bottom-up method. The relevant WPs and ideas are collected and listed. This can be carried out in the course of team meetings (brainstorming). Afterwards, an analysis and assignment of the WPs to defined project tasks must be made. A project structure (hierarchy) is the result of this iterative process. The structuring is performed according to the different functions and objects. From this structure, a horizontal and vertical gradation can be derived (see Section 6.1.5).

WBS is a dynamic management instrument. New knowledge derived from the environmental analysis and additional information necessitate a cyclical revision of WBS during project realization. WBS must always represent the latest version of project performance.

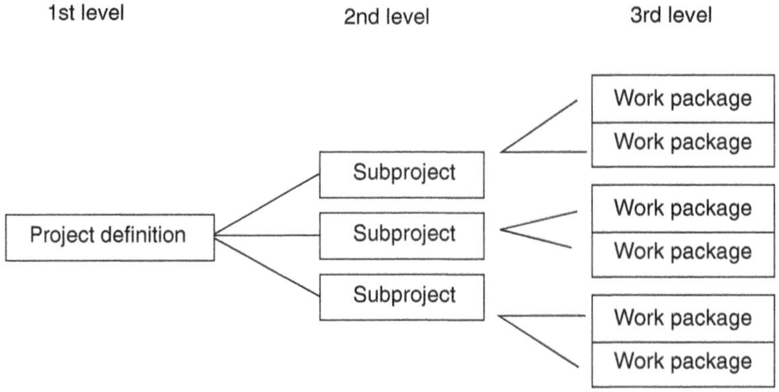

Fig. 6.8. WBS top-down method. Example AU03 Hydroelectric power station (1st level). Subprojects: fieldwork, power house, etc. (2nd level). Work packages of subproject fieldwork: setting up of water resources (see Fig. 6.9, no. 1), pressure pipeline (see Fig. 6.9, no. 9), etc. (3rd level). Work packages could be subdivided into more checkable units, e.g. pressure pipeline: measurement, cutting, etc. (see Fig. 6.9, nos 10, 11, etc.).

Work package specification. The WPs are listed in the WBS. This list represents the real basis for all further management tasks. The sum of all WPs represents the complete extent of project performance. Due to the importance of a WP description, standardized forms have been developed (see Madauss, 1991, p. 188; Patzak and Rattay, 1996, p. 174). The following aspects must be considered for an explicit description of a WP:

- WP title, organizer, responsibility, duration (start, end).
- Detailed definition of contents and performance.
- Documents required.
- Resources required.
- Specification of the WP results.
- Progress measurement (meetings, reporting system, etc.).

The results of the WP specifications should be documented in the project handbook.

PROGRESS AND TIME PLANNING. For the process of a project, the temporal factor (duration, the beginning, end, time reserves) and the logical coherence of all activities must be defined (see project evaluation of case-study NL15 – problems occur because of poor scheduling). The complete duration of the project can be derived, and it is possible to define the end of the project. Finally, the schedule serves performing and controlling purposes, as well as reviewing each package to assess whether realization is practicable in time. Rough and detailed planning can be distinguished. If the WPs of the WBS are neither subdivided further nor provided with additional details, it is only a rough plan. If the single WP is subdivided into events, a very high degree of detail can be reached. This could be a disadvantage for communication and planning purposes.

Process and time planning can be carried out for the complete project or only for single phases (subtasks). There are three planning methods available: list of processes, bar chart (Gantt chart) and the critical path diagram (network plan; see Altrogge, 1996).

List of processes/timetable. Depending on the extent of the project, it could be helpful to picture all activities and their dependencies. Several tasks could be performed in parallel and independently, while some tasks will imply prerequisites (pre-activities). The logical dependence of all subtasks must be derived from this planning tool. The representation of the plan must be decided by the project leader. Regarding RES-related applications, Table 6.12 gives a

Table 6.12. Combined list of processes and schedule.

List of Processes

Project: .... Name: .... Date: .... No.: .... Page: ....

No.	WP	Process	Required processes		Calculated duration (days)	Schedule		Time reserve
			Before	After		Start	End	

Process, Breakdown of each WP in separate activities (lowest plannable level); Required processes, logical connections between the listed processes are represented; the number of the WP or process should be filled in in the relevant column; Time reserve, for several activities it is not possible to define the time frame exactly; reserves concerning start/end and depending activities must be calculated (buffer times); partial tasks are perfectly independent.

combined list of processes and schedule. (For further examples of a more simple as well as a more detailed kind, see Gareis (1991, p. 77) and Jenny (1995, p. 224). An example of a time schedule is illustrated in Section 6.1.5.)

On the basis of a detailed representation of the performance of each process and the respective time frame, the next planning step concerning the bar chart can be done in an easy way. An additional and helpful instrument that is easy to communicate is a list of milestones (see Section 6.1.5). Milestones are defined deadlines and imply a controlling mechanism for each team member.

Bar chart (Gantt chart). The bar chart is the graphical presentation of relevant WPs. It shows the tasks according to the schedule and the duration. Due to visualization, it is an essential instrument of communication within PM. Temporally overlapping WPs indicate that various activities could be done in parallel. However, with increasing overlapping of tasks, the risk must be calculated in more detail.

The bar chart of Fig. 6.9 was made with the software MS-Project. The theoretical background of how to work with it is documented in Berning and Mehrmann (1992). The WPs 'setting up of water resources and pressure pipeline' are pictured as an assembly procedure (black bar). The duration of each package derives from the sum of the processes. Necessary resources are shown on the right-hand side of the bars. The inter-dependence of processes is represented by arrows. It is thus possible to recognize the critical path for project realization.

No.		Duration (days)	From	To
1	**Construction of the well**	**13.5**	**Mon 04.08.97**	**Thur 21.08.97**
2	Delivery of building materials	2.5	Mon 04.08.97	Wed 06.08.97
3	Excavation	2	Mon 04.08.97	Tues 05.08.97
4	Sides of the well	4	Wed 06.08.97	Tues 12.08.97
5	Supply and drawing off	2	Tues 12.08.97	Thur 14.08.97
6	Settling of stones	3	Thur 14.08.97	Tues 19.08.97
7	Build-up devices	2	Tues 19.08.97	Thur 21.08.97
8	Overflow	1	Thur 14.08.97	Fri 15.08.97
9	**Pressure pipe**	**25.5**	**Mon 04.08.97**	**Mon 08.09.97**
10	Survey and marking out	3	Mon 04.08.97	Wed 06.08.97
11	Felling	10	Thur 07.08.97	Wed 20.08.97
12	Wood transportation	5	Mon 18.08.97	Fri 22.08.97
13	Delivery of pipes	2	Mon 11.08.97	Tues 12.08.97
14	Digging and laying	17	Thur 14.08.97	Mon 08.09.97

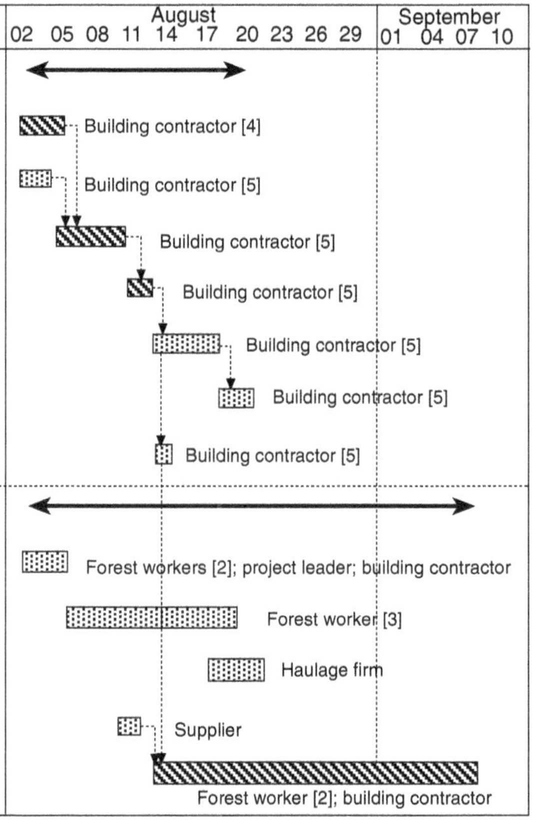

Fig. 6.9. Bar-chart example (construction of a hydroelectric power station, Austria).

USE OF RESOURCES AND PRINCIPLES OF PROJECT COST PLANNING. As mentioned before, the WBS should be the basis for resource and cost planning (Table 6.13). The allocation of resources must be planned for each particular process. In this way, bottlenecks can be recognized in time, and an optimal exploitation of resources can be ensured (Brandstätter and Synek, 1988, p. 395). More attention must be given to resource planning in small and medium projects. Failing to meet deadlines can usually be traced back to lack of capacity or to incorrect planning (Zielasek, 1995, p. 142).

The procedure to determine required resources can be subdivided into the following steps (Mende and Bieta, 1997, p. 83):

1. Definition of relevant resources for particular processes.
2. Selection of 'bottleneck resources'.
3. Assignment to particular WPs.
4. Definition of available resources (internal).
5. Encumbrance diagram.

An encumbrance diagram indicates where capacity compensations must be carried out. This could be done by:

- moving or elongation of non-fixed processes within the buffer time;
- elevation of resources (overtime, etc.);
- moving or elongation of fixed processes; deadlines must be displaced;
- allocation of WPs (processes) to external institutions.

Predefinition of deadlines and capacity planning cannot be carried out separately. The whole planning process is iterative.

The following principles of project cost planning are useful:

- the structure of the project cost calculation should correspond to the WBS (see Table 6.13). Therefore, the individual WPs of the WBS should represent units for controlling predicted and actual costs.
- An adequate degree of detail should be applied to the calculation of project-related costs. A low degree of detail reduces the chances for recognizing deviations in time. A high degree causes high administrative expenditure. Exact calculation should be concentrated on those WPs which are cost-intensive, difficult to monitor and where deviations may occur readily (critical cost factors). Basically, the calculation should be carried out in as detailed a manner as necessary and as roughly as possible.
- The performance structure of external institutions involved in the project should correspond with the WBS. Allocation of external costs to the respective WP is straightforward.
- Actual costs must be recorded in the same way and with the same degree of detail as was the case in cost planning.
- Cost should be assigned to the project directly. Process cost accounting may help to identify cost drivers (unlike department orientation).
- All responsible team members should be integrated into the process of establishing a classification of expenses.

6.1.5 An example – the project handbook of the Hebalm beach volleyball camp

This handbook is handed out to all team members.

HANDBOOK INTRODUCTION. The winter sports centre Malteser Ritterorden Hebalm was

Table 6.13. Resource and cost plan of the project (from Gareis, 1995b, p. 10).

WP number	Work package	Resources (person-days)	Internal costs	External costs
	Total			

developed further by establishing five vacation houses last spring. The enterprise disposes of 60 beds in ten apartments, and the organization was satisfied with the former winter business. As no satisfactory utilization of the apartments could be achieved in the summer season, the forest management tried to find additional possibilities for use.

After the idea to establish a beach volleyball camp was agreed upon, the issue of costs and a realization/marketing concept were discussed.

The tasks of the project (see B, below) comprised the setting up of two volleyball sand courts and also the setting up of a marketing concept. After the end of the project, the project team should retain responsibility for business and marketing activities.

Project boundaries

(A) Objectives. Factual goals and non-goals:

- to set up an attractive beach volleyball court.
- To promote multifunctional management of forestry.
- Use of available infrastructural facilities in the summer months.
- To safeguard/create new employment and to support the company's engagement to care for handicapped persons.
- However, the ecological strategy and forestry targets will not be impaired.
- Under no circumstances will activities be combined with motor sports.

(B) Project definition.

- Planning for setting up sand courts with grandstands.
- Construction and implementation.
- Setting up a marketing concept.

(C) Project roles – social dimension. See Table 6.14.

(D) Time dimension.

- Project start: 1 October 1996.
- End of project: 15 July 1997.

Context analysis

(A) Objective dimension. The construction of the volleyball camp is part of a far-reaching project to support summer tourism. The existing and new infrastructural facilities (holiday accommodation) should be occupied to a greater extent.

In this connection, the following projects have already been realized:

- the construction of five holiday houses.
- The provision of mountain-bike routes.
- Adaptation of a pier and two rocks for climbing activities.

A further project to offer orientation rallies is in development.

(B) Pre-project, start, end and post-project phases. See Fig. 6.10.

(C) Social dimension.

- Project name and logo.
- Relevant project environments (Fig. 6.11).

A detailed analysis of the relationships between the project and the project owner (Malteser Ritterorden) is carried out in D, below.

The meaning of the relationship between the project and its environment is

Table 6.14. Project roles – the Hebalm example.

Project owner	Project manager	Team members	External team members
Souveräner Malteser Ritterorden (forest enterprise)	University assistant	Forest manager Forest assistant Sports instructor Volleyball coach	Advertising agency Austrian Volleyball Association Lawyer Tax consultant

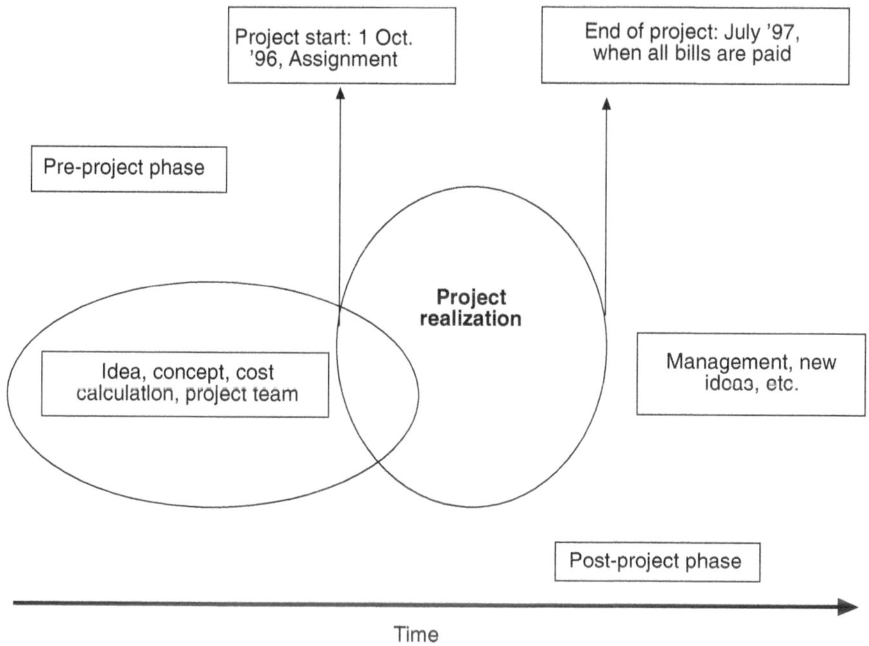

Fig. 6.10. Project phases – the Hebalm example.

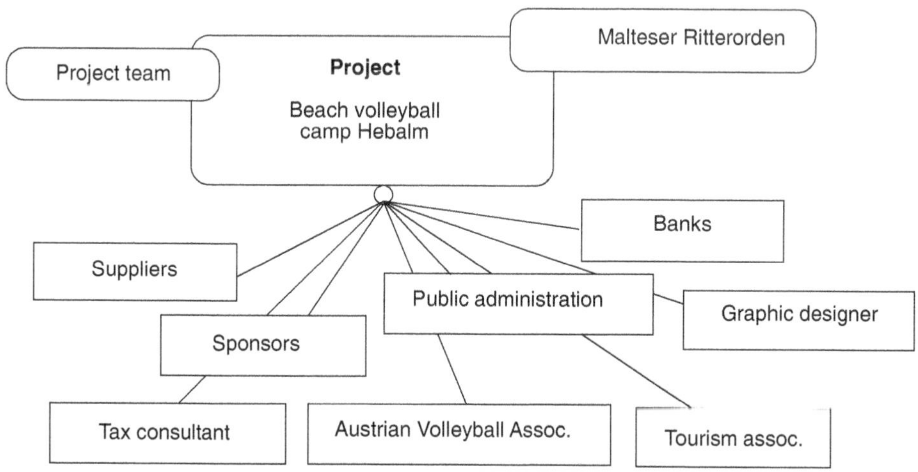

Fig. 6.11. Project environment – the Hebalm example.

highlighted by the distance and by the type size (suppliers ↔ tourism association).

- Influence on other projects (Fig. 6.12).
- Project values.

The special company philosophy was taken into account when setting up the project concept. The following values were defined:

- sport and nature.
- Fitness and relaxation.
- Social engagement in the partner's areas of occupation and to foster handicapped persons.

(D) Relationship between project owner and project (Fig. 6.13). The conflict potential refers mainly to the levels 'control–autonomy' and 'resource provision'.

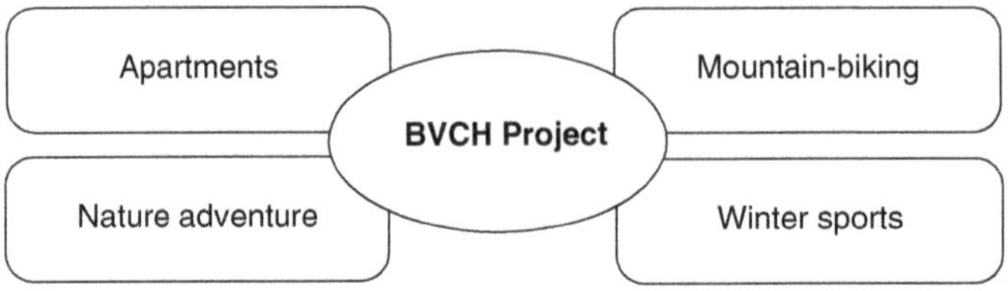

Fig. 6.12. Influence on other projects – the Hebalm example. BVCH, Beach Volleyball Camp.

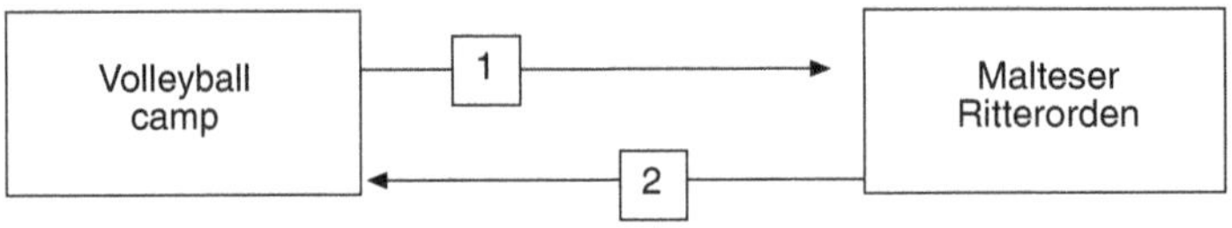

Fig. 6.13. Project–owner relationship – the Hebalm example. 1. Setting-up of the sand courts at minimal expense: organization, management, chronological table, quality guarantee, control possibility. 2. Sharing financing: autonomy, support (other resource provision).

Countermeasure: intensive information exchange between the project owner and the project team:

- regular meetings (once monthly; here the quantitative and temporal allocation of the required resources and the controlling are carried out).
- The contract determines, among other issues, the disposal right of using resources (machines, staff).

WORK BREAKDOWN STRUCTURING. See Fig. 6.14.

FOLLOW UP CHART

(A) Time schedule. See Table 6.15.

(B) Milestones. See Table 6.16.

(C) Bar chart. See Table 6.15.

PROJECT ORGANIZATION – NEED FOR RESOURCES See Table 6.17.

6.2 Potential Analysis – a Multifunctional Forest Management Tool

6.2.1 Introduction

Due to constantly changing market conditions caused by increasing competition (such as cheap imports) and by increasingly demanding/critical customers, forest management is forced to develop new business fields on various markets and simultaneously to use all available potentials. Considering that the market situation requires permanent product diversification, business activities can be realized in a more successful way. The dynamics and uncertainty of new procedures must be seen as a new opportunity. Besides the implementation of new management methods (e.g. PM), it is essential to analyse various infrastructural performances of forests, as well as relevant resources related to any RES activity, in a more detailed way. This is a circumstance which is still insufficiently integrated into internal company planning (Sekot, 1993, p. 439).

In connection with the assessment of an enterprise, PA could be a very helpful procedure. PA could be defined as an investigation of success factors with regard to expected development strategies (prediction). Therefore, a number of strategically important questions should be answered: What medium-term trends can be derived from the social environment? How can these requests be integrated into the existing enterprise concept? What promising internal potentials can be used? How can internal company performances be further differentiated in order to open up new

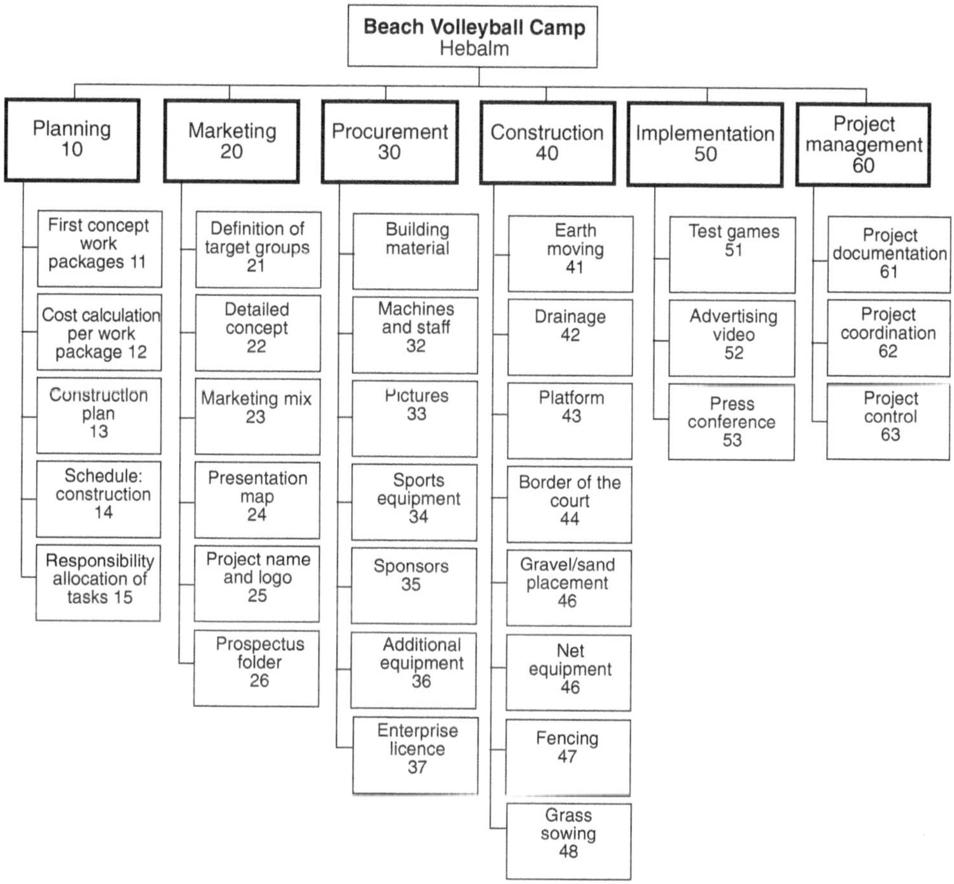

Fig. 6.14. Work structure – the Hebalm example.

market segments? How can the enterprise organization be used as a success potential? What fields of innovation are recognizable? Is it possible to derive new competitive advantages from a European point of view?

Mantsch (1995) showed that PA represents an essential instrument for long-term business planning in the Austrian furniture industry. On the one hand, a high rate of application could be verified (39 of 43 enterprises), while, on the other, successful product diversification could be documented after the implementation of PA. In the field of forestry, the forest management plan represents a specific PA at a more or less standardized level. For niche market activities, additional information related to the product is imperative. With the case-studies, it could be stated that only 34 out of 98 cases applied an additional RES-related PA (e.g. DE24, NL16, AU13, IT27). At the same time, several projects could be documented in which the lack of information related to internal and external potentials caused a broad range of drawbacks (see IT04, 17, AU09, 16, DE23, 27, NL05, etc.).

The objective of this section is to provide principles and guidelines on how to identify strengths which the forest owner can utilize for a specific RES activity. It also discusses weaknesses that need to be overcome to avoid the possibility of failure. For new projects in the field of environmental and recreational activities, forest management has to evaluate the enterprise's capabilities in the light of its

Table 6.15. A time schedule – the Hebalm example.

No.	Work package	Designation	Start	End
10	Planning	Planning of the building, incl. building plan	1 Oct. 1996	30 Nov. 1996
		Time schedule	1 Oct. 1996	15 Nov. 1996
20	Marketing	Project contracts	15 Oct. 1996	3 Dec. 1996
		Marketing concept	1 Oct. 1996	31 March 1997
30	Procurement	Supply contracts	28 Oct. 1996	6 Dec. 1996
		Sports equipment, etc.	17 Feb. 1997	11 April 1997
40	Construction	Process of construction (weather-dependency, therefore, appointment shift calculated)	21 April 1997	16 May 1997
50	Implementation	Inspection, test, film for TV (weather-dependency)	20 May 1997	23 May 1997

Table 6.16. Definition of milestones – the Hebalm example.

1	Project start; placing of order by the project owner	1 October 1996
2	Contracting	14 October 1996
3	Official opening of the vacation apartments and simultaneous presentation of the project (presentation map finished)	3 December 1996
4	Start of construction	21 April 1997
5	Opening of the beach volleyball court	1 June 1997
6	End of project; all bills are paid	*c.* 15 July 1997

Table 6.17. Needs of resources per work package – the Hebalm example.

Work package	Designation	Kind of resources	Quantity	Duration	Total
10/20	Project presentation map	Graphic and design	1 Designer 1 Copywriter 1 Photographer	4 weeks 2 days 1 week	40 days
	Marketing concept	Legal advice	1 Lawyer	3 days	3 days
40	Construction	Work	1 Machinist 2 Workers	2 weeks 3 weeks	50 days
		Material	Drainage		60 m
			Gravel		188 m^3
			Sand		300 t
			Fleece		650 m^2
			Lumber		15 m^3
			Windbreak		50 m^2
			Net-equipment		2 pieces
			Anchor		6 pieces

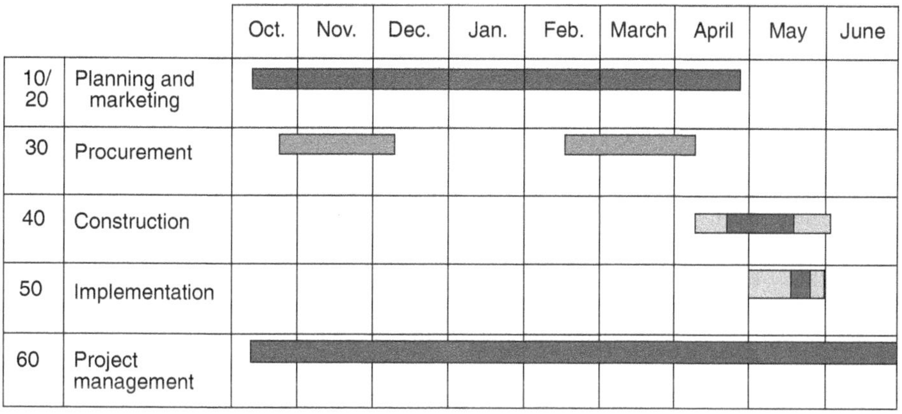

Fig. 6.15. A bar-chart table for project planning – the Hebalm example. Light grey = time buffer.

resources. Considering internal/external circumstances and the specific strategic view, the forest management should be able to allocate niche markets for successful business fields.

The related aspects of information management will be discussed first. The tasks and methods of PA will be presented in the next step and then the internal and external appraisal will be discussed. Finally, after the introduction of general methods for the assessment of recreational and environmental forest functions, a practical procedure of how to evaluate water resources is documented.

6.2.2 Strategic management

Definition and instruments of strategic analysis

Strategic management deals with the question of what paths to take in order to accomplish the operational aims of the enterprise in the long run. The defined business goals represent the basis of any strategic orientation (Gaiser, 1992, p. 10). The goal of the strategic planning is to find the best possible way (strategy) to realize defined business goals. This is an essential part of strategic management, in which the implementation and the evaluation are included. Attention should be paid to the explicit definition of an enterprise strategy, which promotes the process of market-orientated development, as well as specific controlling. Apart from a few exceptions, in forestry no strategies are defined for traditional (timber production, hunting) or niche markets (RES).

The necessity of permanent further business development is caused by the market. Especially in the tertiary production sector, forestry should be aware of the positive forest functions in regard to RES and, in consequence, of its own economic benefit.

In principle, a strategy in the field of business management deals with the creation and preservation of operational potentials. With the increasing instability of the markets, the past analysis (cost accounting) is of minor relevance. At the same time, relevant developments (trends, European Union (EU) programmes) should be considered more intensively by the analysis of relevant indicators (Staehle, 1987, p. 354). The spectrum of specific indicators is related to the individual task and must be defined as the case arises.

Indicators of tourism could be travelling behaviour, overnight stays, quantity and quality of accommodation in the surrounding area, etc. Environmental indicators could be investment value, environmental behaviour in general, legal changes and modifications (e.g. Austrian Water Law).[1] The compulsion for operational innovation is evoked by many causes, such as diminishing net income from timber production, changing moral concepts of society, newcomers (see national park management) and the necessity to customize faster, more flexibly and to individual satisfaction.

The strategic analysis supports the process of decision-making by providing relevant information to the management. The following instruments/methods could be helpful (Eschenbach, 1993, p. 219):

- catalogue of strengths and weaknesses.[2]
- Analysis of chances and risks (opportunities/threats).
- Portfolio analysis.
- PA.
- FMP.
- GAP analysis.[3]
- Definition of strategic success factors.
- Corporate identity, company profile.
- Definition of responsibilities.

RES projects, in particular, require strategic decisions. The successful implementation of a new product/service depends on an individual, but methodical and strategic, procedure. Due to incomplete external and internal assessments, a lack of strategic formulations and mistakes in the realization process, many good ideas have failed. Specific circumstances (e.g. ideology of the forest landowner, traditional borderlines, etc.) should be taken into account to ensure a better chance of realization. In the following, a guideline and relevant tasks for project planning will be presented, which could be adapted by any FLO (farm forestry as well as forest enterprises).

Procedure/strategic business unit (SBU)

Strategic management requires a straightforward business policy. Thus, creativity, innovation and vision on the side of the decision-makers are necessary. An adequate information level, which is generated mainly from the business and environmental analysis, should be provided. Expected internal and external conflict potentials, risks and elements of uncertainty should be taken into account.

Therefore, a prerequisite for the active layout of new business fields is structural adaptability (adaptation), which is ensured by coordination ability, realistic goals and ability to anticipate. After an internal and external analysis, the relevant and available potentials of the forest enterprise should be identified. These resources are the basis of the definition of a strategy (Fig. 6.16). Referring to the resources, possible business fields and special product market combinations should be decided on. In the next step, internal policy guidelines and basic conditions must be marked out. Structural as well as business organization is to be developed further and adapted to this setting by reason of a better execution and control of the strategy.

Feedback on the level of analysis and strategy definition is particularly impor-

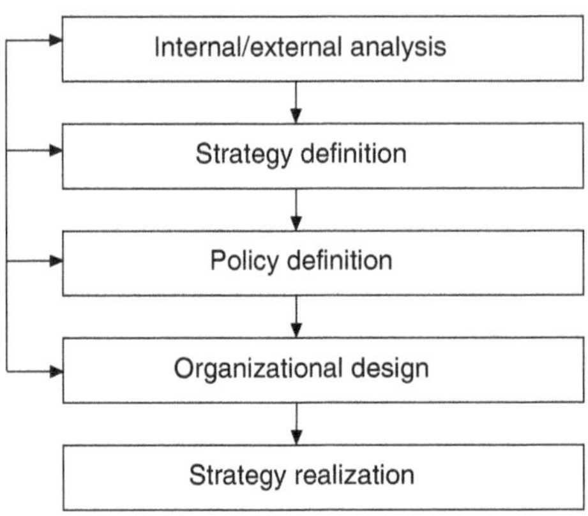

Fig. 6.16. Phases of strategic management (from Sekot, 1992, p. 26).

tant. In the field of forestry, the definition of a strategy and an enterprise model is often restricted by the focus of the FLO itself. The reconnaissance and assessment of market gaps can be missed and, in consequence, these new success potentials are not recognized (Sekot, 1992, p. 25).

As mentioned before, new business fields and the decision about how to open up the market will trigger the necessity of organizational changes within the enterprise. As a practical approach, it is helpful to define specific business units. These so-called strategic business units (SBUs) for which separate strategies are developed always represent a designated product market combination. The main criteria for the delimitation of SBUs are listed in Table 6.18. By structuring the enterprise in SBUs, the business processes for special markets should be easier and more transparent with regard to the decision and leadership competence (Sekot, 1993, p. 76). The segmentation of the market (according to products, customers, sales regions) corresponds to a specialization. For each product market combination, the relevant information must be acquired separately.

CRITERIA FOR SBU STRUCTURING. Basically, SBUs can be established at any business level, from farm forestry enterprises to large forest enterprises (such as the Austrian Federal Forest Administration (ÖBf AG): see Fig. 6.17). For instance, AU05 (Hochreiter eco-park, farm forestry) could be discussed here. The scope of business concerns five areas: agriculture, forestry, eco-park, forest tavern and biological rubbish dump (forest land base 110 ha). It is obvious that each area has to fulfil a specific market task and that the responsibility could be divided between the family members. However, the possibility for monitoring the economic conditions and profitability in the different business areas could be promoted by the explicit definition of SBUs.

The essential features of an SBU can be summarized as follows (Hinterhuber, 1992, p. 147):

- the business unit has to fulfil a specific market task, independently from other production systems (e.g. timber production).
- Organizational independence.
- To achieve relative competitive advantage, operational potentials relevant for the specific market segment must be available.
- Relative independence of the decision-making process in other business units is required for the realization of strategic plans.
- The SBU presupposes a corresponding leadership efficiency.

The administration of the ÖBf AG is located in Vienna, and about 25 forest enterprises are distributed all over Austria. The organizational structure shown in Fig. 6.17 is established at both levels. Thus, in each forest enterprise, a forest manager should be responsible for the forestry busi-

Table 6.18. Criteria of SBU delimitation.

Products or services	In any case where the production of goods or services is not heavily interrelated, a differentiation of SBUs can be considered (e.g. holiday apartments, NL18; hydroelectric power, AU03; outdoor events, DE18)
Target groups	Different target groups of potential customers of a product/service, e.g. in the field of drinking-water production: institutional (NL01, AU19) and private customers (AU16)
Sales regions	Regional and international marketing, several corporate locations (e.g. ÖBf AG: AU01, 19)
Specific marketing strategy	When specific marketing strategies for different products already exist, business organization should be adapted correspondingly (e.g. AU20: pound management)

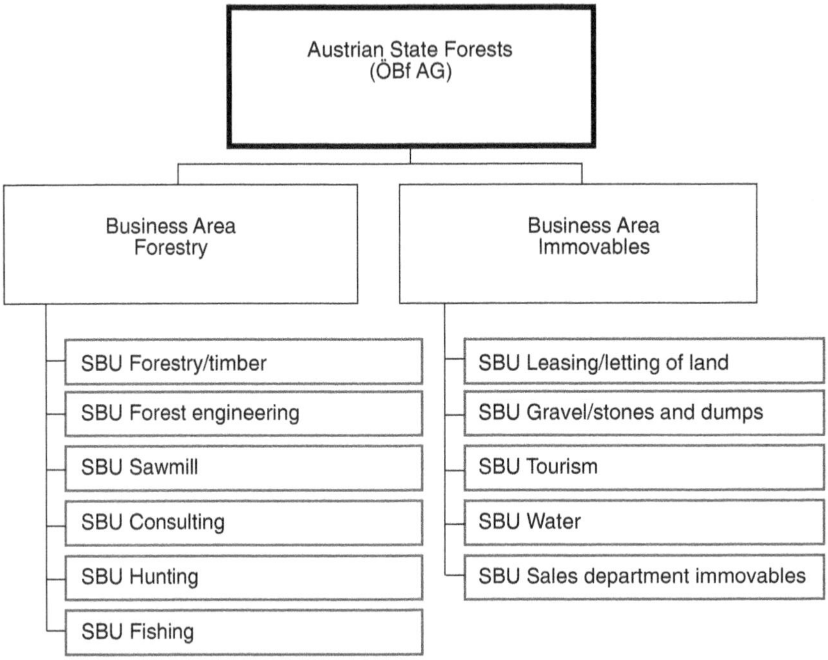

Fig. 6.17. Organizational structure of ÖBf AG (modified from ÖBf AG, 1997).

ness area and a second one should be responsible for immovables (reorganization still not finished).

Strategic information management

INFORMATION MANAGEMENT (IM). In addition to the traditional factors of production – property, work and capital – the importance of access to the relevant information is increasing (Gazdar, 1989, p. 11). The acquisition, evaluation and arrangement of information are becoming a decisive competition factor. Due to the increasing complexity of business relations, the uncertainty of the decision-making process increases (Rota, 1997, p. 5). In principle, additional information reduces the uncertainty. Even so, in individual cases, additional information can cause a delay in the decision-making process through additional uncertainty (Hanf, 1991, p. 65).

The significance of specific information could be stated with the 98 case-studies. About one-fifth of the respondents stated that lack of information was responsible for several drawbacks during realization of the RES project. Additional factors of drawbacks, such as misjudgement or lack of motivation of the personnel, were mainly caused by the lack of information (e.g. DE11, IT17). From this point of view, more than 30% of all documented RES cases were, to a greater or lesser degree, negatively influenced by missing information (multiple answers possible).

The necessity of a consistent processing of strategic information (primarily concerning the economic environment) arises from the increasing diversification of the market conditions (globalization, e.g. the EU). Typical side-effects are overcapacities, aggressive competition behaviour, shorter product life cycles and changing consumer behaviour.

According to a survey, 57% of all managers concede that strategic information supply has the highest priority. But, after accounting for all operative jobs, the decision-makers have less than 10% of their working time left over for strategic planning (Becker, 1994, p. 147). For the opening up of niche markets and for long-term market success, the current provision of information must be ensured (OECD, 1995, p. 22). Consequently, IM is likely to be of major importance, along with the task of implementing RES marketing activities. What tasks does IM have to fulfil?

The most important goal of IM is the provision of correct information in the right quantity, in corresponding quality, at the right time with as little effort as possible and to provide information at the right place (decision-maker, etc.), to ensure an optimal aim achievement (Herget, 1995, p. 27). However, the provision of information is just one task of IM. Internal information systems or knowledge resources must be built up to be at every employee's disposal (Grudowski, 1996, p. 354). By the use of modern information and communication techniques, employees will learn to work decisively with information resources (additional motivation of innovative employees). Therefore, an efficient IM includes the opening up of various information resources through adequate training (e.g. on-line information sources).

The obtaining and processing of information (= trade of information) which is performed daily by every employee represents a social activity (Grudowski, 1996, p. 355). Due to the ever-changing information demand, the trade of information must be organized in a more efficient and effective way. The tasks of planning, controlling and organizing must be taken into account. The various aspects are represented in Fig. 6.18.

Referring to the reorganization of ÖBf AG (see Fig. 6.17), the internal transfer of information is secured by frequent and regular meetings between the administrator of the current business area and the responsible SBU managers at forest enterprise level.

INFORMATION RESOURCES. Concerning multiple forest production systems, the increasing information demand can be covered by various sources. Personal information sources represent the most essential resource for small and medium-sized enterprises (SME) (Gazdar, 1989, p. 48). This statement could be verified by 98 RES case-studies. More than 40% of all respondents specified that internal personal knowledge is the most important information source (see Chapter 4, section 4.2).

In the context of a written interview, 750 SME of Nordrhein-Westfalen (Germany) were asked about their information behaviour (Staudt *et al.*, 1992). The remarkable result of this report is the meaning of external information sources for SME. About 85% of the information is obtained from external sources (see relevant information sources in Table 6.19).

With respect to forestry, the internal information source takes on a greater meaning than the report of Staudt *et al.* (1992) demonstrated (more than 40%, instead of 15%). This circumstance could be explained by still existing forest

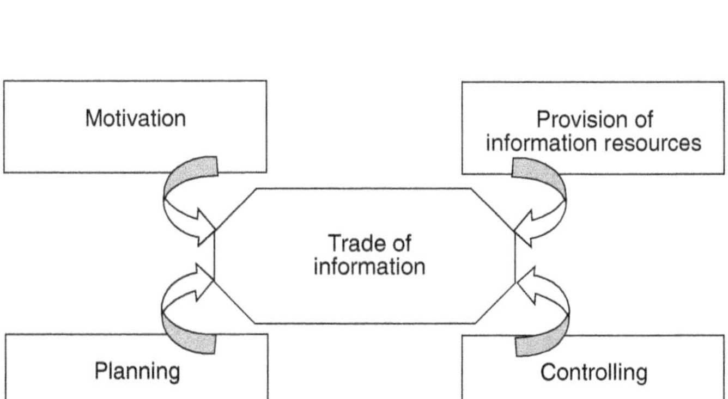

Fig. 6.18. Management aspects of information processing (from Grudowski, 1996, p. 356).

Table 6.19. Structuring of external information resources (from Staudt *et al.*, 1992, p. 1000).

External information resources	%
Personal communication	52
Print media	18
Databases	9
External consultation	21

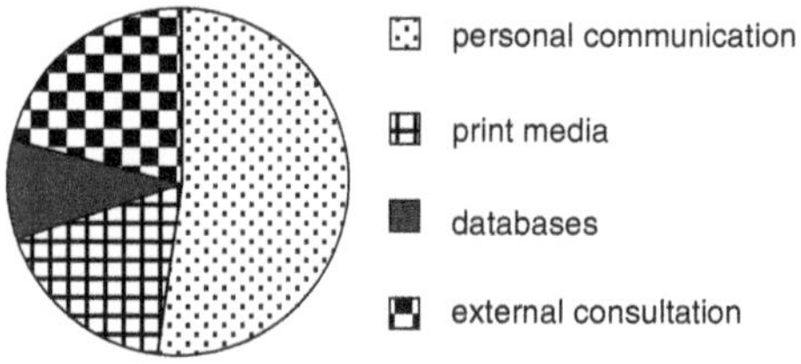

management plans (Table 6.20), which could not be ignored:

- of 98 interview partners, 46% derived relevant information from forest maps.
- Inventories were significant in about 28% of all documented cases.
- The significance of still existing plans in about one-third of all cases.

In addition to the forest management plan, one-third of the FLOs interviewed performed an RES-related potential inventory (e.g. involving the use of geographical information systems (GIS): AU03).

The importance of databases as information resources has increased strongly in the last years. On the one hand, the internet can compensate for the disadvantages of rural locations with regard to information access (provider service), while, on the other, price and quality differences between overcrowded and rural areas are still evident. However, a short overview of internet addresses that could provide relevant information in regard to RES activities is given here:

- Directorate General VI at the EC (Agriculture) (http://europe.eu.int/comm/dg06)
- Recreational and environmental services (www.rrz.uni-hamburg.de/holzw/res/)
- International Timber Exchange (www.timber-exchange.com)
- European Forestry Institute (www.efi.fi)
- Geographical information systems (http://info.er.usgs.gov/research/gis/title.html)
- *Nature* – international weekly journal of science (www.nature.com/)
- Federal Ministry of Agriculture and Forestry (www.bmlf.gv.at)
- Bayerische Staatsforstverwaltung (www.forst.bayern.de)
- Wilderness and Survival Forum (http://survival.imweb.com)
- Swiss Forest Agency (www.admin.ch/buwal/forst)
- ProHolz Österreich (www.proholz.at)
- The Austrian Country Market (www.country.co.at)
- Austrian Conference on Spatial Planning (www.oerok.gv.at)
- Austrian Central Statistical Office (www.oestat.gv.at)
- Federal Environment Agency (www.ubavie.gv.at)
- Municipia (www.municipia.at)
- Chair for World Forestry (www.rrz.uni-hamburg.de/biologie/weltf.html)

Table 6.20. Existence of a forest management plan. (Question 6.1.1: Is the project area covered by a forest management plan?)

Forest management plan	*n*	%
Yes	85	86.7
No	12	12.2
Missing	1	1.0
Total	98	100.0

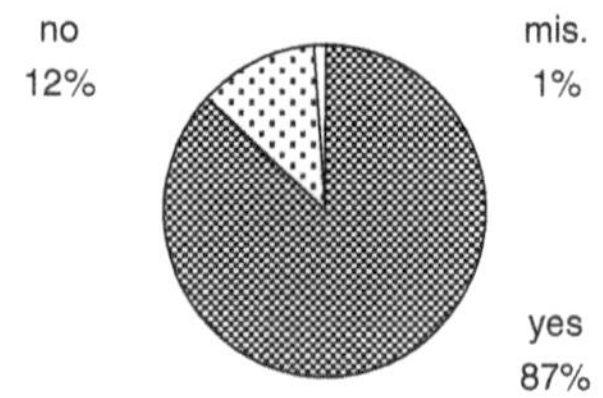

- Institute for and Chair of World Forestry (www.dainet.de/bfh/ins1/index.html)
- Institute for Forestry and Nature Research (www.ibn.dlo.nl/index-eng.html)
- Universita Degli Studi di Padova (www.unipd.it)
- Austrian National Tourist Office (www.austria-tourism.at)
- University of Agricultural Sciences (www.boku.ac.at)
- International Timber Exchange (www.timber-exchange.com)
- Chair of Forest Policy and Forest Economics (www.waho.ethz.ch/ppo/e_welco.html)
- Department of Forest Economics (www.sekon.slu.se/eng/default.html)
- VivaWALD (www.vivawald.at)
- Forest Economy Association (www.wvs.ch)
- Unit of Forestry (www.flec.kvl.dk/forestry/)

Austrian forest enterprises:

- Stiftung Fürst Liechtenstein (www.sfl.at)
- Souveräner Malteser Ritterorden (www.hebalm.at)
- Bistum Gurk (www.bistum-gurk.at)
- Fürst Esterhàsy'sche Privatstiftung (www.esterhazy.at)
- Österreichische Bundesforste (www.oeb f.at)
- Metternich'sche Forstverwaltung Grafenegg (www.grafenegg.com)
- Traun'sche FV Wolkersdorf und Rappotenstein (www.traun.vivawald.at)
- Stiftsforstbetrieb Klosterneuburg (www.vivawald.at/klosterneuburg)
- Forstgut Weissenegg (www.vivawald.at/weissenegg)

Internal knowledge is playing an increasing role. IM requires, in addition to general basic knowledge, a corresponding specialized knowledge and a vocationally orientated and enterprise-specific experience knowledge. The continuation of one's education is a strategic success factor in this context. A planning and further educational programme, which includes the employees, increases the acceptance, minimizes the start-up problems and promotes readiness for assuming self-responsibility, quality assurance and readiness for permanent further education of each employee (Gaiser, 1992, p. 126).

The internal support of interdisciplinary knowledge promotes ordinary information exchange, especially since the information transfer due to 'embodied knowledge' (staff transfer, advice performances, cooperations) becomes most important. Derived from the case-studies, several fields of education relevant for RES activities are listed below:

- meaning and relevance of PR.
- Knowledge about forest-related business fields (especially tourism and nature conservation associations).
- Tools of marketing.
- Internal and external PA.
- Use of PCs and the internet.
- Tasks of project management, teamwork.
- Communication systems (meetings, briefing, etc.).
- Multifunctional (networked) way of thinking.
- Forest policy, lobbying.
- Use of GIS, RES-related accounting methods.
- Kinds of cooperation.
- Legal aspects.
- Knowledge of regional, national and European Community (EC) subsidies.

This list could be seen as a checklist. What information resources are still available? In which fields of education should the management invest?

In the past, forest rationalization measures were mainly confined to modernization of logging procedures and to personnel policy. Therefore, employees and workers were hardly available for the development and realization of RES projects. Thus, in several documented RES cases, outsourcing was essential in the realization process (Table 6.21).

Outsourcing is already a very common management instrument in forestry (e.g. FMP, timber harvesting). The decision on outsourcing tasks of strategic information depends on several considerations (Michels, 1995, p. 91):

Table 6.21. Examples of documented outsourcing cases.

RES case-study	Subject	Done by
NL01 Drinking-water	Hydrological inventory	Forest extension service, drinking-water association
NL06 Sponsoring	Potential analysis	Private consultants
NL13 Tree-crown path	Forest analysis, information centre	Architects, business manager
DE11 Survival centre	Inventory of outdoor potentials	Forest student
DE14 Drinking-water	Soil inventory, definition of protection zones	Non-profit organization (Klimaschutz durch Wald)
DE15 Day-trips	Inventory of forest functions and tour description	Private consultant
AU03 Hydroelectric power	Technical aspects	Building enterprise
AU10 Larch meadows	Square dimension by aerial photograph	Forest and agricultural extension service, department of nature conservation
AU13 Cross-country skiing	Panoramic sites, environmental aspects	Nature conservation association, regional tourism office
IT13 Certification	Forest inventory	Private consultant

- lowering of the extent of work for a professional and internal handling: the potential of human labour could be too low, particularly in SME, to assign the enterprise's own employees or to continue the education of the qualified employees available. Due to missing specialization in definite technical areas, it can only be covered by external competence.
- Avoidance of capacity bottlenecks: the reasons can be (seasonal) peak loads, accidents, wind breakage, short-term capacity failure through illness, etc.
- Use of special know-how: outsourcing can help to compensate for the missing knowledge in the field of IM (global data networks: Häfner, 1996, p. 43), as well as to use new services offered by specialized information agencies.
- Reduction of capacities: mainly financial considerations are in the foreground with outsourcing. As a result of outsourcing, an internal reduction of resources is possible (personnel, machines). Primarily a reduction of fixed costs is ensured. However, the transfer to variable costs should be evaluated by a detailed cost–benefit analysis.

Besides the 'in-house outsourcing', which is confined to a few large forest enterprises, according to Michels (1995, p. 92) two fundamental forms for forest owners could be derived:

- cooperative outsourcing: combining several enterprises with the same interests and a common information institution – for example: in cooperation with the municipality (cost advantage by fixed cost arrangements) or the transfer to forest interest representatives (Chamber of Commerce, engineering office, etc.).
- External outsourcing: the tasks of the information collection will be submitted to a third party. The exact definition of the scope of work is necessary for the quality of the information. In consideration of the dependence, the partner should be known with regard to reliability and stability.

6.2.3 Tasks and methods of PA

PA tasks

The aim of strategic planning is the preservation and creation of success factors. However, the identification and use of the available enterprise potentials makes a strategy of concentration on the strengths of the enterprise possible (Hopfenbeck, 1997, p. 456). On one hand, the factors of production (infrastructure, staff, etc.) and special knowledge of employees are relevant assets. On the other hand, opportunities which could be derived from market trends must be seen as relevant success factors.

The PA starts up with a strengths-to-weaknesses analysis of an enterprise. After the uncovering of risk and opportunities, conclusions are derived. Thereby the management is able to see and eliminate weaknesses and to push the strengths simultaneously. The competition situation in the market can be improved. At this point, it should be mentioned that, in almost one-quarter of all documented RES cases, the inventory of potentials should have been done in a better way (see Table 6.8).

As a result of a PA, strategies and measures can be defined. The PA is concrete and related to an individual case. Usually, several fields must be taken into account and should be valued individually (integrated way of thinking). In contrast to the PA, the portfolio analysis allows only trivial conclusions because of the concentration on the actual situation (Voigt, 1990, p. 18). For example the ÖBf AG uses portfolio analysis merely for a rough derivation of their main strategic objectives (see ÖBf AG, 1997, p. 54).

PA methods

The PA is based on the analysis of the actual state (description of strengths and weaknesses, success factors). After considering potentials, the chances and risks of a required state can be derived. Unlike the statement or ratio analysis, the PA is, in principle, a qualitative method for the operational and environmental analysis (Sagl, 1994, p. 125).

All kinds of qualitative estimation methods can be used:

- induction: derivation of the individual case based on general legitimacies.
- Deduction: general valid legitimacies are broken down to the individual case.
- Intuition: activity which is not based on facts (inspiration).
- Interpolation: enquiry of a mean average value with the help of related data.
- Extrapolation: enquiry of a prospective value from past results by time-series analysis.

The execution of the PA is based on quantitative and qualitative statements. In principle, a monetary assessment is carried out, with direct effect on profit and price expectation. Depending on task complexity, PA requires (more or less) an interdisciplinary cooperation of different experts (e.g. gathering of water potentials). However, the fundamental ideas of PA can be applied at any level (from farm forestry to large forest enterprises).

An essential information source could be customers and employee interviews (Becker, 1994, chapter 4). In most cases, one gets more information from these surveys than from the analysis of balance sheets and profit-and-loss statements. Nevertheless, the PA cannot ignore past results. However, unlike balance sheets and profit-and-loss statements, the interviews represent a more accurate early warning system.

The representation and statements of the PA must be checkable by third persons. The explanations and conclusions resulting from the analysis must be plausible, comprehensible, consistent and professional.

EXAMPLE: POTENTIAL ANALYSIS FOR THE TOURISM SBU BY THE ÖBF AG (see Fig. 6.17; ÖBf AG, 1997, p. 49)

Situation and development of the tourism branch. At present, Austrian tourism is in a structural crisis. The offers and the suppliers have lost international competitiveness. Global market segments have been developed by cheap long-distance flights (decreasing overnight stays). Growth opportunities are seen in the areas of town, culture, adventure and quality tourism. New kinds of sport (snowboarding, mountain-biking) require specific offers (chances of diversification). Tourism regions are working on special-offer packages. For this purpose, partnerships between various suppliers will gain importance (catering trade, hotel business, leisure-time facilities, etc.).

Internal analysis

The tourism area produces regular proceeds. The tourism performance of the ÖBf AG is of great significance for the respective regions. Opening the forests causes areas of conflict with hikers, nature activists and hunters. The market potentials of the previous sources of income are limited. An opportunity can be seen in the area of experience and adventure vacations. The following products could be offered: guided tours in the forests, package tours with overnight stays in huts, survival training, game shootings, children's camps, package offers in cooperation with local tourism organizations.

The policy discussion about the general opening of forest areas for tourism purposes which should be free of charge could lead to income losses in the tourism SBU.

Strategies and activities

- Increased engagement in the regional and local tourism field.
- Additional supply (cycle tracks, bridlepaths, cave tours, etc.).
- Prevention of cost increase and fixed cost risk (lease of buildings).
- Definition of local priorities in utilizing the forest resources to prevent conflicts between hunting, tourism and nature conservation.

6.2.4 Operational and environmental analysis related to RES activities

Judgement of the initial position

Environmental analysis (market research) and the internal analysis of potentials are essential for the definition of a marketing strategy (see Fig. 6.16) in relation to a specific product or service. Both fields of research clarify the market situation. A knowledge of internal strengths and weaknesses, as well as the definition of opportunities and risk sharing, are preconditions for the acquisition and delimitation of products and services, especially in the field of niche markets.

With respect to the initial position of the potential analysis, it is possible to distinguish three levels from which the analysis of RES potentials could start out (three-step approach).

GENERAL INVESTIGATION OF INTERNAL AND EXTERNAL POTENTIALS. In this case, the FLO wants to check possibilities for any diversification. Consequently, the PA is performed at a general level, thereby encompassing various categories of potential products. The PA consists of an operational and an environmental analysis. The internal identification of success factors is concentrated on performance-orientated and product-related determining factors (for example, in the course of FMP). All areas examined are dependent on each other and must be brought together after the individual analysis. In the next step, a profile of the strengths and weaknesses should be made. Considering an additional environmental analysis, opportunities and risks can be defined. Considering the corporate objectives and internal philosophy, the decision-making process on a new product/service development could be concluded (Fig. 6.19).

RESOURCE-SPECIFIC ANALYSIS WITHIN A PRODUCT CATEGORY. The PA is to explore the market potentials associated with a preselected resource. The FLO is already aware of product potentials in a specific product category, which could be the field of investigation of the PA. The knowledge of a niche market could be derived from the state of the market or from surveys. The definition of the success potentials has to consider various products/services within the category. As a result of the investigation, a catalogue of potential products with chances of success should be presented (e.g. category of water resources: drinking-water, fishing, pond management, water-power, etc.; see NL01, DE14, AU03, 06, etc.).

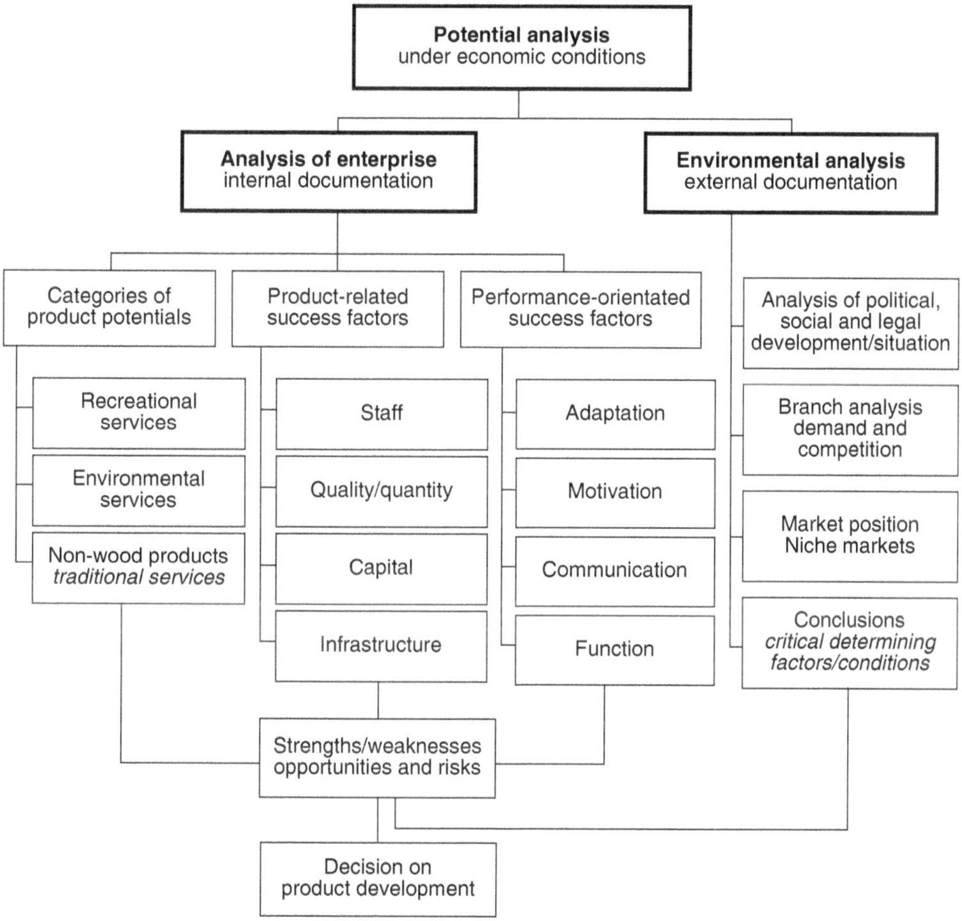

Fig. 6.19. Structure of a general potential analysis.

PRODUCT-SPECIFIC ANALYSIS ABOUT INTERNAL AND EXTERNAL RESOURCES. The product is defined before the execution of the PA (e.g. due to the decided operational objectives or due to market knowledge still available). The investigation of the internal and environmental potentials is related exclusively to the specific product/service. If relevant product-specific resources are available, this approach should present a strategy on how to successfully establish a product on the market (definition of SBUs, etc.).

The gathering of information at strategic as well as operative levels should be concentrated on internal and external factors. Basically, the PA is concentrated on providing strategic information as a decision basis (Krizek, 1994, p. 325).

Positioning of potential products/services

To obtain a clear picture of the strengths and weaknesses of a potential product, significant success factors must be defined and compared with the respective competing product. It is therefore only possible to assess whether a resource is a strength or weakness in relation to identified features of the business. The objective of this process is to identify strengths which the forest owner can utilize in his/her strategy and weaknesses which need to be overcome to avoid the possibility of failure (Bowmann and Asch, 1987, p. 96). With regard to successful market implementation, the list of essential criteria given in Fig. 6.20 should help to evaluate a potential product/service.

	Positioning of the product or service in comparison with possible competitors:		
Success factors:	Poorer	Similar	Better
• quality	☐	☐	☐
• quantity	☐	☐	☐
• attractiveness	☐	☐	☐
• variety of supply	☐	☐	☐
• additional performances	☐	☐	☐
• costs	☐	☐	☐
• cost transparency	☐	☐	☐
• individual solutions	☐	☐	☐
• motivation	☐	☐	☐
• service	☐	☐	☐
• reliability	☐	☐	☐
• infrastructure	☐	☐	☐
• resources	☐	☐	☐
• delivery time	☐	☐	☐
• know-how	☐	☐	☐

Fig. 6.20. Checklist for the preselection of potential products/services.

By the analysis of strengths and weaknesses in the context of PA, the FLO is early on forced to define the new business field analytically, especially in regard to potential competitors, his/her own product advantages, etc. In comparison with timber market conditions, the FLO should be aware that, in the field of niche markets, one is confronted with different market preconditions and strategies.

A comparative analysis of the enterprise's own strengths and weaknesses with those of competitors may identify areas in which strengths provide an advantage capable of sustaining a competitive strategy (niche market) or weaknesses which competitors may be able to exploit. Additional relevant information could be derived to qualify relevant potentials and to reduce the risk of over- or underestimation. The advantages and disadvantages have to be checked also with regard to the operational (internal) objectives. Success factors that were classified as worse relative to those of competitors must be judged and improvement measures (increasing the value) must be taken into account. This comparison allows a first overview of investment fields to be expected (e.g. acquisition of external business know-how necessary, etc.).

6.2.5 Aspects of the acquisition and assessment of RES

Resources overview (quick check)

Considering internal and regional aspects, a superficial evaluation of relevant success factors could present a helpful overview. When preparing an RES project, the FLO should be aware of still existing and/or similar products/services in general and especially in his/her own region. If relevant information is available and comparable with the RES project, this information should be taken into account. However, when considering a defined project, the FLO should first perform an internal check of relevant resources. The factors of success should be discussed with employees (brainstorming). For this purpose, a catalogue of questions could be helpful, e.g. concerning forest guided tours:

- infrastructure (car-park areas, footpaths)? Adequate landscape (forest stands)? Natural highlights (waterfalls, biotopes, etc.)?
- Personnel resources (willingness, training, time, etc.)? Partners?
- Equipment (field-glasses, boots, etc.)?
- Financial resources (investment)?
- Catchment area? Target groups? Informal contacts with several institutions (schools, clubs, trade associations, etc.)?

If relevant resources are available, the FLO should think about organizational, technical and financial aspects, e.g. how to integrate the new tasks in daily business; a first cost–benefit analysis could be done, a rough definition of (non-material) goals, possible procedure of implementation, etc. In case of insufficient resources, the idea/concept could be adapted, relevant factors could be purchased (cooperation with partners, leasing/buying of missing resources) or the project must be rejected.

If the project is still attractive, additional information and planning procedures should be considered (depending on the project). Otherwise, if the planning level is reached, the project could be initiated (implementation).

Several methods of valuation concerning recreational and environmental projects should be discussed. If a detailed planning procedure is necessary, the FLO should have an overview of relevant parameters which could be taken into account. Additionally, the introduced methods and instruments should present guidelines for a systematic and objective realization process. Furthermore, relevant and exemplary criteria should be pointed out.

Forest management planning (FMP)

A vast majority of all interviewed FLOs are using forest management plans (see Table 6.20). However, high significance of maps, inventories and plans with respect to RES activities is the exception rather than the rule.

To support the operational (strategic) decision-making process, the information content of conventional FMP does not suffice. Besides the consideration of ecological, technical and economic factors, the social aspects of the environment must be integrated into the spatial information. Conscious consideration of social claims in forest operational planning ensures the continuance of the area competence for integral forestry (Sekot, 1991, p. 433). In the course of reorientation, the forest manager has to consider regional and infrastructural services of the forest stand (see IT01, mushroom-picking permits: a section is explicitly dedicated to the identification of mushroom allocations). Considering existing/potential markets, new strategies must be developed. Besides the FMP as an essential internal information resource, the case-study research showed that one-third of all interview partners used additional information stemming from inventories of RES-related potentials (Table 6.22). However, the lack of specific information in the field of RES activities is obvious, and

Table 6.22. Additional inventory of RES-related potentials. (Question 6.1.2: Did you undertake any inventory of RES-related potentials (e.g. potential for recreation)?)

RES inventory	*n*	%
Yes	32	32.7
No	66	67.3
Total	98	100.0

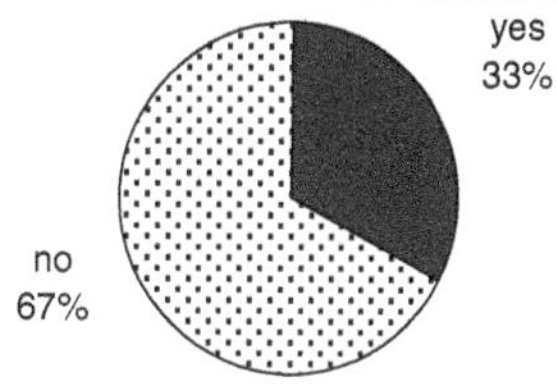

this could be confirmed by the RES questionnaire (relevant drawbacks in the RES realization process: see Table 6.7).

The representation of forest conditions as a main task of the FMP can be carried out with a sample survey, complete measurement of the stands or a composite method. The decision as to which method to use will depend on the extent of relevant information and available resources and must be evaluated for the individual forest (Sekot, 1993, p. 108). The explicit elaboration of infrastructural performance concepts requires a methodical expansion and interdisciplinary specialized knowledge for the acquisition of operational potentials. Interdisciplinary cooperation should take into account the knowledge deficits in the areas of market evaluation, technical, botanical/zoological disciplines, etc. (Schwanecke, 1991, p. 62; Bitter, 1994, p. 164).

Recreational valuation parameters of forest land

Recreation is defined as the keeping, regeneration and increase of the physical and psychological strengths of a person. In the medical meaning, recreation is defined as the regeneration of physical and emotional strengths through quiet, calm (balance) and pleasant change. Also social and individual aspects can be brought into connection with recreation (e.g. prestige).

The data of modified FMP should be able to provide additional information on the various forest functions and services. An adequate representation should make it possible to derive new concepts and decision bases for internal and external (forest public relations) RES activities in an easier and more helpful way.

ELEMENTS OF RECREATION ON FOREST LAND. First, the attributes and elements of the forest land that are relevant for recreational effects must be defined. Broadly outlined, these indicators are examined differentially during the assessment. The forest recreational indicators, as given in Table 6.23, have to be compared with the relationships of an open area (e.g. meadows).

The relevant indicators of the landscape for recreation purposes must be examined and judged. For instance, for environmental protection purposes, the proximity of nature will be an essential indicator for the value of the forest land, but, for the recreational value, this indicator can be insignificant.

Consequently, several valuation parameters are introduced which could be relevant for the assessment of recreational suitability. The valuation can be carried out either by the FLO him/herself (or his/her staff) or outsourced to relevant institutions. The main focus lies on the extensive listing of the single assessment parameters of the landscape. The practical applicability and usefulness in accordance with specific parameters of recreational (e.g. mountain-biking) and environmental (e.g. forest nature reserve, water) products/services should be pointed out.

If the assessment is carried out due to public interest, the formation of a recreational forest according to Austrian forest law (Wohanka and Stürzenbecher, 1991, § 36) is a possible consequence. But the forest owner him/herself may also place an application for the definition and delimitation of a recreational forest area.

Table 6.23. Indicators of recreation in wooded area (modified from Haubenberger, 1984, p. 42).

Indicators	Effects
Radiation and light	Balance of radiation, light–shade effects
Colour contrasts	View of landscape in the annual cycle
Visibilities	Emotional perception (confined, mysterious, etc.)
Thermal effect complex	Thermal level, wind speed, temperature, humidity of the air
Hydrometeorological effects	Precipitation, frost, thaw
Wood's air	Air quality (smells, load of dust, etc.)
Acoustics	Absorption and spread of the sound energy, natural noises
Aesthetics	Subjective feeling (areas, plant species)

SURVEY OF RECREATIONAL PARAMETERS. In the following, the focus should be concentrated on possible parameters that are relevant for a recreational project. In consequence, a specific or individual procedure for the judgement of the recreational value could be derived. If the FLO has to choose between several forest areas for recreational purposes, the method of Scamoni and Hofmann (1969) represents a basis of comparison for investigation at the highest level. For the assessment, three different qualities are examined (parameters with an asterisk are added):

1. Parameters of the nature equipment:

- forest roads (m ha^{-1}) and edge effect.
- Area versatility.
- Variety of wood species.
- Value of various vegetation types.
- Relief factors.
- Climatic factors.
- Natural monuments.*
- Brooks, waterfalls.*
- Pastures.*

2. Air hygiene and recreation-diminishing factors:

- air cleanness.
- Noise disturbance.
- Insect annoyance.
- Reduction of the aesthetic value.

3. Infrastructure in terms of recreation facilities:

- state of forest roads.
- Shelter removal.
- Benches, playgrounds and berths.
- Car-park areas.
- Gastronomic facilities.
- Nature trails, game parks.
- Sanitary facilities.
- Vantage points.*
- Cultural sites.*
- Hunting-lodges.*
- Accommodation.*

This list, or parts of it, could be used to fill the checklist given in Table 6.24. Considered from a general point of view, relevant parameters and the degree of availability should be defined. In the next step, relevance (minor, moderate, major) could be seen in context to a specific project. In combination with the degree of availability, the parameter could be weighted and, finally, factors of competition/synergism should be taken into account.

The parameters will be different for each project/enterprise. One example could be the selection according to Scamoni and Hofmann (1969). Each of the parameters listed could be subdivided into more or less detail. With regard to a concrete project, the parameter 'forest stand' could be of less relevance. In the case of a large recreational project, additional indicators, such as the following, should also be considered:

- geographical distribution.
- Forestry regime (clear-cutting system, group selection system, single tree selection system).
- Age-class distribution.
- Occurrence of brushwood.

Table 6.24. Check list of recreational parameters.

Internal/external recreational parameters		Degree of availability: Low	Medium	High	Relevance	Note
Forest (nature)	Stands					
	Roads					
	Variety of vegetation					
	...					
Infrastructure	Accessibility					
	Restaurants					
	Car-park areas					
	...					

- Change between forest stands and meadows.
- Edge of the forest and its location.
- Special habitats.
- Nature conservation areas.
- Protection forests.
- Percentage of dead wood.
- Flowers, herbage, berries, etc.
- Climatic and relief factors.
- Factors of disturbance (traffic, high-voltage power lines, public roads, neighbours, etc.).

Environmental valuation methods for forest land

The review of economic performances by forestry, particularly for the conservation areas and the preservation of ecological sustainability, requires an evaluation of relevant conservation parameters. In principle, special performances and preconditions which serve the preservation or restoration of natural forest stands should be appropriately paid for (ecological economics: Hampicke, 1991, p. 80).

For the determination of the environmental performance spectrum of forest areas, biotope mapping serves an important database. The factors of appraisal concerning the conservation value of definite forest areas cover the nature proximity (closeness to natural circumstances) and variety (biodiversity), as well as the rarity and endangering of a biotope. Furthermore, the following targets should be taken into account for the evaluation and judgement of biotopes:

- decision base for forest operational planning.
- Basis of the forest development planning and the regional/supraregional planning.
- Derivation of silvicultural management concepts (stand conversion, protection of species, protection and care measures).
- Consideration of forest PR.
- Basis of several enterprise activities (natural paths, forest guided tours, eco-sponsoring, using of water resources, etc.)

Without emphasizing the financial target explicitly, the assessment of the environmental and conservation functions of forest lands are introduced. The relevant appraisal factors will be pointed out in relation to the PA.

THE FOREST ECOLOGICAL EVALUATION SYSTEM (FEES). This assessment scheme is based on the comparison of the current forest land with the potential natural forest stand (PNFS). From the actual-versus-target comparison, so-called 'eco-points' can be derived. If the actual forest stand corresponds with the PNFS, the maximum number of eco-points can be obtained. The set conditions are defined by Frank and Hinterleitner (1994, p. 5):

1. Composition of tree species which is derived from the PNFS.
2. Natural regeneration corresponding to the PNFS.
3. Structures and texture.
4. The ecological sustainability and biodiversity.
5. Measures for the minimization of impairments.

Thus the following criteria must be taken into account for the award of eco-points.

Composition of tree species of the current forest stand. The FEES has been worked out for the assessment of the forest stand with regard to the natural proximity of its composition. The species of the bush, herb and moss strata are used as indicators to define the PNFS. The PNFS was especially worked out in the form of a catalogue representing the targets of the FEES for Lower Austria (Frank and Starlinger, 1994). The assessment only considers the deviation of the current from the potential forest stand.

Structure and variety. With a distinctive age diversity, a structured forest stand is given automatically. Both features are to be judged positively. The ratio of tree species in view of the natural site and the form of the spatial distribution are relevant for the judgement. The occurrence of natural regeneration under the crown and the group selection system are judged very positively.

Special structure elements that increase the value of a biotope are old crop trees and particularly valuable edges of the forest, etc.

Impairments. Damage to forest soil leads to a reduction of the ecological value, due to its long-term damage and limited reversibility. Damage of the stock is taken into account if the forest stand or the distribution ratios of the forest species are endangered. A further reduction of the ecological value is caused by forest roads. As an evaluation criterion, the slope and not the width of the roads is relevant (hillside water flow). Considerate implementation and optimal technical elaboration can compensate for the negative aspects.

Multipliers. An increase of value for the forest stand is taken into account by consideration of social claims. Protection, welfare and recreational performances can best be fulfilled by natural and vital forest stands. Therefore, two parameters are relevant for the derivation of the multiplier. One is the significant value number of the forest development plan and the other is the forest equipment of the municipality (= share of forest area in the entire municipality area).

The extent of the significance of the forest to the public is expressed by the respective figures. The significance of the classified forest increases with the decrease of municipal forests. The use of these parameters for the assessment causes only a little extra work since these parameters are easily available and no additional fieldwork is necessary (Frank and Hinterleitner, 1994, p. 20).

The Austrian case-study AU 12 (Neckenmarkt forest nature reserve) should be mentioned here. In the course of inventory, the PA was exclusively carried out according to the operational valuation criteria of the FEES. Due to a rough structuring of the relevant parameters, a visual judgement was possible and no measurements were necessary. At good sites, the investigation will be repeated every 5–10 years, while, at sites with bad qualities (low growth), the period of investigation can be extended up to 15 years.

Forest biotope mapping (FBM). An FBM considers all the biotopes in the forest and will be judged by the criteria which have already been introduced: nature proximity, variety, rarity and endangering. With this broad view, the principles of FBM are quite different in respect of the view of various nature conservation organizations, where only the criteria of rarity and endangering are taken into account (Volk, 1988, p. 1289).

The goal of FBM is the acquisition and judgement of the conservation function of a defined forest area. This information forms a basis of and decision support for forest economic measures (Kurth, 1993, p. 277). The valuation method is organized with the FMP (Ammer and Utschick, 1982, p. 68; Hanstein, 1992, p. 532; Schirmer, 1992, p. 40). Forest owners should be able to present specialized information on the conservation values of their forests based on objective and verifiable examinations, instead of intuitive statements.

For the execution of FBM, the following sources could provide important information (Arbeitskreis Forstliche Landespflege, 1996, p. 7):

1. Forest technical information sources:

- growth zones, growth regions.
- Forest site survey: derivation of PNFS within the growth region.
- Mapping of forest functions: areas with special ecological meaning can be examined specifically (documented in the Austrian Forest Development Plan).
- FMP: data about the crop establishment, ratio of tree species, management objectives, silvicultural system, etc.

2. Other information sources:

- other biotopes mapped by various institutions: available examinations of nature conservation institutions should be taken into account.
- Austrian catalogue of biotope types.

- Vegetation mapping: derivation of the nature proximity and rarity.
- Red Lists: the criteria of species' rarity and endangered status must be considered for the stating of biotopes.
- Various assessments of fauna and flora.
- Programmes of nature conservation.
- Aerial photographs.

In principle, two alternative attempts are to be considered for the FBM:

1. Selective forest biotope mapping (sFBM) means that only those biotopes which correspond to defined biotope types are cartographically included and written up. In Baden-Württemberg, the following biotope types can be distinguished (Volk, 1990, p. 153):

- rare and natural forest stands (floodplain forests, ravine forests, etc.).
- Forest stands with rare and particularly valuable plants.
- Forest stands with rare and particularly valuable animals.
- Dry biotopes in the forest.
- Successive forest areas (pre-forest stands).
- Damp biotopes (fen wood, ponds, damp meadows, etc.).
- Edges of the forest.
- Remains of historical management (former brushwood management).
- Rare natural rock formations, caves, ravines, etc.
- Structured forest stands (forests of variable ages, diversity of habitats, etc.).

The biotopes are documented in the forest management plan. The biotope type and indications for the management should be described. Biotope areas of approximately 0.5 ha are defined as separate forest plots if the structure and the planned measures of the surrounding forest stands are clearly different (Arbeitskreis Forstliche Landespflege, 1996, p. 27).

2. Full-coverage forest biotope mapping (fFBM) should just be mentioned. Unlike sFMB, the whole forest area is examined, in principle simultaneously with the tasks of the FMP. In this way, prerequisites are defined from considering the claim of nature conservation institutions for an integral concept of nature conservation of the total forest area. Thereby the fFBM offers a proof and control system for the ecological behaviour of forest management. As an additional source of information, the fFBM can also be used as an essential forest policy instrument (Volk, 1993, p. 803). Using objective and checkable appraisals, the forest owners can give information on the conservation value of their forests at any time (Hanstein, 1992, p. 531).

6.2.6 RES-related potential analysis: practical examples

Parameters of the concept of a mountain-bike route

As shown in the previous section several recreational parameters could be used with regard to a specific project. As an example, the relevant planning indicators for the opening up of a mountain-bike route are listed in Table 6.25 and Fig. 6.21, which could be derived from the analysis of the documented case-studies DE23, AU01 and AU17. In coordination with the defined operational and strategic targets, the forest owner has to weigh up and judge the single indicators.

In addition to the estimation of suitability for the functional fulfilment, it is also essential to explicitly carry out an assessment of use conflicts and possible sources of danger in the context of the examination. Finally, regional and national aspects must be taken into account (see Fig. 6.21).

Relevant cooperation partners could include several sponsorships for signposting and events (benches, etc.) and costs of preparation and maintenance could be shared with tourism organizations, municipalities and sports clubs. In case-study AU17, the regional forest authority achieved a solution (Tyrolese mountain-bike concept) between public and landowner interests. Routes of FLO which are of best quality are financed by the Tyrolese government and Tyrolese tourism association. Furthermore, well-known

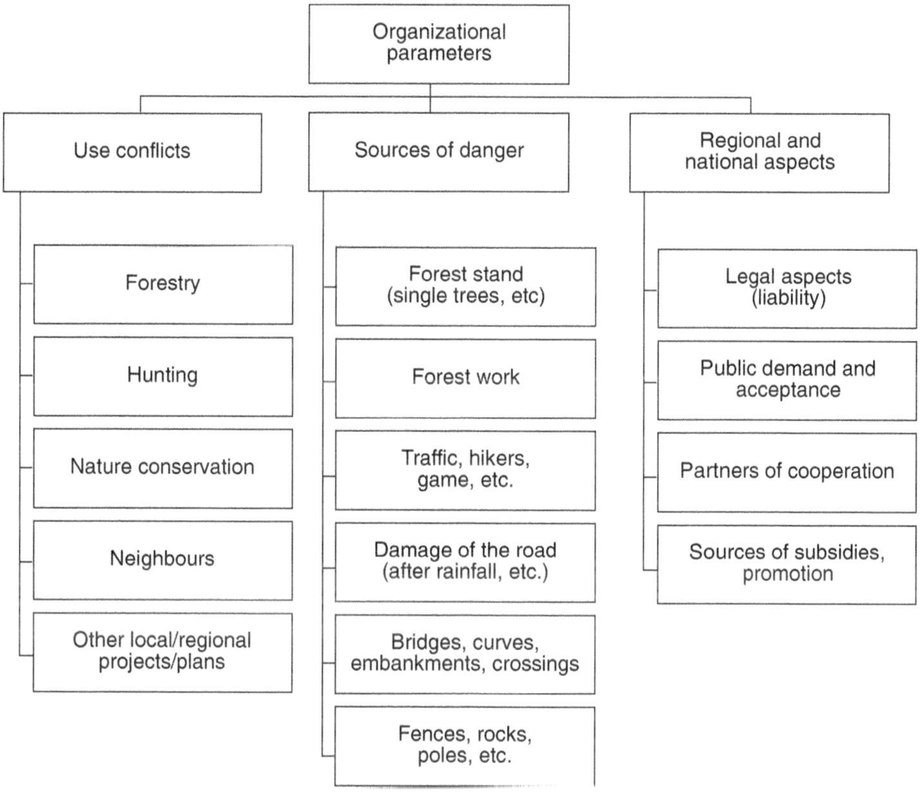

Fig. 6.21. Organizational parameters of a mountain-bike route.

Table 6.25. Technical parameters of a mountain-bike route.

Road parameters	Parameters of environment
Height profile, % share of the grade classes	Terrain (structure variety, sources of danger etc.)
Levels of difficulty	Starting-point (accessibility, parking possibilities, etc.)
Length of the course	Infrastructure (restaurants, etc.)
Surface composition	Use priorities
Drainage system	Landscape variety
Embankment condition	Vantage points
Quality of bridges, fords	Catchment area
Clarity of intersections, curves	Other attractions (ruins, rocks, peaks, alpine huts, etc.)
Signposting	
Site of the routes (circuit, dead end, accessibility)	

sportsmen or institutions/persons, which/who are able to test/plan or to carry out a commission could be relevant cooperation partners.

Acquisition of water resources

Due to the Austrian Water Law, the property of water resources is strictly connected to the ownership of land (Rossmann, 1993: §§ 3,5,9 Wasserrechtsgesetz). Thus the groundwater is defined as a private good. Each landowner with water resources is authorized to take water for his/her own requirements (household), and farmers can also use this water for animals without any authorization procedure. But the property rights are restricted, e.g. irrigation for plantation purposes is dependent on legal proceedings. The borders of property rights are not clearly defined. However, water is partly a public good for common use (§§ 2,4,8 WRG). This double position causes a big conflict potential in the domestic discussion (Pernthaler, 1998, p. 3).

In addition, the basic conditions of water management are changing, which can be shown by several developments:

- on the one hand, the restriction of national sovereignty caused by the European integration (Sager, 1997, p. 509).
- On the other hand, the rules of the market economy and the globalization of the economy.

The establishment of European water networks is already under consideration. Since Austria joined the EU in 1995, the EC Treaty has come into force. Thereby, the EU has the right to determine measures for the management of water resources. Such measurements require a concordant decision in law (Art. 130s (2) EC Treaty).

Owners of water resources must be prepared for possible legal changes. For this, it is necessary that the owner is informed on water potentials (quality, quantity, use restrictions, etc.). It is conspicuous that almost one-quarter of all documented Austrian cases deal with water resources (AU03, hydroelectric power, AU16 and 19, drinking-water, AU06 and 20, fishing and pond management; in the Netherlands (NL01) and in Germany (DE14), drinking-water case-studies are also documented). Furthermore, strategic precautions should be taken into account to save the properties' own water resources (opening up a reservoir, securing the right of several water uses, civil law arrangements, etc.).

For instance, case-study AU03 could be mentioned at this point. In connection with a hydroelectric power project, the FLO was able to secure additional water resources for future drinking-water purposes. Otherwise, there would be no legal opportunity to secure the drinking-water resources because of no actual public demand.[4]

An efficient PA presupposes a clearly defined goal. However, to conciliate a broad perspective of the varied fields examined, general questions should be answered: 'What distribution rights and what water resources can the management dispose of? What products could be derived from this?'

An internal (operational analysis) and external (environmental analysis) potential evaluation serves as a basis for organizational behaviour. There has to be an individual decision as to which properties should be analysed at which time. If the enterprise pursues an offensive strategy, the main attention will be on internal potential evaluation first. The question is: 'What products could be offered?' The environment analysis has essentially to investigate the legal situation. The strategic decision as to what products should be produced and marketed must be the next step. Two alternatives are possible:

1. Due to established demand (market and competition analysis).
2. Product definition without present demand (considerable additional expenses to marketing costs).

With the defensive strategy, the enterprise tries to react to customer requests. The question is: 'What products could be offered by the enterprise due to the existing demand?' The environment analysis gives decisive impulses for any diversification activities.

Internal potential analysis. The properties to be examined can be defined clearly and openly, as shown in Fig. 6.22. In accordance with the enterprise philosophy, the available potentials of water use should be defined. As a good example, the strategic orientation of the ÖBf AG in regard to the water business field should be mentioned here (ÖBf AG, 1997, p. 51):

- opening up of available water potentials (at present 3300 springs have been documented).
- Securing the right of utilization.
- Active marketing of various water products.

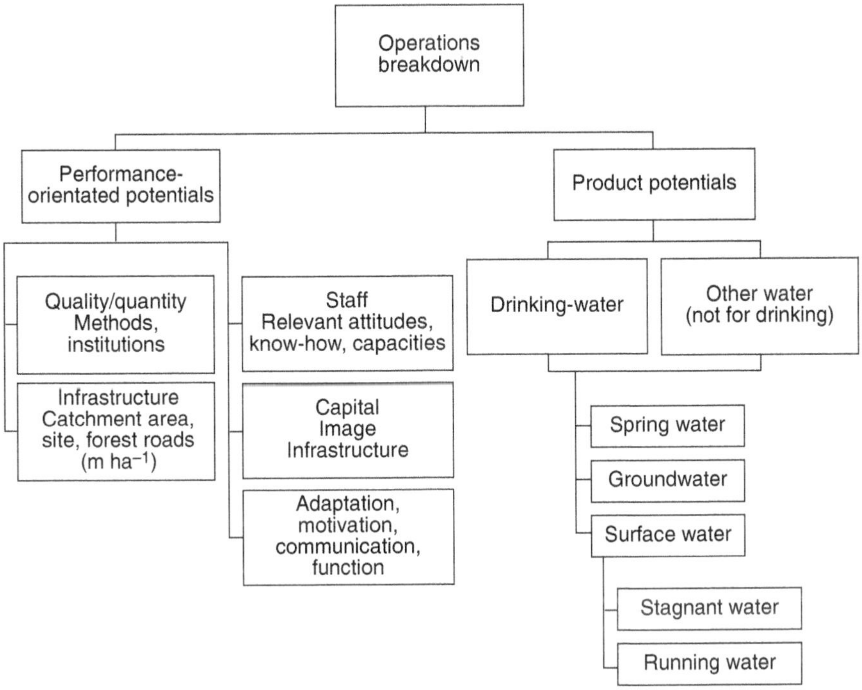

Fig. 6.22. Acquisition of internal water resources.

Either the product is already defined by the management (e.g. pond management) or the definition of the product is carried out after the acquisition of all internally available water resources. Simultaneously, the qualitative, quantitative and infrastructural features must be taken into account. For further planning steps, it is helpful to show the water locations on forest maps. By the visualization of spatial relations, the resources can be seen more distinctly.

The acquisition of water resources has to include research on quantitative variation throughout the year. An initial summary of the available resources arises during mapping, which should already be connected with an evaluation of productiveness. For the final calculation, at least an annual monitoring data set with short intervals (weekly) should be available. The project extent is limited by the minimum water quantity, so that the behaviour of the water flow during dry periods must be known.

Detailed instructions for the assessment of water quantities are obvious in rule sheet no. 205 of the Österreichischer Wasserwirtschaftsverband (ÖWWV; Austrian association for water management) (ÖWWV, 1990, p. 53). As a rule of thumb and with special prerequisites (quality), even a water flow of 0.1 $l\ s^{-1}$ can be of economic relevance (Plaimer, 1997a, p. 14).

The extent of the qualitative examination will depend on the intended application. If drinking-water use is planned (qualitative extreme), a very strict check is necessary. In the annual cycle, the consequences of weather extremes have to be observed (thaw, strong precipitation, dry periods). The assessment has to consider the whole catchment area to recognize potential problem sources in the planning phase. The chemical, physical and bacteriological examinations may only be carried out by state or state-recognized institutions.

The protection of the drinking-water provision within the catchment area must be ensured (see DE14: definition of protection zones). Danger potentials which could affect the quality and quantity of water must be identified. Corresponding precautions should be considered which prevent pollution of the groundwater.

The second area of performance-orientated internal potentials could be defined by staff, operational organization and, furthermore, the business politics and culture. The basis of a service company is readiness for duty and the identification of the employees with the enterprise.

EXTERNAL POTENTIAL ANALYSIS. Depending on operational targets and strategic procedures, the timing of internal and external analysis must be defined individually. However, no examination area can be investigated independently. On one hand, solely the knowledge about the global market situation without having a detailed environmental analysis could be the reason for a PA. On the other hand, it could be necessary to investigate the market chances of a defined product before the management is willing to invest in an internal PA. The disadvantage of the last variant is that, at this time, no concrete ideas about the internal quality, quantity and possible diversification of the defined product exist. However, a quick check, as discussed before, could deliver a first project overview and a decision basis.

Water related environmental analysis referring to economic aspects can roughly be structured in three areas (Fig. 6.23):

1. Due to costs, a discussion of legal conditions must refer to a specific project. In the case of water which is available to each district authority, the relevant and actual legal aspects are obvious. Maps and the overview of all water cooperative societies could be taken from the register of water. Water economic plans define general objectives for future water economic developments. These basic conditions apply for a definite region without any immediate legal consequences. When judging concrete projects, the legal authority must take these basic conditions into account, together with the public interest.

An essential examination parameter concerns the water requirement check (§ 13 WRG). The extent to which the available water resources may be used depends, on the one hand, on the water quantity (ecological function must be preserved) and, on the other, on the kind and extent of the planned use. Also prospective or possible demands of water consumption are taken into account by the authority.

2. In the field of forestry, the market

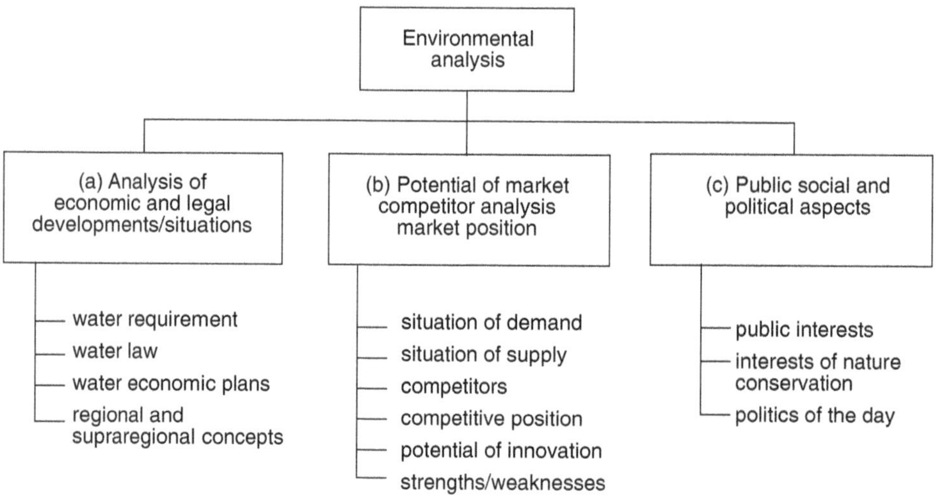

Fig. 6.23. Environmental analysis related to water resources.

analysis will be confined to observations, interviews and expert interviews. With sound business planning, these information sources will suffice to be able to realize opportunities, developments or trends. A concrete note on the procedure of a forest enterprise, according to Pototschnig (1994, p. 42), could be represented by a case-study. However, with increasing project extent, interdisciplinary cooperation with competent partners becomes a decisive success factor.

A detailed market analysis will contain the following aspects:

- demand: definition of products and services, stability, market requirements, position in the market life cycle, negotiation strength of customers, etc.
- Supply: which products and services are offered, field analysis (exploitation, capacities, etc.: see ÖBf AG, 1997, p. 50), labour costs, cost of materials, disturbance susceptibility, distribution channels, negotiation strength of the suppliers, etc.
- Market share of the enterprise, alternative structures of products and services.
- Competition analysis and competition situation: identification of the main competitors, number, size, financial power, analysis of the competition differences and competition instruments (price, quality, design, etc.), strengths/weaknesses, established clientele, reaction capacity of the competitors, leadership systems of established competitors, threat by additional competitors, business strategy of competitors, etc.

3. In the analysis, public interests and the regional/national political situation should also be taken into account. Water is considered by the vast majority of citizens as a public good, but water in Austria is explicitly defined as the property of the landowner (Pernthaler, 1998, p. 3). Due to importance for life, management of this private economic good is in the interest of all. Additionally, on the one hand, forestry is confronted with an increasing conservation claim and, on the other, forestry is regarded with a certain suspicion in the field of nature conservation (Wildburger, 1997, p. 188). Therefore, the forest owner must combine economic goals with ecological principles. Water-use projects which ignore ecological principles have no chance of realization, either in the legal or in the public area.

Acquisition scheme of water potentials. If the management exclusively concentrates on the internal potential elevation of water resources (this would concern the performance-orientated potential quality, quantity and infrastructure), the analysis can be further differentiated at various examination levels.

The recommended procedure for the elevation of water resources (Plaimer, 1996, p. 81) is as follows:

- Step 1: environmental analysis – influences further procedure (e.g. potential customers yes/no).
- Step 2: recording and mapping of all water resources (groundwater, spring water, flowing/still waters, special locations: waterfalls, moist biotopes, etc.); of internal success factors.
- Step 3: qualitative and quantitative investigation of financial resources and, simultaneously, a detailed analysis of the associated field.

The immediate relevant work steps are represented in an acquisition scheme (Fig. 6.24). Starting from a general basis, all relevant planning features of possible products or services are taken into account. The experiences of the OBf AG with regard to the recording of water potentials could be considered in this scheme. The goal of the ÖBf AG is the construction of a computer-supported spring-water register, which will contain the most essential features of about 4000 springs.

The acquisition of the initial position, the basic evaluation, analysis of the technical parameters and the assessment of all influential factors represent the four essential evaluation levels (see Fig. 6.24, right-hand column).

The presentation of the initial position

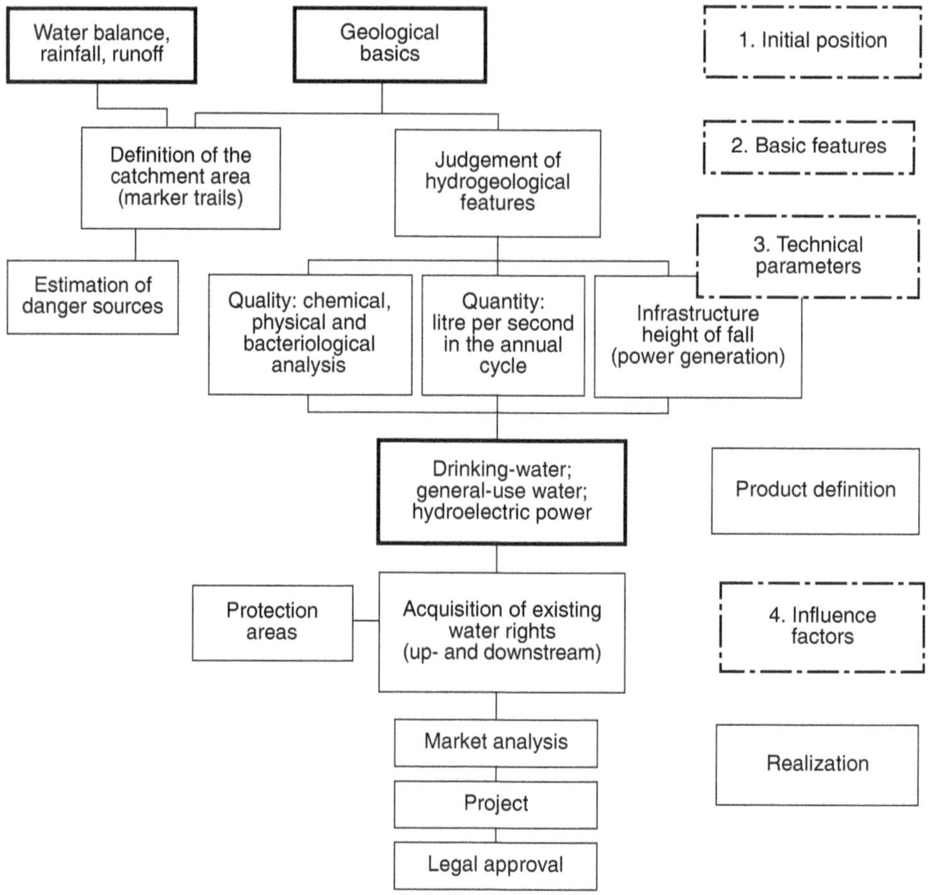

Fig. 6.24. Acquisition of water resources (hierarchical structure) (modified from Plaimer and Putzgruber, 1998, p. 12).

could be done by studying available maps and data materials. For the examination of the catchment area and the hydrogeological qualities, specialists must be consulted (official and state-approved authorities or institutions). Following exact evaluation guidelines, the technical parameters (quantity, water sampling for the qualitative analysis) can be carried out.

At the level of product definition, all possible products should be subdivided into potential use categories (see Fig. 6.22). On the basis of a cost–benefit analysis, the management finally has to decide which product will be produced.

After the decision on the products, the legal aspects and the consequences (e.g. water protection area) must be evaluated.

For the execution of an efficient PA, the knowledge of relevant planning parameters is essential. For the exploitation of water resources for drinking-water purposes, Plaimer (1996, p. 81) created a checklist, the relevant items of which are illustrated in Figs 6.23 and 6.24 (internal and external approach). Further clues can be found in the series of guidelines of the ÖWWV.

Aspects of establishing forest apartments

INTERNAL AND EXTERNAL PARAMETERS. Based on Mischkulnig (1996) and the RES case-studies NL18, DE02, DE27 and AU11, relevant parameters for renting a forest apartment/building should be discussed. A rough overview of internal and external preconditions could be helpful in finding a niche market for 'forest-house holidays'. Starting from the willingness and acceptance by the FLO/staff in principle, the following parameters could be taken into account:

1. Internal parameters:

- quantity of accommodation (number of single/double rooms, apartments, forest houses, etc. which are involved).
- Quality of accommodation (building material, façade, cleanness, heating, furniture, kitchen, bathroom/toilet, warm water, additional services).
- Infrastructure (forest roads, natural and sports facilities (mountain-bike/hiking routes, water surfaces, game park, etc.)).
- Location (altitude, tranquillity, panoramic view, surrounding area (alpine pastures, old forest stands, rocks, etc.), accessibility, neighbourhood).
- Personnel resources (maintenance, cleaning, additional services (forest guided tours, hunting, etc.)).
- Temporal use (from ... till ..., all the year).
- Organizational aspects (booking system, information service, marketing).
- Conflict potentials (disturbance of hunting activities, restrictions of forest management tasks (cutting, transport, etc.)).

2. External parameters

- Legal aspects (law of rent, income from property (tax burden), liability, waste water (environmentalism), quality of drinking-water, municipal rates (tourism), insurance aspects).
- Market situation (see below (determining factors), competition).
- Regional preconditions (image of the region/municipality; recreational, natural, sports, shopping and cultural facilities, marketing activities).
- Climate (temperature, rainy/sunny days).
- Infrastructure (see regional preconditions; traffic lines (highway, train, airport)).
- Cooperation (travel agencies, farmhouse holidays, forest enterprises, representatives of forestry and commerce, restaurants, sports clubs).

DETERMINING FACTORS. The determining factors can be subdivided into general aspects, factors related to forest houses/apartments and those related to the regional/global market situation. The following factors could be underlined:

1. General:

- motivated personnel, commitment.
- The enterprise's own forest land base in the surrounding area.
- Existing and varied infrastructure and additional services (a good example could be given by the Malteser Ritter Orden forest enterprise, AU 11: www.hebalm.at).
- Philosophy of the forest enterprise (certified enterprise model, multifunctionality, risk-taking, etc.).
- Satisfaction of personal needs and personal contact with guests.
- Concept in accordance with hunting and forestry.

2. Related to the accommodation:

- loneliness, quietness.
- Good roads – easy access also in winter.
- Long-term lease, availability all the year.
- Well-tended appearance and general cleanliness.
- Heating in all rooms.
- Bathroom and toilet *en suite*.
- Adequate furniture (living-room, bedroom, kitchen).

3. Related to the market situation: It could be stated that forest accommodation completes the tourism options in Austria. A niche market still exists because of:

- self-catering accommodation is mainly concentrated on mountain pastures;
- forest houses/apartments are not comparable with other accommodation, especially with respect to location in wooded areas (Mischkulnig, 1996, p. 85).

FARMHOUSE HOLIDAYS. In considering different categories of accommodation, a practical example and one that is very close to forestry could be given by the successful initiative of farmhouse holidays in Austria (see: www.lisa.at/urlaub).

A specialist jury put each participating farmhouse in Austria into one of three categories, using 150 different criteria (for farmhouses specializing in cyclist guests, more than 25 criteria are listed below). The accommodation was then awarded two, three or four flowers accordingly. The high standards expected of each farm are represented by the fact that, in three out of four cases, a toilet and bath/shower are available *en suite*. Among other things, the categorization is also based on the surroundings, the style of building and environmental criteria (this system has also been established by forest enterprises, e.g. www.weissenegg.vivawald.at).

Comfortable rooms with adequate facilities. Shower or bath/toilet *en suite* or on the corridor. Cosy accommodation in natural surroundings.

Very good facilities in the room and on the farm in general. Toilet and bath/shower usually *en suite*. A farm environment you can feel really comfortable in.

Excellent facilities all round. *En suite* toilet and bath/shower. A farm holiday for the most discriminating of guests.

Costs: bed-and-breakfast accommodation in 'three-flower' farmhouses costs from Euro 90.- per week, with 'four-flower' accommodation for as little as € 112.- per week.

FARMHOUSE HOLIDAYS FOR CYCLISTS (PARAMETERS) (see Bundesverband Urlaub am Bauernhof, 1996, pp. 26–27)

1. Postulated criteria.

- General aspects: (i) the local situation of the farmhouse must be described; (ii) inspection by social insurance institution (risk of accident); (iii) member of the hire community; (iv) categorization (flowers); (v) obligatory participation in exchange of experience (once a year); (vi) features of specialization must be defined.
- Facilities: (i) bikes for lending and bikes for children are available (three bikes minimum); (ii) seats for children available; (iii) bike shed; (iv) maintenance room and adequate tools; (v) laundry room for drying clothes; (vi) bicycle stands; (vii) cycling helmets.
- Additional services: (i) guided tours (at least once a week); (ii) tips for cycling tours (insider tips); (iii) adequate breakfast (fruit, home-made juice, etc.); (iv) lunch bag for day-trips; (v) information on marked mountain-bike routes; (vi) liability insurance; (vii) additional information concerning bicycles and cycling; (viii) one overnight stay is assumed.

2. Possible criteria (suggestions)

- Bicycle–train taxi service.
- Bonus for guests arriving with bicycles (free drink, etc.).
- Fitness room.
- Awards for cycling tours (king of the tour, etc.).
- Taxi service in cooperation with other enterprises.
- Alternative offers in case of rainy weather (alternative programme).

STATUS QUO. At present, forest accommodation is mainly rented as hunting lodges or as weekend houses by long-term contract (20 years). Adequate buildings are also used as residential houses for approximately 10 years, as well as official residences for forest staff. In several cases, students can use these buildings free of charge or forest landowners provide the accommodation exclusively for their friends. But it is also common for forest enterprises to give up this market by selling or demolishing unprofitable or old buildings.

According to the statements of about 70 forest managers, the main problems in dealing with forest apartments/houses was from legal regulations (leasing law, tax regulation, liability, etc.). A second focus is on the tasks of hunting and the restriction of economic 'elbow-room'. Problems with access, rubbish and restrictions in forest management are negligible (Mischkulnig, 1996, p. 91).

However, the opportunities to establish an additional economic source in an attractive market segment are quite good. Starting from the internal resources (quantity and quality of the buildings), the forest management has to choose between long-term and short-term strategies. If the forest enterprise has at its disposal good accommodation (two, three or four flowers) and personnel resources, the best advantage will come from short-term leasing (up to 3 weeks). In practice, about 90% of all investigated forest enterprises dealing with apartments/houses use long-term contracts, because of the bad order and condition of the buildings (which the leaseholder is obliged to maintain) and a lack of forest personnel at their disposal (Mischkulnig, 1996, p. 92).

6.3 Land-use Potential Evaluation and Multiple Land-use Planning

6.3.1 General principles

In many European countries, there exists a long tradition of spatially referenced management planning for larger forest holdings, both publicly owned and private, which has also influenced those forestry traditions which in one way or another were inspired by European traditions. Forest managers are thus used to working with management maps, plans and regular management planning, and the production of related map materials has become an integral part of forestry curricula (see Table 6.20).

In the context of private land management, planning is seen as a tool for internal control of management measures. By linking objective physical information (e.g. site quality, tree-stand information) and the subjective goals of the manager/landowner (e.g. maximization of financial profit), as well as external influences (e.g. neighbouring communities), specific measures which have to be taken in a given time frame are identified and a potential state of the planning unit at the end of the planning period can be projected. By comparing the actual development with the projected outcome, it is then possible to evaluate the success of implemented management actions.

The identification of land-use conflicts can also be of interest if only one landowner is involved. Even in the traditional concept of multipurpose management, several ways of using land have been integrated into the running of a forest enterprise. Hunting is probably the most prominent example of this, at least in forest estates in Middle and Central Europe (Table 6.26, especially the first column: the conflict between hunting and the RES product in question, which was addressed by all of the 98 RES case-studies).

Damage from deer-browsing is a problem in many European countries (Ottitsch and Weiß, 1998), even in those where hunting rights are linked to landownership. Conflicts between these two activities arise, for example, from the simultaneous allocation of damage-prone succession stages (e.g. regeneration areas, young stands) and game-attracting hunting infrastructure (e.g. winter feedings) (Reimoser, 1993). By outlining the potentials for different land-use activities, it is possible to realize possible negative or positive interactions in advance and to integrate this knowledge into management decisions. Besides hunting, conflicts with the RES business could be caused by several forest functions, e.g. timber production, recreation, ecology and protection (Table 6.27). Relevant conflicts could be documented in about one-quarter of all RES cases (e.g. DE05, NL05, AU13, etc.).

For private land-use management, one value of a well-established land-use planning system is to have access to decision-relevant information down to the level of an individual management unit. This may be of interest for being able to react to changing demands of the market – for example, if specific timber dimensions are required in larger amounts than usual.

Table 6.26. Interdependencies of the RES business. (Question 6.1.4: Please indicate the degree of competition/conflict between the forest uses listed below and your RES product within the surrounding area. (Total basis: 98 RES case-studies.))

Competition in %	Hunting	Timber	Recreation	Ecology	Protection
No	52.0	64.3	78.6	64.3	81.6
Low	21.4	19.4	5.1	18.4	12.2
Moderate	14.3	8.2	9.2	9.2	3.1
High	8.2	6.1	5.1	6.1	1.0
Missing	4.1	2.0	2.0	2.0	2.0

However, when pointing to such potential advantages, one argument frequently encountered in discussions with land managers or owners, who are naturally reluctant to invest in the establishment of such a system, is that the responsible forest personnel would in any case have at their disposal a good knowledge of the area so that such tools would not really be necessary. In about one-third of all documented RES cases, additional information stemming from inventories of RES-related potentials is available. However, about 20% of all respondents would like more information.

As shown above, there are conflicting goals in planning. For the individual landowner, this means having to decide which combination of goals will be preferred in the final land-use plan. If more than one stakeholder is involved in the process, strategies of conflict management will have to be considered.

The result of this is a land-use concept which is applied to the situation. The implementation of this concept has to be evaluated within a given time limit (which is determined in accordance with the chosen temporal element of sustainability planning), and possibly adaptations of the concept will have to be made using the same basic sequence of planning steps:

1. Land-use PA.
2. Land-use conflict analysis.
3. Conflict solution strategies.
4. Land-use concept.
5. Land-use concept implementation.
6. Land-use concept evaluation.

6.3.2 One landscape, many images – land-use potential evaluations

Elements of land-use decisions

Though seemingly a process only of interest to professional land-users, such as foresters, farmers or real-estate developers, land-use potential evaluation is in fact an everyday process, which is performed by every human being more or less continuously in choosing his or her position in the environment. The decisions on which way we choose to go to work, or which place we seek out to rest at a public picnic area or beach, the table we look for in a restaurant, are ultimately based on more or less

Table 6.27. Significance of timber production, hunting, recreation, ecology and protection. (Question 6.1.3: Please indicate the significance of the surrounding area regarding the following forest functions. (Total basis: 98 RES case-studies.))

Significance in %	Timber production	Hunting	Recreation	Ecology	Protection
Low	28.6	29.6	8.2	7.1	24.5
Medium	24.5	36.7	18.4	42.9	48.0
High	44.9	24.5	71.4	48.0	25.5
Missing	2.0	9.2	2.0	2.0	2.0

consciously made evaluations of the respective environments that we encounter. Evaluations of the very same environment will differ not only between different individuals but also for the same individual, depending on the planned activity and also on varying physical or emotional conditions. While in the morning we might choose the shortest way to go from home to work, in the evening we might choose a longer, but perhaps more scenic or otherwise entertaining, route.

It is therefore appropriate to state that there are two types of factors that influence our land-use planning decisions. On the one hand, physically existing factors, which can be either 'natural' or man-made, define the range of alternatives. Values and interests, on the other hand, are the factors determining human decisions for a certain alternative.

Physical factors may be assessed by means of objective measurement; therefore, statements about them are open to falsification and are thus suitable for scientific debate (AU12 – a special check list was used).

In contrast to this, values and interests cannot be proved either wrong or right. Any judgement about values is only determined by other, possibly contradicting, values, as is any statement about the legitimacy of interests.

The potential offered by a given landscape element to perform a certain type of land use may thus be calculated using the following formula (Eastman *et al.*, 1993):

$$P = S \times W_i \times X_{i1}$$

where:

P = land-use potential value
X = indicator value for a land-use-relevant physical factor
W = weight assigned to each land-use factor
S = selection (1, 0)

The values for X_n in this formula represent objective field measurements (e.g. slope, elevation, soil fertility), whereas values for W_n represent user-specific subjective assessments of each factor's relevance for the land use for which a potential is to be calculated. The other subjective element is the choice of which factors are to be used for the calculation.

Whether such an evaluation is performed explicitly or is manifested implicitly in expressed preferences, such as real-estate prices, depends on the circumstances of an evaluation. On the market, a potential buyer will try not to reveal his/her specific preferences for a certain object for fear of encouraging the seller to demand an even higher price.

Explicit evaluation methods are usually applied if the process of evaluation has to be revealed to other actors. A possible scenario for this could be a court hearing to define compensation rates. Another common application is the appraisal of land value for taxation purposes. The Austrian system of standard unit value (*Einheitswert*) for real-estate taxation uses a system of indicators, based on a variety of physical factors (e.g. geographical location, altitude, steepness of terrain, etc.), which ultimately influence the potential revenue from land management. These indicators are assigned values in the form of raise ratios (*Hebesätze*), depending on how greatly the conditions on the specific site are considered to differ from a regional average. While physical factors can be measured on site, the related value assignments are actually the result of negotiations between authorities and landowners' organizations.

Yet, even in the context of private land-use planning, the use of explicit evaluation methods will be useful for improving the rationality of a land-use decision or at least for revealing which preferences actually led to a certain decision.

Indicator qualities and units of assessment

As far as the choice of indicators is concerned, three main qualities are of interest. These are relevance, reliability and efficiency. In selecting appropriate indicators, it will thus be necessary to make trade-offs between the three qualities.

An indicator's relevance is judged according to the extent to which it supplies information on the criterion in question.

Frequently, the criterion of interest itself cannot be easily measured directly and it is therefore necessary to assess it via one or more easily obtainable indicators. One common example in forestry is the assessment of standing timber volume. Once the site-characteristic relationships between diameter and volume are known (using the appropriate growth curve and a combination of diameter and height measures of several trees), it is possible to assess the comparatively difficult factor 'volume' with the more easily obtainable factor 'diameter at breast height'. Certainly, the relevance of this indicator depends on the accuracy of the assumed relationship, i.e. the accuracy of the applied yield curves and tables. Different land-users will apply different criteria to assess the respective attractiveness of land for their purposes. This is even true for different forms of the same general type of land use.

An indicator's reliability describes the accuracy of its measurement. This depends on the quality of the instruments used for individual measurements, the measurement procedure (e.g. control measurements) and sampling design (in case of a statistical approach), but also on the quality of work of the persons taking the measurements.

The efficiency of an indicator describes the relationship between its potential information value and the efforts necessary to assess it. Frequently, the higher the relevance and the reliability of an indicator, the higher will also be the efforts needed to acquire it and the lower may be its efficiency.

The land-use potential can be calculated for each individual landscape element. The definition of 'landscape element' will certainly vary depending on the intended form of land use. In forestry, individual forest stands are a commonly used unit of planning decision, though their size will also vary depending on local forest practices.

For recreation activities, it becomes apparent that the preferences of the intended customer group are of interest. While some people do not mind or even seek out crowded conditions for recreation activities such as sunbathing or picnicking, others seek out more isolated locations. Once again, this depends on their respective attitudes towards recreation. The allocation of infrastructure is also of interest with regard to the degree to which people will want to combine different activities. In the allocation of potential facilities, the types of customers one wants to attract has to be taken into account. If a service is aimed at different groups, it has to be taken into account that their activities may not necessarily be free of conflict.

The landscape potential of a given landscape is determined by ecological factors (climate and terrain), which are limited for any spatial entity within a specific period of time, as has been illustrated before. On the other side of this concept there is demand by the socio-economic system for various forms of output. This demand is determined by the land-use interests of different groups and individuals. If the accumulation of these different land-use interests becomes larger than the limited landscape potential, land-use conflicts result.

6.3.3 Principles of land-use conflict solution

There are two basic strategies for the solution of land-use conflicts: segregation (see AU11: a mountain-bike area was defined; NL02, DE10) and adaptation (IT13: for certification of sustainability, the forest management plan was adopted for the whole forest area).

In the segregation approach, conflicting forms of land uses are either assigned to different land-use elements or scheduled for different time periods (AU11: mountain-biking versus hunting). The adaptation approach tries to allow for the simultaneous performance of different forms of land use by adapting the intensity of conflicting land uses to a compatible level. The decision for one of these two strategies will depend on the individual conflict situation. It may also be appropriate to apply a mixture of both strategies, termed 'priori-tary-determined adaptation'. In this approach, multiple land-use combinations are possible, but first priority land uses are assigned, to which other land-use forms then have to adapt.

Spatially exclusive allocation of land-use priorities may be performed either by ranking land uses in accordance with their importance for the planning individual or group or by applying a weighting system (see the section on potential analysis: the forest ecological evaluation system, which was applied in case-study AU12). RES case-study examples of spatial exclusive allocation include: NL01, DE14, AU16 and AU19, drinking-water production; AU04, military training area; AU14, car-testing course; IT11, camping area; IT12 and IT16, parking places.

In contrast to the above, land may exist which is of little or no value to a specific form of land use, but may be of high interest for other activities. A potential low-cost solution to allocating competing land uses may thus consist in identifying those areas, which are of little or no use to a primary objective. Obviously this approach was assumed in some countries for the allocation of conservation areas, which abound in areas of less economic value, such as high mountain areas in Middle Europe or sub-arctic regions in Scandinavia.

In the context of management planning at enterprise level, it is thus worth looking for areas which, for whatever reason, are of lower interest for the activity that has been defined as the primary objective and to consider alternative uses.

Another way of avoiding land-use conflicts is the temporal segregation of potentially conflicting activities. Examples of this are agreements on temporal restrictions for recreation activities in favour of wildlife management-related issues, such as breeding periods or hunting seasons.

With temporal segregation within each period, each activity is performed at full intensity (RES case-study examples: NL05 and NL07, horse riding; NL18 and DE02, forest apartments; AU07 and IT10, guided tours). The reader is referred to the literature for more information about GIS functionalities, data sources and data preparation, as well as organizational aspects (see, for example, Eastman *et al.*, 1993; Ottitsch, 1996). The following examples demonstrate the potential of GIS in multiple land-use planning.

6.3.4 Practical examples of land-use potential evaluation and multiple land-use planning

Achenkirch land-use analysis project – the land-use information system

THE PROJECT. Environmental studies in various disciplines have been performed in an area of about 7000 ha near Achenkirch, funded by the Förderungsfonds für Umweltstudien (FUST) – Tirol (the Tirol Fund for Environmental Studies). Due to this long-term activity, a vast amount of already existing environmental data could be used as the main source of input for the Achenkirch land-use potential analysis project. The goal of this project was the design of an integrated land-use concept for this area. Nature conservation, agriculture, protection against disasters (floods and avalanches), tourism, forestry and hunting were the land-use forms considered in this project.

The area is situated about 70 km east of Innsbruck, the capital of Tyrol. Elevations are between 900 and 2200 m above sea level. Winter and summer tourism are of great importance for the local economy. Forests are managed for timber production, by both public (Austrian Federal Forests) and private owners (small-area farmers). Like many other areas in the Tyrolese Alps, the region is also known for excellent hunting opportunities. Red deer, chamois and roe-deer are the most important game species. Livestock grazing is the main agricultural activity, grazing rights exist within the forests and large areas below the natural timberline have been cleared of forest cover in the past in order to gain grazing land. Nowadays, tourism is the most important source of income for the region, which is in the convenient situation of being able to offer both summer (hiking, water sports) and winter (downhill and cross-country skiing) recreation opportunities. The multitude of interests in relation to the use of natural resources is the main challenge for land-use managers and politicians alike in alpine regions (Bätzing, 1991; see also Table 6.27).

According to the RES case-studies, it could be stated that different forest functions of the surrounding area are potential prerequisites for RES activities. Besides timber production, hunting and protection, medium and high significance in respect to recreational and ecological functions prevails (see Table 6.28).

CONTENTS OF LAND-USE INFORMATION SYSTEM. Apart from the actual land-use concept, the established land information database is one result of the project, which is also of value for other scientific work in the area. The development of information systems for multiple land-use inventories is also strongly suggested by international land-use-related organizations (UNEP, 1993; Lund, 1998). The example presented here may serve as an encouragement in this direction.

Tables 6.28–6.30 give an account of this information, which is organized on three levels. They also show which type of information may be used in the establishment of a land-use information system even at the enterprise level. In this example, information is organized in several logical levels. The first level contains information that has been gathered in the field or has been obtained from existing sources without any additional processing. The second level

Table 6.28. Land-use information system, Achenkirch – first level.

Information layer	Description of contents
Data collected from field surveys	
Topography	Digital elevation model (Austrian Federal Office of Survey)
Vegetation	Potential natural vegetation (site mapping)
	Actual vegetation
Site information	Site mapping (soil and vegetation type)
	Site descriptions (Austrian Federal Forests)
Hydrology	Stream network
Fauna	Observation data for ungulates
Infrastructure	
General infrastructure	Road network
Hunting infrastructure	Hunting lodges and stands, feeding areas
Agricultural infrastructure	Agricultural buildings and roads
Forestry infrastructure	Forest roads
Tourist infrastructure	Lodges, restaurants, marked paths
Legal information	Ownership structure, land-use rights, protected areas
Land-use information	
(This describes already existing information on different forms of land use)	
Timber production	Austrian Federal Forests' management information
	Management information from biotope descriptions
Agriculture	Information about the use of grazing rights
	Assessment of grazing intensity by local informants
Tourism	Assessment of tourism activities by local informants
Hunting	Information about game kills
Conservation	Distribution of disturbance sources
Disaster protection	Distribution of forest cover, protection installations
Land-use interactions	Damage to forest vegetation by game and cattle

Table 6.29. Land-use Information system, Achenkirch – second level.

Information layer	Description of contents
Combinations of first-level data	
Timber production	Structure and texture
	Assessment of forest access
Agriculture	Suitability of land for grazing
Tourism	Assessment of access possibilities
Hunting	Habitat assessment
	Time-series analyses of observations and kills
	Influence of disturbances
Conservation	Biotope quality assessments (e.g. diversity)
Disaster protection	Avalanche and landslide risk assessments

Table 6.30. Land-use information system, Achenkirch – third level.

Information layer	Description of contents
Interpretations and valuations derived from 1st- and 2nd-level data	
Land use	
Timber production	Land-use potential evaluation
Agriculture	Land-use potential evaluation
Tourism	Land-use potential evaluation
Hunting	Land-use potential evaluation
Conservation	Land-use potential evaluation
Disaster protection	Land-use potential evaluation
Land-use conflicts	Conflict potentials from overlapping land-use potential evaluations
Land-use concept	Concept for the solution of actual land-use conflicts

contains combinations of different first-level information. The third level contains information that has been gathered by combinations and interpretations of first- and second-level data.

META-INFORMATION. Since one of the advantages of integrating land-use-relevant data into a GIS is the possibility to make them available for further use, it is also necessary to maintain information on the input data, since their qualitative and quantitative features are necessary for deciding upon their potential use for further applications.

In the demonstrated examples, such descriptions were developed specifically for the first-level category of data. This is sufficient since second- and third-level data have been derived from them. They are also most likely to be used in further projects, potentially by different personnel.

The following scheme was developed for data description:

- type of spatial representation: this describes the way in which spatial features are represented (point/lines/polygons/raster cells).
- Accuracy/resolution: this describes the geometric accuracy of vector data or the grid size of raster data.
- Type of data gathering: this describes how data have been acquired.
- Digitizing scale: for information digitized from analog sources, this describes the scale of the original map.
- Year of last actualization.
- Revision period: this describes the period within which information updates are performed.

- Covered area: this describes which parts of the project area are covered by the information.
- Ownership of data: copyright ownership of information.
- Thematic descriptions: explanation of information content.

This system is illustrated by the description for the information layer 'potential natural vegetation' below:

- type of spatial representation: polygon information.
- Accuracy/resolution: 0.1 ha.
- Type of data gathering: field surveys.
- Digitizing scale: 1 : 10,000.
- Year of last update: 1989.
- Revision period: – .
- Covered area: Achental-Welst and Pitz-Dollmannsbach hunting districts (two-thirds of the project area).
- Ownership of data: FUST-Achenkirch.
- Thematic descriptions: for mapping potential natural forest associations, 22 different site types have been defined and charted in the study area. Soil type, geology, morphology and hydrology were recorded for use for site characterization. In addition, there is information on ground vegetation and current and potential tree species composition and advice for silvicultural treatment. Information on site indicator values (Ellenberg) and growth-yield data complete the information.

Land-use potential evaluation – Achenkirch

Land-use potential analyses were performed for the land uses analysed in the project. The potentials were evaluated using multiple criteria and weights (see Section 6.3.2). The individual evaluations have been published in Ottitsch (1995).

A closer description of the evaluation algorithm is given for recreation to illustrate the process, while only short summaries are presented for other land uses. Maps 1 and 2 (Figs 6.25 and 6.26) show the results for the potential evaluations for nature conservation and recreation, respectively.

Diversity, rarity and closeness to natural condition were the factors used to evaluate nature conservation potential in the project area. Diversity was assessed using habitat diversity (number of different habitats within a reference area) as an indicator. Rarity was determined by the presence of biotopes of special value (riparian zones, bogs, sites with rare vegetation species). The distance from human infrastructure (buildings, roads etc.) was used to assess closeness to natural conditions (see Section 6.2 on potential analysis and AU12: evaluation of a forest nature reserve).

The potential for recreation is determined by two groups of factors. On the one hand, there are site-specific natural factors, such as terrain and climate, which determine the overall habitat type of the area. On the other hand, there are human-made factors, such as tourism infrastructure, but also infrastructure originally erected for other land uses (e.g. forest roads, agricultural buildings; see also Section 6.2 on potential analysis). Ecosystem diversity, exposure, intensity and duration of sunshine and distance from riparian areas were used to determine the natural components of recreation potential. The infrastructure was used as an indicator to evaluate the human component of the distance from recreation. Infrastructural elements were classified into point features (lodges, restaurants), linear features (hikes, trails) and area features (ski-slopes).

The evaluation process for site-specific potential is described in detail here to illustrate the technique. The calculation was done using habitat diversity, solar radiation input and closeness to streams or lake shores.

FACTORS TAKEN INTO CONSIDERATION. For the assessment of habitat diversity, an existing biotope description of the area was used, which gives information on the number and spatial extent of different habitats in 250 × 250 m raster cells. In order to be considered in this evaluation, though, a minimum share of 10% had to be achieved by a habitat type to be counted as being of relevance in the respective raster cell.

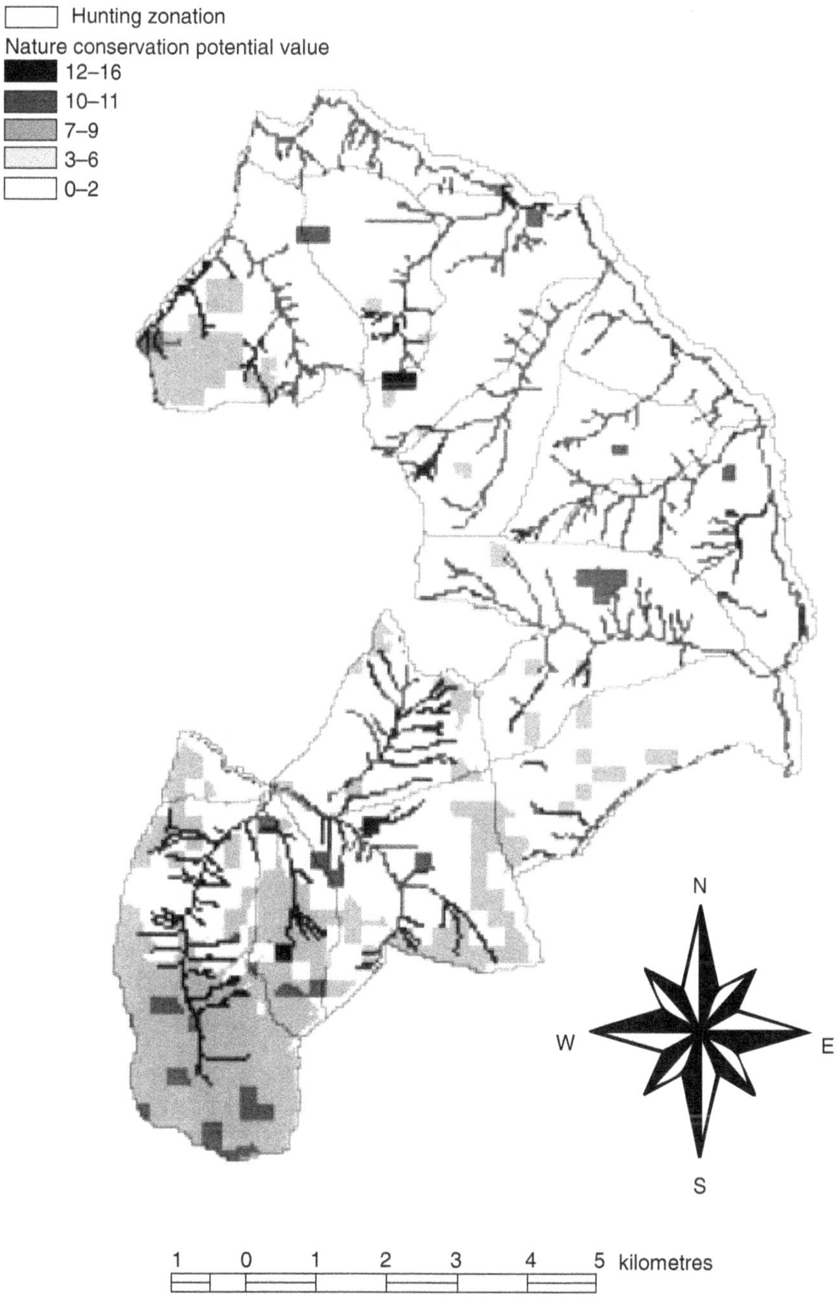

Fig. 6.25. Map 1: nature conservation potential.

The different habitat-type areas were:

- rock or gravel areas.
- Grassland.
- Alpine dwarf pine (*Pinus mugo*) and heathers.
- Shrubs.
- Forest areas.

Built-over land was excluded from this evaluation.

Relative solar radiation input was

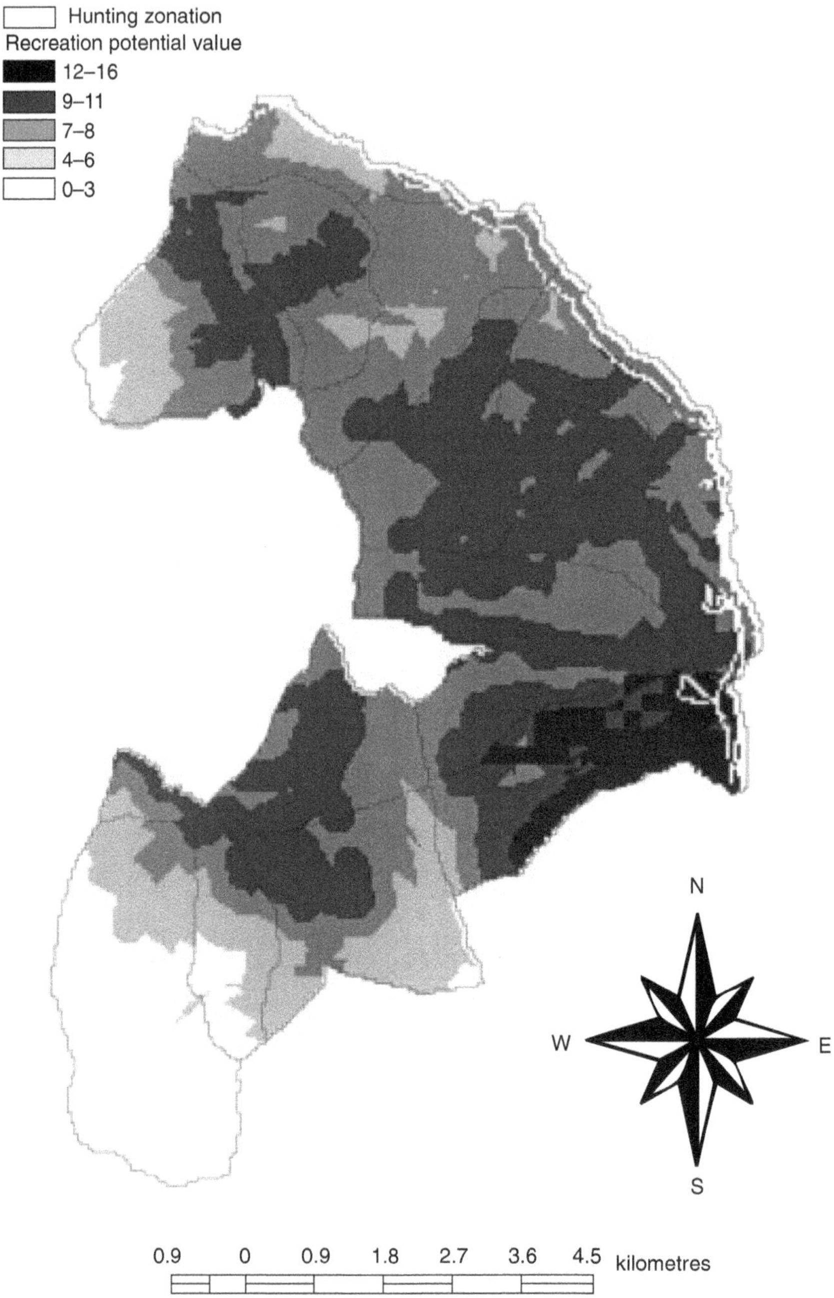

Fig. 6.26. Map 2: recreation potential.

calculated based on the position of the sun over the horizon at 9.00, 12.00 and 15.00 hours at midsummer, midwinter and both spring and autumn equinoxes, as well as the slope, aspect and potential shading effects of surrounding features. The digital elevation model served as base information input in this process.

The distance to the nearest body of water was calculated using the digitized surface-water network as input into the GIS's distance calculation routines. Slope data were used to convert horizontal distance to real terrain distance.

THE PROBLEM OF UNITS. The example shows that a variety of factors with no comparable units are considered. In order to integrate them into the potential evaluation, it is necessary to convert these different values to a single indicator unit. This can be done quite simply by projecting the actual values to one range of indicator values. For example, in this project, the number of habitats in a given grid cell could take on values from 1 to 5. For this project, it was decided to use an indicator range of 1–15. Correspondingly, if the reading for habitat number was 5, the indicator value would be 15. If the reading was 3 (the average value of the original scale), the indicator value would be 7.5 (the average value of the indicator scale).

INTEGRATION OF FACTOR VALUES AND WEIGHTS. For the overall value for site-specific potential, weights were assigned to the individual indicator values. For every evaluation in the project, a total of 10 weight points was available. Thus, the site-specific potential for recreation was calculated, as follows:

$$PRS = 5DS + 3DR + 2DW$$

where:

PRS = site-specific recreation potential value
DS = habitat diversity
DR = solar radiation input
DW = influence of water bodies

It has to be stated here that this example illustrates only one of a myriad of possible combinations of factors that could be taken into account to assess the recreation value of different landscape elements. Once the input data are entered into the GIS, it is up to the users to make their selection of factors and weights. Thus, it is possible to produce different images of the same landscape which illustrate how different groups of potential users might see it.

PROTECTION AGAINST NATURAL CATASTROPHES. The risk of erosion and avalanches was evaluated using slope, exposure and elevation as main indicator elements.

LIVESTOCK GRAZING. Vegetation productivity, which was determined using site survey data and the quality of infrastructure (distance from agricultural roads and buildings), was used to evaluate the potential for cattle grazing. This evaluation could only be performed in areas where site survey data were available.

TIMBER PRODUCTION. Site survey data were also used to determine the natural potential for timber production. In addition, the distance from forest roads was used as an indicator value in this evaluation. This evaluation could only be performed in areas where site survey data were available.

HUNTING. Habitat quality and the quantity of game stands were used to evaluate the potential for hunting. Habitat quality was assessed by combining slope, exposure, ecosystem diversity, camouflage potential and disturbance factors. Records on game sightings, referenced to a system of raster squares, were used to evaluate the quantity of game.

Land-use allocation concept – Achenkirch

The goal of the project Achenkirch land-use concept was not only to perform land-use potential analyses, which has already been described above, but, even more importantly, to assess potential and actual land-use conflicts in the area and to suggest potential concepts for conflict solutions.

ASSESSMENT OF POTENTIAL LAND-USE CONFLICTS. An overlay of different land-use potential evaluations was used to assess potential land-use conflict areas. It is assumed that areas of high interest for different land uses (darker areas in the maps) are also prone to become the object of land-use conflicts.

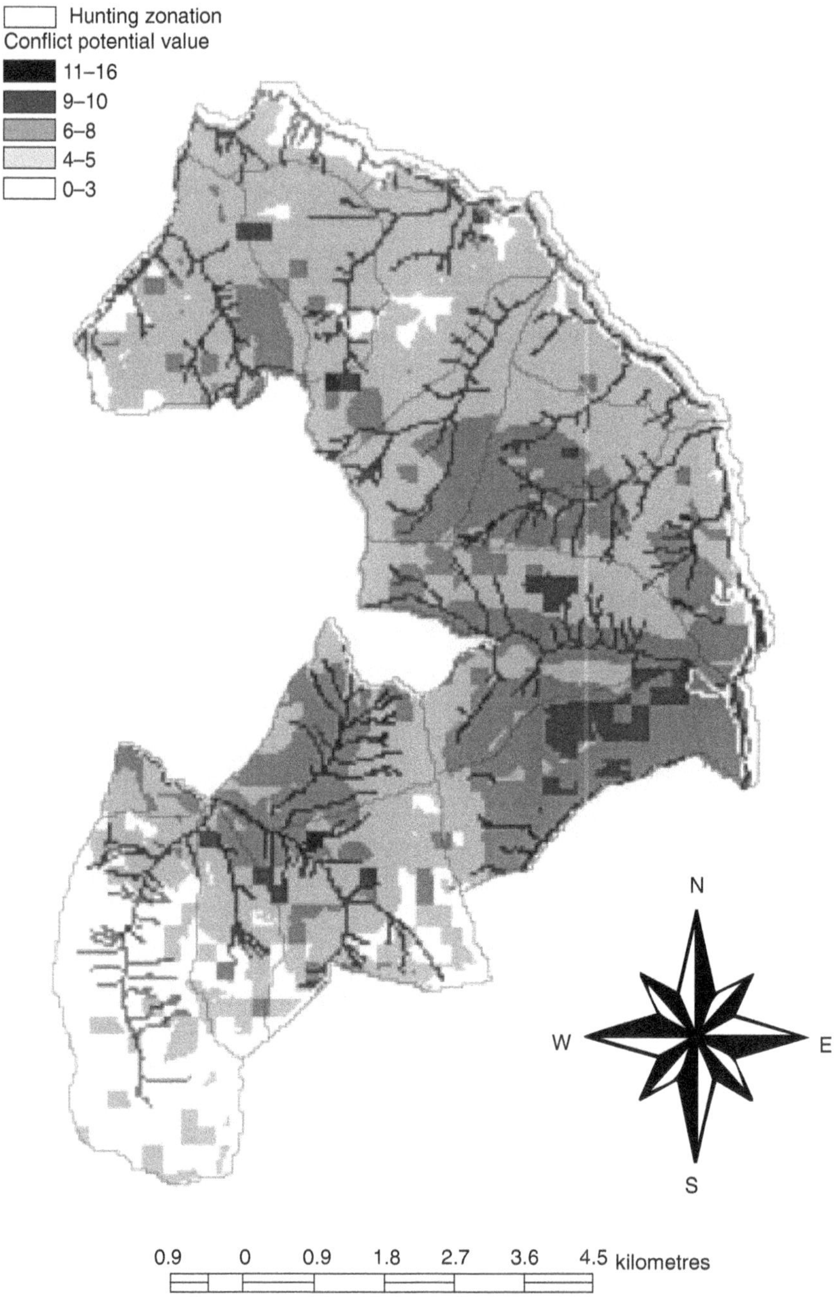

Fig. 6.27. Map 3: Recreation/conservation conflict potential.

Map 3 (Fig. 6.27) shows an example of the conflict potential between recreation and conservation. High conflict potential exists in those areas where there is high land-use potential for both land uses. In the project area, this is the case in areas of high habitat diversity and on specifically sensitive sites (e.g. bogs), which are situated either close to the village or in the surroundings of high-intensity recreation infrastructure (e.g. restaurants, lodges). River-banks, too, are a type of habitat of interest for both land uses.

Assessment of actual land-use conflicts. The comparison of conflict potentials with the actual conflict situation helped to identify compatible intensities of potentially conflicting land uses. For the solution of the remaining land-use conflicts, a land-use concept was developed which consisted of a mixture between the two basic strategies of adaptation and segregation.

Combining spatial segregation and adaptation. On the one hand, segregation is suggested for such activities where only very low levels of intensity are compatible with one or more of the other land-use forms. Adaptation, on the other hand, is suggested for land-use forms where compatibility problems occur only at rather high levels of use intensity.

Interviews with representatives for all the land-use forms that were subject to the project, together with official records, led to the assessment given in Table 6.31 on the situation of land-use conflicts and possible conflict management strategies in the project area. The whole land-use planning process was performed in cooperation with the FUST-Lenkungsausschuß, an advisory board to the FUST-Achenkirch. This advisory board allows representatives of all interested groups (landowners, tenants with various land-use rights, administration agencies, the community, etc.) to participate in hunting management and research activities in the area.

For those land-use forms between which there is no known conflict within the project area, it may be assumed that they are at compatible intensity levels.

In order to regulate actual and potential stress from recreation activities, priority zones for recreation are suggested. These are areas where the infrastructure-dependent recreation potential indicator value is higher than the conservation potential indicator (the latter is not discussed in this chapter). This zonation is the result of weighting recreation and conservation with a 1 : 1 ratio. New recreation infrastructure is to be established only in areas that are already easily accessible, whereas remoter areas are held to be suitable for conservation and game-reserve purposes. In addition to this weighted distribution choice, a buffer value was introduced, so that a certain distance between the two conflicting land-use priorities can be assured.

Though a separation of cattle and grazing is considered desirable for the entire project area, the rather large area with existing grazing rights shows that this goal will not be easy to achieve. Therefore, priorities for this separation have to be decided. The potential, or rather the demand for disaster protection, was chosen as the most relevant factor for this decision. Areas in which the indicator value for disaster protection is higher than a given threshold value (12) are always assigned protection as a priority or as part of a priority combination. Map 4 (Fig. 6.28) illustrates this concept.

Table 6.31. Land-use conflicts and conflict management strategies.

	Hunting	Timber production	Conservation	Recreation	Disaster protection	Grazing
Hunting		C/I	+	C/P	C/I-P	C/I
Timber production	C/I		C/I-P	+	+	C/P
Conservation	+	C/I-P		C/I-P	+	C/I
Recreation	C/P	+	C/I-P		+	+
Disaster protection	C/I-P	+	+	+		C/P
Grazing	+	C/P	C/I	+	C/P	

+, No current conflict within the project area; C, conflict; I, conflict management by intensity adaptation; P, conflict management by priority assignments; I-P, conflict management by small-area priority assignments and wide-area intensity adaptations.

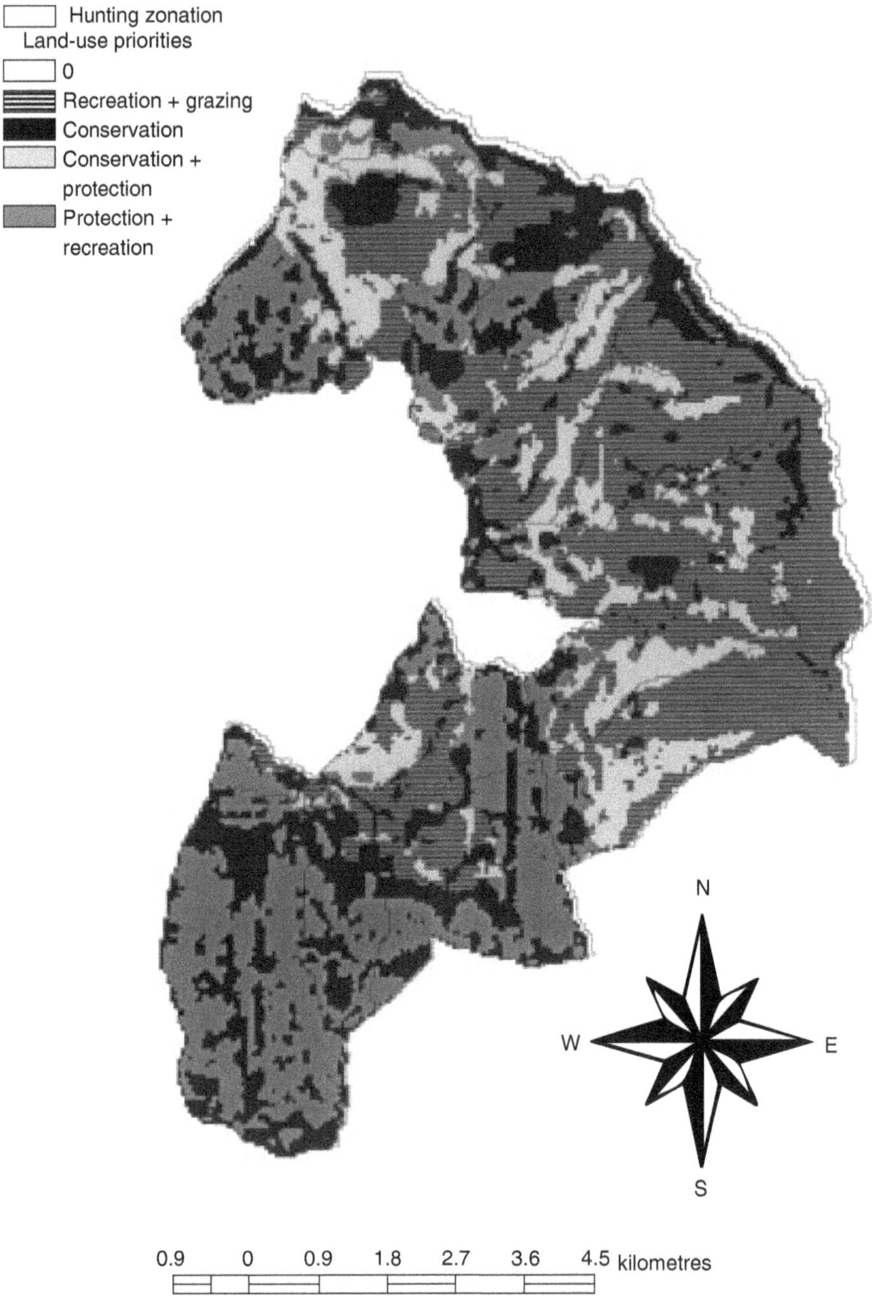

Fig. 6.28. Map 4: land-use allocation – Achenkirch.

A concept of temporal segregation for optimizing hunting activities

A temporal segregation-based approach was suggested for the improvement of hunting activities in the area, as well as lower conflict potentials between hunting and other land-use activities, specifically tourism and timber harvest (see the conflict potentials shown in Table 6.27).

For a better understanding of the considerations that led to this approach, it is necessary to give a short explanation on

Austria's district-based hunting system. All land, except for urban areas and other settlements, is divided into hunting districts. In general, a hunting right is part of a landownership right, but, in order to be zoned as a hunting district, an individual landholding has to exceed an area of 115 ha. In this case, the landowner may either go hunting him/herself, or he/she may rent out the hunting rights to gain additional revenue. Smaller landholdings are integrated into larger hunting districts, the landowners of which are members of a hunting association, which then rents out the hunting right for the hunting district. The revenues of a hunting association are divided up between the associates according to the share of their landholding compared with the size of the hunting district. Whoever wants to hunt, whether on his or her own land or in a rented district, has to hold a valid hunting licence, which can be obtained from the hunting authorities after taking the appropriate examination. The hunting season in Austria is quite long: depending on the detailed regulations in the individual state, it starts around the middle of May or beginning of June and lasts until the end of December. During this rather long period, different regulations apply as regards game species and, within each species, the sex and age of animals that may be hunted. In addition, there may be voluntary associations, called *Hegegemeinschaften* (stewardship associations), between neighbouring hunting districts. The *raison d'être* of these associations is to regulate management for species with individuals roaming areas of considerably larger size than single hunting districts. Specifically these associations form agreements on whether certain individuals will be considered free for hunting in the respective year. This is common practice for the management of red deer, as, due to trophy bearing, mature males of this species are considered to be the most valuable domestic hunting prey in Austria.

At the time of the development of this concept (1994), the 7000 ha of FUST's project area for the purpose of hunting management had been divided into three hunting districts, each about the size of 2300 ha. It has to be mentioned here that in 1997 part of this area (the middle part, west of the village of Achenkirch) was taken out of the FUST project due to a change in the tenure of hunting rights; thus, the regulations described in detail have only been applied there between 1994 and 1996.

This rather lengthy description was necessary to illustrate the consequences this may have for hunting practices. During the long hunting season, the game-keepers may feel obliged to keep up a constant watch on the whole area in order to know where the various herds and individuals of game, specifically the trophy bearers that may be of interest for the holder of the hunting rights and his/her guests, have their favourite places. In addition, the potential to take potential prey by surprise may be seen as a factor enhancing the chances of the hunter. Yet this unpredictability may also result in a general disturbing effect for the whole district, since animals that live under the constant threat of a hunter will adjust their behaviour accordingly (i.e. become more active at night, while they hide in the undergrowth by day).

Scientists (e.g. Reimoser, 1993) have thus suggested a new hunting strategy, called interval hunting, which tries to avoid this disturbance effect. In this approach, the whole district is divided into separate interval zones and the hunting seasons is divided into intervals. During each interval, hunting activities take place only in one zone, while any activity, including monitoring, is suspended in the other zones during that time. In the next period, hunting activities are shifted to the next zone and so on. Thus, in a hunting district with three interval zones and an interval length of 2 weeks, each interval area will be without any activity for 4 weeks.

Interval zones have to be chosen so that activities in one zone have as little influence as possible on any other zone. Ideally this would be adjoining watersheds, separated by a mountain range. These considerations have to be made not only for the actual hunting activities but also for all related activities, such as driving on access roads.

In addition, other land uses may also be taken into account in this zonation, as well as the scheduling of the intervals. For example, tourism can be taken into account by adjusting zones and intervals in such a way that hunting is avoided in favour of recreation in certain areas during known seasonal peaks of tourism. In addition, it is possible to coordinate hunting management with timber harvest activities, so that, for example, a harvest operation is not scheduled immediately before the beginning of the hunting interval in a zone, thus disrupting the intended effect of keeping an area as undisturbed as possible in the time before the hunting interval.

These considerations have been taken into account during the zonation of interval zones for the three districts. According to Tyrolese hunting laws, the hunting season there starts on 1 June. At this time, tourism has not yet fully started; thus, the most tourism-intensive area was chosen to be the first interval zone for that district. These are the mountainsides located west of the village Achenkirch, where several restaurants are within a 60- to 90-min walking distance from potential starting-points (car-park areas). Similar considerations were also made for the other two districts.

In addition, a coordination process with the locally responsible management of the Austrian Federal Forests (Österreichische Bundesforste) was initiated for the coordination of hunting and timber harvest activities. It should be noted here that this cooperation was based on mutual voluntary activities.

Map 5 (Fig. 6.29) shows the different interval zones, whereas the series of maps that are compiled as Map 6 (Fig. 6.30) show how hunting activities shifted between the different zones during the hunting season in 1994.

Even though this system in the beginning required a good deal of adaptation and self-restraint from the professional personnel, it was finally accepted and eventually seen as an improvement on former practice.

For smaller hunting districts, which, due to their size and geographical conditions, cannot be divided up into several interval zones, the corresponding practice would be to shorten the length of the hunting season in the area to only a few weeks per year. In this short period, then, lively hunting activity will enable the fulfilment of the prescribed hunting quotas. Interestingly, such an approach is also favoured by nature conservationists, since it minimizes human influence on wildlife over the year. The World Wildlife Fund (WWF)-Austria, which shares ownership of a medium-sized forest estate about 40 km east of Vienna, applies this approach in their area (Lutschinger, 1996).

In addition to this interval system, several zones have been defined in the FUST project area where hunting pressure is kept constantly high, even making exceptions from certain seasonal regulations. Tyrolese hunting laws allow such exceptions for areas which it is considered desirable to keep as free of game as possible, since the situation of game damage to forest vegetation requires it. On the other hand, potential game retreat areas have also been zoned which are intended as winter habitat. For these areas, the goal is to keep them as undisturbed as possible during the winter period, which is done by restricting tourist access to these zones. These two types of zones are examples of both temporal and spatial segregation, since they do not apply constantly and only for comparatively small parts of the whole planning area. They are depicted in Map 5 (Fig. 6.29).

Identification of land-use potential niches in the context of utility-line tracks

Utility lines are necessary infrastructural elements of technology-dependent societies. In highly forested areas, they will inevitably be present also on forest land; moreover, there exists a certain preference for establishing utility-line tracks on forest land for reasons of landscape aesthetics (Hartl, 1997), given the potential view-shading effects of surrounding forest stands.

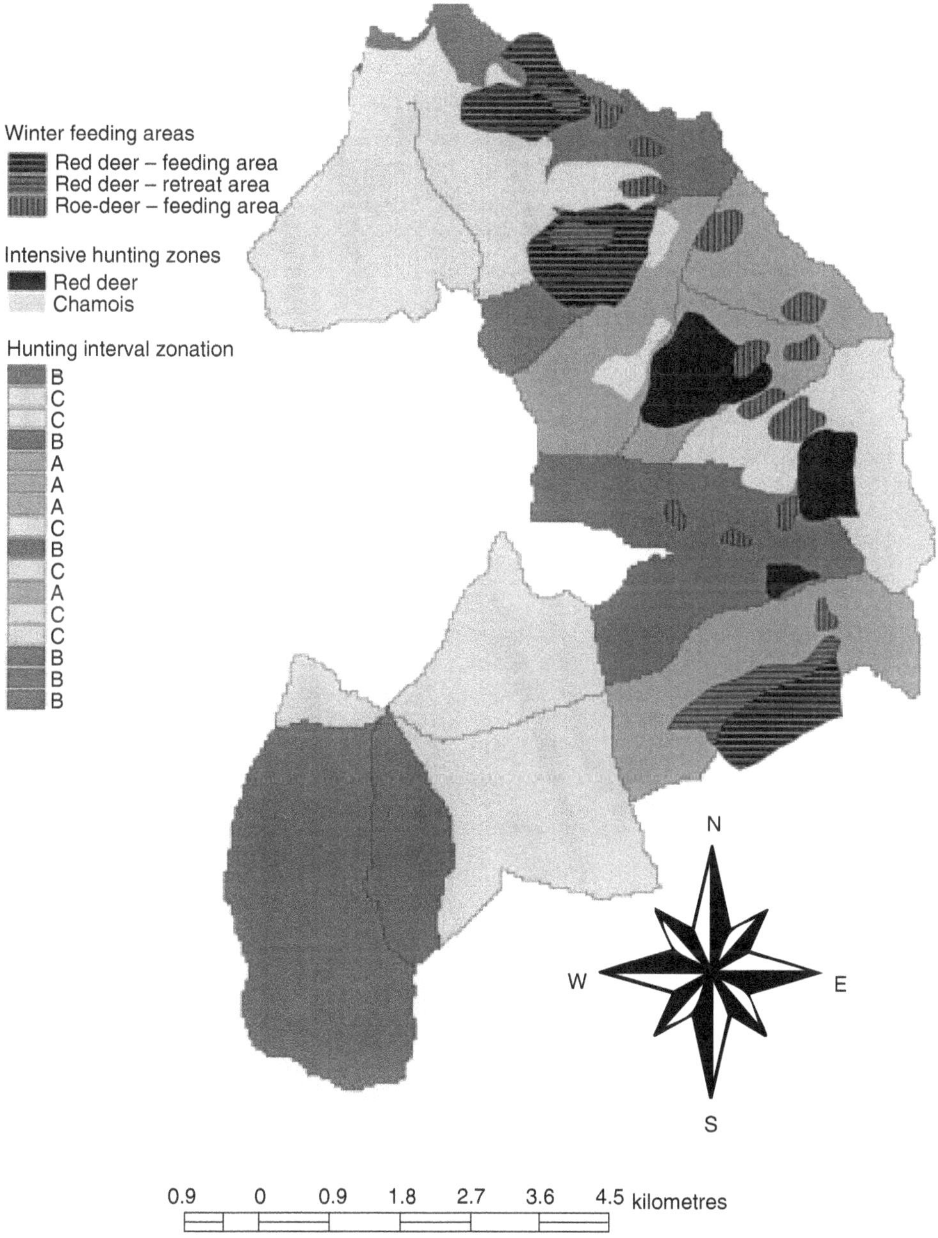

Fig. 6.29. Map 5: hunting zonation and specific measures.

In order to avoid accidents, forest vegetation below and neighbouring utility lines must not exceed certain height limits, which are defined by safety regulations. These regulations represent considerable restrictions on conventional forest management on the tracks. In a project funded by the Austrian Electricity Company Verbund-AG, the options that exist for the management of high-voltage power-line tracks (110 kV, 220 kV, 380 kV) were examined by a multidisciplinary research team in the

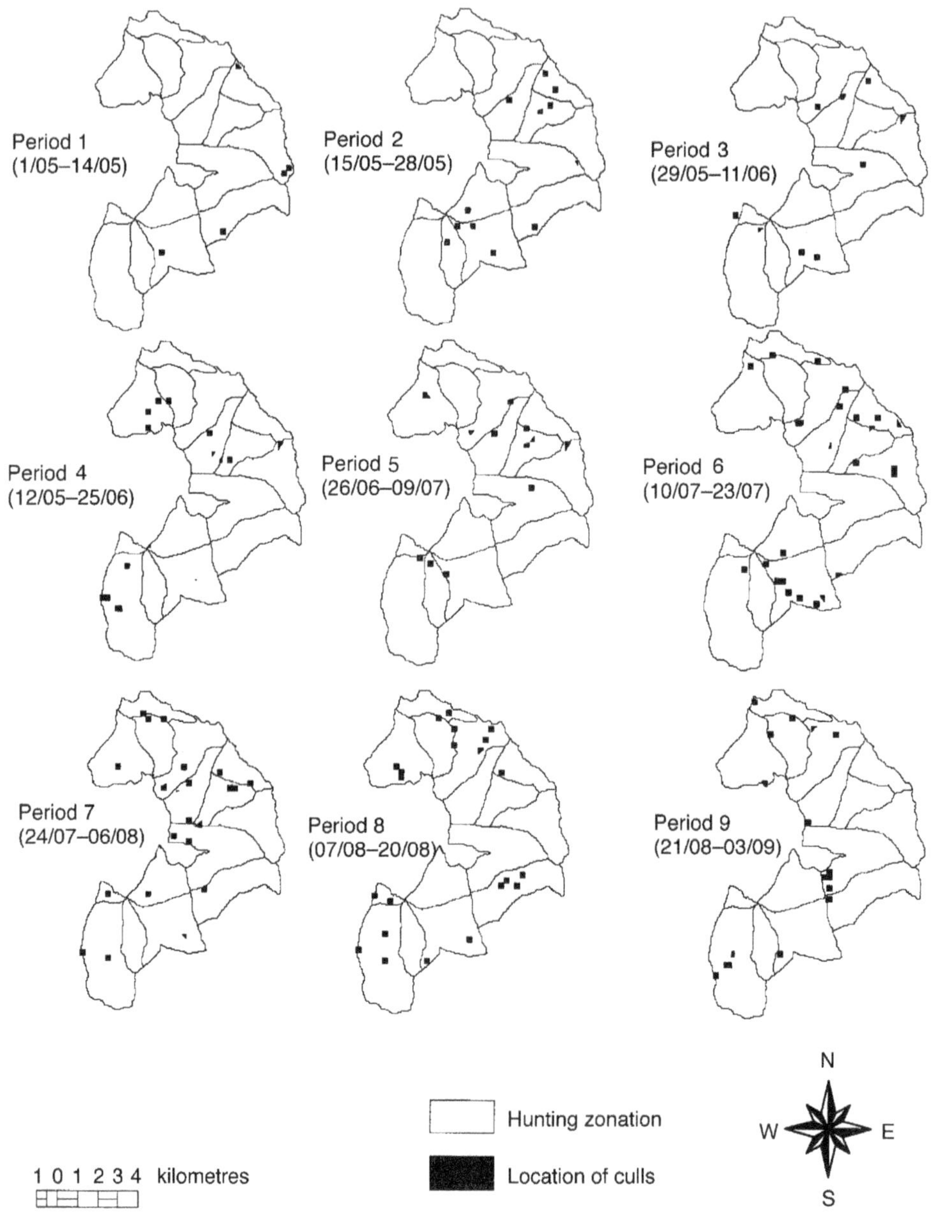

Fig. 6.30. Map 6: time series – hunting intervals.

context of safety regulations, economic considerations and the realization of ecological potentials, using three study sites in Austria as examples (Draxler, 1998). In this project, the main goal was to identify forms of land use for which the specific conditions on a power-line track can be regarded as an additional potential or even as a prerequisite. Specifically, by allocating land uses that may require the removal of tree vegetation to areas where timber production is hampered through safety constraints, it is possible to gain land for other activities without losing too much wood production capacity.

A land-use information system was also established for this project, based on official cadastral maps, digital orthoimagery and documentation material provided by the utility company. GIS allows for an overlay of these different information sources. The variation between the different information sources clearly points to the difficulties that can be encountered in the establishment of a basic information system. In the course of this project, it was then decided to rely mainly on the digital orthophotographs as the main reference.

From the viewpoint of nature conservation, fauna and flora were examined paying specific attention to the potentials for rare and endangered habitats and species. In this context, utility-line track areas are viewed mainly as areas of locally rare habitats (grassland and shrubby areas amidst closed forest land), and recommendations were made to maintain this character and to increase the closeness to natural conditions as regards stand compositions, especially by bringing in a deciduous tree species wherever this is appropriate to natural site conditions.

From a silvicultural point of view, the importance of closeness to natural conditions with regard to species composition and structures (e.g. decaying wood) is also pointed out. High regard is given to site-specific growing conditions and the complex of forest fringe treatment. The latter is also seen in close relation with the treatment of neighbouring stands.

Wildlife ecology sees such tracks as areas that may be used for habitat improvement, as well as for facilitating improvements in hunting conditions (free areas for hunting, construction of hunting stands, increase in forage). Measures that alter habitat conditions have to be coordinated with the larger surroundings to avoid unwanted effects of redirection of game (e.g. closeness to damage-prone regeneration areas and younger stands, traffic routes).

Alternative land uses on tracks are mainly seen as those which will ultimately result in the removal of tree vegetation from the track in the interest of other land uses. On the one hand, such activities may be related to forest land use as defined by the Forest Act in Austria, so that no special permits for land-use change will be required (e.g. forest roads, timber-storage areas). Other activities that are in close connection with traditional forestry, though, may require specific provisions, such as Christmas tree production or forest nurseries.

On the other hand, the track areas also offer potential for other land uses, especially tourism (e.g. cross-country ski-tracks, restaurants, car-park areas), or agricultural activities, such as livestock grazing. Also special forms of land use, such as construction waste dumps, can be taken into consideration.

6.3.5 Conclusions

Multipurpose land-use planning is about visualizing the different angles from which a landscape is viewed by different potential users. By combining these different views, it is also possible to assess where different requirements will interfere with each other.

The type of approach which is suggested in this contribution is based on a rational approach to land-use planning. However, human emotions govern a large part of our behaviour. This is taken into account by contributing both rational facts and value statements to the process.

By viewing a property through the eyes not only of one but of many potential land-users it is also possible to assess the potential for new products or services. Yet the suggested approach also draws attention to possible difficulties in widening the production range.

The use of GIS technology is nowadays widespread in land-use management planning. Yet, in many cases, only a small range of the variety of functionalities which this technology offers is used. By deciding between different possible levels of involvement in the technology, a land-user may choose the one best suited to his or her qualitative and quantitative requirements.

By closer identification of the spatial distribution of land-use potentials, land-use managers are also able to detect niches in which to locate specific uses without any production loss for the main production goal.

Yet the introduction of new planning tools itself will not solve any land-use problems or lead to success in marketing new products or services. The bait ultimately has to be accepted by the potential prey and not by the person hunting or fishing. This means that, as in other industrial product developments, it will be necessary to assess the users' preferences first. Thus, some elements of participation in planning may rather be seen as implementing techniques of PR and market research than as an unwanted intrusion into private management.

6.4 Business Organization and Accounting

6.4.1 Introduction

Any RES project can and should be regarded as a kind of business undertaken to better achieve the goals of the forest owner. As far as financial goals are of major importance, respective monetary measures are key indicators for the success of the project. Even when non-monetary considerations are the main motivation for an RES activity, the provision of the respective services requires the input of productive agents of some kind, thereby competing with other activities for scarce resources. As a leading principle of any economic activity, the economic principle states that either the maximum utility should be achieved at a given expense or a given utility is to be achieved at minimal expense. Consequently, efficiency is a key issue of any RES activity, whereas the significance of profitability is determined by the particular goals, threshold levels or limits established by the forest owner.

Accounting provides a framework for evaluating the input and output of economic activities in monetary terms. Thus accounting tools are prerequisites for monitoring and controlling the profitability, as well as the efficiency, of RES projects. Accounting also comprises the documentation of activity-related input and output in physical terms. Relevant records are a valuable source of data for planning and controlling purposes, especially as regards the provision of productive agents and capacities.

The significance of accounting with respect to RES activities is discussed under two major headings. In the first place, the formal requirements of financial accounting have to be considered. In practice, specific tax regulations are likely to be of some importance. Furthermore, managerial accounting provides a variety of tools for investigating the efficiency of any such project. Whereas the concept and instruments of managerial accounting may be adapted to the individual needs, the legislation on taxation requires financial accounting to comply to certain standards. As these regulations differ between countries and certain details, such as threshold values, may even change from year to year, only general aspects of at least potential significance are discussed here. (For an international comparison of accountancy in forestry see, for example, Hogg and Jöbstl, 1997.) Where necessary, reference is made to the currently valid regulations in Austria, the particular details serving illustrative purposes. On the national or subnational level, there may be quite specific literature available providing relevant hints in great detail, such as the brochure *Kostenrechnung und Finanzierung für 'Urlaub am Bauernhof'-Anbieter* (*Managerial Accounting and Financing for Offerers of 'Vacancies on the Farm'*) by Gattermayer *et al.* (1996) or the book entitled *Altgebäude als Einkommensquelle* (*Old Farm Buildings as a Source of Income*) by Damm and Grahlmann (1996).

6.4.2 Formal delimitation of forestry and the RES business

Indicators for and consequences of a formal delimitation of RES activities

Rendering recreational and/or environmental services may be part of the activities of a forest enterprise, at least up to a certain extent. Alternatively, it may be either necessary on legal grounds or advisable for certain reasons to have a separate legal entity, be it an association or a company of some kind, running the RES business. When establishing RES activities, one should therefore, on the one hand, check the relevant legal framework and, on the other, consider the additional advantages and disadvantages of founding some kind of RES-specific organization. General information on this topic may be available, e.g. from a forest extension service or the chamber of commerce. Finding the optimal solution for bigger projects will usually require specialized advice provided by lawyers or consultants and thus increase the transaction costs of the service in question.

Depending on the national legislation, relevant indicators for a legal obligation to treat the RES activities as a separate business could refer, for instance, to one or several of the following items, the most relevant regulations referring to tax law and trade regulations:

- type of business/activity. According to tax legislation, on the one hand, and trade regulations, on the other, forestry activities may be defined in a more or less strict sense, allowing for specific RES activities only up to a certain degree or even not at all. Depending on the type of activity, special preconditions may have to be fulfilled, such as a certificate of qualification of the particular manager or personnel.
- Size of the business. Turnover or income in absolute values or in relation to the forestry income may serve as indicators in the sense that minor values or shares may be related to forestry, whereas a greater economic significance necessitates a formal separation between forestry and RES activities.
- Regularity of activities. Single events may not be regarded as a separate business, whereas regular activities may be an indication for services in question constituting an individual business on their own.
- Means of living. A further criterion may be the intention and/or the possibility to earn an income from the kind of activity in question. If the revenues associated with the service are just reimbursements more or less covering additional costs in forestry or compensating for forest income foregone, providing the service will hardly be regarded as a business in itself. Real market revenues, on the other hand, are a strong indication for a separate line of business.

Apart from the legal regulations, there may be different reasons for establishing an RES-specific organization, financing, liability and organization being the main aspects in this context.

- Cooperation. When the RES business is not to be run by one forest owner exclusively but involves other forest owners or companies, it might be advisable or even necessary to found a company or an association for running the business. Especially when the business is to be developed on a larger scale and/or necessitates the integration of many stakeholders, contracts between all of the parties involved will possibly not be sufficient in the long run. The role of cooperation is stressed by the fact that within the case-studies up to 200 enterprises may contribute to the product. On average, five forest enterprises are involved in the product. Examples are an Italian consortium of FLOs that sells chestnuts (IT27), the foundation of a consortium to sell certified mushrooms (IT02) and an FLO organization that takes care of the maintenance of forest roads for several owners (AU18). Although product management lies with the forest owner in about half of all cases, consortia, cooperatives and associations were also reported as being in charge of the RES business.
- Distribution. Founding an association may be advisable also in terms of distribution of the service. In some cases, the users or the ones paying for the service in question are not willing or even not allowed to pay money to other than non-profit organizations. Especially when the money paid for the service stems more or less directly from public

budgets, direct payments to the forest owner may turn out to be problematic, if acceptable at all. In such cases, an intermediate non-profit organization has to be established, so that it can be argued on the political stage that public interests are furthered without those paid for the service making a profit out of it. For instance, in Austria the Land und Wirtschaft association (AU01) was founded and acted as an intermediate organization between the National Forests providing mountain-bike trails, on the one hand, and the Ministry of Economic Affairs paying for the provision of this service, on the other. In some cases, it may even be possible that the forest owner him/herself is a member of the non-profit organization. This applies to the BIOSA association (Biosphere Austria; AU09), which mediates services for nature conservation between forest owners and sponsors. A second example is the non-profit-making association Klimaschutz durch Wald e.V. (DE08), which was founded by foresters to acquire sponsoring money for afforestation. Strategies for communication and public relations, too, may be affected by the organizational background for rendering the service. Customers may prefer specialized companies to forest enterprises for reasons of expected professional skill and experience or even object to the social or economic status of a landowner.

- Liability. Certain services in which forest owners may engage themselves are associated with high levels of financial risk, such as running a waste dump or the provision of drinking-water. According to the legal status of the forest enterprise, most forest owners are personally fully and unlimitedly liable for any damage arising from their business. By founding a limited company, the risks associated with liability can be considerably reduced. The more risky the business, the less capital should be left with the company, be it in terms of assets or funds.
- Creditworthiness. There is some relationship between liability, on the one hand, and creditworthiness, on the other. Creditworthiness is highly dependent on the value of the property available as cover for a loan. Within the forest business, the forest itself usually constitutes the major part of the property, real estate being a good security for the creditor. If the RES business is run by an independent firm, the value of the property and consequently also creditworthiness may be limited. When the company running the RES business is backed by several partners, however, the capital and therefore also the securities may well exceed the potential of an individual forest owner.

Apart from the motives discussed above, the following possible consequences of establishing the RES activities as a separate line of business should be taken into account.

1. The relevant authorities are likely to be different ones, so that the forest owner has to deal not only with the forest authorities but possibly also with others.

2. Different legislation has to be applied, including regulations on trade and labour. This may cause some organizational difficulties if workers or employees are working not only within the forest enterprise but also for the company running the RES business. It might even be necessary to create appropriate part-time jobs. Alternatively, to some extent it may be permitted to hire out labour of the forest enterprise to the company running the RES business, the forest enterprise being reimbursed accordingly. A relevant example is provided by case-study DE03. The forest owner founded an association, and this association is running the RES business. Buildings and the forest property are rented from the forest enterprise. Conversely, the forest enterprise hires out labour to the association.

3. The transfer of assets from the forest enterprise to another legal entity which is to run the RES business may trigger tax payments on the part of the forest owner. For instance, a house formerly inhabited by a forester is no longer occupied and is let to tourists on a regular basis. In this case – according to the Austrian regulations – this asset no longer belongs to the forest enterprise but has to be transferred to the letting business. Usually, the book value will be lower than the market value, which has to be applied for that purpose. The difference in value, which has been a secret reserve of the forest enterprise so far, enhances the forestry income for the relevant period.

4. By means of contracts between the forest enterprise and the company running the RES business, it is possible to shift their costs and revenues to some extent. An example is NL20, where the forest landowner founded a separate business to let apartments, and this separate company has to hire the forest property from the forest enterprise. This aspect is of importance especially when the two enterprises do not belong to the same owner or different tax regulations are to be applied (e.g. lump-sum taxation of the forestry income). For example, the forest owner may derive an income not from directly selling the RES on the market, but from letting parcels of land to the company running the RES business or from being reimbursed by such a company for higher forestry costs or diminished revenues.

5. The founding of an enterprise or the incorporation of a company is usually associated with substantial costs. Additional costs may constantly occur every year for preparing separate balance sheets and profit-and-loss accounts.

Roughly two-thirds of the RES case-studies were fully integrated into a forest enterprise. Thirteen projects, however, were run by a separate company and, in some 16 cases, the RES business was organized as a cooperation of different enterprises. These empirical results indicate that, although the majority of the projects were hitherto realized within a forest enterprise, special organizational arrangements are to be considered as well. Product management is closely related to the type of enterprise running the RES business. About half of the documented RES projects are managed by the forest landowner, but there are also examples of various other kinds of management structures, including consortia, cooperatives and associations.

Choice of the type of business unit

As mentioned above, it may be advantageous for several reasons to found an RES specific organization even if there is no formal necessity to do so. Trade regulations usually allow a choice between several types of companies, partnerships and associations, so that it is most important to evaluate their respective advantages and disadvantages (see, for example, Juen, 1997). Depending on the size of the project, it is advisable, if not indispensable, to consult experts for the evaluation process, as well as for the formal realization.

In general, the following types of business unit may be applicable:

- sole proprietorship. In this type of business, one person is solely responsible for providing the capital and for bearing the risks of the enterprise. The sole trader is subject to unlimited liability.
- Association. A more or less loose grouping of people established for mutually beneficial purposes and held together by an orderly system of management.
- Consortium (management group, syndicate). This is an association of a number of firms working together on a specific project that is too large or too capital-intensive for a single firm.
- Cooperative society. This is a self-governing enterprise collectively owned by those with a direct interest in the firm. Ownership and control rest with those who work in the enterprise and profits are shared.
- Ordinary partnership. An association of two or more persons carrying on a business in common with a view to profit. The partners provide the finance for the organization and the profits and losses will normally be shared in an agreed proportion depending on the individual's contribution to the partnership. All the partners are responsible for the debts of the firm.
- Limited partnership. In this case, the liability of some of the partners is

restricted to the amount of capital they have invested in the business, whereas at least one general partner is unlimitedly liable.

- Private limited company. A group of persons associated together for the purpose of undertaking some form of business activity, the liability of each being limited to the nominal value of the respective shares held. Shares can be sold only with the permission of the other shareholders.
- Public limited company. Such a company must have a minimum authorized and allotted share capital. The shares may be sold to anyone and may be quoted on the stock exchange.

Since the type of company triggers various legal, organizational and financial consequences, various criteria have to be evaluated when determining the kind of enterprise that is to run the RES business. The following list indicates some of the main aspects to be considered.

- The company's own legal existence. Some types of business units have their own legal existence, whereas with others all respective rights and duties rest with the persons building the company.
- Management and representation. The owner, partner or shareholder may or may not be an employee of the company. The company may be represented by one person alone or it may be necessary for all the partners to act together.
- Organizational regulations. Some types of companies are legally bound to have specific bodies, such as a board of directors or a supervisory board.
- Intensity of cooperation with partners. Depending on the intended degree of cooperation, a more or less loose cooperation of independent persons or firms may be established or a common company may be founded.
- Time frame of the cooperation with partners. The shorter the time frame of the intended cooperation, the more relevant are the specific costs for the establishment and liquidation of the company.
- Distribution of profits and bearing of losses. Whereas a sole trader has to bear all the losses him/herself, a shareholder will not usually have to compensate for losses of the company.
- Bearing of risks. As indicated above, the degree of personal liability may vary from the nominal value of the shares held to all a person's private possessions in case of unlimited liability.
- Costs of establishment and liquidation of the company. They are an important element of a financial appraisal.
- Requirements for founding the company. Usually, some kind of registration will be necessary. If the business necessitates a certificate of qualification, the owner(s), the general manager or an employee may be required to provide one.
- Financial requirements of the business. The possibilities for fund-raising are also interrelated with the legal type of the company and with the company's assets.
- Requirements for bookkeeping. Different threshold levels and regulations have to be observed.
- Requirements for publishing the company's accounts. For some kinds of company, there is a legal obligation to publish the company's accounts annually.
- Taxation. Different regulations may apply, the field of taxation encompassing such elements as corporation tax, income tax, property tax, value-added tax, gift tax and death duties.
- Changes in partnership and liquidation of the company. There may be legally binding causes for the liquidation of the company along with changes in partnership.
- Passing on the company or shares in it.
- Social insurance of the owner(s).

When establishing a company, various authorities, offices and institutions have to be contacted for licences, permissions, registrations and so forth. Special regulations – for example, stating deadlines or preconditions – have to be observed. Information and support may be provided by syndicates, (tax) consultants and lawyers. Additional contacts with banks, insurance companies, suppliers, the refuse collection department or a notary, to name just a few, might be necessary as well.

Referring to Austrian regulations, the following steps are necessary for the establishment of sole proprietorship (Wirtschaftsförderungsinstitut, 1996):

1. General, legal and financial advice.
2. Licensing of the trade.
3. National insurance of the employees.
4. National insurance of the entrepreneur.
5. Registration at the tax office.

In the case of a private limited company, some additional steps are necessary:

1. General, legal and financial advice.
2. Partnership agreement.
3. Appointment of the general manager.
4. Confirmation of a bank concerning the availability of the minimum share capital.
5. Registration of the company.
6. Trade licensing.
7. National insurance of the employees.
8. National insurance of the entrepreneur.
9. Registration at the tax office.

As documented by the case-studies, some 40% of the RES projects necessitated specific licences or permits. The structural organization of the RES business is not directly related to the question of approval by the authorities. Nevertheless, the empirical data indicate a certain correlation, such that there is, to some degree, a coincidence between product management by a separate company, on the one hand, and the necessity for specific licences, on the other. (For further details on special permits, see Section 5.4.1.)

6.4.3 Financing and taxation of RES activities

Finance and investment

The financial significance of an RES activity is measured, like any other business, in terms of expenditure, proceeds and profits. Once the business is established and profitable, the current expenditure should, on average, be more than compensated for by the revenues, so that the business can more or less sustain its liquidity on its own.

The establishment of a new business, however, may be associated with substantial investments, such as for the investigation of its potentials, the procurement of productive resources or market research. It is highly recommended to make a detailed plan at an early stage of the project, specifying all the investments and the associated financial burden. Objects of investment are not only fixed assets, such as buildings or machinery, but also current assets, such as material. Ultimately, capital has to be provided not only for financing the investments, but also for covering incidental costs, such as fees and interests. Further financial means are necessary for covering the current expenditures, such as wages and costs for energy, as long as they are not counterbalanced by revenues.

In about 60% of all investigated case-studies, no substantial investment was needed for establishing the RES project. This finding is a strong indication that a great variety of activities in the field of RES can be implemented just by utilizing the resources already available in the forest enterprise. Apart from questions of liability, there is hardly any financial risk in such cases and financial means are not the bottleneck for project realization. The identification and utilization of available resources are clearly a question of the spirit of enterprise in the first place. Forty cases were associated with substantial investments. As far as documented, the sum invested amounted, on average, to about 2.5 times the expected yearly turnover of the RES business.

Although the empirical basis is not sufficient to investigate specific hypotheses, several cross-tabulations performed indicate some general relationships, at the same time providing examples for exceptions from simplistic rules. For one thing, the share of RES related income is, to some extent, related to investments. In eight out of 13 cases where RES-based revenues contribute more than 30% to the forestry income, a substantial investment was made. Conversely, where RES products scarcely contribute to the forestry income, eight out of 11 cases were not based on a substantial investment. When a net profit was aimed for, half the cases involved substantial investments, as compared with a share of 30% in those cases where no financial goals were specified. Combined products, especially traditional products, tend to necessitate considerable investments. Also the structural organization seems, in some respects, to be related to the issue of investments: when substantial investments are reported, separate companies running the RES business are more frequent compared with when there has been no substantial investment, where a cooperation of different enterprises occurs more often.

Various sources may be taken into account for financing the investments. In so far as the RES business is to be established within the forest enterprise, the financial potential of the forestry business may be directly relied upon. As an alternative to such internal financing, the forest owner may invest part of his/her personal capital. The company capital can be increased by taking in additional partners. Outside capital can be provided, especially by banks. There are several types of bank advances, differing in terms, interest rates and conditions of loan repayment. Leasing (e.g. of vehicles, machinery or buildings) is a means of lowering the actual need for capital at the very start of the business. Subsidies may be granted by European, national, regional or local bodies and may help to lessen the financial burden. Especially in less favoured areas, RES projects of significance for the local economy may benefit from subsidies. Relevant enquiries are, therefore, a key element of financial planning in general and for bigger RES projects in particular.

Bank loans, sponsors and incentives were obviously not the predominating sources of capital for the investigated projects. For instance, only every ninth project benefited from incentives of some kind. One out of seven projects was backed by sponsors, indicating a relatively high significance of sponsorship, whereas only one out of 16 relied on bank loans. Bank loans occurred only in connection with substantial investments but were not exclusively connected to the financial goal of making net profits.

The higher the investments and hence also the financial risk, the more relevant will a comprehensive investment appraisal be (for a general description of the methodologies of investment appraisal, see a standard textbook on this topic, such as that by Holmes (1998)). The logic of an investment appraisal underlines the fact that a certain project is not the only way of action, but that alternative possibilities are to be investigated and evaluated, including the option not to invest at all. The core of any financial appraisal is the identification of all costs and revenues associated with each option. It is essential not to miss any major cost item; thus costs for detailed planning, costs for the establishment of a company or costs of consulting have also to be taken into account.

The investment appraisal has to cover three aspects, providing answers to the relevant questions:

- rentability of the project (Is the project the best way to invest the money?).
- Liquidity (Will there be enough means constantly available to be able to cover the current expenses?).
- Risk (What are the likely effects of changes in costs, prices, demand or other frame conditions?).

When calculating the net present value (NPV) of an investment, all future costs and revenues triggered by the investment are to be discounted, the NPV being the total discounted benefit from an investment project minus its total discounted cost. Clearly, the discount rate used for the calculations may well influence the results, so that some sensitivity analyses will also be advisable in this respect. If the options are associated with different time periods, this aspect is also to be taken into account. Additional measures besides NPV are, for example, the annuity or the internal rate of return.

The *Investment Appraisal Handbook* of the Forestry Commission provides a checklist (Table 6.32) for conducting an investment appraisal (see Insley *et al.*, 1987, appendix 1).

In the cases where no substantial investment was needed, more than half of the respondents addressed by the empirical study stated that they did not conduct a financial appraisal for their project and that it was not at all necessary. However, about 25% of the interviewees confirmed that there had been a formal investment appraisal and that it proved necessary. The potential significance of a formal financial appraisal is underlined by the fact that roughly every fifth case was associated with shortcomings of this instrument, the manager in charge suggesting respective improvements in the light of experience.

The stated significance of an investment appraisal is clearly related to the fact of a substantial investment. However, only one in two of those who invested substantially stated that an investment appraisal in fact proved necessary. In one out of three cases of investment, an appraisal was judged not to be necessary at all. On one hand, the cross-tabulations do not indicate any direct relationship between the financial goals and the stated significance of the financial appraisal. On the other, there is a clear indication that, when banks or the external management unit contributed to the finance of the project, an investment appraisal proved necessary in most cases. Also, other sources of external financing, such as incentives or sponsorships, are related to a higher frequency of investment appraisals. This finding clearly underlines the general expectation that external investors tend to be interested in the value as well as in the risk of a project.

The outgoings and the income associated with an RES activity influence the liquidity of the particular management unit. Especially when the RES business is not integrated in other businesses in one firm, a specific financial plan is of utmost importance for sustaining the liquidity of the enterprise. In general, the recommendation is to establish a monthly budget, specifying the planned financial expenditures and

Table 6.32. Checklist for conducting an investment appraisal.

Step	Content
1. Objectives	What is the objective of the investment? Quantify the level and scope where possible
2. Options	What options have been considered but not analysed in detail? Why were they dismissed? Is a 'do nothing' option included? Are options treated as inclusive packages when they can be split?
For each option	
1. Costs	What is planned expenditure and when does it occur? (How long are capital items expected to last? Explain why you have chosen this time length)
2. Benefits	What are the revenue benefits or financial savings and when do they occur? Are there other monetary benefits which will result? To what extent do they meet or surpass the objective?
3. Unmarketed costs and benefits	Are there any other costs and benefits which can be identified but not quantified?
4. Relative price	Are all costs and benefits likely to increase by the same amount with inflation? (If so, use today's prices throughout. If not, adjust for relative price changes)
5. Present value of costs and benefits	Discount all values to the present for comparison. Do options have different time lengths? If so, annualize the NPVs
6. Uncertainty – sensitivity analysis	Are any values of costs and benefits uncertain? Revise the assumptions using optimistic and pessimistic estimates
7. Conclusion	Clearly indicate the preferred option and the extent to which its performance exceeds that of its rivals

receipts, as shown in Table 6.33 (Wirtschaftsförderungsinstitut, 1996, p. 20):

Taxation of RES activities

Although the RES-related activities of a forest owner may be based on different motivations, the main emphasis in the context of this research lies on projects which are to contribute to the income of the forest owner. The case-studies indicate, however, that, at least for the time being, such a financial goal is predominant only in relative terms. Only less than half of all investigated cases were meant to achieve a net profit and 30 out of all 98 cases had no financial goals specified at all. Various aspects of taxation are none the less potentially relevant items to be taken into account when planning an RES project.

The net revenues of an RES activity may be allocated to different types of income, such as income from forestry, income from a trading company, income from letting and leasing. Such a differentiation can be of some importance, especially where different tax rates or different tax regulations are to be applied (e.g. lump-sum taxation of forestry). Also, the type of business unit may influence the tax burden, as discussed by Mantau and Mertens (1997) in regard to one of the RES case-studies. When the RES business does not allow for profits, even in the long run, it might even be considered as a private occupation from the taxation point of view. Consequently, relevant losses must not be included in the tax return.

The case-studies indicate that the majority of projects are not necessarily to be considered as a separate business from the taxation point of view. (For the criteria for delimitation and its consequences in terms of taxation in Austria, see, for example, Kleinszig, 1994.) Conversely, more than a quarter of all cases are definitely outside the field of forestry and have to be treated as separate businesses. A delimitation might be of great importance for an indi-

Table 6.33. Cash budget planning.

Cash budget item	1st month	2nd month	3rd month
1 Opening balance			
2. Planned receipts			
Transactions (incl. VAT)			
Loans			
Private-asset contributions			
Other receipts (incl. VAT)			
Sum of receipts (1 + 2)			
3. Planned payments			
Costs to be paid (incl. VAT)			
Payments for investments (incl. VAT)			
Private withdrawals			
Payments to the tax authorities			
Interest to be paid			
Repayments of borrowed capital			
Other payments (incl. VAT)			
Sum of payments			
Surplus/deficit (1 + 2 − 3)			
Coverage of the deficit			
Cheque account loan			
Private-asset contributions			
Other			
Use of surplus			

vidual project. National tax regulations will not immediately provide definite answers in all cases, usually leaving some space for individual judgement by the manager, but especially by the tax authorities. One has, therefore, to take into consideration the possibility that a project might at first be regarded as pertaining to forestry but will have to be treated as a separate business later on, especially when the size of the business increases. Within reasonable limits, it could be possible to influence this allocation by designing the contracts in such a way that the forest enterprise is formally just being reimbursed, instead of deriving an income from rendering services. This has been discussed to some extent in Section 6.4.2. Case DE03 is referred to as a relevant example.

On the one hand, purely environmental products hardly account for the RES activity to be regarded as a separate business from the taxation point of view. On the other hand, five out of the nine projects providing products classified as environmental and traditional constitute a separate item for taxation. Also, the financial goal pursued in the project seems to be only partly related to the question of taxation. Only one-third of the respondents striving for a net profit stated that the RES business is to be treated separately. A somewhat closer relationship can be observed between investment and treatment in taxation: in one in two cases where there has been a substantial investment, separate taxation applies, as compared with a share of 17% when there was no substantial investment involved in establishing the project. The case-studies do not provide any evidence as to a correlation between the age class of a project and its fiscal treatment.

Regulations pertaining to income tax usually state certain threshold levels as to the requirements of bookkeeping for taxation purposes. (Further requirements of accounting may be associated with national regulations on social security, however.) Although financial accounting is primarily to provide information and evidence for taxation purposes and has to comply with respective rules, it is in principle also a major source of management information. The balance sheet and the profit-and-loss account, as well as various ratios, such as ratios of liquidity, solvency and profitability, and especially the return on capital, characterize the financial and economic situation of the business in question. In general, the greater the economic significance of a project or business, the more important management tools based on information stemming from financial accounting will be for controlling purposes.

Once more, the empirical evidence provided by the case-studies indicates the great variety of practical implementations of RES projects. According to the responses received, financial accounting does not provide any relevant information for efficiently running the RES business in about half of all cases. Either there is no specific financial accounting at all or it just serves taxation purposes. Roughly 50% of all respondents state the contrary, claiming that financial accounting does provide them with relevant and significant management information.

There is, however, a significant tendency revealing that the more ambitious the financial goals, the more frequently financial accounting is applied and esteemed as a management tool. Whenever the RES business is to be regarded as a separate business from the taxation point of view, some kind of financial accounting is obviously required and at least in 50% of those cases it is regarded not only as a formal obligation but as highly significant for running the business. Furthermore, the case-studies indicate a certain relationship between financial accounting and the share of RES-related income. The higher the share of project-related income, the more frequently financial accounting is applied and the more often a high significance of financial accounting is attested to.

As to the product categories, financial accounting has been significant when selling traditional tangible products (product group (PG 2.1)), when offering accommodation (PG 3.1), with complex recreational facilities (PG 4.1) and with complex seminars (PG 5.1). Finally, every sixth respondent using RES-related financial accounting stated that, in the light of experience, financial accounting should have been performed in a better way.

Usually, all goods and services delivered to the market are subject to value-added tax (VAT). Consequently, VAT will apply to most RES products. As far as the RES activities are regarded as part of the forestry business, specific regulations, such as lump-sum taxation or tax exemption, will also apply to the RES-related part of the business. Within a forest enterprise, however, a rate of VAT different from the one for unprocessed timber may have to be applied, the tax rates being specified for the different kinds of products and therefore not necessarily being equal for all products of one enterprise. In Austria, for instance, the current rate for VAT is 10% for timber and other forestry and agricultural products, whereas the rate to be applied for most of the other goods and services is 20%.

Besides income tax and VAT, other taxes, such as local taxes, property tax and land tax, may also be affected by an RES project. Furthermore, various fees and duties may become due, so that bigger projects will necessitate professional advice from a tax consultant.

6.4.4 Dealing with RES in managerial accounting

Tasks and means of managerial accounting for controlling RES activities

Irrespective of the question as to whether the RES business can remain integrated into the financial accounts of the forest enterprise or is seen as a separate business from the viewpoint of taxation, the management in charge of the business has to rely on specific information for controlling purposes. Whereas various ratios pertaining to the whole company, such as ratios of profitability, can be derived from the financial accounts, further management information, such as measures of efficiency or product profitability, can be provided only by means of managerial accounting. The main purpose of managerial accounting being the provision of management information for efficiently running a business, there are no legal rules or regulations to be observed. Consequently, this part of accounting is not subject to any distortions due to tax regulations or any tax-avoiding measures. Cost accounting can, therefore, be seen as a tool-box, and it is up to the management to decide which tools are to be applied and to what extent. (For a general outline of the principles and methods of management and cost accounting, see textbooks such as those by Drury (1996) and Glynn *et al.* (1998).)

The economic principle postulates that the maximum utility is to be achieved at given expenses or a given utility is to be achieved at minimum expenses. In order to assess efficiency, respective measures of utility and expenses are required. In managerial accounting, expenses are measured in terms of costs and utility is referred to in terms of revenues. Obviously, a clear delimitation of the respective costs and revenues of a certain business is a precondition for the application of any accounting tools. Supplementary to the monetary information, managerial accounting should also document the input and output in physical terms – that is, the quantities of productive agents, on the one hand, and the number of products, on the other. By combining physical and monetary data, one can derive, for example, ratios of productivity.

Information from managerial accounting is of potential relevance for various management decisions, encompassing topics such as the pricing of products or the analysis of the impact changing levels of activity may exert on the economic result. Cost accounting is of special importance for the following tasks:

- determination of the total production costs of a product.
- Investigation of the profitability of individual products.
- Determination of the minimum price to be asked from the customer.

- Make-or-buy analysis.

Ultimately, relevant data and accounts are a prerequisite for the strategic analyses and planning of forest enterprises that are engaged in several lines of production, as pointed out, for example, by Hostettler (1996).

In most cases, the analysis is based on the data of the previous period, investigating the results that have been achieved (for an RES-related example, see, for example, Dupasquier, 1997). Besides, business planning can be improved by means of standard cost accounting. Generally speaking, managerial accounting strives to model the interrelationships between the input of productive agents in terms of value, on the one hand, and the value of the output, on the other, thereby providing hints on how to improve and sustain efficiency. Cost accounting encompasses a variety of concepts, each being of special significance for a certain task. Before implementing any kind of cost accounting, one should therefore specify the kinds of management information needed. Only when the question to be answered is clearly posed can the right tool be selected. For instance, the full costs indicate the long-term price floor, whereas the marginal costs are a more accurate measure for deciding whether to increase the output at a given price (sell an additional unit of the product) or not. Consequently, different tools, such as full costing, contribution margin accounting or standard cost accounting, need to be distinguished.

Cost analysis is a central element of cost accounting. Costs may be classified according to various characteristics. The most important features for categorizing are the following:

- type of cost.
- Dependence on the level of activity (fixed or variable costs).
- Relation to payments (actual or imputed costs).
- Tangibility (tangible or intangible costs).

A breakdown according to the types of cost shows the composition of the input of productive agents in terms of value. The differentiation between fixed and variable costs is of utmost importance for assessing the total costs at different levels of activity. Fixed costs are related to the offer as such and comprise items such as depreciation, interest on capital (borrowings as well as equity), insurance premiums and various duties and fees. On the one hand, such costs occur irrespective of the number of product units marketed. Variable costs, on the other hand, change according to the level of activity or output. Typical items are wages and costs for maintenance or energy. Once the total costs can be broken down accordingly, it is possible to calculate the level of activity (the number of product units) necessary to just achieve a zero balance, the revenues covering all the costs. The result of such a calculation, the so-called break-even point, is a most important figure for planning purposes (see, for example, Gattermayer *et al.* (1996), who provide an example of a break-even calculation for the letting of rooms on a farm). Capacity, break-even point and the actual or planned level of use are most significant parameters for resource management.

Since they are a monetary measure for all the input into a business, costs are not necessarily related to payments. Equity capital and family labour are typical examples of a kind of input where there are no actual costs connected with payments, so that costs have to be imputed. Imputations are necessary to derive full costs and hence for assessing real efficiency. Consequently, a sufficient level of liquidity may be associated with a poor performance in terms of efficiency.

Especially those real costs directly related to certain cost units are tangible ones. Contribution margin accounting or direct costing concentrate on these costs. The contribution margin indicates to what extent a certain product contributes to covering the overheads which cannot be adjoined from one product to another. In the short run, the contribution margin should be at least zero, whereas in the long run it should well exceed the overheads and allow for some net profit.

The notion of controlling implies the goal-orientated integration of feedback and feed-forward in management. Consequently, tools for planning and tools for *ex post* analysis have to be integrated to establish a controlling system. As a management philosophy, controlling is of special importance, along with PM, where there is little experience and hence high levels of uncertainty have to be coped with. Consequently, the accounting tools for managing an RES project should be designed and integrated in such a way as to best fulfil the requirements of controlling the relevant process of innovation (see, for example, Littkemann, 1997; Holmes, 1998, pp. 145–148).

According to the case-studies, cost accounting is not regarded as essential for all of the documented RES activities. More than half of the projects are not managed on the basis of cost accounting. In almost two-thirds of the cases currently rendering a net profit, the managers judged cost accounting as being unnecessary. One out of four respondents, however, stated a high significance of cost accounting for managing the project. Major deficiencies of cost accounting were reported in eight cases, whereas ten interviewees stated potentials for improvements in terms of controlling.

The case-studies provide some empirical evidence as to the significance of cost accounting in relation to the financial goals of the project. Where there are no financial goals specified, 71% of the respondents judged cost accounting as being unnecessary, as compared with a share of 46% when there is a financial goal to be achieved. Conversely, when a net profit or at least a positive contribution margin is aimed for, one-third of the respective interviewees gave a high significance to the existing cost accounting. There is also a clear indication for a correlation of investments and the significance of cost accounting: in 72% of the cases where no substantial investment was undertaken, cost accounting was considered unnecessary. Conversely, in more than half of the projects involving substantial investments, cost accounting was judged highly significant. The general consideration that managerial accounting has to rely on both instruments – cost accounting as well as financial accounting – is also supported by the empirical data. The cross-tabulation shows a significant coincidence as to the respective judgements of these two instruments. It is a further, very remarkable, finding that, in 21 out of the 25 cases where cost accounting was considered highly significant, the manager could still see potential improvements in this field.

In terms of product categories, cost accounting has been significant when selling traditional tangible products (PG 2.1), when offering accommodation (PG 3.1), with complex recreational facilities (PG 4.1) and with complex seminars (PG 5.1).

Furthermore, 85 out of 98 respondents were able and willing to describe the economic performance of the RES project in terms of contribution margin. In most cases, a positive contribution margin could be achieved, roughly half of those also showing a net profit. For those seven projects showing a negative contribution margin (DE12, 19, 24, 27, IT05, 11, 18), the database does not indicate any common characteristics.

Only 61% of those heading for a net profit had achieved this goal at the time of the interview. Two out of three striving for a positive contribution margin also reported such a result, the others either exceeding or missing this goal. Even where no financial goals were specified, the vast majority (91%) reported at least a positive contribution margin of the RES business.

Whereas the notion of contribution margin seems to be very familiar to most respondents, the most significant result for the questions addressing the break-even point is the high rate of missing answers. Two-thirds of the interviewees were either not able or not willing to report the number of product units necessary to achieve a balanced result. It has to be taken into consideration that, for some kinds of products, such as management agreements in favour of nature conservation, it is by no means straightforward to specify a certain unit of measurement in such a way that the physical quantities are more or less proportional to the revenues. When there is only one contract of a certain kind (e.g. with an intermittent organization and referring to a network of mountain-bike trails or concerning a sponsorship for some set-aside area), it is hardly possible, if not totally meaningless, to consider a relationship between the length of the trails (in km) or the acreage and respective revenues. In such cases, just an overall assessment of profitability is relevant, and some alternatives may be negotiated, but there is no sense in estimating any minimum quantity to achieve a net profit. Conversely, where there is not one overall contract but the products are marketed successively and there is hence a multitude of transactions, the number depending on marketing success, the break-even point would be a valuable indicator for management purposes.

Due to the low response in regard to this item, the cross-tabulations performed on the database of the case-studies did not reveal any significant insight.

Organizational provisions in accounting

In order to be able to readily assess the RES-related costs and revenues, it is advisable to provide relevant accounts within the system of accounts. Depending on the magnitude of the RES business, as indicated, for example, by the turnover or the number of entries in bookkeeping, one additional account may suffice for all the RES business or only a number of accounts can cope with the task. In practice, the degree of differentiation within the system of accounts will also depend on the technology available. Whereas computer programs for accounting allow for a high degree of differentiation with hardly any additional costs, the number of accounts that can be handled reasonably will be quite limited when there is only manual bookkeeping. However, financial documentation of some kind is highly recommended for all RES activities. Besides the monetary documentation, registering respective input and output in physical terms (e.g. number of working hours; quantities of the services rendered) is an important task of managerial accounting.

The basic approach is to just add one or several RES-specific cost centres to the system of accounts. This allows for registering the direct costs of the RES business, so that one can judge the financial input exclusively related to the RES activity. When summing up the respective revenues and subtracting the direct costs, the resulting contribution margin shows whether those activities can bear part of the overheads or not.

Along with structural organization and personal responsibility, the RES business can be established as an individual profit centre within the company, especially when profitability is the major indicator of success, the business is of a significant size and there is a clear personal responsibility. The concept of profit-centre organization may help in concentrating on the profitability of any measure, on the one hand, and further the motivation of the personnel, on the other. Although the concept was originally developed for large enterprises, one person may well be in charge of two or even more profit centres. Profitability of a profit centre being the main issue, the accounting system has to provide relevant information relating costs and revenues associated with the line of business in question (see, for example Binder and Berger, 1998).

Since establishing an RES business is a strategic decision in terms of products and markets, any kind of related activities should be regarded as an SBU. An SBU is defined by the combination of a certain product and the specific market where it is sold. It is, therefore, a concept related more to marketing and marketing goals than to the production process. Supplementary to strategic indicators, such as market shares, specific financial measures derived from managerial accounting are none the less of fundamental interest for controlling such a unit. Ultimately, strategic considerations should be based on indicators of market success and on profitability.

Calculating the full costs of an RES activity

Cost accounting implies the valuation of all kinds of input and output in monetary terms, irrespective of whether there are payments involved or not. Consequently, some cost items that do not correspond to respective expenditures in financial accounting have to be imputed. Without such an imputation of costs, the calculation would be incomplete and give a wrong impression of the efficiency of the production. Typical items that have to be imputed are depreciation, costs of equity capital, private consumption and family labour. Any such valuation has to rely on data describing the kind and quantity of the input, on the one hand, and the value per unit of input, on the other. As there are no binding regulations for cost accounting but just some general recommendations, the determination of the value per unit to be applied leaves some space for individual judgement. This has to be taken into account when interpreting and comparing the results. Some imputations are just a revaluation of items stemming from the financial accounts, where tax regulations require a certain way of calculation that does not comply with the objectives of cost accounting. This is the case with private consumption, which is usually to be revalued at market prices, and also with depreciation, where different methods and rates may be applied. Other imputed costs have no counterpart in financial accounting at all, such as the interest for equity capital and family labour. Especially in farm forestry, family labour tends to be the predominant cost item, its valuation being of utmost significance for the results (see, for example, Sekot, 1998). Keeping a special record of working hours is, therefore, indispensable when family labour is involved and cost accounting is to be performed. Guiding principles for valuation are the market price for substitutes, on the one hand, and opportunity costs of the input factors, on the other. For instance, the interest for equity capital might be valued according to the interest to be paid on bank loans or alternatively at the rate obtainable when putting the money into a savings deposit account. The same applies for family labour: one may price the working hour according to the hourly rate of workers or contractors or derive the opportunity costs from the average hourly income from agricultural work on the farm. The valuation of family labour is a prerequisite for differentiating the imputed earned income, on the one hand, and the net profit of the business, on the other.

As regards the financial analysis of projects aiming at the provision of forest services within a forest enterprise, the main problem usually concerns the sound allocation of overhead costs. The determination of respective full costs is of increasing importance, many people objecting to pay a price for a service that is seemingly provided without any substantial cost. Basic concepts of full costing rely on a general burden rate for adjoining a respective share of overheads to an individual product (for a general consideration of dealing with overheads in accounting, see, for example, Drury, 1996, pp. 83–109; Glynn *et al.*, 1998, pp. 255–286). However, such an approach is not satisfactory when the overheads account for a substantial share of the full costs, as there is no unanimous rationale for this or any other way of cost allocation. Some kinds of RES products typically show only minor direct costs in terms of running expenses and do not even involve substantial investments, but the main costs are overheads associated with management and administration. Alternative approaches are needed here to derive reliable information on the full costs of providing the service (see, for example, the scheme for allocating administration costs proposed by Bitter *et al.*, 1994). A Swiss proposal for cost-centre accounting in respect to forest services relies on several stages of cost accounting (see Blum, 1994, pp. 16–21; Sekot, 1999). An exact record of working hours for each cost centre encompassing the whole personnel is a basic requirement. (In practice, however, many foresters may at first object to keeping such a record for themselves.) Referring to the financial accounts, an expense classification allows the differentiation between individual costs and overheads. The overheads are further classified into those which can be adjoined to certain cost centres and general overheads. Those cost centres which are not to be charged with a certain kind of overhead are identified by means of a negative selection before allocating the remaining overheads by applying a burden rate. Finally, imputed costs are derived from cost accounting, the sum of all the items giving the full costs for each line of production.

Drawing on the case-studies, the calculation of overheads seems to be rather the exception than the rule so far. About 70% of the valid answers state that no overheads were considered. Only 8% of the valid responses refer to a calculation of respective overheads. However, the data available on these cases (DE12, 18, IT19, AU03, NL10, 18, 19) do not indicate any further common characteristics. In about one out of five cases overheads were at least estimated.

Although one could expect a higher interest in respective overheads when financial goals are specified and especially when a net profit should be achieved, the empirical data do not support such a relationship. Also the structural organization of the RES product is not clearly related to the way overheads are dealt with. Furthermore, 20% of those not considering any overheads state possible project amendments in terms of cost accounting, but this is true also for one out of the seven (14%) already calculating overheads.

RES products are typically an additional output of a forest enterprise. Since provision of the service at least partly relies on the same resources as the other line(s) of business, interrelationships may cause opportunity costs of production which have to be accounted for. This is the case, for example, when the establishment of a mountain-bike trail reduces the potential income from hunting licences, as the hunting experience is affected negatively by such activities and some interesting species of game may become less abundant due to disturbances. Similarly, management agreements in favour of nature conservation may also affect the profitability of wood production in those parts of the forest that are not subject to the agreement. This might be due to necessary detours in wood transportation, a remaining excess in harvesting capacity or the like. There might also be complementary effects, one line of production furthering the success of others, e.g. by providing an additional contribution margin for covering the overheads. Any such interrelationship may considerably affect the overall profitability, so that it ought to be investigated.

The majority of the respondents were not aware of negative effects exerted by the RES business on other businesses or forest functions. The most sensitive business seems to be hunting, almost one in four cases showing a moderate or even high competition between the RES activity and hunting.

Multifunctional management not only has to coordinate the different uses of the forest but should also encompass a trade-off analysis in monetary terms. In addition to ordinary investment appraisal, such a calculation determines whether the company as a whole may profit from establishing the new line of business. In essence, discounted contribution margins associated with the different options are compared. Since the period of the investigation and the interest rate influence the results a great deal, it is advisable to perform appropriate sensitivity analyses.

According to Reiterer (1995), the following items are to be evaluated when calculating the economic impact of management agreements on nature conservation:

1. At stand level:

- increase or decrease in costs;
- increase or decrease in revenues.

2. As regards the soil:

- deterioration or improvement.

3. At company level:

- investments;
- loss of working income;
- higher levels of remaining fixed costs;
- additional costs for administration;
- financial impacts on hunting and other nontimber products.

The guidelines for evaluating the magnitude of reimbursements to be paid to forest owners voluntarily participating in a newly established national park in Upper Austria postulate the following general scheme for pricing related management agreements (see Landwirtschaftskammer für Oberösterreich, 1995, p. 6):

reimbursement = additional costs for administration
+ difference of contribution margins (without and with management agreement in the fields of forestry, hunting and non-timber production)
+ diminution of the market value of the forestry asset
+ negative impacts on the remaining forest enterprise

Accounting for trade-offs has been the exception with the case-studies investigated, a relevant analysis existing for only about 10% of all projects. Another 10% of the respondents consider trade-offs as probably relevant without having analysed them. No relevance was stated in one-third of the cases. The most significant result in this respect, however, is that the majority admitted to having no insight into the economic interrelationship between the RES product and other lines of production. This seems to indicate a substantial lack of know-how, information, guidance and advice in this respect.

Of those striving for financial goals, 16% had relevant trade-offs analysed, whereas another 38% had no idea about financial interrelationships of that kind. Where no financial goal was specified, trade-offs were not analysed at all, and 56% of the respondents had no notion in this respect. Trade-offs would be of special interest where the RES business is fully integrated into the forestry enterprise. In 10% of these cases, trade-offs were analysed and, in another 40%, they were considered to be definitely not relevant. Thus, there remain another 50% of cases where the manager either considered trade-offs as probably being of relevance (7%) or had no idea at all (43%). However, there is no statistical indication that the awareness of financial trade-offs is related to stated deficits as regards cost accounting.

One might expect a significant correlation between the level of conflict between the RES project and other lines of production, on the one hand, and the awareness of financial trade-offs, on the other. This assumption can be supported by the empirical findings in terms of a general tendency only. Including multiple answers, there were 21 situations of high conflict between the RES business and one other forest function reported (in total, five functions were considered). The four categories as to the awareness of trade-offs were evenly addressed, which means that in every fourth case of high conflict an analysis was performed and with the same frequency the manager had no idea about financial trade-offs. Where there was no conflict situation, an analysis took place in 7% of the cases, the managers being ignorant of trade-offs in another 46% of all relevant cases. The limited number of high conflict situations does not allow for a further differentiation as to the lines of production affected.

Complementarity between the RES project and another forest function could be another reason for investigating respective financial trade-offs. Relevant analyses of the data provided by the case-studies could not reveal any such relationship, however.

The cross-tabulation with product categories indicates that the calculation of trade-offs has been comparatively significant for long-term contracts for drinking-water and energy production (PG 1.1), when offering accommodation (PG 3.1), for complex seminars (PG 5.1), for contracts about environmental services (PG 6.2) and in cases of sponsoring contracts (PG 6.3).

Practical calculations

The bigger the project is in monetary terms, the more necessary is a sound calculation of associated expenses and revenues. Where the financial volume is low, however, a rough consideration of the monetary effects will suffice. Nevertheless, one should always try to evaluate and manage the associated risks, keeping in mind that liability may well trigger a financial burden exceeding the capital invested. Irrespective of the attractiveness of the idea concerning the establishment of a certain RES activity, a financial appraisal of some kind is indispensable to justify any investment or risk.

The calculation of an RES project is typically characterized by the novelty of the activity envisaged, a lack of respective experience and information being the bottleneck for a sound investment appraisal. Whereas the forest owner may be supposed to be quite familiar with an investment appraisal concerning wood production and relevant connections being more or less readily available (see, for example, the *Investment Appraisal Handbook* of the Forestry Commission, 1987), such guidance is missing in the new and diverse field of RES. Consequently, a first step should be to gather information on sources of information. References may be derived from pilot studies but also from various non-forestry sources, and usually it will be well worth while to look for those. In the following, some corrections for the calculation of RES products as documented in literature are discussed.

Quotations of market prices for certain types of services may provide a first impression as to the magnitude of reasonable price expectations, as exemplified in Table 6.34. However, they represent only more or less significant experiences (averages or ranges) derived from a limited set of cases and do not necessarily apply to a specific case in question. Consequently, they cannot substitute for relevant market research, cost analysis and pricing.

More specific hints may be available in terms of tabulations of standard net return. Such tables are used, for example, in Austria by the agricultural extension service of the chambers of agriculture. Under the heading of agricultural and forestry activities, they also refer to certain non-agricultural activities open to farmers. In essence, the calculation sheet for lodging services provides the following information (see Bundesministerium für Land- und Forstwirtschaft, 1997a, pp. 221–229), which may be of interest also for forest owners considering whether to establish such a business:

- working hours per room, depending on the length of the season.

Table 6.34. Example of reported prices of forest services (adapted from Grill, 1996).

Service	Reported prices
Short-term storage of material	€0.36–0.73 m^{-2} $month^{-1}$
Long-term use of area	4% of the market value of the area per year
Short-term use of roads by lorries	€2.91–3.63 t^{-1} km^{-1}; €1.45–2.91 m^{-3} km^{-1}
Short-term use of roads by car	Minimum: €3.63 km^{-1}
Long-term use of roads	€109–363 $year^{-1}$
Supply networks (underground)	€1.45–2.91 m^{-2} (once)
Pipes for natural gas	About €10.17 m^{-1}
Local water pipes	€1.09–2.18 m^{-1}
Regional water pipes	€10.90–29.07 m^{-1}
Drinking-water and springs	€182–363 l^{-1} s^{-1} $year^{-1}$; €0.04–0.07 m^{-3} $year^{-1}$
Gravel	€1.09–2.18 m^{-3}
Ski-slopes	€0.02–0.22 m^{-2} $year^{-1}$
Tracks for cross-country skiing	€0.07–0.22 m^{-2} $year^{-1}$
Horse-riding	€145–218 ha^{-1} $year^{-1}$; €109–145 $horse^{-1}$ $year^{-1}$
Bridle-paths	€0.07–0.87 m^{-1} $year^{-1}$
Cycling trails	€0.29–0.51 m^{-1} $year^{-1}$
Letting of huts	€1453–2907 $year^{-1}$

- Investments and the respective economic life of the assets.
- Breakdown of variable costs related to the length of the season.
- Additional overheads independent of the season.

The tables allow the estimation of the respective costs for single and double rooms, as well as apartments, as determined by the length of the season. By introducing a price per day, one can derive the respective contribution margins and, when pricing the working hours and calculating the depreciation and the interest on capital invested, also net profits. The example in Table 6.35 illustrates the scheme referring to an apartment for four persons, the invested capital for furnishing the apartment being €12,750.-.

Such tables may help the forest owner to judge the economic significance of the service in general and thus provide the basis for a first project evaluation during the planning process. However, they are not sufficient either for pricing or for an investment appraisal as they are based on the generalized data of a previous year (in this case 1995). In contrast, the structure of the calculation may well be used for the forest owner's own specific computations, using the actual prices of the productive agents employed. It has to be stressed that the example quoted does not include either labour costs as part of variable costs or depreciation and interest on capital invested as part of the fixed costs. The terms 'variable cost' and 'contribution margin' are, therefore, somewhat misleading.

For some kinds of activities, detailed calculation sheets may be readily available, as, for example, is the case in respect to lodging services provided by farmers (see Gattermayer *et al.*, 1996; similar documentation is available in Germany). The textbook of Gattermayer *et al.* (1996) provides a concise background information covering the most relevant aspects of the business, as well as calculation schemes and examples of pricing, profitability assessment and cost analysis. Various checklists help the forest owner not to miss any relevant item.

A further reference is provided, for instance, by the database for measures of nature conservation and landscape protection, as compiled by the Bavarian Agency for the Protection of the Environment (Bayerisches Landesamt für Umweltschutz, 1998). This database comprises cost calcu-

Table 6.35. Calculation of standard net return of lodging (adapted from Bundesministerium für Land- und Forstwirtschaft 1997a, p. 229; values per apartment per season).

	40-day season	120-day season
Variable costs depending on the season		
local tax	€174	€523
detergents	€103	€308
Sum	€277	€831
Costs independent of the season		
Maintenance	€191	€191
Charge of tourist agency	€70	€70
Insurance	€38	€38
Sum	€299	€299
Total of 'variable' cost	€576	€1130
Contribution margin at a price of €43.60 day^{-1}	€1168	€4102
Contribution margin at a price of €50.87 day^{-1}	€1458	€4974
Contribution margin at a price of €58.14 day^{-1}	€1749	€5846
Working hours per year	45	127

lations for a variety of measures. The individual measures are categorized according to the general type of the measure, the individual measure and the technology applied, as well as special conditions. A system of surcharges and discounts allows for adapting the calculation to the specific case. The database allows for an estimation of the direct costs of each measure, whereas overheads and profit surcharge are to be calculated separately. Consequently, the price to be determined on this basis covers the costs for carrying out the measure only, but does not include any reimbursement of the landowner. The example given in Table 6.36 illustrates this calculation scheme. The calculation refers to a detailed description of the procedure as well as of the machinery and further conditions, providing hints as to the source of data and the approximate season for carrying out the work, as well as the literature.

Further hints for cost calculation and the pricing of some services may be derived from general agreements, such as those concerning the reimbursement of forest owners affected by newly established national parks or those participating in national programmes, such as the Austrian network of natural forests.

In pursuing the goal of establishing a network of natural forests throughout Austria, the Ministry of Agriculture and Forestry negotiated the reimbursement of the voluntarily participating forest owners with forest owners' associations and thus developed the following general scheme for calculation (Bundesministerium für Land- und Forstwirtschaft, 1997b). The yearly reimbursement per ha is the sum of a fixed basic fee of €47.24 (ATS 650), the production value of the forest and the VAT. The basic fee is meant to cover the following items:

- administration concerning the protected area.
- Additional difficulties and increased costs of forest production on adjacent woodland.

Table 6.36. Calculation sheet for pricing the measure 'removal of trees and bushes' (adapted from Bayerisches Landesamt für Umweltschutz, 1998, p. 75).

Favourable conditions:			
Density of stocks: 0.06 m^{-2}; slope: 0%; distance for transport: 0 m; size of the plot: 0.5 ha			
(1) Labour costs	€13.29 h^{-1}	201 h	€2672.01
(2) Chain-saw	€5.11 h^{-1}	86 h	€439.71
(3) Rope winch (6 t)	€8.18 h^{-1}	25 h	€204.52
(4) Tractor (70 kW)	€25.05 h^{-1}	83 h	€2079.42
(5) Other machinery	€3.83 h^{-1}	59 h	€226.25
(6) Tools	€0.51 h^{-1}	83 h	€42.44
(7) Travel costs (10 km)	€0.46 km^{-1}	120 km	€55.22
Costs ha^{-1}			€5719.57
Surcharges for unfavourable conditions (multipliers/percentages)			
Density of stocks	0.1 m^{-2}: 1.35; 0.2 m^{-2}: 2.20; 0.3 m^{-2}: 3.00		
Slope	25–40%: 1.12; 41–55%: 1.30		
Distance for transport	50 m: 3%; 100 m: 5%; 150 m: 8%; 200 m: 12%		
Size of the plot	0.2 ha: 1.10		
Sum of surcharge percentages		%	
Product of multipliers			
Adapted costs ha^{-1}		€	
Surcharge for overheads and profit		€	
Price of the offer (ha^{-1}; VAT not included)		€	

- Contribution to the protection of biodiversity.
- Utilization of the area for research, teaching and education.

The production value is calculated according to the following formula:

$$V = YC \times (1 - LOY) \times SD \times SV$$

where:

V	= value
YC	= yield class (in terms of mean total increment)
LOY	= loss of yield
SD	= stand density
SV	= stumpage value

An escalator clause states that, when the general index of wood prices exceeds +5%, a respective correction of the production value is due. Additional measures asked from the forest owner are calculated separately.

Wood production, on the one hand, and nature conservation, on the other, are alternative ways of utilizing woodland, thereby competing for the forestry resources. Consequently, services of nature conservation are likely to affect wood production in economic terms, triggering either higher costs or diminished revenues or both. Thus, the net revenues of wood production forgone are an important item within the economic assessment of such services.

In connection with the establishment of a national park in Upper Austria, a team of specialists worked out a general framework for evaluating a broad range of respective services provided by forest owners (see Landwirtschaftskammer für Oberösterreich, 1995). The guidelines cover the following items, which are of potential relevance regarding services in nature protection:

- diminution of income, revenue foregone.
- Diminution of the value of hunting.
- Loss or diminution of revenues from non-timber products.
- Loss of fixed assets, such as buildings and forest roads.
- Taxes and fees still to be paid for the area (e.g. real property tax).
- Increased costs for administration and surveillance.
- Loss of working income.
- Transaction costs for purchasing supplementary wood lots.
- Higher levels of remaining fixed costs.
- Negative effects on the remaining part of the forest enterprise.
- Higher costs for management and forest protection.
- Diminution of the market value of affected areas.

However, one should notice that, contrary to the calculation scheme for the natural forests presented above, these guidelines merely refer to the various kinds of impacts on ordinary forestry production and do not reflect any benefits on the side of the customer brought about by such services. Consequently, the guidelines are helpful for deriving the necessary amount for reimbursement but are not sufficient for pricing a respective service. (A general list of items with potential relevance for calculating reimbursements, as well as specific hints for calculation, is provided, for example, by Sagl, 1995, pp. 181–237.) Although such calculations allow the estimation of the specific costs of production associated with the service, thereby providing a lower limit for price negotiations (see, for example, Brabänder, 1992), potential uses of the forest associated with higher returns than the present ones are to be considered additionally, as stressed by, for example, Moog (1992, p. 690).

Moog and Brabänder (1992) propose tables like Table 6.37 for calculating discounted contribution margins associated with environmental services directed at nature conservation. Relevant tables should be used for different discount factors to allow for sensitivity analyses.

6.4.5 Summary

In terms of strategic management, an RES project is a matter of business diversifica-

Table 6.37. Calculation sheet for estimating discounted contribution margins (adapted from Moog and Brabänder, 1992).

Year	Measure	Revenue	Expense	Contribution margin	Discount factor $P = 2\%$	Discounted contribution margin
0					1.000	
1					0.985	
2					0.971	
3					0.956	
4					0.942	
5					0.928	
10					0.862	
15					0.800	
...					...	
Total						

tion, in that a new product is offered to already existing or new markets from the point of view of the forest enterprise. The specific combination of product and market constitutes an SBU. It is at the level of the SBU that strategic goals are to be specified and market strategies are to be developed and applied. The task of managing an SBU asks for clearly defined personal responsibilities within the company. Furthermore, specific management tools are to be designed.

A quick scan performed at an early stage of project development helps to identify the respective requirements. Such an analysis has to take into account legal regulations, on the one hand, and necessary measures and instruments for controlling the RES-related activities, on the other. Depending on the significance of the project in monetary as well as non-monetary terms, the necessary steps to take may vary to a great extent, ranging, for example, from negotiating a single contract to founding a complex company for running the RES business.

A formal delimitation of the RES activities from the forestry business may turn out to be necessary on legal grounds or just advisable in terms of, for example, liability or taxation. The choice of the type of business unit triggers various consequences in different fields, so that usually specialized advice provided by lawyers or consultants will be worthwhile.

Tax laws or trade regulations may ask for specific provisions in terms of financial accounting. Besides, an investment appraisal can serve to evaluate the prospective profitability of the project and provide essential information for financial planning. Managerial accounting can be regarded as a tool-box, offering a great variety of tools for controlling the business activities. Such tools, such as full costing or break-even point analysis, are to be selected and customized specifically. The allocation of overhead costs and the calculation of opportunity costs are two of the main problems to be dealt with where production costs are to play a major part in the price negotiations. In practice, costing and pricing may be supported by specific aids such as formulas for calculating reimbursements on the basis of opportunity costs.

Notes

1 As a source of relevant information, the Austrian Central Statistical Office may be mentioned here.
2 A source of relevant information could be the EU studies of regional programmes: for instance, LEADER II (Amt der Kärntner Landesregierung, p. 8), INTERREG II, etc.
3 In the course of the realization process, the deviation between the aims of planning and the forecast always becomes bigger. This process is caused by basic conditions (strengths/weaknesses, chances/risks), which continually change. The difference can be corrected by strategic measures.
4 In Austria, the use of drinking-water resources is subject to permission. A new project concept must be able to prove a public demand (customer). Otherwise, it would be very hard to get a concession from the water authority. The use of drinking-water and its future opportunities are furnished with the highest priority.

References

Altrogge, G. (1996) *Netzplantechnik.* Oldenburg Verlag, Munich/Vienna, 279 pp.

Ammer, U. and Utschik, H. (1982) Methodische Überlegungen für eine Biotopkartierung im Wald. *Forstwissenschaftliches Centralblatt* 101(1–6), 60–68.

Anon. (1998) Streit ums Grundwasser – Ziviltechniker als Alternativ-Zertifizierer. *Holzkurier* 53(4), 6–7.

Arbeitskreis Forstliche Landespflege (ed.) (1996) *Waldlebensräume in Deutschland – Ein Leitfaden zur Erfassung und Beurteilung von Waldbiotopen.* Ecomed, Landsberg, 124 pp.

Balck, H. (1994) Projects as elements of a new industrial pattern: a division of project management. In: Cleland, D. and Gareis, R. (eds) *Global Project Management Handbook.* McGraw-Hill, New York, pp. 2-1–2-11.

Bätzing, W. (1991) *Die Alpen: Entstehung und Gefährdung einer europäischen Kulturlandschaft.* Beck, Munich, p. 216.

Bayerisches Landesamt für Umweltschutz (ed.) (1998) *Kostendatei für Maßnahmen des Umweltschutzes und der Landschaftspflege.* Merkblätter zur Landschaftspflege und zum Naturschutz 5, Munich, 176 pp.

Becker, J. (1994) *Strategische Ausrichtung der Informations- und Organisationsstruktur des Unternehmens.* Forkel, Wiesbaden, 286 pp.

Berning, U. and Mehrmann, E. (1992) *Microsoft project für Windows – Einsteigen leicht gemacht.* Vieweg, Braunschweig, p. 168.

Binder, B. and Berger, W. (1998) Profit Centre: Industriestandard für die Forstwirtschaft. *Österreichische Forstzeitung* 109(12), 7–8.

Birker, K. (1995) *Projektmanagement,* 1st edn. Cornelsen, Berlin, 175 pp.

Bitter, A. (1994) Aufgaben der Forsteinrichtung im Rahmen einer mulitfunktionalen Forstwirtschaft. *Forst und Holz* 49(6), 162–165.

Bitter, A.W., Johann, M. and Brabänder, H.-D. (1994) Weiterentwicklung des Betriebsabrechnungsbogens zur Verteilung der Verwaltungskosten. *Forstarchiv* 65, 58–64.

Blum, A. (1994) *Zur Marktfähigkeit infrastruktureller Leistungen des Waldes.* Report 94/4, Professur für Forstpolitik und Forstökonomie, Zurich, 67 pp.

Bowman, C. and Asch, D. (1987) *Strategic Management.* Macmillan, London, 407 pp.

Brabänder, H.-D. (1992) Vertragsnaturschutz – eine Chance für die Forstwirtschaft? *Forstarchiv* 63, 41–45.

Brandenberger, G. and Ruosch, E. (eds) (1996) *Projektmanagement im Bauwesen,* 4th edn. Baufachverlag, Dietikon, 231 pp.

Brandstätter, G. and Synek, H. (1988) Fehler im Projektmanagement: Planung, Kontrolle, Organisation. *Zeitschrift Führung und Organisation* 57(88)(6), 394–398.

Bundesministerium für Land- und Forstwirtschaft (ed.) (1997a) *Standarddeckungsbeiträge und Daten für die Betriebsberatung 1996/97/98.* Ausgabe Ostösterreich, Vienna, 229 pp.

Bundesministerium für Land- und Forstwirtschaft (ed.) (1997b) Vereinbarung über ein Naturwaldreservat. Vertragsmuster, unpublished.

Bundesverband Urlaub am Bauernhof (ed.) (1996) *Spezialisierung der Höfe.* Vienna, 34 pp.

Damm, T. and Grahlmann, G. (1996) *Altgebäude als Einkommensquelle*. Landwirtschaftsverlag GmbH, Münster-Hiltrup, 232 pp.

Draxler, R. (ed.) (1998) *Neue Lösungsansätze zur Gestaltung und Nutzung von Freileitungstrassen nach ökologischen und ökonomischen Gesichtspunkten. Zwischenbericht 1998*. Österreichischen Elektrizitätswirtschafts-Aktiengesellschaft (Verbund), Vienna.

Drury, C. (1996) *Management and Cost Accounting*, 4th edn. International Thompson Business Press, London, 928 pp.

Duncan, W.R. (1996) *A Guide to the Project Management Body of Knowledge*. Project Management Institute, Upper Darby.

Dupasquier, P. (1997) Was kosten die Erholungsfunktionen des Waldes? *Wald und Holz* 6, 12–15.

Eastman, J.R., Kyem, P.A.K., Toledano, J. and Jin, W. (1993) *GIS and Decision-Making*. Explorations in Geographic Information Systems Technology, Vol. 4, United Nations Institute for Training and Research (UNITAR), Genf.

Eid, M. (1986) *The Project Cycle*. Technical Paper 4, FAO Investment Centre, Rome, 16 pp.

Eschenbach, R. (ed.) (1993) *Privatisierung öffentlicher Leistungen: Beurteilung – strategische Optionen – volkswirtschaftliche Konsequenzen*. Manz, Vienna, 258 pp.

Fiedler, S., Rabl, W. and Weidinger, B. (1994) Management von Krisen und Chancen im projektorientierten Unternehmen, 2nd edn. In: Project Management Group (ed.) *Project Journal 1994*. Project Management Group, Vienna, pp. 12–16.

Forestry Commission (ed.) (1987) *Investment Appraisal Handbook*. Edinburgh.

Frank, G. and Hinterleitner, F. (1994) *Wald-Öko-Punkte-System*. Forstliche Bundesversuchsanstalt, Vienna, 24 pp. + appendix.

Frank, G. and Starlinger, F. (1994) *Katalog der natürlichen Waldgesellschaften in Niederösterreich*. Forstliche Bundesversuchsanstalt, Vienna, 28 pp.

Gaiser, H. (1992) *Erfolgreiche Strategieplanung und moderne Führungstechniken*. VDI, Düsseldorf, 136 pp.

Gareis, R. (1991) *Projektmanagement im Maschinen- und Anlagebau*. Manz, Vienna, 407 pp.

Gareis, R. (1993) Projektmanagement. Lecture Notes, University of Economics and Business Administration, Vienna, 23 pp.

Gareis, R. (1994) Management by projects: specific strategies, structures and cultures of the project-orientated company. In: Cleland, D. and Gareis, R. (eds) *Global Project Management Handbook*. McGraw-Hill, New York, pp. 4-1–4-22.

Gareis, R. (1995a) Controlling in project-orientated enterprises. In: Project Management Group (ed.) *Project Journal 1995*. Project Management Group, Vienna, pp. 4–6.

Gareis, R. (1995b) Design of project organization – systemic-constructivistic project management. Paper presented in the Project Management Seminar, University of Economics and Business Administration, 16 pp. + appendix.

Gareis, R. (1997) Organizational design of projects. Unpublished, PM Seminar, Vienna.

Gattermayer, F., Burgstaller, M. and Wöber, G. (1996) *Kostenrechnung und Finanzierung für 'Urlaub am Bauernhof'-Anbieter*. Leopold Stocker, Graz, 119 pp.

Gazdar, K. (1989) *Informationsmanagement für Führungskräfte: Konkrete Perspektiven für Wissenschaft, Verwaltung und Politik*. FAZ, Frankfurt-on-Main, 301 pp.

Glynn, J., Perrin, J. and Murphy, M. (1998) *Accounting for Managers*. International Thompson Business Press, London, 477 pp.

Grill, F. (1996) Entgelte für Dienstleistungen. *Österreichische Forstzeitung* 107, 15.

Grudowski, S. (1996) Begriffsverständnis 'Informationsmanagement' aus der Sicht von Information und Dokumentation. *Nachrichten für Dokumentation* 47, 351–360.

Häfner, C. (1996) Globale Datennetze: Chancen oder Risiko für unsere Gesellschaft? *Nachrichten für Dokumentation* 47, 43–44.

Hampicke, U. (1991) *Naturschutz-Ökonomie*. Ulmer, Stuttgart, 342 pp.

Hanf, C.H. (1991) *Entscheidungslehre: Einführung in Informationsbeschaffung, Planung und Entscheidung unter Unsicherheit*, 2nd edn. Oldenburg, Munich, p. 193.

Hansel, J. and Lomnitz, G. (1993) *Projektleiter-Praxis: Erfolgreiche Projektabwicklung durch verbesserte Kommunikation und Kooperation*, 2nd edn. Springer, Berlin, 224 pp.

Hanstein, U. (1992) Vollflächige Waldbiotopkartierung als Teil der Forsteinrichtung. *Forst und Holz* 47(17), 531–533.

Hartl, G. (1997) *Factors influencing the allocation of utility-line planning – oral interview*. Vienna.

Haubenberger, G. (1984) *Erholungswaldbewirtschaftung.* Vorlesungsskriptum, Wien, p. 187.
Herget, J. (1995) Das betriebliche Informationsmanagement vor einer Neuorientierung – Perspektiven und Konsequenzen. *Nachrichten für Dokumentation,* 46, 25–32.
Hinterhuber, H. (1982) *Strategische Unternehmensführung,* Vol. 2, 5th edition. de Gruyter, Berlin.
Hirzel, M. (1988) Miteinander stark: Linien- und Projektmanagement als Partner. Wie lassen sich Fehler im Projektmanagement vermeiden. *Zeitschrift Führung und Organisation,* 57(1988), No. 4, pp. 261–264.
Hogg, J.N. and Jöbstl, H.A. (1997) Zum Stand des forstlichen Rechnungswesens in einigen europäischen Ländern. *Centralblatt für das gesamte Forstwesen* 114(2/3), 73–95.
Holmes, P. (1998) *Investment Appraisal.* International Thompson Business Press, London, 160 pp.
Hopfenbeck, W. (1997) *Allgemeine Betriebswirtschafts- und Managementlehre: das Unternehmen im Spannungsfeld zwischen ökonomischen, sozialen und ökologischen Interessen,* 11th edn. Moderne Industrie, Landsberg/Lech, 1056 pp.
Hostettler, T. (1996) Produkte- und Dienstleistungsportfolio – ein neues Führungshilfsmittel. *Wald und Holz* 16, 15–17.
Insley, H., Harper, W. and Whiteman, A. (1987) *Investment Appraisal Handbook.* Forestry Commission, Edinburgh.
Jenny, B. (1995) *Projektmanagement in der Wirtschaftsinformatik.* Hochschulverlag, Zurich, 494 pp.
Juen, O. (1997) *Rechtliche Aspekte zu gemeinsamen Projekten der Land- und Forstwirtschaft.* Fernschule der Landwirtschaft, Innsbruck, 16 pp.
Kleinszig, M. (1994) Forstwirtschaftliche Nebenbetriebe im Steuerrecht. *Österreichische Forstzeitung* 105(5), 75–77.
Krizek, I. (1994) Informationsversorgung als Erfolgsdeterminante strategischer Unternehmensführung. *Zeitschrift Führung und Organisation* 63(5), 324–329.
Kurth, H. (1993) *Forsteinrichtung – Nachhaltige Regelung des Waldes.* Deutscher Landwirtschaftsverlag, Berlin, 592 pp.
Landwirtschaftskammer für Oberösterreich (ed.) (1995) Nationalpark Oberösterreichische Kalkalpen. Bewertungsrichtlinien für forstliche Grundflächen und Ödflächen. Unpublished, 12 pp.
Litke. H.D. (1995) *Projektmanagement,* 3rd edn. Hansaverlag, Munich, p. 114.
Littkemann, J. (1997) Erfolgreiches Innovationscontrolling. *Zeitschrift für Betriebswirtschaft* 67(12), 1309–1331.
Lund, G. (ed.) (1998) *IUFRO Guidelines for Designing Multipurpose Resources Inventories.* International Union of Forest Research Organizations, Vienna, p. 216.
Lutschinger, W. (1996) Naturschutz und Ökosystemmanagement. Presentation at workshop of Austrian Forestry Association, 24 April 1996–26 April 1996, Ossiach, Austria.
Madauss, B.J. (1991) *Handbuch Projektmanagement,* 4th edn. Poeschl, Stuttgart, p. 454.
Mantau, U. and Mertens, B. (1997) Forstbetriebliche Umweltagenturen öffnen den Weg zur Vermarktung von Umweltleistungen des Waldes. *Allgemeine Forstzeitung/Der Wald* 17, 918–922.
Mantsch, W. (1995) Strategische Unternehmensführung in der Österreichischen Möbelindustrie. Dissertation, Institut für Sozioökonomik, Universität für Bodenkultur, Vienna, 220 pp.
Mende, W. and Bieta, V. (1997) *Projektmanagement.* Oldenburg, Munich, p. 114.
Michels, M. (1995) Outsourcing: Gestaltungsalternative zur Optimierung der Informationsversorgung. *Nachrichten für Dokumentation* 46, 91–94.
Mischkulnig, G. (1996) Marktchancen für Forsthäuser als Ferienhäuser. Eine angebotsbezogene, qualitative Untersuchung über Marktgröße und Marktpotential. Dipl.-Arb. an der Wirtschaftsuniversität, Vienna, 148 pp. + appendix.
Moog, M. (1992) Vertragsnaturschutz im Wald. *Holz-Zentralblatt* 118(43), 690–692.
Moog, M. and Brabänder, H.-D. (1992) *Vertragsnaturschutz in der Forstwirtschaft.* Sauerländer's, Frankfurt-on-Main, 199 pp.
ÖBf AG (1997) *Unternehmenskonzept 97.* Österreichische Bundesforste, Vienna, 84 pp.
OECD (1995) *Niche Markets and Rural Development – Workshop Proceedings and Policy Recommendations.* Paris, 142 pp.
Ottitsch, A. (1995) *Nutzungspotentialanalyse Achenkirch: Projektendbericht.* Institut für forstliche Betriebswirtschaft und Forstwirtschaftspolitik, Universität für Bodenkultur, Vienna, Förderungsfonds für Umweltstudien (FUST) Tirol, Vienna.
Ottitsch, A. (1996) *Analyse und Optimierung waldbezogener Raumnutzungen – Entwicklung der*

rauminformationsgestützten Nutzungspotential- und Konfliktanalyse als Instrument rationaler Landnutzungsplanung (Analysis and Optimization of Forest Land Uses – Development of Spatial Information-based Land Use Potential and Conflict Analysis as an Instrument for Rational Land Use Planning). Institut für Sozioökonomik der Forst- und Holzwirtschaft, Universität für Bodenkultur, Vienna.

Ottitsch, A. and Weiß, G. (1998) *A Comparison of Mountain Forest Policies in Europe: Proceedings of COST-E3 Final Meeting in Villa Real, Portugal.* EFI Proceedings, European Forest Institute, Joensuu.

ÖWWV (1990) *Nutzung und Schutz von Quellen in nicht verkarsteten Bereichen.* ÖWWV Regelblatt 205, Österreichischer Wasserwirtschaftsverband, Vienna, pp. 4–6.

Patzak, G. and Rattay, G. (1996) *Projektmanagement: Leitfaden zum Management von Rrojekten, Projektportfolios und projektorientierten Unternehmen.* Linde, Vienna, 592 pp.

Pernthaler, P. (1998) *Wem gehört das Waser? Nationale und internationale Wassernutzungen als Rechtsproblem.* Ziviltechnikertagung, Niklasdorf/Leoben, 11 pp.

Pichler, F. (1992) Strategien des Kostenmanagements bei unterschiedlichen Projektarten. In: Gareis, R. (ed.) *Projekte und Strategien.* Service, Vienna, pp. 33–42.

Plaimer; P. (1996) Rahmenanalyse zur Vermarktung des Gutes Wasser durch Waldeigentümer. Diplomarbeit an der Universität für Bodenkultur, Vienna, 121 pp. + appendix.

Plaimer, P. (1997a) Wasser – ein forstliches Wirtschaftsgut? *Österreichische Forstzeitung* 108(6), 13–14.

Plaimer, P. (1997b) Projektmanagement: Beachvolleyball auf der Alm – eine Erfolgsstory. *Österreichische Forstzeitung,* 108 (11), 11–13.

Plaimer, P. and Putzgruber, N. (1998) Mit der Potentialanalyse Diversifikationsmöglichkeiten erkunden. *Internationaler Holzmarkt* 89(4), 12–13.

Pototschnig, G. (1994) Marketing in einem Forstbetrieb des Weinviertels. Diplomarbeit, Universität für Bodenkultur, Vienna, 115 pp.

Reimoser, F. (1993) *Schalenwildmanagement Achenkirch.* Planungsunterlage des FUST-Tirol, Vienna.

Reiterer, F. (1995) Vertragsnaturschutz im Wald und seine Bewertung. *Österreichische Forstzeitung* 106 (3), 29–31.

Rinza, P. (1985) *Projektmanagement, Planung und Steuerung von technischen und nichttechnischen Vorhaben,* 2nd edn. VDI, Düsseldorf, 169 pp.

Rossmann, H. (1993) *Das österreichische Wasserrechtsgesetz,* 3rd edn. Österreichische Staatsdruckerei, Vienna, 837 pp.

Rota, F (1997) *Informationsmittel des Unternehmens: Wege und Formen effizienter Marktinformation.* C.H. Beck, Munich, p. 244.

Sager, W. (1997) Wasser: Welle von Konfrontation und Kooperation – Eine überlebenswichtige Ressource wird knapp. *Österreichische Militär Zeitschrift* 35(5), 507–518.

Sagl, W. (1993) *Organisation von Forstbetrieben: Grundlagen, Organisation, Führung, Analyse.* Paul Parey, Hamburg/Berlin, 205 pp.

Sagl, W. (1994) Betriebsanalyse – eine Einführung in Aufgaben und methodische Grundlagen. *Centralblatt für das gesamte Forstwesen* 111(2), 109–127.

Sagl, W. (1995) *Bewertung in Forstbetrieben.* Blackwell Scientific, Berlin, 306 pp.

Scamoni, A. and Hofmann, G. (1969) Verfahren zur Darstellung des Erholungswertes von Waldgebieten. *Archiv für Forstwesen* 18(3), 283–300.

Schirmer, C. (1992) Verfahren und Ergebnisse der Waldbiotokartierung. *Allgemeine Forstzeitung* 47(1), 38–41.

Schwanecke, W. (1991) Waldbiotopkartierung. *Der Wald* 41(2), 62–63.

Sekot, W. (1991) *Stand und Entwicklungsmöglichkeiten der Forsteinrichtung als Führungsinstrument im Forstbetrieb.* Institutes für forstliche Betriebswirtschaft und Forstwirtschaftspolitik, Vienna, 545 pp. + appendix.

Sekot, W. (1992) Strategische Analyse der Dienstleistungsproduktion im Wald. *Österreichische Forstzeitung* 103(6), 24–27.

Sekot, W. (1993) *Studienunterlagen zur Forsteinrichtung,* Vol. 19. Institut für forstliche Betriebswirtschaft und Forstwirtschaftpolitik, Vienna, p. 508.

Sekot, W. (1998) Methodological issues of cost accountancy in farm forest enterprises – Austria. In: Hyttinen, P. and Kallio, T. (eds) *Cost Accountancy in European Farm Forest Enterprises.* EFI Proceedings No. 20, European Forestry Institute, Joensuu, pp. 13–21.

Sekot, W. (1999) Challenges associated with the integration of forest services into forest management planning and accounting. In: Forestry Commission (ed.) *The Living Forest. Non-Market Benefits of Forestry.* Stationery Office, London, pp. 373–379.

Staehle, W. (1987) *Management – Eine verhaltenswissenschaftliche Einführung*, 3rd edn. Vahlen, Munich, 823 pp.

Staudt, E., Bock, J. and Mühlemeyer, P. (1992) Informationsverhalten von innovationsaktiven kleineren und mittleren Unternehmen – Ergebnisse einer empirischen Untersuchung in Nordrhein-Westfalen. *Zeitschrift für Betriebswirtschaft* 62(9), 989–1008.

Thamhain, H. and Nurick, A. (1994) Project team development in multinational environments. In: Cleland, D. (ed.) *Global Project Management Handbook.* McGraw-Hill, New York, pp. 19-1–19-24.

Tuman, J. (1994) Cultural strategies for global project management. In: Cleland, D. (ed.) *Global Project Management Handbook.* McGraw-Hill, New York, pp. 11-1–11-24.

UNEP (1993) *Developing Large Environmental Databases for Sustainable Development: UNEP Workshop 14–16 July 1993.* United Nations Environmental Programme, Nairobi.

Voigt, J. (1990) *Unternehmensbewertung und Potentialanalyse: Chancen und Risiken von Unternehmen treffsicher bewerten.* Gabler, Wiesbaden, 159 pp.

Volk, H. (1988) Auswirkungen des Biotopschutzes, der Erholung und anderer Schutzaufgaben des Waldes auf die Zukunft der Forstbetriebe. *Allgemeine Forstzeitung* 43(47), 1288–1289.

Volk, H. (1990) Waldbiotopkartierung – ökologische Bestandsaufnahme in den Wäldern. *Allgemeine Forstzeitschrift* 44(6/7), 152–154.

Volk, H. (1993) Ziele und Aufgaben der flächendeckenden Waldbiotopkartierung in Baden Württemberg. *Schweizerische Zeitung für Forstwesen* 144(10), 803–816.

Wildburger, C. (1997) Naturschutzkompetenz der Forstwirtschaft in Österreich. Dissertation, Universität für Bodenkultur, Vienna, 204 pp.

Wirtschaftsförderungsinstitut (ed.) (1996) *Leitfaden für Betriebsgründer.* Vienna, 67 pp.

Wischnewski, E. (1993) *Projektmanagement auf einen Blick – Der unentbehrliche Ratgeber.* Vieweg, Braunschweig, 116 pp.

Wohanka, E. and Stürzenbecher, K. (1991) *Forstgesetz 1975.* Österreichische Staatsdruckerei, Vienna, 263 pp.

Zielasek, E. (1995) *Projektmanagement – Erfolgreich durch Aktivierung aller Unternehmensebenen.* Springer, Berlin, 215 pp.

7

Delimitation of Property Rights

This chapter deals with the main legal aspects of public law that influence the disposal rights of a forest landowner seeking new types of use for his/her forest. In the following pages, answers will be given to questions that are raised when analysing the documented recreational and environmental services (RES) cases of this project. These questions mainly refer to access rights, specific provisions concerning the recreational use of the forest, conversion of forests and particular questions of planning and nature conservation law. However, forest law is the most important matter in this context. But, first of all, the relevant basic rights of the German constitution will be considered, since they describe the frame conditions of legislation in the area of forest law. For practical reasons, this research mainly focuses on German law. Only one aspect – the right of access – has been investigated in all countries of the project partners involved.

7.1 Basic Rights Relevant for Forest Land Use in Germany

The German constitution (Basic Law, *Grundgesetz*, GG) provides different basic rights (*Grundrechte*) which can be asserted by the forest visitors, on the one hand (Section 7.1.1), and by the forest owner, on the other (Section 7.1.2). The rights of the general public are of interest for the forest owner because his/her rights might be influenced and restricted by those of others. The rights of both groups are delimited as follows.

7.1.1 Public rights of forest use

The German Basic Law does not contain an explicit right of recreation or mobility; only the Bavarian constitution grants a right of enjoyment of natural beauties (*Recht auf Genuß der Naturschönheiten*) and recreation in natural surroundings, especially access to forests and mountain pasture, just as boating and the gathering of forest fruits, to the local customary extent, is allowed in Bavaria (Bayern (Bay)).

However, according to the ruling of the Federal Constitutional Court (*Bundesverfassungsgericht*), riding and walking in forests is protected by Art. 2 I GG as part of the general freedom of action (*allgemeine Handlungsfreiheit*). This basic right also applies to other recreational or leisure-time activities, such as playing, resting and participation in organized or commercial events in the forest.

Basic rights other than that of Art. 2 I GG do not serve as a legal basis to claim a right of recreation or mobility. Thus, in Art. 2 II 1st sentence GG – the right to life and physical integrity (*Recht auf Leben und körperliche Unversehrtheit*) – it is mainly a state and not an activity that is protected. This basic right is not involved until the physical health is impaired by refusing recreation in forests, so that a comprehensive protection is not granted by this basic right. Art. 11 GG protects the freedom of movement (*Freizügigkeit*), but it does not substantiate a public right of forest use. Although the liberty to move and settle everywhere within the country is granted by this right, neither the particular types of movement, the use of particular vehicles nor the access to particular areas is included. Finally, Art. 12 GG, which protects the free choice of occupation or profession (*Berufsfreiheit*), could be affected by restrictions on access to forests, e.g. where the interests of a riding-school are concerned. But a recent decision (VerfGH SN, NuR 1998, 248) states that such restrictions do not closely relate to the practice of a profession nor do they objectively influence the profession itself. Therefore, even a business can only refer to Art. 2 I GG for a claim to use other people's forest land.

As the general freedom of action is the main legal basis of a right of recreation and mobility, it will be considered in more detail in respect of the rights of forest visitors. First, the scope of protection (*Schutzbereich*) of Art. 2 I GG will be looked at. This will help to explain which activities this basic right applies to and the problems of application that are raised in respect of forest uses.

The scope of protection in Art. 2 I GG

Art. 2 I GG grants a right of 'free development of personality' (*freie Entfaltung der Persönlichkeit*), which has been interpreted by the Federal Constitutional Court as general freedom of action (*allgemeine Handlungsfreiheit*) since 1957. Accordingly, every type of human action is protected, irrespective of the importance the action has for the individual personality. As stated by the Federal Constitutional Court, Art. 2 I GG applies, for example, to the feeding of doves, falconry and riding in forests (BVerfGE 54, 143; 55, 159; 80, 137). Consequently, the scope of protection of Art. 2 I GG is very large and generally includes any type of recreation and mobility, though its application is excluded when a specific basic right, e.g. the freedom of assembly (*Versammlungsfreiheit*), is concerned.

However, it has to be noted that the basic rights are addressed to the state and usually serve as defensive rights (*Abwehrrechte*) of the citizens against interference by the state. Thus, forest visitors may not directly refer to the basic rights when the interference is carried out by private landowners: the citizens may only claim that the state has neglected its duty of protection.

Thus, it is generally accepted that the basic rights exert an influence on the relationship between individual citizens pertaining to different groups, e.g. between forest owners and recreationists, because they represent an objective order of value (*objektive Wertordnung*). Accordingly, the state generally has to enable the citizens to practise recreation and mobility. But it is mainly left to the discretion of the state – in particular, of the legislature – how this duty will be realized. Interference in the general freedom of action of a citizen can only be claimed when the state has obviously acted insufficiently, which rarely actually happens.

In this respect, it is of importance that the right of access provided by forest law entitles recreationists to enter not only public forests but also private ones. Furthermore, the duty to tolerate the construction of recreational facilities can be imposed on the forest owner by public bodies in designated recreational forests (see Section 7.2.7). Consequently, the legislator obviously has not neglected his/her duties as far as public use of privately owned land is concerned. Moreover, forest law provides a duty of approval of conversion of forests in order to preserve forest areas as a prerequisite for recreation.

In addition, the administration could obviously neglect its duties of protection when forest law is applied. Thus, the effectiveness of the legal instruments might be questioned when the appropriate authorities admit or tolerate enclosures of big private forest areas or conversion of forests in a very extensive way. If forest areas of primary importance for recreation, in particular those close to cities, are withdrawn from recreational use by the general public, it might be possible that a citizen is entitled to challenge such measures with reference to the general freedom of action (OVG Bln, DVBl. 1977, 901; see Section 7.6 for abbreviations of German state names). But, in practice, this will only be relevant in very exceptional cases.

As far as recreation and mobility in the forest are concerned, forest visitors may refer to the general freedom of action protected by Art. 2 I of the Basic Law. By this, citizens can challenge interference by the state – in particular, the legislator. In so far as interference is caused by private landowners, recreationists may only refer to the state's duty of protection, which is deduced from the Basic Law. But, in this context, recreationists can claim an interference in their basic right only when the state has acted insufficiently in a very obvious way, which usually is not the case.

Interference in Art. 2 I GG

Even though the Basic Law only provides a few duties of the state to enact protective regulations for recreationists, the main importance has to be seen in the defensive function of the basic rights. If the state deliberately restricts the legal position of forest visitors, an interference in the general freedom of action can be claimed. This might also be of importance for the forest landowner, as the legal rules on recreation in the forest have an effect on the admissibility of RES products.

However, it has to be borne in mind that affecting the scope of protection of a basic right by an interference does not in itself imply an infringement of the constitution. It rather signifies that justification of the interference has to be given by proving the limits (*Schranken*) set up in Art. 2 I GG. Thus, it might comply with the Basic Law if the legislature restricts recreation and mobility in forests in another way than is at present the case in Germany (Fig. 7.1).

First of all, any legal provision restricting recreation and mobility in forests, e.g. by a statutory ban on motor sport or a duty of approval for commercial events, imposes restrictions on the respective activity without touching the question of access to a particular site. These provisions impose restrictions on forest visitors even if they act with the consent of the landowner. As access is generally restricted by public bodies, an interference in the general freedom of action can be claimed.

Furthermore, the removal of the right of access according to the German Federal Forest Act (§ 14 BWaldG), as well as restrictions on this right, would be an interference, because a formerly given right is totally or partially withdrawn by the legislator. In this context, the Federal Constitutional Court has stated that the property rights of the forest landowners cannot be regarded as a general limitation for application of the basic right of Art. 2 I GG (BVerfGE 80, 137). But, in so far as specific duties are imposed on landowners, it has to be investigated if these restrictions are reasonable and if the landowner may claim compensation in the scope of his/her property right. Apart from this, it must be mentioned that public bodies may not refer to the property right of Art. 14 GG at all; only communes can at least assert their

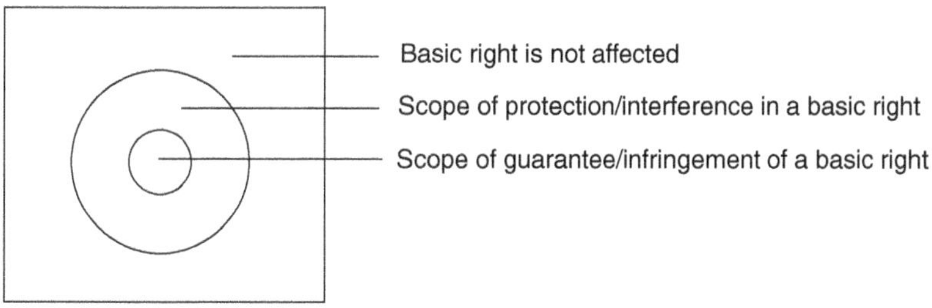

Fig. 7.1. Relation between interference and infringement of a basic right.

right of self-government of Art. 28 II GG, though this right grants less protection than the property right.

Furthermore, most federal states allow riding on designated lanes only, which implies a ban on other, non-designated lanes. In respect of § 14 BWaldG, which generally allows riding on (all) forest lanes, this regulation is a restriction of the states on the rights of the riders and thus an interference in the general freedom of action. However, the Federal Constitutional Court has already stated that such an interference is justified, so that the basic right is not infringed (BVerfGE 80, 137).

On the contrary, the restriction on the recreational purpose of the right of access (§ 14 BWaldG) is no interference, as this provision does not order any exclusion. Rather, it is stated that forest visitors do not have a further right to use forests, e.g. for organized events or commercial purposes. Thus, private law rules that the landowner has exclusive disposal rights (§ 903 BGB), as long as he/she does not contravene other rules. In general, forest visitors may not enforce specific rights with reference to the Basic Law, because of their defensive function.

Finally, it has to be questioned whether the construction of barriers by public bodies in (public and private) forests should be regarded as an interference. Even though a particular type of action is not generally prohibited, barriers impose restrictions on the extent of the recreational use of forests. Thereby, access to particular sites is questioned. But it is not quite clear if the basic right generally implies a right of access to non-public areas, in particular when they are private. However, a right of access for recreational purposes could be justified, at least to the extent where a claim to particular services, such as the construction of lanes, is not raised. This issue has to be assumed on the grounds that recreation is of prime importance for physical and psychic health, protected by Art. 2 II GG, and because it is impossible for every forest visitor to acquire his/her own ground for recreation. Thus, any recreationist who can prove significant restriction on exercising his/her right of access by barriers may uphold a right to use private land. To this extent, a comprehensive constitutional protection against unjustified barriers is granted. But, in so far as barriers are erected by private landowners, a recreationist may usually only refer to the right of access of the forest law. The basic law would only be applicable additionally if the state evidently neglects its duty of protection, as already described above.

Justification of interference

The general freedom of action is only granted within the limitations (*Schranken*) of Art. 2 I GG. These limitations provide that everybody is free to act as he/she wants to as long as he/she does not violate the rights of others or offend against the constitutional order or the moral code. Consequently, the general freedom of action can be restricted if adequate justification is given. Only an essential sphere of the individual way of life is absolutely protected, which refers back to the close relation to the unconditional basic right of human dignity (*Menschenwürde*, Art. 1 I GG) and the guarantee of the essential content of a basic right (*Wesensgehaltsgarantie*, Art. 19 II GG). As far as recreation in forests is concerned, even the right of Art. 2 II 1st sentence GG – the right to life and physical integrity – requires a minimum of recreation for everybody.

The 'constitutional order', in the sense of Art. 2 I GG, comprises the entirety of rules that comply with the constitution in a formal and material sense. In this context, the Federal Constitutional Court always proves if the general constitutional principles, to which, in particular, the principle of proportionality (*Verhältnismäßigkeitsgrundsatz*) and the principle of the reliability of law (*Vertrauensschutz*) belong, are fulfilled. Furthermore, the legislative competences and the rules of procedure (*Verfahrensvorschriften*) have to be complied with.

In respect of the constitutional order, the other limitations of Art. 2 I GG – 'the rights of others' and 'the moral code' – are not of great importance. On the one hand, the rights of others are already included in the constitutional order and are only mentioned separately for clarity. On the other, the moral code is not relevant in the context of the recreational use of forests.

As a result, the general freedom of action can be restricted by law because of overriding public interests or because of the property right of the forest owner. In general, this implies that the more the general freedom of action is restricted, the more important the overriding public or private interest has to be. Public interests that justify a restriction on recreation and mobility in forests may, in particular, consist of the interests of nature conservation and regulation of recreational traffic. In respect of nature conservation, it has to be noted that the protection of the environment is an aim of state (*Staatsziel Umweltschutz*) according to the German constitution (Art. 20a GG). Consequently, both legislature and administration have to give due weight to the ecological concern.

Furthermore, conflicts between different user groups can necessitate restrictions on recreational use, e.g. by separating the user groups. As all recreationists may refer to the same basic right (Art. 2 I GG), a balance between the different interests is required. The order of preference of the different uses has to be judged by the quantity of users and the degree of damage caused by them. Accordingly, severe restrictions can be imposed on the rather small group of riders in favour of pedestrians' security. But it would offend the constitution if a type of recreation generally admitted by the Federal Forest Act is entirely banned by state law. Neither would it be acceptable that access to forests is generally made subject to entrance fees.

Besides, the prevention of damage to the forest, the forest lanes and the vegetation lies in the interest of the forest owner as well. Moreover, the legislator might be entitled to regulate access to private forests differently from the present case in Germany, e.g. he/she might enact provisions – similar to the Dutch situation – that grant subsidies to a private forest owner for opening his/her forest to visitors; otherwise the landowner would be entitled to deny access or to charge entrance fees. By doing so, the disposal rights of the owners would be stressed and a compensation would be granted according to the services rendered by the owners, which also implies a change in the system of subsidies. However, such a tendency in legislation is not being discussed at present.

> All interference of the state in recreational use of the public requires justification. Justification might be given when recreation is restricted for reasons of nature conservation, regulation of public traffic or because of property rights of landowners. Nevertheless, unreasonable restrictions infringe the Basic Law and therefore are not admissible.

7.1.2 Property right

Subject-matter of property right

Art. 14 I 1st sentence GG provides that property is guaranteed. But the term 'property' is not defined in the Basic Law. Rather, the subject-matter of the property right has to be ascertained by the objects and the function of property right under consideration of its importance in the constitution as a whole. In the jurisdiction of the Federal Constitutional Court, every right allotted to an individual by the legal system for private disposal and benefit deserves the protection of Art. 14 GG. Thus, the term 'property' is mainly defined by statutory law and other legal rules, which include not only private law but public law as well. Due to cultural and economic changes, the definition of property may also be altered. However, the far-reaching definition of property does not mean that the scope of application of this basic right is unlimited. Not every aspect influencing the legal position of an owner is included in the constitutional guarantee of property.

First of all, private property of land and all derived rights are protected by Art. 14 GG. In this way, silvicultural, agricultural and recreational use of one's own land is included. Restrictions on this legal position, as well as dispossession, affect the property right. Thus, landed property is fairly well protected by Art. 14 GG. Even though only realized uses are protected in any case, continuance of a once acquired property right may be a matter of this basic right as well. Only potential uses that might possibly be realized in the future are generally excluded.

At times, the question has been raised as to whether development of land for building purposes is still part of the property right, since the construction of buildings is subject to far-reaching legal provisions. But it has to be taken into account that everybody has a right to development as far as admitted by law. To this extent, property right also applies to the freedom of development (*Baufreiheit*).

Furthermore, a forester can refer to Art. 14 GG if he/she has established a business (*Gewerbebetrieb*) or when he/she is only the leaseholder (*Pächter*) instead of being the owner of land. But, in these cases, the property rights are restricted to an extent which is not yet clarified. In any case, a business cannot claim a right of continuance of an advantageous legal situation, in so far as it has merely made use of a favourable opportunity. The rights of a leaseholder may be limited by law (leasehold interest – *Pachtrecht*) or by a contract with the landowner. However, the Federal Constitutional Court has recently stated that a leaseholder has to be treated like an owner in so far as he/she has the same disposal rights of a property (BVerfGE 89, 1).

Limitation of determination of property by legislator

According to Art. 14 I 2nd sentence GG, the legislature has to determine the 'substance and limits' (*Inhalt und Schranken*) of property. Even though this instruction is given unconditionally, the legislature is not entirely free. Art. 14 II GG provides that property rights are subject to social obligation (*Sozialbindung*), which is considered to be a 'guiding principle and limitation' (*Richtschnur und Grenze*) for the legislature determining the substance and limits of property. Consequently, the position of proprietors may only be restricted for public interests. Thus, for example, the social obligation of real property is defined by legal provisions permitting access to private land for recreationists or by those restricting uses that damage the ecosystem. Moreover, the social obligation of landed property reaches further than in any other case of valuable goods. This has to be concluded from the fact that land is indispensable and may not increase (BVerfGE 87, 114). Therefore, landowners have to accept far-reaching restrictions on land use, in particular by the forest law, nature conservation law and planning law.

Furthermore, every restriction on property rights has to pay due regard to the institutional guarantee (*Institutsgarantie*) and the guarantee of individual rights (*Individualrechtsgarantie*) of property, which also is called the guarantee of legal status (*Rechtstellungsgarantie*) or guarantee of continuance (*Bestandsgarantie*). The constitutional guarantee of property is characterized by the private benefit of property (*Privatnützigkeit*) and freedom of disposition (*Verfügungsfreiheit*). But this does not mean in general that every matter or every good must be placed at private disposal. Private goods can be subject to administrative regulations and single rights can be excluded from private disposal without infringing the constitutional guarantee of property.

As a result, the legislature has to delimit the different opposite interests (private and public ones) in such a way that every position concerned is preserved appropriately. Thus, the general constitutional principles have to be taken into account, to which belongs, above all, the principle of proportionality (*Verhältnismäßigkeitsgrundsatz*). This principle provides that laws, actions and measures of public bodies are permitted only to the extent necessary to achieve the legal goal. Besides, the principle of reliability of law (*Vertrauensschutz*), the principle of legal certainty (*Bestimmtheitsgebot*) and the right of equality (*Gleichheitssatz*) in particular have to be considered when property rights are delimited.

Compensation for determination of property
If determination of a property right complies with the Basic Law, there is not only a question of compensation for otherwise unfair restriction. Compensation is not only payable in the context of expropriation (*Enteignung*), although this was discussed formerly. Since the so-called '*Naßauskiesungsbeschluß*' of the Federal Constitutional Court (BVerfGE 58, 300) from 15.7.1981, a strict distinction between determination of property and expropriation has to be made. Expropriation is defined as a deliberate withdrawal of private property for the realization of public purposes. On the contrary, the abstract definition of the rights and duties of an owner – e.g. by forest and nature conservation law – always involves a determination of the substance and limits of property (BVerwGE 94, 1). Thus, general rights and duties referring to forest land use always have to be regarded as the determination of the substance and limits of property, even though particular uses might be generally excluded. A determination of property does not turn into expropriation when property rights are restricted unreasonably.

In addition, another decision of the Federal Constitutional Court (BVerfGE 58, 137, 147) provides that a claim to compensation can arise from a determination of property. Nowadays, this is described by the term 'determination of property with the obligation of compensation' (*entschädigungspflichtige Inhaltsbestimmung*). However, the former discussion on the distinction between the social obligation of property (always without compensation) and expropriation (always with compensation) is now applied to differentiate determination of property with or without compensation. But, in any case, every restriction on property rights has to be justified by the social obligation of property.

CRITERIA FOR ASSESSMENT OF COMPENSATION. Even though an owner has to bear a loss, not every determination of property entitles him/her to a compensation. The basic right of Art. 14 GG does not grant a claim to the largest profit possible on private goods. Restrictions on landowners, e.g. by the right of access or by designating special protected areas, to a certain extent can be imposed uncompensated in the scope of the social obligation of property. At present, there is a discussion on the issue of the cases in which compensation has to be given, with reference to the terms '*Situationsgebundenheit*' (≈ situation linkage) and '*Situationsberechtigung*' (≈ situation privilege) established by the courts in the context of the quite far-reaching restrictions that are increasingly imposed on land use in protected areas (BGHZ 123, 242; 121, 328; BVerwGE 94, 1).

According to jurisdiction, the 'situation' of landed property is determined by its location and its condition, as well as by its relation to the surrounding environment. Hence there follows an immanent restriction on the user rights of the landowner that is merely written down in forest or nature conservation regulations. On the one hand, compensation is not regarded as necessary in so far as a 'sensible and reasonable owner who has not lost sight of the common weal would renounce use' (BGHZ 126, 379; 123, 242). On the other hand, landed property must not become entirely worthless; it must still be possible to make use of the land in at least some respect.

Restrictions on land use generally have to be accepted uncompensated in the scope of '*Situationsgebundenheit*'. This is the case if only the existing use is fixed and a status quo has to be preserved. In respect to other uses which have not already been carried out, the Federal Court of Justice (*Bundesgerichtshof*) usually investigates if permission could be granted. If a legal claim to a particular use cannot be raised, generally no compensation has to be paid. In other cases, reasonableness of restriction without any compensation has to be proved carefully.

On the contrary, compensation has to be granted in the scope of '*Situationsberechtigung*' for otherwise unreasonable restrictions of property. Any grave and unfair restriction imposing a 'special sacrifice' (*Sonderopfer*) on a landowner necessitates compensation. The previous use which has been realized legally raises a guarantee of continuance, in particular when the landowner has invested money and/or effort and if he/she may trust in the legal situation in respect of Art. 14 GG. Thus, compensation will usually be required when landowners are placed under particular duties, e.g. to tolerate construction of recreational facilities on their land by public bodies, which compels them to bear significant losses of income from their formerly implemented forest management. However, it also has to be noted that the interruption of use for several years can also imply removal of reliability in law.

In particular cases, even a use that has not already been realized may raise a claim of compensation if it is prohibited in the future. Such a claim is assumed if a use has been legal up to this time and, according to the particular situation, it 'almost urges an owner to get performed'. Compensation, for example, has to be granted when development of a building site within Inner Areas (*Innenbereich* – § 34 BauGB) or within a binding land-use plan (*Bebauungsplan*) is refused. On the contrary, a landowner generally has to accept that development of Outer Areas (*Außenbereich* – § 35 BauGB) is not permissible. Afforestation of fruit orchards or excavation in forest areas may also be prohibited without compensation, even though a more profitable utilization is restricted thereby. Moreover, changes between agricultural and silvicultural use or between silvicultural and cultural use are not privileged by law; thus, no claim may arise therefrom. However, an owner may not be forced to bear long-lasting and unreasonable losses, e.g. an average annual loss of about DM 20,000 for preservation of a forest area. If the existence of a forest landowner is risked, this urges him/her to change use so that a claim to change use can be asserted or, in the case of refusal, compensation has to be paid. This is of particular importance for the conversion of forests. Nevertheless, the landowner has to substantiate his/her losses. As the requirements imposed on claims to compensation are difficult to comply with, a landowner will usually not get any money for restrictions imposed on him/her unless he/she has not already carried out a particular use.

If a general amendment of the law is adopted which imposes general restrictions on land use and which does not refer to single areas, the legislature does not impose a 'special sacrifice' on an individual landowner. Provisions introducing restrictions on land use – e.g. on the construction of barriers or on commercial events in the forest – apply to all landowners in the same way. Generally, these restrictions are not subject to compensation, as the legislature is empowered to make a legal amendment. But every restriction on land use has to be justified by public interest, in respect of the principle of proportionality, which places a ban on excessive measures. Grave and unreasonable restrictions have to be mitigated by exemptions or transitional provisions or even by granting compensation.

AMOUNT OF COMPENSATION. Full compensation for loss of profits is not necessary in the light of the social obligation of property. Only when a so-called 'threshold of sacrifice' (*Opferschwelle*) is crossed does compensation have to be granted (BGHZ 105, 15). This threshold has to be defined with respect to duration, type and intensity of property infringement, in order to assess the individual concern. Therefore, the actual loss of value has to be determined at least approximately. If profits had not been achieved by the former use, compensation will not be paid. But investments effected, for example, in the construction of buildings that may not be used any more or that have to be removed, are subject to reimbursement. In any case, assessment of compensation has to pay due regard to the loss of value created by the proprietor's own work.

As a 'basic amount' (*Sockelbetrag*) has not to be compensated in respect of the social obligation of property, no compensation is payable at all in the case of insignificant losses. This might be of particular relevance in the context of the recreational use of forests, because in this area frequently only low profits are made. It is generally admissible to restrict land use to the extent compatible with environmental requirements. In order to balance private economic and public ecological aspects, transitional provisions can be enacted, so that no compensation is payable to landowners. However, in some cases, the legislature has even granted further compensation, which is done for reasons of equity (*Billigkeit*), not because of binding constitutional law.

> The rights of a forest landowner are protected by Art. 14 of the German Basic Law, the property right. In this case, the legislature has to determine the substance and limits of private property. Existing property rights can be restricted for the public interest, due to the social obligation of property explicitly stated in the Basic Law. Accordingly, property rights in forests are restricted for the recreational interests of the general public, in particular by the right of access, and for reasons of nature conservation. However, any restriction must be reasonable in order to comply with the constitution. If grave or unfair restrictions have to be imposed on a landowner for the public interest, compensation must be granted by the state.

7.2 Relevance of Forest Law for RES Products in Germany

7.2.1 Main structure of German forest law

German forest law is the principal legal foundation for forest uses on which to build new economic user concepts concerning RES products. According to German constitutional provisions, forest law is mainly enacted by the federal states. For better comprehension of the framework, the constitutional background of German forest law will be outlined first.

Legislative competences of the federal legislature and the states

According to Art. 74 No. 17 of the German Basic Law (*Grundgesetz*, GG), promotion of forestry production and the import/export of forestry products is part of concurrent legislation. Thus, the federal states (*Bundesländer*) are allowed to enact a law only in so far as the federal legislature has not exercised its rights. This implies the precedence of federal legislation. But the federal legislature may pass a law only if a need for centralized federal regulation exists. According to Art. 72 II GG, with the latest amendments from 27 October 1994 (Federal Law Gazette, *Bundesgesetzblatt*, 1994 I, p. 3146), such a need is recognized when and in so far as the establishment of equivalent conditions in the whole area of the Federal Republic of Germany or the preservation of legal or economic unity necessitates a federal rule.

Besides concurrent legislation, forest law is related to framework legislation in Art. 75 No. 3 (hunting, nature conservation and landscape protection) and No. 4 GG (federal area-wide planning). In the area of the framework legislation, the competence of the federal legislature also depends on the requirement imposed under Art. 72 II GG. But, contrary to concurrent legislation, the federal legislature is not allowed to enact complete regulations, but only a framework law, which leaves substantial competences to the federal states. The constitutional amendment of 1994 has instituted further restrictions on the federal competences: only in exceptional cases may the federal legislature pass detailed regulations or provisions having a direct effect. Nevertheless, law enacted before 15 November 1994 remains valid.

The Federal Forest Act

The federal legislator has made use of his/her competence by enacting the Law on the Conservation of Forests and the Promotion of Forestry (Federal Forest Act, *Bundeswaldgesetz* – BWaldG: Federal Law Gazette 1975 I, p. 1037, amended 1984 I, p. 1034). This law is based on the competence of the federal legislature before the amendment of the constitution in 1994. It consists of framework provisions (chapter 2: Preservation of Forests, § 5 to § 14

BWaldG), which are only addressed to the states, and of others having direct effects. But even the provisions with direct effects, in particular chapter one (General Rules), leave the main competences to the federal states.

Forest law of the states

Almost every federal state (except Bremen) has enacted a state forest law (*Landeswaldgesetz*). Thus, 15 state laws exist. Due to the frame character of the Federal Forest Act, the state laws differ considerably. This can be important for approvals or other official permits necessary for RES products, as the state law determines the legal requirements.

Relationship between forest and nature conservation law

Nature conservation law is a matter of federal framework legislation as well. Thus, a Federal Nature Conservation Act (*Bundesnaturschutzgesetz* – BNatSchG: Federal Law Gazette 1987 I, p. 889, last amended 1998 I, p. 2994) and 16 state laws have been passed. Whereas forest law only refers to forests, the concern of nature conservation law are nature and landscape as a whole. On one hand, nature conservation law contains regulations that may apply to forests (e.g. protection of endangered species and areas, intrusions), while, on the other hand, it is replaced by forest law to some extent (right of access). In conformity with the Federal Nature Conservation Act (§ 27 BNatSchG), right of access to agriculture areas is usually subject to the respective state nature conservation law, which does not apply to forests; to this extent, forest law is *lex specialis*. On the other hand, few states (Bay, Niedersachsen (Nds), LSA) have enacted uniform access regulations that apply to the whole of nature, including forests as well as meadows, heath and pastures. These regulations are contained in the nature conservation law (Bay) or in another (third) separate law (*Feld- und Forstordnungsgesetz* in Nds and LSA). In Nordrhein-Westfalen (NW), uniform rules provided by the nature conservation law apply only to riding, whereas walking and cycling in forests are subject to the state forest law.

However, there are further interrelations between forest and nature conservation law, especially on the state law level, concerning particular uses, such as camping, gathering mushrooms and the construction of recreational facilities.

7.2.2 Access to forests in Germany

The right of access to forests (§ 14 BWaldG and corresponding state law) is of primary importance for the development of RES products. This right as upheld by recreationists implies the duties of the forest landowner to tolerate particular public uses. This means a restriction of the disposal rights of the landowner. Therefore, a delimitation of the right of access is necessary.

Access to forests for recreational purposes

§ 14 BWaldG allows access to forests for recreational purposes. In the sense of the forest law, 'access', on the one hand, means walking, hiking and running. Resting, playing and other activities are also included, in so far as they are done mainly for recreational purposes. Consistent with recreational purpose are scientific interests, individual playing of games and long-distance skiing. Even a rather small group of persons who join in the forest for common recreation may refer to the right of access. On the contrary, intensive types of sporting, e.g. a ski-run and overnight stays, are not regarded as recreational activities. The same applies to commercial or professional use of forests and organized or public events and meetings (VGH BW, NuR 1995, 462; VG Arnsberg NuR 1995, 485). Corresponding to that, races and festivals taking place in the forest are generally not allowed by the right of access.

This limitation of access to forests is to prevent damage to the forest and inconvenience to the forest owner; the forest ecosystem must be kept intact. The public has to conduct itself in such a way that other forest visitors and forest plants and animals are not disturbed. These rules of conduct are laid down in more detail in the so-called *Wohlverhaltensregeln* (rules of good conduct) or *Gemeinverträglichkeitsklauseln* (clauses of compatibility with the common interest) of the state laws.

It has to be emphasized that access by pedestrians is not confined to roads and lanes in the forest: the public is permitted to enter the whole forest. Only the state Schleswig-Holstein (SH) has enacted a different rule; here, access is allowed only on roads and lanes. But, in the general discussion on legal theory, there is controversy over whether this state rule contradicts federal law.

Cycling, driving in wheelchairs and riding in forests

Cycling, driving in wheelchairs and horse-riding are permitted in forests on roads and lanes only (§ 14 I BWaldG). This provision refers solely to private roads and lanes in the forest; public roads designated for public traffic are exclusively subject to traffic regulations. The function of the traffic authorities is to ensure public safety and traffic order. As far as these aspects are involved, the traffic authorities are allowed to take suitable measures in forest as well (VGH BW, NuR 1995, 264). But they must not regulate access to forests in any further way.

Access to forests in general, as well as cycling, driving in wheelchairs and riding, is permitted for recreational purposes only. Cyclists, drivers and riders have to pay attention to pedestrians and the property of the landowner. In general, narrow lanes and hiking tracks are not considered suitable for riding. In many states, riding is only allowed on designated and marked roads and lanes to prevent conflicts with other forest visitors. According to the Federal Constitutional Court, this rule is consistent with the Federal Forest Act (BVerfGE 80, 137).

Transport other than wheelchairs is not permitted. In particular, driving by motorized vehicles, such as cars and motorbikes, without authorization is prohibited (VG Freiburg NuR 1992, 94).

Other types of use allowed by the federal states

Other types of forest use can be equated, entirely or in part, with access by the states (§ 14 II BWaldG). But only a few kinds of uses are accordingly permitted. In two states (Hessen (Hess), Thüringen (Thür)), horse-driving is generally permitted on fixed lanes broader than 2 m. In some other states (Bay, Baden-Württemberg (BW), Brandenburg (Bbg), Mecklenburg-Vorpommern (MV), LSA, Thür), it is permitted to gather forest products, such as berries, herbs, nuts, mushrooms, flowers and branches, in small quantities. These permissions do not comprise commercial or professional use, as they are given only for recreational purposes.

Legal classification of the right of access

Access to forests is allowed to the public without the necessity of any particular agreement of the forest landowner. For this reason, it is occasionally spoken of as the 'common use' (*Gemeingebrauch*) of forests (OVG NW, NuR 1983, 122; VGH BW, NuR 1995, 264; 1995, 462). Common use in the sense of administrative law is linked to public goods or to the dedication of use for public purposes. This is the case as far as road traffic law and water rights are concerned. But forests are not declared to be public goods, nor are they dedicated to public use by forest law. The forest owner basically retains disposal rights over his/her forest: he/she is allowed to manage the forest and to admit further use as far as this does not contravene forest or other law. Therefore, the term 'common use' is inappropriate in connection with forests even if they are owned by public bodies (Kolodziejcok and Recken, 1977/1999, ref. no. 4553, margin no. 12).

The right of access implies the duty of the landowner to tolerate particular uses. His/her statutory claims based on private law are accordingly restricted. If the forest owner wants to exclude the public from his/her forest areas, he/she has to apply to the administration (see below). Only illegal uses can be averted by referring to private law.

On the other hand, the forest visitor holds an individual right of access that he/she can enforce under certain conditions when the forest owner restricts use of his/her forest illegally. In several cases, legal action has been taken (BVerwGE 71, 324 = BVerfGE 80, 137; VGH Bay, NuR 1991, 184; VGH BW, NuR 1995, 264). The forest visitor has to prove an individual interest deserving protection to have a *locus standi*. He/she is generally not authorized to use self-help to obtain access.

The right of access does not include a claim to particular recreational services or recreational facilities in the forest (VGH Bay, NuR 1991, 184). The permission relates to the forest in its actual state. The forest owner is obliged neither to keep up given conditions nor to maintain roads and lanes. But he/she is not allowed to take action only to impede access or to prevent damage on occasion of access (OVG NW, NuR 1993, 240; VGH BW, NuR 1993, 142).

No charge for recreational use

The recreational use of forests admitted by law has to be tolerated free of charge by the forest owner (OVG NW, NuR 1986, 215; VG Arnsberg NuR 1995, 485). Some state laws spell this out explicitly. In other states, free use relies on the toleration of the forest owner.

However, the forest owner may permit further uses or uses for other than recreational purposes, e.g. organized and commercial use, in so far as he/she respects the law in force. Furthermore, particular recreational facilities may be placed at the disposal of the public under the same condition. These uses do not have to be permitted without charge, as they are not part of the imposed toleration duties. Nevertheless, the collection of charges will be a problem in many cases, as the forest owner is not free to build enclosures at will in order to exclude other people. In so far as recreational use by the general public is restricted, the requirements of forest law (or nature conservation law in the case of uniform access rules) have to be complied with.

Restrictions on the access to forests

According to § 14 II BWaldG, the states can impose restrictions on the access to forests on the basis of important reasons. As mentioned by the law, such reasons may concern:

- forest protection;
- forest or game management;
- the safety of forest visitors;
- the prevention of substantial damage;
- the safeguarding of related interests of other forest owners which deserve protection.

The restrictions may refer not only to pedestrians, but also to other permitted uses of forests, such as riding and cycling, and to other uses equated with access.

Public access to forests is restricted in very different ways by the federal states when federal law is translated into state law. In particular cases, access is generally limited by state law, e.g. by placing a ban on entering forest or game facilities and juvenile cultures. Moreover, forest owners are allowed to enclose juvenile cultures to a certain extent by fences without any authorization in some states (Bay, Hamburg (Hmb), NW). But, apart from this, every hindrance of access to forests is subject to approval according to state forest law. Consequently, establishing enclosures for recreational facilities in the forest and the charging of entrance fees also need permission of the forest (or nature conservation) administration, as free recreation is restricted or excluded in particular areas.

Admissibility of entrance fees and barriers for additional services

Enclosures and entrance fees for additional services, e.g. picnic sites, ski-tracks or events, offered by the forest owner are only admissible if the permission of the forest authorities is granted for restriction on access for recreational purposes. But forest law does not explicitly provide justification in such cases. According to § 14 II BWaldG, only 'other forest owner-related interests deserving protection' (this means other than forest management or forest protection) might be put forward as the legal basis of a claim to restrict access. But some states (Berlin (Bln), Hess, Nds, Rheinland-Pfalz (RP), LSA, SH, Thür) have restricted the rights of the landowners even further, as they do not provide justification for barriers. In this case, barriers may only be erected for reasons of 'prevention of substantial damage', which might be of particular importance when an event is to take place in the forest and requires some safety precautions. Bavaria (Bayern) is the only state where it is explicitly stated that barriers may be erected for sports contests for a certain limited time.

Enclosures and barriers will almost always have disturbing effects, even if they are of small size (VGH Hess, NuR 1990, 472). Moreover, the summation of several small encroachments can lead to significant deterioration. Therefore, enclosures for recreational facilities cannot be protected by forest law for commercial interests. The courts have decided in several cases that these interests cannot be accepted as adequately important reasons to justify a restriction on access (OVG NW, NuR 1979, 125; 1986, 215; VG Arnsberg NuR 1995, 485). Thus, not only fences but even the prevention of free access by psychological impediments, such as signboards, are illegal when they exclusively serve for charging entrance fees. Consequently, recreational facilities, e.g. a track for long-distance skiing, cannot be offered by charging money on the spot (VG Arnsberg NuR 1995, 485). In general, access must be tolerated free of charge, even when additional services are offered!

Apart from this, it has not yet been clarified, either by statutory law or by judicature, to what extent enclosures and charges for recreational facilities might be permitted in so far as these facilities are offered mainly for the interests of the general public. According to the objectives of forest law, only facilities serving nature-compatible recreation can be considered for this purpose, e.g. an open-air museum in the forest. Besides, substantial impairments of the forest ecosystem or of free recreation must not be caused. This has to be judged in consideration of the affected locality and by weighing the different interests against and among one another. The same conditions apply to public events in so far as access to forests is restricted thereby. But restrictions on access for events generally raise fewer problems, because recreationists have to be excluded for only a short time. Nevertheless, public interests can be subservient to the interests of a private forest owner only in exceptional cases.

In contrast to this, it must be pointed out that the legal restriction provisions are not applicable if remuneration for additional services is paid by contractual agreement with particular interest groups, as in this case no barriers restricting recreational use have to be erected and no money is charged in the forest. To this extent, there are at least some possibilities to establish an enterprise where remuneration for additional services offered by the forest owner is allowed (Table 7.1).

However, numerous provisions concerning particular uses of the forest, e.g. riding, driving and organized events, do exist. Section 7.2.3 describes the extent to which the rights of the forest owner may be restricted by these specific provisions. But, first, another provision on access to forests will be considered.

Specific provisions on access for animal enclosures

A specific provision of the nature conservation law (§ 24 BNatSchG) refers to the construction and management of animal enclosures. According to § 24 BNatSchG, access to the forest must not be restricted inappropriately when animal enclosures are erected. As access to the forest is explicitly regulated in this provision, the access right of forest law is not applicable, because nature conservation law is *lex specialis* and, to this extent, derogates forest law.

Thus, there is a collision of interests between the free access of recreationists and the interests of the managers of animal enclosures. It is necessary to investigate, by analysing the factual situation, whether unreasonable restrictions are imposed on recreationists. In general, it will be necessary for recreationists to be able to go round the enclosure without any great difficulty. Furthermore, freely accessible areas outside the enclosure must be left for

Table 7.1. The right of access and possible RES products.

Free forest uses (no RES products)	Examples of potential RES products*
Right of access for recreational purpose Walking Riding on roads and lanes Cycling on roads and lanes	Granting further access rights for Commercial use, e.g. by a riding school Participation in organized events, e.g. hunting, outdoor games, nature excursions
In some federal states also Coach driving on specific roads and lanes Gathering of mushrooms and picking of berries in small amounts	Use (rent) contracts for Bridle-paths Ski-tracks
	Offering additional goods and services Christmas-tree markets Sponsoring of recreational facilities

* Under consideration of the respective legal provisions.

recreation. Under these conditions, it will also be admissible to charge entrance fees, at least when they are within the means of the general visitor (Kolodziejcok and Recken, 1977/1999, ref. no. 1170, margin no. 11; Dipper *et al.*, 1976/1999, § 34 margin no. 8).

However, § 24 BNatSchG also requires that the ecosystem and the landscape must not be adversely affected. In so far as boar, red deer or roe-deer is kept in enclosures in the forest, damage to the forest can become a considerable problem because of grazing and trampling by the game. Therefore, it will be extremely difficult to establish new animal enclosures in the forest.

> The German right of access to forest areas allows recreational use by pedestrians, riders and cyclists. These uses have to be tolerated free of charge by the forest owner. However, he/she may still dispose of his/her land as far as non-recreational uses are concerned, e.g. commercial uses or organized events. But construction of fences or other kinds of barriers is severely impeded by the right of access.

7.2.3 Restrictions on specific forest uses by German state law

In so far as forest visitors do not possess a subjective right to access to forests for recreational purposes, they require the consent of the forest owner. To this extent, the forest owner is entitled to free disposal of his/her forest areas by the Civil Code (§ 903). Thus, he/she can permit use for other than recreational purposes by contract, e.g. for commercial use or organized events, and charge money for this service. But, to a certain extent, he/she has to consider further provisions referring to particular uses and differing in each state. These provisions are enacted for reasons of nature conservation and protection of the forest owner's interests, as well as those of recreationists. Thus, usually no questions about their compatibility with the constitution are raised. But sometimes it is not quite clear as to how far the forest owner him/herself is affected by the provisions, in particular when they refer to recreational uses already permitted by law. Then, it must be checked whether even the forest owner's disposal rights or only those of recreationists are restricted. It has to be assumed that the forest owner has to pay attention to provisions that mainly relate to the protection of public interests. In this respect, he/she may not free others from legal restrictions.

In this section, the provisions of forest and nature conservation law applying to specific recreational uses of the forest are described. Moreover, general rules of forest law are dealt with that are of prime importance where no specific provisions are enacted by the state. Though the legal requirements concerning a permit for forest conversion and a building permit are explained in separate sections, they might also apply. Apart from this, it should be noted that further restrictions can be imposed in protected areas and that application of the intrusion provision of nature conservation law has often to be proved in addition.

Riding

Horse riding is subject to a number of restrictions imposed by the states. These restrictions may be relevant for particular RES projects, such as riding events, as well as contracts with riders concerning the use of bridle-paths. In general, it has to be taken into consideration that particular zones or lanes must not be used for riding, as described below. But RES projects on riding are not generally prohibited by these restrictions.

First of all, in a number of German states (BW, Bay, Bln, Bbg, Hmb, MV, Saarland (SL), Sachsen (SN)) riding is explicitly confined to roads and lanes for reasons of forest protection. Thus, the forest owner in principle is not authorized to consent to cross-country riding, e.g. in the context of a riding event. In other states, such a restriction is not explicitly stated, but in general it applies that – according to the 'clauses of compatibility with the common interest' of the states – severe damage of the forest ecosystem must not be caused. As cross-country riding is likely to damage the forest when it is practised intensively, the forest owner has to be cautious when he/she wants to give further consent. Otherwise, an infringement of forest law is risked.

Moreover, riding in forests is prohibited on marked hiking lanes and footpaths, as well as on sporting and educational paths, in a number of states (BW, Hmb, Hess, RP, LSA, SL, Thür). These rules are aimed at the protection of pedestrians and, thus, are important even by the landowner. In addition, riding is confined to appropriate (*geeignete*) lanes according to their width and condition in Bayern and Sachsen-Anhalt. Protection of pedestrians is intended by this measure also. Furthermore, conservation of lanes will be guaranteed, which serves not only the interests of the forest owner but also those of recreationists. Accordingly, the forest owner may not allow riding when the lanes are significantly damaged by riders, e.g. on soft ground, or where other recreationists are endangered, e.g. on narrow paths.

Moreover, riding is often restricted to designated and marked bridle-paths only. This refers to the total area of some states (Bln, Bbg, MV, Nds, NW, SN, Schleswig-Holstein (SH)) and to others only in so far as particular areas, e.g. nature conservation areas or recreational forests, are concerned. In addition, particular procedural provisions (*Verfahrensvorschriften*) often refer to the method of designating bridle-paths: in general, the forest or the local authorities have the duty of designating bridle-paths. Forest owners, riders, hunters, other recreationists and nature conservation organizations may participate in the process of designation to a certain extent, differing between states. These procedural provisions ensure sufficient designation for the riding purposes of the general public, on the one hand, and due consideration of other affected interests, on the other.

With respect to the implications involved, a differentiation needs to be made between states in which riding is generally confined to designated paths and those where this refers only to particular, designated areas. In the latter case, the disposal rights of the forest owner are generally restricted: he/she may not consent to riders leaving officially designated bridle-paths, for reasons of nature conservation or the safety of forest visitors (Dipper *et al.*, 1976/1999, § 37 margin no. 12). Thus, he/she may not make 'individual' designations for riding purposes nor may he/she allow riding on paths that are not officially designated, as far as assigned areas are concerned.

But these restrictions on disposal rights of the forest owner do not apply when riding is generally confined to roads and lanes within a state, because there is no general justification or need for it. The provisions referring to the method of designation are rather aimed at safeguarding the interests of the riders, which is necessary because of the far-reaching restrictions imposed on riding by state law. Therefore, a forest owner must be allowed to permit riders to leave officially designated bridle-paths, e.g. by individual contracts with riders or by making a contract with a riding-school. However, the forest owner has to bear in mind that he/she may not legally exclude unauthorized riders on contract paths, because of the right of access. Apart from that, he/she has to consider that marked hiking lanes and footpaths, as well as sporting and educational paths, must not be assigned for riding in order to ensure the safety of forest visitors and in order to avoid conflicts. This is stated explicitly in some state laws (MV, NW).

Moreover, the forest owner may not free the riders from obligations imposed by other rules of forest law. Thus, a number of states empower the appropriate administrative bodies with a duty to mark riding horses, which is subject to administrative discretion (Bay, Bbg, Hess, Nds). These provisions serve to identify riders damaging the forest or forest lanes or disturbing other recreationists.

Other states have enacted strict obligations to mark riding horses, combined with the charge of a levy (BW, NW, Saarland (SL), Sachsen (SN); provided by law but not yet enacted in SH). These riding levies (*Reitabgaben*) are spent on the repair of damage caused by riders and (only in NW, SH) on the construction and maintenance of bridle-paths (Fig. 7.2).

As the riding levies are paid for particular services and are not imposed exclusively for riding, which has to be tolerated free of charge, they can be regarded as legal (BVerwGE 71, 324). Consequently, the forest owner is also entitled to charge money for additional services, in particular by entering into private contracts. The levies are mostly imposed only for the elimination of damage. Therefore, the forest owner may still ask for further reimbursement for the construction or maintenance of lanes. Where the levy is also spent on providing lanes (only NW, SH), the forest owner will probably not get any additional public money provided by the riding levy for his/her private-contract paths. But, even in this case, he/she may still enter into contracts concerning the construction and maintenance of bridle-paths, in particular as contracting always depends on the voluntary contribution of the riders. In this respect, participation of riders is not generally unreasonable. Another problem lies in the fact that riders are not willing to pay two different duties at the same time for riding in the forest. In this respect, further considerations are of interest.

On one hand, in Baden-Württemberg and Sachsen, the riding levy exclusively applies to officially designated bridle-paths

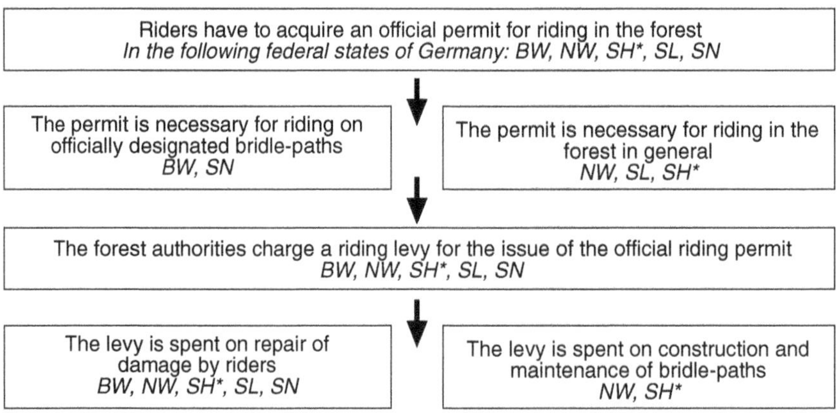

Fig. 7.2. Charge and spending of the riding levy. * In Schleswig-Holstein (SH) the riding levy is not yet enacted.

(Dipper *et al.* (1976/1999), § 39 margin no. 6). This relates not only to the charge of the levy, but also to its spending. In this case, the forest owner may never claim official money for elimination of damage caused by riders on non-designated lanes. However, any conflict with private contracting is avoided on privately offered bridle-paths, as an additional public levy cannot be imposed there.

On the other hand, the levy is charged for riding in general in Nordrhein-Westfalen and Saarland (and SH if a state regulation is enacted). In this case, the official riding levy always has to be paid by riders. Shunning of officially designated lanes may not free them from this duty. But the question might be raised as to whether the levy is still applicable when riders only use those lanes that are placed at their disposal solely by making a contract with the forest owner. In this respect, it has to be considered that the forest owner may still claim official money provided by the riding levy. Accordingly, he/she may not exempt the riders from paying the duty. Therefore, it will be necessary for the state legislature to enacts an exceptional clause by forest law providing that the riding levy does not have to be paid in particular areas where the forest owner enters into private contracts with riders. Moreover, the legislature will also ensure that a forest owner making use of such an exception will not receive any public money for bridle-paths or the repair of damage, even if this damage is caused by non-contractors. Such a regulation will be fair in that it avoids an unreasonable burden on those riders who are actually paying the duty.

Cycling

In general, cycling in forests is regulated slightly less restrictively than riding on horseback. Nevertheless, cross-country cycling is explicitly prohibited in some states (BW, Hmb, MV, SL, SN) as well and thus may not be permitted by the forest owner. In those states where there is no such restriction, the forest owner must take into consideration the fact that severe damage to the forest ecosystem must not be caused by cross-country cycling, due to the clauses of compatibility with the common interest.

Furthermore, in some states cycling is not allowed on top of banks (Bln), on footpaths (SN, LSA), paths narrower than 2 m (BW) or sporting and educational paths (BW, SN, Thür). Moreover, it is confined to appropriate (*geeignete*) lanes according to the width and condition of the lane in Bayern and Sachsen-Anhalt. In Hessen and Thüringen, cycling is restricted on firm (*feste*) roads and lanes, exclusively comprising surfaced lanes broader than 2 m. In addition, particular areas can be designated by so-called segregation plans (*Entmischungs-/Entflechtungspläne*) in both states, in which restrictions on cycling can be imposed. However, such designations always require the consent of the forest owner. But, where particular areas have been designated by segregation plans, cycling can be prohibited on particular lanes in Hessen or it can be confined to designated lanes in Thüringen.

The forest owner may not free the cyclists from the above-mentioned rules of conduct, as they are mainly enacted in favour of pedestrians' safety.

Driving and motor sport

Cars and other motorized vehicles are not allowed by forest law. However, the forest owner may permit driving, as well as parking, in so far as this is not subject to a specific restriction provision of a state. Thus, in Mecklenburg-Vorpommern, driving is strictly prohibited in the forest with no possibility of permission being granted. Exemptions only refer to driving on designated public roads and to driving by the forest owner and his/her commissioners (*Beauftragte*). In Brandenburg, driving by recreationists is subject to authorization by the forest administration without mentioning the conditions of granting approval. But it has to be assumed that consent will be given when driving by car is justified by a particular reason (e.g. access to a rented house inside the forest) and when the forest ecosystem and recreationists are not endangered. In Hamburg, driving is only permitted on officially marked lanes. If the forest owner intends to approve driving in his/her forest, he/she has to mark his/her lanes with official signs. Apart from this, it has to be borne in mind that the traffic authorities (*Straßenverkehrsbehörde*) may restrict access if public traffic is admitted that damages forest lanes or endangers recreationists. But, most probably, the traffic authorities will not take action against the express wishes of the landowner unless public interest requires such a measure.

Furthermore, motor sport is forbidden with no possibility of dispensation either by the forest owner or by the administration in several states (Bbg, MV, LSA, Thür). As motor sport disturbs recreationists with noise and exhaust fumes, and causes severe damage to forest lanes, the forest ground and forest animals and plants, a strict ban on this action is justified. Therefore, it may also be prohibited by the forest authorities within the scope of the danger defence if severe impairments are caused and even if no specific legal ban is stated.

Camping

Camping is not part of the access right and thus always depends on permission from the forest owner. Furthermore, camping with tents and caravans and other mobile homes is subject to the approval of the forest authorities in two federal states, Brandenburg and Mecklenburg-Vorpommern. According to the objectives of the provisions, approval will only be granted when endangering of the forest, e.g. by fire, severe damage of forest plants and animals, pollution and disturbance of scenic aspects, is excluded. In any case, permission can only be granted for a single overnight stay, as provided by the nature conservation law of these states.

In Schleswig-Holstein, camping with tents and caravans is even strictly prohibited by a state regulation (*Landesverordnung*), which is mainly enacted to guarantee protection of the forest against fire. Considering this factor and the other negative effects of camping, the ban does not imply an excessive and illegal restriction on the disposal rights of the forest owner or on the general freedom of action of forest visitors. However, camping is not generally excluded: it is still permitted on camp-sites and in other places outside the forest.

In the other states, no specific duties apply on obtaining a permit. However, the forest authorities of Hamburg are explicitly entitled to forbid camping where forest plants are damaged, the forest is polluted or wild animals are disturbed. It has to be assumed that the appropriate authorities of the other states have the same competence if similar damage is caused.

Furthermore, it remains to be proved whether local police decrees (*Polizeiverordnungen*) apply to camping (VG Freiburg NuR 1992, 94). These decrees may place a ban on camping outside camp-sites if sanitary systems are not offered. In any case, police decrees apply to camp-sites in all states. This means that requirements concerning the supply of potable water, sanitary systems and fire protection have to be fulfilled. According to the state decrees, a camp-site already exists when a place is used for camping with some regularity during particular seasons and with more than three tents or caravans. Apart from this, authorization for the construction of camp-sites is required (see below in this section and Section 7.2.6).

Events and meetings in forests

Although organized events and meetings (*organisierte Veranstaltungen*) in forests are not permitted by forest law, nor are they strictly prohibited and thus they may be allowed by the forest owner. But, in some states, these activities are subject to additional approval of the appropriate forest authorities, which implies a restriction on the disposal rights of the forest owner. However, application of these provisions always presupposes some extent of organization, though this will generally be fulfilled when commercial aims are pursued. Besides, the duty of approval applies when impairments of nature and landscape or disturbances of recreationists might be caused.

A prerequisite of granting approval is, first, the fulfilment of specific legal provisions where these exist and are applicable. An organized tour with mountain bikes for example may only be permitted if the legal restrictions on cycling are observed. Furthermore, the general interests of nature preservation and of recreationists have to be taken into account. In this connection, the clauses of compatibility with the common interest (*Gemeinverträglichkeitsklauseln*) of the states are of particular relevance, as they describe the rules of conduct in the forest in a general way. Accordingly, disturbances and damage likely to be caused by an event or meeting have to be evaluated in every single case. The negative effects, on the one hand, and the interests of the forest owner and the participants, on the other, have to be weighed against one another. Consent has to be refused if significant encroachments would otherwise occur. This will be the case when, for example, a large group wants to walk or ride cross-country, which would disturb wild animals and damage biotopes (VGH BW, NuR 1995, 462), or when events and meetings are intended to take place during the rearing season of the game. In addition, severe disturbances of recreationists by noise and disruption of landscape have to be avoided (VGH Hess, NuR 1989, 86; VG Darmstadt NuL 1998, 38) and recreationists must not be endangered by participants in the event, in particular when vehicles are used. The forest authorities have to guarantee that the rights of the recreationists are upheld, as they have a subjective right of access (§ 14 BWaldG).

It should be mentioned that the forest owner (or the organizer) has to apply for additional consent if recreationists are excluded by barriers or if they are expressly kept out of the forest in another way, e.g. by signs or fees. In general, barriers will not be necessary for the execution of an event. However, they might be required for the safety of forest visitors, in particular when a race is taking place.

Picking of mushrooms and berries

The gathering of mushrooms and the picking of berries, nuts and other forest products (*Walderzeugnisse*) in small amounts is explicitly permitted in some German states. Accordingly, the forest owner has to tolerate picking to a certain extent, and he/she may not charge a fee for it. However, the legal permission does not refer to commercial or organized picking or picking in large amounts, which always requires the consent of the forest owner. Furthermore, the forest owner may demand a remuneration in the latter cases. But, where the forest owner allows picking, it also has to be proved that it is admissible in respect of specific legal provisions in the forest or nature preservation law.

First of all, the forest owner may not grant permission for picking of special protected species that are submitted to provisions of nature conservation law. These species, which include, for example, various mushrooms, such as *Boletus edulis* and all species of *Cantharellus*, *Morchella* and *Tuber*, must not be picked in total or in part and they must not be possessed or traded. However, according to the Federal Regulation of Species Protection (*Bundesartenschutzverordnung*), some exemptions are made for particular species. But, even then, these species may only be picked in small amounts, so that a forest owner cannot charge additional fees for their picking at all if he/she generally has to tolerate picking according to state forest law. Besides, the exemptions only refer to particular species: of those mentioned above, *Tuber* is not included.

In most states (BW, Bln, Bbg, Hess, MV, NW, RP, SL, SN, LSA, SH, Thür), general provisions on forest subsidiary use (*forstliche Nebennutzung*) have been enacted, which also apply to the picking of forest products, such as mushrooms and berries. These provisions mainly consist of general rules demanding that orderly forest management or the functions of the forest must not be adversely affected. But it has to be assumed that the forest authorities will be allowed to intervene when significant impairments of the ecosystem are caused by intensive types of picking. A preventive duty to apply for approval for intensive types of subsidiary use only exists in Rheinland-Pfalz. Accordingly, organized and commercial picking of forest products always requires the consent of the forest authorities in this state. Following from the relevant state forest law, approval has to be refused if negative effects cannot be avoided by attaching conditions. Otherwise, the forest authorities have to assent, if need be under the condition of appropriate restrictions on the intensity of picking.

Construction and marking of roads and lanes
The construction of new forest roads and lanes and their marking (*Kennzeichnung*) may be in the interest of the forest owner if he/she enters into a contract with user groups that agree to pay a remuneration. But, to a certain extent, provisions of forest law have to be considered which apply specifically to the construction and marking of forest roads and lanes.

First, the forest owner has to seek approval of the forest authorities for the construction of forest lanes in several states. In Thüringen, this refers to ski-tracks and lanes that are used for hiking, riding or cycling, whereas mere management lanes (*Wirtschaftswege*) are not affected. Brandenburg has constituted a duty of approval merely for the construction of 'hiking' lanes (*Wanderwege*), but it has to be assumed that riding and cycling paths are also included, as they have the same effect as hiking lanes. Moreover, this point of view is based on considerations of idiom, as the term '*Wanderwege*' can be considered as an overall term (*Reitwanderweg*). In Mecklenburg-Vorpommern, only the construction of hiking lanes which are possessed by more than one proprietor is subject to a permit by the forest authorities. On the contrary, only the construction of sealed (*versiegelte*) roads requires consent of forest administration in Sachsen-Anhalt.

Pursuant to the provisions of state forest law, consent to the construction of lanes has to be granted when the lane is 'required' (*erforderlich*) for the accessibility of the forest. This criterion implies that lanes must not occupy more than an adequate proportion of the forest, which has to be judged in respect of the real demands of recreationists. Furthermore, the interests of nature conservation and landscape protection have to be regarded as explicitly expressed by the legal provisions. Accordingly, there should be as few forest lanes as possible, and due regard has to be paid to the sensitivity of the forest ecosystem according to the individual circumstances. Besides, lanes should only be constructed when the ground is appropriate for the permitted use, e.g. bridle-paths are not convenient on soft or wet ground because of ground solidification, as well as for reasons of safety of the users.

In another two states, a duty of announcement has been constituted for the construction of lanes. This refers to every construction of forest management lanes (*forstwirtschaftliche Wegebaumaßnahmen*) in Nordrhein-Westfalen. In this context, it is not quite clear whether the duty of announcement solely applies to management lanes or if mere recreational paths are affected as well. It has to be assumed that the announcement applies to the two types of lanes in the same way, as the effects of construction on the forest ecosystem are also similar. Conversely, in Bayern the construction of lanes (all types) has to be announced only in the alpine area. Both announcements enable the responsible authorities to take suitable action if necessary, but the explicit consent of forest administration is not required in advance. Suitable measures of interference include the application of the intrusion provision (*Eingriffsregelung*) of the nature conservation law, which often presupposes official approval or announcement. But this provision also applies in cases where a permit of the forest authorities is required or when its application is ensured by the states in another way. In this case, further duties can be imposed on the construction of forest lanes. It must be mentioned that the construction of roads and lanes may require an additional building permit (see Section 7.2.6), at least when they are surfaced.

Construction of recreational facilities
The construction of recreational facilities will be of increasing interest for the forest

owner if a business can be developed thereby. As the charging of entrance fees is usually not allowed because of the right of access (see Section 7.2.2), it is necessary to find a sponsor or other sources of revenue. If this is achieved, the construction of picnic sites, huts, car-park areas, sports facilities or educational paths may be considered. However, some states have enacted specific provisions of forest law referring to the construction of recreational facilities in the forest.

Sachsen-Anhalt and Thüringen are the only two states where a strict duty of approval by the forest authorities has been introduced for the construction of recreational facilities. In Baden-Württemberg, such a duty only applies when recreational grounds larger than 1 ha are built.

According to state forest law, approval has to be refused when discharge of the forest functions is endangered. These functions refer, first, to the economic value of the forest (economic function), secondly, to the environmental significance of the forest (protective function) and, finally, to the recreational importance of forests (recreational function). The protective function includes the significance of the forest for the ecosystem, climate, soil fertility, water resources, filtration of the atmosphere and scenic aspects (§ 1 BWaldG).

In order to take account of the protective function of the forest, modification and removal of forest soil, e.g. by the construction of large buildings or planting of non-native plants, should be avoided. Furthermore, appropriate sites should be chosen for the construction of recreational facilities. Areas that are already accessible to recreational traffic, e.g. close to car-park areas, restaurants or villages, are preferred. Besides, various facilities should be concentrated in one place as far as possible. In this way, disturbances of the game by additional noise and traffic can be limited and particular areas of the forest are reserved for the quiet recreation of visitors.

Moreover, impairment of scenic aspects should be avoided and wild animals and plants, in particular endangered ones, must not be adversely affected. Considerable deterioration will lead to refusal of approval. But, where only minor facilities are built, such as a few tables and benches, signs for an educational path or equipment for a playground, the consent of the forest authorities will be granted if no substantiated objections are raised.

In addition, it has to be mentioned that the construction of camp-sites and golf-courses is subject to authorization according to nature conservation law in Schleswig-Holstein. Consent will only be granted if the interest of nature conservation is not impaired. Accordingly, construction of these facilities will not be permissible at all if conversion of the forest is required. But they might be built in agrarian areas of low ecological value.

With regard to the building of fireplaces, most federal states have enacted particular provisions intended to prevent forest fires. These provisions constitute a duty of approval in almost all states when fireplaces or barbecue sites are built. In order to obtain consent, appropriate safety measures have to be taken. Thus, fireplaces must be fixed so that fire cannot spread to the forest. In areas of increased fire risk, permission may generally be refused.

Finally, it should be mentioned that additional permits might be required. If trees are felled or if the forest ecosystem is seriously impaired in any other way, it must be proved whether a conversion of forests has been realized, which is always subject to approval (see Section 7.2.5). Furthermore, the landowner has to seek an additional building permit when buildings are erected or when the soil is significantly modified in any other way (see Section 7.2.6).

Restrictions on access by forest authorities

In particular cases, access can be restricted totally or in part by instruction (*Anordnung*) of the forest authorities. By this means, access can be excluded to a certain extent, which implies a limitation on the disposal rights of the forest owner.

Administrative instruction of restrictions on access always has to be justified by grave and substantial reasons, mentioned in § 14 BWaldG. Accordingly, access can be restricted by the forest authorities in particular for the following reasons: forest protection, game management, safety of forest visitors or prevention of substantial damage. These reasons will be discussed here.

The term 'forest protection' has to be interpreted to refer not only to the preservation of forest trees but also to the general interests of nature conservation. This is concluded from § 1 BWaldG, in which the protective function of the forest is explained and the ecosystem (*Naturhaushalt*) is explicitly mentioned. Therefore, the forest authorities may restrict access, in particular when endangered or special protected species and biotopes are disturbed or damaged by forest visitors. This might be especially relevant during the rearing season of game, when visitors should not leave forest lanes. If more strict and durable restrictions on access are justified on the grounds of nature conservation, protected areas can be designated; in this context, the imposition of further restrictions on access is possible.

In addition, suitable measures have to be taken to avoid substantial damage of the forest, e.g. during the summer season, when significant risks of forest fires are threatening. But, in this respect, it will usually be sufficient to confine access to lanes and to prohibit smoking in the forest. Furthermore, game management is aimed at the prevention of damage to the forest. Accordingly, restrictions on access might be required for hunting in exceptional cases when this is necessary for the safety of forest visitors. Moreover, where intensive recreational traffic takes place in the forest, the establishment of refuge zones for game can be considered. Finally, conflicts between recreationists can require restrictions on particular user groups to prevent damage. Where different types of recreational traffic are not already separated due to specific legal rules, riders and pedestrians, in particular, can be directed on to different lanes when accidents are likely to happen.

In any case, forest authorities may only restrict access as far as this is actually necessary and justified by grave and substantial reasons. Apart from this, the actual scope of competence of the forest authorities depends on state forest law, as the Federal Forest Act only provides the frame for state legislation. But some state laws do not mention all the grave and substantial reasons provided by the Federal Forest Act. In this way, the concrete administrative competence is sometimes limited in comparison with the federal law. As the state laws differ significantly, the legal situation always has to be considered in regard to the respective state law.

An analysis of the state laws shows that the forest authorities of some states are only entitled to restrict access for reasons of prevention of forest fires (Hess, RP). In other cases, access may only be confined to forest roads and lanes and only during particular times of the day (NW). Moreover, in some states it is necessary to come to an agreement with the forest owner (Hess). These provisions show that the forest authorities are required to restrict access very prudently, in particular when the measure is implemented against the express will of the owner. General restrictions on access will usually only be justified when a specific clash of interests occurs, as mentioned above.

Where conflicts with other recreationists or with nature conservation are caused by the forest owner permitting further uses of the forest, specific provisions of forest or police law apply, which give the appropriate administrative bodies additional power to intervene.

General restrictions by rules of good conduct

The 'rules of good conduct' (*Wohlverhaltensklauseln*), also called 'clauses of compatibility with the common interest' (*Gemeinverträglichkeitsklauseln*), enacted by the federal states serve to prevent damage to the forest ecosystem, the forest owner's property and the forest visitors making use of their right of access for recreational purpose. Accordingly, the forest owner may not exempt forest visitors from these rules, as they were passed in consideration of not only his/her own interests but those of the public as well. However, free disposal of the forest owner has to be conceded in so far as the regulation is aimed mainly at the protection of his/her property. This means that damage has to be tolerated to some extent when the forest owner consents to particular uses. But it has to be ensured that serious damage of nature and endangerment of recreationists is avoided. Widespread destruction of forest plants (not only of trees but of herbs as well) – for example, caused by sports activities or by other events – must be considered as significant. Therefore, the forest owner should be cautious when promoting recreational uses. However, insignificant disturbances of recreation and minor impairments of forest lanes and the ecosystem have to be tolerated by recreationists and by the administration in favour of the interests of the forest owner, as long as specific legal provisions are not infringed.

Infringement of the rules of good conduct can be punished by a fine (Bußgeld) in most German states. The provisions of state forest law constitute a legal basis for fining pollution, disturbance of recreationists by noise or other impediments on recreational use, as well as damage to the forest ecosystem. It has to be noted that fines can even be addressed to the forest owner in cases where he/she is regarded as a 'participant' (*Beteiligter*) in the sense of the infringement law (§ 14 *Ordnungswidrigkeitengesetz*). If the forest owner deliberately promotes uses that constitute an offence, he/she has to be regarded as a participant even if he/she has not taken part in the action itself (Rebmann *et al.*, 1988/1998, § 14 margin no. 14; Göhler § 14 margin no. 6). Furthermore, realization of an infringement indicates that the forest authorities are also entitled to intervene against forest uses. But it has to be taken into account that forest uses can only be punished by fining in so far as this is explicitly expressed by state law. As not every state has enacted provisions imposing a fine on particular actions, fining is limited accordingly.

As a result, the rules of good conduct are of prime importance for additional uses permitted by the forest owner. In so far as no specific regulations referring to particular uses exist – as is the case in a number of states – the rules indicate the cases in which the forest authorities are entitled to intervene. Moreover, the rules of conduct specify the conditions of the administration granting consent where duties of approval have been constituted, as discussed above. Although applying for approval requires the initiative of the forest owner (or the forest users admitted by him/her), the advantage of official approval granted in advance lies in the certainty about the legality of forest uses permitted by the owner. Accordingly, it would also be advisable in other doubtful cases where no official permit is required for the forest owner still to clarify the admissibility of questionable uses with the forest authorities. In this way, he/she will avoid unexpected interference by the forest authorities.

General restrictions by the duty of sustainable forest management

Pursuant to § 11 BWaldG and the corresponding state provisions, the forest has to be managed in an orderly and sustainable manner (*ordnungsgemäß und nachhaltig*). The term 'orderly' implies that the forest has to be utilized in respect of its determined purposes and functions and according to acknowledged forestry principles (*anerkannte forstliche Grundsätze*). These principles are the ordinary rules of management, which are mainly defined by forestry science and the literature. The minimum requirement – which rather goes without saying – is that legal provisions must not be infringed. The term 'sustainable' points to the fact that the forest has to be managed in such a way that the forest functions – including economic, protective and recreational functions (§ 1 BWaldG) – are guaranteed even for future times.

As a result, the forest owner has to pay due regard to the protective and recreational functions of the forest when utilizing the forest. In this context, some state legislators have published explicit restrictions on clear-cutting and on felling juvenile cultures up to 50 (conifers) or 70–80 (broad-leaved trees) years. Moreover, clear-cut forest areas or thinned-out forest stands always have to be reafforested or supplemented within a reasonable period of time (§ 11 BWaldG). Furthermore, some states specify as a principle that forest soil and its fertility (*Bodenfruchtbarkeit*), as well as biological variety, should be preserved and that the forest should be made accessible in a controlled manner.

Accordingly, on the one hand, the obligation of sustainable management can influence the recreational use of the forest, because sustainable forest management must not be adversely affected. On the other hand, due regard must be paid even to the recreational function of the forest when managing the forest. However, jurisdiction has shown that the obligations arising under the management principle cannot be taken too far. The administrative court of Nordrhein-Westfalen (OVG NW, NuR 1988, 303) has stated that a clear cut of 50 ha cannot be regarded as an infringement of the management principle unless the state legislator has enacted specific provisions that impose restrictions on clear-cutting. Thus, the ordinary procedures of forest management – including clear-cutting, for example – cannot generally be restricted by the forest authorities. In addition, the construction of forest lanes is lawful when the lanes are required for forest management (VGH Bay, NuR 1981, 209). A judgement has to be made in respect of forestry and economic requirements, as well as the type of structure, whether a lane is actually 'required' for management. In any case, the fewest possible lanes must be constructed and surfaced.

As far as construction of recreational facilities (including lanes) and implementation of recreational uses, such as events in the forest, are concerned, it is to be assumed that the management rules are opposed to them when significant encroachments on the natural environment are involved. In this case, the forest authorities will be entitled to intervene because permitting intensive types of recreational activities does not correspond to the ordinary and acknowledged way of utilizing the forest. But as recreational interests have to be considered as part of forest management as well, even recreational activities are admissible to a certain extent, depending on recreational demand and the impacts of use. As a result, the limit of admissible intervention by the forest authorities will not go very much beyond that already defined by the rules of good conduct. The most important fact is that the forest authorities are explicitly entitled to give instructions on reafforestation if trees are felled in an excessive way. But, as far as mere recreational use is concerned, rather more problems are raised with regard to whether intervention is allowed at all.

Administrative intervention in case of offences

When legal provisions are infringed by the forest owner or by forest visitors, the forest authorities are generally entitled to intervene within the scope of the danger defence (*Gefahrenabwehr*). In most states, the competence of the forest authorities is explicitly stated by the forest law. Otherwise, the forest authorities have to refer to the general clause (*Generalklausel*) of police law, which comes within the legislative competence of the states. Accordingly, a legal basis for intervention is given in any case. Some important rules will be described that apply to implementation of danger-defence measures.

Where an infringement of a legal provision – either a specific or a general one – is imminent or has actually occurred, the forest authorities can take action against unlawful forest uses. In terms of police law, a specific danger (*konkrete Gefahr*) is cited, which justifies interference. But, to some extent, it is subject to the discretion (*Ermessen*) of the forest authorities whether action is taken at all and as to the type of intervention chosen. However, the forest authorities are obliged to act when the lives of people are endangered or serious damage of the forest is caused. In cases of illegal recreational use of the forest, such significant dangers do not usually occur, so that the forest authorities have to exercise discretion. In this respect, it is also of importance that administrative intervention might contradict the ban on excessive measures (*Übermaßverbot*), especially when there have been only minor impairments. This will be particularly questioned where only general rules, e.g. the rules of good conduct or the duty of sustainable management, are infringed. But discretion can be further limited by the principle of equality (*Gleichheitssatz*), provided in Art. 3 GG, or by administrative guidelines (*Verwaltungsvorschriften*) that are binding on the forest authorities, although they are not statutory law but only internal instruments of administration.

Furthermore, most German states provide that the forest administration – in general, the forest ministry – is empowered to enact regulations (*Verordnungen*) on recreational use of the forest. These may impose duties of approval or a ban on particular forest uses. These regulations make up statutory law and are externally binding on any person. When forest law does not provide such a power, police law is applicable instead, if forest law does not explicitly exclude application of police law in exceptional cases. It has to be taken into account that issuance of regulations restricting forest uses will probably increase the more conflicts are caused by admitting further forest uses by the forest owner.

In general, measures of danger defence have to be addressed to the person responsible for the danger. If the forest owner explicitly allows a forest use that is illegal, not only the admitted actors but even the forest owner him/herself bears responsibility. The forest owner can be regarded as a so-called occasioner (*Zweckveranlasser*) even though he/she does not directly take part in the illegal action. In this case, measures of danger prevention can also be addressed to him/her. When the forest authorities have to choose between different disturbers of public order, the one who can eliminate the danger in the fastest and most effective way has to be chosen. In general, this will be the person directly causing the danger (*Verursacher*), when a concrete danger is threatening. However, explicit bans on illegal uses for the future will be appropriately addressed to the forest owner if he/she has permitted these uses before.

> A number of specific legal provisions impose restrictions on particular uses, such as riding, camping, motor sport and events in the forest. But, in general, such actions are not absolutely banned; rather, authorization of the appropriate authorities is required in some federal states of Germany. As the main result, it can be stated that specific uses can be permitted as long as no significant damage of the forest ecosystem and no disturbances of or dangers to recreationists are caused.

7.2.4 Afforestation

Afforestation of formerly agricultural or other areas can be of relevance in context with environmental products that are aimed at enlarging forest areas. Even though afforestation is usually desirable for ecological reasons, afforestation is subject to authorization according to § 10 BWaldG. However, authorization may only be denied when the requirements of country and landscape planning are opposed. Moreover, a number of federal states have provided that a permit is not necessary in particular areas that are designated for afforestation.

As a result, it is not normally difficult to get a permit for afforestation, if such a permit is required at all. Problems may occur, in particular, when ecologically valuable sites, e.g. humid meadows, are concerned or when the afforestation does not fit into the surrounding landscape. But, even then, the permit still has to be granted if the encroachments can be dealt with by attaching appropriate conditions to the permit.

7.2.5 Conversion of forest areas according to German forest law

Where implementation of recreational services necessitates the construction of recreational facilities or clear-cutting of forest areas, forest law referring to conversion of forests might be applicable. According to § 9 I BWaldG and the respective state law, conversion of forest areas is subject to approval by the forest authorities. However, application of this provision mainly depends on the fact that forest is cleared in order to carry out another type of use. Therefore, a judgement has to be made, in the light of the legal definition of forest areas, as to when a conversion is realized.

Furthermore, § 9 BWaldG determines the minimum requirements of granting approval when a conversion of forest is intended. But, as § 9 BWaldG is only a framework provision addressed to the federal states, state law has to be considered additionally to clarify the restrictions imposed on a forest owner seeking a permit.

Definition of forest areas

According to § 2 I BWaldG, forests are defined as areas planted with forest trees. Clear-cut or thinned-out forests, forest roads and lanes, forest meadows, game feeding points, timber-storing places, as well as other areas connected with forests and ancillary to them (*mit dem Wald verbundene und ihm dienende Flächen*), also have to be rated as forests. However, smaller lots planted with clusters of trees or hedges or serving as forest nurseries and being situated on farmland or within built-up areas are not regarded as forests (§ 2 II BWaldG). Moreover, the states may assign other tracts of land as forests, and they may exempt parks within residential areas and Christmas-tree plantations from the legal definition of forests (§ 2 III BWaldG).

In this context, it is of importance that the construction of forest lanes does not fulfil the definition of a conversion. Furthermore, areas connected with forests and ancillary to them are still regarded as forests. The term 'ancillary' is interpreted to hint at a relation to the forest functions mentioned in § 1 BWaldG, namely economic, protective and recreational functions (Kolodziejcok and Recken, 1977/1999, ref. no. 4529, margin no. 10; VG Kassel NuR 1981, 71). Therefore, even recreational areas can be considered as 'forests' to a certain extent. As all three functions of the forest are equivalent, a balance between the different functions has to be drawn when the term 'forest' is interpreted and defined in respect of the law. Accordingly, modification of areas by construction of rather small recreational facilities that are open for the general public, such as benches and huts, are not subject to approval for conversion of forest even where a few trees have to be felled. But this has to be judged differently when rather big clear cuts are made or when the facilities are built for private or economic reasons (OVG Nds, NuR 1987, 36; Klose and Orf, 1998, § 9 margin no. 43; Dipper *et al.*, 1976/1999, § 4 margin no. 10).

Finally, it should be mentioned that a conversion can be realized when the forest is intensively used for a particular purpose although none of the trees is felled. This might be the case when parking or camping sites or enclosures for game are built. If a typical forest flora cannot be durably preserved, a conversion is awarded (Klose and Orf, 1998, § 9 Rn. 43). Impairments of the forest flora can be caused by driving or trampling damage (*Trittschäden*), by forest visitors as well as by game, biting by game (*Wildverbiß*) or pasturing of cattle (*Waldweide*). On the contrary, a forest area is preserved if significant impairments of the forest are avoided by limitations of the intensity of use, e.g. by appropriate density of game in an enclosure (VGH BW, NuR 1995, 357).

It also has to be borne in mind that the state may define other areas (apart from those mentioned in § 2 II BWaldG) as forests. Some states have made use of this power by defining recreational areas and car-park areas (BW, Bln, Bbg, Hmb, MV, Nds, SL, LSA) or restaurants (Bln) as forest. As far as recreational areas are concerned, the statement merely serves for clarification because these areas are already treated as forest pursuant to the Federal Forest Act. In this respect, there is no hint that state law goes beyond federal law except in Baden-Württemberg, where recreational areas can explicitly be even larger than 1 ha (§ 9 VII LWaldG BW) without fulfilling the fact of a conversion. Nevertheless, another kind of permit is then required. But the determination of car-park areas and restaurants as 'forests' always goes beyond federal law. However, it has to be assumed that these areas may only be rated as forests if they were established in order merely to satisfy the needs of recreationists and if they, to a greater or lesser degree, fit into the forest. Therefore, construction of paved and very large car-parks still requires consent of the forest authorities for conversion in these states.

Christmas-tree plantations usually fulfil the definition of forests, because these areas are planted with forest trees. However, they are exempt from the definition of forests in a few states (BW, Hmb, SH), according to § 2 III BWaldG. Other states only exclude them when they are situated outside the forest (Bay, Bbg, MV, Nds, NW, LSA). Consequently, it depends on state law whether a change of forest management in favour of the cultivation of Christmas trees requires the consent of the forest authorities.

Requirements of the Federal Forest Act

When passing a decision on a conversion, the rights, obligations and economic interests of the forest owner, on the one hand, and the interests of the general public, on the other, must be weighed against and among one another (§ 9 I 2nd sentence BWaldG). Authorization will be denied if conservation of forest land is predominantly in the public interest, especially if forest land is of essential importance for the ecosystem, for silvicultural production or for recreation (§ 9 I 3rd sentence BWaldG).

Although it is stated in the Federal Forest Act that the interests of the forest owner have to be taken into consideration, the general economic interests of the forest owner do not normally justify a conversion (Kolodziejcok and Recken, 1977/1999, ref. no. 4543, margin no. 16). The forest owner must rather be obliged to make the conversion due to imperative reasons of his/her actual situation. This will usually only be the case when his/her existence is actually endangered if consent is denied (OVG Nds, NuR 1997, 100). But, in general, the forest owner will not be able to prove such a risk of existence so that these assumptions are fulfilled only in exceptional cases.

Nevertheless, the construction of recreational facilities might be admissible in so far as they are built predominantly in the public interest. But not every recreational facility is desired by the general public, in particular when significant encroachments are made into the protective function of the forest. Therefore, the different public interests – recreation, on the one hand, and nature preservation and silvicultural production, on the other – have to be weighed against one another. Approval will not be granted when only a very particular group of recreationists may have the usufruct, e.g. holiday houses, golf-courses or camp-sites, or when economic aspects are dominant, e.g. construction of a shop, a museum or a viewpoint. Where a permit is required for rather big picnic places, playgrounds or other recreational facilities that are placed at the free disposal of the general public, approval can be granted if the facility concerned meets the demand of recreationists and if it does not cause significant impairments of the ecosystem or scenic aspects. Significant impairments are caused, for example, when ecologically valuable woodland or small residual forest areas are destroyed or when typical and impressive sections of the landscape are damaged. Furthermore, the designation of special protected areas indicates that forest should not be converted. But exceptions can refer, first of all, to recreational forests, where construction of recreational facilities might be particularly desired. If a conversion is permitted it must be restricted to the minimum size necessary to achieve its purpose.

In individual cases, the forest authorities can deviate from these principles, as the Federal Forest Act only requires that consent 'shall' be denied in a case of predominant public interests; the forest authorities are not strictly obliged to do so. However, specific reasons must be given to justify approval when predominant public interests are opposed. Such a justification might be given when only minor deterioration of the forest functions is caused or when compensatory measures mitigate impairments. Nevertheless, further restrictions in this area are imposed by a number of states, as described below.

Apart from this, the prevailing opinion is that approval has to be granted if private interests (which may be supported by public ones) on conversion are predominant. In this case, the forest owner has a claim to approval (*Rechtsanspruch*) that he/she can enforce before court; administration is not left any discretion in this respect. If approval is denied despite the fact that private interests are predominant, compensation has to be paid to the forest owner for otherwise unreasonable restriction of his/her property rights.

Requirements of the state forest laws

The states can pass regulations providing that a permit is not required if the forest area concerned has been designated for another use in a legally binding manner by other regulations of public law. On the other hand, the states may impose further restrictions on a conversion and they may even strictly prohibit any conversion, in particular when protected forests or recreational forests are involved (§ 9 III BWaldG).

Exceptions from approval particularly apply to conversions that are assigned by a land-use plan in some states (Bay, Hmb, Nds, NW, SL). In this case, the interest of forest conservation has to be regarded when the land-use plan is passed. To this extent, the exemption of a formal permit for conversion is a simplification of procedure provisions, rather than a facilitation of material requirements. However, the state legislator runs the risk of the forest being less protected when authorities other than the forest administration have to decide upon conversion.

On the contrary, some states constitute stricter requirements on conversions in comparison with the Federal Forest Act by providing that authorization is necessary even when a conversion of a forest area according to its legal definition is not fulfilled (Bay, BW, Bbg, LSA, SN). However, this provision is mainly relevant for forestry facilities (*forstliche Einrichtungen*) larger than 1 ha in Brandenburg and Sachsen, but it does not apply to recreational facilities in these two states. In contrast, a permission even for recreational areas is required in Baden-Württemberg and Sachsen-Anhalt when woodland is cut but not converted. Additionally, in Sachsen-Anhalt the construction of carpark areas and Christmas-tree nurseries in the forest is subject to approval, even though these areas are also rated as forests. But, in Baden-Württemberg, approval is only required if recreational areas larger than 1 ha are concerned. Besides, forest lanes are explicitly excluded in this state so that they can always be built without a permit for conversion. In Bayern, a permit for conversion is required when forest lanes or areas connected with the forest and ancillary to it are built in protected forests (*Schutzwald*). As conversion is not realized until forest pursuant to its legal definition is removed, a second permit is required when the above-mentioned areas are again changed into other than forest areas.

Moreover, additional restrictions on conversions are imposed by a number of states. These restrictions refer first of all to protected forests and recreational forests. Thus, a strict ban on any conversion in these areas has been introduced in Schleswig-Holstein and Thüringen, as well as in particular types of protected forests (called *Bannwald*) in Hessen. In other states (Bay, Hmb, Hess, NW), particular requirements have to be fulfilled when conversion is to take place in protected or recreational forests. In these cases, compensatory reafforestation is strictly required and justification of conversion is made more difficult.

Besides, some states specify that authorization 'shall' not be granted in areas with a low proportion of forests (Nds, Bbg) or when forest function plans (*Waldfunktionspläne*) are opposed to it (Bay). However, in exceptional cases, adaptations might be admissible when a strict order has not yet been formulated.

Furthermore, conversion always 'has to' be denied (even away from protected areas) in numerous states (Bln, MV, LSA, SH, Thür) if public interests of forest conservation are predominant. In Brandenburg, this is the case when conversion is not compatible with the objectives of federal area-wide planning, as well as those of state-wide planning (*Ziele der Raumordnung und Landesplanung*). In these cases, the forest authorities may never grant approval when public interests of forest conservation are predominant, not even in exceptional cases. Thus stricter provisions in comparison with the Federal Forest Act have been introduced, as federal law only requires that consent 'shall' be denied in these cases.

When the forest authorities consent to conversion of forest, a compensatory measure will usually be asked of the forest owner gaining by the conversion. Such a measure usually consists of reafforestation. If it is not possible to carry out compensatory measures, a contribution (*Abgabe*) might be demanded. These requirements are stated by forest law in almost every federal state. Where forest law is not applicable in the other states, nature conservation law provides similar rules by the intrusion provision (Section 7.2.8).

Finally, it has to be mentioned that approval is generally bound to a time limit, which is usually indicated in the permit. Some states determine a fixed limit of 2 years (Hess, Thür) or of 5 years at the most (MV, SH) in their state forest law.

> Conversion of forest areas is subject to approval of the forest authorities and cannot be admitted if conservation of forest land is predominantly in the public interest. However, conversion for recreational areas can be admissible where they are to be established predominantly in the public interest. In addition, some state laws impose further restrictions in specific cases.

7.2.6 Construction of recreational facilities according to German planning law

In the case of construction of recreational facilities in the forest – such as huts, holiday houses, viewpoints, restaurants, shops, car-park areas, roads and golf-courses – planning law might be applicable as well as forest law. Where forest law does not state any requirements at all, only planning law (and additionally nature conservation law (see Section 7.2.8) has to be regarded.

Furthermore, forest owners sometimes have at their disposal additional non-forest land, in particular agricultural land, which they can use as a building site. In this case, forest law – which is rather restrictive – is not applicable at all, so that the realization of some projects might be facilitated. Therefore, in the following research, the right of development will be looked at even when recreational facilities are, to some extent, built outside the forest.

Development of buildings is subject to various special laws. Planning law (*Bauplanungsrecht*) is regulated by the Federal Planning Act (*Baugesetzbuch*, BauGB; Federal Law Gazette 1997 I p. 2141). Building control from the point of view of safety (*Bauordnungsrecht*) is state law, but its content is, to a greater or lesser degree, equivalent in all of the states. Whereas state law sets up an obligation of applying for approval, planning law formulates the main requirements of admissibility for the construction of buildings. Furthermore, state law defines which types of buildings are subject to approval and which are exempt.

Necessity of approval for buildings and recreational facilities

Any building operation requires the planning permission of the local planning authorities (building permit). According to the respective state building law (*Landesbauordnung*), the term 'building' (*bauliche Anlage*) usually includes any structure or construction made of building material and connected – directly or indirectly – to the ground. Fillings (*Aufschüttungen*) and diggings (*Abgra-*

bungen), parking and storage places, as well as playgrounds, are usually also regarded as building operations. Exemptions of approval refer, first of all, to minor alterations to a building, but change of use is not exempt. Accordingly, the conversion of a former hunting hut into a holiday hut is subject to approval, even when the outside view is not modified.

Movable structures (e.g. caravans) are not normally classified as buildings, unless they are attached to the land or made permanent in some way. Furthermore, smaller structures, such as fences, advertisement boards, lanes and recreational facilities, are exempt from approval to a certain extent, differing in each state. Thus, fences are usually not subject to approval if they serve forestry business. Then, only forest law is applicable. Fences are designated as having a forestry purpose when their function and design is determined by the forest's management. But fences with pavements, closed fences or rather high ones are not usually exempt.

Moreover, fillings and diggings are generally subject to approval. Therefore, even construction of unpaved forest lanes is subject to approval if forest ground is moved or filled in order to obtain some kind of foundation. However, fillings and diggings up to a certain extent, e.g. up to 300 m^2 in Niedersachsen and 1000 m^2 in Schleswig-Holstein, are exempt so no permit is then necessary. In contrast, paved roads are always made of building material and are not only a 'filling', so a building permit is required.

Finally, playing facilities (*Spielgeräte*) for playgrounds are sometimes exempt from approval. But – if at all – this refers only to typical facilities, such as a swing or a climbing pole, or to playgrounds up to a certain size (e.g. 100 m^2 in RP). In general, playgrounds, camp-sites and car-park areas are explicitly subject to approval.

Admissibility of recreational facilities in forests

A building permit, according to the respective state building law, can only be granted in so far as the requirements of the law itself are fulfilled and other law is not opposed to it. Consequently, the requirements imposed by other laws have to be proved. In this regard, first of all planning law has to be considered that does not relate to a single authorization body.

In exceptional cases, planning law is also applicable when a building permit is not required. This is the case when development, e.g. construction of fences, is subject to approval pursuant to a landscape protection regulation or when it is an intrusion in nature and landscape pursuant to nature conservation law (VGH Hess, NuR 1995, 296).

According to the Federal Planning Act, admissibility of a building depends on the planning area in which the building is to be erected. The law differentiates between three different types of areas: first, the area within a binding land-use plan (*Bebauungsplan* – §§ 30–33 BauGB) that determines the admissible development, secondly, sites within developed areas but outside a land-use plan, the so-called Inner Areas (*Innenbereich* – § 34 BauGB), and, thirdly, the Outer Areas (*Außenbereich* – § 35 BauGB), which are neither within a land-use plan nor within the Inner Area.

Forest and agricultural land usually belongs to the Outer Area in so far as no land-use plan is formally launched by the local community (*Gemeinde*). In the following, only Outer Areas are considered, because a land-use plan will not usually exist for forest areas or farmland. But it has to be taken into account that the community can take advantage of its competence to enact land-use plans. This is of particular relevance when the community is also the owner or when the community is interested in a building operation on private land. However, any person still has to obtain a permit for conversion of forest in order to get an additional building permit, in so far as no exemptions are made by state forest law for building operations assigned by a land-use plan (see Section 7.2.5). A land-use plan is illegal when conflicts of forest conversion cannot be solved in the process of granting approval for the building operations (BVerwG NuR 1998, 134).

The Outer Area (§ 35 BauGB) generally must not be developed. Therefore, the Federal Planning Act imposes far-reaching restrictions on the construction of a building in this area. However, some mitigations refer to so-called privileged operations (*privilegierte Vorhaben* – § 35 I BauGB), whereas non-privileged buildings can only be realized in individual cases when 'public interests are not impaired' (§ 35 II BauGB). But it has to be assumed that public interests are almost always impaired if forest or agricultural land is developed, in particular because nature conservation, the particularity of the landscape (*Eigenart der Landschaft*) and its recreational value are impaired (§ 35 III no. 5 BauGB). Furthermore, a building operation must not contradict the preparatory land-use plan (*Flächennutzungsplan*) of the community, and fragmentary settlement in Outer Areas must be avoided (§ 35 III no. 1, 7 BauGB). However, in the case of rather small facilities, it might be possible to avoid impairments so that they can be admissible, in particular when they are erected within areas that are already developed for recreational purpose.

PRIVILEGED OPERATIONS. Nevertheless, admissibility of a building operation in the Outer Area mainly depends on whether the operation is privileged or not. Relevant privileged operations are those which:

- serve (*dienen*) a forestry business and require only a minor part of the business area (§ 35 I no. 1 BauGB);
- serve a business that is bound to a particular place (§ 35 I no. 3 BauGB);
- must be realized in Outer Areas because of their particular purpose (§ 35 I no. 4 BauGB).

A serving relation to a forestry business according to § 35 I no. 1 BauGB is defined when the building is indispensable for forest management, e.g. a dwelling-house occupied by a forester. Besides, a non-forestry business can take part in the privilege when it is subordinated to forestry (a so-called *mitgezogener Betrieb*). This might be the case when a few holiday houses, apartments or caravan sites or a shop where mainly the products of the forestry are sold are to be built. But, in any case, it is necessary that the Outer Area is spared as much as possible and that the subordinate relation is externally visible. Accordingly, it will generally be presupposed that the buildings are erected within the precincts (*Hofstelle*) of the forest base. Moreover, it must be intended that the subordinated business runs profitably (*Gewinnerzielungsabsicht*), though its income must not exceed that of forestry so as to keep up the subordinate relationship.

Secondly, a business is privileged when it is bound to a particular place (*ortsgebundener gewerblicher Betrieb*). This is the case when it may only be erected at one particular place for geographical or geological reasons, which might apply, for example, to a quarry, a mine or even to ski-lifts. But mere advantages of position (*Standortvorteile*) do not constitute a privileged position within the Outer Area (BVerwG NuR 1996, 251).

Finally, particular building operations are privileged because they must be realized in Outer Areas due to their particular purpose. This is of prime importance for recreational facilities. However, not every recreational facility 'shall' be erected in Outer Areas. The main preconditions are that the facilities are open to the general public, and that they do not serve individual interests (Smollich, 1993, p. 67; Krautzberger, 1998, § 35 margin no. 41; BVerwG UPR 1992, 28). Therefore, golf-courses, holiday houses, camp-sites, rest centres (*Erholungsheime*) and educational centres (*Bildungszentren*) are not privileged. The same is valid for fences established for commercial purposes (Ernst *et al.*, 1987/1992, § 35 margin no. 62). In contrast, viewpoints, game enclosures and protective huts that are open to the general public free of charge may benefit from the privilege. In addition, other recreational facilities, such as playgrounds, picnic-sites and car-park areas, are privileged in so far as they are demanded for recreation by the general public. Hunting huts are included when they are required for hunting, which has to be decided in respect of their position, size and features (BVerwG NuR 1997, 141). Even restaurants might be privileged if they are indispensable for the supply of recreationists and if they are erected within areas that have already been developed, e.g. for stopping off at an inn (*Einkehrzonen*).

Construction of recreational areas involving considerable planning problems because of their size and the extent of affected interests are subject to formal development planning (*Bauleitplanung*). This refers in particular to golf-courses, holiday-house areas and camp-sites. The same has to be assumed where entrance fees are demanded for bigger recreational facilities.

ADMISSIBILITY OF PRIVILEGED BUILDINGS. Even when a building is privileged, a building permit has to be denied 'when public interests are opposed' (§ 35 I BauGB). Public interests are particularly opposed when other provisions of public law are contradicted. Thus, a permit has to be denied when a building is erected in the forest without approval for conversion or when protected areas are affected without permission according to the respective protective regulation (*Schutzverordnung*) (see Sections 7.2.5 and 7.2.7). Moreover, forest law often stipulates a minimum distance (50–100 m) to the forest, in particular when the building includes a fireplace. Besides, intrusions on nature and landscape must not be caused (see Section 7.2.8).

Furthermore, it must be ensured that the site to be developed is made accessible in a sufficient way (*ausreichende Erschließung* – § 35 I BauGB). Thus, supply with electricity and water and accessibility to fire brigades and ambulances are normally required. But the extent of necessary measures depends on reasons of security and environmental protection according to the individual circumstances. Thus, the necessity of water-supply will not apply, for example, to a playground.

Moreover, a building is not admissible when other interests of planning law are infringed. Although the criteria listed in § 35 III BauGB only refer to non-privileged buildings, they are to some extent applicable for privileged ones as well (Krautzberger, 1998, § 35 margin no. 62–67). However, these criteria are dealt with in a less restrictive way in the case of privileged operations in accordance with the intention of the legislator. The crucial point is that – pursuant to § 35 III no. 5 BauGB – the landscape (*Landschaftsbild*) must not be disfigured (*verunstaltet*) and the peculiarity of the landscape (*Eigenart der Landschaft*) and its recreational value must not be impaired. But a building has to be rather obviously unpleasant or unsuitable to contradict these criteria. However, recreational facilities must not be unreasonably big in size, and they must more or less fit into the surroundings. It would, for example, not usually be admissible to establish recreational facilities in the middle of the forest. On the contrary, public interests will normally not be opposed when the facilities are erected within areas that are already developed to some extent.

CHANGE OF USE. Change of use is not generally privileged. Moreover, change of use may imply cease of privilege, e.g. when a former hunting hut in the Outer Area is converted into a holiday hut. As a consequence, approval will not be granted in such a case because public interests – in particular the principle of avoidance of fragmentary settlement (Splittersiedlung) in Outer Areas according to § 35 III BauGB – are impaired.

However, some preferences refer particularly to change of use of buildings that have formerly 'served a forestry business' pursuant to § 35 I no. 1 BauGB. In particular, fragmentary settlement and impairments of the peculiarity of the landscape cannot contradict such a change of use (§ 35 IV BauGB). But some conditions are imposed in order to benefit from the legal preference. First of all, it is presupposed that the shape of the building is preserved in the main. Therefore, significant alterations, e.g. developing a former cattle-shed into a holiday house, are not admissible. Furthermore, the building must be situated in spatial and functional connection to the forestry administration centre (§ 35 IV no. 1 lit. e BauGB). In order to meet the legal conditions, in particular a shop, a restaurant or holiday apartments may be constituted in buildings that have formerly been used as dwelling-houses or for the management of the forestry.

On the contrary, buildings that have formerly served a business that is bound to a particular place or that have to be realized in Outer Areas because of their particular purpose, pursuant to § 35 I no. 3 and 4 BauGB, e.g. a hunting hut, may not benefit from these mitigations. As a result, a forest owner merely seeking to change the use of his/her buildings is still subject to a number of restrictions imposed by planning law, although a permit for conversion of forest is not required in these cases.

A building permit is required when building operations take place. Construction of buildings in Outer Areas, to which forest areas usually belong, is privileged by law in specific cases. Thus, approval is usually granted for specific projects that are erected close to the forestry centre or that are accessible free of charge to the general public. On the contrary, it is rather difficult to get a permit for buildings that are erected merely for private commercial interests.

7.2.7 Restrictions on protected areas

According to forest and nature conservation law, habitat protection is pursued through designation of specific key sites. By this means, additional restrictions can be imposed on management, as well as on the recreational use of forests.

In general, designation is made by enacting a protective regulation (*Schutzverordnung*) determining the rights and duties of every person concerned in respect of the objectives of protection. The regulation is passed by the appropriate authorities of forest or nature conservation administration. In particular cases, protection is granted otherwise, as described in the relevant context below.

The restrictions imposed on protected areas depend on the category of protection. Even though administration includes a certain degree of discretion (*Ermessen*), an area always has to be worthy of protection (*schutzwürdig*) according to the definitions of the chosen category. It should be remembered that many designations are cumulative.

Protected areas pursuant to forest law

PROTECTED FOREST (SCHUTZWALD). According to § 12 I BWaldG, woodland can be designated as protected forest, especially if protection is needed against:

- damaging environmental impacts pursuant to the Federal Immission Control Act;
- soil and water erosion;
- desiccation;
- harmful run-off of precipitation;
- avalanches.

Any clear felling or any clearing activity producing the same effect requires the authorization of the forest authorities. Consent can be granted with legal conditions attached where this is necessary for preservation of the forest functions (§ 12 III BWaldG).

The term 'protected forest' is not uniform in the federal states. Terms used in some states include 'soil-protection forest' (*Bodenschutzwald*), 'ban forest' (*Bannwald*), 'nature forest reserve' (*Naturwaldreservat*) or 'forest protection area' (*Waldschutzgebiet*). The different terms characterize various types of protected forests aiming at distinct objectives of protection. In addition, differences refer to the method of designation, which can be achieved not only by enacting a protective regulation, but also by a contract with the landowner, or protection is directly granted by law to particular types of woodland described in the particular state law. In Berlin and Hamburg, the whole forest is declared to be protected forest by state law.

Furthermore, intensity of protection varies to a considerable degree in different types of protected forests. Sometimes, merely the minimum requirements imposed by the Federal Forest Act apply. But conversion of forests is explicitly made more difficult in numerous states (see Section 7.2.5). Apart from this, access to the forest for recreational purpose must not be excluded in total, but restrictions on recreational use can be imposed as far as they are necessary in respect of the pursued aim of protection. However, this is usually not the case in practice. Besides, protection and scientific research of particular forest plants, animals and biotopes can be a matter of designation, which might require further restrictions on recreational use and a strict ban on any forest management. But these objectives of protection raise some problems because of overlapping with nature conservation law and because empowerment by the Federal Forest Act is missing in this respect. Therefore, objectives of nature conservation are usually assigned only to state forests or in agreement with private forest owners. However, additional designation by nature conservation law can be made, which might impose further restrictions on forest use.

RECREATIONAL FOREST (*ERHOLUNGSWALD*). Pursuant to § 13 BWaldG, woodland can be designated as recreational forest if general public interests require the protection, care and shaping of the forest for recreational purposes. In particular, provisions can be enacted that regulate:

- the type and extent of forest management;
- restraints on hunting for protection of the forest visitors;
- the obligation of the forest owner to tolerate the construction of lanes and other recreational facilities, as well as the removal of intrusive installations;
- the conduct of forest visitors.

Designation of recreational forest presupposes a general public interest. Therefore, individual interests of particular groups, e.g. a hiking club, do not justify designation. Usually protection is only required close to urban areas or health resorts where there is a high demand for recreation. In general, the federal states ensure that forests owned by public bodies are preferred for designation. By this means, restrictions on private owners are avoided as far as possible.

On the one hand, forest owners can be placed under certain obligations, including, first of all, the prescription to execute particular measures of forest management, such as the choice of species to be planted or restrictions on the application of pesticides. Furthermore, the forest owner is subject to toleration duties imposed for recreational interests. On the other hand, forest visitors cannot be entirely excluded as recreational forests are designated particularly for recreational purposes. However, restrictions may be necessary if intensity of use causes conflicts between different user groups or with forest protection. The permissible restrictions to be imposed depend on individual circumstances.

In Berlin, the forest as a whole is defined as recreational forest by state law. The same refers to state forests in Hamburg and to private forests in district-free towns (*kreisfreie Städte*) in Niedersachsen. In other states, designation is usually made by enacting protective regulation.

Protected areas pursuant to nature conservation law

NATURE PROTECTION AREAS (*NATURSCHUTZGEBIETE*). In accordance with § 13 I BNatSchG, nature protection areas are defined as territories in which it is necessary to organize special protective measures for the countryside as a whole or for certain parts of it. These territories:

- serve for the conservation of living communities or biotopes of particular species of wild plants and animals;
- serve for scientific, natural historical or physiographical interest;
- have to be protected because of their rarity, uniqueness or extraordinary beauty.

All activities which might destroy, damage or alter the nature protection area or parts thereof are prohibited. As far as the objectives of nature conservation are not opposed, the area can be made accessible to the public (§ 13 II BNatSchG).

Nature protection areas are the sites best-protected by German law. In general, a strict ban on any alteration is imposed. Sometimes, even uses that have already been carried out are prohibited for the future, although this implies compensation for the losses incurred by the landowner. Moreover, the areas may even be closed to the public as far as this is necessary to achieve the aim pursued. However, this does not usually occur in practice. But the landowner and the visitors are placed under far-reaching restrictions, as any action that is likely to damage the site will be prohibited according to the particular protective regulation. Accordingly, promotion of recreational activities is extraordinary difficult, if not even impossible, in these areas.

It has to be noted that an average forest will not be part of a nature conservation area unless special protection is justified by particular reasons. This might refer, for example, to alluvial or moor forests and to specific stands (*Sonderstandorte*) in other forests.

NATIONAL PARKS (*NATIONALPARKS*). In accordance with § 14 I BNatSchG, national parks are safeguarded by law on a uniform basis and meet the following specific requirements:

- they are of considerable size and have specific features.
- They satisfy the definition of nature protection areas in large parts.
- They are situated in a region that has suffered little or no human influence.
- They ensure protection of the indigenous fauna and flora and serve to retain the greatest possible variety of species.

With respect to their size, national parks have to be protected like nature protection areas. As far as compatible with the objectives of protection, the national parks are to be made accessible to the public (§ 14 II BNatSchG).

The main difference in comparison with nature protection areas is the bigger size of national parks, which might also imply less intensity of protection. In general, national parks are divided into different protective zones according to their ecological sensitivity. Only the core zones are strictly protected, e.g. by excluding the public, whereas outer zones may also include developed or other relatively intensively used areas. Nevertheless, a number of legal restrictions refer to national parks, so that recreational activities will be significantly impeded. It should be mentioned that a few German national parks are typically characterized by forest areas.

The German category national park can be traced back to the international regulations of the International Union for Conservation of Nature (IUCN). However, most German national parks do not meet the criteria of the IUCN, because of lack of naturalness due to human influence.

LANDSCAPE PROTECTION AREAS (*LANDSCHAFTSSCHUTZGEBIETE*). Pursuant to § 15 I BNatSchG, landscape protection areas are assigned:

- for preservation or restoration of the natural potential;
- because of their variety, specificity or natural beauty;
- because they are particularly suited for recreation.

Actions that change the character of such a site or that contradict the particular objectives of protection are not admissible (§ 15 II BNatSchG).

Significant parts of the Federal Republic of Germany are designated as landscape protection areas, and these areas have a considerable share in forests. Therefore, this category is probably the most important one dealt with in this context. Although no strict ban on any alteration is usually imposed and already existing uses are normally not restricted, development of buildings will often contradict the objectives of landscape protection even in case of 'privileged' operations (see Section 7.2.6). Exceptions particularly refer to areas that are already partly developed or to recreational facilities that are open to the general public. Apart from this, it depends on the rights and duties determined in the particular protective regulation by which uses are restricted. The actual intensity of protection differs very greatly in different sites. But any restriction imposed has to be justified by particular reasons of nature preservation. Accordingly, recreational use that is likely to damage the protected site may be prohibited. But recreational traffic must not be totally excluded, as landscape protection areas are also designated for recreational aspects.

NATURE PARKS (*NATURPARKS*). Pursuant to § 16 I BNatSchG, nature parks are defined as territories that are to be managed in a uniform way and which fulfil the following specific requirements:

- they extend over a large surface area.
- They include nature or landscape protection areas.
- They are particularly suitable for recreational use.
- They are assigned for recreation and tourism in country-wide planning.

In addition, nature parks must be planned, structured and made accessible in accordance with their recreational objectives (§ 16 II BNatSchG).

In contrast to the categories mentioned above, nature parks are not always designated in a legally binding manner, so that protective regulation constituting restrictions on land use is not always enacted. In this case, protection is granted by designating the whole or parts of the area as a landscape or nature protection area. The declaration of an area as a nature park is mainly aimed at the administrative management and recreational planning. However, in some federal states, a formal designation is provided, so that equivalent restrictions to those for nature and landscape protection areas can be imposed, at least for parts of the nature park.

OTHER PROTECTED SITES. Nature monuments (*Naturdenkmäler*) and protected parts of the landscape (*geschützte Landschaftsbestandteile*) only comprise rather small sites. This means that, for example, a single, very old oak may be designated as a nature monument, and that small woodlands or hedges that characterize the landscape can be formally protected. The main goal is to preserve an object rather than a site. Therefore, these categories are not of further interest in this context.

The category of biosphere reserves (*Biosphärenreservate*) has just recently been defined in the Federal Nature Conservation Act and had already been introduced before, especially in the laws of the new (eastern) federal states. This category refers back to a programme of the United Nations Educational, Scientific and Cultural Organisation (UNESCO) called 'Man and the Biosphere'. Designation of biosphere reserves pursues preservation of large-sized areas of cultural use where ecologically compatible management activities are to be developed. German law mainly grants the same protective regime as in case of nature or landscape protection areas. Thus, a totally new category is not constituted.

The federal states are empowered by § 20c BNatSchG to enact provisions granting protection to particular biotopes of extraordinary value for nature conservation. This refers particularly to bog and alluvial forests, as well as to dry and warm forest stands. Protection is granted directly by law, as a protective regulation does not exist. But not every state has made use of the federal empowerment, so that statutory protection of biotopes is not applicable throughout the country. Where legal provisions have been enacted, the biotopes must not be destroyed or impaired, e.g. by a riding event (VGH BW, NuR 1995, 462). By this means, rather strict protection is achieved, but owners or recreationists cannot be placed under particular duties, due to lack of an individual protective regulation. Furthermore, only small sites usually fulfil the legal criteria of biotope definition.

Protected areas pursuant to European law

Two European provisions are of prime importance as far as habitat protection is concerned: the Birds Directive (79/409/EEC) and the latterly enforced Habitats Directive (92/43/EEC). Whereas the Birds Directive is aimed at the protection of habitats for wild bird species, the Habitats Directive relates to the protection of habitats for wild plants, animals and biotopes according to the annexes of the directives. In these annexes, numerous species and biotopes of the forest are listed, e.g. alluvial forests, bog woodland and different types of beech and oak woods. In addition, some species, such as lynx (*Lynx lynx*), wolf (*Canis lupus*), capercaillie (*Tetrao urogallus*) and different beetles (e.g. *Osmoderma eremita, Rosalia alpina*), are of importance in the context of forest protection.

According to the Birds Directive, special protection areas (SPAs) have to be designated by the member states of the European Community. Under the Habitats Directive, the member states are required to designate special areas of conservation (SACs) and to make up a coherent European ecological network, known as Natura 2000, including the SPAs. It has to be mentioned that – in contrast with German law – discretion of designation can be condensed to a strict obligation of designation. The European Court of Justice condemned Spain and Britain for having failed to classify particular areas as SPAs (C-355/90, C-44/95). Consequently, all sites that meet the criteria for designation as a SPA should be treated as if they had been formally designated. Although the Habitats Directive mitigated the provisions of the Birds Directive, the rules for designating SPAs are not changed by the Habitats Directive. But, for some reason, it is not yet clarified whether a strict obligation of protection also applies to areas yet to be classified (but not proposed or designated) under the Habitats Directive.

The protective regime launched by the Birds Directive is now replaced by the obligations arising under Art. 6 of the Habitats Directive: the member states are required to take appropriate steps to avoid deterioration of the sites and significant disturbance of the species for which the areas have been designated. In order to achieve this goal, management plans have to be outlined. In this context, restrictions can be imposed on specific uses of a site. Moreover, any plan or project – thus a RES project as well – which is likely to exert a significant effect on the site must be subject to an appropriate assessment of the implications; the national authorities can agree only if this will not 'adversely affect the integrity of the site concerned' (Art. 6 III). However, a plan or project may be carried out if there are 'imperative reasons of overriding public interest' (Art. 6 IV). It is of prime importance that these obligations refer not only to areas formally designated, but even to those proposed by the member states and adopted by the European Commission as sites of community importance (SCIs) according to the Habitats Directive.

In Germany, both directives are transformed by designating areas according to the German categories of nature conservation law. However, the protective regime to be applied is stricter than any German category. In a number of European countries, including Germany, legal transformation, as well as designation of areas, is not yet accomplished, so that effects on RES projects will not occur for a few years to come.

> Further restrictions on RES projects can be imposed if the project is to be realized within protected areas. German nature conservation and forest law provides a number of categories of protected areas. Admissibility of a project depends on the category of protection concerned and the individual requirements for protection of the specific site.

7.2.8 Intrusions in nature and landscape by German nature conservation law

As far as nature and landscape are adversely affected by recreational use or by construction of recreational facilities, it must be assessed whether an 'intrusion' is recognised (*Eingriff*) according to nature conservation law. Intrusions are subject to numerous restrictions that might impede their implementation.

Definition of intrusions

Pursuant to § 8 I BNatSchG and the consequent provisions of state law, intrusions are defined as a change of the shape or use of land which might adversely affect the ecosystem or the landscape in a significant or sustainable manner.

The orderly (*ordnungsgemäße*) forestry and agricultural use of the ground (*Bodennutzung*) is exempt from application of the provisions relating to intrusions (§ 8 VII BNatSchG). However, this privilege only refers to the mere use of the ground and the ordinary and everyday practice of management. In this connection, the setting up of enclosures for the protection of juvenile forest cultures still benefits from the privilege. But change between forestry and agricultural use or between arable ground and pasture, as well as the construction of buildings for management use, is not included (BVerwGE 67, 93; VGH Hess, NuR 1992, 86; BVerwG NuR 1989, 84).

Accordingly, change of the shape of land by the construction of lanes, recreational facilities and other buildings, positioning of advertisement signs or caravans in the Outer Area (§ 35 BauGB (see Section 7.2.6)), conversion of forest and hindrance of free access for non-forestry reasons are usually regarded as 'intrusions'. This is often spelt out explicitly by listing rule examples (*Regelbeispiele*) in the state laws. On the contrary, rather small structures, such as single benches, will not usually lead to significant impairments and thus are not intrusions. As far as significance is concerned, it will be of particular importance if protected areas or endangered species are affected.

Furthermore, mere change of use can fulfil the conditions of an intrusion. This is the case when, for example, a meadow is used as a flying field for model planes (VGH BW, NuR 1992, 126). But the temporary use of the forest for events and meetings will not usually be regarded as an intrusion. However, this has to be judged differently if significant impairments are caused, e.g. by a motocross race (VGH BW, NuR 1987, 129), an organized hunt on horseback (VGH BW, NuR 1995, 462) or organized games in the forest (VG Darmstadt NuL 1998, 38).

On the contrary, federal law presupposes that intrusions are subject to announcement or approval by administrative authorities (§ 8 II BNatSchG). In this way, a number of actions are effectively exempt, even though they impair the ecosystem or the landscape. Therefore, most states have introduced a subsidiary duty of authorization if no specific permit or announcement is required. In this case, the provisions referring to intrusions are always applicable when significant impairments are caused, irrespective of specific duties of authorization or announcement applying to particular uses.

Duties imposed on intrusions

If an intrusion is accomplished, four different obligations arise. First, the person carrying out the project has to avoid any encroachment that is not indispensable to achieve the pursued aim (obligation of avoidance – *Vermeidungsgebot*). Thus, the alternative causing the least possible impairments has to be chosen; for example, a lane that is required for forest management usually must not be paved (VGH Bay, NuR 1981, 209).

Secondly, compensatory measures (*Ausgleichsmaßnahmen*) have to be carried out by the person causing the intrusion. The destroyed or impaired functions and structures of the ecosystem and the landscape are to be replaced by new ones of the same kind in the vicinity of the place of intrusion. If forest is converted, an equivalent reafforestation will be an appropriate compensation. As far as an equivalent duty of reafforestation is already attached to a permit for conversion of forest, no further compensatory measures have to be carried out due to nature conservation law. An intrusion is compensated for when no significant or sustainable impairments remain and when the landscape is restored or newly shaped (§ 8 II, 3rd sentence, BNatSchG).

Thirdly, all requirements concerning nature and landscape have to be weighed against one another (obligation of weighing – *Abwägungsgebot*). An intrusion has to be forbidden when impairments cannot be avoided or compensated for and when precedence is given to the interests of nature conservation and landscape protection (§ 8 III BNatSchG). However, in practice, intrusions are not often prohibited if severe impairments are avoided and compensation is afforded.

Finally, measures of substitution (*Ersatzmaßnahmen*) have to be implemented as far as compensatory measures are not possible or not convenient to carry out. This duty is imposed by state law, but it refers back to an empowerment pursuant to § 8 IX BNatSchG. In contrast to compensatory measures, substitution is not bound to a close relation between the impaired functions and those to be replaced. The spatial relation can be wider as well. However, some functional and spatial relationship still has to be maintained, and substitution must still be equivalent in value. If it is not possible to realize even substitutional measures, most states require compensatory payment (*Ausgleichsabgabe*).

Intrusions in nature and landscape are subject to specific provisions of German nature conservation law. Thus, the forest ecosystem must not be impaired more than necessary and encroachments must be compensated for by appropriate measures, e.g. afforestation. In specific cases, the intrusion, and consequently the whole RES project, may even be prohibited.

7.2.9 Identification of relevant legal provisions in Germany

As this research focuses on the frame conditions for RES projects in general, it is not always simple to identify the legal problems in a specific case. Therefore, some checklists and tables will be presented in order to give some orientation to the reader.

It has to be noted that establishing RES projects in the forest can be affected by many legal provisions. Their application depends not only on the type of project, but on the place of realization as well. Thus, provisions of state law are only applicable in that particular state, and provisions on protected areas only apply if a designated site is concerned. According to individual circumstances, several provisions can be relevant and even several permits can be required. Table 7.2 gives a rough overview of the legal provisions that have to be checked and that can be cumulatively applicable for specific types of projects.

As a next step, the relevant legal provisions have to be considered in detail in order to evaluate the admissibility of RES projects. However, in some cases it might turn out to be a problem to identify the relevant laws that contain the respective legal provisions. Unfortunately, the provisions are divided between a number of different laws and, moreover, the legal situation is not uniform throughout Germany, as the most important laws in this context – nature conservation and forest law – are mainly subject to the legislative competence of the federal states. Besides, the legislature sometimes entitles the appropriate administrative bodies to enact specific provisions. Therefore, an overview of the different laws and their scope of application is given (Table 7.3).

Going into further detail, the provisions of the states on specific types of forest uses turn out to constitute a very broad and complex field. Nevertheless, it seems necessary to review some of the details, because of their primary importance. Therefore, the main important provisions of forest and nature conservation law of the federal states are shown in Table 7.4. Regarding this table, it has to be borne in mind that it is not complete. Further provisions on specific uses exist and other provisions might be additionally applicable, in particular regulations of police law or provisions of planning law. In addition, Table 7.4 contains some simplifications; further details are described in the different sections of

Table 7.2. Checklist to identify relevant legal provisions for RES projects.

Type of project	Relevant legal provisions
Mere access to the forest (e.g. events, camping, riding, etc.)	Right of access Specific provisions on the respective activity
Recreational facilities	Conversion of forest areas Building permit Specific provisions of state law Construction of fences Animal enclosures
Environmental projects	Afforestation
All projects (additionally)	Intrusion in nature and landscape Protected areas

Table 7.3. Relevant German laws for RES-projects in the forests.

Relevant law	Important provisions
Federal Forest Act (federal law)	Right of access to forests Conversion of forest areas Afforestation Protection of specific forest areas
Federal Nature Conservation Act (federal law)	Right of access to nature except forests Protection of specific areas Protection of species Animal enclosures Intrusions in nature and landscape
Forest and/or nature conservation law (state law)	Construction of fences Restrictions on riding, cycling, driving, camping and other activities in the forest Construction of recreational facilities Details on application of federal law
Traffic law (federal law)	Traffic regulations (order, security, events, fences)
Street law (state law)	Access to public roads Access to private roads (only Bre)
Building law (state law: order, security) (federal law: planning)	Construction of recreational facilities
Regulations of police law (state or local law)	Camping Organized events
Local decrees (empowerment by state law)	Gathering of mushrooms (Nds) Riding (NW, LSA)

this research. Apart from this, it has to be noted that a lack of specific provisions does not imply that the respective activity is generally admissible. Other provisions may still impose restrictions in individual cases and, furthermore, the appropriate authorities may prohibit specific uses when disturbance, damage or danger is caused.

Table 7.4. Provisions on specific forest uses by German state laws.

Type of use	BW	Bay	Bln	Bbg	Bre	Hmb	Hess	MV	Nds	NW	RP	SL	SN	LSA	SH	Thür
Riding permitted on lanes in general	F	N				F	F			L	F	F		F/N		F
Riding permitted only on marked lanes			F	F				F	F/N	N			F	L	F	
Riding prohibited on footpaths	F					F					F	F		F/N		
Riding prohibited on sport and nature paths	F					F	F	F		N						F
Riding prohibited on marked hiking lanes	F					F		F		N	F	F				
Cycling prohibited on footpaths	F												F	F/N		
Cycling prohibited on sport and nature paths	F												F			F
Cycling prohibited on raised paths			F													
Riding prohibited away from lanes	F	N	F	F		F		F				N	F			
Cycling prohibited away from lanes	F					F		F				N	F			
Duty to mark riding horses	F	N		F			F		F/N	N		F	F		(F)	
Riding levy	F									N		F	F		(F)	
Driving by car				F, a				F, p								
Motor sport				F, p				F, p						F/N, p		F, p
Camping				F, a				F, a							F, p	
Caravanning				F, a				F, a							F, p	
Organized sport events	F, a			F, a				F, a						F/N, a		F, a
Organized meetings	F, a													F/N, a		
Organized gathering of mushrooms	F, a						N, a		L		F, a			F/N, a	N, a	
Construction of sale stalls								F, a								
Construction of advertising installations								F, a								
Construction of lanes		N, n		F, a				F, a		F, n						F, a
Construction of recreational facilities	F, a													F, a	N, a	F, a
Construction of fireplaces	F, a	F, a	F, a	F, a		F, a		F, a		F, a	F, a		F, a			F, a

a, approval; n, announcement; p, prohibited; F, provided by forest law; F/N, provided by common forest and nature conservation law; N, provided by nature conservation law; L, empowerment by forest or nature conservation law for provisions to be enacted by the local authorities; (F), provided by forest law, but not enacted.

Finally, it has to be pointed out that the admissibility of a specific RES project can only be ascertained in individual cases according to the legal and factual situation of the case. This research can only present a rather rough overview on the most important legal provisions and their general interpretation.

7.3 Comparative Study of Access Rights in Different Countries

As the legal situation differs in each country of the project partners, at least one aspect should be examined and compared in all countries. This ought to be the right of access, because it describes the different starting-points of the countries for the development of RES projects. However, a problem in this context is that access to forests is not influenced only by a single legal rule or a single law; a rather large number of rules and laws of each country are concerned. Furthermore, different kinds of legal interpretation apply in the different countries, which are difficult to investigate because of language problems. In this respect, it has to be noted that laws, comments and jurisdiction are often only available in the native language and not in English. Therefore, foreign co-workers have been engaged in describing the right of access in the Netherlands and Italy (Caruso, 1998; Hertoghs, 1998); their results are merely summarized in the following section. As this examination is focused on the German situation, legal problems are only regarded from a rather general point of view as far as the other countries of the European partners involved are considered.

This examination is a descriptive comparison of the different legal rules in each country, rather than a scientific work dealing with comparative law (*Rechtsvergleichung*). As comparative law is aimed at the valuation of national rules, the following steps are usually taken (Rheinstein, 1987, p. 12; Zweigert and Kötz, 1996, p. 7):

- a rather short description of national rights.
- Comparative reflections.

Comparative reflections lead to a review of legal policy and allow the drawing of conclusions for interpretation of national law. Another result may consist of a supranational systematology of national law and a study of the role of law in society.

In the following, a synoptic description of the different national laws will be presented and major differences will be pointed out. The role of access right is examined only in respect of the significance for the forest owner offering recreational services. Thus, only a very small aspect of law is considered. This has to be taken into account when conclusions are drawn to evaluate the different national laws.

7.3.1 Austrian right of access

Pursuant to Art. 10 I no. 10 Austrian Basic Law (*Bundes-Verfassungsgesetz*, B-VG), the legislation and execution of forest matter (*Forstwesen*) are part of the federal competence. Corresponding to that, the Federal Forest Act of Austria (*Forstgesetz*, ForstG: Federal Law Gazette 1975, p. 440, last amended 1996, p. 419) comprises much more detailed provisions than the German Federal Forest Act. Moreover, the legal situation in Austria is more uniform than in Germany as the federal states of Austria have less competence in forest legislation. On the other hand, nature conservation law is only enacted by the states; a Federal Nature Conservation Act does not exist. Accordingly, nature conservation law is much more split up in the different Austrian states than it is in the German ones.

It must be mentioned that only the nature conservation laws of the following states have been analysed for the following research: Burgenland, Kärnten, Oberösterreich, Salzburg, Vorarlberg and Wien. As not every state could be considered, there might exist even more state provisions that are of relevance in this context.

Access to forests for recreational purposes

According to § 33 I ForstG, access to forests is permitted for recreational purposes. 'Access' includes walking and resting by day; long-distance skiing is allowed 'under application of necessary prudence'. By contrast, riding and cycling are not permitted by law. In this respect, there is a significant difference from the German situation. In addition, camping, resting overnight and driving motorized vehicles are not generally permitted; but all of these activities can be allowed by the forest owner (§ 33 III ForstG). To prevent damage to young trees, ski-runs in forests leaving designated pistes and ski-routes in the area of ascent facilities is strictly prohibited (§ 33 III, § 174 IV lit e ForstG) (Gatterbauer, 1993, p. 15). In the latter case, the forest owner may not give further approval.

As the right of access is granted only for recreational purposes, commercial or organized activities are not permitted thereby (Wohanka, 1983, p. 63). In this respect, the same situation prevails as in Germany.

Restrictions of the right of access

Access to forest facilities, e.g. buildings and storage places for wood, as well as to young trees up to 3 m of height (§ 33 II ForstG), is not allowed. In addition, the forest owner may erect barriers under certain conditions (§ 34 ForstG). Barriers limited in time are permitted to ensure forest or scientific works. Permanent barriers may be erected for protection of particular cultures, such as Christmas trees, for recreational facilities and for residential areas. The forest owner has to apply for official approval for restrictive barriers that are kept for longer than 4 months and for permanent barriers that enclose forest areas bigger than 5 ha. Approval has to be granted when the barrier is essential in order to reach the pursued aim (§ 34 IV ForstG). In this case, the forest owner has to make it possible for people to go round permanently closed areas (§ 34 VIII ForstG).

In contrast with German law, the construction of fences for recreational facilities in forests is explicitly regulated and is allowed under particular conditions. Furthermore, enclosures smaller than 5 ha are generally not subject to authorization. Accordingly, the disposal rights of the forest owners are less restricted. But it has to be borne in mind that the charging of entrance fees for particular areas can depend on further provisions. The forest owner will need additional permission in most cases for the construction of the enclosed facilities. This can be a permit for conversion of forests (§ 17 ForstG) and, if the need arises, an additional building permit. In contrast to German law, a permit for forest conversion is more frequently required, as recreational areas are not normally rated as forests (§ 1 ForstG). However, some privileges refer to recreational forests (§ 35 ForstG): in these areas, obtaining approval is facilitated for conversion of forest and construction of recreational facilities. In this respect, recreational forests in Austria fulfil a different function from those in Germany.

Furthermore, the nature conservation laws of the Austrian states often require authorization for the construction of roads, car-park areas, fences and various recreational facilities, such as golf-courses, skiing pistes, camp-sites and motocross sites. Approval will only be granted when the landscape and ecosystem are not negatively affected. This will always be a considerable problem as far as forests are concerned. Therefore, nature conservation regulations are of prime importance with respect to the development of recreational areas.

Acquisition of forest products

The acquisition of small quantities of forest products, such as berries and mushrooms (less than 2 kg), is no infringement (*Verwaltungsübertretung*) of forest law. Only the implementation of and the participation in organized picking events (§ 174 IV lit d ForstG) is strictly forbidden. This rule serves to prevent the extinction of the plants and mushrooms (Kalss, 1990, p. 143). But this does not generally mean that forest products are removed from the control of the forest owner. He/she may still prohibit collecting or may demand money for individual picking permits as long as state law is not contravened (for a contrary view, see Podlipnig and Stock, 1998, p. 36). Thus, some federal states have restricted the gathering of mushrooms in a further way by specific nature conservation regulations (*Verordnungen*) that prohibit commercial picking or picking in larger amounts or that introduce a duty of approval for these activities (see *Pilz(schutz)verordnungen* Tirol LGBl. 30/1992, Salzburg LGBl. 47/1994 and Kärnten LGBl. 79/1992). Moreover, the same restrictions on picking are sometimes imposed by the particular nature conservation law of a state.

Restrictions on forest uses by nature conservation law

All nature conservation laws contain various provisions that are aimed at the protection of nature and recreational areas. In this respect, nature conservation law comprises more rules than forest law. They have to be applied not only to agricultural areas but to forests, as well. The Federal Forest Act cannot be considered as the final (*abschließend*) regulation in this respect, even though this is not explicitly expressed by any law. On the contrary, the constitutional principle of consideration (*Berücksichtigungsprinzip*) demands that the right of access granted by the federal legislature is only restricted for essential reasons and not completely abolished (Giese, 1996, p. 196; Podlipnig and Stock, 1998, p. 16). In any case, this is not the fact if only particular uses are restricted in forests as realized by the nature conservation law of the Austrian states.

Thus, numerous provisions have to be considered that might be of importance for the implementation of recreational services in forests. This refers in particular to the setting up of sale stalls and advertising signboards, to leaving designated areas, to the exercising of bicycle or motor sport beyond designated areas, to camping outside camp-sites or camping with caravans and to the chemical preparation of ski pistes. These activities are subject to authorization or announcement or they are even strictly prohibited. As nature conservation law is only enacted by the states, the legal situation in the various Austrian states differs considerably and has to be properly checked in every single case.

7.3.2 Access to forests in Italy

The distribution of powers between the state and the regions (the regions mainly correspond to the German federal states) is provided for in Art. 117 of the Italian Constitution (Const.), which lists the subjects of legislative and administrative competence that are delegated to the regions. These subjects are, among others, forestry, agriculture, hunting, fishing and planning regulations. Nature conservation is not mentioned, so that the national legislator has primary powers in this area. However, further competence is delegated to the regions by national law on nature conservation.

Art. 117 Const. refers to regions that are ruled by an 'ordinary statute' (15 out of 20). Their competence is granted within the 'general principles set by state laws' and 'the interest of the state and other regions'. Other regions enjoy so-called 'special statute self-government'. Their legislative powers are limited only by the 'general principles of the Italian legal system' (not the principles expressly laid down in laws) and 'international duties'. Accordingly, these regions possess greater competence than the regions of 'ordinary statute'.

It may be observed that a national law on forestry does not exist, as this subject is delegated to the regions' competence. However, several national laws explicitly refer to forests and set a framework for regional legislation. Thus, restrictions are imposed on forest use due to hydrogeological reasons by L 3267/1923 or due to protection of the environment and landscape beauty by L 431/1985. Other national laws refer to particular types of use, such as the gathering of mushrooms (L 352/1993) or truffles (L 752/1985). Moreover, particular forests can be designated as parks or reserves, and in this context they can be placed under further duties and restraints (L 394/1991). It should be noted that nature conservation law is split between numerous national laws. Furthermore, regions have enacted specific provisions to implement national law. But only some of the regions have passed complex laws on nature conservation (e.g. Regional Law (LR) Piemonte 32/1982) or forestry in general (e.g. LR Lombardia 8/1976).

Access rights to parks and reserves

Even though national laws impose restrictions on particular matters of forest management, a right of access to forests for the general public is not granted. A general principle of free access is only stated for natural parks and reserves of the state, according to Art. 1 of *Decreto Ministeriale* (DM) 1984–12–15, which has similar effects to national laws. But it is also provided that access can be restricted due to reasons of nature conservation and that access for groups of a particular number is always limited to predetermined paths under the authorization and surveillance of competent authorities.

According to Art. 11 of L 394/1991 (frame law on natural protected areas), access to forests in more detail is up to regulations issued by the *Ente Parco* (park government authorities); in particular they concern 'stay and circulation of the public by all means of transportation', thus covering all types of access – even by cars and motor cycles. Specific attention must be paid to providing adequate means of access for handicapped and old people. Park regulations can specify visiting hours, and they can permit particular types of use, such as for exercise, as well as recreational and educational activities. Particular regulation of access is left to the discretion of the park authorities. But, as far as prohibitions are concerned, national law sets directly applicable principles, e.g. it is forbidden to use open fires or to advertise without authorization.

As far as regional and local protected areas are concerned, L 394/1991 sets out only general principles and delegates legislative and administrative competence to the regions, which entitles them to regulate public access to these areas.

As a result, access to forests is explicitly regulated for particular areas designated as parks or reserves – even though access rights may differ considerably in different areas due to the respective regional law. The rules on parks and reserves impose specific duties on landowners to tolerate access to such areas and to bear restrictions on their rights of use. It is of interest that the law does not provide compensation for the landowners for toleration of access, but only for restrictions on agro-forestry and other activities (Art. 15 L 394/1991).

Access to non-protected areas

Public forests are run by regions or local authorities and are subject to their regulations on access. Usually, regional laws explicitly or implicitly take account of public access to forests and permit (or do not prohibit) particular types of access. On the other hand, restrictions can be imposed on access, e.g. by a ban on using motorized vehicles or off-road driving. These restrictions refer either to forests in general or only to particular forests, depending on the regional law. But they may also apply to private forests. However, the private landowner him/herself will not be subject to such restrictions on access, because of the principles of private property. Furthermore, regional laws cannot explicitly permit access to private forests, as this – according to the prevailing opinion – would infringe the principles of private property laid down in the constitution. Notwithstanding, it is assumed that access is accepted by the owner (or at least it is not forbidden by criminal law) unless he/she has made use of his/her right to ban other persons from his/her grounds by enclosures (see below).

However, the disposal rights of a private forest owner are limited to some extent by the access rights of others granted by civil or other law.

Access rights by civil law

The Italian Civil Code (CC) expressly allows single cases of access to areas (not specifically forests) of other proprietors, in particular for neighbours when there is a real necessity to do so, e.g. to repair a wall (Art. 843 CC), for owners of animals or bees (Art. 942 CC) and for hunters holding a hunting permit (Art. 842 CC). Hunting is only excluded when there are cultivated areas capable of being damaged or when land is closed pursuant to hunting law, which usually requires the construction of fences higher than 1.8 m or water ditches broader than 3 m (L 968/1977, L 157/1992 and most regional laws). Access for any other purpose is subject to the consent of the owner, in particular access in order to take photos or to observe animals for scientific or artistic reasons (Constitutional Court sentence n. 57, 1976–3–25).

Acquisition of natural products

Another important source of access rights is constituted by the so-called '*usi civic*', civic uses, which are defined as 'rights of common enjoyment of lands' (L 1766/1927). These traditional rights are of two categories: on the one hand, 'essential' uses necessary for basic needs, such as the right of grazing and the gathering of wood or fruits, and, on the other hand, uses for economic or profit purposes exceeding personal or family needs. According to the main law on forests, L 3267/1923, civic uses for personal and family needs (as defined in Art. 1021 CC) are explicitly admitted, while uses for economic or profit purposes cannot be allowed. But it has to be noted that this law refers only to particular forests that are designated and protected for hydrogeological reasons.

A forest owner can prohibit civic uses by contracts or – if contracting is not possible – by a special procedure of winding up, where a share of the forest property is transferred to the holder of the civic rights. In so far as civic uses are maintained without winding up, they are run according to economic plans (Art. 130, 135 of L 3267/1923) and they are taken into consideration by laws on parks and reserves (L 394/1991) and on landscape protection (L 431/1985). Besides, regional laws and park regulators take account of them. In this context, it is of prime importance that specific national and regional laws refer to the gathering of natural products, such as mushrooms and fruits.

The national legislature has enacted specific provisions on mushrooms (L 352/1993), medicinal plants (L 39/1931) and truffles (L 752/1985). These laws are completed by regional laws on mushrooms, truffles and plants. In addition, regions have passed further laws on natural products and the protection of plants. The different laws describe permitted and forbidden picking of natural products by setting conditions, time and quantity limits. Several regions have introduced picking permits especially for mushrooms, but the requirements for obtaining such a permit differ greatly. According to LR Emilia Romagna 6/1996 and LR Piemonte 32/1982, a fee is charged for the permit, while other regions provide free permits. The Constitutional Court asserted that it is no violation of private property principles when picking is allowed without any form of compensation for owners, except for professional cultivation (sentence 328, 1990–7–13). However, the regional law of Piemonte, for example, explicitly states that the money from the picking permits is transferred to private landowners. As Art. 2 of L 352/1993 empowers the regions to regulate the ways of picking mushrooms without further specification, regional legislature are given wide discretion.

When restrictions on picking are set, ordinarily residents can be favoured. This principle is based on the necessity of giving precedence to civic uses of residents in contrast to the mere entertainment of visitors from outside. Apart from this, it has to be noted that the picking of natural products is generally prohibited in private gardens and forests near houses or when cultivation is professional and licensed. Moreover, restrictions on picking do not usually apply to the forest owner, except for protected plants. Thus, the landowner may still pick unlimited quantities without a permit or he/she may obtain a permit free of charge.

Power of enclosure

Art. 841 CC expresses the general power of enclosure of one's own land, according to the character of exclusiveness typical for private property. However, it is not entirely clear whether and in which cases other legal provisions impose restrictions on enclosures by landowners.

According to L 3267/1923, authorization is required for alterations to forests protected for hydrogeological reasons. L 431/1985 (Landscape and Environment Protection Act or *Legge Galasso*) generally qualifies forests as goods of public interest and subjects them to the provisions of L 1497/1939 (law on landscape protection). According to these laws, particular activities are admitted without explicit licence: ordinary and extraordinary maintenance that does not alter the site, as well as agro-forestry works that definitely do not modify the condition of a site and do not alter the hydrogeological structure of land. Any other activity altering the landscape is subject to authorization of the competent authority pursuant to Art. 7 of L 1497/1939. Accordingly, if enclosures impair or alter the landscape and do not fit into the category of agroforestry works, they are subject to an express licence (L 431/1985). Moreover, owners cannot use their competence of enclosure in opposition to civic uses. But it also has to be mentioned that regional laws sometimes explicitly allow the erection of signs (LR Lazio 58/1989) by the owners in order to reserve the picking of natural products for themselves. On the contrary, mere custom, in particular access to private land for recreational purposes, cannot derogate the right of enclosure.

Furthermore, it is questionable whether local council plans or planning regulations could restrain owners from fencing in. As walls and fences do not normally have to be authorized, it sometimes happens that councils or local authorities are interested in imposing restrictions on enclosures, e.g. due to skiing or other sports activities. It is not entirely clear if it is necessary to compensate owners for such restrictions, unless regional or national laws provide prohibitions or place licences on enclosures.

Finally, it should be mentioned that mere trespass is a typical and historical tort. However, the effectiveness of this provision for protecting private property may be questioned because of the lack of a civil remedy, as mere trespass does not usually cause specific damages. But Art. 637 of the Criminal Code prohibits abusive access to lands when they are enclosed by hedges, ditches or stable fences. No violation occurs when enclosure is interrupted or is only temporary or when it results from signs and oral or written warnings.

7.3.3 Access to forests in the Netherlands

The Dutch nation is a so-called decentralized unitary state (*gedecentraliseerde eenheidsstaat*). This means that the primacy of legislative and administrative powers rests with the central government. The decentralized bodies, such as the provinces or communities, have a limited legislative and administrative competence within their territory. These competencies are limited in the sense that the decentralized bodies have to respect the regulations of the central level. As forest and nature conservation law is mainly regulated by national law, the legal situation is more uniform in the Netherlands than in any other country referred to above. However, provinces and communities may enact additional provisions, and they may additionally grant specific subsidies. Besides, it has to be noted that the Dutch Constitution contains no specific provisions on forest law or on nature conservation law.

The main legal provisions concerning forests can be found in the Forest Act (*Boswet*, 1961–7–20, last amended 1997–12–17). This law regulates forestry in order to safeguard the existence of a reasonable area of woodland. Furthermore, the Forest Act provides the legal basis for the policy on state subsidies concerning forestry. Another important regulation is the Nature Conservation Act (*Natuurbeschermingswet* (Nbw), 1998–5–25). The main objective of this law is to protect specific natural reserves, including woodlands. Besides, the Landscape Act (*Natuurschoonwet* (Nsw), 1928–3–15, last amended 1997–12–18) provides tax deductions for the forest owner admitting forest visitors to his/her land.

Furthermore, other laws and regulations can influence particular uses of the forest, e.g. the Open Air Recreation Act (*Wet op de openluchtrecreatie*) of 1994–3–25, last amended 1997–12–17. Moreover, the policy of outdoor recreation is mainly laid down in a governmental policy plan that is periodically updated (*Tweede Kamer* 1992–1993, 22 990, nr. 2). This plan does not impose direct restrictions on outdoor recreation, but it has to be observed in the decisions of the competent authorities, e.g. when licences are granted. Finally, the government master plan based upon the Town and Country Planning Act (*Wet op de Ruimtelijke Ordening* 1962–7–5, last amended 1997–12–4), the Green Space master plan (*Structuurplan Groene*

Ruimte, Tweede Kamer 1992–1993, 22 880), has to be considered.

Neither forest nor nature conservation law provides an explicit right of access to forests. However, the Dutch government stimulated the acceptance of public access to private land by granting state subsidies and tax deductions for the landowner.

Financial incentives for opening of forests

Theoretically, the private forest owner is free to exclude forest visitors from his/her property. He/she is not obliged to give access to the forest. Trespassing without permission of the owner is penalized in § 461 of the Dutch Penal Code (*Wetboek van Strafrecht*). However, the relatively low cost-effectiveness and the subsidies of the government give an incentive to open the forests to the public. Moreover, open-air recreation in forests is very popular and public access does not usually influence the output of forestry in a negative sense. An exception can be made for the access of cars or other motor vehicles (government policy plan, p. 70). The incentives that exist to stimulate forest owners to open their property to the public differ for private owners, for nature preservation organizations and for public bodies. Only about 14% of the entire forest is not accessible to the public (*Uitvoeringsprogramma Meerjarenplan Bosbouw, Tweede Kamer* 1989–1990, 21 671, nr. 2, p. 40).

PRIVATE FOREST OWNERS. The earliest act that provides a legal basis for a financial incentive to open the forest to the public is the Landscape Act of 1928 (*Natuurschoonwet*/Nsw). This law is aimed at the preservation of rural estates. According to the law, an estate is prescribed as 'real estate on the Dutch territory that is wholly or partly covered with forest or timber' (§ 1 Nsw). Rural estates that are accessible for the public are not subjected to property taxes (§ 7 Nsw). In order to benefit from the tax deduction, walking and cycling on paths and roads must not be denied (§ 7 Rangschikkingsbesluit Natuurschoonwet 1928 RN, 1990–11–23, last amended 1996–6–14). Apart from this, the owner can allow other forms of extensive outdoor recreation. Nevertheless, a rural estate that is used for motor sport or for intensive outdoor recreation is not classified as a rural estate (§ 4 RN). Specific provisions refer to camping, stating that the size and number of camp-sites on an estate is limited (§ 6 RN). Thus, the size of a camp-site must not exceed 1 ha, and on an estate of 25 ha only one camp-site is allowed; on an estate of 100 ha two camp-sites are permitted, and on an estate of more than 250 ha not more than three camp-sites may be built. The owner who does not open his/her property to the public but preserves his/her rural estate in accordance with the provisions of the Landscape Act is taxed over half of the estimated value of his/her estate.

In general, the landowner cannot charge entrance fees to his/her estate. An exception is made for estates that are situated near big cities. The owner can charge a very small fee in order to limit the number of visitors. The fee must not exceed Dfl. 0.50 for a 1-day visit and Dfl. 2.50 for an annual pass. An entrance fee can also be charged for special gardens when the costs of maintenance exceed the income. For both forms of charging entrance fees, the owner must have the permission of the Ministry of Agriculture, Nature Preservation and Fisheries.

Furthermore, the Forest Act contains provisions that are important for the access to the forests. This act gives a legal basis for subsidy schemes addressed to the private forest owners. There is a regulation with respect to compensation for certain functions of forests or nature reserves (*Regeling functiebeloning bos en natuurterreinen* 1994–8–18, last amended on 1997–12–15). This regulation grants certain subsidies to the forest owner for the preservation of his/her property. Among the subsidies, there is an annual amount per hectare. The owner who opens his/her forest to the public receives a sum that is four times the amount than that given to an owner excluding visitors. Thus, an owner may receive Dfl. 140 per hectare instead of Dfl. 35. For very valuable forests, an extra sum of Dfl. 40 can be granted for both categories. The access must meet certain requirements: the grounds must cover at least 5 ha and the entrance(s) must be accessible from a public road; finally, there must be several entrances depending upon the size of the area. The length of the open paths must be at least 50 m per hectare, and the forest must be opened throughout the year from sunrise till sunset. The owner is not allowed to charge entrance fees.

NATURE PRESERVATION ORGANIZATIONS. Nature preservation organizations play an increasingly important role in preserving Dutch forests. About 11% of the Dutch forest is owned by such organizations. These organizations are generally subsidized by the state, and they pursue their goals by purchasing forests and other nature reserves which they consider to be valuable. For each piece of land that these organizations acquire, the government pays half of the acquisition price. Furthermore, the organizations can obtain an annual sum to cover their management costs. This sum is individually determined for each organization (§ 16 *Regeling subsidies particuliere terreinbeherende natuurbeschermingsorganisaties*, 1997–12–15). In addition, the provinces give some financial support. The objective of acquiring nature reserves is to realize a network of nature reserves in order to allow migration of fauna and flora. The territories of the nature preservation organizations, in general, are open to the public from sunrise to sunset. The subsidies do not depend on the accessibility of the territory. In some areas, access is restricted, e.g. during the breeding periods of certain animal species. In most cases, restriction is done by zoning. The restricted area is surrounded by accessible areas. In general, Dutch policy tries to combine the preservation of valuable natural domains with the stimulation of open-air recreation. Nature preservation organizations cannot receive subsidies on the same basis as private owners, in so far as they are based upon the *Regeling functiebeloning bos en natuurterreinen* (§ 5).

PUBLIC BODIES. Since the beginning of this century, the management of forests owned by the state is carried out by Staatsbosbeheer, a service from the Department of Agriculture, Nature Preservation and Fisheries. This organization has recently been privatized. The costs of managing the state forest by Staatsbosbeheer are paid by the state and are part of the budget of the above-mentioned department.

When other legal bodies, such as communities, possess forests, they can obtain a contribution to the costs of forest management on the condition that their property is accessible to the public. According to Art. 4 *Regeling functiebeloning bos en natuurterreinen*, the amount is Dfl. 70 per hectare and can be Dfl. 20 more when very valuable forest is involved.

Other specific subsidies

The creation of a network of long-distance footpaths through the forests is encouraged by the government. Certain sums are granted to facilitate the accessibility of roads and lanes. The same applies to a network of cycle-tracks. The provinces administer the subsidies. Certain private organizations cooperate with the government to establish such networks.

Restrictions on access in natural monuments

The Dutch Nature Conservation Act (*Natuurbeschermingswet*/Nbw) only provides a single category of protected sites, the so-called natural monuments (*natuurmonumenten*). As far as these sites are concerned, access can be restricted for reasons of nature conservation. If necessary, some areas can be entirely fenced. However, restrictions on access cannot be imposed on the proprietor or the tenant and his/her relatives without their consent (§ 20 Nbw). Apart from this, other restrictions apply, such as a prohibition on picking flowers or disturbing animals. Even further restrictions can be imposed by the authorities that are competent to indicate certain sites as natural monuments (§§ 19, 24 Nbw). Due to the serious influence on the property rights of the owner, restrictions are necessary for the protection of the specific site in order to be lawful. The Nature Conservation Act allows financial compensation for restrictions on property rights.

Camping

Camping is generally restricted to campsites, and the exploitation of a camp-site is prohibited without a permit from the local authorities (§§ 8, 15 *Wet op de openluchtrecreatie* (Open Air Recreation Act)). In order to preserve forests or other valuable natural sites, the provincial authorities can restrict camping in certain areas. In these areas, no camping permit may be given. An even more specific regulation applies to camping in the forest with respect to nature preservation (*Regeling natuurkampeerterreinen* 1995–5–15). Camping in nature reserves, including forests, is only allowed under restricted conditions. The size of the camp-sites must not be larger than 1 ha, and there must not be more than 40 sites for tents or campers. Visitors are not allowed to stay longer than 6 weeks.

7.3.4 Comparative reflections

A comparison of the legal situation in Austria, Italy, the Netherlands and Germany shows that the right of access to forests is regulated very differently. However, some similarities can be observed between the countries.

Even though there are still a number of differences between Germany and Austria, both countries have enacted an explicit right of recreationists to enter other persons' land. Thus, the position of forest visitors – at least of pedestrians – is more or less the same, as they uphold a subjective right. In respect of the rights of the landowner offering RES products, an advantage of the Austrian situation may be seen in the fact that it is stated rather clearly by law which conditions have to be fulfilled in order to erect barriers for recreational facilities.

In contrast, Italy and the Netherlands do not provide a right of access. In Italy, at least a principle – not a subjective right – of access is granted; however, this only refers to state nature reserves. As a consequence, the property rights of landowners are less restricted than in Germany and Austria; the landowner is generally entitled to exclude other people (even though there might still occur some problems of enforcement, especially in Italy). However, access to forests is actually not totally prohibited: numerous landowners tolerate the access of the public, although they are not obliged to do so. But, in particular for Italy, general statements on the admissibility of recreational and environmental services in the forest are difficult to make because of the importance of provincial legislation.

The Dutch situation is very specific, as access is mainly determined by subsidies granted by the state. Similar subsidies do not exist in any other country concerned. The main consequence of restricting public access in order to offer RES products is the removal of subsidies, which is not directly a legal problem. The Dutch subsidy system merely does not encourage a landowner to develop RES projects that might conflict with public benefits. In any case, the policy strongly depends on the support of the general public and can probably not be realized in a similar way in any other country.

As a final result, it has to be pointed out that the right of access does not encroach very much upon the property rights of landowners, although their disposal rights are certainly restricted. The German and Austrian right is limited to recreational purposes and, thus, only mere trespass is admitted. In the area of recreational and environmental services, rather complex and specific products may be developed that are usually not in conflict with the subjective right of recreationists. However, in some cases, the access right causes problems because it is difficult to exclude unauthorized people, e.g. on contract paths. But, in general, it cannot be concluded that RES products are easier to develop in Italy and the Netherlands, where no right of access is enacted. In respect of RES products, specific provisions imposing restrictions on particular types of forest use by forest, nature conservation and planning law are of prime importance. As these provisions have not been investigated in detail in the other countries, a final statement cannot be made.

Comparing the legal situation of Austria, the Netherlands and Italy with the German situation shows that only Germany and Austria provide a right of access to the forest for recreational purposes. Whereas in Austria only walking is permitted, in Germany horse-riding and cycling on forest lanes are also allowed. On the contrary, a right of access does not exist in Italy and the Netherlands. In this respect, the rights of forest landowners are less restricted. However, the admissibility of RES projects mainly depends on specific provisions referring to particular uses. As these provisions have not been investigated in detail in the other countries, a final statement cannot be made.

7.4 Final Reflections and Recommendations for German Legislation

In respect of the legal situation in Austria, Italy and the Netherlands, some recommendations for the German legislature should be made.

The comparative study on the legal situation in different countries shows that the right of access has not turned out to be a general impediment for the implementation of RES products. Therefore, it should be maintained in the interest of recreationists. Even a regulation based on financial incentives for toleration of access, similar to the Dutch law, would imply a significant deterioration in recreational rights. Besides, this would necessitate very far-reaching legal changes and a considerable financial burden on public bodies in financial respect, which probably would not be accepted by German policy. But, in order to facilitate realization of specific RES products, some legal amendments on the specific uses of the forest would be helpful.

First of all, the legislature should specify the right of access with regard to the construction of fences and other types of barriers in the context of recreational uses. In this respect, mainly the state legislatures are asked to fill the legal frame provided by the Federal Forest Act. Thus, the state legislatures could specify that fences and even the charging of entrance fees may be admissible in the forest when the respective recreational facility or event is offered for the overriding public interest. However, this would still imply that significant impairments of free access and of the ecosystem must not be caused, so that only particular RES products would benefit. But an even greater change in the legal situation in favour of fences would no longer comply with the right of access provided by the Federal Forest Act.

Nevertheless, an amendment of federal law could be taken into consideration in order to determine the conditions of fences for recreational facilities in a general way throughout the country, similar to Austrian law. However, the German state legislatures would still have to transform the federal law by adaptation of their state laws, because the federal legislature of Germany is usually not allowed to pass provisions of direct effect within the area of forest law, which is subject to framework legislation.

Besides, enlargement of the right of access by the state legislatures should be done prudently in respect of the interests of landowners. In particular, it could be stated that the gathering of mushrooms and berries in small amounts is only permitted free of charge when the landowner does not introduce picking permits. Thus, a forest owner would still be entitled to implement RES products in this field.

Another problem of RES products is the riding levy which impedes the making of contracts with riders or riding organizations if riders have to pay two duties at the same time: the public levy and, additionally, a private fee. However, only a few federal states are actually concerned (NW, SL; SH only if the levy is enacted). In this respect, adoption of legal provisions is necessary in order to introduce an exemption clause for the riding levy as far as private contracts on riding are effected on forest areas by the landowner.

Apart from that, it should be mentioned that riding is not included in the Austrian right of access at all, which generally facilitates implementation of RES projects in this context. However, exclusion of riding from the German right of access will probably not be possible for political reasons; besides, it is not really necessary if the above-mentioned amendments on the riding levy are adopted. The other restrictions on riding imposed by the state legislatures do not have similar impeding effects on RES products.

Finally, it has to be conceded that the German state legislatures have enacted a number of restrictions on specific recreational activities. Moreover, most of these provisions have been passed in recent years, due to the increasing demand for recreation by the general public and the various conflicts caused thereby. Even though these provisions are also aimed at the protection of private property, they can impose restrictions on those landowners who want to implement RES projects. Therefore, it is necessary for the legislature to pay due regard to the rights of the landowners when new provisions on recreation are enacted. Until now, RES products that are implemented by forest landowners are probably not considered by policy leaders, so that some publicity in this market field is necessary.

In general, it can be recommended that a permit of the landowner for minor recreational projects in the forest, such as events with small groups, should be sufficient. Additional authorization by the forest authorities should only be required when the particular activity is dangerous, e.g. for large, disrupting events. Besides, a strict ban will only be necessary in very exceptional cases, e.g. for motor sport in the forest. It has to be borne in mind that further restrictions can still be imposed on protected areas in order to comply with the requirements of nature conservation. On the contrary, it should be explicitly stated by law that the appropriate authorities may prohibit recreational uses in individual cases when disturbance, damage or danger is caused, in order to protect nature, as well as providing safety to other recreationists. By this means, public interests can be sufficiently safeguarded while RES projects are impeded as little as necessary.

> Some recommendations for the German legislature can be made. The right of access should be upheld in the interest of recreationists. But some legal amendments on specific uses of the forest would be helpful in order to stress the disposal rights of private landowners and, thus, to facilitate implementation of RES products.

7.5 Summary

First, the provisions of the German Basic Law that apply to forest uses were described. In this context, the basic rights of recreationists as well as those of private landowners were determined. As far as recreation and mobility in the forest is concerned, forest visitors may refer to the general freedom of action protected by Art. 2 I of the Basic Law. In this respect, citizens can challenge interference by the state – in particular, by the legislature. Where interference is caused by private landowners, recreationists have to refer to the state's duty of protection, which is deduced from the Basic Law. But, in this case, recreationists can claim an interference in their basic right only when the state has acted insufficiently in a very obvious way. However, all interference in recreational uses requires justification. Justification might be given when recreation is restricted for reasons of nature conservation, regulation of public traffic or because of the property rights of landowners. Nevertheless, unreasonable restrictions infringe the Basic Law and, therefore, are not admissible.

The rights of a forest landowner are protected by Art. 14 of the German Basic Law, which is the property right. In this case, the legislature has to determine the substance and limits of private property. Existing property rights can be restricted for public interests due to the social obligation of property, explicitly stated in the Basic Law. Accordingly, property rights within forests are restricted for the recreational interests of the general public, in particular the right of access, and for reasons of nature conservation. However, any restriction must be reasonable in order to comply with the constitution. If grave or unfair restrictions have to be imposed on a landowner for the public interest, compensation must be granted by the state.

In the next step, the simple law, mainly forest and nature conservation law, which contains numerous provisions on different types of forest uses, has been dealt with. These provisions are looked at in detail in so far as they impose restrictions on the disposal rights of a forest landowner in Germany seeking implementation of RES products.

First of all, the German right of access to forest areas has been examined, which allows recreational use by pedestrians, riders and cyclists. These uses have to be tolerated free of charge by the forest owner. However, he/she may still dispose of his/her land as far as non-recreational uses are concerned, e.g. commercial uses or organized events. But the construction of fences or other kinds of barriers is severely impeded by the right of access. Moreover, many specific provisions impose further restrictions on particular uses, such as riding, camping, motor sport and events. But, in general, such actions are not absolutely banned; rather, authorization by the appropriate authorities is required in some federal states. As a main result, it can be stated that specific uses can be admitted as long as no significant damage of the forest ecosystem and no disturbance or danger to recreationists are caused.

Moreover, conversion of forest areas is subject to approval of the forest authorities and cannot be admitted if conservation of forest land is predominantly in the public interest. However, conversion for recreational areas might be admissible in so far as they are established predominantly in the public interest. Additionally, a building permit is required when building operations take place for recreational facilities. Construction of buildings in Outer Areas, to which forest areas usually belong, is privileged by law in specific cases. Thus, approval is usually granted for specific projects that are erected close to the forest headquarters or that are accessible free of charge to the general public. On the contrary, it is rather difficult to get a permit for buildings that are erected merely for private commercial interests.

Further restrictions on RES projects can be imposed if the project is to be realized within protected areas. German nature conservation and forest law provides a number of categories of protected areas. Admissibility of a project depends on the category of protection concerned and the individual requirements of protection of the specific site. Apart from this, intrusions in nature and landscape are subject to specific provisions of German nature conservation law. Thus, the forest ecosystem must not be impaired more than necessary and encroachments must be compensated for by appropriate measures, e.g. by afforestation. In specific cases, the intrusion, and consequently even the entire RES project, may even be prohibited.

Analysing the legal situation in Germany leads to the conclusion that disposal rights of landowners are mainly affected in connection with recreational products. Environmental projects on ecological management or afforestation are either not at all or at least much less restricted by legal provisions. In this case, it is mainly the problems in the context of contracting that are concerned. But these problems are not dealt with in this chapter as they do not belong to the area of public law, but rather to civil law. In this respect, they are mentioned in Chapter 5 of this volume.

Another field of investigation has been the legal situation in Austria, the Netherlands and Italy as far as access to forests is concerned. Comparing the legal situation of these countries with that in Germany shows that only Germany and Austria provide a right of access to the forest for recreational purposes. Whereas in Austria only walking is permitted, in Germany horse-riding and cycling are also allowed on forest lanes. By contrast, a right of access does not exist in Italy and the Netherlands. Therefore, the rights of forest landowners are less restricted. However, admissibility of RES projects mainly depends on specific provisions referring to particular uses. As these provisions have not been investigated in detail in the other countries, a final comparative statement cannot be made.

Finally, some recommendations for the German legislature have been made. The right of access should be upheld in the interest of recreationists. But some legal amendments on specific uses of the forest would be helpful in order to stress the disposal rights of private landowners and thus to facilitate implementation of RES products.

Glossary

Basic Law (*Grundgesetz*): name of the German constitution (*Verfassung*).

Basic rights (*Grundrechte*): rights of individuals in relation to governmental organs, listed in the constitution of a state.

Constitution (*Verfassung*): fundamental provisions at the apex of ranking of national law; these provisions include the basic rights and, furthermore, refer to the organization of the state.

General freedom of action (*Allgemeine Handlungsfreiheit*): common term for the German basic right of Art. 2 I GG granting 'free development of the personality'.

Inner Area (*Innenbereich*): term of German planning law (§ 34 BauGB), describing areas outside a binding land-use plan but within developed areas.

Intrusion (*Eingriff*): term of German nature conservation law (§ 8 BNatSchG), describing actions that might impair nature and landscape; intrusions are subject to specific legal provisions.

Outer Area (*Außenbereich*): term of German planning law (§ 35 BauGB), describing areas outside a binding land-use plan and outside the Inner Area.

Regulation (*Rechtsverordnungen*): executive legislation according to empowerment by a formal law (*Gesetz*).

Common abbreviations of the German federal states:

Bay	Bayern (Bavaria)
Bbg	Brandenburg
Bln	Berlin
Bre	Bremen
BW	Baden–Württemberg
Hmb	Hamburg
Hess	Hessen
LSA	Sachsen–Anhalt (Saxony–Anhalt)
MV	Mecklenburg–Vorpommern (Mecklenburg–Western Pomerania)
Nds	Niedersachsen (Lower Saxony)
NW	Nordrhein–Westfalen (North Rhine–Westphalia)
RP	Rheinland–Pfalz (Rhineland–Palatinate)
SH	Schleswig–Holstein
SL	Saarland
SN	Sachsen (Saxony)
Thür	Thüringen (Thuringia)

References

Caruso, A. (1998) Access to forests. Italian discipline. Unpublished research, Catania, Italy.

Dipper, H., Ott, W., Schlessmann, H., Schröder, H.-W. and Schumacher, W.H. (1976/1999) *Waldgesetz für Baden-Württemberg*. Kommentar, Kohlhammer, Stuttgart.

Ernst, W., Zinkahn, W. and Bielenberg, W. (1987/1992) *Baugesetzbuch (BauGB)*. Kommentar, Beck, Munich.

Gatterbauer, H. (1993) *Uneingeschränkte Erholung in der Natur – Ein Rechtsanspruch?* Diskussionspapier, Universität für Bodenkultur, Vienna.
Giese, K.J. (1996) Forstrecht. In: Bachmann, S. *et al.* (eds) *Besonderes Verwaltungsrecht.* Springer, Vienna, pp. 193–212.
Göhler, E. (1995) *Ordnungswidrigkeitengesetz,* 11th edn. Beck, Munich.
Hertoghs, M. (1998) Dutch law concerning the access to forest. Unpublished research, Maastricht, the Netherlands.
Kalss, S. (1990) *Forstrecht. Walderhaltung und Umweltschutz.* Springer, Vienna.
Klose, F. and Orf, S. (1998) *Forstrecht.* Kommentar, 2nd edn. Aschendorff, Münster.
Kolodziejcok, K.G. and Recken, J. (1977/1999) *Naturschutz, Landschaftspflege und einschlägige Regelungen des Jagd- und Forstrechts.* Schmidt, Berlin.
Krautzberger, M. (1998) § 35. In: Battis, U., Krautzberger, M. and Löhr, R.P. (eds) *Baugesetzbuch.* Kommentar, Beck, Munich.
Podlipnig, C. and Stock, W. (1998) *Wegefreiheit im Wald. Umwelt im Interessenkonflikt, Informationen zur Umweltpolitik 131.* Bundeskammer für Arbeiter und Angestellte, Vienna.
Rebmann, K., Roth, W. and Herrmann, S. (1988/1998) *Gesetz über Ordnungswidrigkeiten.* Kommentar, Kohlhammer, Stuttgart.
Rheinstein, M. (1987) *Einführung in die Rechtsvergleichung.* Beck, Munich.
Smollich, T. (1993) *Der Konflikt Sportanlagen und Umweltschutz.* Nomos, Baden-Baden.
Wohanka, E. (1983) Skifahren im Wald – Zulässigkeit und Durchsetzbarkeit von Verboten in Österreich. In: Carlsen, C. (ed.) *Natur und Recht (NuR).* Blackwell, Berlin, pp. 63–64.
Zweigert, K. and Kötz, H. (1996) *Einführung in die Rechtsvergleichung auf dem Gebiet des Privatrechts.* Mohr, Tübingen.

8

Forest Policy Analysis – Evaluation Studies of Public Acceptance

8.1 Public Acceptance of RES Products in the Netherlands

8.1.1 Motivation

Problem exploration

In Europe, there is little experience with pricing for recreation in forests. There is a widespread view that recreation should be and can only be provided as a public service and, therefore, be paid for by tax revenues. This view originates from the characteristics of many recreational goods, being non-marketable because exclusion is not possible or too expensive and rivalry is not greatly felt. The introduction of individual owner or user rights, which is one way of establishing a market for this kind of goods, is regarded as undesirable, because in the view of many people there should be as few barriers to recreation as possible. Another argument against pricing is that recreation is considered to be an intrinsically motivated and relatively unstructured experience (Leuschner *et al.*, 1987). On the one hand, this recreational experience might be negatively altered through the introduction of facilities for collecting the fees or the notion of not being 'free to go wherever you want'. On the other hand, a result of having to pay might be that less people will visit the area. This can enhance the quality of experience for some groups seeking peace and tranquillity in outdoor recreation (Frost and McCool, 1988).

In the Netherlands, attitudes towards pricing are changing. As a result, opportunities to introduce pricing in government-subsidized recreation areas have increased on the condition that public access is guaranteed. Especially the more common (basic) activities, such as walking, cycling and stationary recreation, are considered not to be subject to pricing in any form. Pricing is considered to be appropriate only for services and facilities provided for specific recreation activities that are not associated with the general recreational use of public land (Van der Aalst and Van der Straaten, 1992; LNV, 1996). The fact that public access for basic activities has to be guaranteed limits the opportunities for pricing of (semi-)public goods, because it makes the exclusion of other activities difficult or even impossible. There are possibilities for specific situations (special features)[1] and for individual goods, but the importance in economic terms is small (Hekhuis and De Baaij, 1997). However, there are some experiences with entrance fees in coastal areas (dunes) and national parks. As many people arrive by car, a more or less common form of pricing is possible and allowed, namely pricing for parking. Other activities suitable for pricing are horse-riding and mountain-biking (both on specially designed and managed tracks). These activities are considered to be 'not basic' for forests (Segeren *et al.*, 1997). Special services sometimes subject to pricing are visitor centres and excursions (Van Setten, 1989).

Research objectives

The research started with a literature study on economic valuation and pricing experiences. The objectives of the literature study were:

1. To determine factors and variables that influence the attitude towards pricing (acceptability of pricing, price levels and changes in price levels; acceptability of different means of payment).
2. To determine the major consumer groups for the Dutch situation.

The results of the literature study have led to the third objective:

3. To develop a conceptual framework for future research on determining public acceptance of pricing for recreation in forests.

The results of the literature study and the conceptual framework have been used in two surveys about determining public acceptance – visitors of forests and the general public – leading to the fourth research objective:

4. To test the relevant factors and variables from the literature study on relevant consumer groups and the general public in the Netherlands.

8.1.2 Literature research and conceptual framework

Introduction

The research on public acceptance of pricing for recreation in forests started with a literature research on economic valuation and pricing experiences. The aim of this literature research was to find:

1. Factors that influence the acceptability of pricing for recreation in forests;
2. Hints for the acceptability of (changes in) price levels;
3. Factors that influence the acceptability of means of payment;
4. results that define the important consumer groups.

The literature research reports on the psychological aspects of pricing (attitude development theory), and the economic aspects of pricing have been described briefly. The psychological aspects of pricing are considered more important in the context of trying to find factors that influence acceptability. 'One problem evident with the economic approach is that, while it offers a sophisticated predictive tool, it offers little explanation for why people respond to prices in a given manner' (Kerr and Manfredo, 1991).

Pricing principles and tips

The literature outlines a number of pricing principles (psychological dimensions) and practical clues that can be of use for individual forest owners when determining a price.

Crompton (1982) explores psychological dimensions of pricing for leisure services (pricing principles) because: 'The rationale for establishing a specific price must consider the psychological impact which that price will have on potential clients.' He argues that a group's reactions to a new price or a price change are not always logical or rational. Crompton distinguishes eight different psychological dimensions:

1. Expected price threshold. Clients have a lowest and a highest price they want to pay for a particular programme or service. On the one hand, if a price is set too high, the consumer is likely to find it too expensive. On the other hand, if a price is set too low, the consumer might think the quality is not good enough. The price setting at the first introduction of a programme is important, because the initial price is likely to be the midpoint of the expected price threshold. 'The first pricing decision may have a strong determining impact on the level of price which can be charged for that programme throughout its life.'
2. Price tolerance zone. Consumers have a limited zone in which a price increase will be accepted. Tolerance of gradual price increases will prevent influence on change in participation: 'A series of small incremental increases in price that fall within the tolerance zone, over a period of time, are less likely to meet client group resistance than a single major increase.'

3. Client adjustment period. The period to psychologically adjust to the new price after a price change is called the client adjustment period: 'Provide client groups with an advanced warning ... at least some of the client adjustment period will have taken place by the time the price change occurs.'

4. Price–quality relationships. Price is an indicator for the quality of the programme. Higher prices indicate higher quality, better service, etc.: 'This is not an irrational action by consumers because in most cases a higher quality necessitates higher costs and, hence, a higher price has to be paid to meet those higher costs.'

5. Anchor pricing. Marketing literature suggests that the lowest and highest prices charged for a programme or service are likely to be the most noticeable (Monroe, in Crompton, 1982). The 'end prices' may accentuate the perceived value for a service: 'Because the anchor prices are more visible and noticeable than prices spaced between the lowest and highest prices, changing them is likely to arouse more clientele resistance than changing any of the intervening prices.'

6. Change the perceived value of services. By providing a point of reference for the price (e.g. information on service, facilities, other suppliers and other prices), client resistance to a price increase can be reduced. 'If they [the clients] perceive the quality level to be higher, then they are more likely to expect the price for the service to be higher.'

7. Customary pricing. If traditional or customary prices exist for a programme, client groups expect a certain price to be charged: 'It is difficult for a manager to ignore them ... this puts the emphasis on cost control.'

8. Odd pricing. Odd or uneven pricing (e.g. $9.95 instead of $10.00) is thought to create the illusion of low prices, but: 'There is little support to this contention ... and there appears to be no good reason why it should be [adopted].'

McCarville (1992, 1993) and McCarville *et al.* (1996) summarize and describe three pricing principles (fairness, value and choices), illustrated by a number of pricing clues:

1. Participants seek fairness in pricing. Although there is considerable agreement on what is fair and what is unfair in price levels, the tolerance may vary from programme to programme. The following tips can help to establish 'fair' price levels:

- Price is considered most appropriate for those activities that clearly benefit only the participants.
- Use other providers' prices as a guide when developing new price levels.
- If you have to increase your prices, do so in small increments on a regular basis (see price tolerance zone; client adjustment period).
- Tell consumers how much it costs to provide the programme they are about to enjoy.
- Recent payment of fees at similar sites will encourage acceptance of first-time fees. Respondents with recent payment experience at similar day-use areas will report higher 'fair' price levels.
- Information suggesting increased service levels will increase reported 'fair' price levels.
- Local users will report lower 'fair' price levels.

2. Consumers seek value. Value is the positive result of the difference between perceived benefits and perceived costs. Both can be influenced to increase the perceived value of consumers.

- Focus on benefits to be enjoyed through purchase.
- Assign programme names that focus on the benefits of participation.
- Ensure your clients know how they benefit from paying a fee.
- Compare new programmes with well-established and valued alternatives.
- Tell the world how wonderful you and your staff are.
- Stress the convenience element in all of your programmes.
- Offer liberal refund policies to reduce uncertainty.

3. Consumers seek choices. Any threat to users' sense of freedom or choice can result in psychological reactions. Response to such threats ranges from an increased desire for the lost behavioural option to hostility and aggressive behaviour (Iso-Ahola, 1980, cited in McCarville, 1993). Complaints, vandalism and displacement are manifestations of this phenomenon. It is advisable to make sure that people can exercise control over their own decisions to avoid psychological reaction.

- Always provide price alternatives.
- Give the consumer choices as to the kind of price to be paid.

The psychological dimensions of Crompton and the pricing principles of McCarville lead to a better understanding of acceptance of price levels: the reference price. McCarville and Crompton (1987a, b) have also described the 'role of reference prices in the context of public recreation services'. In their articles, they come to nine propositions resulting from a literature research. These nine propositions have been grouped into three groups of consumers' perceptions of reference price:

1. The roles of perception and reference price.

- The acceptance of an increased price for a public leisure service is likely to be influenced by the existing reference price (consumers' notion of prices felt to be appropriate) for that service.
- Consumers expect prices charged by public-sector agencies to be different from those charged by private-sector suppliers for a similar service.

2. The influence of information on reference price.

- Acceptance of a price increase would be enhanced by informing those affected by it of:
  - the prices charged by other suppliers for a similar service;
  - the justification for raising the price including the cost of delivering the service and the level of subsidy involved;
  - the projected benefits emerging from the price increase.
- The order in which comparative price and/or cost information is introduced to consumers may influence eventual beliefs, attitudes and expectations toward the price charged for the service.
- 'Non-consuming' taxpayers, when given the choice, will favour higher user fees and lower levels of tax support for park and recreation services.
- Consumers are more likely to support user fees when such fees are used to maintain and improve the resource for which they are collected.
- Both non-consuming taxpayers and consumers are more likely to support fees for a variety of public recreation and park services than administrators and elected officials.

3. The role of equity in reference price perception.

- Residents will expect the proportion of a programme's costs covered by revenues generated from fees to vary according to the type of recreation programme (differential pricing fees).
- Different groups of residents will support different price structures for any selected recreation service.

Gratton and Taylor (1995) also describe the reference price. They argue that, for a good understanding of consumer reactions to prices, not only is the actual price asked for a service important, but also the perception and expectation of that price. Reference prices are levels of prices felt to be appropriate by consumers and result from a variety of stimuli, including the following:

1. Psychological determinants, the manner in which an individual evaluates and stores stimuli, such as values held by the consumer.
2. Contextual determinants, objective external stimuli such as other prices and facilities and information on current subsidy or cost levels.
3. Immediate determinants, such as price and the physical quality of the facility.

In their view, information, experiences and perceptions informing the reference prices are in a continual state of change and can, for example, be altered through managerial initiatives, such that the reference prices fluctuate. The division in determinants by Gratton and Taylor has been used for the construction of the research model and the conceptual framework.

Price levels and means of payment

The following means of payment have been found in the literature:

- Entrance fee: money which one pays to enter (e.g. the facility or area, mostly paid at the check-in point).
- Permit: an official written statement giving someone the right to do something.
- Licence: permission to do something in return for a fixed payment.
- Membership: member's ticket.
- Excise tax: government tax on certain goods (e.g. related to recreation, for example equipment) produced and used inside a country.
- Rescue insurance: in some areas the highest costs are made for operating rescue services. A compulsory insurance can cover those costs.
- Donation: non-compulsory amount of money to support the provider of a service.

The means of payment have been used for different user groups. Several examples are known. Entrance fees are quite common in national parks and other public lands (Betz and Cordell, 1989). Permits are used for camping (Bamford *et al.*, 1988). A trail licence has been used for cross-country skiers (Berrier and Carlson, 1984). The excise tax has been used on hunting and fishing equipment (Prosser, 1984). The rescue insurance has been used for backpackers and climbers in mountains.

There are a number of differentiations of types with their own implications:

1. Differentiation in money collection.

- Voluntary programmes. People pay for the use of a facility voluntarily, for example in a money collection box. There is more support for money collection in this way (Agee, 1984; Powell, 1984; Martin, 1986).
- Payment through user organizations. The members pay through their user organization for the use of the facility. Research among anglers in the Netherlands shows more support for this type of payment (Bakker *et al.*, 1988).

2. Differentiation in tariff construction.

- All-inclusive admission charges vs. separate charges. All-inclusive admission charges are considered inappropriate for sites with important attractions (Bovaird *et al.*, 1984; De Bruin, 1988).
- Only paying for extra opportunities. There is more support for payments providing extra opportunities (Driver, 1984a, b).
- Differential pricing. Meaningful choices within a pricing strategy. Differential pricing may eliminate much of the psychological reaction arising from new price levels (McCarville, 1993).
- Two-part pricing. Guldin and Kroon (1986) apply a two-part tariff to recreation, with user fees set equal to marginal cost (part one) and a fixed charge on any user for the right to enjoy the initial unit of use (part two), set to cover the costs of management not covered by marginal costs.
- Multi-part pricing. Guldin and Kroon (1986) describe multi-part pricing as 'a pricing system to be used to obtain a quasi-optimal allocation of resources – a second best solution'. They call it 'a realistic compromise for funding recreation' which 'serves both the benefits-received and the ability-to-pay principles of equity'. One part of the price in their view is, for example, collected through taxes and given to the management as a subsidy. The subsidy should cover the same proportion of the costs as the external benefits created are to the benefits perceived by the user. The second part can be a user fee. The user fee should be based upon the marginal costs of providing one extra unit of use. The third part, for example an annual

fee, should cover the costs remaining after the subsidy and the user fee.

3. Differentiation in provision and use of a service.

- Provision through user involvement. If a group of users is involved in, for example, the construction of a track (information!), there is more support for paying a fee to support this work (Kemsley, 1984; Kerr and Manfredo, 1991).
- Exclusive use. People or a group of people have the exclusive use of the facility when paying. There is more support for pricing in the case of exclusive use (Allen, 1981).
- Public vs. private provision of a service. Consumers expect prices of a similar service charged by public-sector agencies to be lower than those charged by private-sector suppliers. Jarvis (1984) argues that maximum fees should be compared with private-sector fees, but minimum fees should not.

4. Differentiation in certificate or ticket.

- Stamp. Proof in addition to official paper to show that tax (stamp duty) has been paid.
- Windshield sticker. Sticker for the car to show payment (e.g. entrance fee) has been paid.

Research model and conceptual framework

This section describes the research model (Fig. 8.1) and the conceptual framework that has been used for the surveys. The results of the literature research show that 'past use' and 'past paying', together with the construction of a 'reference price' in people's minds, are the basis of attitude development. The research is about attitude development. The resulting change in behaviour is not part of the research and the model that will be constructed (suggested reading for research on behaviour change: Fishbein and Ajzen, 1976).

Consumers have reference price structures that influence the way in which they evaluate pricing options. The reference price is considered to be the key factor in this model, determining attitude and thus behaviour. Psychological determinants must be considered in every pricing decision because they influence reference price structures. Other important determinants of reference prices are contextual and immediate determinants. Altogether, this forms the basis (reference price) on which new price information is processed and attitudes are formed. The resulting behaviour (paying) is strongly influenced by attitude (towards paying). However, it is important to stress that a change in attitude does not automatically lead to a change in

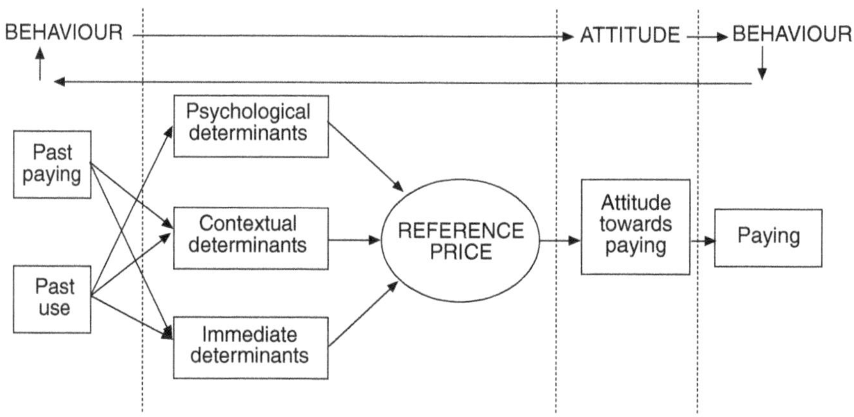

Fig. 8.1. Research model (based on Kerr and Manfredo, 1991).

the behaviour of users (Reiling *et al.*, 1988; Boersema *et al.*, 1991; Van der Aalst and Van der Straaten, 1992). A host of factors influence this step (including social values). Once behaviour has changed, the new behaviour does not automatically stay that way. New information can, for example, change attitude and thus change behaviour again (RMB, 1995).

Behaviour (paying) influences the attitude (towards paying) the next time because current paying and use is 'past paying' and 'past use' for the next visit. This is in accordance with the view from the attitude development research field that causal models that incorporate two views – 'attitudes cause behaviours' and 'behaviours cause attitudes' – provide the best explanation of behaviour.

The research model has been made operational in the conceptual framework (Fig. 8.2). Personal (sex, age, education, etc.) and use variables (time, distance, expenditure, etc.) have been added. Past use and past paying are important because they determine the accessibility of pre-existing information about the present fee (contextual determinants: information, subsidy, etc.), the price and quality of the facility (immediate determinants: quality, price, etc.), the types of beliefs – values – a person already possesses about paying fees and the polarity of belief strengths (psychological determinants: values, expectations, etc.). The conceptual framework has been used to design the questionnaires for on-site and household surveys.

8.1.3 On-site and household surveys

Research method and research areas

RESEARCH METHOD. The on-site survey took place on two research days in the spring of 1998. Two different areas were selected: Berg en Bos, where visitors are expected to be familiar with paying because in one part of the area visitors have to pay a fee, and Edese Bos, where visitors are less likely to be familiar with paying because the whole area is free of charge. Two areas were selected because the literature study showed that previous experiences with pricing are a major influence on attitude development towards pricing. A total of 1499 visitors were asked to cooperate. In total, 1261 questionnaires were handed out to visitors (16 years and older) on the five research points in both

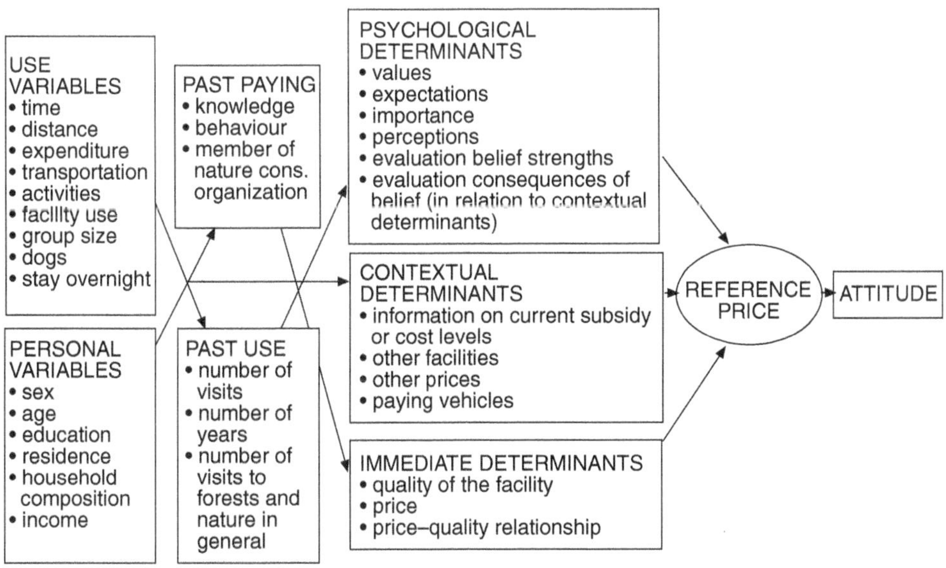

Fig. 8.2. Conceptual framework.

areas – 592 in Berg en Bos and 669 in Edese Bos. The 'next to pass' method was used to make sure that the sample was representative for the visitors on the research days. A total of 827 (66%) of these questionnaires were returned in the reply envelopes – 382 from Berg en Bos and 442 from Edese Bos (three questionnaires could not be coupled with a research point). This makes a net response of 55% in both areas.

The sample is not representative for all visitors throughout the year because data were collected on only two research days. As it is difficult to approach horse-riders and mountain-bikers, an extra attempt was made to approach them through their clubs and riding-schools. However, only three of the separate questionnaires that were sent to riding-schools were returned. As a result, only two groups (walkers and cyclists) will be taken into account when discussing the different user groups.

A selection of the questions from the on-site survey were submitted to a representative sample of 501 people of the Dutch population (16 years and older). This household survey was conducted to approach non-users as well and was carried out in the summer of 1998. The household survey was executed by a market research bureau specializing in interviewing by telephone, with random telephone-number selection. The response of the household survey was 22% (because of refusals, not at home, incorrect age or sex, illness and language problems). In fact slightly more than 2000 people were approached to arrive at a sample size of 501 persons. This may have affected the results of the survey. For the questions that were part of the household survey, the results have been included as a separate category in the following tables and figures.

GENERAL DESCRIPTION OF THE RESEARCH AREAS. The areas Berg en Bos (396 ha) near the city of Apeldoorn and Edese Bos (435 ha) near the city of Ede are both part of what is called the Veluwe, one of the largest natural areas in the Netherlands. Both forests are part of the province of Gelderland. For one part of Berg en Bos, visitors have to pay an entrance fee or they can buy an annual licence. Inside this part, there is a zoo with a separate entrance fee. The visitors to Berg en Bos are expected to be more familiar with paying compared with the visitors to Edese Bos, which is free of charge. Both areas are more or less comparable on other characteristics, such as area, type of forest, other facilities and distance from residential areas.

Description of the sample groups

Personal variables might have an influence on the acceptance of pricing. For the comparison of Berg en Bos with Edese Bos, it is important to have two similar groups. The personal variables in the questionnaire can be used to define new target groups for marketing and promotion. The results have been compared with results for other forests in the Netherlands (Segeren and Visschedijk, 1997). The following personal variables are part of the questionnaire.

SEX. About 56% of the respondents in both areas are male and 44% female. This is comparable to other forests near residential areas in the Netherlands. Other forests relatively far from residential areas show a higher proportion of male visitors. The household survey consists of a stratified sample of male and female respondents, both 50%.

AGE-GROUPS. The two areas have a similar age-group composition (Table 8.1). The average age in both areas is 47. This is comparable to other forests in the Netherlands.

EDUCATION. The educational level of the respondents in Berg en Bos is somewhat higher than in Edese Bos (Table 8.2). The educational level of visitors to both forests is high compared with the general public from the household survey.

NET MONTHLY HOUSEHOLD INCOME GROUP. The net monthly household income in Berg en Bos is high compared with Edese Bos (Table 8.3). The relatively high net monthly household income in Berg en Bos is linked to the type of houses in the city district that borders on the forest.

Household composition. The number of persons per household in the two areas is comparable.

Residence. Most of the visitors to both forests live in the cities near the forest: Ede (53%) in Edese Bos and Apeldoorn (63%) in Berg en Bos. Table 8.4 shows the distance from the residence to the research area in four groups. This table is a result of

Table 8.1. Age-groups of respondents (%) in both areas and household survey.

	% in the sample		
Age	Berg en Bos	Edese Bos	Household
17–31 years	12	12	26
31–45 years	34	35	30
46–60 years	41	38	23
61 years and over	13	15	21
Total	100	100	100

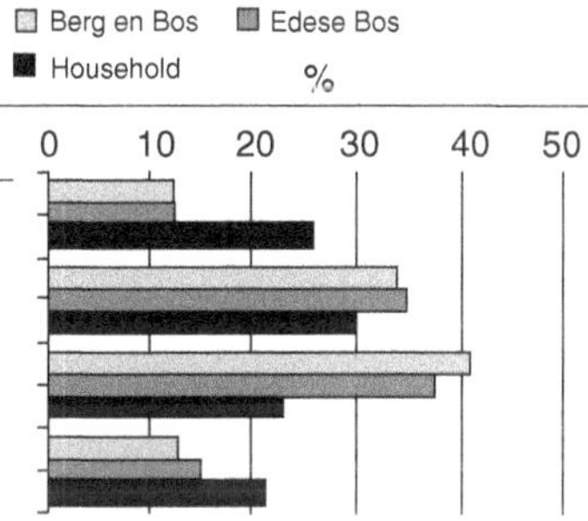

Table 8.2. Education of respondents (%) in both areas and household survey.

	% in the sample		
Education	Berg en Bos	Edese Bos	Household
Primary	6	9	21
Secondary	26	32	40
High school	20	15	9
University	48	44	29
Total	100	100	99

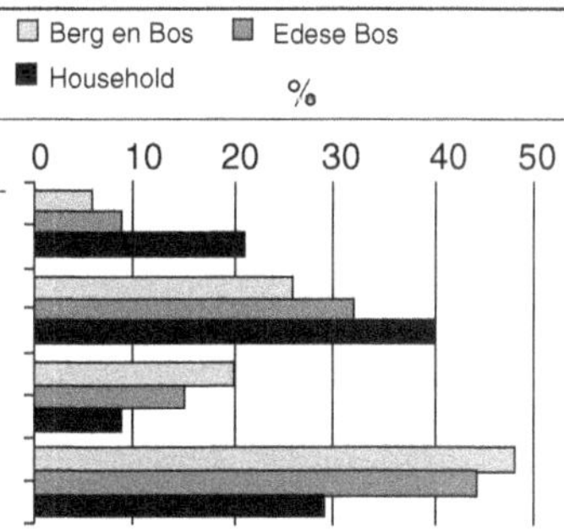

Table 8.3. Net household income of respondents (%) in both areas.

	% in the sample	
Income	Berg en Bos	Edese Bos
Less than 2500 guilders	9	9
2500–5000 guilders	40	51
More than 5000 guilders	32	22
Missing	19	19
Total	100	101

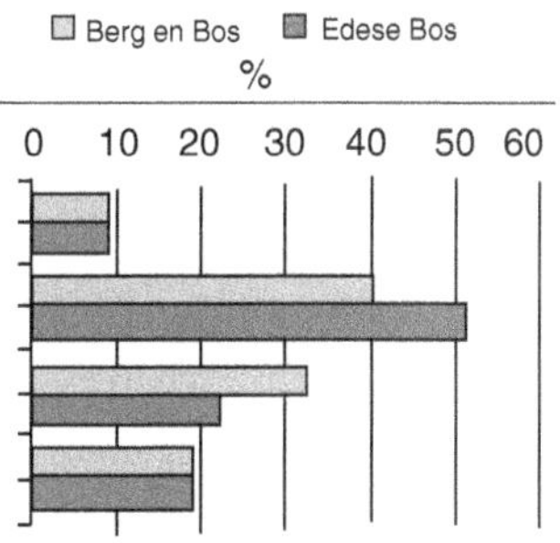

Table 8.4. Distance of the residence of the respondents (%) from the research area.

	% in the sample	
Distance	Berg en Bos	Edese Bos
< 5 km	56	44
5–10 km	10	13
10–25 km	3	11
> 25 km	32	32
Total	101	100

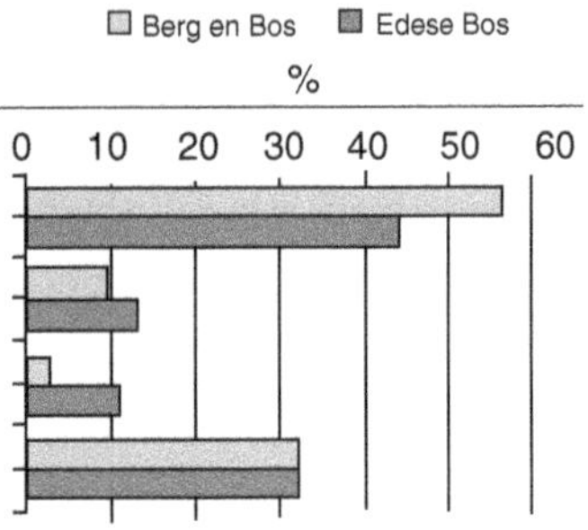

linking the centre of the postal area to the centre of the research area. In Berg en Bos, a higher percentage of the visitors live near the forest.

The results in both areas show that the research groups in the two areas are comparable as to personal variables.

The use of the research areas

The way in which visitors use the research area might have an influence on acceptance of pricing as well.

TOTAL GROUP SIZE. The total group size is comparable in the two research areas. Most people come to the forests alone (average 28%) or in pairs (average 46%). This is comparable with other forests in the Netherlands. Only 15% of the visitors have children younger than 12 years in their group. An average of 40% of the groups bring a dog. This percentage is particularly high because both research areas lie close to the cities, and both forests have specially assigned places where dogs are allowed to run free.

PLACE OF DEPARTURE. The variable 'place of departure' is about whether people came from their home or from a temporary residence to the forest. In Berg en Bos 25% of the visitors on both days were staying in a temporary residence, while in Edese Bos this was 29%. This might be higher than normal because of the long weekend most people in the Netherlands had during the research days.

TIME OF ARRIVAL AND TIME OF DEPARTURE. Arrival and departure times are almost the same in the two areas and are comparable with other research in the Netherlands (Table 8.5).

EXPENDITURE. Expenditure by visitors in Berg en Bos and Edese Bos is, in general, low. Apart from the entrance fee and travel costs, an average of only 27% of the visitors spend money in other ways (beverages, food, etc.). An average of only 11% spend more than 10 guilders per person on expenditure other than entrance fee or travel. The average expenditure per person (Dfl. 18.55 in Berg en Bos and Dfl. 10.92 in Edese Bos) is greatly influenced by this group of respondents and should be considered with caution.

MEANS OF TRANSPORTATION TO THE AREA/MEANS OF TRANSPORTATION WITHIN THE AREA. Most people come to the research areas by car (average 56%). Other means of transportation that score high are touring bicycles and walking. Most people who arrive by car go walking. With an average of 67% for both areas, 'walking' is the largest group (Table 8.6).

ACTIVITIES. Walking and bicycle rides are the most important activities in both forests. The higher percentages for Berg en Bos for 'visiting café/restaurant' and 'other activities' is probably due to the zoo inside the park. The respondent has spent time on more than one activity. A question about

Table 8.5. Arrival and departure times of the visitors (%; both areas together).

Time	% of forest visitors Arrival	Departure
Before 9.00	3	1
9.00–10.00	8	2
10.00–11.00	12	5
11.00–12.00	16	8
12.00–13.00	14	12
13.00–14.00	16	11
14.00–15.00	14	16
15.00–16.00	10	18
16.00–17.00	5	19
17.00–18.00	0	8
After 18.00	0	1
Total	98	101

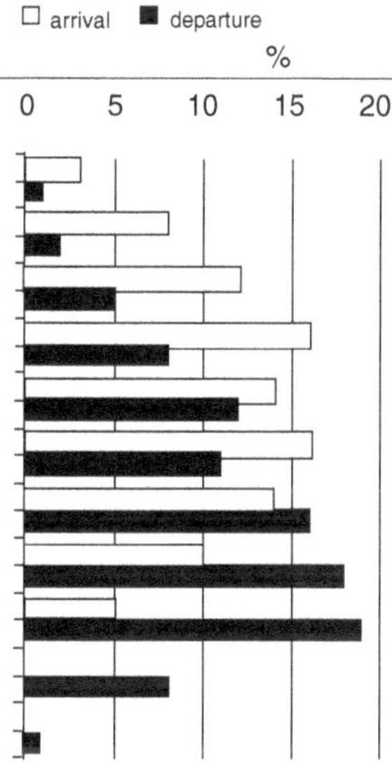

Table 8.6. Means of transportation (%) to and within the research area.

Means of transportation	Transportation to ... (% in the sample)		Transportation in the area (% in the sample)	
	Berg en Bos	Edese Bos	Berg en Bos	Edese Bos
Walking	11	14	69	66
Touring bicycle	27	28	26	31
Racing bicycle	1	1	1	1
Mountain bike	3	1	3	1
Car	58	55	1	1
Horse	0	1	0	1
Other	0	1	0	0
Total	100	101	100	101

which of the activities most time was spent on has been included to see which activity was most important to the respondent (Table 8.7). Walking (long and short walks) is the activity visitors spent most time on, followed by cycling. Again, visiting the zoo in Berg en Bos is the reason for the high percentage of 'other activities'.

FACILITY USE. In Edese Bos the car-park and in Berg en Bos the recreation area for dogs is used more often (Table 8.8).

THE IMPORTANCE OF RECREATION FACILITIES IN THE FOREST. Twenty items about the importance of different facilities were presented. The respondents could rate them as abolutely not important (1), not very important (2), neutral (3), important (4) and very important (5). Table 8.9 shows the average value of importance for all facilities. The most important were 'absence of noise' (4.57) as well as 'enough walking paths' (4.00). Almost totally unimportant was being able to 'reach everything by car'

Table 8.7. Activities (%) on the research days/on which most time was spent on the research days.

	% in the sample		Activities on which most time was spent (% in the sample)	
Activities	Berg en Bos	Edese Bos	Berg en Bos	Edese Bos
Long walk	26	32	24	29
Short walk	40	34	26	29
Jogging/exercise	6	5	4	5
Bicycle ride	30	32	29	30
Horse-riding	1	1	1	1
Sitting/lazing	7	9	1	2
Day camping	1	0	0	1
Sport and games	3	2	2	1
Nature study	3	1	1	0
Organized excursions	0	0	0	0
Visiting café/restaurant	8	4	1	1
Other activities	10	2	8	2
Missing	1	0	3	1
Total (basis all respondents)	136	122	100	102

(1.80), while 'sufficient car parking' ranked much higher (3.27).

Attitude towards paying

PAST PAYING. The literature shows that experience of paying is an important variable. The past paying behaviour can be divided as follows:

- Knowledge.
- Past paying behaviour.
- Membership of nature conservation organizations.

Whether visitors know of different means of payment ('knowledge') is important for acceptance of that paying vehicle. A question about the knowledge of

Table 8.8. Facility use (%) on the research days.

	% in the sample	
Facility use	Berg en Bos	Edese Bos
Recreation area for dogs/dogs can run free	38	29
Car park	34	47
Marked cycle paths	21	19
Marked walking trails	13	14
Bench/picnic tables	13	15
Café/restaurant	10	5
No answer/no facilities	10	14
Observation hut/viewing point/hide	8	2
Recreation area/playing field	5	1
Bike park	2	0
Marked mountain-biking trails	1	0
Marked bridle-paths	0	1
Total (basis all respondents)	155	147

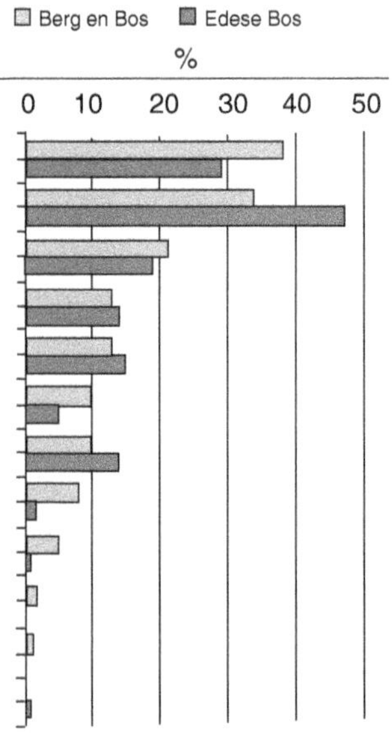

Table 8.9. The importance of recreation facilities and other aspects.

Facility use	Value, both areas: Importance	Value, both areas: Res. area
Absence of noise	4.57	3.31
Enough walking paths	4.00	3.95
Elements of special natural interest	3.98	3.56
Enough cycle tracks	3.96	3.67
Varied landscape	3.89	3.82
Free access	3.88	3.31
Maintenance of facilities	3.68	3.54
Size of the area	3.66	3.59
Few people in area	3.63	2.98
Supervision and regular enforcement	3.60	3.05
Signposting	3.58	3.58
Road and path maintenance	3.49	3.83
Provision of benches	3.45	3.17
Local information and education	3.31	3.29
Sufficient car-parking area	3.27	3.62
Playing areas and picnic meadows	2.69	3.35
Basic water and lavatory facilities	2.56	2.83
Removal of dead branches	2.43	3.32
Refreshment facilities	2.12	3.33
Reach everything by car	1.80	2.44
Mean	3.4	3.4

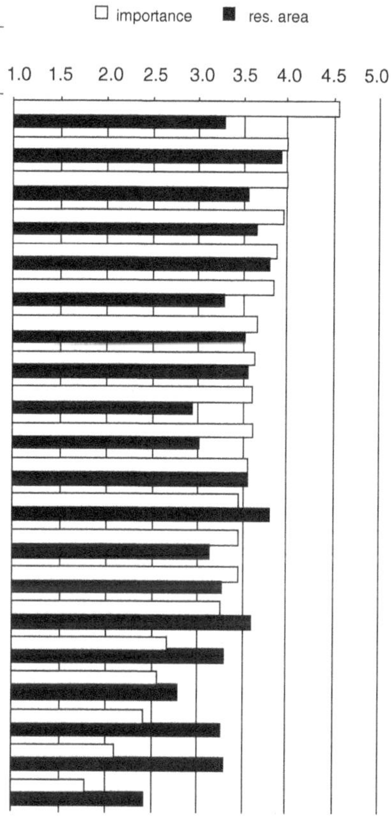

Table 8.10. Past paying: knowledge and past paying behaviour of eight different means of payment – on-site survey (%).

Means of payment	Knowledge (% in the sample)		Past paying behaviour (% in the sample)	
	Berg en Bos	Edese Bos	Berg en Bos	Edese Bos
Annual licence	55	37	17	11
Entrance fee	51	39	24	16
Parking fee	42	26	16	9
Entrance fee for visitor's centre	39	36	11	10
Excursion fee	38	42	4	7
Rider's pass	18	22	2	3
Mountain-bike pass	5	4	0	0
Payment for special events	28	17	7	4
Total (basis all respondents)	276	223	81	60

(familiarity with) eight different means of payment is part of the questionnaire. Whether people actually paid for one of them ('past paying behaviour') is important for acceptance as well. Again, a question about eight different means of payment is part of the questionnaire (Table 8.10). The knowledge and past paying behaviour of most means of payment is higher in Berg en Bos than in Edese Bos (except 'excursion fee' and 'rider's pass'). The paying vehicle 'entrance fee' is well known (53%) and accepted by 26% of the persons asked within the household survey, who paid an entrance fee when visiting a forest or nature conservation area during the past 12 months.

Membership of nature conservation organizations can be seen as paying for forest and nature and, in some cases, for recreation. Some of the forests and nature conservation areas owned by these organizations are open for members only. A question about the membership of eight different organizations is part of the questionnaire. The results showed that 57% of the respondents are members of one of these organizations, with 42% of the respondents being members of Vereniging Natuurmonumenten or of one of the Provinciale Landschappen. These organizations own areas that sometimes are open for members only.

PAST USE. Past use, or use experience, was divided into number of visits, number of years and number of visits to forest and nature conservation areas in general.

The attitude towards paying has been divided into:

1. General paying: attitude towards general principles and consequences.
2. Specific paying: attitude towards more specific means of payment and paying for specific facilities.
3. Willingness to pay (WTP): how much visitors say they want to pay.

This results in the attitude towards paying in both areas.

GENERAL PAYING. General paying (Table 8.11) has been divided, as follows:

- Is recreation in forests and nature conservation areas regarded as a public good and should the government pay?

Table 8.11. The attitude to five statements concerning payment for forest services.

	Agree		Neutral		Disagree	
Statement	BeB	EB	BeB	EB	BeB	EB
Should forests and nature always be accessible free of charge (% of answer categories)?	65	80	15	10	20	11
Should the government pay for recreation in forests and nature conservation areas (% of calculated score of two questions)?	71	80	18	11	11	9
Should the user pay for recreation in forests and nature conservation areas (% of calculated score of two questions)?	17	8	21	15	63	78
It is a good idea that the government plans to introduce entrance fees (% of answer categories).	8	3	12	10	81	88
It is a good idea that private owners are allowed to charge entrance fees (% of answer categories).	29	24	26	24	45	52

BeB, Berg en Bos; EB, Edese Bos.

- Does recreation have private components and is it fair to let the user pay as well?
- Other statements on safety, private owners and lower income groups.

The household survey contained only one statement on the 'government pays' principle: 'It is the responsibility of the government to provide recreation facilities in forests and nature conservation areas.' The answers of the general public from the household survey were: 2% strongly disagree, 15% disagree, 6% are neutral, 63% agree and 15% strongly agree. The opinion of the respondents from the household survey is rather different from that of the visitors of both research areas.

The household survey contained only one statement on the 'user pays' principle: 'As there are so many opportunities for recreation in forests and nature, I am prepared to pay for my visit.' The answers of the general public from the household survey on this question were: 7% strongly disagree, 29% disagree, 7% are neutral, 52% agree and 5% strongly agree. The general public from the household survey seem to be less reluctant to pay than the respondents from the on-site survey.

Other statements related to paying are statements on private ownership, lower income groups and safety. The visitors think differently about private owners introducing entrance fees compared with public owners. The respondents think that it is fairer for private owners to charge entrance fees than for the government. In the opinion of the visitors of Berg en Bos and Edese Bos, it would be unacceptable if entrance fees were to prevent low-income groups from visiting (an average of only 5% disagrees). Hardly anybody feels unsafe in forests and nature conservation areas as a result of everyone being able to visit free of charge (only 3% feel unsafe).

> It can be concluded that the respondents strongly oppose the 'user pays' principle. In their opinion, the government should pay for recreation opportunities in forests and nature conservation areas. This conclusion is stronger in Edese Bos than in Berg en Bos. In the opinion of the general public, recreation in forests and nature is the responsibility of the government, but a considerable number do not mind paying for it, because they value possibilities for recreation highly. Whether they actually pay is somewhat different, however. Private owners can count on more support from the respondents than public owners when they want to introduce an entrance fee.

SPECIFIC PAYING. Specification of the opinion towards paying has been divided into:

- different means of payment;
- paying for additional facilities.

More specific forms of paying are different means of payment for different user groups. A question about the attitude towards eight different means of payment, which are still in use or have been used in the Netherlands was part of the questionnaire (Table 8.12). Results show that the attitude towards all the different means of payment is more negative in Edese Bos than in Berg en Bos. This was to be expected because experience with paying is lower in Edese Bos. Visitors are most negative about paying if they do not get something special in return (e.g. general entrance fee or a season ticket).[2]

Another more specific attitude towards paying is whether respondents want to pay for additional facilities. Results show that an average of 91% of the respondents in both areas agree or strongly agree that the forests should be freely accessible for walking and cycling. An average of 49% of the respondents in both areas agree or strongly agree that a forest owner should be able to ask money for additional facilities, such as mountain-bike and horse-riding tracks.

More usually, the respondents reply as shown in Table 8.13 to the question: 'I would rather pay an entrance fee if the

Table 8.12. Attitude towards means of payment (% of all answers) – on-site survey and household survey.

	Attitude (% of all answers)								
	Positive			Neutral			Negative		
Payments	BeB	EB	HH	BeB	EB	HH	BeB	EB	HH
Season ticket	33	18	36	20	10	21	48	72	43
Entrance fee	10	5	43	17	8	21	74	88	37
Excursion fee	65	63	69	21	21	18	14	16	12
Parking fee	26	24		20	15		54	62	
Entrance fee for visitor's centre	47	37		25	29		28	34	
Rider's pass	60	53		22	22		18	25	
Mountain-bike pass	55	48		22	22		23	30	
Payment for special events	64	52		16	22		19	26	

BeB, Berg en Bos; EB, Edese Bos; HH, households.

Table 8.13. Answers (%) to the questions whether visitors would be happier to pay if the entrance fee is used for additional recreation facilities, nature conservation measures or information facilities.

	Attitude (% of all answers)								
I would be happier to	Positive			Neutral			Negative		
pay for better ...	BeB	EB	HH	BeB	EB	HH	BeB	EB	HH
Recreation facilities	37	21	59	24	20	9	39	59	32
Nature conservation measures	61	52		20	18		20	30	
Information facilities	41	30		32	30		27	41	

BeB, Berg en Bos; EB, Edese Bos; HH, households.

money is used for additional ... '. Nature conservation measures are the most 'popular' additional facility. Again, the attitude towards paying is more positive in Berg en Bos than in Edese Bos.

Another conclusion is that, if the respondents know that their money is spent on additional facilities, their acceptance of paying is higher. A customer-orientated information policy on the activities undertaken will reduce price resistance. This was also concluded by McCarville and Crompton (1987a): 'Consumers are more likely to support user fees when such fees are used to maintain and improve the resource at which they are collected.'

Willingness to pay (WTP)

The WTP of the respondents has been divided into the amount people want to pay:

- per visit to the area (entrance fee);
- for a season ticket for the area;
- for a season ticket for all forests and nature conservation areas in the Netherlands.

The amount people want to pay per visit to the area depends on the research area (Fig. 8.3). The amount of WTP per visit for Berg en Bos is much higher than that for Edese Bos. This is mainly caused by the fact that 80% of the visitors of Edese Bos do not want to pay at all. In Berg en Bos, this is 'only' 51%.

As a result, the potential for the introduction of an entrance fee is higher in Berg en Bos (Table 8.14). The potential for the introduction of an entrance fee can, for example, be analysed by comparing the number of visits with the WTP. Four different groups and strategies have been constructed:

- 'non-payers' do not visit very often and do not want to pay. This group has the least potential. Strategy for introduction of a fee: new products to persuade people to visit and to pay.

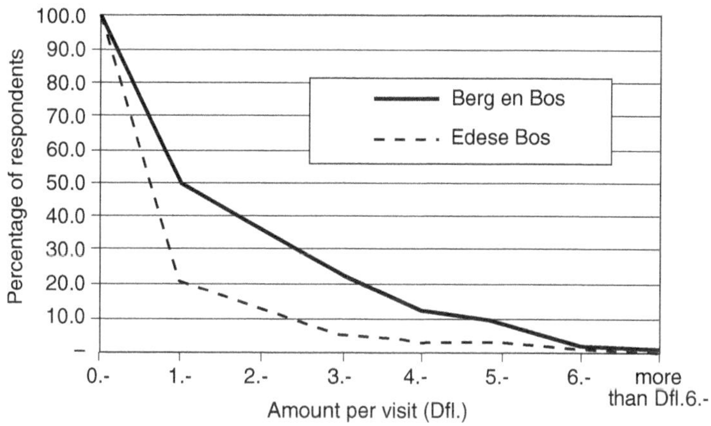

Fig. 8.3. Willingness to pay per visit (% of respondents who would pay the amount).

Table 8.14. Potential for the introduction of an entrance fee.

	Berg en Bos		Edese Bos		Description	
	Low number of visits	High number of visits	Low number of visits	High number of visits	Low number of visits	High number of visits
Does not want to pay entrance fee	20%	31%	29%	51%	Non-payers	Rejecters
Is willing to pay an entrance fee	25%	24%	12%	8%	Potentials	Payers

- 'Rejecters' visit very often but do not want to pay. This group has some potential because of the high number of visits. Strategy for introduction of a fee: the attitude might be altered through the pricing principles as outlined above.
- 'Potentials' do not visit very often but do not mind paying. This group has potential if they can be persuaded to visit more often. Strategy for introduction of a fee: quality improvement, exhibitions, discounts, etc.
- 'Payers' visit very often and do not mind paying. This group has the highest potential. Strategy for the introduction of a fee: maintain service and quality and focus on the optimal price level.

It can be concluded that the potential for the introduction of an entrance fee is very low in Edese Bos but higher in Berg en Bos.

What people are disposed to pay for a season ticket for the area depends on the research area as well (Fig. 8.4). Again, respondents in Edese Bos would pay less, and 65% of the respondents do not want to pay at all for an annual season ticket, compared with 39% in Berg en Bos.

Again, the potential has been analysed by comparing the number of visits with the WTP, this time for an annual season ticket. The description of the four different groups is the same as before, and the results are given in Table 8.15. From the figures in the table, it can be concluded that the potential for introduction of an annual season ticket is doubtful in Edese Bos but very promising in Berg en Bos.

The amount of money the respondents are willing to pay for an annual season ticket for all forests and nature conservation areas in the Netherlands depends on the research area too, but not as much as in either of the other WTP examples (Fig. 8.5). The WTP for an annual season ticket for all forests and nature conservation areas is also higher in Berg en Bos. So there is still a difference,

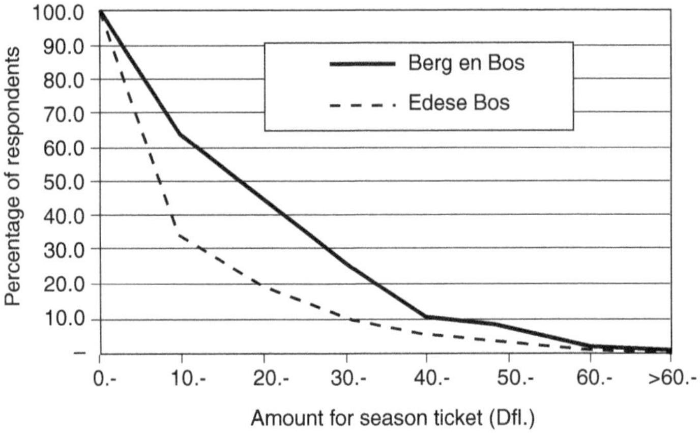

Fig. 8.4. Willingness to pay for a season ticket (% of respondents who want to pay the amount).

Table 8.15. Potential for the introduction of a season ticket.

	Berg en Bos		Edese Bos		Description	
	Low number of visits	High number of visits	Low number of visits	High number of visits	Low number of visits	High number of visits
Does not want to pay for a season ticket	18%	21%	26%	39%	Non-payers	Rejecters
Is willing to pay for a season ticket	26%	36%	14%	21%	Potentials	Payers

even when the season ticket is valid for all areas, regardless of where the respondent lives. It can be concluded that the potential for the introduction of an annual season ticket for all forests and nature conservation areas is promising in Edese Bos and very good in Berg en Bos (Table 8.16).

Another way of seeing whether visitors want to pay an entrance fee is their reaction to the introduction of an entrance fee

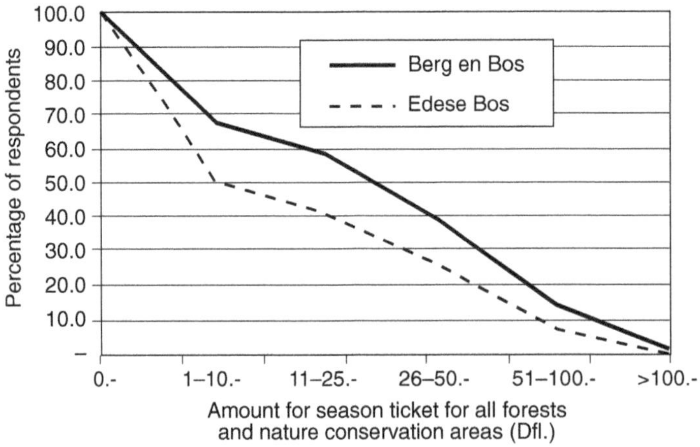

Fig. 8.5. Willingness to pay for a season ticket for all forests and nature conservation areas (% of respondents who are disposed to pay the amount).

Table 8.16. Potential for the introduction of a season ticket for all forests and nature conservation areas.

	Berg en Bos		Edese Bos		Description	
	Low number of visits	High number of visits	Low number of visits	High number of visits	Low number of visits	High number of visits
Does not want to pay for a season ticket	7%	24%	13%	37%	Non-payers	Rejecters
Is willing to pay for a season ticket	21%	48%	14%	36%	Potentials	Payers

Table 8.17. Reactions to introduction of entrance fee of 2.- guilders per person.

Reaction to an entrance fee (Dfl. 2.- per person)	% in the sample	
	Berg en Bos	Edese Bos
Visit more often	1	0
Visit as often as I do now	25	13
Visit less often	35	33
Never visit again	39	54
Total	100	100

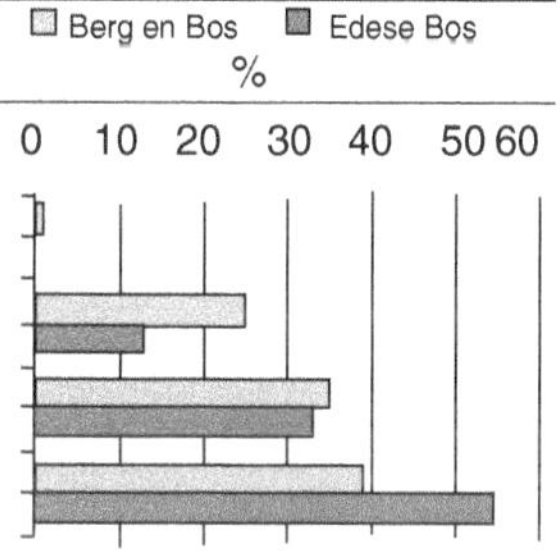

> For each forest or type of forest, marketing management should provide the optimal mix of strategies as different strategies might influence each other.

(Table 8.17). The reaction of most people is that they would visit less often. Again, in Edese Bos the reaction is more negative.

When forest owners try to get more income out of their property, this might have some consequences for the visitors. Nine statements about these possible consequences were part of the questionnaire (Table 8.18). The results show that the respondents are very negative about 'restricted access', 'more cafés and restaurants', 'more riding-schools and camp-sites', 'prohibition of fruit-picking, etc.' and 'building in forests'.

Table 8.18. Attitude towards possible consequences of pricing (%).

Possible consequences of pricing	Attitude (% of all answers)					
	Positive		Neutral		Negative	
	BeB	EB	BeB	EB	BeB	EB
Ability to buy green products	58	53	30	35	12	12
More supervision by a ranger	54	62	34	28	13	9
More advertising for the area	42	36	31	37	27	26
Signposted paths carry a sponsor's name	31	26	32	37	37	37
Prohibition of free fruit-picking, etc.	20	16	18	20	62	63
Restricted access to certain areas	8	5	17	15	75	82
More cafés and restaurants in the area	8	10	24	23	68	68
Certain new buildings in forests	8	5	15	18	78	77
More riding-schools and camp-sites	4	5	20	22	77	72

BeB, Berg en Bos; EB, Edese Bos.

8.1.4 Attitude dependencies

Overview of possible dependencies

This section gives the results on attitude dependencies from the different variables of the research model and the conceptual framework. The attitude to the introduction or increase in entrance charge was asked directly in the questionnaire. The answers to each of the other variables from the conceptual framework have been grouped into two-answer (e.g. yes/no; did/did not; high/low; much/little) categories.

Whether variables from the conceptual framework influence attitude development can be tested with the χ^2 test. With the Pearson χ^2 test, it is possible to see whether there is a statistically significant relation between two variables. A high χ^2 value means that it is more likely that there is a relation between the two. A high χ^2 value also depends on the degrees of freedom (df = number of rows − 1 × number of columns − 1). The probability (P) is the chance of a wrong conclusion (saying that there is a relation, when this is actually not true). In this research, the conditions for a statistically significant relation between two variables are: $P < 0.05$; $\chi^2 \geqslant 9.488$ (df in all tables = 4); the minimum expected count for each cell (combination of values) is 1; a maximum of 20% of the cells (combinations of values) has an expected count less than 5.

Table 8.19 summarizes the probabilities for possible dependencies.

Influence of personal and use variables

Age, sex, education and household income have no statistically significant relation to attitude towards paying.

MEANS OF TRANSPORTATION. Respondents in Edese Bos who are walking show a more negative attitude towards paying than respondents who are cycling. There is no statistically significant relation between these two variables in Berg en Bos or for the group as a whole.

DOGS. Respondents with a dog in the group report a more negative attitude towards paying than respondents without a dog (Table 8.20).

STAYING OVERNIGHT. Respondents who were staying in temporary accommodation report a less negative attitude than day-visitors. The relation is not statistically significant for Berg en Bos.

Influence of past paying

KNOWLEDGE. Knowledge is about whether visitors know of the different means of payment used in forests or nature conservation areas in the Netherlands. The total sample has been divided into two groups: respondents who know of at least one of the means of payment and respondents who do not know of any paying vehicle. The results show that there is no statistically significant relation between knowledge and attitude towards paying.

Table 8.19. Significance of several factors towards the attitude to paying.

Test: Attitude towards paying by ...	Significance in the sample								
	Berg en Bos			Edese Bos			Total		
	Sig.	P	χ^2	Sig.	P	χ^2	Sig.	P	χ^2
Means of transportation	n.s.	0.352	3.266	*	0.016	10.385	n.s.	0.211	4.517
Dog in the group	*	0.019	9.896	**	0.002	14.581	***	0.000	18.205
Staying overnight	n.s.	0.611	1.817	***	0.000	17.830	*	0.011	11.172
Knowledge of pricing	n.s.	0.089	6.508	n.s.	0.741	1.249	n.s.	0.062	7.317
Payment in the past	**	0.005	12.662	n.s.	0.078	6.820	**	0.002	14.434
Membership nature cons. org.	n.s.	0.671	1.548	n.s.	0.621	1.774	n.s.	0.330	3.426
Number of visits (res. area)	*	0.028	9.072	***	0.000	26.830	***	0.000	34.996
Number of years	n.s.	0.384	3.049	***	0.000	18.392	***	0.001	16.751
Number of visits in general	n.s.	0.095	6.377	n.s.	0.286	3.780	n.s.	0.062	7.331
Mark of the area	n.s.	0.353	3.261	n.s.	0.712	1.372	n.s.	0.876	0.688

sig., Significance; n.s., not significant; *, significant; **, high significance; ***, maximum significance.

Table 8.20. Dog in the group (yes or no) and attitude towards paying.

	Berg en Bos				Edese Bos				Total			
	No dogs		Dogs		No dogs		Dogs		No dogs		Dogs	
Attitude	*n*	%	*n*	%	*n*	%	*n*	%	*n*	%	*n*	%
Very negative	57	27	66	41	148	55	122	72	205	43	188	57
Negative	91	43	53	33	72	27	34	20	163	34	87	26
Neutral	57	27	38	24	39	15	11	7	96	20	49	15
Positive	9	4	3	2	9	3	2	1	18	4	5	2
Very positive	0	–	0	–	0	–	0	–	0	–	0	–
Total	214	100	160	100	268	100	169	100	482	100	329	100
	$\chi^2 = 9.896$; $P = 0.019$				$\chi^2 = 14.581$; $P = 0.002$				$\chi^2 = 18.205$; $P = 0.000$			

PAST PAYING BEHAVIOUR. Past paying behaviour is about whether visitors have actually paid one of the different means of payment for visiting forest or nature before. The total sample has been divided into two groups: respondents who actually paid by at least one paying vehicle and respondents who did not pay by any paying vehicle before. The results show a statistically significant relation between past paying behaviour and attitude towards paying for the total sample. Respondents with recent payment experience report a less negative attitude towards paying. However, for the subsamples, the relation is only statistically significant for Berg en Bos and not for Edese Bos. This was to be expected because the Berg en Bos area was specially selected on the assumption that people who paid before have a more positive attitude towards paying.

MEMBERSHIP OF NATURE CONSERVATION ORGANIZATION. The total sample has been divided into two groups: respondents who answered 'yes' and respondents who answered 'no' to the question: 'Are you a member of a nature conservation organization?' The results show that there is not a statistically significant relation between membership and attitude towards paying.

Influence of past use

NUMBER OF VISITS. Respondents have been divided into two groups by the number of visits to the research areas: respondents with a high number of visits a year (12 and higher) and respondents with a low number of visits (up to and including 11). The results show a statistically significant relation between the number of visits and the attitude towards paying (Table 8.21). Respondents who visit more often report more negative attitudes.

NUMBER OF YEARS. Respondents have been divided into two groups by the number of years they have been visiting the research area: respondents who were visiting for the first time or for the first year

Table 8.21. Number of visits (low and high) and attitude towards paying.

	Berg en Bos				Edese Bos				Total			
	Low		High		Low		High		Low		High	
Attitude	*n*	%	*n*	%	*n*	%	*n*	%	*n*	%	*n*	%
Very negative	41	25	81	40	86	49	188	72	127	37	269	58
Negative	67	41	73	36	53	30	51	20	120	35	124	27
Neutral	48	29	46	22	30	17	20	8	78	23	66	14
Positive	7	4	5	2	8	5	3	1	15	4	8	2
Very positive	0	–	0	–	0	–	0	–	0	–	0	–
Total	163	100	205	100	177	100	262	100	340	100	467	100
	$\chi^2 = 9.072$; $P = 0.028$				$\chi^2 = 26.830$; $P = 0.000$				$\chi^2 = 34.996$; $P = 0.000$			

and respondents who have been visiting for 2 years or longer. The results show a statistically significant relation between the number of years visitors have already been visiting the research area and their attitude towards paying. Respondents who have been visiting for a longer time report more negative attitudes. However, caution is necessary: this statistically significant relation does not exist for both subsamples. The relation is only statistically significant for Edese Bos (and not for Berg en Bos).

NUMBER OF VISITS TO FOREST AND NATURE CONSERVATION AREAS IN GENERAL. Respondents have been divided into two groups by the number of visits to forests and nature conservation areas in general: respondents with a high total number of visits a year (12 and higher) and respondents with a low total number of visits (up to and including 11). There is no statistically significant relation between this variable and the attitude towards paying.

Influence of psychological, contextual and immediate determinants

PSYCHOLOGICAL DETERMINANTS. The influence of psychological determinants is a result of other variables and difficult to research.

CONTEXTUAL DETERMINANTS. Questions about information on current subsidy or cost levels and on other facilities were not included in the questionnaire. The questionnaire tried to find out whether respondents wanted to pay less or more for other forests or nature reserves. This question was not stated or understood correctly (many respondents made comments). Therefore, the results are not valid. Respondents are very negative about 'entrance fee', 'annual season ticket' and 'parking fee'. There seems to be more support for means of payment with a clear compensation, such as extra facilities. There is more support for payment for extra nature conservation measures than for information or recreation facilities.

IMMEDIATE DETERMINANTS. The price and therefore the price–quality relationship are not relevant in this research because both research areas are free of charge. The only immediate determinant is the (perceived) quality of the facility. Respondents have been divided into two groups by the mark they gave for the total area: respondents with a high mark (8 or higher) and respondents with a low mark (up to and including 7). There is no statistically significant relation between this variable and the attitude towards paying.

8.1.5 Conclusions

Conclusions on description of the sample groups

THE VISIT ON THE RESEARCH DAYS

- The results show that at the moment there are only two relevant user groups in both areas: walkers (67%) and cyclists (29%).
- There is no significant difference between Berg en Bos and Edese Bos on personal variables apart from the net household income groups. There are more high incomes in Berg en Bos.
- Compared with the household survey, fewer young people (between 17 and 30 years old) and fewer older people (61 years and over) visit forests and nature conservation areas. Visitors from the on-site survey on average have a higher educational level compared with the general public from the household survey.
- The perceived quality of both areas is roughly the same. Edese Bos scores better on free access and cycle tracks. Berg en Bos scores better on playing areas and meadows, basic water and lavatory facilities (though also negative) and supervision and regulations enforcement.
- There is no significant difference between Berg en Bos and Edese Bos on the use of the research areas on the research days, apart from facility use. More people in Edese Bos use the car-

parks in the research area. In Berg en Bos, more people use the day recreation area and a viewpoint. There are more facilities of this kind in Berg en Bos.
- There is a significant difference on knowledge and experience with paying (past paying). Most means of payment score higher on knowledge and experience in Berg en Bos. This was expected beforehand, and it was one of the reasons for selecting these research areas.
- The general public from the household survey score higher on knowledge of and experience with an entrance fee than the visitors of, especially, Edese Bos.
- Past use and membership of an environmental or nature conservation organization are not significantly different in the two areas.

ATTITUDE TOWARDS PAYING: GENERAL PAYING

- The opinion that the government should pay for recreation in forests and nature is strongly felt by visitors in both areas and by the general public in the household survey. The visitors of Edese Bos are more distinct in their opinion compared with the visitors of Berg en Bos and the general public.
- There is hardly any support for the introduction of the 'user pays' principle for general visits.
- Public acceptance for the introduction of an entrance fee by private owners is somewhat higher compared with its introduction by public owners, but still most visitors disagree.
- The majority of the general public (household survey), however, do not mind paying, considering the possibilities for recreation in forests and nature.

ATTITUDE TOWARDS PAYING: SPECIFIC PAYING

- The attitude towards the introduction of all means of payment is more negative in Edese Bos than in Berg en Bos (again: this was expected beforehand).
- Visitors in both forests have a strongly negative attitude towards the introduction of an annual season ticket, introduction of an entrance fee and charging of a parking fee.
- Visitors are positive about an excursion fee.
- Visitors are also positive about a licence for mountain bikes, horse-riding and events.[3]
- When the money from an entrance fee is used for the improvement of the area, visitors of Berg en Bos are significantly more positive than visitors of Edese Bos. Additional nature conservation measures score better, followed by additional information facilities and additional recreational facilities.
- Most visitors, especially in Edese Bos, regard additional recreational facilities as undesirable.
- The majority of the general public, however, would rather pay if the money is used for additional recreational facilities. Additional nature conservation measures and additional information facilities were not part of the questions in the household survey.

ATTITUDE TOWARDS PAYING: WILLINGNESS TO PAY

- The WTP per visit and for an annual season ticket is much lower in Edese Bos than in Berg en Bos. This is mainly due to the fact that most people in Edese Bos are reluctant to pay any amount at all.
- The number of visitors who want to pay for an annual season ticket for all forests and nature conservation areas in the Netherlands is higher compared with an annual season ticket just for the research area, but still 50% of the visitors of Edese Bos do not want to pay at all.

ATTITUDE TOWARDS THE CONSEQUENCES OF PRICING

- Consequences of pricing that are felt to be positive are more supervision by the manager and the possibility of buying ecological products.
- Visitors do not mind or are positive about extra sponsoring on road signs

(64%) or extra advertising in information materials (73%).

- Visitors are negative about partly closing the forest for specific user groups, the opening of extra catering facilities, riding-schools or camp-sites, building in forests and licences for fruit-picking.
- It can be concluded that the visitors consider accessible forests and nature conservation areas without any more buildings important. For the majority of the general public, access to forests is important as well (56%).

Conclusions and recommendations on attitude towards pricing

In general, visitors regard recreation in forests and nature conservation areas as a public service to be paid for by tax revenues. Especially the more common activities, such as walking and cycling, are considered to be 'basic' and not to be subject to pricing in any form. The non-visitors from the household survey are less unanimous in their refusal of pricing.

Pricing is considered most appropriate for those activities that clearly benefit the participants. Means of payment with the emphasis on something 'extra' (facilities, services) have higher support. The emphasis in information materials should be on the investment in extra facilities and services which could not have been provided without pricing.

The government should make clear what is basic (and thus subsidized) and what is not (and thus possibly subject to pricing). At present, the status of many activities is vague in the Netherlands, even an activity like cycling or the provision of benches. Private owners get a subsidy when the forest is opened to the general public, but there is no subdivision in subsidy levels for different activities and facilities.

RELATION BETWEEN VARIABLES AND ATTITUDE TOWARDS PRICING. There is a statistically significant relation between some variables and the attitude towards pricing. Visitors who visit more frequently (E + B + T),[4] local users (E + B + T), visitors with a dog (E + B + T), day-visitors (E + T), walkers (E) and visitors who have been coming for many years (E + T) show lower support for pricing. They probably do so because they regard the area as 'their own'. In other words, the reference price formed by these groups of visitors is lower compared with that of other visitors.

The results support the statement that 'recent payment at other similar sites will encourage acceptance of first-time payment'. Respondents with experience with pricing show higher support for pricing. Respondents with recent payment experience at similar sites also report higher 'fair' price levels (WTP). In other words, their reference price has changed because of recent payment at other similar sites.

No statistically significant relationship was found for personal variables (age, sex, education and income), apart from the residence of the respondent, for knowledge of different means of payment, for membership of a nature conservation organization, for frequency of visits to forest and nature in general (in addition to the research area) or for evaluation of the quality of the research area. Especially for this last variable, one would expect to find a difference.

Special arrangements for local and frequent visitors might change the more negative attitude of local users. User involvement in part of the management and organization or provision of recreational facilities through user involvement might have the same effect. 'If a group of users is involved in for example the construction of a track, there is more support for paying a fee to support this work.' Examples are the introduction of a 'friends of the Edese Bos' society or giving more authority (exclusive use for some activities) for local user groups. For small individual forest owners, this might be difficult to organize, but forest owners might be more successful as a group. Cooperation within the forest sector is important.

PRIVATE OWNERS AND NATURE CONSERVATION MEASURES. Respondents show higher support for payment for areas of private owners (compared with public owners). This supports the statement that 'consumers expect prices charged by public sector agencies to be different (lower) from those charged by private sector suppliers for a similar service'. Respondents also show higher support for payment for nature conservation measures (compared with recreation or information facilities).

Private owners can use information materials to emphasize the fact that the visitors are guests on private grounds and that the money that has been collected is used for nature conservation measures.

DIFFERENT MEANS OF PAYMENT. Visitors do not support entrance fees, season tickets or parking fees. These are the more general means of payment (many visitors arrive by car). When these means of payment are introduced, visitors only have the choice between visiting and not visiting and not on different ways of visiting. As said before, visitors seek choices and a sense of freedom and want to exercise control over their own destinies.

USE OF INFORMATION. The effect of information has not been part of the research, but, from the literature, it is known that information for and communication with the visitor can improve public acceptance of pricing. The sort of information that can best be used is:

- information on possibilities and the benefits of a visit (facilities, etc.);
- information that the introduction of pricing is used to improve the resource at which it is collected, especially nature conservation measures;
- information on the possibility of better supervision and regulation enforcement;
- information on the special natural and cultural elements that have to be preserved for coming generations;
- information on current subsidy and cost levels;
- information on ownership (private owners).

OPPORTUNITIES ACCORDING TO HEKHUIS AND DE BAAIJ (1997). Hekhuis and De Baaij describe a 'top ten' list of opportunities for pricing that have the best chance in terms of profit for forest owners. These top ten include five opportunities for recreation (5–9) and two related items (2 and 10). The results from our research give support to their conclusions, but some remarks have to be made.

- A visitor's pass combined with overnight stays. (Potential because visitors pay indirectly. The cooperation of the camp-site or hotel operator, who must collect the money, is not certain.)
- A season ticket combined with extra services. (Potential because of extra facilities and services.)
- Recreational arrangements. (Potential because of extra facilities and services.)
- Regional mountain bike, horse-riding, camping or canoe permits. (Non-users do not object and there are potential paying vehicles because of the extra facilities and services which are involved. Specific users of these means of payment are not sufficiently represented in our sample to draw conclusions.)
- Sponsoring. (There is a potential for sponsoring on signposted paths and information material about the area because visitors do not object to these places for commercial outlets.)
- Payments for forest and nature because of the quality factor for investments in buildings (living and working). (Not part of this research, but visitors object if this means that more area is built upon.)
- 'Green' products (for example, organic meat). (Visitors do not object, but the market potential was not part of the research.)

DISCUSSION ON TAXES. Though it has not been part of the research, there should be a discussion on the possibilities of 'single-purpose taxes'. Should the user pay directly to the owner or indirectly in the form of (local) taxes, which again are used to provide the service? In this way, negative effects of pricing, such as substitution by and crowding in other (public) forests, might be prevented.

MARKETING. For each (type of) forest, marketing management should provide the optimal mix of strategies. Each forest is different and has different (potential) user groups, and different strategies might influence each other. The results of this research, including the results of the literature research, can be of help when implementing a strategy.

REFLECTION.

> Fortunately, most management policies are not irreversible. This characteristic distinguishes managerial attributes from biophysical attributes. Mistakes can be corrected, policies refined, effectiveness monitored, changes made. But only through an understanding of visitor attitudes can this be accomplished successfully.
> (McCool and Lime, 1989)

8.2 Public Acceptance of RES Products in Germany

8.2.1 Introduction

The recreational and environmental services (RES) project deals with the investigation and representation of marketable products from the RES of the forest. Thereby, on the one hand, the population with its attitude towards the marketing of such forest services defines the frame conditions of society. On the other hand, the potential customers for recreational forest products can be found among the part of the population which spends a certain amount of time in the forest for recreational purposes. For successful marketing of RES offers, it is essential to gain some knowledge about the structure and social attitude towards the forest, particularly of the forest visitor's attitude towards the financing of these services. Therefore, the answers to the following questions – based on a household survey carried out in Germany – need to be analysed:

- What do forest visitors think about recreational offers if they are obliged to pay for them?
- Is it possible to estimate the acceptance for RES products?

It is frequently stated, on first consideration of the development of new RES offers, that there is no acceptance for paid forest offers and therefore no customers will be found. It is an objective of the RES project to examine these rapid judgements with regard to concrete offers and the different influences and attitudes of very different forest visitors. Accordingly, it is valid to consider the acceptance of possible RES offers in connection with different sociodemographic features and the behaviour of forest visitors, with special regard to the reasons for a forest visit. By means of concrete product examples, reactions to specific offers can finally be estimated. There is no question that a 'general admission charge' for a forest visit is refused socially within the former Federal Republic of Germany (FRG). Nevertheless, paying for the use of forest services that can be understood as 'a right of the proprietor' gains broad acceptance within the German population.

8.2.2 Material and methods

The results to be presented are based on a household survey entitled 'Forest visit behaviour and acceptance of RES offers', which was carried out by the Moelln INRA institute under the aegis of the University of Hamburg.[5] For this, within the framework of a multitopic survey, personal questioning of 1096 households in Germany was carried out in autumn 1998. The method included representative random samples and the results were further subjected to a proportionate population weighting. The questionnaire used was divided into the three parts 'standard demography', 'questions on the forest and forest visit behaviour' and 'questions on the "forest backpack"' as an example of a concrete RES product.

Standard demography can be important for the description of different segments of society. Therefore, some simple sociodemographic and geographical segmentation variables were taken from the great number of household features:

- structure of the household according to number of persons, family structure and number of earners.
- Age and sex of the respondent.
- Size and area of the habitation, afforestation status of the area.
- Education, profession and salary of the respondent.

These features can influence the behaviour of forest visitors and therefore lead to different user segments.

A similar division was also carried out for the features of forest visits and the attitude towards RES products:

1. Use of forest for visits.
2. Frequency of forest visits.
3. Duration of the forest visit.
4. Motives for the forest visit.
5. Wish to receive information about the forest.
6. Knowledge about 'forest admissions'.
7. 'I want to go to the forest free of charge.'
8. 'The forest owner should benefit.'
9. 'I would pay something for special forest offers.'
10. Financing sources.
11. Modes of payment.

In the following section, the answers to these points are represented both in detail and in comparison with each other.

The methods of data investigation and data evaluation correspond to the usual practice of empirical social investigation based on questioning. It is a representative random sample of the entirety of German private households.[6] Thus, the fundamental preconditions for carrying out random samples of any kind were met. Determined by a systematic random key, one person in every target household over the age of 14 years was interviewed personally. The present investigation was carried out by 210 interviewers, who were informed in detail about ascertaining the target person as well as the carrying out of the interview. The fieldwork was guided and supervised by the INRA institute. All the interview questionnaires were tested after having been returned, and methodological errors were excluded by machine control. The investigation occurred within the framework of a multitopic survey carried out between 13 and 19 October 1998. The random sample contained 1400 households, whereas the number of interviews actually carried out was 1069. The influence of these failure rates on the total results cannot be accurately verified.

Of the 1069 respondents, 664 persons indicated that they had visited the forest at least once in the last 12 months in Germany or in neighbouring states. These 'forest visitors' were asked about the forest itself and about the 'forest backpack'.[7] Thus, the net number of respondents for this study was $N_{\text{forest visitors}} = 664$. Within the present investigation, usually occurring whilst carrying out random samples, deviations from the official evaluation of the population (information on the population of the Federal Republic of Germany within the current statistical annual) were to be observed. An attempt was made to eliminate these deviations by iterative weighting. According to the computational transformation of household into person random sample – replacing the equal opportunity of the households by equal opportunity of the persons – an approximation is executed according to the features 'state' and 'local political area', as well as 'age' and 'sex'. The three-digit factor resulting from this iterative weighting was included in the data material as an individual emphasis of every respondent. If not characterized differently, all numerical data shown in the following are to be seen as weighted percentages.

8.2.3 Forest visit behaviour in Germany

Sociodemographic features of German forest visitors

In this section, the socio-demographic features of the forest visitor are described, as well as features of a forest visit. 'Forest visitors' are defined as respondents who visited the forest at least once during their leisure time within the last 12 months. About two-thirds of the popula-

Table 8.22. General use of forest for visits. (Question 2.1: Have you been to the forest at least once in your leisure time during the last 12 months?)

Forest visit during the last 12 months	*n*	%	% w.
Yes	664	62.1	61.7
No	379	35.5	35.9
Don't know	26	2.4	2.4
Total	1069	100.0	100.0

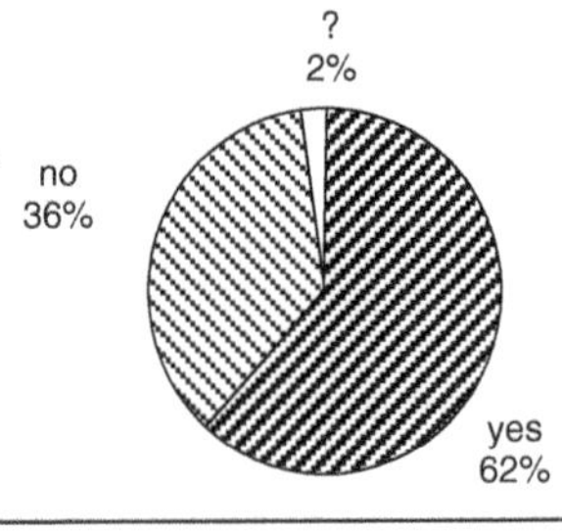

% w., weight %.

tion of the FRG use the forest according to this definition as a place for recreation (Table 8.22).

How can these forest visitors be distinguished according to sociodemographic characteristics from the non-forest visitors? A simple sociodemographic feature is the household structure, with regard to the number of persons, the number of earners and the number of children, as well as the age and sex of the respondent. Figure 8.6 represents five separate household structures from the entirety of the respondents and from the 'forest visitor' part of the entirety.

It can be seen that persons who are in a one-person household or a two-person household with only one earner visit the forest more rarely than would correspond to their proportion of the total population. From these household groups, only half have visited the forest at least once during the last 12 months. On the other hand, from among the respondents, about two-thirds of the 'forest visitors' live in 'several-person households' (families) or in dual-income families in two-person households (DINKs). Persons from these groups visit the forest slightly more often than the average. This result is supported by observations that respondents whose household contains at least one child under the age of 14 years visited the forest slightly more often than households without any children.

Furthermore, connections, though not very strong ones, can be recognized between age and the use of the forest as a resort (Fig. 8.7). It is obvious from the figure that it is above all the age classes

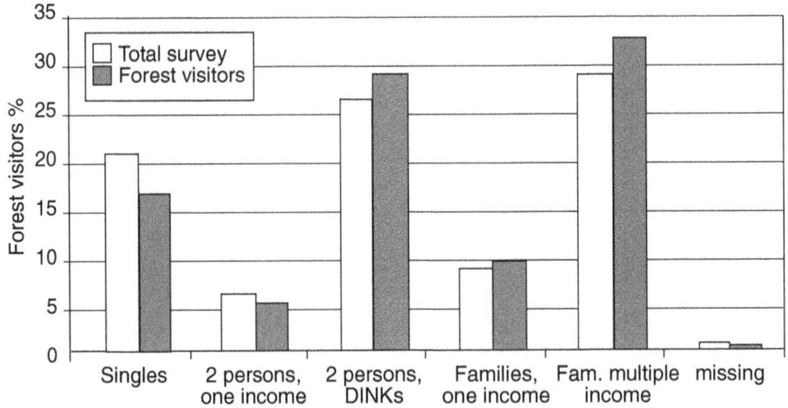

Fig. 8.6. Division of forest visitors by their household structure (no. total survey = 1069, no. forest visitors = 664 (61,7%); scale in weight %; DINKs, double income, no kids).

between 36 and 65 years who visit the forest. Young people between 19 and 25 years and older persons of more than 65 years of age visit the forest more rarely than expected from their share of the entirety of the respondents. This can be explained simply by typical leisure-time interests, on the one hand, and impaired physical mobility, on the other. With regard to the sex of the forest visitors, only a weak trend could be recognized, men visiting the forest a little more often than women.

Geographical influences such as the size of the home town, the region within Germany and the forest richness of the *Bundesland* (state), are considered to be important. With regard to the size of the home town, there is a greater tendency for inhabitants of medium-sized cities to visit the forest. However, the influence of the town size is strong enough to be able to draw the conclusion of a fundamental dependence on forest use for recreation purposes upon the size of the place of residence.

However, quite striking differences appear if the region is considered, as shown in Fig. 8.8. Whilst North Germany, which is poorly covered with forests, is represented to a noticeably low degree within the entirety of forest visitors, inhabitants of the new (East German) states constitute a particularly high proportion of the forest visitors. Fewer than half of the respondents from the north specified having been in the forest at least once during the last 12 months, whereas this question was answered positively by more than three-quarters of the respondents from East Germany. Forest visits were also rarer in the west and in the middle of Germany, whereas the forest in the southern states was used more often for recreation.

It can be presumed that, of the new states (including Berlin), only Thuringia and Brandenburg were classified as 'rich in forest area', and thus there is only a weak trend for the forest in forest-rich states to be visited by more respondents than in forest-poor states. The strikingly weak nature of the expected trend can then be evaluated as a reference to the fact that the forest in forest-poor areas has an attraction which balances its difficult accessibility.

As a third field of the sociodemographic features, education, profession and income will be considered. These three features are interconnected, which is also reflected in the forest visit behaviour. Thus, Fig. 8.9, concerning the effect of education on forest visits, represents the entire range of social groups. It can be seen from this figure that the educational level of the forest visitors is non-significantly higher than in the entirety of the respondents. The tendency that visiting the forest for recreation purposes increases with level of education is repeated in consideration of the part of

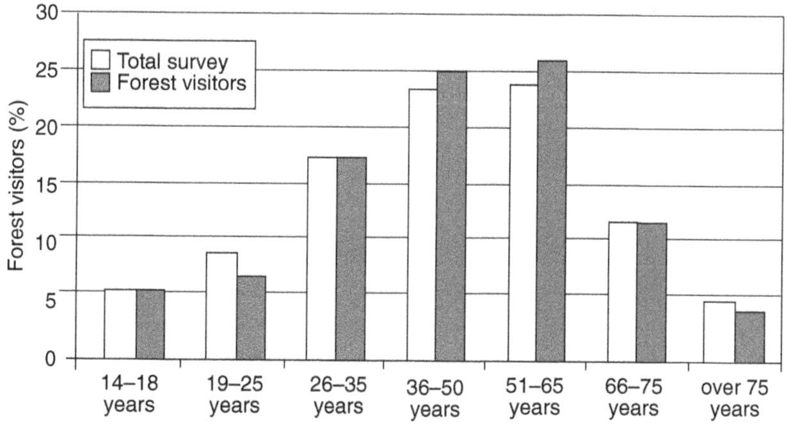

Fig. 8.7. Division of forest visitors by age (no. total survey = 1069, no. forest visitors = 664 (61.7%); scale in weight %).

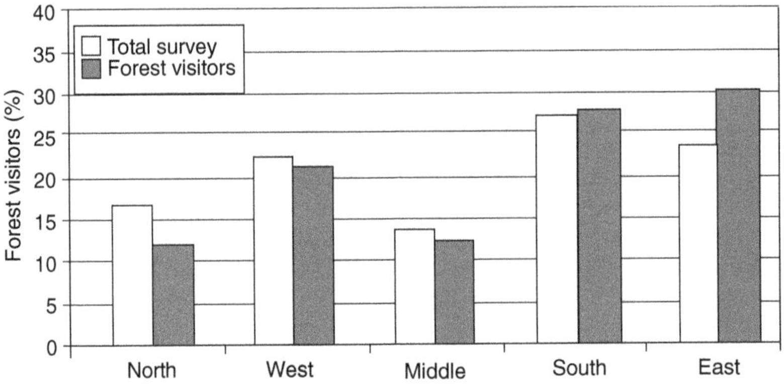

Fig. 8.8. Division of forest visitors by region (no. total survey = 1069, no. forest visitors = 664 (61.7%); scale in weight %).

forest visitors in different occupational groups. Officials and the self-employed, but also unemployed persons, visited the forest a lot more often than pensioners, pupils, students, apprentices and workers. Similarly, an increase in forest visits becomes obvious with increasing salaries. Nevertheless, it has to be stated that also in the lower earnings group more than half of the respondents visited the forest. Therefore, forest visits are not at all the privilege of the upper social classes.

Typical forest visitors are persons – with a non-significantly greater proportion of males – who often derive from family households – rarely from single- or two-person households – and whose age is between 25 and 65, mostly above 50 years of age. They often come from medium-sized towns in the south and east of Germany. In addition, they are often more highly educated, have a slightly higher salary and belong to the upper social class, e.g. as officials or the self-employed.

Typical forest visitors, according to this study as well as according to a Hamburg region survey (Elsasser, 1994), could be described as in the box below.

Frequency and duration of forest visits

The frequency of forest visits represents a basic variable and a parameter for different question formulations with regard to recreation in the forest. While up to now those respondents who were at least once in the forest during the last year have been treated equally as 'forest visitors', Table 8.23 shows the actual frequency of the forest visits. While the results shown in Table 8.23 record a generally smaller forest visit frequency than in preceding investigations, the frequencies are nevertheless quite comparable, as shown in Table 8.24.[8] Here the results of the forest visit frequency from six regional enquiries are represented comparatively. According to the Sample-Institut (1995), approximately 33 annual forest visits per registered German citizen can be calculated, leading to a total sum of approximately 1.6 thousand million annual forest visits in the Federal Republic. These results may be compared with 29.3 forest visits (without 'missings') per citizen derived from the presented RES study, which leads to a total of approximately 1.4 thousand million annual forest visits.

As a primary sociodemographic variable, age influences the frequency of forest visits. While adolescents very rarely and young adults up to 35 years rarely visit the forest, the frequency of forest visits increases continuously with the increase of age. Above all, the group of over-65-year-old people, who are represented in a low proportion when considering whether forests are generally used for recreation,

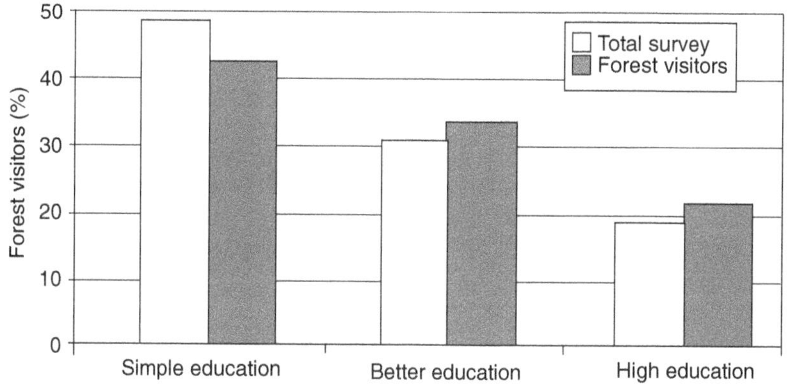

Fig. 8.9. Division of forest visitors by educational level (no. total survey = 1069, no. forest visitors = 664 (61.7%); scale in weight %; the 'education' of the interviewed person was generated from his/her school and professional examination degree).

Table 8.23. Frequency of forest visits. (Question 2.2: How often did you visit a forest area in Germany or in a neighbouring state in your leisure time during the last 12 months (approximately)?)

Frequency of forest visits	*n*	% w.
Up to 10 times a year (< monthly)	357	53.6
10–50 times a year (< weekly)	217	33.0
More than 50 times a year (> weekly)	72	10.3
Don't know	17	3.0
Missing	1	0.1
Total	664	100.0

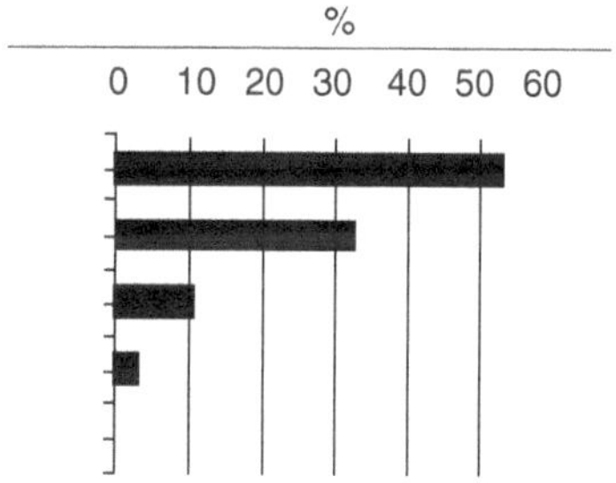

% w., weight %.

Table 8.24. Results from six regional investigations on forest visit frequency.

Source	Number of respondents	Frequency of forest visits (%): Seldom or never	At least once a year	At least once a month	At least once a week	Missing
Dertz and Nießlein (1993), Germany	1777	22	28	48		2
Sample-Institut (1995), Germany	1414	33	28	28	9	2
RES (1998), Germany	1069	35	33	20	7	5
Rozsnyay (1972b), Bremen	(?)	24	25	38	13	
Elsasser (1994), Hamburg	806	20	23	29	28	
Braune (1998), Lübeck	100	34		31	35	

shows an especially large frequency of forest visits if this possibility of recreation is available. This observation is supported by various earlier investigations (Rozsnyay, 1972b; Dunkel *et al.*, 1994; Elsasser, 1994).[9]

With regard to geographical influence variables, there was a tendency for respondents from big cities to visit the forest either very rarely or very often. Among the social population features, only the profession shows an influence on forest visit frequency: the self-employed, officials, pensioners and annuitants belong to the 'group of frequent forest visitors' who are under-represented in the range of up to ten visits and over-represented in the range of more than 50 visits. Workers, apprentices, pupils and students can be found more often in the range of up to ten visits and more rarely in the range of over 50 annual forest visits ('group of rare forest visitors'). The groups of employees and of unemployed persons tend to make forest visits more rarely, but their behaviour is more or less average. This result is also supported by earlier investigations (Rozsnyay, 1972b; Dertz and Nießlein, 1993; Dunkel *et al.*, 1994).

Besides the frequency, the duration of an individual forest visit determines how much time the individual forest visitor spends in the forest within a specific period. As shown in Table 8.25, forest visits of between half an hour and 2 h (56%), followed by stays of up to 4 h (32%), dominate. Short stays of less than half an hour and a duration of more than 4 h are rarer (5–6%). An average forest visit duration of 1–2 h, with visits at weekends a little longer than on workdays, is also described in other investigations (Rozsnyay, 1972b; Dunkel *et al.*, 1994; Elsasser, 1996).

It was shown, among other things, that, with regard to the respondents' connections with geographical features, either short visits of up to 30 min or longer visits of 2 h or even more than 4 h are popular among people from big cities. Furthermore, inhabitants of forest-rich states tend to stay a little longer in the forest. While Dunkel *et al.* (1994) determined that respondents who visit the forest frequently stay there for a rather short time (and vice versa), the present study showed no fixed connections between the frequency and the duration of a forest visit.

Motives for the forest visit and the wish for information

There are very different motives and reasons for forest visits. The historical importance of the forest as a place of physical use in order to secure human existence has over time almost completely disappeared, being replaced by the non-material recreational use of the forest. About 1% of the respondents indicated that they visited the forest to work. But approximately 62% of the German population visits the forest for various forms of recreation. By far the most common form of forest recreation is hiking and walking (84%) and enjoying nature through impressions (34%). Table 8.26 shows some motives for forest visits and their frequency in the 'at least occasionally forest-visiting' population.

Sports recreation in the forest is above all carried out by bike (19%), whereas a use of mountain-bikes and all-terrain bikes is far rarer (1.5%). Another frequent form of forest sports recreation is jogging (11%).

Table 8.25. Duration of forest visits. (Question 2.3: How long did your last forest visit take (approximately)?)

Duration of the last forest visit	*n*	% w.
Less than half an hour	38	5.4
Half an hour to 2 h	364	55.7
2–4 h	219	32.1
More than 4 h	36	5.6
Don't know	6	1.1
Missing	1	0.1
Total	664	100.0

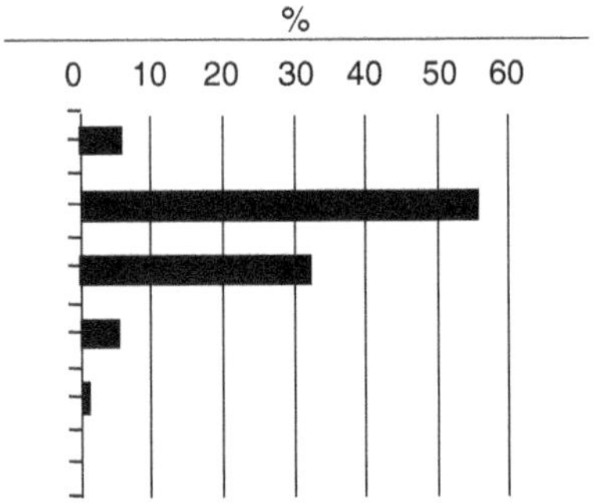

% w., weight %.

Table 8.26. Motives and reasons for forest visits. (Question 2.4: Why do you visit forest areas (multiple answers possible)?)

Reasons for forest visits	*n*	% w.
Silent recreation (walking, hiking, relaxing)	602	88.2
Sport activities (jogging, riding, skiing, cycling)	194	31.4
Playing with the children	83	12.0
Observing nature	130	19.0
Collecting forest fruits (mushrooms, berries)	231	32.4
Hunting	7	1.0
Work	8	1.0
Other	1	0.2
Total (basis: 664 respondents)	1256	185.2

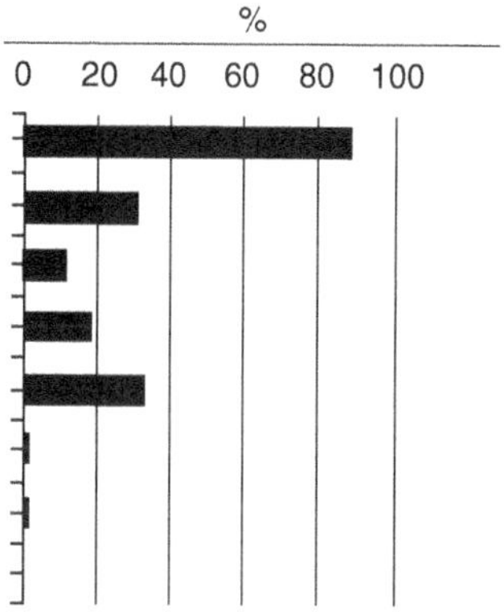

% w., weight %.

Cross-country skiing (4%) and riding (3.5%) were mentioned quite rarely as reasons for a forest visit.

About 12% of German citizens visit the forest in order to play there with their children. About 19% try to learn in the forest by nature observations. The 'free use' of physical forest products, such as berries and mushrooms, can be observed with 32% of German citizens. Finally, the forest is visited by approximately 1% of the population in order to hunt.

The significance of the forest as a place of 'quiet recreation' for walks, excursions and enjoying nature was already obvious in various studies dealing with reasons and motives of forest visits (see, for example, Hanstein, 1967; Rozsnyay, 1972a,b; Dunkel *et al.*, 1994; Elsasser 1994, 1996).

To what extent do the reasons for a forest visit depend on sociodemographic features? As expected, there is a clear connection between the household structure and the motive for a forest visit 'to play (with the children)'. Approximately 80% of the respondents indicating these motives come from families with one or several earners. Furthermore, the age of the respondents obviously has an influence on the motive for a forest visit, as shown in Table 8.27.

Table 8.27. Change of reasons for forest visits with increasing age (weight %).

Reasons for forest visit	All	Age: 14–18 years	19–35 years	36–50 years	51–65 years	Over 65 years
	Weight:	6.1	25.1	25.5	26.5	16.9
Going for a walk/hiking	83.5	47.1	77.6	81.1	93.1	93.8
Enjoying nature through impressions	33.6	26.8	27.7	32.3	40.3	36.3
Serious observation of nature	19.0	5.8	21.5	18.2	23.7	13.9
Collecting forest fruits	32.4	18.2	28.2	36.1	37.9	29.4
Hunting	1.0	3.0	0.4	1.2	1.0	0.7
Playing outdoors (with the children)	12.0	10.0	19.0	18.0	7.0	0.9
Cross-country skiing	4.3	2.1	2.8	6.6	4.0	4.3
Cycling	18.5	27.0	24.2	21.2	13.5	10.8
Riding	3.6	14.8	1.3	8.8	0.5	0.0
Cross-country running/jogging	11.2	27.4	15.9	11.3	5.9	6.4
Mountain-biking	1.6	12.9	1.7	0.2	0.0	1.7

The table shows that the forest is often used by adolescents for athletic purposes, such as cycling, riding, jogging and mountain-biking. Between 19 and 35 years of age, the importance of these sporting activities recedes a little and motives like 'to play outside with the children' and 'serious nature observations' gain importance. In the middle age field between 36 and 50 years, the events of skiing and riding become more important, as well as playing with the children and collecting forest fruits. This occupation, going back to handed-down knowledge and habits (from postwar times), is seen as a motive for visiting the forest for the group of the 50–65-year-old people (nearly 40%). In this age stage, the aspects of 'quiet forest recreation' deriving from nature observation and pleasure from impressions, as well as going for a walk and hiking, have a special importance. Within the over-65 age-group, walking and simply enjoying nature are especially important reasons for a forest visit. Nature studies or the collecting of forest fruits is reduced here and playing with children and athletic activities decrease to a minimum, as expected. However, it should not be overseen that senior citizens also indicate some sporting activities, e.g. in the field of skiing, cycling or jogging. According to the results of Elsasser (1994), a clear shift from 'sporting activities' to 'a walk with the dog' goes along with increasing age.

Geographical connections appear with regard to the collecting of forest fruits. This reason for forest visits was, above all, stated by respondents from small towns and rarely specified by people from big cities. The 'popular sport of collecting mushrooms' is evident in East Germany: whilst only 17% and 9%, respectively, of persons from the north and the west of Germany specified visiting the forest for collecting berries and mushrooms, 60% of respondents in the east did so.

With regard to the social features of the respondents, a quite homogeneous picture emerges from the level of education and the salaries. Respondents with a simple education and/or small salaries generally show less interest in forest visits and so indicated fewer reasons also. Above all, they mentioned 'playing with children in the forest' below the average, as well as the different sports opportunities in the forest. On the other hand, these sports opportunities were more often mentioned reasons for a forest visit for persons with a high education and/or high salaries, whereas in this group collecting forest fruits had quite a low representation. Finally, respondents with a basic education visited the forest more often for reasons of nature observation and in order to play outside with their children. This motive emerged in addition to collecting forest fruits in the case of respondents with middle salaries. In this group, sports activities are clearly reduced.

It seems that, with regard to 'forest sportsmen', there is a connection between the frequency and duration and the reasons for forest visits. These people often indicated frequent forest visits (more than 50 times per year) but also short stays of up to half an hour. Forest visits with the purpose of playing there (with children) often last between 2 and 4 h. Nature observations and the collecting of forest fruits show a homogeneous picture towards time-increasing forest visits of up to more than 4 h.

Wishes to receive information. Visiting a forest always reflects an interest in specific qualities of the forest. Where it is a question of a conscious interest, an information request is often combined with it. The information requests of German forest visitors are illustrated in Table 8.28. This table shows the widespread interest in receiving information about the forest. About 80% of the respondents stated at least one subject area about which they would like to receive information. Questions of nature conservation (33%)[10] and nature trails in the forest (25%), as well as information about dangers for the forest and the problems of 'forest decline' (37%), are of special importance. Next comes information about animals and plants (39% and 34%) and information about socio-economic questions (21%), wood use and forest cultivation, as well as the forest laws and prohibitions in the forest (16% each),

Table 8.28. Wishes to receive forest information. (Question 2.5: Concerning which aspects of the forest and forestry do you wish to receive more information (multiple answers possible)?)

Information wishes concerning ...	*n*	% w.
Ecology, protection of nature, forest decline	350	49.9
Animals of the forest	255	38.9
Plants of the forest	234	34.0
Forestry (technical, economical, etc.)	168	23.8
Laws, prohibitions in the forest	106	16.3
History, international forestry	178	24.8
Hunting	55	0.1
High knowledge available	11	1.7
No interest in information	117	18.4
Other	4	1.0
Missing	11	1.7
Total	1489	218.6

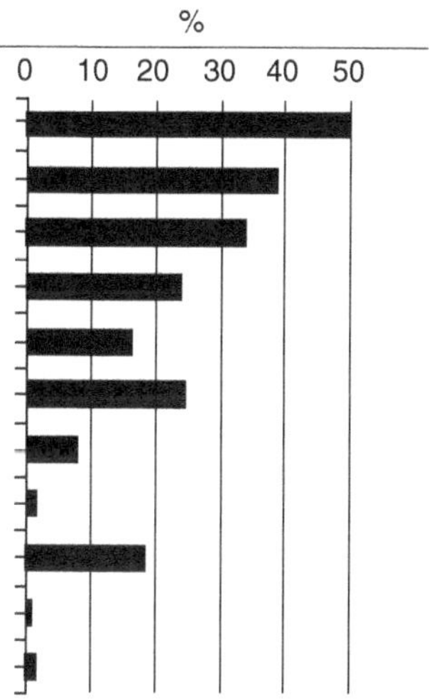

% w., weight %.

information about forests in other countries (20%) and forest history (17%). About 8% of the respondents were interested in questions of the hunt.

Discussion of forest visit behaviour

Along with the increasing necessity for the infrastructure services of the forest to no longer be seen as a free gift of 'forest functions' given by the forest owner to society, and the need to search actively for market chances (see, among others, Mantau, 1993a,b, 1995), forest opinion surveys have led to more. The representative questioning of German households carried out within the framework of tho RES project, as presented here, emphasizes questions of financing forest services. Conclusions should be drawn from this as to the potential acceptance of offers for which a charge is made.

However, the use of earlier reported surveys can, as a secondary market investigation, also be a valuable contribution to ascertaining possible markets and, above all, lead to customer segments for RES offers.[11] At this point, the methods of questioning will not be explained further, but reference should be made to Rozsnyay (1972a,b), Elsasser (1994), Henze (1994) and Oesten and Roeder (1994). The method of household questioning used in the present study can, with regard to the selection and the extent of the random sample, be judged as sufficient to derive general statements on the behaviour and the opinion of German forest visitors. It remains without question that these statements, particularly if they are based on cross-evaluations with a small field allocation, can only be designated as significant within a broad framework. Nevertheless, within the limits of limited financing, this study shows sufficiently probable tendencies about as many aspects concerning the marketing of infrastructure services of the forest as possible.

FOREST VISITORS AND FOREST VISITS IN GERMANY. About two-thirds of the inhabitants of Germany from the age of 14 years upwards occasionally or regularly visit the forest for recreation purposes, on the average about 30 times per year. Although these numbers are slightly below the results given in earlier investigations, they identify the broad importance of the forest as a place for recreation. On the one hand, it is evident, that considering the sociodemographic structure of the forest visitors, forest visits are used for recreation by all parts of the population. On the other hand, demographic shifts can be recognized between the 'forest visitors' and the average population. These shifts can be used for segmenting forest visitors as customers for RES recreation offers. As already stated above, forest visitors can be distinguished, among other things, by the following sociodemographic features:

- they come more often from families than from one- or two-person households.
- They are often between 25 and 65 years old (and, non-significantly, are more often male).
- They are more frequently represented in medium-sized small towns in the south and east of Germany.
- They are often more highly educated, have a slightly higher salary and belong to the upper social class, e.g. as officials or self-employed persons.

However, as the forest is used for recreation, to different degrees, by the whole society, too narrow a concentration on this demographic customer segment would certainly not be reasonable.

Based on the frequency of forest visits, the considerable 'customer potential' of approximately 1.5 thousand million annual forest visits by Germans over 14 years of age can be recognized, which possibly would be extended still further by attractive offers in the forest. However, the frequency of forest visits only becomes interpretable by combination with the reasons and motives. This also applies to the duration of forest visits. The average duration time of 1–2 h may give a clue for the organization of recreation offers such as guided tours in the forest. The duration of forest visits might also be increased by specific offers. To show this, the stay duration in the forest at the 71 recreational RES case-studies (see Chapter 2) was divided as shown in Table 8.29. From this table, it is obvious that the use of specific recreation offers in the forest often extends the stay duration. This leads to the next section, in which the different reasons for a forest visit are considered as a part of the 'forest visitor demand'.

Demand – reasons, motives and wishes of forest visitors. The reasons and motives for forest visits, as well as the wishes of the forest visitors as 'customers' of recreation offers in the forest, from the central components of the 'forest visit behaviour' regarding the qualitative demand aspects of German forest visitors. Based on the results of the present study, as well as of all the other investigations discussing this subject, there is no question that the forest is, above all, used for 'quiet recreation' on foot. Motives for this forest use are primarily rest, relaxation, fresh air and health promotion (see, among others, Rozsnyay, 1972a; Elsasser, 1996). Special attractions are not normally a primary motivation for a forest visit: 'the forest itself apparently serves as a particular attraction' (Dunkel *et al.*, 1994).

The image of the forest as a place of 'quiet, natural' recreation is very non-

Table 8.29. Duration of forest visits at the RES case-studies. (Question 2.2.10 RES questionnaire: What is the usual length of the stay in the forest of your most important target group?)

Length of stay in the forest (most important customer group)	*n*	%
0–4 h	24	33.8
4 h–1 day	28	39.4
> 1 day	12	16.9
Missing	7	9.9
Total	71	100.0

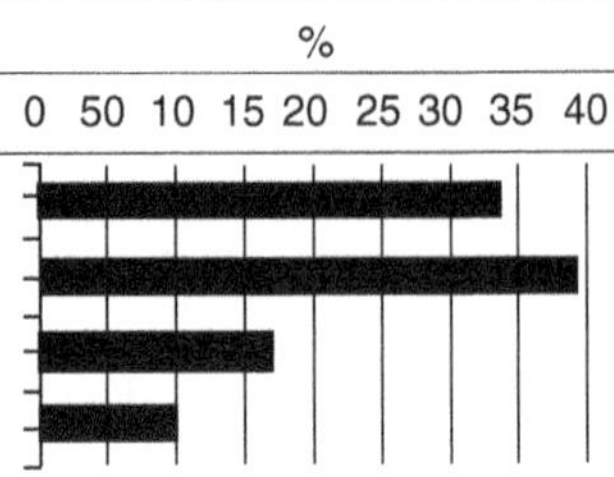

specific, comparable with the image of a 'mixed forest', without determining more closely its characteristic qualities, e.g. the tree species composition (Rozsnyay 1972a, b; Elsasser, 1996; Lehmann and Schriewer, 1999). The forest track network and its condition in the forest are, indeed, often judged subjectively as of lesser importance by recreation searchers; however, those same persons are often severely critical about the lack of volume, fixing and ease for walking on (Rozsnyay, 1972b; Elsasser, 1996; Braune, 1998; Lehmann and Schriewer, 1999). Sometimes, a lack of recreation equipment was also lamented in the past (Hanstein, 1967; Rozsnyay, 1972a, b), but this should under no circumstances lead to 'furnishing the forest' (Dunkel *et al.*, 1994; Elsasser, 1996).

Consequently, Elsasser (1996) concludes that 'equipment similar to that of a leisure park' in the forest 'would privilege minorities at the expense of the majority' (normally refusing such offers), but reduces this problem to the free use of additional recreation equipment in the forest. Since such options would anyhow be expected to only a limited extent in the forest, the problem is not settled for the predominant part of forest visitors, who concentrate on the 'nature experience' as 'strollers' (see Chapter 7).

These forest visitors, who see active recreation offers in the forest with hardly any interest or with outright rejection, will not be won as direct customers of specific RES offers. However, certain offers of the forest owner would be possible and worth considering for these visitors too, such as a focusing of forestry measures on the forest as a place for recreation which is as 'natural' as possible, 'genuine', characterized by a great variety of species or an undisturbed nature experience. This could also be marketable for certain interest groups ('contract recreation forest').[12] None the less, a most elastic demand is to be expected in this field. This becomes obvious from a consideration of different forest opinion surveys. Hanstein (1967), Rozsnyay (1972b), Dertz and Nießlein (1993) and Dunkel *et al.* (1994) came to the conclusion that a 'usual forestry' is generally accepted by recreation searchers. A WTP for changing this kind of forestry is thus improbable.

In the RES questioning presented here, however, reasons for a forest visit are stated which indicate an existing demand for specific offers in the forest (Table 8.30). The central topic of the present RES project is the practical implementation of these 'possible RES offers'. On the basis of his results of questioning Hamburg forest visitors, discussed above, Elasser (1996) also recommends that additional recreation possibilities should only be marketed on a private law basis and offered to limited user groups. In this case, he emphasizes the advantages of a demand-orientated offer, of the financial release of the forest owners and of the development of new income possibilities (see also Dunkel *et al.*, 1994).

Furthermore, from the priorities of forest visitors determined by Dunkel *et al.* (1994), the potentially high acceptance for 'contract nature/water protection' becomes obvious. Elsasser (1996) also determined that 90% of the Hamburg forest visitors he surveyed would agree with a partial closing of the forest for recreation searchers in order to be able to operate specific nature conservation measures.

Before dealing with the question of who should pay and in what manner for which services of the forest, the wishes of forest visitors according to the kind of information they ask for should be mentioned. A wish, for example, for guided forest tours is manifest in various earlier investigations (Rozsnyay, 1972a; Dertz and Nießlein, 1993; Elsasser, 1996). Different approaches to 'forest information products' can be found here, which, however, should not be confused with 'forest public relations (PR)'.[13] They could be adapted to the requirements of those forest visitors who are neither rarely nor especially frequently visit forest to be better informed about diverse topics of the forest through the increased offer of the corresponding products, such as guided tours, information centres or a 'forest information parcel' in the form of a 'forest backpack'.

Table 8.30. Forest utilizations, user groups and suitable RES offers.

Reason for forest visit (% respondents)	Typical customer segments for this forest use	Possible RES offers
Cycling (19%)	Young (dynamic), higher education and income, frequent, rather short forest visits	Cycle-routes
Jogging (11%)		Jogging paths
Cross-country skiing (4%)	Young or middle-aged, higher education and income, frequent, rather short forest visits	Cross-country ski-runs
Riding (3.5%)		Bridle-paths
Mountain-biking (1.5%)	Young, dynamic, frequent, rather short forest visits	Mountain-bike tracks
Playing (with children) (12%)	Families, middle to higher education and income, average forest visits	Special events, forest games
Intensive nature observation (19%)	Middle-aged, longer forest visits	Guided tours, nature trail, observation stands, identification aids
Collecting forest fruits (32%)	Middle-aged or older, from small towns, esp. from the east of Germany, elementary education and lower income, longer forest visits	Picking permits, excursions, identification aids
Hunting (1%)	High income	Hunt possibilities

Now that the general features of forest visit behaviour have been discussed and characterized, the attitude towards offers in the forest for which the user is liable to pay, as a crucial question of the RES project, may be discussed.

8.2.4 Acceptance of forest offers the user is liable to pay for

Knowledge and use of forests where the user is liable to pay

First, the question should be considered to what extent 'admission fees' for forest areas are known about in Germany and to what extent such forests are used for recreation (Table 8.31). Much the greater part of the 'forest visitors' indicated that they never before heard anything of 'forests where the user is liable to pay' (62%). Only 8% had stayed in forests where they had to pay an admission fee for its use. This information must be evaluated under the reservation that the definition of the term 'forest' is understood differently among the respondents. Some respondents, for example, identify a wildlife park as 'forest', while others regard it as a 'zoo', for which an admission fee is natural.

Knowledge of and visits to forests with an admission demand are shown by the following combination with geographical variables and other aspects of forest visit behaviour:

- forests where the user is liable to pay are known better to people from big cities than to inhabitants of small towns.
- Respondents from forest-poor states have, to a much greater extent, already paid an admission fee for specific forests.

Furthermore, the number of those who have knowledge of forests where they are liable to pay grows with increasing level of education, increasing earnings and increasing social position. This corresponds with a generally increasing interest in the forest.

Beyond that, only poor connections are detectable between the knowledge of forests where the user is liable to pay and/or its use and the reasons for forest visits. Only persons who use the forest for natural observations indicated a little more often that they have already used forests where they paid, e.g. nature parks, etc.

Table 8.31. Knowledge about 'forest admissions'. (Question 2.6: Have you ever heard about an admission fee for particular forests, in which, for example, special animals or plants exist? If yes, which answer fits for you?)

Knowledge of charge on forests	*n*	% w.
No, never heard about it	398	62.1
Yes, but never visited	166	24.2
Yes, visited and paid admission	56	7.6
Don't know	36	4.8
Missing	8	1.3
Total	664	100.0

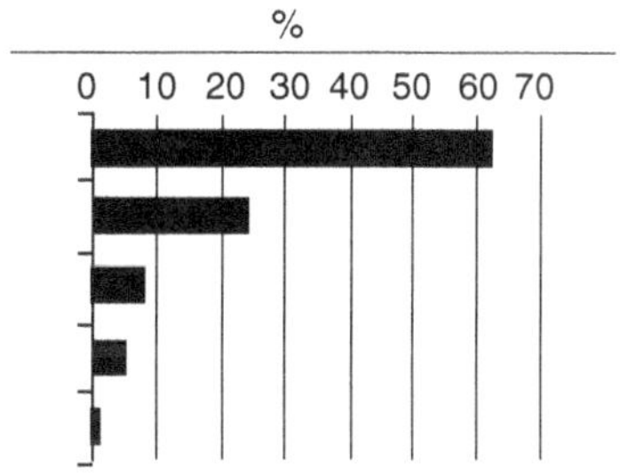

% w., weight %.

The personal attitudes towards financing forest services

Table 8.32 shows personal attitudes towards financing forest services. It gives the agreeing, neutral or refusing attitude towards three statements concerning the forest use and its financing. The table shows that unobstructed admission to the forest for recreation use is of great importance for almost all forest visitors. But, at nearly 45%, many more persons would promote financial support for the forest owner than those who are against it (28%). And, even though about half of the respondents would refuse to pay for use of the forest, almost a quarter of them approve such a WTP.[14]

By a combination of the answers to these three questions, it becomes evident that only 9% of those respondents who find support for the forest unimportant expressed their own WTP, whereas 38% of the respondents favouring support for a forest owner expressed their personal willingness. The educational status also shows the expected effect: Along with increasing education goes an increasing number of respondents who consider support of the forest owners as reasonable. This could once more derive from the increasing interest in the forest and, furthermore, from an increasing problem consciousness regarding economic factors.

With regard to their own readiness to pay, dependencies of age and salaries can be recognized. The WTP for 'special' offers in the forest first increases with the age, and then falls again at the same rate. Maximum approval and minimum refusal is in the age stage between 36 and 50 years. The strong negation of their own readiness to pay in the case of respondents under 19

Table 8.32. Attitude to the supply and financing of forest activities. (Question 2.7: To what extent do you agree with these statements personally?)

	Attitude in % w.			
Statement	Positive	Don't know	Negative	Missing
Free access to the whole forest is very important for me	89.5	6.8	3.2	0.5
The forest owner should gain from the environmental and recreational services of their forest	44.8	27.0	27.9	0.2
I would pay something for special environmental/ recreational forest offers	23.5	26.0	49.9	0.6

% w., weight %.

and over 65 years of age can be explained by the generally smaller interest in the forest, the financial situation of these age groups and, at least with the senior citizens, also by habituation, to a large extent, to free forest use. Although it is alone not sufficient, it is obvious that the financial situation has an influence on the readiness to pay: respondents from households with a monthly net income of more than DM 4000 are over-represented in the group of the ready-to-pay supporters. Here too, an intercorrelation with increasing interest in the forest is probable.

There are also connections to be recognized between other features of forest visits and the attitudes towards forest financing. The respondents' own willingness to pay slightly decreases in tendency along with an increasing frequency of forest visits. This statement appears to be logical if one imagines that increased forest visits could involve increased payments. However, this contradicts the connections found earlier between the willingness to pay and the interest in the forest. This could be a kind of buffering of inhomogeneous effects. A clear result is gained if the two positions knowledge of forests where the user is liable to pay and the respondents' own WTP are opposed. Respondents who do not know any forests where the user has to pay refuse making a payment to 56% and only agree with a such contribution to 18%, while respondents who had already used forests where they were liable to pay the costs refuse a payment only to 41% and admit to being ready for payment to 36%.

Finally, the questions on forest financing should again be considered in connection with the different reasons for a forest visit. Above all, those respondents who use the forest for sport should be mentioned, because a consistent picture resulted in this group: free entry of the forest was considered to be of slightly below-average importance, financial support for the forest owner was more often held to be important and the number of respondents who would be ready to pay for specific forest offers was higher among these 'sportsmen'. In addition, 'collectors of forest fruits' more often expressed their WTP and their approval for forest owners being supported financially. However, unrestricted entry to the forest was especially important for these 'collectors'. The remaining user groups answered more or less on the average.

The question as to the extent to which the answer behaviour of the users of concrete forest services towards the financing of these offers is consistent leads to the next question: By whom and in what way should offers in the forest be financed?

Attitudes about who should pay for special forest services

Different financing sources apply to different uses of the forest. Therefore, the opinion of the German forest visitors was sought. For this purpose, 16 different possible uses of the forest, from sports offers to environmental services, were asked about, as to whether the corresponding service from the point of view of the respondent should be paid for by the user him/herself, by the tax-paying society or by the forest owner him/herself. A remarkable picture, represented in Table 8.33, resulted.

Sports offers, hunting possibilities and recreation services were judged by the majority of respondents as offers which the user him/herself would have to pay for. Financing by the taxpayer was refused to a large extent. On the other hand, financing environmental qualities, services in the environmental field and a basic provision of the forest with recreation facilities was seen as the task of the taxpayer. However, a considerable number of the respondents stated that the forest owner also has a duty. As a service to be supported by the forest owner him/herself, the larger number finally stated the 'simple recreation uses guaranteed by law and to be tolerated', such as the use of forest tracks and the free collecting of forest fruits.

This result indicates two different motives for choosing the answer. On the one hand, various respondents based their answers on the present state of affairs, above all with regard to the present financing by the forest owner. On the other hand,

Table 8.33. Opinion about different possible financing sources for forest services. (Question 2.8: Please express your opinion (not the actual situation) as to which of the following possibilities and advantages of forest utilization should be paid for by the users themselves, which should be financed by the public through taxes and which should be offered by private forest owners free of charge.)

	Financing of forest service (attitude in % w.): payment by:				
Kind of forest use	User	Taxes	Forest owner	Don't know	Missing
Silent recreation	4.1	27.7	59.6	7.5	1.0
Sport activities, hunting	67.5	10.6	12.0	8.7	1.2
Recreation facilities	25.8	41.5	24.0	7.2	1.6
Recreation services	48.6	16.0	24.2	10.0	1.1
Collecting forest fruits	13.5	12.5	63.1	9.8	1.1
Environmental qualities	5.3	50.3	24.7	18.0	1.7
Environmental services	5.7	39.5	32.9	20.3	1.6

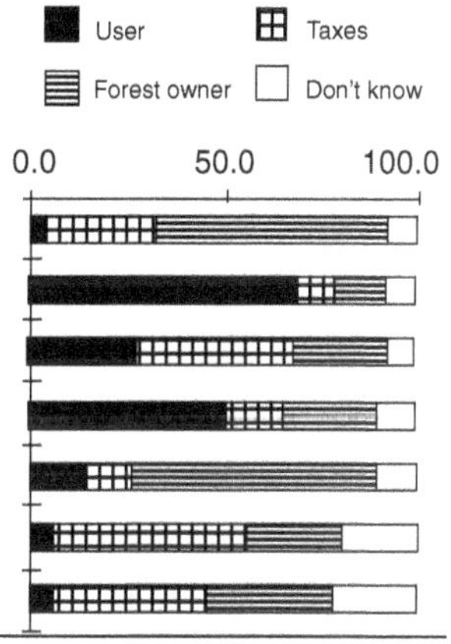

% w., weight %.

this interview result expresses a very fundamental feeling of justice within the population. Services which are to be assigned directly to the individual user should also be paid by him/her. Services for the general public are to be paid for by the taxpayers. Only services of the forest established as 'natural' and 'genuine' in society's life are expected to be a free service of the forest owner.

Considering the importance of mushroom-picking permits in Italy (RES case-studies IT01, 02, 08, 19), the financing of this forest use will be explained briefly. That the collection of berries and mushrooms – clearly limited physical forest components – should be paid for by the collector is stated by only approximately 13% of the respondents, whereas approximately 65% state that it would be a task of the forest owner to promote this service without being paid for it. In this case, regional differences appear: whilst in the north and south of Germany one-quarter of the respondents stated that the collector of forest fruits should him/herself pay for it, only 2% of the respondents from the middle of Germany stated that this forest service should be financed by the user. The proportion of persons who would impose the financing of collecting forest fruits on the forest owner is 69% in the west and 86% in the east of Germany. This result again shows the special importance of collecting mushrooms in the new states. In addition, persons from households with low salaries surprisingly often see these services as a duty of the forest owner. Altogether, it is evident that a marketing of licences for collecting mushrooms would be difficult due to an especially low acceptance in Germany, in addition to legal and organizational problems.

Finally, the attitude towards guided forest tours is also interesting. Although the majority of those asked are of the opinion that the users should pay for this service, 24% nevertheless see this service as a task of the forest owner to be made available without any payment. This refers to the fact that the corresponding free offers of 'forest PR' have already negatively influenced the payment consciousness within the population.

Interesting connections are to be expected, however, especially with regard to the reasons for a forest visit and the financing of specific uses. It could be presumed that the users of specific, concrete forest services do not indicate the user as a financing source for this service. Such a tactical behaviour of answering can actu-

Table 8.34. Attitude to financing of forest services by the reasons for forest visits (total 100%).

Forest service	User group	n	Financing by: User	Taxes	Forest owner	Don't know	Missing
Hunting							
	All forest visitors	664	81.9	4.4	6.3	6.6	0.8
	Hunters	7	67.6	9.3	23.1	0.0	0.0
Bridle-paths							
	All forest visitors	664	67.9	8.9	13.8	8.0	1.3
	Rders	15	78.8	16.4	1.4	3.4	0.0
Mountain-bike tracks							
	All forest visitors	664	61.3	13.2	12.4	11.5	1.5
	Mountain-bikers	8	3.8	30.7	56.3	9.1	0.0
Cross-country ski-runs							
	All forest visitors	664	58.9	16.0	15.5	8.7	1.0
	Cross-country skiers	22	59.9	34.7	5.4	0.0	0.0
Guided forest tours							
	All forest visitors	664	48.6	16.0	24.2	10.0	1.1
	Nature connoisseurs	239	54.6	11.2	23.5	10.5	0.3
	Nature observers	130	45.9	19.4	26.1	8.6	0.0
Cycling paths							
	All forest visitors	664	31.0	36.1	25.2	6.8	1.0
	Cyclists	124	22.8	40.3	34.2	2.8	0.0
	Mountain-bikers	8	0.0	40.7	50.2	9.1	0.0
Collecting of mushrooms and berries							
	All forest visitors	664	13.5	12.5	63.2	9.8	1.1
	Mushroom (berry) collectors	231	4.5	18.2	68.7	8.0	0.6
Walking on forest paths							
	All forest visitors	664	4.1	27.7	59.6	7.5	1.0
	Cross-country runners, joggers	71	8.5	34.4	49.9	6.5	0.8

ally be seen in different areas of application, as shown in Table 8.34.

The different recreational services of the forest in Table 8.34 are arranged according to the question, 'What proportion of all persons being asked in the survey think that these services should be paid for by the user?' It is obvious from the table that, above all, mountain-bikers, but also hunters, cyclists, mushroom collectors and nature observers, would prefer the forest owner or the taxpayer to be responsible for paying the costs rather than the user him/herself.[15] Very high proportions of riders, joggers and people who just enjoy nature state that bridle-paths, hiking tracks and guided forest tours, respectively, should be paid for by the users. A greater number of respondents in these user groups want to support the forest owner financially. This is also valid for skiers, who want the taxpayers to finance their tracks, as with some riders for their bridle-paths and joggers for the installation of forest tracks. All in all, Table 8.34 shows a differentiated picture, from which a strict refusal of the user of specific forest services to pay a financial contribution for these forest services themselves cannot be derived.

The end of the question complex concerning forest financing deals with possible ways of financing forest services by the user.

Attitudes as to how special forest services should be paid for

In accordance with the examples of use chosen beforehand, the respondents were asked about the different kinds of forest services and whether these could be paid for preferably 'per use', in the form of an 'annual lump sum', as a 'voluntary donation' or in the course of another mode of payment. Although it was assumed in the question formulation that the service

Table 8.35. Opinion about different possible modes of user payment for forest services. (Question 2.9: For the (hypothetical) case that every forest user has to pay for the advantages of the use of the forest, please give your opinion as to which mode of payment would be the best for each forest service.)

	Mode of payment for forest services (attitude in % w.):						
Kind of service	Per use	Annual lump sum	Voluntary donation	Other	No payment	Don't know	Missing
Bridle-paths, cross-country ski-runs, mountain-bike tracks	46.3	23.6	5.8	1.7	14.2	7.0	1.5
Barbecue places, game parks, forest education paths	37.5	9.9	14.0	5.5	27.6	4.6	0.8
Guided tours, seminars	51.6	9.2	10.6	3.2	17.7	6.1	1.4
Collecting of forest fruits	11.9	7.1	6.3	2.6	66.0	4.8	1.3
Protection of drinking-water	7.8	17.0	9.2	8.8	41.1	15.2	0.8

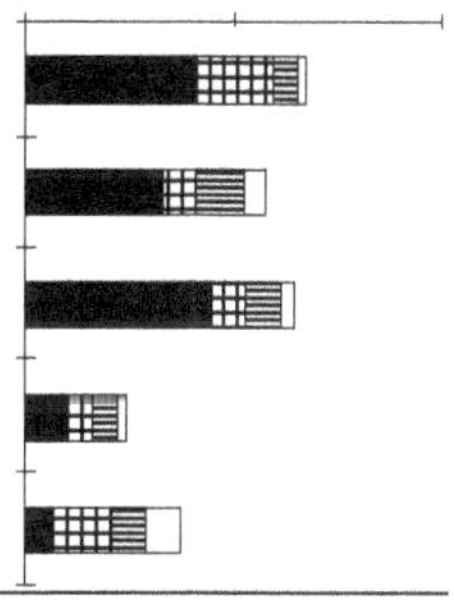

% w., weight %.

would have to be paid for by the users themselves, 'no payment' as a variable for conflict avoidance was offered. This resulted in the picture given by Table 8.35.

Payment 'per use' was judged as the most suitable and apparently fairest method of financing. One-fifth of the respondents stated that annual lump sums for sport equipment and environmental services would be quite suitable. Voluntary donations and other modes of payments were only exceptionally judged as suitable. The high occurrence of the answer 'no payment' for collecting berries and mushrooms' (66%) and the environmental service 'water protection' (41%) is striking.

Finally, connections between the mode of financing 'guided forest tours' and 'sports equipment in the forest' and some features of the forest visit behaviour should be considered more closely. Rare forest visitors (fewer than ten visits per year) tend to suggest lump-sum and 'other' modes of payment. They were represented as relatively low in number concerning the alternative 'no payment', whereas persons visiting the forest once a week or more frequently specifically rejected lump-sum payments and gave the 'escaping answer' 'no payment' instead. This picture is also reflected in the answers to the question as to how important it is for the respondent to be able to enter the forest at any point without any admission fee. Persons who found that free entry to the forest is less important (3%) stated various modes of payment, except donations. Those to whom free admission to the forest was important (89%) tended to give the answer donations or 'no payment'.

Considering the proposals for the mode of payment for specific sports offers in the forest by persons who use the forest for athletic purposes, it is striking that a payment per use in the case of mountain-bikers, riders and skiers is not very popular. Whilst the mountain-bikers (1.5% of all respondents) refuse a payment categorically, riders and skiers propose financing via an annual lump sum. It must be taken into consideration that the question formulation hypothetically presumed that a payment would have to be made.

Conclusions and discussion concerning the financing of forest services

It is not possible and would not be socially or politically achievable by the German system of jurisprudence to burden the broad mass of people looking for 'quiet recreation' with a general 'forest admission fee'. Any experiments or even the appearance of such a request for marketing for products from the recreation services of the forest would be not only unrealistic but even damaging.[16]

This guaranteed public freedom of being in the forest is typical for Germany but also for Austria and Scandinavia, but cannot be found in some other European countries. Whereas in Germany any forest demanding an admission fee must contain, for example, a game park or a comparable offer, every private forest owner in the Netherlands can combine admission to his/her forest with a demand for an admission fee. Table 8.36 compares these two countries concerning knowledge about forests for which fees are demanded and their use (see Section 8.1).

Within this comparison it must be noted that in Germany only those people who had been in the forest at least once during the last year were questioned while in the Netherlands an unfiltered cross-section of the population was asked. With this assumption, the comparison between the two states is quite obvious: more than half of the inhabitants of the Netherlands know of the existence of admission fees in the forests, whereas this is the case for only 40% of the German forest visitors. Visit behaviour is distinguished even more clearly in forests with an admission fee: 26% of the Dutch population but only 8% of forest visitors in Germany had used a forest where the visitor had to pay before. Table 8.36 explains the fact that the consideration of a 'fictional' forest admission fee is far away from the specific understanding of the forest within the German population. However, it also became obvious that respondents from cities and forest-poor regions are indeed more familiar with forests with an admission fee. This shows the fundamental connection between the scarcity of a certain good and its marketability.

The question complex relating to the wish for a 'free forest visit', the financial support of forest owners and the personal WTP something for specific products from the environmental and recreation services of the forest leads to the marketing of new forest products via the basic principle of offer and demand. This is the crucial question of the European Union (EU) RES project, whose empirical basis consists of 98 case-studies of successful offers in this economic field.

Successful marketing supposes the acceptance of the customer and his/her WTP. Thus, the products examined within the framework of the RES case-studies

Table 8.36. Comparison of forest admission fees in Germany and the Netherlands.

Statement regarding forests accessible only by admission fee	Germany *n*	Germany %	Netherlands *n*	Netherlands %
I never heard about such a forest	398	59.9	234	46.7
I heard about it, but I never visited such a forest	166	25.0	138	27.5
I have already visit such a forest	56	8.4	129	25.7
Missing/don't know	44	6.6	0	0.0
Total	664	100.0	501	100.0

Table 8.37. Customers' acceptance of documented products in the RES case-studies. (Question 7.3.3 RES questionnaire: How did the process of consumer acceptance develop over time?)

The offer was ...	*n*	%
Immediately and totally accepted	57	58.2
Irregularly but totally accepted	32	32.7
Not accepted	1	1.0
Not initiated yet	5	5.1
Missing	3	3.1
Total	98	100

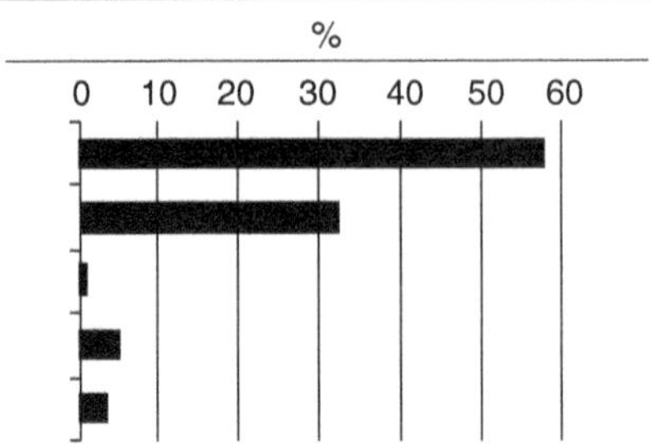

showed a high acceptance within the group of customers (Table 8.37). Concerning this table, it has to be noted that, naturally, these case-studies consciously concentrate on successful offers. However, it also shows that customer-orientated designed products in the forest are mainly accepted immediately and completely.

It is a fact that not every forest visitor can also be gained as a customer for RES offers. This is obvious from a cautious interpretation of the answers concerning the wish for free forest admission (89% positive), the financial support of the forest owners (45% positive) and the respondents' WTP (24% positive). The unobstructed admission to the forest for recreational use is of great importance for forest visitors. However, the opportunity for forest owners to gain a financial advantage from this recreation function and the environmental services is also evaluated positively by a considerable part of the forest visitors. Different reasons could be stated for the refusal of financial support. On the one hand, in the case of many respondents questions concerning a financial improvement for somebody else cause a more or less conscious refusal. The thoughts leading thereto are, for example: 'Why should I grant money to someone I do not know?' or 'Here is somebody who wants me to pay for others.' On the other hand – again more or less consciously – many people grant payments for such services only if the product derives from the receiver, while the forest is often considered globally as an existing landscape and natural resource of the entire society. Finally, knowledge about the forest ownership structures in the population is quite poor. It is normally presumed that the forest is a public, government estate from which no gains should be achieved.

The denial of their own WTP stated by the predominant part of respondents was to be expected due to tactical answer behaviour. Nevertheless, about 23% of forest visitors (approximately 14% of the population) agreeing to pay something for a visible improvement of the recreation possibilities and environmental services of the forest suggests a respectable market potential for products from the RES of the forest. Possible customers for accepted RES offers can indeed be found in this field. From the consideration of various possible factors influencing a positive attitude towards RES offers from the forest owner results in the following population segments, in which customers for such products can be found with raised probability, namely those respondents:

- from mid-life;
- from southern and eastern Germany;
- with higher education and salaries;
- who visit the forest neither rarely nor far too often;
- who already know and have used forests with an admission fee;
- who use the forest for athletic purposes.

Presumably, the most important result of the present study is the attitude towards financing forest services.[17] It shows that, first, with regard to the financing propositions the respondents are strongly differentiated by the different forest uses. Based on this, specific uses of individual groups should be paid for exclusively by these users themselves. At present, this is the usual practice for hunting; in the case of bridle-paths or mountain-bike trails, it is still the exception. Certainly, great possibilities exist for new products deriving from recreational services of the forest for which there is at least a broad social acceptance. It is also obvious from the results presented above that such an acceptance would also be expected by at least a part of the affected users of the corresponding services. Based on these results, the marketing of mountain-bike trails would face some difficulties; on the other hand, the marketing of bridle-paths may be expected to be realizable.

Considering the economic possible uses of forest fruits, a further aspect becomes obvious. Payment by the user is refused, particularly in the middle and in the east of Germany and in the case of respondent households with small salaries. Therefore, a conceivable renaissance of selling 'mushroom-picking permits', currently representing an important RES product, for example, in Italy, and also commonly practised in Germany in former times, cannot be expected, due to legal and technical control problems, as well as problems of acceptance. Nevertheless, it would be conceivable to use the forest for mushroom-collecting by selling collecting baskets and identification aids or by 'mushroom study trips'.

Basically, it seems likely and is also supported by the survey results that persons who themselves indicated a WTP for special forest services also agreed with the payment of such services by the direct user.

Services summarized as 'simple, general recreational services', such as the provision of forest tracks or simple recreation equipment, for which 'specific' user groups are hard to determine, are seen as the task of the forest owner and also, partially, of the state. The marketing possibilities in this field are certainly limited on account of questions of excludability, the legal situation and social acceptance. But some possibilities have already been mentioned above in the context of reasons for forest visits.

As became obvious within the scope of the survey results, the financing of general and specific environmental services of the forest is seen more and more as a public financing measure, although a contribution is also expected by the forest owner. Problems concerning payment by the user could appear because it seems to be difficult to determine, as with general recreational services, the concrete user of an environmental service. The proportion of the respondent households without a decision on the financing of forest environmental services was correspondingly high.[18] Considering the answer that 'the taxpayer should pay for environmental services' as a basis of an acceptance of contract nature conservation by public institutions and the answer that the forest owner might be supposed to finance these services him/herself as a refusal of such marketing approaches, the following picture results. Contracts for groundwater protection (56% for payment by taxes and 21% for payment by forest owners), biotope care measures (47% to 31%) and 'ecological forestry' (42% to 32%) appear to be especially promising. Whilst more respondents would support the retention of carbon dioxide (CO_2) by financing solutions carried out by public authorities (36% to 30%), the 'increase of dead-wood in the forest' is obviously not often judged as useful for society (29% to 36%). It seems to be certain that there is a communication problem with the term 'dead-wood' and its benefits.

Summarizing what has been stated above, considerable financing potentials from external sources, above all with regard to specific forest services, can be recognized, i.e. considerable socially accepted income possibilities for the forest owner. Therefore, as already suggested by Elsasser (1996), single claims of specific groups have to be seen in a wider context and the universality of such interests (e.g. in bridle-paths) have to be considered.

In addition to the questions of acceptance of the marketing of RES products and the legal general conditions, which were not discussed in further detail, concrete questions about the technical achievement of admissions have to be solved. One question is the mode of payment, which should be carried out 'per use' for specific forest uses of specific groups, according to the point of view of almost every respondent. Provided that this means of charging is technically possible and reasonable, it should be used for the sale of RES products.

A second, often reasonable, possibility for obtaining admission fees consists of lump-sum offers, e.g. annual calculations for specific offers. Such lump-sum calculations are often indicated, for example, by riders and skiers for bridle-paths and loipes. This appears to be obvious, as these offers are already marketed, e.g. in the form of a pass which is valid for the whole year. Finally, the possibility of a voluntary payment in the form of donations should not be ignored either, even if this mode of payment generally encountered little approval within the survey. Voluntary 'admission fees' are raised in Great Britain for bird-watching plots and the *Traufsteinloipe* located in Hessia, Germany, is partly financed via the voluntary acquisition of a ski-sticker. Voluntary payments always need a higher communication expenditure in order to motivate the user to pay; however, they help to overcome specific problems of excludability.

As a concrete, unlimitedly marketable RES example product, the Schleswig-Holstein forest backpack and its marketing possibilities are investigated in the third part of the evaluated household survey.

8.2.5 An RES example product – market chances of a 'forest backpack'

The development of the Schleswig-Holstein forest backpack

A concrete possibility for using the recreational services of the forest in a marketable product is the offer of a forest backpack, whose content enables the user to make his/her forest visit more intense and to increase his/her personal benefit from this forest visit on account of specific information or increased enjoyment. Such a forest backpack was developed and offered in Schleswig-Holstein as a possible experimental project of the Schleswig-Holstein RES workshop.[19] The design and contents of this product have already been described (see Section 8.2.2, note 8). Within the scope of the marketing launch, the 'Trappenkamp forest experience' (forest pedagogic centre of the Schleswig-Holstein forest administration) and the intensely used tourist regions of Albersdorf and Kellenhusen were chosen. The final product design and manufacture, as well as the central marketing of the backpack, was carried out by the 'Trappenkamp forest experience'.

In order to meet the wishes of backpack customers for the introduction and further development of the forest backpack, the household survey on forest visit behaviour in Germany carried out within the framework of the RES project was complemented by some questions about the forest backpack. The results of this part of the survey are described in 'Basic decisions for marketing a "forest backpack"' by Welcker and Mantau (2000).

Potential customers

At the beginning, potential customers for a forest backpack were determined. Therefore, the question was posed as to whether a potential interest would exist in the Schleswig-Holstein prototype backback, as described. Furthermore, such an interest was derived from a question about possible motives of product acquisition. As shown in Table 8.38, the percentage of the part of potential customers among 'forest visitors' is rated at approximately 63%. Referring the number of statements to the total group of respondents, including the 'non-forest visitors', 39% of the population of Germany over 14 years of age showed an interest in the forest backpack.[20] This share must be recognized as an enormous market potential.

Based on the sociodemographic features determined beforehand, it could be stated, in a strongly simplified manner, that above all young people (women) from/with families with a high education level and higher

Table 8.38. Share of potential backpack customers among the forest visitors (number of potential customers calculated from the forest visitors).

Potential backpack (rucksack) customers	*n*	% w.
Yes	417	63.5
No	247	36.5
Total	664	100.0

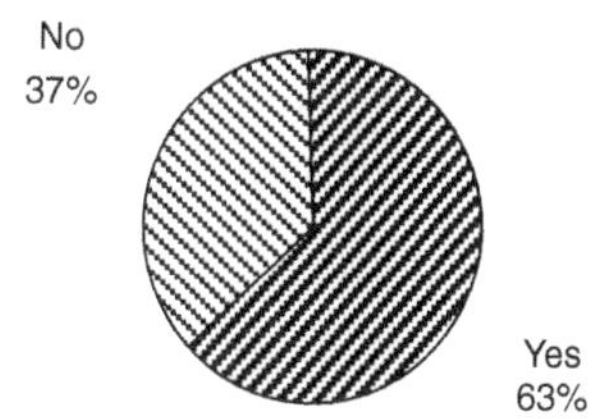

% w., weight %.

salaries would express an interest in a forest backpack. With regard to the forest-visit behaviour of 'potential backpack customers', it was shown that especially regular forest visitors who stay there for diverse reasons (longer-acting use) for a relatively long time are interested in the forest and forestry (information requirement) and therefore notice the economic problems of the forest owners and accept realistic implementation approaches. These survey segments do, indeed, come into consideration as typical backpack customers.

Results for the optimization of the marketing mix

In order to optimize the marketing mix for the forest backpack, questions were asked about acquisition motives, expected purchase places, the favoured sales method, WTP, favoured price-fixing measures and the expected advertising media.

Hints for the product design could be gained from the acquisition motives, shown in Table 8.39, and from the wishes for information about the forest, which were requested earlier (see Section 8.2.3).

By looking at the determined segment of the potential backpack customers, the following can be stated. The contents of the Schleswig-Holstein forest backpack prototype model were confirmed as a possible acquisition motive by from a quarter to half of all the respondents in the household survey. Above all, the importance of identification aids and information about the forest as a component of the product became obvious. The desired information emphasis changes along with an increasing stay duration in the forest, from information about the closer environment to information about forests and forestry in general. With regard to the information about forests and forestry, biological–ecological subjects are of great interest, but a notable number of forest visitors would also like to be informed better about questions on the economy and forest policy. Persons who visit the forest in order to play (with their children) show great interest in the product component 'game instructions' as expected. This group, as a part of the 'families' group, is an important and relatively specific customer segment

Table 8.39. Motivations for the acquisition of a forest backpack. (Question 2.10: With what prerequisites could you imagine buying such a forest backpack or borrowing it?)

Prerequisites for interest in the rucksack	*n*	% w.
Rucksack components		
Information about nature, forest, forestry	166	23.9
Aids for recognition of plants, animals	187	26.9
Information about the region, surroundings	165	24.0
Material for simple experiments	71	11.9
Games for children	98	14.4
Instruction for use of rucksack contents	94	16.0
Other		
A favourable price	144	22.3
Generally no interest	296	43.7
Don't know	45	8.1
Missing	6	0.9
Total (basis: 664 respondents)	1272	192.1

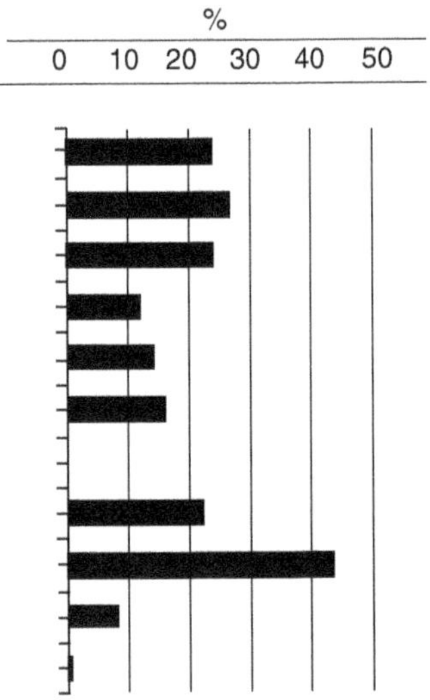

% w., weight %.

for the forest backpack. Therefore, it should receive special attention when designing the product.

A motivation for acquisition which is determined by the price was expressed by nearly one-quarter of the respondents. Persons from family households, also a major target group of the forest backpack, were less concerned with prices than the average. Especially for respondents from the new states and from households with high salaries, a low price was of great importance.

Information on the optimization of distribution could be derived from the question on the expected place of acquisition of the backpack. Further references result from the question as to whether these customers would favour a purchase of the product or a loan, and, if appropriate, what lending duration would be requested. Table 8.40 shows the variety of possible marketing places for a forest backpack. In order to summarize this table, the following can be stated with regard to the optimal distribution of a forest backpack: next to spa resort administrations and tourist information centres, forest offices are suitable as places for marketing the backpack. In particular the important customer segment 'families' and respondents with an interest in the forest backpack prototype model would expect a forest backpack here. Further expected marketing places are accommodation institutions, such as hotels, youth hostels or camp-sites. This applies especially to the important customers segment of families with several earners. However, the wide range of possible methods of distribution is obvious from the variety of free answers.

Provided that an interest in the acquisition of a backpack exists, one-third favours the purchase and two-thirds the loan of the product; the desired duration of the loan varies strongly between some hours and a longer vacation time.

The price design of a product is closely associated with distribution policies. The WTP as a price basis is difficult to determine

Table 8.40. Expected points of purchase and places with information for a forest backpack. (Question 2.11: Where would you expect to receive both information about this forest backpack and the possibility of acquiring it?)

Expected points of purchase	*n*	% w.	%
Given answer possibilities			
Restaurants, cafes	31	3.6	
Hotel, pension, camping sites, YH	141	19.8	
Spa authorities, tourist information	378	56.1	
Ranger station, forest office	294	44.8	
Events	55	7.3	
Free answers			
Sports shops	5	0.6	
At forest/direct in the forest	4	0.4	
Communities	6	1.1	
Schools	3	0.8	
Private clubs/associations	5	1.0	
Other	4	0.6	
Don't know	141	21.2	
Missing	10	1.5	
Total (basis: 664 respondents)	1046	155.2	

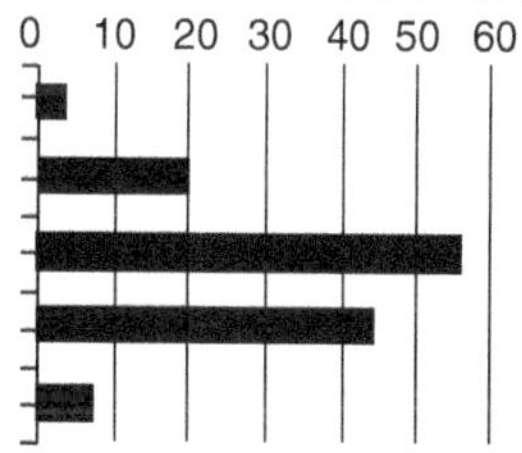

% w., weight %.

from direct questioning and therefore can only be estimated with considerable reservations. Tactical answer behaviour normally leads to the refusal of information on the respondent's financial situation or price ideas are indicated consciously lower than in reality. In addition, WTP is strongly dependent on the general conditions of the respective acquisition situation, such as, in the case of the loan of a product, the duration of its use.

The following picture concerning the forest backpack resulted from the corresponding questions of the RES household survey: The potential customer's WTP for a forest backpack goes up to about DM 100.- in the case of a purchase. In the case of a loan, the trend is that the accepted lending price increases with the desired lending duration; however, it is to be found in the range of DM 5.- to DM 25.- in particular. Proposals for a price differentiation received broad approval from the respondents, above all for lending and for families with several earners. The proportion of the respondents who requested a generally low price was no higher than the proportion who would pay more for higher quality. This quality-dependent price increase was above all accepted in the important target group 'families' and in the case of respondents who would be ready to pay something for special forest services or who expressed an interest in a backpack. The WTP a little more for good quality is also represented slightly more strongly in the case of a desired longer lending duration and in the case of a decision to purchase the backpack.

Finally, the distribution is complemented by communication. Communication includes information on the customer and his/her motivation for acquisition. The survey also supplied information on suitable means of communication (Table 8.41). It was shown that the usual methods of advertising, such as announcements, handbills or posters appeared suitable according to the expectations of the potential customers of the backpack, while 20% of the forest visitors thought of popular but expensive advertising measures, such as TV or broadcasting spots.

GENERAL CONCLUSIONS WITH REGARD TO THE FOREST BACKPACK. The apparently considerable marketing potential of the RES example product, the forest backpack, shows how far attractive offers for recreational purposes in the forest would be accepted. A proportion of 39% potential backpack customers within the population

Table 8.41. Expected advertising media for a forest backpack. (Question 2.12: Where and how should such a forest backpack be advertised in your opinion or where else would you expect information about it?)

Expected communication	*n*	% w.
Advertisements in newspapers, magazines	233	36.5
Advertisements in booklets	305	45.4
Leaflets, prospectuses	216	28.5
Posters, handbills	141	20.0
Broadcast and television spots	136	19.8
Other	1	0.3
Don't know	137	21.7
Missing	21	3.2
Total (basis: 664 respondents)	1190	175.4

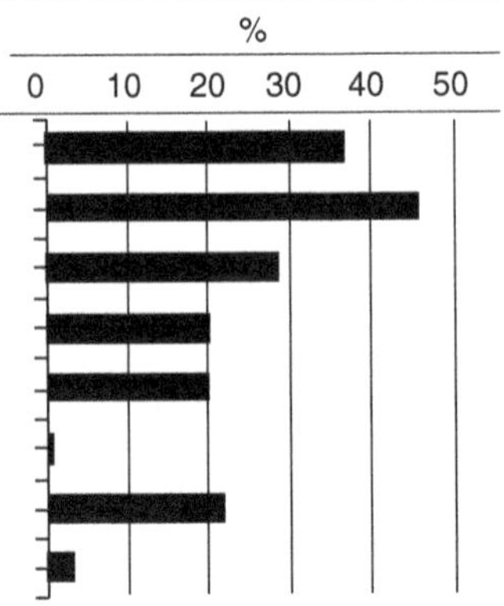

% w., weight %.

of Germany over 14 years of age was calculated, which would correspond to 20–30 million persons in the FRG. As a matter of fact, it would never be possible to serve this 'customers pool' completely with the concrete offer of one special product. However, by means of corresponding marketing strategies, some demographically and regionally limited market segments can be formed. As derived from the survey results of the RES household study, the most important customer segment is the (young) family from a better background. However, the relatively weak expression of these features also shows that too narrow a market segmentation for the forest backpack does not appear to be either necessary or suitable.

8.2.6 Summary

This section is concerned with the forest visiting behaviour of German households, their attitude towards financing specific forest services and the forest backpack as a possible RES product example. The results of the household questioning carried out are therefore discussed with regard to the possibilities of potential products being derived from the environmental and recreational services of the forest. The study represents, as part of the RES project, the German contribution to the investigation complex 'evaluation studies of public acceptance in the participating countries'. A representative study of 1069 German households, which was carried out within the survey of the University of Hamburg in autumn 1998 by INRA Germany, serves as the data basis.

About two-thirds of the German population over 14 years of age use the forest as a place for recreation at least once a year. Though there are some shifts towards specific sociodemographic segments, it can be determined that forest visits are used for recreation by all parts of the population. On average, the forest is used by forest visitors for recreational purposes for 1–2 h on 30 days of the year. The main motive for these forest visits is quiet recreation on foot. However, the forest is also regularly used for various athletic purposes, for playing (with children), for nature studies and for collecting forest fruits by different sociodemographic population groups. There is a high information requirement among the population with regard to the forest, its components and its social importance.

Forests where an admission fee is liable to be paid are relatively unknown in Germany and, also on account of legal impediments, are only poorly distributed within the country. Nevertheless, a basic attitude towards financing specific forest services can be observed within the population, which appears to make the marketing of these RES products quite realizable. Although free admission to the forest is important to about 90% of German forest visitors, a predominant proportion agrees with financial support for the forest owners and a considerable number state their own WTP for specific offers in the forest. It turns out that these individual forest services are respected by the population in a very differentiated manner. Therefore, with regard to their marketing possibilities, they have to be handled in a differentiated manner. In the view of the German households, specific forest uses of individual groups, such as riders or skiers, should be financed by these users themselves, and, if possible, payment should be made per use. Financing specific environmental services of the forest is predominantly imposed on the community of taxpayers, which supports market chances of corresponding cultivation/protection contracts with public authorities. The forest owner's duty is mainly seen only as providing the basic environmental and recreation services of his/her forest.

This result – encouraging for marketing – leads to a discussion of the market opportunities for different RES products. Some knowledge of forest-visit behaviour allows basic decisions to be made about the product design of recreation offers in the forest and market segmentation into different forest-visit customer categories. Reasons and motives for forest visits generate – along with the wishes of the population concerning the forest as a place for recreation – the qualitative demand of forest visitors for forest recreation services. The task of forestry will be to implement these in suitable offers. The huge information requirement of forest visitors implies the conclusion that forest 'information products' would meet a receptive market.

The forest backpack is one possible RES product for marketing forest recreation, with special regard to the demand of forest visitors for information about the forest. Such a backpack was developed and has already been marketed within the framework of the Schleswig-Holstein RES workshop as a forest backpack prototype model. From the questions to households, a considerable customer potential appeared for a forest backpack, as well as first approaches to a reasonable market segmentation into specific sociodemographically limited target groups. The target group 'families', which was particularly important within the course of developing the Schleswig-Holstein forest backpack, could be confirmed as a significant market segment. Furthermore, the content-related construction of the developed backpack as an information product with interactive components met with great approval in the household investigation. The broad range of possible distribution channels and forms became obvious, along with the great number of possible advertising instruments. Although a problem appeared with the WTP for the offer corresponding to the price of the offer, this concrete RES product example did indeed show the potentially high acceptance of attractive RES offers.

According to the results of this study, a broad acceptance for specific customer-orientated forest recreation offers can be expected, both within society and by potential customers. The assumption that forest owners would have to produce and finance the entirety of all recreation services of the forest, along with the wood production, can, based on this study, be considered from the start with regard to more intelligent business and forest policy solutions.

Notes

1 Probably the best-known example in the Netherlands is the De Hoge Veluwe National Park, where art and nature are combined and the entrance fee includes visits to both the museums and the park. There is also a separate admission to the park and the visitor centre only.

2 The negative attitude towards parking fees (in return for a safe place for the car) might be explained by the general aversion to parking fees in city centres, but this is merely an assumption and was not part of the research.

3 Please note that visitors who actually undertake these activities might be less positive.

4 E = Edese Bos, B = Berg en Bos and T = Total group.

5 The following statements on the survey are based on the explanations of method 'Set-up and method of the investigation in the Federal Republic of Germany (persons)' of the INRA Deutschland GmbH.

6 The basis is therefore an ADM (Arbeitskreis Deutscher Marktforschungsinstitute) master sample (information from INRA Deutschland GmbH).

7 The Schleswig-Holstein forest (experience) backpack (rucksack) is a concrete RES-product example, which was developed as a practically marketable offer by the forest backpack workshop within the Schleswig-Holstein RES circle. The forest backpack is a regionally flexible offer and should serve to make the forest visit more interesting, less monotonous and more precious for its user. It consists of a robust, medium-size backpack with a built-in compartment in which five or six plastic cases of varying content can be found. These bags contain information on the forest, the components and the use, aids for identification, material for easy-to-handle natural scientific investigations, ideas for 'forest experience games', regional information and also small wood samples.

8 Unlike the RES study presented here, both the survey of Dertz and Nießlein (1993) and the three surveys in specific cities were characterized by similar questions concerning the frequency of visits and therefore will not be discussed in further detail.

9 The partially contrary results of the survey of Dertz and Nießlein (1993) could again be explained by the different questions concerning the frequency of visits.
10 The values in parentheses indicate the weighted percentage of the answers concerning specific topics which are partly summarized in the table.
11 A selection of corresponding publications in the German language (some of which are quoted in this study) are: Fischer (1965); Hanstein (1967); Bichlmaier (1969); Weimann (1969); Rozsnyay (1972a, b); Weber (1976); Hertig (1979); Ott (1980); Glück (1984); Volk (1978, 1989, 1992); Dertz and Nießlein (1993); Dunkel *et al.* (1994); Elsasser (1994, 1996); a European link can be found in the publication of Rametsteiner (1999).
12 Here, a link is indicated with the marketing of infrastructure services in the city forest department of Baden, Switzerland by G. Schoop (Blum, 1992).
13 See also Welcker and Sierk (1999); there, the term forest information product (environmental education) and forest public relations are explained in detail.
14 Here, no decisive differentiation between 'forest effects' and 'services of forestry' is made, as this is a methodological problem and, from the point of view of possible marketing, not efficient (see Mantau, 1997, 1998).
15 Based on this explanation, the low absolute number of mountain-bikers and hunters limiting a profound statement has to be considered.
16 The interruption of a target region survey (Pfaelzerwald) can be given as an example. Based on a press release speculating that an actual, real introduction of forest admission fees (hypothetically stated in the survey questionnaire) could be realized soon, the refusal and protest rate of the forest visitors increased dramatically and continuation of the investigation did not seem to make any sense. Similar experiences were had by Löwenstein in the Harz (Dunkel *et al.*, 1994).
17 See the prepublication of this aspect in Mantau *et al.* (1999).
18 The proportion of the answer 'I don't know' was rated – depending on the environmental service asked about – between a minimum of 9% and a maximum of 30%.
19 See Mantau, contributing session to the workshop 'Markets for recreational and environmental services of the forest in Schleswig-Holstein', 2nd session of the workshop, November 1996, and 3rd session of the workshop, April 1997, University of Hamburg, unpublished; further participants in the Schleswig-Holstein RES workshop and the 'forest rucksack' work shop were – in addition to the University of Hamburg and the state forest department of Schleswig-Holstein – various institutions of forest property and tourism (see Welcker and Mantau, 2000).
20 This statement seems to be permissible, although the 'non-forest visitors' were not asked the question; people who do not visit the forest frequently will have no interest in the forest backpack.

References

Agee, P. (1984) A new excise tax to support non-game habitat. In: Martin, B.H. and Taylor, D.T. (eds) *Fees for Outdoor Recreation on Lands Open to the Public: Conference Proceedings.* Research Department Appalachian Mountain Club, Gorham, pp. 101–104.

Allen, F.D. (1981) Perceptions of recreation and park user fees in Austin, Texas. *Dissertation Abstracts International* 42(3), 1311.

Bakker, J.G., Kugel, E. and Zwam, C.R. (1988) Privatiseren Maasplassen kost sportvisser geld. *Recreatie & Toerisme* 20(12), 363–366.

Bamford, T.E., Manning, R.E. and Koenemann, E.J. (1988) Differential campsite pricing: an experiment. *Journal of Leisure Research* 20(4), 324–342.

Berrier, D. and Carlson, D.M. (1984) A trail license for cross-country skiers. In: Martin, B.H. and Taylor, D.T. (eds) *Fees for Outdoor Recreation on Lands Open to the Public: Conference Proceedings.* Research Department Appalachian Mountain Club, Gorham, pp. 134–135.

Betz, C.J. and Cordell, H.K. (1989) Trends in recreation participation on public lands. In: *Outdoor Recreation Benchmark 1988: Proceedings of the National Outdoor Recreation Forum, Tampa, Florida, 13–14 January, 1988. No. SE-52.* Southeastern Forest Experiment Service, USDA Forest Service, Athens, Georgia, pp. 128–146.

Bichlmaier, F. (1969) *Die Erholungsfunktion des Waldes in der Raumordnung.* Paul Parey, Hamburg, Berlin.

Blum, A. (1992) Zur Marktfähigkeit infrastruktureller Leistungen des Waldes: Schlußfolgerungen einer Analyse des Stadtforstamtes Baden/Schweiz. Diplomarbeit erstellt am Institut für Forsteinrichtung und Forstliche Betriebswirtschaft der Forstwissenschaftlichen Fakultät an der Albert-Ludwigs-Universität Freiburg im Breisgau.
Boersema, J.J., Copius Peereboom, J.W. and De Groot, W.T. (eds) (1991) *Basisboek Milieukunde.* Boom, Meppel, 583 pp.
Bovaird, A.G., Tricker, M.J. and Stoakes, R. (1984) *Recreation Management and Pricing: the Effect of Charging Policy on Demand at Countryside Recreation Sites.* Gower, Vermont, 182 pp.
Braune, U. (1998) Meinungen der Lübecker Bürger zu ihrem Wald. Diplomarbeit, erstellt am Institut für Forstökonomie und Forsteinrichtung der Fakultät Forst-, Geo- und Hydrowissenschaften der Technische Universität Dresden.
Crompton, J.L. (1982) Psychological dimensions of pricing leisure services. *Recreation Research Review* 9(3), 12–20.
De Bruin, Th. (1988) *De recreatieve betekenis van het Nationale Park De Hoge Veluwe.* Mededelingen 14 van de Werkgroep Recreatie, Landbouwuniversiteit Wageningen, Wageningen, 206 pp.
Dertz, W. and Nießlein, E. (1993) Die Bevölkerung akzeptiert die Waldbewirtschaftung. *Holz-Zentralblatt* 119(46), 693–696.
Driver, B.L. (1984a) Recreation on public lands: should the user pay? *American Forests* 90(3:11), 51–53.
Driver, B.L. (1984b) Public responses to user fees at public recreation areas. In: Martin, B.H. and Taylor, D.T. (eds) *Fees for Outdoor Recreation on Lands Open to the Public: Conference Proceedings.* Research Department Appalachian Mountain Club, Gorham, pp. 45–48.
Dunkel, K., Elsasser, P., Oesten, G. and Roeder, A. (1994) Wertschätzung des Waldes aus der Sicht der Waldbesucher – Ergebnisse einer Zielgebietsbefragung im Pfälzerwald. In: Oesten, G. and Roeder, A. (eds) *Zur Wertschätzung der Infrastruktleistungen des Pfälzwaldes.* Mitteilungen der Forstlichen Versuchsanstalt Rheinland-Pfalz 27/94, Trippstadt, pp. 1–72.
Elsasser, P. (1994) *Waldbesucher in Hamburg – Ergebnisse einer Bürgerbefragung im Hamburger Stadtgebiet.* Arbeitsbericht des Institutes für Ökonomie 94/4, Bundesforschungsanstalt für Forst- und Holzwirtschaft, Hamburg, 11 pp.
Elsasser, P. (1996) *Struktur, Besuchsmotive und Erwartungen von Waldbesuchern – Eine empirische Studie in der Region Hamburg.* Arbeitsbericht des Institutes für Ökonomie 96/1, Bundesforschungsanstalt für Forst- und Holzwirtschaft, Hamburg, 19 pp.
Fischer, F. (1965) Der åetliberg als Erholungsgebiet. *Zeitschrift für Forstwesen, Schweiz,* 467.
Fishbein, M. and Ajzen, I. (1976) *Belief, Attitude, Intention and Behaviour: an Introduction to Theory and Research.* Addison-Wesley, Reading.
Frost, J.E. and McCool, S.F. (1988) Can visitor regulations enhance recreational experiences? *Environmental Management* 12(1), 5–9.
Glück, P. (1984) Die Einstellung Österreichischer Stadtbewohner zum Wald. *Centralblatt für das gesamte Forstwesen* 101(3), 142–159.
Gratton, C. and Taylor, P. (1995) From economic theory to leisure practice via empirics: the case of demand and price. *Leisure-Studies* 14(4), 245–261.
Guldin, R.W. and Kroon, H.J.J. (1986) Economic and social aspects of user fees. In: *18th IUFRO World Congress, Division 6, General Subjects, 7–21 September, Ljubljana, Yugoslavia.* IUFRO, Vienna, pp. 386–396.
Hanstein, U. (1967) Über die Gewohnheiten, Ansichten und Wünsche der Waldbesucher. *Allgemeine Forstzeitung* 78, 465–467.
Hekhuis, H.J. and De Baaij, G. (1997) *Toepassing van het profijtbeginsel voor de financiering van bos- en natuurbeheer.* IBN-rapport 254, DLO-Instituut voor Bos- en Natuuronderzoek, Wageningen, 146 pp.
Henze, A. (1994) *Marktforschung: Grundlage für Marketing und Marktpolitik.* UTB für Wissenschaft 1792, Ulmer Verlag, Stuttgart.
Hertig, H.P. (1979) Die Einstellung der Bevölkerung zu Problemen des Waldes und der Waldwirtschaft: Ergebnis einer Meinungsumfrage. *Zeitschrift für Forstwesen, Schweiz* 130(8), 591–618.
Jarvis, D. (1984) Towards a rational fee policy. In: Martin, B.H. and Taylor, D.T. (eds) *Fees for Outdoor Recreation on Lands Open to the Public: Conference Proceedings.* Research Department Appalachian Mountain Club, Gorham, pp. 2–7.

Kemsley, W., Jr (1984) Concluding remarks. In: Martin, B.H. and Taylor, D.T. (eds) *Fees for Outdoor Recreation on Lands Open to the Public: Conference Proceedings.* Research Department Appalachian Mountain Club, Gorham, pp. 150–153.

Kerr, G.N. and Manfredo, M.J. (1991) An attitudinal based model of pricing for recreation services. *Journal of Leisure Research* 23(1), 37–50.

Lehmann, A. and Schriewer, K. (1999) Oh, du schöner deutscher Wald. *Forschung – Magazin der DFG* 3–4(99), 20–23.

Leuschner, W.A., Cook, P.S., Roggenbuck, J.W. and Oderwald, R.G. (1987) A comparative analysis for wilderness user fee policy. *Journal of Leisure Research* 19(2), 101–114.

LNV (1996) *Structuurschema Groene Ruimte; toelichting Planologische Kernbeslissing.* Ministerie van Landbouw, Natuurbeheer en Visserij, The Hague, 112 pp.

McCarville, R.E. (1992) Successful pricing: in the eye of the beholder. Revenue management. *Parks and Recreation* 27(12), 26–50.

McCarville, R.E. (1993) Prices and pricing for outdoor recreation opportunities. In: Van der Stoep, G.A. (ed.) *Proceedings of the 1993 Northeastern Recreation Research Symposium.* Michigan State University, Park and Recreation Resources Department, East Lansing, Michigan, pp. 38–39.

McCarville, R.E. and Crompton, J.L. (1987a) Propositions addressing perceptions of reference price for public recreation services. *Leisure Sciences* 9, 281–291.

McCarville, R.E. and Crompton, J.L. (1987b) An empirical investigation of the influence of information on reference prices for public swimming pools. *Journal of Leisure Research* 19(3), 223–235.

McCarville, R.E., Reiling, S.D. and White, C.M. (1996) The role of fairness in users assessments of first time fees for a public recreation service. *Leisure Sciences* 18(1), 61–76.

McCool, S.F. and Lime, D.W. (1989) Attitudes of visitors toward outdoor recreation management policy. In: *Outdoor Recreation Benchmark 1988: Proceedings of the National Outdoor Recreation Forum, Tampa, Florida, USA, 13–14 January 1988.* Nor. SE-52. Southeastern Forest Experiment Service, USDA Forest Service, Athens, Georgia, pp. 401–411.

Mantau, U. (1993a) Die Leistungen des Waldes zwischen ökonomischem und ethischem Wertdenken. *Forst und Holz* 48(11), 299–302.

Mantau, U. (1993b) Die forstliche Produktlücke: Infrastrukturleistungen des Waldes marktfähig machen. *Allgemeine Forstzeitung* 4, 186–190.

Mantau, U. (1995) Von der Waldfunktionenlehre zur Waldproduktlehre. *Holz-Zentralblatt* 26, 445, 451–453.

Mantau, U. (1997) Funktionen, Leistungen, Wirkungen oder ganz was Neues? – Vom Wert der Waren und Dienste. *Allgemeine Forstzeitung/Der Wald* 15, 826–827.

Mantau, U. (1998) Zu: Funktionen, Leistungen, Wirkungen oder ganz was Neues? *Allgemeine Forstzeitung/Der Wald* 16, 855.

Mantau, U., Mertens, B. and Welcker, B. (1999) Produktinnovationen für Umwelt- und Erholungsleistungen des Waldes. Block C/I, 32 S. In: *Zusammenstellung der fachlichen Beiträge, präsentiert am Jahrestreffen der deutschsprachigen Lehrstühle für Forstpolitikwissenschaft 1999, Siders, Schweiz.* Departement Wald- und Holzforschung – ETH Zürich, Zürich.

Martin, B.H. (1986) Hikers' opinions about fees for backcountry recreation. In: Lucas, R.C. (ed.) *Proceedings – National Wilderness Research Conference: Current Research.* General Technical Report INT-212, United States Department of Agriculture, Forest Service, Intermountain Research Station, Ogden, USA, pp. 483–488.

Oesten, G. and Roeder, A. (eds) (1994) *Zur Wertschätzung der Infrastrukturleistungen des Pfälzerwaldes.* Mitteilungen der Forstlichen Versuchsanstalt Rheinland-Pfalz 27/94, Trippstadt.

Ott, W. (1980) Wald und Forstverwaltung im Spiegel der Öffentlichen Meinung: Ergebnisse einer Meinungsumfrage in Baden-Württemberg. *Allgemeine Forstzeitung* 15, 397–399.

Powell, B. (1984) National park gift catalogs and donation boxes. In: Martin, B.H. and Taylor, D.T. (eds) *Fees for Outdoor Recreation on Lands Open to the Public: Conference Proceedings.* Research Department Appalachian Mountain Club, Gorham, pp. 119–120.

Prosser, N.S. (1984) A successful excise tax: the Dingell–Johnson program. In: Martin, B.H. and Taylor, D.T. (eds) *Fees for Outdoor Recreation on Lands Open to the Public: Conference Proceedings.* Research Department Appalachian Mountain Club, Gorham, pp. 115–118.

Rametsteiner, E. (1999) The attitude of European consumers towards forests and forestry. *Unasylva 196* 50, 42–48.

Reiling, S.D., Criner, G.K. and Oltmanns, S.E. (1988) The influence of information on users' attitudes towards campground user fees. *Journal of Leisure Research* 20(3), 208–217.

RMB (1995) *Draagvlak voor het milieubeleid.* RMB 95–05. Raad voor het milieubeheer, The Hague, 62 pp.

Rozsnyay, Z. (1972a) Ziele, Wünsche und Vorstellungen der Waldbesucher – Ergebnisse forstlicher Meinungsumfragen. *Forst und Holz* 1(72), 7–9.

Rozsnyay, Z. (1972b) Ergebnisse eines Jahrzehnts forstlicher Meinungsumfragen. *Forstarchiv* 43(8), 149–159.

Sample-Institut (1995) Einschaltung in die Sample-Mehrthemenumfrage 'Waldnutzung in der Freizeit'. Data sample, intended for publication in: Elsasser, P. (ed.) *Der Erholungswert der Wälder in Deutschland: Eine 'Benefit-Funtion-Transfer-Schätzung' für Tagesbesucher.*

Segeren, A.H.J. and Visschedijk, P.A.M. (1997) *Het recreatief gebruik van SBB-terreinen in de regio Brabant-West.* IBN-rapport 264, DLO-Instituut voor Bos- en Natuuronderzoek, Wageningen, 79 pp.

Segeren, A.H.J., Hekhuis, H.J. and Van Wijk, M.N. (1997) Recreatie in bossen: overheidsfinanciering of marktwerking? *Recreatie en Toerisme* 7(10), 16–19.

Van der Aalst, P. and Van der Straaten, J. (1992) *Profijtheffing bij recreatievoorzieningen: praktijkbeschrijvingen en principiële keuzes.* Katholieke Universiteit Brabant, Tilburg, 87 pp.

Van Setten, T.A. (1989) *Retributieheffing bij het Staatsbosbeheer in bezoekerscentra en tijdens ekskursies.* Dorschkamp-rapport 575, Instituut voor Bosbouw en Groenbeheer 'De Dorschkamp', Wageningen, 129 pp.

Volk, H. (1978) Gewohnheiten beim Waldbesuch: Ergebnisse einer Meinungsumfrage in Freiburg. *Forstwissenschaftliches Centralblatt* 97, 192–208.

Volk, H. (1989) Ergebnisse der Erholungswaldstudie 1988: Der Schönbuch und seine Waldbesucher. *Allgemeine Forstzeitung* 51–52, 1378–1382.

Volk, H. (1992) Neue Entwicklungen bei der Walderholung in Südwestdeutschland. *Forstwissenschaftliches Centralblatt* 111, 282–292.

Weber, H. (1976) Einstellungen und Verhaltensweisen von Waldbesuchern im mittleren Neckarraum. *Der Forst- und Holzwirt* 17, 356–360.

Weimann, J. (1969) Wald und Erholung. *Der Forst- und Holzwirt* 2, 42–45.

Welcker, B. and Mantau, U. (2000) Entscheidungsgrundlagen zur Vermarktung eines 'Waldrucksacks'. Unpublished.

Welcker, B. and Sierk, K. (1999) Ein Forstamt auf neuen (Wald)Wegen. *Allgemeine Forstzeitung/Der Wald* 14, 749–752.

Index

Note: page numbers in *italics* refer to figures and tables.